# SCHAUM'S SOLVED PROBLEMS SERIES

## 700 SOLVED PROBLEMS IN

# VECTOR MECHANICS FOR ENGINEERS Volume II: DYNAMICS

by

**Joseph F. Shelley, Ph.D., P.E.**
Trenton State College

**McGRAW-HILL, INC.**
New York   St. Louis   San Francisco   Auckland   Bogotá   Caracas
Hamburg   Lisbon   London   Madrid   Mexico   Milan   Montreal
New Delhi   Paris   San Juan   São Paulo   Singapore
Sydney   Tokyo   Toronto

Joseph F. Shelley, Ph.D., P.E, *Professor of Mechanical Engineering at Trenton State College*.
Dr. Shelley earned the Ph.D. at The Polytechnic University of New York. He has authored a three-volume set of textbooks in Engineering Mechanics, published by McGraw-Hill in 1980.

*To*
*Gabrielle*
*and*
*Stefanie, Suzanne,*
*Matthew, and Meredith*

*Your sweet love*
*such wealth brings,*
*I scorn to change*
*my state with kings*

Project supervision was done by The Total Book.

**Library of Congress Cataloging-in-Publication Data**
(Revised for volume 2)

Shelley, Joseph F.
  800 solved problems in vector mechanics for engineers.

  (Schaum's solved problems series)
  Vol. 2 has title: 700 solved problems in vector
mechanics for engineers.
  Contents: [1] Statics – v. 2. Dynamics.
  1. Mechanics, Applied – Problems, exercises, etc.
2. Vector analysis – Problems, exercises, etc.
I. Eight hundred solved problems in vector mechanics for
engineers. II. Title: 700 solved problems in vector
mechanics for engineers. III. Title: Seven hundred
solved problems in vector mechanics for engineers.
TA350.7.S54  1990        620.1'04'076        89-7951

**Library of Congress Cataloging-in-Publication Data**

ISBN 0-07-056582-1
ISBN 0-07-056687-9 (v. 2)


1 2 3 4 5 6 7 8 9 0 SHP/SHP 9 5 4 3 2 1 0

ISBN 0-07-056687-9

# CONTENTS

# TO THE STUDENT

Engineering mechanics is the study of the effects that forces produce on bodies. It has two major subdivisions: statics, in which the bodies are at rest or are moving with constant velocity; and dynamics, in which the bodies may possess any type of motion. Thus, acceleration is a necessary part of the description of dynamics problems. It is the absence of acceleration effects that distinguishes statics from dynamics.

This book is the second volume of a two-volume set. The first volume, in Chapters 1–13, treats the subject of statics. Statics is one of the beginning courses in the fields of aeronautical, civil, and mechanical engineering and is required of all engineering students. A thorough understanding of its fundamental principles is a prerequisite for further study in dynamics, strength (or mechanics) of materials, structural engineering, stress analysis, and mechanical design and analysis.

This second volume treats the subject of dynamics. The mastery of the principles of dynamics has direct, useful application in itself and is a prerequisite for further study in vibrations, dynamics of machinery, fluid mechanics, and mechanical design and analysis.

This book is a completely self-contained treatment of dynamics. All the fundamental definitions, concepts, and problem-solving techniques of dynamic force and motion analysis are introduced through questions. All groups of problems with numerical solutions are preceded by presentation of the particular definition, concept, or technique required for the solution of those problems. The material in each chapter is arranged in grouped sections of topics, and, within each section, the problems are arranged in a generally increasing order of difficulty.

The last question in each chapter is a review of the fundamental definitions, concepts, and techniques introduced and used in that chapter. The final chapter in this book is a self-study review of all the earlier questions on fundamental definitions, concepts, and techniques. This chapter contains 251 review questions, which are referenced by problem number to the original question in order to make it easy for the reader to refer to the answer. These review questions are not counted as part of the 700 Solved Problems.

Many problems are presented as easily recognizable mechanical or machine systems. The intent is to give a physical, real-world flavor to these problems with which the reader can readily identify. Additional commentary on the solutions is frequently provided at the end of problems to clarify or point out a particular characteristic or limitation of the solution. In many problems a comparison is made between the solutions when certain conditions of the original problem are varied. This gives a general engineering design flavor to these problems. The units used in this text are equally divided between U.S. Customary (USCS) units and International System (SI) units.

There is a carefully developed index, by problem number, at the end of the text. All problems involving definitions, concepts, or techniques are cross-referenced by topic. All problems in the text are listed in this index, and those which are more advanced, or have unusually lengthy solutions, are identified. The reader is encouraged to review this index and become familiar with its use, and thus be able to rapidly identify specific problems in any desired area of dynamics.

This book may be used with any textbook in dynamics. It may also be used by itself. A cross-reference of this book, by topics, with the three leading textbooks in dynamics is included in the appendices. These three texts are Beer and Johnston, *Vector Mechanics for Engineers: Statics and Dynamics*, 5th ed.; Hibbeler, *Engineering Mechanics: Statics and Dynamics*, 5th ed.; and Meriam and Kraige, *Engineering Mechanics*, volume 2, *Dynamics*, 2d ed.

Preparation of a work such as this is a very subjective exercise in creativity. It reflects many judgments on the part of the author with respect to organization of material and emphasis of topics. As with any other book, it receives its ultimate review by the readers. The author welcomes comments and suggestions on any matters of content, organization, or emphasis, and such information may be sent to the Schaum Division, McGraw-Hill Publishing Company, 1221 Avenue of the Americas, New York, N.Y. 10020. Every effort will be made to reply to this correspondence.

As a final note, the author wishes to thank Meredith Ann Shelley for her yeoman service in helping to perform the myriad tasks required to bring this work to its final form.

# LIST OF SYMBOLS

| | |
|---|---|
| $\mathbf{a}$ | Acceleration |
| $a$ | Magnitude of $\mathbf{a}$, acceleration in rectilinear translation |
| $a, b, \ldots$ | Constants, lengths points |
| $a_1, a_2, \ldots$ | Constants, lengths |
| $a_{avg}$ | Average acceleration |
| $\mathbf{a}_a, \mathbf{a}_b$ | Acceleration of points $a$ and $b$ |
| $a_a, a_b$ | Magnitudes of $\mathbf{a}_a$ and $\mathbf{a}_b$ |
| $\mathbf{a}_{ab}$ | Relative acceleration of point $a$ with respect to point $b$ |
| $a_{ab}$ | Magnitude of $\mathbf{a}_{ab}$ |
| $\mathbf{a}_n$ | Normal acceleration in curvilinear translation |
| $a_n$ | Magnitude of $\mathbf{a}_n$ |
| $\mathbf{a}_t$ | Tangential acceleration in curvilinear translation |
| $a_t$ | Magnitude of $\mathbf{a}_t$ |
| $a_x$ | Rectilinear acceleration in $x$ direction |
| $a_x, a_y$ | $x$ and $y$ components of $\mathbf{a}$ |
| $a_0, a_1, \ldots$ | Magnitudes of acceleration |
| $a_{an}, a_{bn}$ | Magnitude of normal acceleration of points $a$ and $b$, respectively |
| $a_{ta}, a_{tb}$ | Magnitude of tangential acceleration of points $a$ and $b$, respectively |
| $\mathbf{A}$ | General vector quantity |
| $A$ | Magnitude of $\mathbf{A}$, projected area of body on plane normal to direction of motion |
| $A, A_1, A_2, \ldots$ | Plane areas |
| $\mathbf{a}_A, \mathbf{a}_B, \ldots$ | Translational acceleration of bodies $A, B, \ldots$ |
| $a_A, a_B, \ldots$ | Magnitudes of $\mathbf{a}_A, \mathbf{a}_B, \ldots$ |
| $\mathbf{a}_c$ | Absolute acceleration of center of mass of a body |
| $a_c$ | Magnitude of $\mathbf{a}_c$ |
| $c$ | Viscous drag force constant |
| $c_A, c_B$ | Viscous drag force constant of dampers $A$ and $B$, respectively |
| $C_D$ | Drag coefficient |
| CM | Center of mass of a body |
| $d, d_1, d_2, \ldots$ | Diameter of circular path, cylinder, gear, or sphere; axial separation distance between rotating point masses |
| $d_i, d_0$ | Inside and outside diameters of cylinder or disk |
| $d_x, d_y, d_z$ | Distance between center of mass of body and $x$, $y$, and $z$ axes, respectively |
| $D$ | Length |
| $d_{0y}$ | Distance between center of mass of an element and $y_0$ axis |
| $d_{x1}, d_{x2}, \ldots$ | Distance between center of mass of elements $1, 2, \ldots$ and $x$ axis |
| $d_{y1}, d_{y2}, \ldots$ | Distance between center of mass of elements $1, 2, \ldots$ and $y$ axis |
| $d_{z1}, d_{z2}, \ldots$ | Distance between center of mass of elements $1, 2, \ldots$ and $z$ axis |
| $d_{x_1}, d_{y_1}, d_{z_1}$ | Distance between center of mass of a body and $x_1$, $y_1$, and $z_1$ axes, respectively |
| $d_{x_2}$ | Distance between center of mass of a body and $x_2$ axis |
| $d_{x_{13}}$ | Distance between center of mass of element 3 and $x_1$ axis |
| $d_{y, y_1}$ | Distance between $y$ and $y_1$ axes |
| $e$ | Coefficient of restitution |
| $f$ | Function, such as $y = f(x)$ |
| ft | Foot (USCS unit of length) |
| $\mathbf{F}$ | Force, friction force, resultant force acting on particle or body |
| $F$ | Magnitude of $\mathbf{F}$ |
| $F_a, F_b, F_c$ | Forces acting on pins $a$, $b$, and $c$, respectively |
| $F_c, F_d, F_e$ | Centrifugal force of material which would occupy holes $c$, $d$, and $e$, respectively |
| $\mathbf{F}_i, i = 1, 2, \ldots$ | Forces acting on particle $1, 2, \ldots$, or body $1, 2, \ldots$ |
| $F_i, i = 1, 2, \ldots$ | Magnitude of $\mathbf{F}_i$, $i = 1, 2, \ldots$, centrifugal forces acting on mass elements $1, 2, \ldots$ |
| $F_x, F_y$ | $x$ and $y$ components, respectively, of $\mathbf{F}$ |

| | |
|---|---|
| $\mathbf{F}_n$ | Normal force acting on particle or body |
| $F_n$ | Magnitude of $\mathbf{F}_n$ |
| $\mathbf{F}_t$ | Tangential force acting on particle or body |
| $F_t$ | Magnitude of $\mathbf{F}_t$ |
| $F_{in}, F_{it}$ | Normal and tangential components, respectively, of force acting on mass element $m_i$ |
| $F_A, F_B, \ldots$ | Centrifugal forces acting on mass elements $A, B, \ldots$ |
| $F_{12}, F_{21}$ | Pair of action-reaction forces on mass particles $m_1$ and $m_2$ |
| $F_{12}, F_{23}$ | Tangential force transmitted between teeth on gears 1 and 2, and on gears 2 and 3, respectively |
| $F_D$ | Drag force |
| $F_{NC}$ | Nonconservative force |
| $F_s$ | Spring force |
| $\mathbf{g}$ | Acceleration of gravitational field |
| $g$ | Magnitude of $\mathbf{g}$ ($g = 32.2 \, \text{ft/s}^2$ or $386 \, \text{in/s}^2$ in USCS units, and $9.81 \, \text{m/s}^2$ in SI units) |
| $h$ | Vertical displacement of a body, height of a body above the ground or above a reference point |
| $h_1, h_2, \ldots$ | Height of cylinder $1, 2, \ldots$; height of body $1, 2, \ldots$ above the ground |
| $h', h'', \ldots$ | Rebound height after first, second, $\ldots$ impact |
| hp | Horsepower |
| $\mathbf{H}$ | Angular momentum of a body |
| $H$ | Magnitude of $\mathbf{H}$ |
| $H_1, H_2, \ldots$ | Magnitude of angular momentum at endpoints $1, 2, \ldots$ of a time interval |
| $\mathbf{H}_0$ | Angular momentum about center of mass of body |
| $H_0$ | Magnitude of $\mathbf{H}_0$ |
| in | Inch (USCS unit of length) |
| $\mathbf{i}, \mathbf{j}, \mathbf{k}$ | Unit vectors in $x$, $y$, and $z$ directions, respectively |
| $i$ | Index of summation, $i = 1, 2, \ldots$ |
| $I_x, I_y, I_z$ | Mass moment of inertia of body about $x$, $y$, and $z$ axes, respectively |
| $I_{0x}, I_{0y}, I_{0z}$ | Mass moment of inertia of body about centroidal $x_0$, $y_0$, and $z_0$ axes, respectively |
| $I_{x_1}, I_{y_1}, I_{z_1}$ | Mass moment of inertia of body about $x_1$, $y_1$, and $z_1$ axes, respectively |
| $I_{x_2}$ | Mass moment of inertia of body $x_2$ axis |
| $I_1, I_2, \ldots$ | Mass moment of inertia of mass elements $1, 2, \ldots$ about a specified axis |
| $I_{0x,1}, I_{0x,2}, \ldots$ | Centroidal mass moment of inertia of mass elements $1, 2, \ldots$ about $x$ axis |
| $I_{0y,1}, I_{0y,2}, \ldots$ | Centroidal mass moment of inertia of mass elements $1, 2, \ldots$ about $y$ axis |
| $I_{0z,1}, I_{0z,2}, \ldots$ | Centroidal mass moment of inertia of mass elements $1, 2, \ldots$ about $z$ axis |
| $I_{x1}, I_{x2}, \ldots$ | Mass moment of inertia of mass elements $1, 2, \ldots$ about $x$ axis |
| $I_{y1}, I_{yz}, \ldots$ | Mass moments of inertia of mass elements $1, 2, \ldots$ about $y$ axis |
| $I_{z1}, I_{z2}, \ldots$ | Mass moments of inertia of mass elements $1, 2, \ldots$ about $z$ axis |
| $I_0, I_{01}, I_{02}, I_{03}$ | Mass moment of inertia of body about center axis or center of mass |
| $I_{xA}, I_{yA}, I_{zA}$ | Moment of inertia of plane area about $x$, $y$, and $z$ axes, respectively |
| $I_{xM}, I_{ym}, I_{zM}$ | Mass moment of inertia of thin, plane body about $x$, $y$, and $z$ axes, respectively |
| $I_{x1,A}, I_{x2,A}, \ldots$ | Moment of inertia of plane area elements $1, 2, \ldots$ about $x$ axis |
| $I_{xl}, I_{yl}, I_{zl}$ | Mass moment of inertia of plane body formed of thin rod shapes, length moment of inertia of plane curve, about $x$, $y$, and $z$ axes, respectively |
| $I_{y1,1}, I_{y1,2}, \ldots$ | Mass moment of inertia of mass element $1, 2, \ldots$ about $y_1$ axis |
| $I_{xy}, I_{yz}, I_{zx}$ | Mass product of inertia of body about $x$, $y$, and $z$ axes |
| $I_{xy,A}, I_{xy,B}, I_{xy,C}$ | Mass products of inertia of elements $A$, $B$, and $C$, respectively, about $x$, $y$, and $z$ axes |
| $I_{xy,C}$ | Mass product of inertia of a cylinder about $x$ and $y$ axes |
| $I_{xy,H}$ | Mass product of inertia of mass, which would occupy a hole, about $x$ and $y$ axes |
| $I_{0,xy}, I_{0,yz}, I_{0,zx}$ | Mass product of inertia of body about centroidal $x_0$, $y_0$, and $z_0$ axes |
| IC | Instant center |
| $I_a$ | Mass moment of inertia of body about point $a$ |
| $I_0$ | Mass moment of inertia of body about center of mass |
| $I_{assy}$ | Mass moment of inertia of assembly about specified axis |

| | |
|---|---|
| $I_{0A}, I_{0B}, \ldots$ | Centroidal mass moments of inertia of bodies $A, B, \ldots$ |
| $I_{01}, I_{02}, \ldots$ | Centroidal mass moments of inertia of bodies $1, 2, \ldots$ |
| $\mathbf{I'}$ | Impulse of a force |
| $I'$ | Magnitude of $\mathbf{I'}$ |
| $I'_x, I'_y$ | $x$ and $y$ components of $\mathbf{I'}$ |
| $J$ | Polar mass moment of inertia of a thin, plane body |
| $\mathbf{J}$ | Joule |
| $k$ | Spring constant |
| kg | Kilogram (SI unit of mass) |
| kn | Knot (speed of one nautical mile per hour) |
| $k_x, k_y, k_z$ | Radii of gyration of body about $x$, $y$, and $z$ axes, respectively |
| $k_{0x}, k_{0y}, k_{0z}$ | Radii of gyration of body about centroidal $x_0$, $y_0$, and $z_0$ axes, respectively |
| $k_0$ | Radius of gyration of a body about center of mass |
| $l$ | Length, length of a plane curve |
| lb | Pound (USCS unit of force) |
| $m$ | Mass of particle or body |
| m | Meter (SI unit of length) |
| $m_A, m_B, \ldots$ | Mass of bodies $A, B, \ldots$ |
| $m_1, m_2, \ldots$ | Mass of elements $1, 2, \ldots$ |
| $M_a$ | Moment about point $a$ |
| $M_a, M_b$ | Dynamic moment about points $a$ and $b$, respectively |
| $m_a, m_b, m_c, m_d$ | Mass of dynamic balance correction weights |
| mi | mile (1 mi = 5280 ft) |
| $M_0$ | Moment about center of mass of body |
| $M_{ai}$ | Moment required to produce $F_{it}$ |
| $M_x, M_y, M_z$ | Moment about $x$, $y$, and $z$ axes, respectively |
| $M_{ce}, M_{df}$ | Moment due to distributed centrifugal forces |
| $M_{AB}, M_{BC}, M_{CD}$ | Torque transmitted through lengths $AB$, $BC$, and $CD$, respectively, of shaft |
| $M_{NC}$ | Nonconservative moment |
| $M_C$ | Moment due to centrifugal forces acting on mass element $C$ |
| $\mathbf{N}$ | Normal force |
| $N$ | Magnitude of $\mathbf{N}$ |
| N | Newton (SI unit of force) |
| $N_1, N_2, \ldots$ | Number of teeth on gears $1, 2, \ldots$ |
| $\mathbf{N}_a, \mathbf{N}_b, \ldots$ | Normal forces at points $a, b, \ldots$ |
| $N_a, N_b, \ldots$ | Magnitude of $\mathbf{N}_a, \mathbf{N}_b, \ldots$ |
| $\mathbf{N}_A, \mathbf{N}_B, \ldots$ | Normal forces acting on bodies $A, B, \ldots$ |
| $N_A, N_B, \ldots$ | Magnitudes of $\mathbf{N}_A, \mathbf{N}_B, \ldots$ |
| $\mathbf{N}_{AB}$ | Normal force transmitted between bodies $A$ and $B$ |
| $N_{AB}$ | Magnitude of $\mathbf{N}_{AB}$ |
| $N_L, N_R$ | Left-side and right-side normal reaction forces, respectively |
| $\mathbf{N}_F, \mathbf{N}_R$ | Normal reaction forces on front and rear automobile wheels, respectively |
| $n, t$ | Normal and tangential axes |
| $N_F, N_R$ | Magnitudes of $\mathbf{N}_F$ and $\mathbf{N}_R$, respectively |
| $N'$ | Tangential impulse |
| $O_a, O_b$ | Center of radii of curvatures $\rho_a$ and $\rho_b$, respectively |
| $\mathbf{P}$ | Force |
| $P$ | Magnitude of $\mathbf{P}$, power, resultant centrifugal force |
| $P_a$ | Actual power requirement |
| $P_t$ | Theoretical power requirement |
| $P_1, P_2$ | Resultant centrifugal force on mass elements 1 and 2, respectively |
| $P_{AB}, P_{BC}, P_{CD}$ | Power transmitted through lengths $AB$, $BC$, and $CD$, respectively, of shaft |
| $P_y, P_z$ | $y$ and $z$ components of resultant centrifugal force |
| $q$ | Factor equal to $\rho t$, where $t$ is thickness of a thin, plane body; distance between center of percussion and axis of rotation |
| $\mathbf{r}$ | Displacement, or position, vector |
| $r$ | Magnitude of $\mathbf{r}$, radius of circular path of motion, radius of cylinder or sphere, radial coordinate of a mass element |

| | |
|---|---|
| $r_1, r_2$ | Radii of circle or cylinder |
| $\mathbf{R}$ | Reaction force, dynamic bearing force acting on shaft |
| $R$ | Gyroscopic force, magnitude of $\mathbf{R}$, radius of circular arc |
| $R_x, R_y$ | $x$ and $y$ components of $\mathbf{R}$ |
| $r_x, r_y, r_z$ | Coordinates of differential mass element |
| $r_c$ | Distance between $z$ and $z_0$ axes and between center of mass and axis of rotation |
| $\mathbf{r}_i$ | Position vector from center of mass to mass particle $m_i$ |
| $r_i$ | Magnitude of $\mathbf{r}_i$, radius of path of mass particle $m_i$ |
| $\ddot{\mathbf{r}}_i$ | Relative acceleration of mass particle $m_i$ with respect to center of mass of body |
| $r_{ix}, r_{iy}$ | $x$ and $y$ components, respectively, of $\mathbf{r}_i$ |
| $r_0$ | Position of line of action of resultant centrifugal force |
| $\mathbf{R}_a, \mathbf{R}_b$ | Dynamic bearing forces, normal bearing forces, reaction forces |
| $R_a, R_b$ | Magnitude of $\mathbf{R}_a$ and $\mathbf{R}_b$, respectively |
| $R_{ay}, R_{az}, R_{by}, R_{bz}$ | $y$ and $z$ components of dynamic bearing force at $a$ and $b$, respectively |
| $r_A, r_B, r_C$ | Radii of cylinders $A$, $B$, and $C$, respectively |
| $\mathbf{R}_C, \mathbf{R}_D$ | Reaction forces on bodies $C$ and $D$, respectively |
| $R_C, R_D$ | Magnitude of $\mathbf{R}_C$ and $\mathbf{R}_D$, respectively |
| $R_{cx}, R_{cy}, R_{Dx}, R_{Dy}$ | $x$ and $y$ components of $\mathbf{R}_C$ and $\mathbf{R}_D$, respectively |
| $\mathbf{s}$ | Displacement |
| $s$ | Magnitude of $\mathbf{s}$, rectilinear displacement, coordinate of length along curvilinear path |
| s | second (unit of time) |
| $\dot{s}, \ddot{s}$ | First and second time derivatives of $s$, respectively |
| $s_0$ | Initial displacement |
| $s_1, s_2, \ldots$ | Displacement at times $t_1, t_2, \ldots$; displacement at points $1, 2, \ldots$ |
| $\mathbf{s}_a, \mathbf{s}_b$ | Displacement of points $a, b, \ldots$ |
| $s_a, s_b, \ldots$ | Magnitude of $\mathbf{s}_a, \mathbf{s}_b, \ldots$ |
| $\mathbf{s}_A, \mathbf{s}_B, \ldots$ | Displacement of bodies $A, B, \ldots$ |
| $s_A, s_B, \ldots$ | Magnitude of $\mathbf{s}_A, \mathbf{s}_B, \ldots$ |
| $\mathbf{s}_{ab}, \mathbf{s}_{ba}$ | Relative displacement of point $a$ with respect to point $b$, and point $b$ with respect to point $a$, respectively |
| $s_{ab}, s_{ba}$ | Magnitudes of $\mathbf{s}_{ab}$ and $\mathbf{s}_{ba}$, respectively |
| $t$ | Time, thickness of plane body |
| $t_m$ | Time for particle acted on by drag force to reach maximum displacement |
| $t_1, t_2, \ldots$ | Time, time intervals |
| $\mathbf{T}$ | Cable tensile force |
| $\mathbf{T}_A, \mathbf{T}_B, \ldots$ | Cable tensile force acting on bodies $A, B, \ldots$ |
| $T_A, T_B, \ldots$ | Magnitude of $\mathbf{T}_A, \mathbf{T}_B, \ldots$ |
| $T_1, T_2, \ldots$ | Cable tensile forces |
| $T_{AC}, T_{BD}, T_{BE}, T_{CD}, T_{DE}$ | Cable tensile force between bodies $A$ and $C$, $B$ and $D$, $B$ and $E$, $C$ and $D$, and $D$ and $E$, respectively |
| $T$ | Kinetic energy of particle or body; magnitude of $T$ |
| $T_1, T_2, \ldots$ | Kinetic energy of particle or body at points $1, 2, \ldots$ |
| $T_{A1}, T_{A2}, \ldots, T_{B1}, T_{B2}, \ldots$ | Kinetic energy of bodies $A, B, \ldots$ at points $1, 2, \ldots$ |
| $\mathbf{v}$ | Velocity |
| $v$ | Magnitude of $\mathbf{v}$, rectilinear velocity |
| $v_x, v_y$ | $x$ and $y$ components of $\mathbf{v}$ |
| $\dot{v}$ | First time derivative of $v$ |
| $v_0$ | Initial velocity |
| $v_{0x}, v_{0y}$ | $x$ and $y$ components of $v_0$ |
| $v_{0a}, v_{0b}, \ldots$ | Initial velocities of points $a, b, \ldots$ |
| $v_{0A}, v_{0B}, \ldots$ | Initial velocities of bodies $A, B, \ldots$ |
| $v_1, v_2, \ldots$ | Velocity at times $t_1, t_2, \ldots$; velocity at points $1, 2, \ldots$ |
| $\mathbf{v}_a, \mathbf{v}_b, \ldots$ | Velocity of points $a, b, \ldots$ |
| $v_a, v_b, \ldots$ | Magnitude of $\mathbf{v}_a, \mathbf{v}_b, \ldots$ |
| $\mathbf{v}_A, \mathbf{v}_B, \ldots$ | Velocity of bodies $A, B, \ldots$ |
| $v_A, v_B, \ldots$ | Magnitude of $\mathbf{v}_A, \mathbf{v}_B, \ldots$ |
| $v_{Ax}, v_{Ay}$ | $x$ and $y$ components of $v_A$ |

| | |
|---|---|
| $v_t$ | Tangential velocity in curvilinear translation |
| $\mathbf{v}_{ab}, \mathbf{v}_{ba}$ | Relative velocity of point $a$ with respect to point $b$, and point $b$ with respect to point $a$, respectively |
| $v_{ab}, v_{ba}$ | Magnitudes of $\mathbf{v}_{ab}$ and $\mathbf{v}_{ba}$, respectively |
| $v_{ba,x}, v_{ba,y}$ | $x$ and $y$ components of $v_{ba}$ |
| $\mathbf{v}_{AB}, \mathbf{v}_{BA}$ | Relative velocity of body $A$ with respect to body $B$, and body $B$ with respect to body $A$, respectively |
| $v_{AB}, v_{BA}$ | Magnitudes of $\mathbf{v}_{AB}$ and $\mathbf{v}_{BA}$, respectively |
| $v_{avg}$ | Average velocity |
| $V$ | Potential energy of particle or body, volume |
| $V_1, V_2, \ldots$ | Potential energy at endpoints $1, 2, \ldots$; volume of elements $1, 2, \ldots$ |
| $\mathbf{v}_A, \mathbf{v}_B$ | Velocity of bodies $A$ and $B$ before impact |
| $v_A, v_B$ | Magnitude of $\mathbf{v}_A$ and $\mathbf{v}_B$, respectively |
| $\mathbf{v}'_A, \mathbf{v}'_B$ | Velocity of bodies $A$ and $B$ after impact |
| $v'_A, v'_B$ | Magnitude of $\mathbf{v}'_A$ and $\mathbf{v}'_B$ |
| $v_{Ax}, v_{Ay}, v_{Bx}, v_{By}$ | $x$ and $y$ components of $v_A$ and $v_B$ |
| $v'_{Ax}, v'_{Ay}, v'_{Bx}, v'_{By}$ | $x$ and $y$ components of $v'_A$ and $v'_B$ |
| $v_{Ax_1}, v_{Ay_1}, v_{Bx_1}, v_{By_1}$ | $x_1$ and $y_1$ components of $v_A, v_B$ |
| $v'_{Ax_1}, v'_{Ay_1}, v'_{Bx_1}, v'_{By_1}$ | $x_1$ and $y_1$ components of $v'_A, v'_B$ |
| $v'_x, v'_y$ | $x$ and $y$ components of velocity of center of mass of body after impact |
| $v_T$ | Terminal velocity |
| $v_T^*$ | Constant defined by $\sqrt{\xi/\zeta}$ |
| $\mathbf{W}$ | Weight force |
| $W$ | Magnitude of $\mathbf{W}$, work done |
| $W$ | Watt |
| $\mathbf{W}_A, \mathbf{W}_B, \ldots$ | Weight force of bodies $A, B, \ldots$ |
| $W_A, W_B, \ldots$ | Magnitude of $\mathbf{W}_A, \mathbf{W}_B, \ldots$ |
| $W_1, W_2, \ldots$ | Weight of elements $1, 2, \ldots$ |
| $W_{12}$ | Work done between points 1 and 2 |
| $W_{NC}$ | Nonconservative work done |
| $x$ | Length |
| $x, y, z$ | Cartesian coordinate axes |
| $\ddot{x}, \ddot{y}, \ddot{z}$ | $x$, $y$, and $z$ components of $\mathbf{a}$ |
| $x_m$ | Maximum displacement of particle acted on by drag force |
| $x_0, y_0, z_0$ | Centroidal axes |
| $x_0, y_0$ | Coordinates of differential mass element in $x_0, y_0$ coordinates |
| $x_0$ | Position of line of action of resultant centrifugal force |
| $x_{ab}, y_{ab}$ | $x$ and $y$ components of $\mathbf{s}_{ab}$ |
| $x_A, y_A, x_B, y_B$ | Coordinates of point on body in positions $A$ and $B$, position coordinates of mass elements $A$ and $B$, respectively |
| $x_c, y_c$ | Coordinates of centroid of plane area |
| $x_c, y_c, z_c$ | Coordinates of centroid of body, or volume |
| $x_i, y_i, z_i$ | Centroidal coordinates of $n$ mass, weight, or volume elements $1, 2, \ldots, i, \ldots, n$ |
| $x_1, y_1$ | Coordinate axes |
| $x'_c, y'_c$ | Centroidal coordinates of plane body formed of thin rod shapes or of plane curve |
| $x_{CP}, y_{CP}$ | Coordinates of center of percussion |

## GREEK SYMBOLS

| | |
|---|---|
| $\boldsymbol{\alpha}$ | Angular acceleration |
| $\alpha$ | Angle, magnitude of $\boldsymbol{\alpha}$ |
| $\alpha_A, \alpha_B, \ldots$ | Angular acceleration of bodies $A, B, \ldots$ |
| $\alpha_{avg}$ | Average angular acceleration |
| $\beta$ | Angle, direction of acceleration in plane curvilinear translation |
| $\beta_A, \beta_D$ | Direction of dynamic balance holes in elements $A$ and $D$, respectively |
| $\beta_a, \beta_b$ | Direction of dynamic unbalance forces $R_a$ and $R_b$, respectively |
| $\beta_f$ | Angle |
| $\delta, \delta_0$ | Spring deflection, initial spring deflection |

| | |
|---|---|
| $\Delta\theta$ | Included angle, change in angular displacement |
| $\Delta$ | Change in a quantity |
| $\Delta s$ | Arc length |
| $\Delta T$ | Change in kinetic energy $T$ |
| $\Delta V$ | Change in potential energy $V$ |
| $\Delta v_n, \Delta v_t$ | Normal and tangential components, respectively, of change in velocity |
| $\gamma$ | Specific weight |
| $\boldsymbol{\theta}$ | Angular displacement |
| $\theta$ | Angle, magnitude of $\boldsymbol{\theta}$ |
| $\theta_A, \theta_B$ | Angular positions of line on body in positions $A$ and $B$, respectively |
| $\theta_A, \theta_B, \ldots$ | Angular displacement of bodies $A, B, \ldots$ |
| $\theta_A, \theta_D$ | Direction of centrifugal force acting on mass elements $A$ and $D$, respectively |
| $\theta_1, \theta_2, \theta_3$ | Endpoints of angular motion, angles, angular displacement of gears 1, 2, and 3, respectively |
| $\theta_0$ | Initial angular displacement |
| $\dot{\theta}, \ddot{\theta}$ | Variable angular velocity and acceleration, respectively |
| $\rho, \rho_a, \rho_b$ | Density, radius of curvature, radius of curvature of points $a$ and $b$, respectively |
| $\rho_0$ | Mass density per unit length of homogenous rod material of constant cross section |
| $\boldsymbol{\omega}$ | Angular velocity |
| $\omega$ | Magnitude of $\boldsymbol{\omega}$ |
| $\omega'$ | Angular velocity after impact |
| $\omega_{avg}$ | Average angular velocity |
| $\omega_{bc}$ | Angular velocity of line $bc$ |
| $\omega_0$ | Initial angular velocity |
| $\omega_A, \omega_B$ | Angular velocity of bodies $A$ and $B$, respectively, before impact |
| $\omega'_A, \omega'_B$ | Angular velocity of bodies $A$ and $B$, respectively, after impact |
| $\omega_1, \omega_2, \omega_3$ | Angular velocity of gears 1, 2, and 3, respectively |
| $\boldsymbol{\omega}_1$ | Angular velocity of axis of shaft, or of direction of vector |
| $\omega_1$ | Magnitude of $\boldsymbol{\omega}_1$ |
| $\omega_C, \omega_D$ | Angular velocity of bodies $C$ and $D$, respectively |
| $\zeta$ | Angle, constant defined by $C_D \rho A / 2m$ |
| $\mu$ | Absolute viscosity, coefficient of friction |
| $\pi$ | 3.14159... |
| $\eta$ | Energy efficiency, constant defined by $c/m$ |
| $\xi$ | Constant defined by $p/m$ |
| $\Delta W$ | Energy dissipated by viscous forces |

## OTHER SYMBOLS

| | |
|---|---|
| $1, 2, \ldots$ | Endpoints of an interval of motion, elements |
| %D | Percent difference |

# Kinematics of Particles

## 14.1 RECTILINEAR MOTION, DISPLACEMENT, VELOCITY, AND ACCELERATION

**14.1** (*a*)  What is the definition of the term *kinematics*?

(*b*)  What is the difference between a *particle* and a *body*?

▮ (*a*)  Kinematics is the science which studies the motions of particles, or points, and bodies, *without concern for the forces which produce these motions*. A particle is a physical body which has mass and whose physical dimensions are assumed to be vanishingly small. The consequence of this assumption is that all rotation effects, of the body which is represented as a particle, about any axis through the body may be neglected. The particle may thus be thought of as a *point* in space.

(*b*)  If a physical element is assumed to not be a particle, it is referred to as a *body*. All the bodies considered in this text are assumed to be rigid. In a rigid body, the distance between any two points is always the same, no matter what type of force system acts on the body. Thus a *rigid body* may be viewed as a body with unchanging dimensions. It is left as an exercise for the reader to show why the assumption of rigidity has no meaning when applied to a particle.

**14.2**  What are the only two possible motions of a particle?

▮ Two general types of particle motion may be identified. The first type is referred to as *rectilinear translation*, and this type of motion is straight-line motion. In Fig. 14.2*a* a particle, shown by the dark circle, is constrained to move along the straight line *ab*. The motion of the particle in this case is described as rectilinear translation.

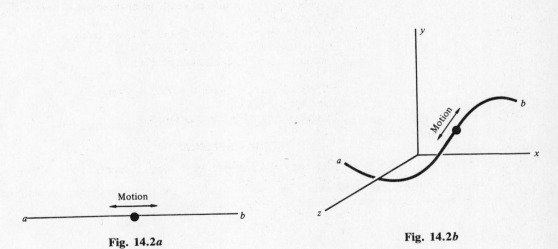

Fig. 14.2*a*

Fig. 14.2*b*

The second general type of particle motion is called *curvilinear translation*. In Fig. 14.2b, the particle is constrained to move along the curved path *ab*. For this case, the motion of the particle is described as curvilinear translation. It is emphasized that, in *both* of the above definitions, the term *translation* implies that no rotation, or angular motion, effects are required for the complete kinematic description of the motion of the particle. The reason for this, from the definition of a particle, is that all dimensions of the particle are effectively zero.

**14.3** (*a*)  What is the definition of the term *displacement*?

(*b*)  Show that displacement is a vector quantity.

▌ (*a*) Figure 14.3 shows a particle which is constrained to move along the straight line *ab*. The *displacement*, or position, of the particle is the location of the particle with respect to a fixed reference point. For the situation shown in Fig. 14.3, the fixed reference point is 0. The coordinate *s* describes the displacement of the particle with respect to this point. The positive sense of the coordinate *s* is the sense of increasing *s*, or to the right in the figure. If a value of *s* is positive, the particle is located to the right of the origin. If this coordinate has a negative value, the particle is located to the left of the origin.

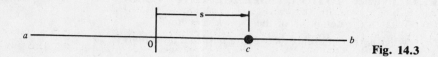

**Fig. 14.3**

(*b*) From the description in part (*a*), it follows that the displacement is a vector quantity. The *magnitude* of the displacement is the value of the *s* coordinate. The *direction* of the displacement is along the line *ab*. Finally, the *sense* of the displacement is determined by the sign of the value of *s*. There is a single, known direction in rectilinear translation problems.

**14.4** (*a*) What is the definition of the term *velocity*?

(*b*) What is the relationship between the positive senses of velocity and displacement?

(*c*) What is meant by the *speed* of a particle?

▌ (*a*) The velocity *v* of a particle is defined to be the first time derivative of the displacement. The velocity may be thought of as the time rate of change of displacement.

In Fig. 14.4, the particle moves from *c* to *c′* during the time interval $\Delta t$. The average velocity $v_{\text{avg}}$ is defined to be

$$v_{\text{avg}} = \frac{\Delta s}{\Delta t}$$

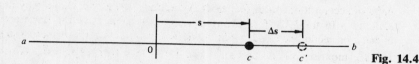

**Fig. 14.4**

The time interval $\Delta t$ is now allowed to decrease without limit. The formal definition of the velocity *v* of the particle is then

$$v = \lim_{\Delta t \to 0} \frac{\Delta s}{\Delta t} = \frac{ds}{dt}$$

Throughout this text, a dot over a quantity will indicate the first time derivative of this quantity, and two dots will represent a second time derivative. Thus,

$$v = \frac{ds}{dt} = \dot{s}$$

The units of velocity are length divided by time. Typical U.S. Customary System (USCS) units for velocity are feet per second or inches per second. In SI units, the velocity is expressed in meters per second.

The direction of the velocity is the limiting direction of $\Delta s$, as $\Delta s \to 0$. Thus, the velocity is a vector quantity.

(*b*) The term $\Delta s$, by definition, using Fig. 14.4, is

$$\Delta s = s_{c'} - s_c$$

Since $s_{c'} > s_c$, $\Delta s$ *is positive in the same sense as s*. $\Delta t$ is always positive, since time may only increase. From consideration of the above equation, it follows that *the velocity is positive in the same sense as the displacement coordinate*. Thus, the choice of a positive sense for the displacement automatically establishes the positive sense of the velocity.

If the velocity of the particle in Fig. 14.4 is positive, then the particle moves to the right. If this velocity is negative, the particle moves to the left. These conclusions are independent of the *positions* of the particle along line *ab*.

(c)  The term speed is a *scalar* quantity used to describe the *magnitude* of the velocity.

**14.5**  (a)  What is the definition of the term *acceleration*?

(b)  What is the relationship among the positive senses of acceleration, velocity, and displacement?

(c)  What is meant by the term *deceleration*?

(d)  Give the form for the average acceleration of a particle.

▮  (a)  Figure 14.5 shows the two velocities of the particle at the locations *c* and *c'*. The acceleration *a* of the particle is defined to be the first time derivative of the velocity. The acceleration may thus be thought of as describing the time rate of change of velocity. The formal definition of the acceleration of the particle is

$$a = \lim_{\Delta t \to 0} \frac{\Delta v}{\Delta t} = \frac{dv}{dt}$$

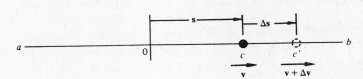

Fig. 14.5

The above results may be expanded further as

$$a = \frac{dv}{dt} = \frac{d}{dt}(v) = \frac{d}{dt}\left(\frac{ds}{dt}\right) = \frac{d^2s}{dt^2} = \ddot{s}$$

where two dots over a variable represent the second time derivative of the quantity. The fundamental units of acceleration, from the above equations, are velocity divided by time, or length divided by time squared. In the U.S. Customary System, typical acceleration units are feet per second squared or inches per second squared. The SI units for acceleration are meters per second squared.

(b)  From comparison of the forms for *a* and *v*, it may be seen that the positive senses of *a* and *v* are the same. As was shown in Prob. 14.4, the positive senses of *v* and *s* are the same. Thus, *the positive sense of the acceleration is always the same as the positive sense of the coordinate s*. It follows from the discussion that the choice of a positive sense for the displacement coordinate automatically establishes the *same* positive senses for the velocity and the acceleration.

(c)  The term deceleration is commonly used to describe an acceleration of negative sense.

(d)  The average value $a_{\text{avg}}$ of the acceleration in the region of motion between *c* and *c'* in Fig. 14.5 is

$$a_{\text{avg}} = \frac{\Delta v}{\Delta t}$$

**14.6**  A particle moves with rectilinear translation. From a prior calculation, the displacement of the particle is known to be

$$s = (100 - 4t^2)$$

where *t* is in seconds and *s* is in meters.

(a)  Find the general forms of the velocity and acceleration of the particle.

(b)  Find the values of the displacement, velocity, and acceleration when  *t* = 0. Show these results on the axis of motion.

(c)  Do the same as in part (b) for the time when the particle passes through the origin.

(d)  Find the value of the average velocity for the interval between  *t* = 0  and the time found in part (c).

(e)  Where is the particle when  *t* = 10 s? What are the values of the velocity and acceleration at this time? Sketch these results on the axis of motion.

**▮**  (a)   The general forms of the velocity and acceleration are

$$v = \dot{s} = \frac{ds}{dt} = \frac{d}{dt}(100 - 4t^2) = -8t \text{ m/s} \qquad a = \dot{v} = \ddot{s} = \frac{d}{dt}(\dot{s}) = \frac{d}{dt}(-8t) = -8 \text{ m/s}^2$$

(b)   When   $t = 0$,

$$s = 100 - 4(0) = 100 \text{ m} \qquad v = \dot{s} = -8(0) = 0 \qquad a = \dot{v} = \ddot{s} = -8 \text{ m/s}^2$$

The results are shown in Fig. 14.6a.

(c)   When the particle passes through the origin,   $s = 0$.   Using the equation for $s$, $v$, and $a$,

$$0 = 100 - 4t^2 \qquad t = 5 \text{ s} \qquad v = -8t = -8(5) = -40 \text{ m/s} \qquad a = -8 \text{ m/s}^2$$

These results are shown in Fig. 14.6b.

(d)   The average velocity between   $t = 0$   and   $t = 5$ s   is

$$v_{\text{avg}} = \frac{\Delta s}{\Delta t} = \frac{s|_{t=5} - s|_{t=0}}{5 - 0} = \frac{0 - 100}{5} = -20 \text{ m/s}$$

At the beginning of this interval the velocity is zero.   At the end of the interval, when   $t = 5$ s,   the velocity has a magnitude of 40 m/s.   It may be seen that the average value of the velocity is representative of *neither* of these two endpoints.   This example illustrates that *caution must be exercised in using average values to describe a problem*.   The average value of velocity will more closely approach the true value of the velocity as the time interval $\Delta t$ decreases and approaches zero.

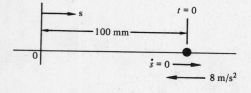

Fig. 14.6a

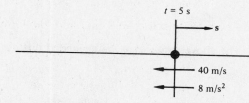

Fig. 14.6b

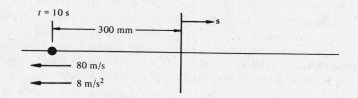

Fig. 14.6c

(e)   When   $t = 10$ s,

$$s = 100 - 4t^2 = 100 - 4(10)^2 = -300 \text{ m} \qquad v = -8(10) = -80 \text{ m/s} \qquad a = -8 \text{ m/s}^2$$

These values are shown in Fig. 14.6c.

**14.7**   A particle moves in rectilinear translation with the displacement-time function   $s = 1 - e^{-0.5t}$,   where $s$ is in meters and $t$ is in seconds.

(a)   Find the general forms of the velocity and acceleration of the particle.

(b)   Find the displacement, velocity, and acceleration when   $t = 0$   and when   $t = 4$ s.   Show these values on the axis of motion.

(c)   Find the average values of the velocity and acceleration in the time interval between   $t = 0$   and   $t = 4$ s.

**▮**  (a)   The velocity and acceleration have the forms

$$s = 1 - e^{-0.5t} \text{ m} \qquad v = \dot{s} = -(-0.5)e^{-0.5t} = 0.5e^{-0.5t} \text{ m/s}$$
$$a = \dot{v} = \ddot{s} = -0.5(0.5)e^{-0.5t} = -0.25e^{-0.5t} \text{ m/s}^2$$

(b) At $t = 0$,
$$s = 0 \qquad v = 0.5e^0 = 0.5 \text{ m/s} \qquad a = -0.25e^0 = -0.25 \text{ m/s}^2$$

At $t = 4 \text{ s}$,
$$s = 1 - e^{-0.5(4)} = 0.865 \text{ m} \qquad v = 0.5e^{-0.5(4)} = 0.0677 \text{ m/s} \qquad a = -0.25e^{-0.5(4)} = -0.0338 \text{ m/s}^2$$

The above results are shown in Fig. 14.7.

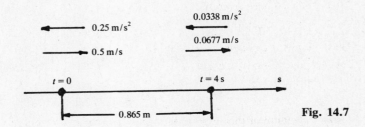

**Fig. 14.7**

(c) In the interval $0 \le t \le 4 \text{ s}$,
$$v_{\text{avg}} = \frac{\Delta s}{\Delta t} = \frac{0.865 - 0}{4} = 0.216 \text{ m/s} \qquad a_{\text{avg}} = \frac{\Delta v}{\Delta t} = \frac{0.0677 - 0.5}{4} = -0.108 \text{ m/s}^2$$

**14.8** Solve Prob. 14.7, if $s = 25 \sin \pi t/2$, where $s$ is in inches and $t$ is in seconds.

▌ (a) The velocity and acceleration have the forms
$$s = 25 \sin \frac{\pi t}{2} \text{ in} \qquad v = \dot{s} = 25\left(\frac{\pi}{2}\right) \cos \frac{\pi t}{2} = 39.3 \cos \frac{\pi t}{2} \text{ in/s}$$
$$a = \dot{v} = \ddot{s} = -25\left(\frac{\pi}{2}\right)\left(\frac{\pi}{2}\right) \sin \frac{\pi t}{2} = -61.7 \sin \frac{\pi t}{2} \text{ in/s}^2$$

(b) At $t = 0$,
$$s = 25 \sin 0 = 0 \qquad v = 39.3 \cos 0 = 39.3 \text{ in/s} \qquad a = -61.7 \sin 0 = 0$$

At $t = 4 \text{ s}$,
$$s = 25 \sin \frac{\pi(4)}{2} = 0 \qquad v = 39.3 \cos \frac{\pi(4)}{2} = 39.3 \text{ in/s} \qquad a = -61.7 \sin \frac{\pi(4)}{2} = 0$$

The above results are shown in Fig. 14.8.

It may be seen that the motion is periodic, and $t = 0 \text{ s}, 4 \text{ s}$ are the endpoints of one cycle.

(c) In the interval $t = 0$ to $t = 4 \text{ s}$,
$$v_{\text{avg}} = \frac{\Delta s}{\Delta t} = \frac{0 - 0}{4} = 0 \qquad a_{\text{avg}} = \frac{\Delta v}{\Delta t} = \frac{39.3 - 39.3}{4} = 0$$

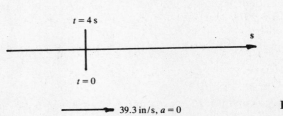

**Fig. 14.8**

**14.9** The displacement of a particle which moves with rectilinear translation is known to have the functional form $s = at^2 + bt + 8$, where $s$ is in feet and $t$ is in seconds, and $a$ and $b$ are constants to be chosen.

(a) Find the required values of $a$ and $b$, if $s=0$ and $\dot{s}=100$ ft/s when $t=2$ s, for the time interval $1.5 \le t \le 2.5$ s.

(b) What are the units of $a$ and $b$?

(c) Find the functional forms of the velocity and acceleration.

(d) Find the average value of the velocity in the time interval between 1.5 and 2.5 s. Would you consider this value to be representative of the velocity in this time interval?

(e) Are there any times in the interval $1.5 \le t \le 2.5$ s when the displacement and velocity are zero?

▌ (a)
$$s = at^2 + bt + 8 \text{ ft} \qquad (1)$$

Using $t=2$ s and $s=0$,
$$0 = a(2)^2 + b(2) + 8 \qquad 4a + 2b + 8 = 0 \qquad (2)$$

The velocity has the form
$$v = \dot{s} = 2at + b \text{ ft/s}$$

Using $t=2$ s and $v=\dot{s}=100$ ft/s,
$$100 = 2a(2) + b \qquad 4a + b - 100 = 0 \qquad (3)$$

Equation (3) is multiplied by $-1$, and added to Eq. (2), to obtain
$$b + 108 = 0 \qquad b = -108$$

Using Eq. (2),
$$4a + 2(-108) + 8 = 0 \qquad a = 52$$

(b) From Eq. (1), the term $at^2$ has the same units as $s$. Thus, $a = s/t^2 = $ ft/s$^2$. The term $bt$ also has the same units as $s$. Thus, $b = s/t = $ ft/s.

(c) The equations of motion are
$$s = 52t^2 - 108t + 8 \text{ ft} \qquad v = \dot{s} = 104t - 108 \text{ ft/s} \qquad a = \dot{v} = \ddot{s} = 104 \text{ ft/s}^2$$

(d) At $t=1.5$ s,
$$s = 52(1.5)^2 - 108(1.5) + 8 = -37 \text{ ft} \qquad v = 104(1.5) - 108 = 48 \text{ ft/s}$$

At $t=2.5$ s,
$$s = 52(2.5)^2 - 108(2.5) + 8 = 63 \text{ ft} \qquad v = 104(2.5) - 108 = 152 \text{ ft/s}$$

$$v_{avg} = \frac{\Delta s}{\Delta t} = \frac{s_{2.5} - s_{1.5}}{t_{2.5} - t_{1.5}} = \frac{63 - (-37)}{1} = 100 \text{ ft/s}$$

The average velocity is not representative of the actual velocity in the interval $t=1.5$ s to $t=2.5$ s.

(e)
$$s = 52t^2 - 108t + 8 = 0 \qquad t = \frac{108 \pm \sqrt{(-108)^2 - 4(52)8}}{2(52)} = \frac{108 \pm 100}{2(52)} = 0.0769 \text{ s}, \quad 2 \text{ s}$$

The time $t=2$ s verifies the problem statement.
$$v = 104t - 108 = 0 \qquad t = 1.04 \text{ s}$$

The displacement is zero when $t=2$ s, and the velocity is not equal to zero in the interval $t=1.5$ s to $t=2.5$ s.

**14.10** Do the same as in Prob. 14.9 if $s=0$ and $\dot{s}=0$ when $t=5$ s, for the time interval $4.5 \le t \le 5.5$ s. $s$ is in meters and $t$ is in seconds.

▌ (a)
$$s = at^2 + bt + 8 \text{ m} \qquad (1)$$

Using $t=5$ s and $s=0$,
$$0 = a(5)^2 + b(5) + 8 \qquad 25a + 5b + 8 = 0 \qquad (2)$$

The velocity has the form
$$v = \dot{s} = 2at + b \text{ m/s}$$

Using $t = 5\,\text{s}$ and $v = 0$

$$0 = 2a(5) + b \qquad 10a + b = 0 \qquad b = -10a \tag{3}$$

Using Eq. (3) in Eq. (2),

$$25a + 5(-10a) + 8 = 0 \qquad a = 0.32$$

Using Eq. (3),

$$b = -10(0.32) = -3.2$$

(b) Using the approach in part (b) of Prob. 14.9, $a$ has the units meters per second squared and $b$ has the units meters per second.

(c) The equations of motion are

$$s = 0.32t^2 - 3.2t + 8\,\text{m} \qquad v = \dot{s} = 0.32(2)t - 3.2 = 0.64t - 3.2\,\text{m/s} \qquad a = \dot{v} = \ddot{s} = 0.64\,\text{m/s}^2$$

(d) At $t = 4.5\,\text{s}$,

$$s = 0.32(4.5)^2 - 3.2(4.5) + 8 = 0.08\,\text{m} \qquad v = 0.64(4.5) - 3.2 = -0.32\,\text{m/s}$$

At $t = 5.5\,\text{s}$,

$$s = 0.32(5.5)^2 - 3.2(5.5) + 8 = 0.08\,\text{m} \qquad v = 0.64(5.5) - 3.2 = 0.32\,\text{m/s}$$

$$v_{\text{avg}} = \frac{\Delta s}{\Delta t} = \frac{s_{5.5} - s_{4.5}}{t_{5.5} - t_{4.5}} = \frac{0.08 - 0.08}{1} = 0$$

The average velocity is not representative of the actual velocity in the interval $t = 4.5\,\text{s}$ to $t = 5.5\,\text{s}$.

(e)
$$s = 0.32t^2 - 3.2t + 8 = 0 \qquad t = \frac{3.2 \pm \sqrt{(-3.2)^2 - 4(0.32)8}}{2(0.32)} = \frac{3.2 \pm 0}{2(0.32)} = 5\,\text{s}$$

$$v = 0.64t - 3.2 = 0 \qquad t = 5\,\text{s}$$

$s$ and $v$ are zero at $t = 5\,\text{s}$, which agrees with the problem statement.

**14.11** The acceleration of a particle which moves with rectilinear translation is given by $a = (t - 2)\,\text{m/s}^2$. At $t = 0$, the displacement and velocity are zero.

(a) Find the velocity and displacement when $t = 2\,\text{s}$ and when $t = 4\,\text{s}$.

(b) Show sketches of $s$, $v$, and $a$ for $0 \le t \le 4\,\text{s}$.

(c) Find the average values of the velocity and the acceleration in the intervals $0 \le t \le 2\,\text{s}$, $2 \le t \le 4\,\text{s}$, and $0 \le t \le 4\,\text{s}$, and discuss the results.

▌ (a) The velocity at $t = 2\,\text{s}$ is

$$v = \int a\,dt = \int_0^2 (t - 2)\,dt = \left(\frac{t^2}{2} - 2t\right)\Big|_0^2 = -2\,\text{m/s}$$

The displacement at $t = 2\,\text{s}$ is

$$s = \int v\,dt = \int_0^2 \left(\frac{t^2}{2} - 2t\right) dt = \left[\frac{t^3}{2(3)} - 2\left(\frac{t^2}{2}\right)\right]\Big|_0^2 = -2.67\,\text{m}$$

At $t = 4\,\text{s}$,

$$v = \int a\,dt = \int_0^4 (t - 2)\,dt = \left(\frac{t^2}{2} - 2t\right)\Big|_0^4 = 0$$

$$s = \int v\,dt = \int_0^4 \left(\frac{t^2}{2} - 2t\right) dt = \left[\frac{t^3}{2(3)} - 2\left(\frac{t^2}{2}\right)\right]\Big|_0^4 = -5.33\,\text{m}$$

(b) Figure 14.11 shows the functional forms of $s$, $v$, and $a$ in the time interval 0 to 4 s.

(c)
$$v_{\text{avg},0-2} = \frac{-2.67 - 0}{2} = -1.33\,\text{m/s} \qquad v_{\text{avg},2-4} = \frac{-5.33 - (-2.67)}{2} = -1.33\,\text{m/s}$$

$$v_{\text{avg},0-4} = \frac{-5.33 - 0}{4} = -1.33\,\text{m/s}$$

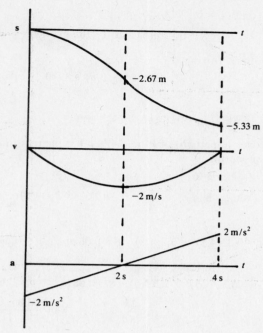

**Fig. 14.11**

$$a_{\text{avg},0-2} = \frac{-2-0}{2} = -1 \text{ m/s}^2 \qquad a_{\text{avg},2-4} = \frac{0-(-2)}{2} = 1 \text{ m/s}^2 \qquad a_{\text{avg},0-4} = \frac{0-0}{4} = 0$$

The range of the velocity is $-2 \text{ m/s} \le v \le 0$, and the range of the acceleration is $-2 \text{ m/s}^2 \le a \le 2 \text{ m/s}^2$. None of the above average values of velocity and acceleration would be considered to be good representations of the actual values.

**14.12** Solve Prob. 14.11, if $a = (4-t) \text{ ft/s}^2$.

❚ (a)  At $t = 2 \text{ s}$,

$$v = \int a \, dt = \int_0^2 (4-t) \, dt = \left(4t - \frac{t^2}{2}\right)\Big|_0^2 = 6 \text{ ft/s}$$

$$s = \int v \, dt = \int_0^2 \left(4t - \frac{t^2}{2}\right) dt = \left[4\frac{t^2}{2} - \frac{t^3}{2(3)}\right]\Big|_0^2 = 6.67 \text{ ft}$$

At $t = 4 \text{ s}$,

$$v = \int a \, dt = \int_0^4 (4-t) \, dt = \left(4t - \frac{t^2}{2}\right)\Big|_0^4 = 8 \text{ ft/s}$$

$$s = \int v \, dt = \int_0^4 \left(4t - \frac{t^2}{2}\right) dt = \left[4\frac{t^2}{2} - \frac{t^3}{2(3)}\right]\Big|_0^4 = 21.3 \text{ ft}$$

(b)  The functional forms of $s$, $v$, and $a$ are seen in Fig. 14.12.

(c)  $v_{\text{avg},0-2} = \dfrac{6.67-0}{2} = 3.34 \text{ ft/s} \qquad v_{\text{avg},2-4} = \dfrac{21.3-6.67}{2} = 7.32 \text{ ft/s} \qquad v_{\text{avg},0-4} = \dfrac{21.3-0}{4} = 5.33 \text{ ft/s}$

$a_{\text{avg},0-2} = \dfrac{6-0}{2} = 3 \text{ ft/s}^2 \qquad a_{\text{avg},2-4} = \dfrac{8-6}{2} = 1 \text{ ft/s}^2 \qquad a_{\text{avg},0-4} = \dfrac{8-0}{4} = 2 \text{ ft/s}^2$

In this problem,

$$0 \le v \le 8 \text{ ft/s} \qquad \text{and} \qquad 0 \le a \le 4 \text{ ft/s}^2$$

None of the above values of average velocity and acceleration are good representations of the actual values.

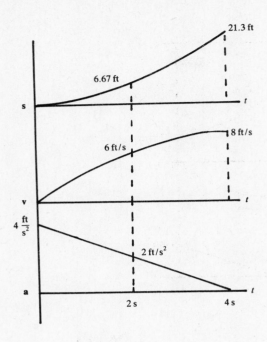

**Fig. 14.12**

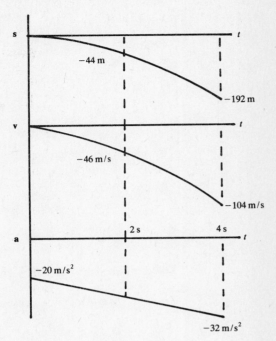

**Fig. 14.13**

**14.13**  Solve Prob. 14.11, if   $a = (-3t - 20) \text{ m/s}^2$.

▌ (*a*)  At   $t = 2 \text{ s}$,

$$v = \int a\, dt = \int_0^2 (-3t - 20)\, dt = \left(-\frac{3t^2}{2} - 20t\right)\bigg|_0^2 = -46 \text{ m/s}$$

$$s = \int v\, dt = \int_0^2 \left(-\frac{3t^2}{2} - 20t\right) dt = \left[-\frac{3t^3}{2(3)} - 20\,\frac{t^2}{2}\right]\bigg|_0^2 = -44 \text{ m}$$

At   $t = 4 \text{ s}$,

$$v = \int a\, dt = \int_0^4 (-3t - 20)\, dt = \left(-\frac{3t^2}{2} - 20t\right)\bigg|_0^4 = -104 \text{ m/s}$$

$$s = \int v\, dt = \int_0^4 \left(-\frac{3t^2}{2} - 20t\right) dt = \left[-\frac{3t^3}{2(3)} - 20\,\frac{t^2}{2}\right]\bigg|_0^4 = -192 \text{ m}$$

(*b*)  The plots of $s$, $v$, and $a$ are in Fig. 14.13.

(*c*)    $$v_{\text{avg},0-2} = \frac{-44 - 0}{2} = -22 \text{ m/s} \qquad v_{\text{avg},2-4} = \frac{-192 - (-44)}{2} = -74 \text{ m/s}$$

$$v_{\text{avg},0-4} = \frac{-192 - 0}{4} = -48 \text{ m/s}$$

$$a_{\text{avg},0-2} = \frac{-46 - 0}{2} = -23 \text{ m/s} \qquad a_{\text{avg},2-4} = \frac{-104 - (-46)}{2} = -29 \text{ m/s}$$

$$a_{\text{avg},0-4} = \frac{-104 - 0}{4} = -26 \text{ m/s}$$

From consideration of Fig. 14.13, it may be concluded that the average velocity values are a poor representation of the actual values, while the average acceleration values are a fair representation of the actual values.

**14.14** Solve Prob. 14.11, if

$$a = -4.8t \text{ in/s}^2 \quad 0 \le t \le 2 \text{ s} \qquad a = -9.6 \text{ in/s}^2 \quad 2 \le t \le 4 \text{ s}$$

▌ (*a*)  At  $t = 2$ s,

$$v = \int a \, dt = \int_0^2 (-4.8t) \, dt = -4.8 \left.\frac{t^2}{2}\right|_0^2 = -9.6 \text{ in/s}$$

$$s = \int v \, dt = \int_0^2 \left(-4.8 \frac{t^2}{2}\right) dt = -\left.\frac{4.8t^3}{2(3)}\right|_0^2 = -6.4 \text{ in}$$

The velocity in the interval  $t = 2$ s  to  $t = 4$ s  is found as

$$v = \int a \, dt = \int_0^2 -4.8t \, dt + \int_2^t -9.6 \, dt = -4.8 \left.\frac{t^2}{2}\right|_0^2 - 9.6t \Big|_2^t = -\frac{4.8}{2}(2)^2 - 9.6(t-2)$$

$$= -9.6 - 9.6(t-2) = -9.6(t-1)$$

At  $t = 4$ s,

$$v = -9.6(4-1) = -28.8 \text{ in/s}$$

The displacement in the above interval is

$$s = \int v \, dt = \int_0^2 v \, dt + \int_2^4 v \, dt = -6.4 + \int_2^4 -9.6(t-1) \, dt = -6.4 - 9.6 \left.\frac{(t-1)^2}{2}\right|_2^4$$

$$= -6.4 - \frac{9.6}{2}(3^2 - 1^2) = -44.8 \text{ in}$$

(*b*)  Figure 14.14 shows the plots of *s*, *v*, and *a*.

(*c*)
$$v_{\text{avg},0-2} = \frac{-6.4 - 0}{2} = -3.2 \text{ in/s} \qquad v_{\text{avg},2-4} = \frac{-44.8 - (-6.4)}{2} = -19.2 \text{ in/s}$$

$$v_{\text{avg},0-4} = \frac{-44.8 - 0}{4} = -11.2 \text{ in/s}$$

$$a_{\text{avg},0-2} = \frac{-9.6 - 0}{2} = -4.8 \text{ in/s}^2 \qquad a_{\text{avg},2-4} = \frac{-28.8 - (-9.6)}{2} = -9.6 \text{ in/s}^2$$

$$a_{\text{avg},0-4} = \frac{-28.8 - 0}{4} = -7.2 \text{ in/s}^2$$

None of the above average values are accurate representations of the actual values.

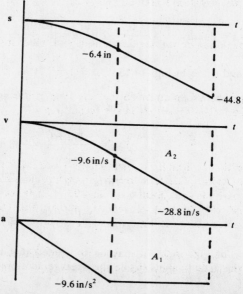

Fig. 14.14

**14.15** What is the graphical interpretation of the definition of velocity as the first time derivative of displacement?

**❚** The fundamental definition of the velocity in rectilinear motion is

$$v = \frac{ds}{dt}$$

It follows from the above equation that the *velocity is equal to the slope of the displacement-time curve.* The above equation may be written in the form

$$ds = v\,dt \qquad \int_{s_1}^{s_2} ds = \int_{t_1}^{t_2} v\,dt \qquad s_2 - s_1 = \int_{t_1}^{t_2} v\,dt$$

where 1 and 2 represent the endpoints of a particular interval of interest along the axis of motion. The term on the right side of the above equation may be recognized as the area under the velocity curve between 1 and 2. Furthermore, this area is equal to the *change*, $s_2 - s_1$, in the *displacement s* over this interval. The graphical interpretation of these results is shown in Fig. 14.15.

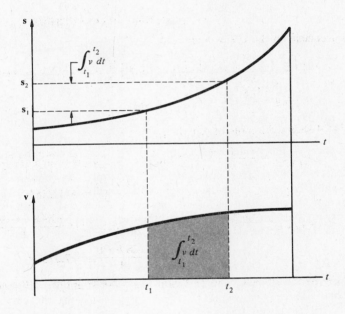

**Fig. 14.15**

**14.16** (*a*) What is the graphical interpretation of the definition of acceleration as the first time derivative of velocity?

(*b*) What is meant by the term *motion diagram*?

**❚** (*a*) The fundamental definition of the acceleration in rectilinear motion is

$$a = \frac{dv}{dt}$$

It may be seen from the above equation that the *acceleration is equal to the slope of the velocity-time curve.* The equation may be written in the form

$$dv = a\,dt \qquad \int_{v_1}^{v_2} dv = \int_{t_1}^{t_2} a\,dt \qquad v_2 - v_1 = \int_{t_1}^{t_2} a\,dt$$

The limits of integration indicated by the numbers 1 and 2 are the endpoints of the interval of interest along the axis of motion. The term on the right side of the above equation is the area under the acceleration-time curve between locations 1 and 2. It may thus be concluded that the area under the acceleration-vs.-time curve between two points 1 and 2 is equal to the change, $v_2 - v_1$, in the velocity of the particle as it moves from the first to the second point. The above effect is shown graphically in Fig. 14.16.

(*b*) The relationships among displacement, velocity, acceleration, and time, for a particle in rectilinear translation, may be plotted to show the variations in these quantities. These curves are referred to as *motion diagrams*.

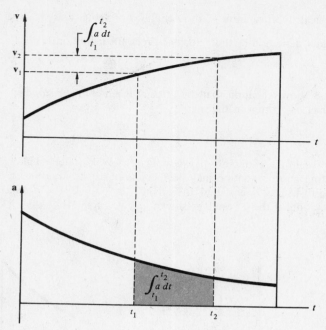

**Fig. 14.16**

**14.17**  (a)  Show how to identify positive and negative slopes, and increasing and decreasing slopes, in motion diagrams.

(b)  Show a summary of the results in part (a).

▌ (a)  Figure 14.17a shows a plane curve referenced to a set of xy coordinates.   The slope of the curve at a given point is found by using the tangent of a typical triangle, such as triangle abc in the figure.   In order to move from point a to point b, one could follow path ac and then path cb.   For this case both $\Delta x$ and $\Delta y$ are positive, since the sense of the travel along both legs of the triangle is in the two positive coordinate senses.   The slope of the curve shown in Fig. 14.17b has the form

$$\frac{dy}{dx} = \lim_{\Delta x \to 0} \frac{\Delta y}{\Delta x} = \frac{+}{+} = +$$

It follows that the slopes of the curves in Figs. 14.17a and b are *positive*.   It may also be seen that these slopes *decrease in magnitude* as one advances along the curve in the positive x coordinate sense.   If the curve were traversed from b to a, along the two legs of the triangle, then both $\Delta x$ and $\Delta y$ would be negative, since the travel would be in the negative coordinate senses.   The above equation would then be the quotient of two negative values; and it would again be concluded that the slope is positive.

Figure 14.17c shows another shape of curve which has a positive slope.   In this case, however, it may be concluded that *the magnitude of the slope increases* as one moves along the curve in the positive x coordinate sense.

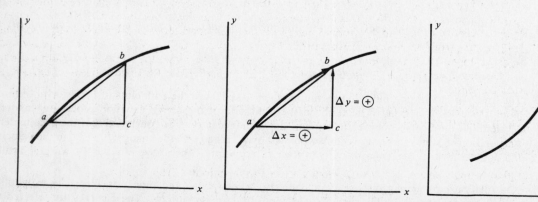

**Fig. 14.17a**　　　　　**Fig. 14.17b**　　　　　**Fig. 14.17c**

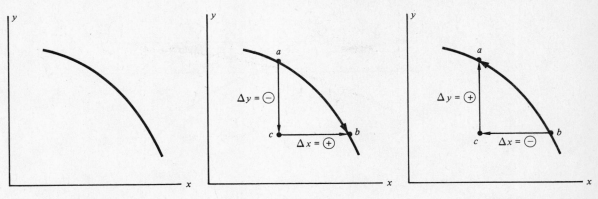

Fig. 14.17d             Fig. 14.17e             Fig. 14.17f

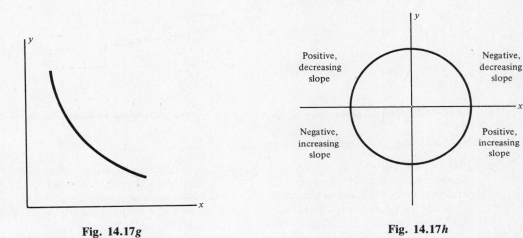

Fig. 14.17g                         Fig. 14.17h

Figures 14.17d, e, and f show a curve and orientation which are unlike those of the previous two curves. If the curve is traversed in the sense a to b, the slope has the form

$$\frac{dy}{dx} = \lim_{\Delta x \to 0} \frac{\Delta y}{\Delta x} = \frac{-}{+} = -$$

The alternative sense of movement along the curve is from b to a. For this case,

$$\frac{dy}{dx} = \lim_{\Delta x \to 0} \frac{\Delta y}{\Delta x} = \frac{+}{-} = -$$

It follows that the slope of the curve shown in Figs. 14.17d, e, and f is negative. It may also be seen from this figure that the *magnitude* of the slope increases as one moves along the curve in the positive x coordinate sense. A rigorous mathematical description says that *the slope of the curve is decreasing*, since the change in the slope is from one negative quantity to a negative quantity of larger magnitude.

A fourth possible curve is shown in Fig. 14.17g, and it may be concluded that this curve has a negative slope. As one advances in the positive x coordinate sense, the value of the slope goes from one negative quantity to a negative quantity of smaller magnitude. Such a change in a negative slope is an *increase*.

(b) The results in part (a) may be conveniently summarized in the diagram shown in Fig. 14.17h. The four quadrants of the circle represent the four general shapes of a plane curve.

**14.18** Using graphical methods, find the solution to parts (c) and (e) of Prob. 14.6.

❚ The displacement, velocity, and acceleration functions were found in Prob. 14.6 as

$$s = (100 - 4t^2)\text{ m} \qquad v = -8t\text{ m/s} \qquad a = -8\text{ m/s}^2$$

These three functions are plotted in Fig. 14.18.

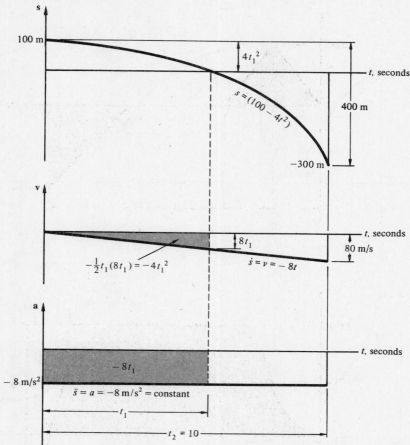

**Fig. 14.18**

When the particle passes through the origin, $s = 0$ and the corresponding time is designated $t_1$. A dashed line, corresponding to $t = t_1$, is drawn through the motion diagrams of Fig. 14.18. The area under the acceleration curve, over the interval 0 to $t_1$, is $-8t_1$. This area is equal to the change in the velocity over the same interval. The area under the velocity diagram over the same interval is $-4t_1^2$, as shown in the figure. When this area is added to the displacement value at $t = 0$, the result is zero, so that

$$100 - 4t_1^2 = 0 \qquad t_1 = 5 \text{ s}$$

For the interval from 0 to 10 s, the area under the acceleration curve is $-8(10) = -80$ m/s. This value is equal to the change in velocity over the same interval. The area under the triangular velocity diagram is then $\frac{1}{2}(10)(-80) = -400$ m. When this value is added to the initial displacement value of 100 m, the final result for the displacement at 10 s is $-300$ m.

**14.19** An experimental vehicle moves along a straight track with the velocity diagram shown in Fig. 14.19a.

(a) Sketch the acceleration and displacement diagrams.

(b) Discuss the location of the vehicle when $6 \text{ s} \leq t \leq 12 \text{ s}$,

(c) Find the position and deceleration of the vehicle when $t = 17$ s.

▮ (a) The velocity diagram is redrawn in Fig. 14.19b. The acceleration diagram is constructed first. Since the velocity diagram consists of straight-line elements, the corresponding acceleration values must be constant over the intervals. The slope of the velocity diagram in the first interval is $60/3 = 20$ in/s², and this is the value of the acceleration. In the second interval, the slope of the velocity diagram is $-60/3 = -20$ in/s². The minus sign is written to recognize that this portion of the velocity diagram has a negative slope. In the last portion of the velocity diagram the slope, which is the acceleration, is $-80/5 = -16$ in/s². The areas under the velocity diagram are computed next, and they are used to sketch the displacement curve.

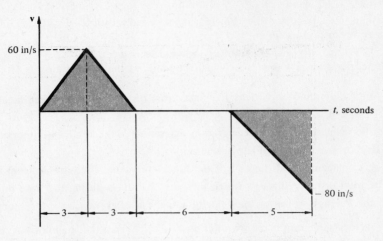

**Fig. 14.19a**

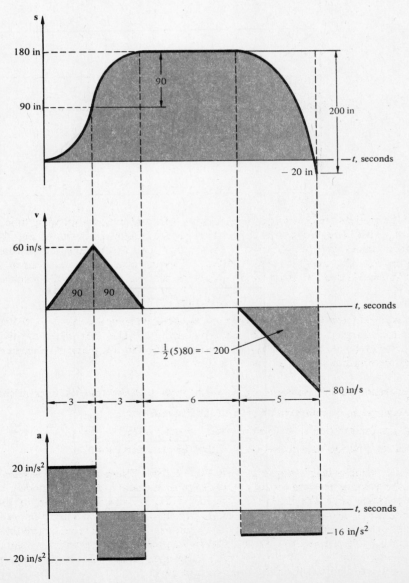

**Fig. 14.19b**

(b) During $6\,\text{s} \le t \le 12\,\text{s}$, the vehicle is at the fixed location $s = 180\,\text{in}$. The velocity and acceleration are zero during this time.

(c) At $t = 17\,\text{s}$, the displacement of the vehicle is given by $s = -20\,\text{in}$. The corresponding value of the acceleration is $-16\,\text{in/s}^2$. Thus, the value of the deceleration is $16\,\text{in/s}^2$.

**14.20** The velocity-time diagram of a particle that starts from rest and moves in rectilinear translation is shown in Fig. 14.20a. All times are in seconds, and $s = 0$ at $t = 0$.

(a) Sketch the displacement and acceleration diagrams for the total time interval shown in the figure.

(b) Find the extreme value(s) of the acceleration.

(c) Find the displacement, at the end of the time interval, of the particle from its starting position.

(d) Find the total distance through which the particle travels during the time interval.

(e) Sketch the average values of the velocity and the acceleration on the respective motion diagrams.

▌ (a) The velocity diagram is redrawn in Fig. 14.20b. From the shape of this diagram, it may be concluded that the acceleration has a constant value over each discrete time interval.

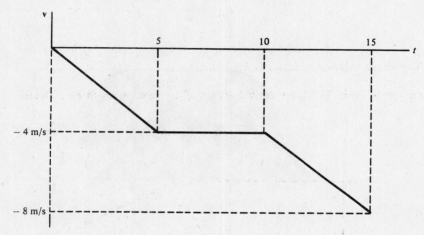

Fig. 14.20a

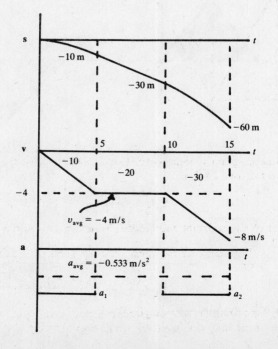

Fig. 14.20b

The values of the accelerations $a_1$ and $a_2$ in Fig. 14.20$b$ are found as

$$a_1 = \frac{\Delta v}{\Delta t} = \frac{-4 - 0}{5 - 0} = -0.8 \text{ m/s}^2 \qquad a_2 = \frac{\Delta v}{\Delta t} = \frac{-8 - (-4)}{15 - 10} = -0.8 \text{ m/s}^2$$

The above values are then used to construct the acceleration diagram in Fig. 14.20$b$. The areas under each portion of the velocity diagram, which are seen in Fig. 14.20$b$, are computed. These values are next used to find the change in displacement over each interval. The final form of the displacement-time curve is shown in Fig. 14.20$b$.

(**b**) From the results in part ($a$), or from inspection of Fig. 14.20$b$,

$$a_{\max} = -0.8 \text{ m/s}^2$$

(**c**) Using Fig. 14.20$b$, at $t = 15$ s, $s = -60$ m.

(**d**) Since the velocity has the same sense over the time interval of interest, the total distance through which the particle travels is the same as the magnitude of the displacement at the end of the interval, with the value 60 m.

(**e**) The average values of the velocity and acceleration over the 15-s time interval are

$$v_{\text{avg}} = \frac{\Delta s}{\Delta t} = \frac{-60 - 0}{15} = -4 \text{ m/s}$$

$$a_{\text{avg}} = \frac{\Delta s}{\Delta t} = \frac{-8 - 0}{15} = -0.533 \text{ m/s}^2$$

The above values are shown in Fig. 14.20$b$.

**14.21**   Do the same as in Prob. 14.20, for the velocity-time diagram shown in Fig. 14.21$a$.

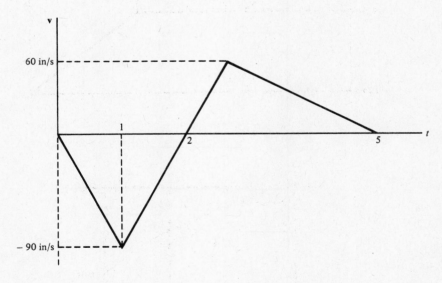

**Fig. 14.21$a$**

▌ (**a**)   The velocity-time diagram is redrawn in Fig. 14.21$b$. Using the similar triangles $A_1$ and $A_2$ in this figure, the time $t$ is found from

$$\frac{90}{2 - 1} = \frac{60}{t - 2} \qquad t = 2.67 \text{ s}$$

The magnitudes of the constant acceleration values $a_1$, $a_2$, and $a_3$ are found as

$$a_1 = \frac{\Delta v}{\Delta t} = \frac{-90 - 0}{1} = -90 \text{ in/s}^2 \qquad a_2 = \frac{\Delta v}{\Delta t} = \frac{60 - (-90)}{2.67 - 1} = 89.8 \approx 90 \text{ in/s}^2$$

$$a_3 = \frac{\Delta v}{\Delta t} = \frac{60 - 0}{5 - 2.67} = 25.8 \text{ in/s}^2$$

The areas under the velocity-time diagrams are next computed. These values are used to construct the displacement-time diagram, with the result shown in Fig. 14.21$b$.

(*b*)  The maximum value of the acceleration is given by

$$a_{max} = 90 \text{ in/s}^2$$

(*c*)  At  $t = 5 \text{ s}$,  $s = 0$.

(*d*)  The particle starts from zero displacement, moves 90 in, in the negative sense, and returns to zero displacement.  The total distance traveled is  $2(90) = 180 \text{ in}$.  It is emphasized that, for this problem, the total distance through which the particle travels during the time interval is *not* equal to the displacement at the end of the time interval.

(*e*)  The average values of the velocity and acceleration over the 5-s time interval are

$$v_{avg} = \frac{\Delta s}{\Delta t} = \frac{0-0}{5-0} = 0 \qquad a_{avg} = \frac{\Delta v}{\Delta t} = \frac{0-0}{5-0} = 0$$

The above results are shown in Fig. 14.21*b*.

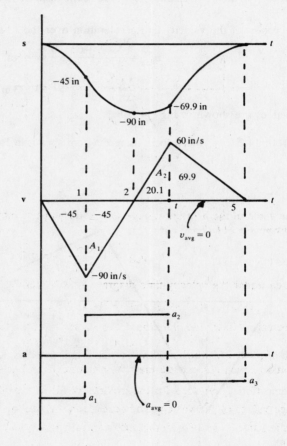

Fig. 14.21*b*

**14.22**  The velocity diagram of a particle that moves in rectilinear translation is shown in Fig. 14.22*a*.

(*a*)  Find the required value of the time $t_1$ if the final displacement of the particle from its starting position is 80 in.

(*b*)  Find the extreme value of the acceleration of the particle.

(*c*)  Do the same as in parts (*a*) and (*b*), if the final displacement of the particle is zero.

▌ (*a*)  The velocity-time diagram is redrawn in Fig. 14.22*b*.  Using the similar triangles $A_1$ and $A_2$,

$$\frac{60}{6-3} = \frac{20}{t-6} \qquad t = 7 \text{ s}$$

The initial displacement is chosen to be zero.  Since the final displacement must be 80 in, the total area under the velocity-time diagram must be 80 in.  Using Fig. 14.22*b*, this area has the value

$$\tfrac{1}{2}(3)(-60) + \tfrac{1}{2}(3)(-60) + \tfrac{1}{2}(1)20 + t_1(20) + \tfrac{1}{2}(1)20 = 80 \qquad t_1 = 12 \text{ s}$$

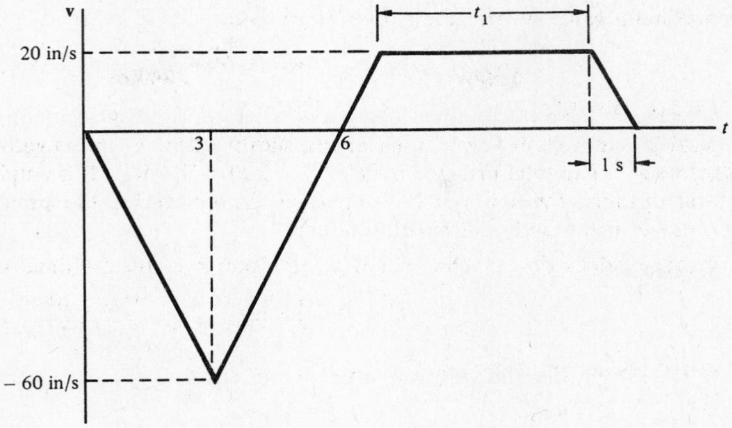

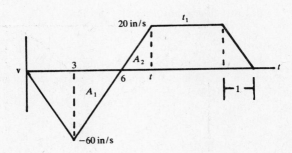

Fig. 14.22a

Fig. 14.22b

(b) The extreme value of the acceleration is equal to the slope of maximum magnitude in the velocity-time diagram. From Fig. 14.22b,

$$a_{max} = \pm \frac{60 - 2}{3 - 0} = \pm 20 \text{ in/s}^2$$

(c) The total area under the velocity-time diagram must be zero. This area has the value

$$\tfrac{1}{2}(3)(-60) + \tfrac{1}{2}(3)(-60) + \tfrac{1}{2}(1)20 + t_1(20) + \tfrac{1}{2}(1)20 = 0 \qquad t_1 = 8 \text{ s}$$

The extreme values of the acceleration are the same as those found in part (b).

14.23 The velocity diagram of a particle which moves in rectilinear translation is shown in Fig. 14.23. The maximum value of the acceleration which the particle may be subjected to is $0.8 \text{ m/s}^2$.

(a) Find the minimum time $t_1$ required for the particle to move through a distance of 3 m.

(b) Find the average values of the velocity and acceleration during the time interval $0 \leq t \leq t_1$.

(c) Do the same as in parts (a) and (b), if the maximum allowable value of the acceleration is $1.3 \text{ m/s}^2$ and the distance is 3.8 m.

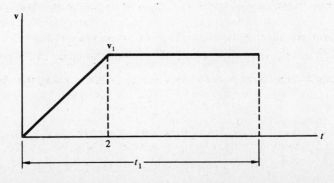

Fig. 14.23

(a)  Using  $a_{max} = 0.8\,\text{m/s}^2$,  the maximum value of $v_1$ is found from

$$a_{max} = \frac{\Delta v}{\Delta t} = \frac{v_1}{2} = 0.8 \qquad v_1 = 1.6\,\text{m/s}$$

The area under the velocity-time diagram is equal to the displacement, so that

$$\tfrac{1}{2}(2)1.6 + (t_1 - 2)1.6 = 3 \qquad t_1 = 2.88\,\text{s}$$

(b) $$v_{avg} = \frac{\Delta s}{\Delta t} = \frac{3 - 0}{2.88 - 0} = 1.04\,\text{m/s} \qquad a_{avg} = \frac{\Delta v}{\Delta t} = \frac{1.6 - 0}{2.88 - 0} = 0.556\,\text{m/s}^2$$

(c)  $v_1$ is found from

$$a_{max} = \frac{\Delta v}{\Delta t} = \frac{v_1}{2} = 1.3 \qquad v_1 = 2.6\,\text{m/s}$$

Setting the area under the velocity-time diagram equal to the displacement results in

$$\tfrac{1}{2}(2)2.6 + (t_1 - 2)2.6 = 3.8 \qquad t_1 = 2.46\,\text{s}$$

The average values of velocity and acceleration are

$$v_{avg} = \frac{\Delta s}{\Delta t} = \frac{3.8 - 0}{2.46} = 1.54\,\text{m/s} \qquad a_{avg} = \frac{\Delta v}{\Delta t} = \frac{2.6 - 0}{2.46} = 1.06\,\text{m/s}^2$$

**14.24**  A vehicle starts from zero velocity and moves along a straight, horizontal roadway.  The rectilinear acceleration of the vehicle is shown in Fig. 14.24a.

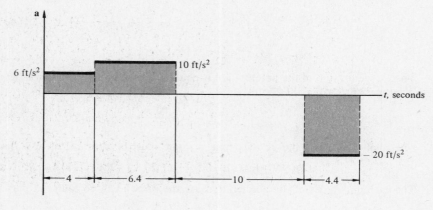

Fig. 14.24a

(a)  Using graphical methods, construct the velocity and displacement diagrams for the motion of the vehicle.

(b)  What is the maximum value of the speed attained by the vehicle?

(c)  How far has the vehicle traveled by the time it comes to rest?

(d)  Find the average velocity of the vehicle over the entire interval of motion.

▌ (a)  The acceleration diagram is redrawn in Fig. 14.24b.  Since this diagram consists of constant values of acceleration, the velocity diagram must consist of curves of constant slope, or straight-line elements. The areas under the three distinct portions of the acceleration diagram are shown in the figure.  These values, then, are the changes in velocity over the corresponding time intervals.

The velocity diagram construction, if one observes that the vehicle starts at zero initial velocity, is shown in Fig. 14.24b.  Since the values of acceleration are positive constants in the first two intervals, the corresponding velocity diagram has positive slopes in these intervals.  It also follows that the slope of the velocity diagram in the second interval is steeper than that in the first interval.  Since the last acceleration value is a negative constant, the corresponding velocity diagram for this interval is a straight line with a negative slope.

The areas under the velocity diagram are computed next, and they are the changes in the displacements over the corresponding intervals.  The slopes and curvature of the displacement curve are determined from the magnitudes of the velocities and from whether these magnitudes are increasing, decreasing, or constant.

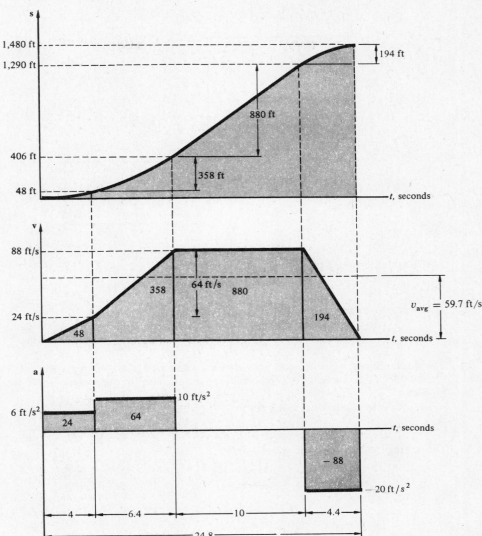

Fig. 14.24b

(b) The maximum value of the speed attained by the vehicle is 88 ft/s.

(c) The total distance traveled by the vehicle is 1,480 ft.

(d) The average velocity $v_{avg}$ of the vehicle over the entire interval of motion is a constant value shown as the dashed line in the velocity diagram of Fig. 14.24b.

$$v_{avg} = \frac{\Delta s}{\Delta t} = \frac{1,480}{24.8} = 59.7 \text{ ft/s}$$

**14.25** A particle starts with zero displacement and moves in rectilinear translation with the acceleration-time diagrams shown in Fig. 14.25a, and with initial velocity $v_0 = 0$. All times are in seconds.

(a) Sketch the velocity and displacement diagrams, and identify the extreme values of velocity and displacement.

(b) Sketch the average velocity and acceleration on the respective motion diagrams.

(c) Solve parts (a) and (b) if the initial velocity $v_0 = 10$ m/s.

(d) Solve parts (a) and (b) if the initial velocity $v_0 = -15$ m/s.

▮ (a) Figure 14.25b shows the areas under the acceleration diagram. These areas are used to construct the velocity diagram, with lines of constant slope. The areas under the velocity diagram are next found, and used to construct the displacement-time diagram shown in Fig. 14.25b.

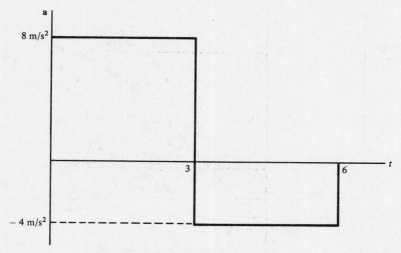

Fig. 14.25a

The extreme values of velocity and displacement are

$$v_{\max} = 24 \text{ m/s} \qquad s_{\max} = 90 \text{ m}$$

(b)  The average values of velocity and acceleration are

$$v_{\text{avg}} = \frac{\Delta s}{\Delta t} = \frac{90 - 0}{6 - 0} = 15 \text{ m/s} \qquad a_{\text{avg}} = \frac{\Delta v}{\Delta t} = \frac{12 - 0}{6 - 0} = 2 \text{ m/s}^2$$

(c)  The displacement, velocity, and acceleration diagrams, for an initial velocity of 10 m/s, are shown in Fig. 14.25c.  The steps used to construct these diagrams are the same as those used in part (a) of this problem.

The extreme values of $v$ and $s$ are

$$v_{\max} = 34 \text{ m/s} \qquad s_{\max} = 150 \text{ m}$$

The average values of $v$ and $a$ are

$$v_{\text{avg}} = \frac{\Delta s}{\Delta t} = \frac{150 - 0}{6} = 25 \text{ m/s} \qquad a_{\text{avg}} = \frac{\Delta v}{\Delta t} = \frac{22 - 10}{6} = 2 \text{ m/s}^2$$

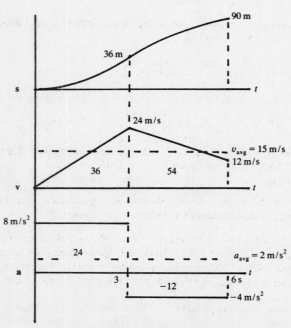

Fig. 14.25b

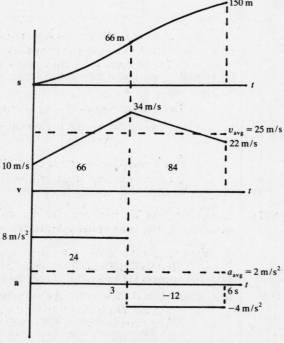

Fig. 14.25c

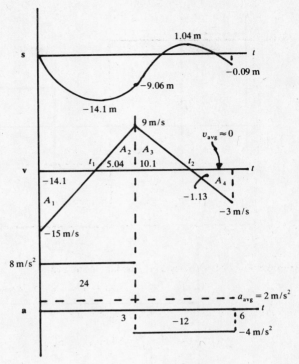

**Fig. 14.25d**

(*d*)  Figure 14.25*d* shows the acceleration and velocity diagrams for the case of the initial velocity equal to −15 m/s.  Using the similar triangles $A_1$ and $A_2$, in the figure,

$$\frac{15}{t_1} = \frac{9}{3 - t_1} \qquad t_1 = 1.88 \text{ s}$$

Using the similar triangles $A_3$ and $A_4$,

$$\frac{9}{t_2 - 3} = \frac{3}{6 - t_2} \qquad t_2 = 5.25 \text{ s}$$

The areas under the velocity diagram, which are shown in the figure, are calculated, and these values are used to construct the displacement diagram.

The extreme values of $v$ and $s$ are

$$v_{\max} = -15 \text{ m/s} \qquad s_{\max} = -14.1 \text{ m}$$

The average values of $v$ and $s$ are

$$v_{\text{avg}} = \frac{\Delta s}{\Delta t} = \frac{-0.09 - 0}{6 - 0} = -0.015 \approx 0 \qquad a_{\text{avg}} = \frac{\Delta v}{\Delta t} = \frac{-3 - (-15)}{6 - 0} = 2 \text{ m/s}^2$$

From comparison of Figs. 14.25*b* or *c* with Fig. 14.25*d*, it may be seen that the shape of the displacement-time diagram is strongly influenced by the values of the initial velocity.

**14.26**  Solve Prob. 14.25, part (*a*), for the acceleration-time relationship shown in Fig. 14.26*a*, for (*a*) initial velocity $v_0 = 0$,  (*b*) initial velocity $v_0 = 1.8$ m/s,  and (*c*) initial velocity $v_0 = -3$ m/s.

▎(*a*)  Fig. 14.26*b* shows the areas under the acceleration diagram.  Since the acceleration values are constants, the velocity diagram has straight lines of constant slope, and these slopes are equal to the magnitudes of the respective acceleration values.  The areas under the velocity diagram are found next, and these values are equated to the changes in displacement.  The final forms of the velocity-time, and displacement-time, diagrams are seen in Fig. 14.26*b*.

The extreme values of the velocity and displacement are

$$v_{\max} = 3 \text{ m/s} \qquad s_{\max} = 5.6 \text{ m}$$

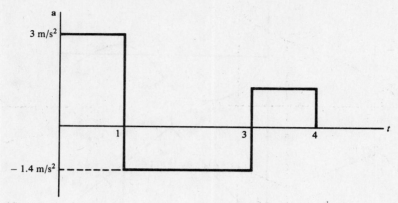

Fig. 14.26a

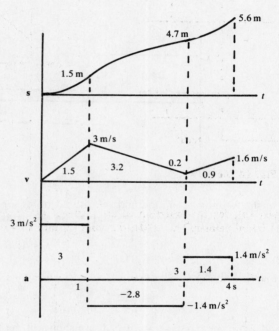

Fig. 14.26b

The average values of the velocity and acceleration are

$$v_{avg} = \frac{\Delta s}{\Delta t} = \frac{5.6 - 0}{4} = 1.4 \text{ m/s} \qquad a_{avg} = \frac{\Delta v}{\Delta t} = \frac{1.6 - 0}{4} = 0.4 \text{ m/s}^2$$

(b) The plots of displacement and velocity, for an initial velocity of 1.8 m/s, are found by using the techniques in part (a). The results are shown in Fig. 14.26c.
  The extreme values of $v$ and $s$ are

$$v_{max} = 4.8 \text{ m/s} \qquad s_{max} = 12.8 \text{ m}$$

The average values of velocity and acceleration are

$$v_{avg} = \frac{\Delta s}{\Delta t} = \frac{12.8 - 0}{4 - 0} = 3.2 \text{ m/s} \qquad a_{avg} = \frac{\Delta v}{\Delta t} = \frac{3.4 - 1.8}{4 - 0} = 0.4 \text{ m/s}^2$$

(c) The plots of displacement and velocity, for an initial velocity of $-3$ m/s, are shown in Fig. 14.26d. These curves are found by using the techniques in part (a).
  The extreme values of $v$ and $a$ are

$$v_{max} = -3 \text{ m/s} \qquad s_{max} = -6.4 \text{ m}$$

The average values of velocity and acceleration are

$$v_{avg} = \frac{\Delta s}{\Delta t} = \frac{-6.4 - 0}{4 - 0} = -1.6 \text{ m/s} \qquad a_{avg} = \frac{\Delta v}{\Delta t} = \frac{-1.4 - (-3)}{4 - 0} = 0.4 \text{ m/s}^2$$

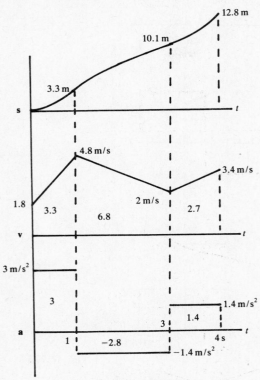

**Fig. 14.26c**

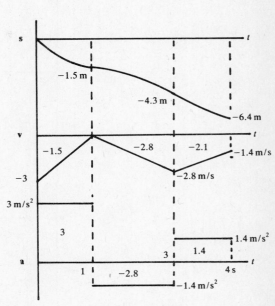

**Fig. 14.26d**

**14.27** Solve Prob. 14.25, part (*a*), for the variation of acceleration with time shown in Fig. 14.27a, for (*a*) initial velocity $v_0 = 0$, (*b*) initial velocity $v_0 = 10 \text{ m/s}$, and (*c*) initial velocity $v_0 = -10 \text{ m/s}$.

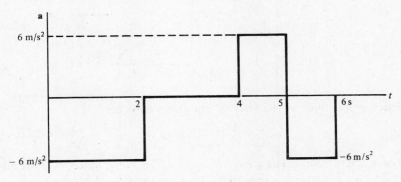

**Fig. 14.27a**

❚ (*a*) The areas under the acceleration diagram in Fig. 14.27*b* are used to construct the velocity diagram, and the areas under the velocity diagram are used to construct the displacement diagram. These results are shown in Fig. 14.27*b*. The extreme values of *v* and *s* are

$$v_{max} = -12 \text{ m/s} \qquad s_{max} = -54 \text{ m}$$

The average values of *v* and *a* are

$$v_{avg} = \frac{\Delta s}{\Delta t} = \frac{-54 - 0}{6 - 0} = -9 \text{ m/s} \qquad a_{avg} = \frac{\Delta v}{\Delta t} = \frac{-12 - 0}{6 - 0} = -2 \text{ m/s}^2$$

(*b*) The velocity diagram, for an initial velocity of 10 m/s, is shown in Fig. 14.27*c*. Using $\int a \, dt = \Delta v$, the three time intervals $a_1$, $a_2$, and $a_3$ in Fig. 14.27*c* are found as

$$a_1(6) = 10 \qquad a_1 = 1.67 \text{ s}$$
$$a_2(6) = 2 \qquad a_2 = 0.333 \text{ s}$$
$$a_3(6) = 4 \qquad a_3 = 0.667 \text{ s}$$

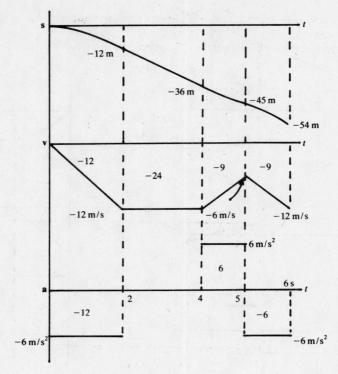

Fig. 14.27b

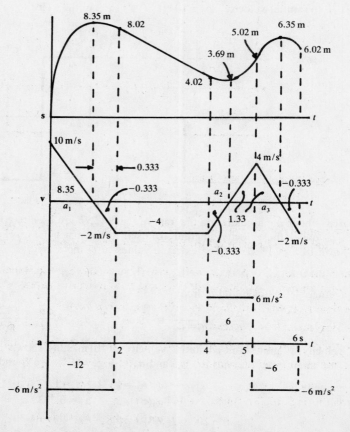

Fig. 14.27c

The areas under the velocity diagram are next computed, and these values are used to construct the displacement curve.

The extreme values of velocity and acceleration are

$$v_{max} = 10 \text{ m/s} \qquad s_{max} = 8.35 \text{ m}$$

The average values of $v$ and $a$ are found to be

$$v_{avg} = \frac{\Delta s}{\Delta t} = \frac{6.01 - 0}{6 - 0} = 1.00 \text{ m/s} \qquad a_{avg} = \frac{\Delta v}{\Delta t} = \frac{-2 - 10}{6 - 0} = -2 \text{ m/s}^2$$

(c)   Figure 14.27d shows the plots of $a$, $v$, and $s$ versus $t$ for an initial velocity of $-10$ m/s.   The extreme values of $v$ and $s$ are

$$v_{max} = -22 \text{ m/s} \qquad s_{max} = -114 \text{ m}$$

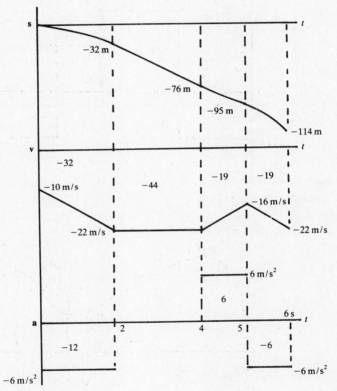

**Fig. 14.27d**

The average values of $v$ and $a$ are

$$v_{avg} = \frac{\Delta s}{\Delta t} = \frac{-114 - 0}{6 - 0} = -19 \text{ m/s} \qquad a_{avg} = \frac{\Delta v}{\Delta t} = \frac{-22 - (-10)}{6 - 0} = -2 \text{ m/s}^2$$

From comparison of Figs. 14.27b and d with Fig. 14.27c, it may be seen that the value of the initial velocity has a significant effect on the shape of the displacement-time curve.

**14.28**   The acceleration diagram of a particle which moves in rectilinear translation is shown in Fig. 14.28a.   The particle starts with an initial velocity of 100 in/s and zero initial displacement.

(a)   Find the time $t_1$ at which the velocity of the particle is zero.

(b)   Find the displacement of the particle at   $t = t_1$.

(c)   Find the total distance through which the particle travels during the time interval   $0 \le t \le t_1$.

(d)   Do the same as in parts (a) through (c), if the initial velocity is   $v_0 = -100$ in/s.

(a)   The velocity at any time is given by   $v = 100 + \int a \, dt$.   Using the areas under the acceleration diagram in Fig. 14.28a, with   $v = 0$,   gives

$$0 = 100 + 2(160) + 1(68) - (t_1 - 3)80 \qquad t_1 = 9.1 \text{ s}$$

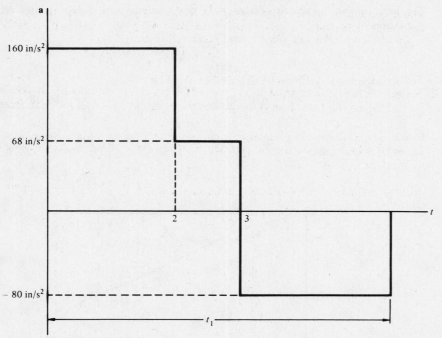

Fig. 14.28a

The displacement, velocity, and acceleration diagrams are sketched in Fig. 14.28b.

(b)  The displacement of the particle at $t_1 = 9.1$ s  is  $s = 2,460$ in.

(c)  The total distance through which the particle travels in the interval 0 to $t_1$ is 2,460 in.  For this case, the results in parts (b) and (c) are the same.

(d)  For an initial velocity of $-100$ in/s,

$$v = -100 + \int a\,dt \qquad 0 = -100 + 2(160) + 1(68) - (t_1 - 3)80 \qquad t_1 = 6.6\,\text{s}$$

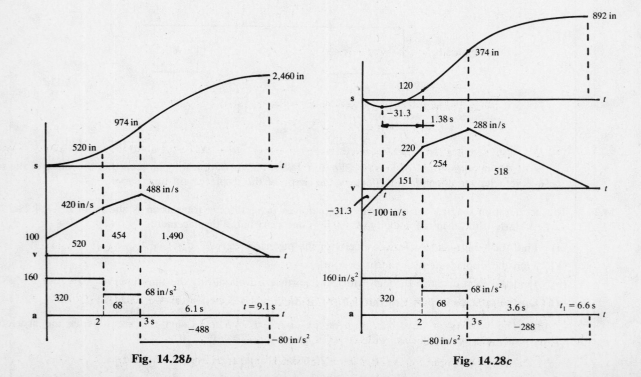

Fig. 14.28b                    Fig. 14.28c

The displacement, velocity, and acceleration diagrams are sketched in Fig. 14.28c. The equation for the time $t$ in Fig. 14.28c is found from the similar triangles in the velocity diagram as

$$\frac{100}{t} = \frac{220}{2-t} \qquad t = 0.625 \text{ s}$$

The displacement of the particle at time $t_1$ is 892 in.

The particle travels from 0 to $s = -31.3$ in, and then comes to rest at $s = 892$ in at time $t_1$. The total distance through which the particle travels during the time interval is thus

$$31.3 + 892 = 923 \text{ in}$$

It may be observed that the displacement at time $t_1$ is *not* equal to the distance traveled by the particle in the time interval 0 to $t_1$.

**14.29** The acceleration diagram of a particle in rectilinear translation is shown in Fig. 14.29. The initial velocity is 12 m/s.

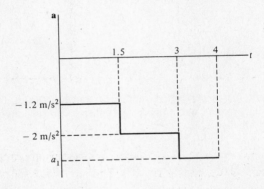

Fig. 14.29

(*a*) Find the value of $a_1$ if the particle comes to rest in 4 s.

(*b*) Find the maximum value of the magnitude of the velocity of the particle.

(*c*) Do the same as in parts (*a*) and (*b*), if the initial velocity is 20 m/s and the vehicle comes to rest in 3.5 s.

❚ (*a*) The initial and final conditions of the motion are

$$t = 0 \qquad v = 12 \text{ m/s}$$
$$t = 4 \text{ s} \qquad v = 0$$

The velocity at any time is given by

$$v = 12 + \int a \, dt$$

Using the area under the acceleration-time diagram in Fig. 14.29,

$$v = 0 = 12 + 1.5(-1.2) + 1.5(-2) + 1(a_1) \qquad a_1 = -7.2 \text{ m/s}^2$$

(*b*) The acceleration is always negative, and the velocity decreases from 12 m/s to zero. Thus, the maximum value of the magnitude of the velocity is 12 m/s.

(*c*) The initial and final conditions of the problem are

$$t = 0 \qquad v = 20 \text{ m/s}$$
$$t = 3.5 \text{ s} \qquad v = 0$$

$$v = 20 + \int a \, dt \qquad v = 0 = 20 + 1.5(-1.2) + 1.5(-2) + 1(a_1) \qquad a_1 = -15.2 \text{ m/s}^2$$

The maximum value of the magnitude of the velocity is 20 m/s.

## 14.2 MOTION WITH CONSTANT ACCELERATION, MOTION WITH GRAVITATIONAL ACCELERATION

**14.30** A particle moves in rectilinear motion with *constant* acceleration.

(*a*) Describe two important physical problems which exhibit this type of motion.

(*b*) Show the relationships among velocity, acceleration, and time.

**▮** (*a*) A very important type of motion is rectilinear translation with constant acceleration. This type of motion describes the case of a freely falling body in the earth's gravitational field, in the absence of any frictional retarding effects. It also describes the motion of a particle which is acted upon by *a resultant force which has a constant value.*

(*b*) The *constant* magnitude of the acceleration will be designated *a*, so that

$$a = \frac{dv}{dt} = \text{constant}$$

The integral form of the above equation is

$$\int_{v_1}^{v_2} dv = \int_{t_1}^{t_2} a \, dt$$

The numbers 1 and 2 represent the endpoints of the time interval under consideration. Since *a* is a constant, it may be moved outside the integral sign:

$$\int_{v_1}^{v_2} dv = a \int_{t_1}^{t_2} dt$$

This equation is integrated, and

$$v\Big|_{v_1}^{v_2} = at\Big|_{t_1}^{t_2} \qquad v_2 - v_1 = a(t_2 - t_1)$$

The above equation is a perfectly general relationship among the velocity, acceleration, and times of the problem. The result will now be put into a more convenient form. The initial time of the problem is $t_1$, with the value

$$t_1 = 0$$

The corresponding velocity $v_1$ will be defined to be the *initial velocity*, designated by $v_0$, so that

$$v_1 = v_0$$

The velocity and time $v_2$ and $t_2$ at the end of the interval of consideration will be written as

$$v_2 = v \qquad t_2 = t$$

With the above changes in notation,

$$v = v_0 + at$$

This equation is the first of three basic equations used to characterize the problem of rectilinear motion with constant acceleration. $v_0$ is the initial velocity at $t = 0$, and $v$ is the velocity at a later time $t$. $v_0$, $v$, and $a$ may have positive or negative values. If $v$ and $a$ are assumed to be positive, the corresponding velocity and acceleration diagrams would have the forms shown in Fig. 14.30*a*.

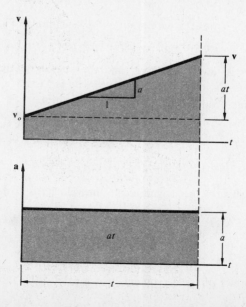

**Fig. 14.30*a***

**14.31** A particle moves in rectilinear motion with *constant* acceleration.

(*a*) Show the relationship among the displacement, acceleration, and time.

(*b*) Show the relationship among the displacement, velocity, and acceleration.

(*c*) Summarize the three equations used to solve problems of rectilinear motion with constant acceleration.

▌ (*a*) The displacement and velocity are related by

$$v = \frac{ds}{dt}$$

The integral form of this equation is

$$\int_{s_1}^{s_2} ds = \int_{t_1}^{t_2} v \, dt$$

The displacement $s_1$, corresponding to $t_1 = 0$, is the initial displacement, designated by the term $s_0$. The terms $s_2$ and $t_2$ will be taken as

$$s_2 = s \qquad t_2 = t$$

The displacement then has the form

$$\int_{s_0}^{s} ds = \int_{0}^{t} (v_0 + at) \, dt$$

$$s\big|_{s_0}^{s} = (v_0 t + \tfrac{1}{2}at^2)\big|_{0}^{t} \qquad s - s_0 = v_0 t + \tfrac{1}{2}at^2 \qquad s = s_0 + v_0 t + \tfrac{1}{2}at^2$$

This result is the second basic equation for the solution of rectilinear motion problems with constant acceleration, where the first basic equation $v = v_0 + at$ was found in Prob. 14.30, $s_0$ is the initial displacement at time $t = 0$, and $s$ is the displacement at a later time $t$. All the terms in this equation, except $t$, may have positive or negative values. In many problems the reference point of the coordinate $s$ may be chosen to make $s_0 = 0$.

If $s_0$, $v_0$, and $a$ are assumed to be positive, the displacement, velocity, and acceleration diagrams would have the forms shown in Fig. 14.31.

(*b*) If $t$ is eliminated between the first and second basic equations, the result is

$$v^2 = v_0^2 + 2a(s - s_0)$$

(*c*) The three basic equations used in the solution of *all* problems in rectilinear translation with constant acceleration are

$$v = v_0 + at \qquad s = s_0 + v_0 t + \tfrac{1}{2}at^2 \qquad v^2 = v_0^2 + 2a(s - s_0)$$

The only requirement for using the above equations is that *the acceleration must have a constant value.*

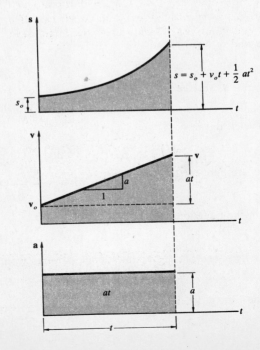

Fig. 14.31

**14.32** A particle slides along a straight track with a speed of 6.5 m/s, as shown in Fig. 14.32. At $t = 0$ a force, shown as the dashed arrow in the figure, is applied to the particle and the resulting deceleration is 1.6 m/s².

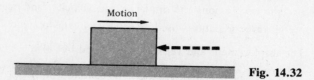

Motion

**Fig. 14.32**

  (**a**)  Find the time when the velocity of the particle is reduced to 50 percent of its initial value and the corresponding distance traveled by the particle.

  (**b**)  Find the time when the particle comes to rest and the corresponding total distance through which the particle travels.

  ▌  (**a**)  Displacement of the particle to the right in Fig. 14.32 is assumed to be positive. The constant acceleration of the particle is $a = -1.6$ m/s² and the initial velocity is $v_0 = 6.5$ m/s. The time to reach 50 percent of the initial velocity is found from

$$v = v_0 + at \qquad 0.5(6.5) = 6.5 + (-1.6)t \qquad t = 2.03 \text{ s}$$

The corresponding distance traveled is found from

$$s = s_0 + v_0 t + \tfrac{1}{2}at^2 = 6.5(2.03) + \tfrac{1}{2}(-1.6)2.03^2 = 9.90 \text{ m}$$

  (**b**)  When the particle comes to rest, $v = 0$. The corresponding time and displacement are found from

$$v = v_0 + at \qquad 0 = 6.5 + (-1.6)t \qquad t = 4.06 \text{ s}$$
$$v^2 = v_0^2 + 2a(s - s_0) \qquad 0 = 6.5^2 + 2(-1.6)(s - 0) \qquad s = 13.2 \text{ m}$$

**14.33** An automobile moves along a straight roadway with a constant speed of 40 mi/h.

  (**a**)  Find the increase in speed if the automobile accelerates at 5 ft/s² for 5 s.

  (**b**)  Find the distance traveled by the automobile during the motion described in part (a).

  (**c**)  At a later time the brakes are applied, causing a deceleration of 10 ft/s². Find the stopping time, and distance, of the automobile.

  ▌  (**a**)  The initial velocity is given as

$$v_0 = 40 \, \frac{\text{mi}}{\text{h}} \left( \frac{5,280 \text{ ft/mi}}{3,600 \text{ s/h}} \right) = 58.7 \text{ ft/s}$$

Using the conditions

$$a = 5 \text{ ft/s}^2 \qquad \text{and} \qquad t = 5 \text{ s}$$

the speed at the end of 5 s is found from

$$v = v_0 + at = 58.7 + 5(5) = 83.7 \text{ ft/s} \qquad v = \frac{3,600}{5,280}(83.7) = 57.1 \text{ mi/h}$$

The increase in speed over the 5-s time interval is

$$\Delta v = 57.1 - 40 = 17.1 \text{ mi/h}$$

  (**b**)  The distance traveled during 5 s is given by

$$s = s_0 + v_0 t + \tfrac{1}{2}at^2 = 58.7(5) + \tfrac{1}{2}(5)5^2 = 356 \text{ ft}$$

  (**c**)  When the brakes are applied, the conditions are

$$v_0 = 83.7 \text{ ft/s} \qquad a = -10 \text{ ft/s}^2$$

Using $v_0 = 0$ in the equation $v = v_0 + at$,

$$0 = 83.7 + (-10)t \qquad t = 8.37 \text{ s}$$

The required stopping distance of the automobile is found from

$$v^2 = v_0^2 + 2a(s - s_0) \qquad 0 = 83.7^2 + 2(-10)s \qquad s = 350 \text{ ft}$$

As a check on the last calculation above,

$$s = s_0 + v_0 t + \tfrac{1}{2}at^2 \qquad s = 83.7(8.37) + \tfrac{1}{2}(-10)8.37^2 \qquad s = 350 \text{ ft}$$

14.34 (a) Two vehicles approach each other in opposite lanes of a straight horizontal roadway, as shown in Fig. 14.34a. At time $t = 0$ the vehicles have the speeds and positions shown in the figure. Find the time and positions at which the vehicles meet if both continue to move with constant speed.

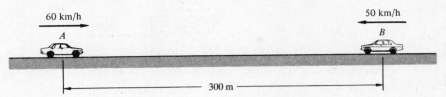

60 km/h          50 km/h

A                B

|←——————— 300 m ———————→|

Fig. 14.34a

(b) Do the same as in part (a) if, at $t = 0$, vehicle A continues to move with constant speed and vehicle B starts to accelerate at $1.6 \text{ m/s}^2$.

(c) Do the same as in part (a) if, at $t = 0$, both vehicles start to accelerate at $1.6 \text{ m/s}^2$.

▌ (a) The displacement coordinates of the vehicles are shown in Fig. 14.34b. The initial velocities are

$$v_{0A} = v_A = 60 \frac{\text{km}}{\text{h}} = \frac{60,000 \text{ m/h}}{3,600 \text{ s/h}} = 16.7 \text{ m/s} \qquad v_{0B} = v_B = 50 \frac{\text{km}}{\text{h}} = \frac{50,000 \text{ m/h}}{3,600 \text{ s/h}} = 13.9 \text{ m/s}$$

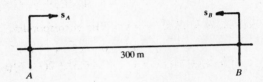

$s_A$              $s_B$

300 m

A                B    Fig. 14.34b

The time $t$ at which the vehicles meet is found from

$$s = v_0 t \qquad s_A + s_B = 300 \qquad 16.7t + 13.9t = 300 \qquad t = 9.80 \text{ s}$$

The positions of each vehicle, at the time they meet, are given by

$$s = v_0 t \qquad s_A = 16.7(9.80) = 164 \text{ m} \qquad s_B = 13.9(9.80) = 136 \text{ m}$$

(b) Vehicle A moves with constant speed and vehicle B accelerates at $1.6 \text{ m/s}^2$. The time $t$ at which the vehicles meet is found from

$$s_A + s_B = 300 \qquad v_{0A}t + (v_{0B}t + \tfrac{1}{2}a_B t^2) = 300$$

Using the initial velocities from part (a),

$$16.7t + 13.9t + \tfrac{1}{2}(1.6)t^2 = 300 \qquad 0.8t^2 + 30.6t - 300 = 0$$

$$t = \frac{-30.6 \pm \sqrt{30.6^2 - 4(0.8)(-300)}}{2(0.8)} = \frac{-30.6 \pm 43.5}{1.6} = 8.09 \text{ s}$$

The corresponding displacement of each vehicle is then found to be

$$s_A = v_{0A}t = 16.7(8.09) = 135 \text{ m} \qquad s_B = v_{0B}t + \tfrac{1}{2}a_B t^2 = 13.9(8.09) + \tfrac{1}{2}(1.6)8.09^2 = 165 \text{ m}$$

(c) For the case where both vehicles accelerate at $1.6 \text{ m/s}^2$,

$$s_A + s_B = 300 \qquad [16.7t + \tfrac{1}{2}(1.6)t^2] + [13.9t + \tfrac{1}{2}(1.6)t^2] = 300$$

$$1.6t^2 + 30.6t - 300 = 0 \qquad t = \frac{-30.6 \pm \sqrt{30.6^2 - 4(1.6)(-300)}}{2(1.6)} = \frac{-30.6 \pm 53.4}{2(1.6)} = 7.14 \text{ s}$$

The positions of each vehicle at the time they meet are found from

$$s = s_0 + v_0 t + \tfrac{1}{2}at^2 \qquad s_A = 16.7(7.14) + \tfrac{1}{2}(1.6)7.14^2 = 160 \text{ m}$$

$$s_B = 13.9(7.14) + \tfrac{1}{2}(1.6)7.14^2 = 140 \text{ m}$$

14.35 (a) Do the same as in Prob. 14.34 if, at $t = 0$, vehicle A continues to move with constant speed and vehicle B starts to decelerate at $0.6 \text{ m/s}^2$.

(*b*) Do the same as in part (*a*) if, at $t = 0$, both vehicles start to decelerate at $0.6\,\mathrm{m/s^2}$.

$\mathscr{l}$ (*a*) The initial velocities are the same as those found in part (*a*) in Prob. 14.34.

$$s_A + s_B = 300 \qquad 16.7t + [13.9t + \tfrac{1}{2}(-0.6)t^2] = 300$$

$$-0.3t^2 + 30.6t - 300 = 0 \qquad t = \frac{-30.6 \pm \sqrt{30.6^2 - 4(-0.3)(-300)}}{2(-0.3)} = \frac{-30.6 \pm 24}{-0.6} = 11\,\mathrm{s}, \quad 91\,\mathrm{s} \tag{1}$$

In order to determine which of the above two values of $t$ to use, the time at which vehicle $B$ comes to rest will be found. This value is found from

$$v_B = v_{0B} + a_B t \qquad 0 = 13.9 + (-0.6)t \qquad t = 23.2\,\mathrm{s}$$

From this result, the time $t = 11\,\mathrm{s}$ in Eq. (1) is used. The displacements of the two vehicles at the time they meet are

$$s_A = v_{0A}t = 16.7(11) = 184\,\mathrm{m} \qquad s_B = s_{0B} + v_{0B}t + \tfrac{1}{2}a_B t^2 = 13.9(11) + \tfrac{1}{2}(-0.6)11^2 = 117\,\mathrm{m}$$

(*b*) For the case where both vehicles decelerate,

$$s_A + s_B = 300 \qquad [16.7t + \tfrac{1}{2}(-0.6)t^2] + [13.9t + \tfrac{1}{2}(-0.6)t^2] = 300$$

$$-0.6t^2 + 30.6t - 300 = 0 \qquad t = \frac{-30.6 \pm \sqrt{30.6^2 - 4(-0.6)(-300)}}{2(-0.6)} = \frac{-30.6 \pm 14.7}{2(-0.6)} = 13.2\,\mathrm{s}, \quad 37.8\,\mathrm{s}$$

It was shown in part (*a*) that vehicle $B$ comes to rest at $t = 23.2\,\mathrm{s}$. Thus, the time $t = 13.2\,\mathrm{s}$ is used. The displacements of the two vehicles, at the time they meet, are found to be

$$s = s_0 + v_0 t + \tfrac{1}{2}at^2 \qquad s_A = 16.7(13.2) + \tfrac{1}{2}(-0.6)13.2^2 = 168\,\mathrm{m}$$
$$s_B = 13.9(13.2) + \tfrac{1}{2}(-0.6)13.2^2 = 131\,\mathrm{m}$$

**14.36** (*a*) If vehicle $A$ in Fig. 14.34a continues to move at constant speed, find the required value of constant acceleration of vehicle $B$ so that the vehicles pass each other at the midpoint of their initial separation distance.

(*b*) If vehicle $B$ in Fig. 14.34a continues to move at constant speed, find the required value of constant deceleration of vehicle $A$ if the vehicles are to pass each other at the midpoint of their initial separation distance.

$\mathscr{l}$ (*a*) The initial velocities are given in part (*a*) of Prob. 14.34.

$$s_A = v_{0A}t \qquad 150 = 16.7t \qquad t = 8.98\,\mathrm{s}$$
$$s_B = s_{0B} + v_{0B}t + \tfrac{1}{2}a_B t^2 = 150 = 13.9(8.98) + \tfrac{1}{2}a_B(8.98)^2 \qquad a_B = 0.624\,\mathrm{m/s^2}$$

(*b*)
$$s_B = v_{0B}t \qquad 150 = 13.9t \qquad t = 10.8\,\mathrm{s}$$
$$s_A = s_{0A} + v_{0A}t + \tfrac{1}{2}a_A t^2 = 150 = 16.7(10.8) + \tfrac{1}{2}a_A(10.8)^2 \qquad a_A = -0.521\,\mathrm{m/s^2}$$

**14.37** An automobile travels along a straight road at 35 mi/h through a 25-mi/h speed zone. A police car observes the automobile approaching. At the instant that the two vehicles are abreast of each other, the police car starts to pursue the automobile and overtakes it 16 s later. For the purposes of this problem, the police car is assumed to move with constant acceleration.

(*a*) Find the acceleration of the police car.

(*b*) Find the total distance traveled by the police car while overtaking the automobile.

(*c*) Find the speed of the police car at the time that it overtakes the automobile.

$\mathscr{l}$ (*a*) The automobile is designated $A$ and the police car is designated $B$, and Fig. 14.37 shows a sketch of the events. The displacements of $A$ and $B$ are

$$s_A = v_{0A}t \qquad s_B = \tfrac{1}{2}a_B t^2 \qquad s_A = s_B$$

Using $t = 16\,\mathrm{s}$,

$$51.3(16) = \tfrac{1}{2}a_B(16)^2$$
$$a_B = 6.41\,\mathrm{ft/s^2}$$

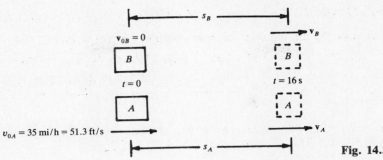

Fig. 14.37

$(b)$  $\qquad s_B = s_A = v_{0A}t = 51.3(16) = 821$ ft

$(c)$  $\qquad v_B = v_{0B} + a_Bt \qquad v_B = 6.41(16) = 103$ ft/s $\qquad v_B = 103\left(\dfrac{3{,}600}{5{,}280}\right) = 70.2$ mi/h

**14.38**  The motorist in Prob. 14.37 observes the police car in his rear view mirror 12 s after the police car started the pursuit. He applies his brakes and decelerates at $10$ ft/s$^2$.

$(a)$ Find the total time required for the police car to overtake the automobile.

$(b)$ Find the total distance traveled by the police car while overtaking the automobile.

$(c)$ Find the speed of the police car at the time it overtakes the automobile.

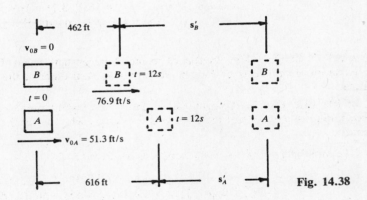

Fig. 14.38

▌ $(a)$ The first step is to find the positions of both vehicles at $t = 12$ s. Using $v_{0A} = 51.3$ ft/s and $a_B = 6.41$ ft/s$^2$ from Prob. 14.37,

$$s_A = v_{0A}t = 51.3(12) = 616 \text{ ft} \qquad s_B = \tfrac{1}{2}a_Bt^2 = \tfrac{1}{2}(6.41)12^2 = 462 \text{ ft}$$

Figure 14.38 shows a sketch of the events. At $t = 12$ s, the velocity of the police car is

$$v_B = a_Bt = 6.41(12) = 76.9 \text{ ft/s}$$

For $t > 12$ s, the police car continues to accelerate at $a_B = 6.41$ ft/s$^2$. Car $A$ decelerates at $10$ ft/s$^2$. From Fig. 14.38,

$$462 + s_B' = 616 + s_A' \qquad 462 + 76.9t + \tfrac{1}{2}(6.41)t^2 = 616 + 51.3t + \tfrac{1}{2}(-10)t^2$$

$$8.21t^2 + 25.6t - 154 = 0 \qquad t = \frac{-25.6 \pm \sqrt{25.6 - 4(8.21)(-154)}}{2(8.21)} = \frac{-25.6 \pm 75.6}{2(8.21)} = 3.04 \text{ s}$$

The total time $t_0$ required for the police car to overtake the automobile is then given by

$$t_0 = 12 + 3.04 = 15.0 \text{ s}$$

$(b)$ The total distance traveled by the police car is given by

$$s_{B,\text{total}} = 462 + 76.9t + \tfrac{1}{2}(6.41)t^2$$

Using $t = 3.04\,$s,

$$s_{B,\text{total}} = 725\text{ ft}$$

As a check on the above value,

$$s_{A,\text{total}} = 616 + 51.3t + \tfrac{1}{2}(-10)t^2 = 616 + 51.3(3.04) + \tfrac{1}{2}(-10)3.04^2 \overset{?}{=} 725 \qquad 726 \approx 725$$

(c) The speed of the police car is given by

$$v_B = v_{0B} + a_B t = 6.41(15) = 96.2\text{ ft/s} = 96.2\left(\frac{3{,}600}{5{,}280}\right) = 65.6\text{ mi/h}$$

**14.39** (a) At $t = 0$ vehicle $A$, traveling at a speed of 60 km/h, passes a road marker on a straight horizontal roadway. Vehicle $B$, traveling with a speed of 90 km/h, passes the marker 2 s later. Find the time when vehicle $B$ overtakes vehicle $A$, and the corresponding distance from the road marker.

(b) Do the same as in part (a) if vehicle $B$ starts to accelerate at the rate of 2 m/s$^2$ at the instant that it passes the road marker.

▮ (a) Figure 14.39a shows a sketch of the events. $s$ is the distance from the marker where $B$ overtakes $A$ at time $t$. From Fig. 14.39a,

$$s = s_A = s_B$$

Using $s = vt$,

$$16.7t = 25(t - 2) \qquad t = 6.02\text{ s} \qquad s = s_A = s_B = 16.7(6.02) = 101\text{ m}$$

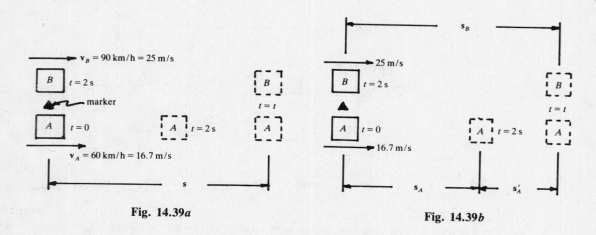

**Fig. 14.39a**        **Fig. 14.39b**

Figure 14.39b shows the detail for use with an alternative solution. The position $s_A$ when $t = 2\,$s is found from

$$s_A = v_A t = 16.7(2) = 33.4\text{ m}$$

Let $t'$ be the time from $t = 2\,$s to the time when $B$ overtakes $A$.

$$s_B = s_A + S_A' \qquad 25t' = 33.4 + 16.7t' \qquad t' = 4.02\text{ s} \qquad t = 2 + t' = 2 + 4.02 = 6.02\text{ s}$$

(b) Using Fig. 14.39a,

$$s = s_A = s_B$$

Using $s_A = v_A t$ and $s_B = v_{0B}t + \tfrac{1}{2}a_B t^2$,

$$16.7t = 25(t - 2) + \tfrac{1}{2}(2)(t - 2)^2 \qquad t^2 + 4.3t - 46 = 0$$

$$t = \frac{-4.3 \pm \sqrt{4.3^2 - 4(1)(-46)}}{2(1)} = \frac{-4.3 \pm 14.2}{2} = 4.95\text{ s}$$

$$s = v_A t = 16.7(4.95) = 82.7\text{ m}$$

**14.40**   Vehicle $A$ in Prob. 14.39 starts to accelerate at the rate of $0.5 \, \text{m/s}^2$ at the instant that vehicle $B$ passes the road marker.   Determine whether vehicle $B$ will overtake vehicle $A$, and the position if this occurs, and discuss the motion of the two vehicles at later times.

▌   From Fig. 14.39$a$,

$$s = s_A = s_B \qquad 16.7 + \tfrac{1}{2}(0.5)t^2 = 25(t-2) \qquad 0.25t^2 - 8.3t + 50 = 0$$

$$t = \frac{8.3 \pm \sqrt{(-8.3)^2 - 4(0.25)50}}{2(0.25)} = \frac{8.3 \pm 4.35}{2(0.25)} = 7.9 \, \text{s}, \quad 25.3 \, \text{s}$$

Using   $t = 7.9 \, \text{s}$,

$$v_A = v_{0A} + a_A t = 16.7 + 0.5(7.9) = 20.7 \, \text{m/s} \qquad v_B = 25 \, \text{m/s} = \text{constant}$$

Since   $v_B > v_A$,   vehicle $B$ will pass vehicle $A$.

$$s_A = s_B = v_{0B}(t-2) = 25(7.9-2) = 148 \, \text{m}$$

Using   $t = 25.3 \, \text{s}$,

$$v_A = 16.7 + 0.5(25.3) = 29.4 \, \text{m/s} \qquad v_B = 25 \, \text{m/s} = \text{constant} \qquad s_A = s_B = v_B(t-2) = 25(25.3-2) = 583 \, \text{m}$$

Since   $v_A > v_B$,   vehicle $A$ will pass vehicle $B$.

If   $t < 7.9 \, \text{s}$,   $s_A = s_B < 148 \, \text{m}$,   and   $A$   is   ahead   of   $B$. If   $7.9 \, \text{s} < t < 25.3 \, \text{s}$,   $148 \, \text{m} < s_A = s_B < 583 \, \text{m}$,   and $B$ is ahead of $A$. If   $25.3 \, \text{s} < t$,   $583 \, \text{m} < s_A = s_B$   and   $A$ is ahead of $B$.

**14.41**   Vehicle $B$ is stopped at a traffic light, as shown in Fig. 14.41$a$.   At the instant that the light turns green, vehicle $B$ starts to accelerate at $3 \, \text{ft/s}^2$.   At this time vehicle $A$ is $300 \, \text{ft}$ behind vehicle $B$, traveling at a speed of $30 \, \text{mi/h}$.

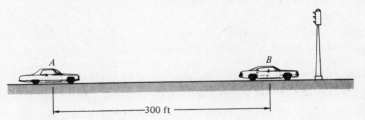

-300 ft-

**Fig. 14.41$a$**

(**a**)   At what distance past the light will vehicle $A$ overtake vehicle $B$?

(**b**)   At what distance past the light will vehicle $B$ overtake vehicle $A$?

(**c**)   For what range of values of constant acceleration of vehicle $B$ could vehicle $A$ not overtake this vehicle?

▌   (**a**)   Figure 14.41$b$ shows a sketch of the events.   The displacement coordinate $s$ is measured from the position of vehicle $A$ at   $t = 0$.   The position when both vehicles are side by side is $s$.

$$s = s_A = 300 + s_B \qquad v_A t = 300 + \tfrac{1}{2}a_B t^2 \qquad 44t = 300 + \tfrac{1}{2}(3)t^2$$

$$1.5t^2 - 44t + 300 = 0 \qquad t = \frac{44 \pm \sqrt{44^2 - 4(1.5)300}}{2(1.5)} = \frac{44 \pm 11.7}{2(1.5)} = 10.8 \, \text{s}, \quad 18.6 \, \text{s}$$

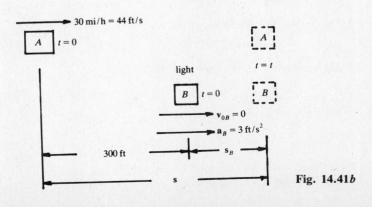

**Fig. 14.41$b$**

Using $t = 10.8$ s,

$$v_A = 44 \text{ ft/s} = \text{constant} \qquad v_B = a_B t = 3(10.8) = 32.4 \text{ ft/s}$$

Thus $v_A > v_B$, and $A$ overtakes $B$.
The distance past the light is given by

$$s_B = \tfrac{1}{2} a_B t^2 = \tfrac{1}{2}(3)10.8^2 = 175 \text{ ft}$$

(b) It is assumed that vehicle $B$ continues to accelerate at 3 ft/s². At $t = 18.6$ s,

$$v_A = 44 \text{ ft/s} = \text{constant} \qquad v_B = a_B t = 3(18.6) = 55.8 \text{ ft/s}$$

Thus $v_B > v_A$, and $B$ overtakes $A$.
The distance past the light is given by

$$s_B = \tfrac{1}{2} a_B t^2 = \tfrac{1}{2}(3)18.6^2 = 519 \text{ ft}$$

(c) The limiting condition occurs when $v_A = v_B$ at the time that $A$ overtakes $B$.

$$s_A = 300 + s_B \qquad 44t = 300 + \tfrac{1}{2} a_B t^2 \qquad (1)$$
$$v_A = v_B \qquad 44 = a_B t \qquad (2)$$

Using Eq. (2) in Eq. (1),

$$44t = 300 + \tfrac{1}{2}t(44) \qquad t = 13.6 \text{ s}$$

Using Eq. (2),

$$44 = a_B(13.6) \qquad a_B = 3.24 \text{ ft/s}^2$$

If $a_B > 3.24$ ft/s², vehicle $A$ cannot overtake vehicle $B$.

14.42 (a) An automobile travels along a straight roadway at a speed of 55 mi/h. The driver sees a disabled vehicle blocking the road ahead and applies the brakes. The resulting deceleration of the car is 18 ft/s². If the driver's visual reaction time is 0.75 s, how far has the automobile traveled between the time the driver first sees the disabled vehicle and the time that the vehicle comes to rest?

(b) Do the same as in part (a) if the initial speed of the automobile is 70 mi/h.

(c) Do the same as in part (a) if, because of some preoccupation, the driver's visual reaction time is 1.5 s.

(d) Do the same as in part (a) if the initial speed is 70 mi/h and the driver's visual reaction time is 1.5 s.

▌(a) The distance required to stop the automobile is found from

$$v^2 = v_0^2 + 2a(s - s_0)$$

Using 55 mi/h = 80.7 ft/s,

$$0 = 80.7^2 + 2(-18)s \qquad s = 181 \text{ ft}$$

The total stopping distance is given by

$$s_{\text{total}} = 80.7(0.75) + 181 = 60.5 + 181 = 242 \text{ ft}$$

It may be observed that the motion during the reaction time is 60.5 ft.

(b) $$v^2 = v_0^2 + 2a(s - s_0)$$

Using 70 mi/h = 103 ft/s,

$$0 = 103^2 + 2(-18)s \qquad s = 295 \text{ ft} \qquad s_{\text{total}} = 103(0.75) + 295 = 372 \text{ ft}$$

(c) Using $s = 181$ ft from part (a),

$$s_{\text{total}} = 80.7(1.5) + 181 = 302 \text{ ft}$$

(d) Using $s = 295$ ft from part (b),

$$s_{\text{total}} = 103(1.5) + 295 = 450 \text{ ft}$$

14.43 Give the three equations of motion for the case of a freely falling particle in the earth's gravitational field, in the absence of any frictional retarding effects.

**▮** For the case of a freely falling body, *the magnitude of the acceleration a is equal to the magnitude g of the acceleration of the gravitational field.* If the coordinate axis *s* has a sense which is upward with respect to the surface of the earth, then

$$a = -g$$

The three equations for motion with constant gravitational acceleration then have the forms

$$v = v_0 - gt \qquad s = s_0 + v_0 t - \tfrac{1}{2} g t^2 \qquad v^2 = v_0^2 - 2g(s - s_0)$$

$v_0$ and $s_0$ in the above equations are the initial velocity and displacement at time $t = 0$. The value of $g$ in USCS units is $32.2 \text{ ft/s}^2$, or $386 \text{ in/s}^2$. In SI units, the value of $g$ is $9.81 \text{ m/s}^2$.

**14.44** A particle initially at rest is dropped from a height of 100 ft. Find the velocity with which it strikes the ground and the time of flight.

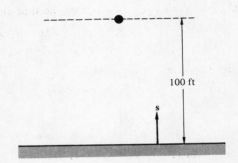

100 ft

s

**Fig. 14.44**

**▮** The definition of the coordinate system is shown in Fig. 14.44. The initial conditions of the problem are described by

$$s_0 = 100 \text{ ft} \qquad v_0 = 0 \qquad t = 0$$

When the particle strikes the ground, $s = 0$. The equation of motion is

$$v^2 = v_0^2 - 2g(s - s_0) = -2g(-s_0) = -2(32.2)(-100) \qquad v = \pm 80.2 \text{ ft/s}$$

From physical considerations, the minus sign is used in the above result. The formal solution for the velocity with which the particle strikes the ground is

$$v = -80.2 \text{ ft/s}$$

The time of flight is found from

$$v = v_0 - gt \qquad -80.2 = 0 - 32.2t \qquad t = 2.49 \text{ s}$$

An alternative way of solving for the time of flight is to use the equation

$$s = s_0 + v_0 t - \tfrac{1}{2} g t^2$$

With $s = 0$,

$$0 = 100 + 0 - \tfrac{1}{2}(32.2)t^2 \qquad t = 2.49 \text{ s}$$

This result serves as a check on the first value obtained for the time of flight.

**14.45** With what minimum value of initial velocity $v_0$ must the particle in Prob. 14.44 be projected downward if the total time of flight is not to exceed 2 s?

**▮** The conditions are

$$s_0 = 100 \text{ ft} \qquad s = 0 \qquad t = 2 \text{ s}$$

The equation of motion is

$$s = s_0 + v_0 t - \tfrac{1}{2} g t^2 \qquad 0 = 100 + v_0(2) - \tfrac{1}{2}(32.2)(2^2) \qquad v_0 = -17.8 \text{ ft/s}$$

The negative sign in the above result is consistent with the physical problem, since the particle must be projected *downward*, in the negative coordinate sense.

**14.46** It is desired to have the particle in Prob. 14.44 be in flight for exactly 3 s. What is the required initial velocity of the particle, if it is launched from the height of 100 ft?

▮ The conditions are

$$s_0 = 100 \text{ ft} \qquad s = 0 \qquad t = 3 \text{ s}$$

The equation of motion has the form

$$s = s_0 + v_0 t - \tfrac{1}{2} g t^2 \qquad 0 = 100 + v_0(3) - \tfrac{1}{2}(32.2)(3^2) \qquad v_0 = 15.0 \text{ ft/s}$$

In this case, the particle would have to be projected *upward* in order to gain the extra time in flight.

**14.47** A person stands on a balcony, as shown in Fig. 14.47, and throws a ball vertically upward with an initial velocity of 8 m/s. At the instant the ball leaves the person's hand, it is 18 m above the ground.

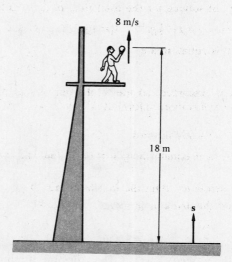

**Fig. 14.47**

(*a*) Find the maximum height reached by the ball during its flight.

(*b*) At what time is this maximum height reached?

(*c*) Find the velocity when the ball passes a point 18 m above the ground.

(*d*) With what velocity does the ball strike the ground?

(*e*) What is the total time of flight?

▮ (*a*) The initial conditions of the ball are

$$s_0 = 18 \text{ m} \qquad v_0 = 8 \text{ m/s}$$

When the maximum height is reached, $v = 0$.

$$v^2 = v_0^2 - 2g(s - s_0) \qquad 0 = 8^2 - 2(9.81)(s - 18) \qquad s = 21.3 \text{ m}$$

(*b*) The time at which the maximum height is reached is found from

$$v = v_0 - gt \qquad 0 = 8 - 9.81t \qquad t = 0.815 \text{ s}$$

(*c*) When the ball passes a point 18 m above the ground, $s = s_0 = 18$ m. Thus,

$$v^2 = v_0^2 - 2g(s - s_0) = v_0^2 - 2g(0) \qquad v = \pm v_0 = \pm 8 \text{ m/s}$$

The positive value of $v$ corresponds to the initial upward velocity of the ball. When the ball passes the 18-m height again, on its way down, its velocity is

$$v = -8 \text{ m/s}$$

The minus sign is consistent with the fact that the ball is moving downward.

(*d*) When the ball strikes the ground, $s = 0$.

$$v^2 = v_0^2 - 2g(s - s_0) = 8^2 - 2(9.81)(0 - 18) \qquad v = \pm 20.4 \text{ m/s}$$

The positive answer above is an extraneous root, and the velocity with which the ball strikes the ground is

$$v = -20.4 \text{ m/s}$$

(e)    The total time of flight may be found by using the result for part (d). The time for the ball to fall from the maximum height to the ground, using $v_0 = 0$ as the initial condition for this part of the problem, is found from

$$v = v_0 - gt \qquad -20.4 = 0 - 9.81t \qquad t = 2.08 \text{ s}$$

The time 0.815 s to reach the maximum height was found in part (b). The time $t_{\text{total}}$ for the entire flight is then

$$t_{\text{total}} = 2.08 + 0.815 = 2.90 \text{ s}$$

An alternative way of solving for the total time of flight is to use the equation

$$s = s_0 + v_0 t - \tfrac{1}{2}gt^2 \qquad 0 = 18 + 8t - \tfrac{1}{2}(9.81t^2) \qquad 4.91t^2 - 8t - 18 = 0$$

The two roots of this equation are

$$t = 2.90 \text{ s}, \quad -1.27 \text{ s}$$

The negative root is discarded, so the total flight time is 2.90 s. It may be observed that the first solution in part (e) yields more information about the problem than does the second equation.

**14.48**    An archer shoots an arrow vertically upward.

(a)    If the arrow ascends to a maximum height of 90 ft, find the required value of the initial velocity of the arrow.

(b)    Find the time for the arrow to attain the maximum height.

(c)    Find the total time that the arrow is in flight.

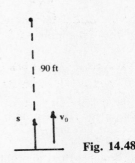

**Fig. 14.48**

▌ (a)    Figure 14.48 shows the definition of the coordinate system. The initial velocity $v_0$ is found from

$$v^2 = v_0^2 - 2g(s - s_0) \qquad 0 = v_0^2 - 2(32.2)90 \qquad v_0 = 76.1 \text{ ft/s}$$

(b)    The time of flight to reach maximum height is found by using

$$v = v_0 - gt \qquad 0 = 76.1 - 32.2t \qquad t = 2.36 \text{ s}$$

(c)    Using $s = s_0 = 0$ and $v_0 = 76.1$ ft/s, the time of flight is found from

$$s = s_0 + v_0 t - \tfrac{1}{2}gt^2 \qquad 0 = 76.1t - \tfrac{1}{2}(32.2)t^2 = t[76.1 - \tfrac{1}{2}(32.2)t] = 0 \qquad t = 0 \text{ s}, \quad 4.73 \text{ s}$$

The first value of $t$ above corresponds to the initial time. The arrow is in flight for 4.73 s.

**14.49**    A ball is dropped from a roof, as shown in Fig. 14.49a. A second ball is dropped from the roof 1 s after the first ball is dropped. Find the position and velocity of the second ball when the first ball strikes the ground.

▌ The displacement coordinate is measured downward from the roof, as shown in Fig. 14.49b. The first ball is designated $A$, and the second ball is designated $B$. The time $t_A$ for the first ball to fall through a distance of 25 m is found from

$$s = s_0 + v_0 t - \tfrac{1}{2}gt^2$$

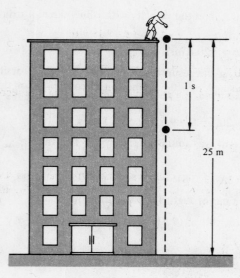

Fig. 14.49a

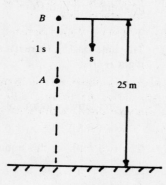

Fig. 14.49b

Using $s_0 = 0$ and $v_0 = 0$ in the above equation

$$s = \tfrac{1}{2}gt^2$$
$$25 = \tfrac{1}{2}(9.81)t_A^2 \qquad t_A = 2.26 \text{ s}$$

The time $t_B$ of flight of ball $B$ is given by

$$t_B = 2.26 - 1 = 1.26 \text{ s}$$

The distance traveled by ball $B$ in 1.26 s is found from

$$s_B = \tfrac{1}{2}(9.81)1.26^2 = 7.79 \text{ m}$$

At this time, the height of $B$ above the ground is given by

$$25 - 7.78 = 17.2 \text{ m}$$

The velocity of ball $B$ at $t = 1.26$ s is given by

$$v_B = gt = 9.81(1.26) = 12.4 \text{ m/s}$$

When the first ball strikes the ground, the second ball is 7.79 m below the roof, or 17.2 m above the ground, moving downward with a speed of 12.4 m/s.

**14.50** At the instant that the ball is dropped from the roof of the building in Fig. 14.49a a second ball is projected upward, from a window at a height of 13 m above the ground, with a velocity of 2 m/s.

(*a*) Which ball strikes the ground first?

(*b*) Find the elapsed time between the impacts of the two balls with the ground.

▌ (*a*) The second ball is designated $B$, and the initial conditions for this ball are shown in Fig. 14.50. The time for ball $B$ to strike the ground is found from

$$s = s_0 + v_0 t - \tfrac{1}{2}gt^2$$

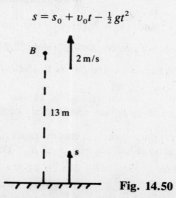

Fig. 14.50

Using $\quad s_0 = 13 \, \text{m} \quad$ and $\quad s = 0,$

$$0 = 13 + 2t - \tfrac{1}{2}(9.81)t^2 \qquad -4.91t^2 + 2t + 13 = 0$$

$$t = \frac{-2 \pm \sqrt{2^2 - 4(-4.91)13}}{2(-4.91)} = \frac{-2 \pm 16.1}{-2(4.91)} = 1.84 \, \text{s}$$

From Prob. 14.49, the time for the first ball to reach the ground is 2.26 s. Thus, the second ball strikes the ground first.

(b) The elapsed time between the impacts of the two balls is given by

$$\Delta t = 2.26 - 1.84 = 0.42 \, \text{s}$$

14.51 A child drops a stone into a well shaft, as shown in Fig. 14.51. Three seconds after the child drops the stone, he hears the sound of the stone hitting the bottom of the well shaft. Find the depth, in feet, of the well. Assume that a sound wave travels with a constant velocity of 1,120 ft/s.

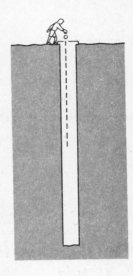

Fig. 14.51

▮ $t$ is the time for the stone to reach the bottom of the well, and $t_1$ is the time for the sound wave to reach the top of the well. For the stone,

$$s = \tfrac{1}{2}gt^2 = \tfrac{1}{2}(32.2)t^2$$

For the sound wave,

$$s_1 = vt_1 = 1,120t_1$$

The two times must satisfy

$$t + t_1 = 3 \, \text{s} \qquad t_1 = 3 - t$$

$s = s_1,$ and $t$ is then found from

$$16.1t^2 = 1,120(3 - t) \qquad 16.1t^2 + 1,120t - 3,360 = 0$$

$$t = \frac{-1,120 \pm \sqrt{1,120^2 - 4(16.1)(-3,360)}}{2(16.1)} = \frac{-1,120 \pm 1,213}{2(16.1)} = 2.89 \, \text{s}$$

Four significant figures are used in the above calculation, to avoid loss of accuracy in the final answer.

$$t_1 = 3 - 2.89 = 0.11 \, \text{s}$$

The depth of the well is then found to be

$$s = \tfrac{1}{2}(32.2)2.89^2 = 134 \, \text{ft}$$

14.52 Figure 14.52a shows an elevator at a construction site. The elevator moves upward with a velocity of 6 ft/s. At the instant that the bottom of the elevator is at a height of 120 ft above the ground, a pebble is dislodged from this surface. Find the position of the bottom of the elevator when the pebble strikes the ground.

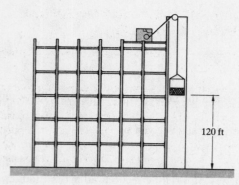

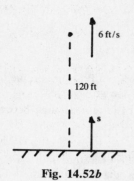

Fig. 14.52a                                      Fig. 14.52b

**❚** Figure 14.52b shows the coordinate system, and the initial conditions of the pebble are $s_0 = 120$ ft and $v_0 = 6$ ft/s. The time for the pebble to reach the ground is found from

$$s = s_0 + v_0 t - \tfrac{1}{2}gt^2$$
$$0 = 120 + 6t - \tfrac{1}{2}(32.2)t^2 \qquad 16.1t^2 - 6t - 120 = 0$$
$$t = \frac{6 \pm \sqrt{(-6)^2 - 4(16.1)(-120)}}{2(16.1)} = \frac{6 \pm 88.1}{2(16.1)} = 2.92 \text{ s}$$

The height of the bottom of the elevator above the ground at this time is

$$s = s_0 + v_0 t = 120 + 6(2.92) = 138 \text{ ft}$$

**14.53** Do the same as in Prob. 14.52 if the elevator moves downward at 6 ft/s and the pebble is dislodged when the bottom of the cage is 120 ft above the ground.

**❚** The coordinate system, and initial conditions, of the pebble are shown in Fig. 14.53. The equation of motion of the pebble is

$$s = s_0 + v_0 t - \tfrac{1}{2}gt^2$$
$$0 = 120 - 6t - \tfrac{1}{2}(32.2)t^2 \qquad 16.1t^2 + 6t - 120 = 0$$
$$t = \frac{-6 \pm \sqrt{6^2 - 4(16.1)(-120)}}{2(16.1)} = \frac{-6 \pm 88.1}{2(16.1)} = 2.55 \text{ s}$$

The height of the bottom of the elevator above the ground, at the time the pebble strikes the ground, is

$$s = s_0 + vt = 120 - 6(2.55) = 105 \text{ ft}$$

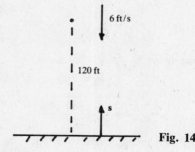

Fig. 14.53

**14.54** Do the same as in Prob. 14.52 if the elevator decelerates at 0.5 ft/s². At the instant that the pebble is dislodged, the elevator velocity is 6 ft/s upward and the bottom of the elevator is at a height of 120 ft above the ground.

**❚** The time for the pebble to strike the ground was found in Prob. 14.52 as $t = 2.92$ s.
The displacement coordinate in Fig. 14.52b is used, with the initial conditions $s_0 = 120$ ft and $v_0 = 6$ ft/s. The acceleration of the elevator is given by $a = -0.5$ ft/s², and the equation of motion of the elevator has the form

$$s = s_0 + v_0 t + \tfrac{1}{2}at^2 = 120 + 6(2.92) + \tfrac{1}{2}(-0.5)2.92^2 = 135 \text{ ft}$$

### 14.3 PLANE CURVILINEAR MOTION, VELOCITY AND NORMAL AND TANGENTIAL COMPONENTS OF ACCELERATION, PLANE PROJECTILE MOTION

**14.55**   Find the magnitude, direction, and sense of the velocity of a particle which moves in *plane curvilinear translation*.

▌ Figure 14.55a shows a particle which experiences curvilinear translation, in the sense of a to b, along a plane curve. The coordinate of length along the curve is s. At time t, the particle is at location a. The *displacement*, or *position*, of the particle at this time is given by the position vector $\mathbf{r}(t)$. This position vector is referenced to the origin of the coordinate system. At a later time $t + \Delta t$ the particle is at the position b on the curve. The displacement of the particle at this new location is given by the position vector $\mathbf{r}(t + \Delta t)$. The *change* in the position, or displacement, of the particle as it goes from a to b is given by the vector $\Delta\mathbf{r}$. From the triangle law of vector addition,

$$\mathbf{r}(t) + \Delta\mathbf{r} = \mathbf{r}(t + \Delta t)$$

The equation may be written as

$$\Delta\mathbf{r} = \mathbf{r}(t + \Delta t) - \mathbf{r}(t)$$

The form of the above equation emphasizes the fact that the change in the displacement of the particle is the *difference* between the position at the *later* time and the position at the *earlier* time.

From the fundamental definition of the velocity $\mathbf{v}$, as the time rate of change of displacement, it follows that

$$\mathbf{v} = \lim_{\Delta t \to 0} \frac{\Delta\mathbf{r}}{\Delta t}$$

As $\Delta t \to 0$, the point b approaches the point a, and the *chord length ab* approaches the *arc length ab*. In the limit, the *magnitude* of $\Delta\mathbf{r}$ is equal to $\Delta s$, and the *direction* of $\Delta\mathbf{r}$ is the direction of a line tangent to the curve at the point a. It follows that the magnitude v of the velocity along the curve is

$$v = \lim_{\Delta t \to 0} \frac{\Delta s}{\Delta t} = \frac{ds}{dt}$$

In summary, when a particle moves along a plane curve, the direction of the velocity is tangent to the curve. The magnitude of this velocity is equal to the speed with which the particle moves along the curve. The positive sense of the velocity is the positive sense of the length coordinate along the curve. Figure 14.55b shows the typical appearance of the graphical representation of the velocity of the particle at point a.

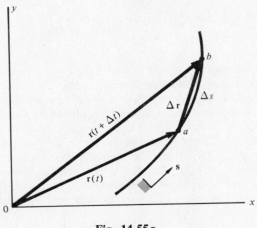

**Fig. 14.55a**

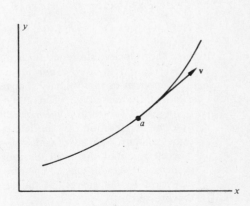

**Fig. 14.55b**

**14.56**   (*a*)   Find the magnitude, direction, and sense of the normal component of the acceleration of a particle which moves in plane curvilinear translation.

(*b*)   What is the name of the normal component of acceleration in part (*a*)?

(*c*)   What is the physical effect which produces a normal component of acceleration of a particle that moves in plane curvilinear translation?

**(a)** Figure 14.56a shows the velocities of the particle at points $a$ and $b$ on the curved path of motion. $\rho_a$ and $\rho_b$ are the two radii of curvature of the plane curvilinear path at these points, with the centers $0_a$ and $0_b$. $\Delta\theta$ is the included angle between $\rho_a$ and $\rho_b$, and $\Delta s$ is the corresponding arc length. The time interval which corresponds to movement of the particle from $a$ to $b$ is $\Delta t$. The magnitude of $\mathbf{v}_b$ in the figure is greater than the magnitude of $\mathbf{v}_a$, to show an increase in this quantity as the particle moves in the positive sense of the coordinate $s$.

The velocity $\mathbf{v}_b$ is now redrawn at point $a$, as shown in Fig. 14.56b. $\Delta\mathbf{v}$ is the *vector difference* between $\mathbf{v}_b$ and $\mathbf{v}_a$, and $\Delta\theta$ is the included angle between these two quantities. The velocity triangle is shown in Fig. 14.56c.

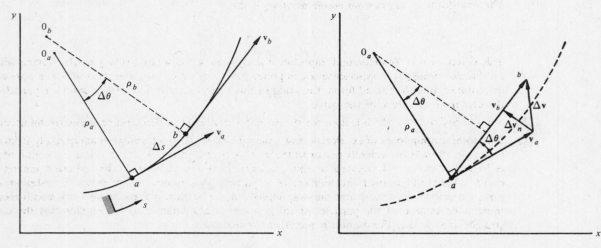

Fig. 14.56a          Fig. 14.56b

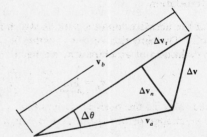

Fig. 14.56c

The velocity difference $\Delta v$ is now represented as the sum of two rectangular components. $\Delta v_n$ will be referred to as the *normal* component of the velocity change, and $\Delta v_t$ is called the *tangential* component of this velocity change. The term *normal* will mean a direction normal to a line which is tangent to the curve. Thus, this direction is that of the radius of curvature of the curve at the point of interest. The term *tangential* will mean the direction of the line which is tangent to the curve.

The acceleration of a particle is defined to be the time rate of change, or the first time derivative, of the velocity. From consideration of Fig. 14.56c, it may be seen that two distinct components of acceleration of the particle along the curve may be identified. The first component is called the *normal acceleration* $a_n$. The direction of this component of acceleration is normal to the line which is tangent to the curve. If the fundamental definition of acceleration is used, the quantity may be expressed as

$$a_n = \lim_{\Delta t \to 0} \frac{\Delta v_n}{\Delta t}$$

The positive sense of $a_n$ must be the same as the positive sense of $v_n$. From Fig. 14.56b, it follows that the normal acceleration $a_n$ is always directed toward the center of curvature of the plane curve. As $\Delta t \to 0$, the corresponding limiting value of the magnitude of $\Delta v_n$ is equal to the product $v_a(\Delta\theta)$. The above equation then appears as

$$a_n = \lim_{\Delta t \to 0} \frac{\Delta v_n}{\Delta t} = \lim_{\Delta t \to 0} \frac{v_a \, \Delta\theta}{\Delta t}$$

As the limit is approached, the value $\rho_a$ of the radius of curvature at point $a$ approaches the general value $\rho$, the term $v_a$ approaches the general term $v$, and $\Delta s \rightarrow \rho \, \Delta \theta$. The normal acceleration then has the form

$$a_n = \lim_{\Delta t \to 0} \frac{v \, \Delta \theta}{\Delta t} = \lim_{\Delta t \to 0} \frac{v}{\Delta t} \frac{\Delta s}{\rho} = \lim_{\Delta t \to 0} \frac{v}{\rho} \frac{\Delta s}{\Delta t}$$

The term $\lim_{\Delta t \to 0} (\Delta s / \Delta t)$ is equal to the magnitude $v$ of the velocity, or the speed, with which the particle moves along the curve, given by

$$\lim_{\Delta t \to 0} \frac{\Delta s}{\Delta t} = \frac{ds}{dt} = \dot{s} = v$$

The final form of the normal acceleration $a_n$ is then

$$a_n = \frac{v^2}{\rho}$$

This equation is of fundamental importance in dynamics. It states that a particle moving with plane curvilinear motion will experience a component of acceleration in a direction which is normal to the direction of motion. In addition, the sense of this acceleration will always be from the particle toward the center of curvature of the curve.

(**b**) An alternative term which is used for the normal component of acceleration is *centripetal* acceleration.

(**c**) The normal component of acceleration of a particle in curvilinear translation arises solely from changes in the direction of the velocity vector as the particle travels along the curve. It is a function of velocity *only* and in no way is related to the acceleration of the particle in the direction tangent to the curve. Thus, the normal acceleration of a particle that moves in curvilinear translation is never zero. It may be observed that the only way for $a_n$ to be zero is for $v$ to be zero, which means that there is no motion of the particle, or for $\rho \rightarrow \infty$. This latter condition implies that the curve is a straight line, so that the motion is rectilinear translation.

**14.57** Find the magnitude, direction, and sense of the tangential component of the acceleration of a particle which moves in plane curvilinear translation.

❚ The second component of the acceleration of a particle which moves in plane curvilinear translation is called the tangential acceleration $a_t$. The direction of this quantity is along a line tangent to the curve. From Fig. 15.46c, and using the basic definition of acceleration, we have

$$a_t = \lim_{\Delta t \to 0} \frac{\Delta v_t}{\Delta t} = \frac{dv_t}{dt}$$

The positive sense of $a_t$ is the same as the positive sense of $\Delta v_t$. The last term in the above equation may be recognized as the time rate of change of the speed $v$ with which the particle moves along the curve.

Using

$$\frac{dv_t}{dt} = \frac{dv}{dt} = \dot{v}$$

the tangential component of acceleration in curvilinear translation has the form

$$a_t = \dot{v}$$

Since $s$ is the coordinate of length along the curve, the velocity $v$ may be written as

$$v = \frac{ds}{dt} = \dot{s}$$

The component of tangential acceleration may then be written as

$$a_t = \dot{v} = \ddot{s}$$

It may finally be observed that if the particle moves along the curved path with constant speed, the tangential acceleration is identically zero.

**14.58** (**a**) Give a summary of the normal and tangential components of acceleration of a particle which moves in plane curvilinear translation.

(**b**) Show the forms for the magnitude and direction of the resultant of the component accelerations in part (*a*).

(**c**) What is the major difference between visualizing the velocity and visualizing the acceleration of a particle which moves in plane curvilinear translation?

**I** (*a*)  The acceleration of a particle in plane curvilinear motion may be described by two components.   The first component, the *normal, or centripetal, acceleration*, is given by

$$a_n = \frac{v^2}{\rho}$$

The direction of this vector is along the radius of curvature $\rho$ of the curve, and the sense is *always* toward the center of curvature.

The second component of acceleration of a particle in plane curvilinear motion is called the *tangential acceleration*, given by

$$a_t = \dot{v} = \ddot{s}$$

The direction of this vector is along a line tangent to the curve, and the positive sense of $a_t$ is the same as the positive sense of $s$, the displacement along the curve.

The two components of acceleration for a particle in plane curvilinear translation are shown in Fig. 14.58.

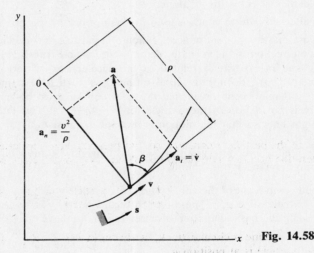

**Fig. 14.58**

(*b*)  The magnitude of the resultant acceleration of a particle which moves in plane curvilinear translation is given by

$$a = \sqrt{a_n^2 + a_t^2}$$

The direction $\beta$ of this acceleration, with respect to the tangent line to the curve, may be found from

$$\tan \beta = \frac{a_n}{a_t} = \frac{v^2}{\rho \dot{v}}$$

These results are seen in Fig. 14.58.

(*c*)  A very important difference between the visualization of the velocity and acceleration of a particle in plane curvilinear translation may be noted.   The velocity of the particle is *always* tangent to the curve, and the magnitude of this velocity is equal to the speed of the particle along the curve.   Thus, a mental picture of this motion may be formed readily.   The direction of the total acceleration of a particle moving along a curved path is *never* tangent to the curve.   The magnitude of the normal component of acceleration is directly proportional to the square of the speed and inversely proportional to the local radius of curvature, and the magnitude of the tangential component of acceleration is proportional to the rate of change of the speed.   It is generally very difficult to form a mental picture of the magnitude and direction of the acceleration of a particle which moves in curvilinear translation.

**14.59**   A particle moves along a curved track that lies in a horizontal plane, as shown in Fig. 14.59.   The speed of the particle has a constant value of 0.15 m/s.   Find the maximum value of acceleration which the particle experiences as it moves from *a* to *g*.

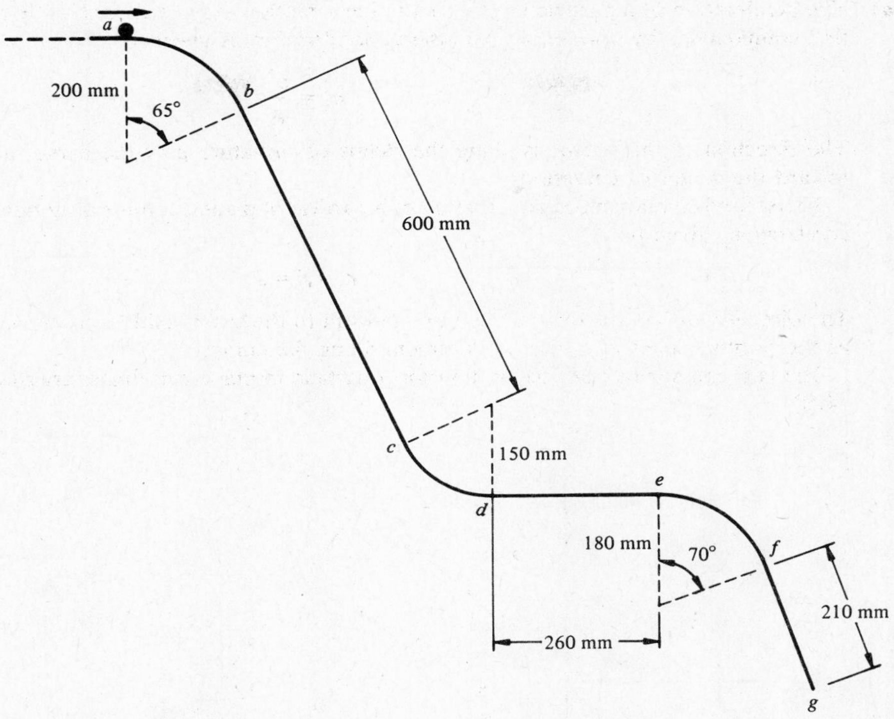

**Fig. 14.59**

▌ The speed of the particle is constant, so that the tangential acceleration component is zero. The maximum normal acceleration occurs where the radius of curvature of the track is minimum, along path *cd*. Using $v = 0.15\,\text{m/s}$,

$$a_{n,\text{max}} = \frac{v^2}{\rho} = \frac{0.15^2}{0.150} = 0.15\,\text{m/s}^2$$

**14.60** A particle travels around a circular track, as shown in Fig. 14.60*a*. The speed of the particle is 2.4 ft/s and, at $t = 0$, the particle is at position *a*.

(*a*) Find the position of the particle, and the *x* and *y* components of the velocity, when $t = 9\,\text{s}$.

(*b*) Find the normal and tangential components of the acceleration when $t = 9\,\text{s}$.

(*c*) Find the *x* and *y* components of the acceleration when $t = 9\,\text{s}$.

(*d*) State the results in parts (*a*) and (*c*) in formal vector notation.

▌ (*a*) The circumference of the track is given by $\pi(8) = 25.1\,\text{ft}$. At $t = 9\,\text{s}$,

$$s = vt = 2.4(9) = 21.6\,\text{ft}$$

The corresponding angular motion is

$$\theta = \frac{21.6}{25.1}\,(360°) = 310°$$

Fig 14.60*b* shows the position of the particle at $t = 9\,\text{s}$.
The velocity components of the particle at $t = 9\,\text{s}$ are shown in Fig. 14.60*c*. The values of the components are

$$v_x = 2.4\cos 50° = 1.54\,\text{ft/s} \qquad v_y = -2.4\sin 50° = -1.84\,\text{ft/s}$$

(*b*) The normal and tangential components of the velocity at $t = 9\,\text{s}$ are

$$a_n = \frac{v^2}{\rho} = \frac{2.4^2}{4} = 1.44\,\text{ft/s}^2 \qquad a_t = 0$$

(*c*) The *x* and *y* components of the acceleration at $t = 9\,\text{s}$ are shown in Fig. 14.60*d*. These components have the values

$$a_x = 1.44\sin 50° = 1.10\,\text{ft/s}^2 \qquad a_y = 1.44\cos 50° = 0.926\,\text{ft/s}^2$$

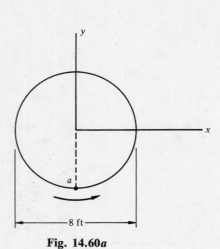

Fig. 14.60a

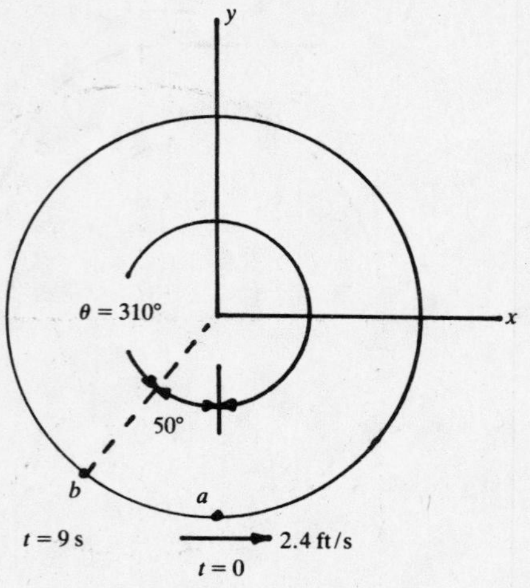

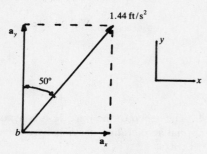

Fig. 14.60b

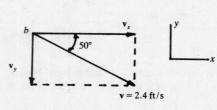

Fig. 14.60c

Fig. 14.60d

(*d*)  The position of the particle is given by

$$\mathbf{r} = (-4\sin 50°)\mathbf{i} - (4\cos 50°)\mathbf{j} = -3.06\mathbf{i} - 2.57\mathbf{j} \qquad \text{ft}$$

The velocity has the form

$$\mathbf{v} = 1.54\mathbf{i} - 1.84\mathbf{j} \qquad \text{ft/s}$$

The acceleration of the particle is

$$\mathbf{a} = 1.10\mathbf{i} + 0.926\mathbf{j} \qquad \text{ft/s}^2$$

**14.61**  A particle moves around a circular track with constant velocity, as shown in Fig. 14.61*a*.  At the instant that the particle is at point *a*, it has tangential deceleration of 80 m/s².

(*a*)  Find the normal acceleration and the magnitude and direction of the total acceleration at point *a*.

(*b*)  Find the *x* and *y* components of the velocity and acceleration at point *a*.

❚  (*a*)  The normal acceleration is

$$a_n = \frac{v^2}{\rho} = \frac{10^2}{0.5(900\text{ mm})(1\text{ m}/1{,}000\text{ mm})} = 222\text{ m/s}^2$$

The sense of this normal acceleration component is from *a* to the center of the track.  Since the particle is described as decelerating at point *a*, the sense of the tangential acceleration at this point must be opposite to the sense of the velocity.

   The two acceleration components are shown in Fig. 14.61*b*.  The magnitude of the total acceleration at *a* is

$$a = \sqrt{a_n^2 + a_t^2} = \sqrt{222^2 + (-80)^2} = 236\text{ m/s}^2$$

Angle $\beta$ is defined in Fig. 14.61*b*, and

$$\tan \beta = \frac{222}{80} = 2.78 \qquad \beta = 70.2°$$

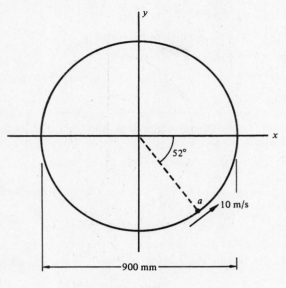

**Fig. 14.61a**

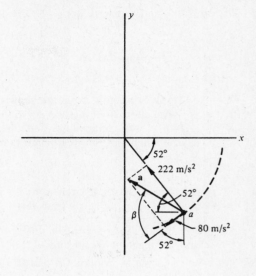

**Fig. 14.61b**

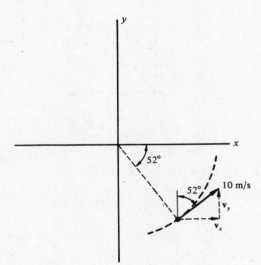

**Fig. 14.61c**

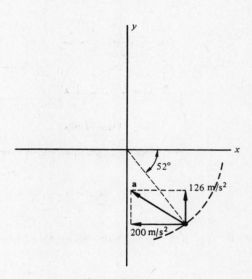

**Fig. 14.61d**

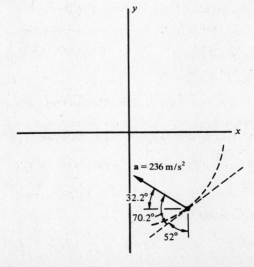

**Fig. 14.61e**

(b)  From Fig. 14.61c, the x and y components of the velocity are

$$v_x = 10 \sin 52° = 7.88 \text{ m/s} \qquad v_y = 10 \cos 52° = 6.16 \text{ m/s}$$

From Fig. 14.61b, the x and y components of the acceleration are

$$a_x = -222 \cos 52° - 80 \sin 52° = -200 \text{ m/s}^2 \qquad a_y = 222 \sin 52° - 80 \cos 52° = 126 \text{ m/s}^2$$

These two components are shown in Fig. 14.61d.

An alternative method for finding the x and y components of the acceleration is shown in Fig. 14.61e. The magnitude and direction of the resultant acceleration were found in part (a). The components of this vector are then found directly as

$$a_x = -236 \cos 32.2° = -200 \text{ m/s}^2 \qquad a_y = 236 \sin 32.2° = 126 \text{ m/s}^2$$

**14.62**  A particle moves along a circular track, shown in Fig. 14.62a, with constant tangential acceleration $a_t = 0.28 \text{ m/s}^2$. The particle starts from rest at point a.

(a)  Find the x and y components of the velocity when the particle first reaches point b.

(b)  Find the magnitude and direction of the acceleration when the particle first reaches point b.

(c)  Do the same as in parts (a) and (b) for the case where the particle is at point b the second time.

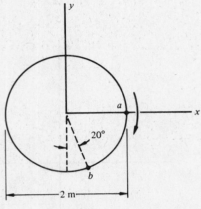

Fig. 14.62a

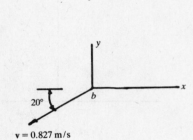

v = 0.827 m/s

Fig. 14.62b

▮ (a)  The displacement coordinate s is measured along the track in a clockwise sense from point a. The position of point b is given by

$$s = r\theta = (1) \frac{70°}{360°} 2\pi = 1.22 \text{ m}$$

The speed of the particle at point b is found from

$$v^2 = v_0^2 + 2a(s - s_0) = 2(0.28)1.22 \qquad v = 0.827 \text{ m/s}$$

The velocity at point b is shown in Fig. 14.62b. The x and y components of this velocity are

$$v_x = -0.827 \cos 20° = -0.777 \text{ m/s} \qquad v_y = -0.827 \sin 20° = -0.283 \text{ m/s}$$

(b)  The components of acceleration at point b are

$$a_n = \frac{v^2}{\rho} = \frac{0.827^2}{1} = 0.684 \text{ m/s}^2 \qquad a_t = 0.28 \text{ m/s}^2 \qquad a = \sqrt{a_n^2 + a_t^2} = \sqrt{0.684^2 + 0.28^2} = 0.739 \text{ m/s}^2$$

The acceleration at point b is shown in Fig. 14.62c. The direction of this vector is found from

$$\tan \theta_1 = \frac{0.28}{0.684} \qquad \theta_1 = 22.3° \qquad \theta_2 = 90° - 20° - \theta_1 = 47.7°$$

(c)  The displacement s when the particle is at point b the second time is given by

$$s = r\theta = (1) \frac{70° + 360°}{360°} 2\pi = 7.50 \text{ m}$$

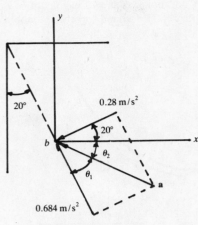

**Fig. 14.62c**

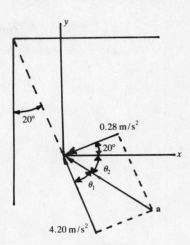

**Fig. 14.62d**

The corresponding value of the speed of the particle is found from

$$v^2 = v_0^2 + 2a(s - s_0) = 2(0.28)7.50 \qquad v = 2.05 \text{ m/s}$$

Using Fig. 14.62b, the components of the velocity are

$$v_x = -2.05 \cos 20° = -1.93 \text{ m/s} \qquad v_y = -2.05 \sin 20° = -0.701 \text{ m/s}$$

The acceleration at point $b$ is found from

$$a_n = \frac{v^2}{\rho} = \frac{2.05^2}{1} = 4.20 \text{ m/s}^2 \qquad a_t = 0.28 \text{ m/s}^2 \qquad a = \sqrt{a_n^2 + a_t^2} = \sqrt{4.20^2 + 0.28^2} = 4.21 \text{ m/s}^2$$

Figure 14.62d shows the acceleration at point $b$. The direction of this quantity is given by

$$\tan \theta_1 = \frac{0.28}{4.20} \qquad \theta_1 = 3.81° \qquad \theta_2 = 90° - 20° - \theta_1 = 66.2°$$

It may be observed that the direction of the resultant acceleration is very nearly equal to the direction of the normal component of this acceleration.

**14.63** (*a*) What is the expression for the *radius of curvature* of a plane curve which is defined by $y = f(x)$?

(*b*) What is a point of inflection on a plane curve?

▌ (*a*) Figure 14.63 shows a plane curve with a shape defined by $y = f(x)$. It is shown in texts on analytic geometry and calculus that the radius of curvature $\rho$ of a plane curve is given by

$$\frac{1}{\rho} = \frac{\pm d^2y/dx^2}{[1 + (dy/dx)^2]^{3/2}}$$

The terms $dy/dx$ and $d^2y/dx^2$ are the first two derivatives of the known function $y = f(x)$. If the equation above is written without the plus or minus sign, the result for $\rho$ will be positive or negative, depending on whether the general orientation of the radius of curvature is that shown in Fig. 14.63 as $\rho_a$ or $\rho_b$. Thus, in using this equation, *the proper sign is chosen to make the radius of curvature a positive value*.

(*b*) The point $c$ in Fig. 14.63 is referred to as a point of inflection. At this location on the curve, $\rho \rightarrow \infty$.

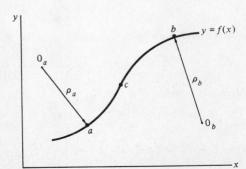

**Fig. 14.63**

**14.64** A particle moves along a track which has the shape of a parabola, as shown in Fig. 14.64a. The speed of the particle along the track has the constant value 100 in/s. The shape of the parabola is given by $y = 5 + 0.3x^2$, where $x$ and $y$ are in inches.

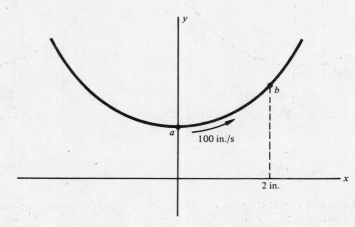

Fig. 14.64a

(*a*) Find the $x$ and $y$ components of the velocity at points $a$ and $b$.

(*b*) Find the magnitude and direction of the normal acceleration of the particle at $a$ and $b$.

▮ (*a*) At point $a$, from inspection of the figure,

$$v_x = 100 \text{ in/s} \qquad v_y = 0$$

The slope of the curve is

$$\frac{dy}{dx} = \frac{d}{dx}(5 + 0.3x^2) = 0.6x$$

At point $b$,

$$\frac{dy}{dx}\bigg|_{x=2} = 0.6x\bigg|_{x=2} = 0.6(2) = 1.2 = \tan\theta \qquad \theta = 50.2°$$

The orientation of the velocity at point $b$ is shown in Fig. 14.64b.

The $x$ and $y$ components of the velocity are

$$v_x = 100\cos 50.2° = 64.0 \text{ in/s} \qquad v_y = 100\sin 50.2° = 76.8 \text{ in/s}$$

(*b*) The first derivative of the curve is given in part (*a*). The second derivative is

$$\frac{d^2y}{dx^2} = 0.6$$

The radius of curvature at point $a$ is then

$$\frac{1}{\rho_a} = \frac{\pm d^2y/dx^2}{[1 + (dy/dx)^2]^{3/2}}\bigg|_{x=0} = \frac{\pm 0.6}{\{1 + [0.6(0)]^2\}^{3/2}} \qquad \rho_a = 1.67 \text{ in}$$

The normal component of acceleration of the particle at point $a$ is

$$a_{n,a} = \frac{v^2}{\rho_a} = \frac{100^2}{1.67} = 5.990 \text{ in/s}^2$$

This acceleration acts upward along the $y$ axis. At point $b$,

$$\frac{1}{\rho_b} = \frac{\pm d^2y/dx^2}{[1 + (dy/dx)^2]^{3/2}}\bigg|_{x=2} = \frac{\pm 0.6}{\{1 + [0.6(2)]^2\}^{3/2}} \qquad p_b = 6.35 \text{ in}$$

The normal acceleration at point $b$ is then

$$a_{n,b} = \frac{v^2}{\rho_b} = \frac{100^2}{6.35} = 1,570 \text{ in/s}^2$$

The direction of the normal acceleration at $b$ is at 50.2° from the direction of the $y$ axis, as shown in Fig. 14.64c.

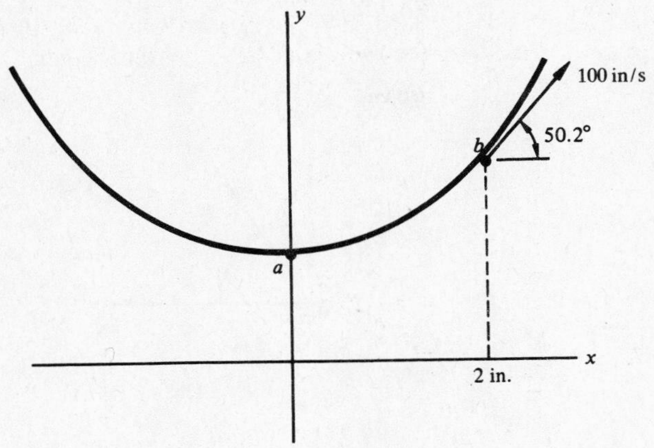

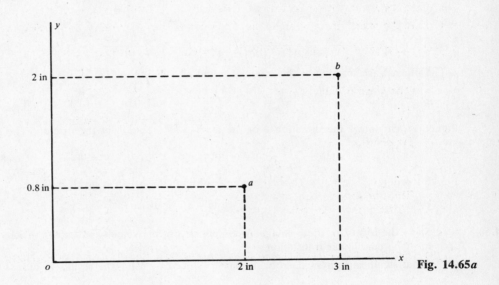

**Fig. 14.64b**

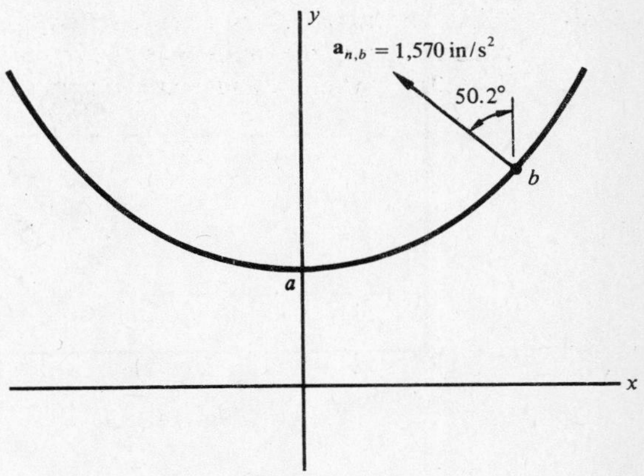

**Fig. 14.64c**

**14.65** A particle moves with constant speed of 22 ft/s along a path of parabolic shape that passes through points $a$, $b$, and the origin, shown in Fig. 14.65$a$. Find the magnitude, direction, and sense of the acceleration at the origin.

**Fig. 14.65a**

▎ The form of the parabolic curve is shown in Fig. 14.65$b$. The general form of the equation which defines a parabola is

$$y = ax^2 + bx + c$$

The constants $a$, $b$, and $c$ will now be found. Using $x = 0$ and $y = 0$,

$$c = 0$$

Using $x = 2$ and $y = 0.8$,

$$0.8 = a(2)^2 + b(2) \qquad 4a + 2b = 0.8 \tag{1}$$

Using $x = 3$ and $y = 2$,

$$2 = a(3)^2 + b(3) \qquad 9a + 3b = 2 \tag{2}$$

Equations (1) and (2) are two equations in the unknowns $a$ and $b$. These two terms are found as

$$a = 0.267 \qquad b = -0.134$$

The equation of the curve is then

$$y = 0.267x^2 - 0.134x$$

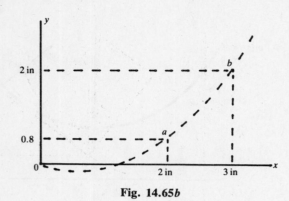

Fig. 14.65b          Fig. 14.65c

The first two derivatives of the curve have the forms

$$\frac{dy}{dx} = 2(0.267)x - 0.134 \qquad \frac{d^2y}{dx^2} = 2(0.267) = 0.534$$

The computation for the radius of curvature at the origin has the form

$$x = 0 \qquad \left.\frac{dy}{dx}\right|_{x=0} = -0.134$$

$$\left.\frac{1}{\rho}\right|_{x=0} = \left.\frac{\pm d^2y/dx^2}{[1 + (dy/dx)^2]^{3/2}}\right|_{x=0} = \frac{0.534}{[1 + (-0.134)^2]^{3/2}} \qquad \rho = 1.92 \text{ in}$$

The normal acceleration at the origin is found as

$$a_n = \frac{v^2}{\rho} = \frac{[22(12)]^2}{1.92} = 36,300 \text{ in/s}^2 = 3,030 \text{ ft/s}^2$$

Figure 14.65c shows the direction $\theta$ of the normal acceleration at the origin. The value of $\theta$ is found from

$$\tan \theta_1 = \left.\frac{dy}{dx}\right|_{x=0} = -0.134 \qquad |\theta_1| = 7.63° \qquad \theta = 90° \qquad -|\theta_1| = 82.4°$$

For this problem, $a_t = 0$. Thus, the acceleration at the origin is given by

$$a = a_n = 3,030 \text{ ft/s}^2$$

**14.66** (*a*) Show the forms for the rectangular components of the velocity and acceleration of a particle which moves in plane curvilinear translation.

(*b*) What significant physical problem may be solved by using the results in part (*a*)?

▌ (*a*) Figure 14.66*a* shows the velocity of a particle moving along a plane curve. Since the velocity is a vector quantity, it may be expressed in terms of two rectangular components. The directions of these components will be taken to be the directions of the coordinate axes. The two components $v_x$ and $v_y$ of the velocity $v$ are shown in the figure.

Angle $\theta$ is related to the slope of the curve by

$$\tan \theta = \frac{dy}{dx}$$

The relationships among the velocity and its components are

$$v_x = v \cos \theta \qquad v_y = \sin \theta$$

Elimination of $\theta$ from the above two equations results in

$$v = \sqrt{v_x^2 + v_y^2}$$

Figure 14.66*b* shows the normal and tangential components of the acceleration of the particle. These components are projected onto the $x$ and $y$ axes, with the results

$$a_x = a_t \cos \theta - a_n \sin \theta \qquad a_y = a_t \sin \theta + a_n \cos \theta$$

$a_x$ and $a_y$ are the $x$ and $y$ components of the total acceleration of the particle. The quantities $a_x$ and $a_y$ are positive in the positive coordinate senses. The magnitude of the acceleration is given by

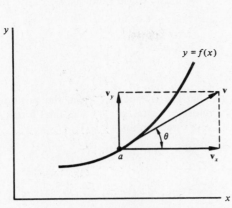

**Fig. 14.66a**

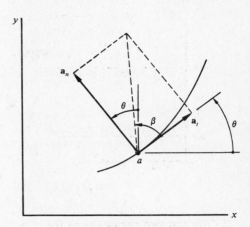

**Fig. 14.66b**

$$a = \sqrt{a_x^2 + a_y^2} = \sqrt{a_n^2 + a_t^2}$$

This equation illustrates that the acceleration may be expressed in terms of two *different* pairs of rectangular coordinates. The direction of the total acceleration with respect to the positive $x$ axis is then given by the angle $\theta + \beta$, where $\theta$ is given by the first equation in part ($a$) and $\beta$ was found in part ($b$) of Prob. 14.58.

The operations presented above are referred to as the description of the velocity and acceleration, in plane curvilinear translation, in terms of rectangular components.

(**b**) The results in part ($a$) may be used to characterize the very important problem of projectile motion in a gravitational field.

**14.67** (**a**) Derive the equations of motion of a particle in a gravitational acceleration field.

(**b**) What is the term used to describe the motion in part ($a$)?

(**c**) Compare the component motions of the particle in the horizontal and vertical directions.

▌ (**a**) Figure 14.67 shows the motion when a particle is launched with initial velocity in a vertical plane. Typical physical situations which would be represented by this motion are a cannon firing a projectile, a player hitting a baseball with a bat, or a pebble being thrown from an automobile tire. The path traced out by the particle is referred to as the *trajectory* of the particle.

The curvilinear translational motion of the particle will be expressed now in terms of *component motions* in the $x$ and $y$ directions. In this model of the actual motion, all air resistance effects will be neglected.

The motion of the particle in the $y$ direction is a case of freely falling motion of a body in a gravitational field. The initial velocity of the particle in this direction is $v_{0y}$. The solution for this motion was given in Prob. 14.43. For the present case, these equations have the forms

$$v_y = v_{0y} - gt \qquad y = y_0 + v_{0y}t - \tfrac{1}{2}gt^2 \qquad v_y^2 = v_{0y}^2 - 2g(y - y_0)$$

The component of the gravitational acceleration in the $x$ direction is zero. Thus,

$$v_x = \text{constant}$$

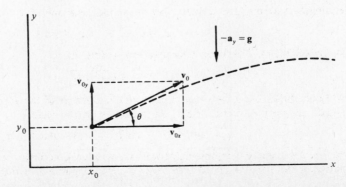

**Fig. 14.67**

The constant value of the $x$ component of velocity must be the same as $v_{0x}$, the $x$ component of the initial velocity $v_0$. Thus,

$$v_x = v_{0x} = \text{constant}$$

This equation is integrated, with the result

$$x = v_{0x}t + \text{constant}$$

The problem commences at $t = 0$ when the particle is launched from the position $x = x_0$, $y = y_0$. Thus, the constant in the above equation must be equal to $x_0$. The final forms of the equations for the component motion of the particle in the $x$ direction are

$$v_x = v_{0x} = \text{constant} \qquad x = x_0 + v_{0x}t$$

It follows from Fig. 14.67 that

$$v_{0x} = v_0 \cos \theta \qquad v_{0y} = v_0 \sin \theta$$

(b) The term used to describe the motion in part (a) is *plane projectile motion*.

(c) The case of projectile motion in a vertical plane exhibits the interesting characteristic that one component of the total motion, the $x$ motion, is rectilinear motion with *constant velocity*, while the other component, the $y$ motion, is rectilinear motion with *constant acceleration*.

**14.68** Show that the trajectory of a particle which moves in a vertical plane without air resistance is a plane parabolic curve.

▮ The initial position of the particle is assumed to be $x = x_0 = 0$ and $y = y_0$. The time $t$ which corresponds to position $x$ is found from

$$x = v_{0x}t \qquad t = \frac{x}{v_{0x}} \qquad (1)$$

The equation of motion in the $y$ direction is

$$y = y_0 + v_{0y}t - \tfrac{1}{2}gt^2 \qquad (2)$$

Using Eq. (1) in Eq. (2),

$$y = y_0 + v_{0y}\left(\frac{x}{v_{0x}}\right) - \tfrac{1}{2}g\left(\frac{x}{v_{0x}}\right)^2 = y_0 + \frac{v_{0y}}{v_{0x}}x - \frac{g}{2v_{0x}^2}x^2$$

The above equation is in the form

$$y = ax^2 + bx + c$$

which is the equation of a parabola.

**14.69** Figure 14.69 shows a shell being fired from an artillery weapon. The muzzle velocity of the shell upon exit from the gun tube is 1,000 ft/s.

(a) Find the maximum height reached by the shell and the time to reach this height.

(b) Find the duration of the flight and the distance from the gun at which the shell strikes the ground.

(c) Find the velocity with which the shell strikes the ground.

(d) Express the result in part (c) in formal vector notation.

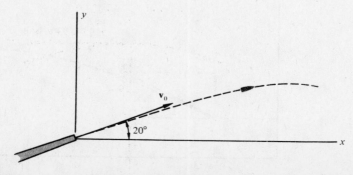

**Fig. 14.69**

**(a)** Since the origin of coordinates is placed at the muzzle,

$$x_0 = y_0 = 0$$

The initial conditions are

$$v_0 = 1{,}000 \text{ ft/s} \qquad v_{0x} = 1{,}000 \cos 20° = 940 \text{ ft/s} \qquad v_{0y} = 1{,}000 \sin 20° = 342 \text{ ft/s}$$

When the maximum height is reached,

$$v_y = 0 \qquad 0 = v_{0y} - gt \qquad 0 = 342 - 32.2t \qquad t = 10.6 \text{ s}$$

The maximum height is found from

$$y = y_0 + v_{0y}t - \tfrac{1}{2}gt^2 \qquad y_{max} = 0 + 342(10.6) - \tfrac{1}{2}(32.2)(10.6^2) = 1{,}820 \text{ ft}$$

The above result may be checked by using

$$v_y^2 = v_{0y}^2 - 2g(y - y_0) \qquad 0 = 342^2 - 2(32.2)(y_{max} - 0) \qquad y_{max} = 1{,}820 \text{ ft}$$

**(b)** When the shell strikes the ground, $y = 0$. The equation of motion is

$$y = y_0 + v_{0y}t - \tfrac{1}{2}gt^2$$

$$0 = 0 + 342t - \tfrac{1}{2}(32.2)t^2 \qquad t(16.1t - 342) = 0 \qquad t = 0, \quad 21.2 \text{ s}$$

The first value corresponds to the initial time. The shell strikes the ground 21.2 s after launch. The $x$ component of motion at this time is

$$x = x_0 + v_{0x}t \qquad x_{max} = 0 + 940(21.2) = 19{,}900 \text{ ft}$$

The shell strikes the ground at 19,900 ft, or 3.77 mi, from the gun.

**(c)** When the shell strikes the ground, $t = 21.2$ s. The corresponding value of the $y$ component of velocity is

$$v_y = v_{0y} - gt \qquad v_y = 342 - 32.2(21.2) = -341 \text{ ft/s} \approx -342 \text{ ft/s}$$

This velocity component has the same magnitude as the initial $y$ component of velocity, but is of opposite sense. The $x$ component of the striking velocity is the same as $v_{0x}$, given in part (a) as 940 ft/s. The magnitude of the striking velocity is

$$v = \sqrt{v_x^2 + v_y^2} = \sqrt{940^2 + (-341)^2} = 1{,}000 \text{ ft/s}$$

This value is the same as the magnitude of the initial velocity. This is an expected result, since there are no air-resistance losses.

**(d)** The velocity of the shell when it strikes the ground is

$$\mathbf{v} = 940\mathbf{i} - 341\mathbf{j} \qquad \text{ft/s}$$

**14.70** Find the radius of curvature of the trajectory of the shell in Prob. 14.69 when the shell is at its maximum height above the ground.

▮ When the shell is at its maximum height above the ground, the acceleration is $g$, acting vertically downward. This condition is shown in Fig. 14.70. The normal acceleration has the form

$$a_n = \frac{v^2}{\rho} = g \qquad \frac{940^2}{\rho} = 32.2 \qquad \rho = 27{,}400 \text{ ft}$$

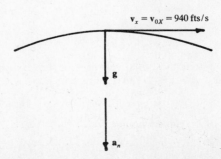

**Fig. 14.70**

**14.71** Figure 14.71*a* shows a ball rebounding from a pavement.

(*a*) Find the maximum height to which the ball will rise.

(*b*) Find the distance from the initial position at which the ball strikes the pavement again.

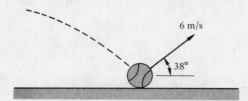

**Fig. 14.71*a***

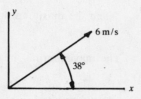

**Fig. 14.71*b***

▌ (*a*) Figure 14.71*b* shows the coordinate system and the initial velocity. The equation of motion is

$$v_y^2 = v_{0y}^2 - 2g(y - y_0)$$
$$0 = (6 \sin 38°)^2 - 2(9.81)y_{max} \qquad y_{max} = 0.695 \text{ m} = 695 \text{ mm}$$

(*b*) When the ball strikes the pavement,

$$v_y = v_{0y} - gt \qquad 0 = 6 \sin 38° - 9.81t \qquad t = 0.377 \text{ s}$$

The corresponding value of the *x* motion is

$$x = v_{0x}t = (6 \cos 38°)0.377 \qquad x = 1.78 \text{ m}$$

**14.72** A ball is struck by a paddle and has the initial velocity shown in Fig. 14.72*a*.

(*a*) Find the maximum height to which the ball will rise, and the time to reach this height.

(*b*) Find the total time that the ball is in flight.

(*c*) Find the velocity with which the ball strikes the ground.

(*d*) How far does the ball travel from the initial position before it strikes the ground?

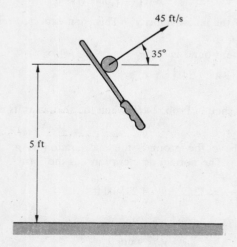

**Fig. 14.72*a***

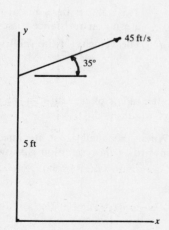

**Fig. 14.72*b***

▌ (*a*) The coordinate system and initial velocity are shown in Fig. 14.72*b*. The initial displacement is given by $x_0 = 0$ and $y_0 = 5$ ft. The equation of motion is

$$v_y^2 = v_{0y}^2 - 2g(y - y_0)$$

At the position of maximum height, $v_y = 0$.

$$0 = (45 \sin 35°)^2 - 2(32.2)(y - 5) \qquad y = 15.3 \text{ ft}$$

When the ball reaches the maximum height, using $v_y = 0$,

$$v_y = v_{0y} - gt \qquad 0 = 45 \sin 35° - 32.2t \qquad t = 0.802 \text{ s}$$

(*b*)   When the ball strikes the ground, using $y = 0$,

$$y = y_0 + v_{0y}t - \tfrac{1}{2}gt^2$$
$$0 = 5 + (45 \sin 35°)t - \tfrac{1}{2}(32.2)t^2 \qquad 16.1t^2 - 25.8t - 5 = 0$$
$$t = \frac{25.8 \pm \sqrt{25.8^2 - 4(16.1)(-5)}}{2(16.1)} = \frac{25.8 \pm 31.4}{2(16.1)} = 1.78 \text{ s}$$

The above value is the total time of flight.

(*c*)   Using the value $t = 1.78$ s, the $y$ component of velocity when the ball strikes the ground is found from

$$v_y = v_{0y} - gt = 45 \sin 35° - 32.2(1.78) = -31.5 \text{ ft/s}$$

The $x$ component of velocity is

$$v_x = v_{0x} = 45 \cos 35° = 36.9 \text{ ft/s}$$

The resultant velocity at impact is

$$v = \sqrt{v_x^2 + v_y^2} = \sqrt{36.9^2 + (-31.5)^2} = 48.5 \text{ ft/s}$$

The direction $\theta$, with respect to the $x$ axis, of the ball at impact with the ground is given by

$$\tan \theta = \left| \frac{v_y}{v_x} \right| = \frac{31.5}{36.9} \qquad \theta = 40.5°$$

(*d*)   The value of the $x$ coordinate when the ball strikes the ground is given by

$$x = v_{0x}t = (45 \cos 35°)1.78 = 65.6 \text{ ft}$$

The ball has moved 65.6 ft to the right, and 5 ft down, at the time it strikes the ground.

**14.73**   Figure 14.73 shows a projectile which is launched from a gun with initial velocity $v_0$.   The range is the distance from the gun to where the projectile strikes the ground.   For what angle of launch will the range be maximum?

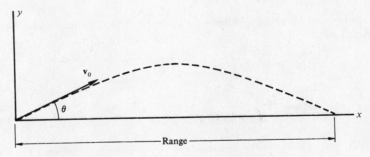

**Fig. 14.73**

❚   The initial displacement has the coordinates

$$x_0 = y_0 = 0$$

The $x$ displacement of the motion is

$$x = v_{0x}t = (v_0 \cos \theta)(t) \tag{1}$$

The $y$ displacement is

$$y = v_{0y}t - \tfrac{1}{2}gt^2$$

When the projectile strikes the ground, $y = 0$, so that

$$0 = (v_0 \sin \theta)t - \tfrac{1}{2}gt^2 \qquad t(v_0 \sin \theta - \tfrac{1}{2}gt) = 0$$

The root $t = 0$ is discarded, since this time corresponds to the initial time of the problem.   Thus,

$$v_0 \sin \theta - \tfrac{1}{2}gt = 0 \qquad t = \frac{2v_0 \sin \theta}{g} \tag{2}$$

Using Eq. (2) in Eq. (1),

$$x = (v_0 \cos \theta)\left(\frac{2v_0 \sin \theta}{g}\right) = \frac{v_0^2}{g}(2 \sin \theta \cos \theta) = \frac{v_0^2}{g}\sin 2\theta \qquad (3)$$

The sine function has its maximum positive value at 90°. The value $\theta_m$ of launch angle for maximum range is then

$$2\theta_m = 90° \qquad \theta_m = 45°$$

**14.74** Figure 14.74$a$ shows a gun emplacement. Find the minimum value of $\theta$ if the shell is to clear point $a$.

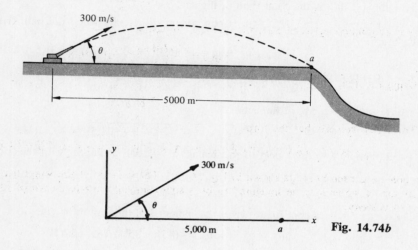

Fig. 14.74$b$

▮ Figure 14.74$b$ shows the coordinate system and initial conditions. Using Eq. (3) in Prob. 14.73,

$$x = \frac{v_0^2}{g}\sin 2\theta \qquad \sin 2\theta = \frac{gx}{v_0^2} = \frac{9.81(5,000)}{300^2} \qquad \theta = 16.5°$$

**14.75** Figure 14.75 shows the observed motion of a cricket. The distances $a$ and $b$ are estimated to be 1.5 m and 400 mm, respectively.

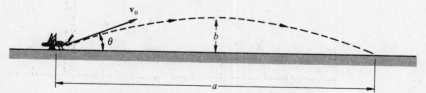

Fig. 14.75

(**a**) Find the values of the magnitude $v_0$ and direction $\theta$ of the initial velocity of the cricket.

(**b**) Express the results in part (**a**) in formal vector notation.

▮ (**a**) The origin of the coordinate system is at the initial position of the cricket. The first equation for the $y$ motion, with $v_y = 0$ at the maximum height, is

$$v_y^2 = v_{0y}^2 - 2gy \qquad 0 = (v_0 \sin \theta)^2 - 2gy = v_0^2 \sin^2 \theta - 2(9.81)0.4 \qquad (1)$$

The second equation for the $y$ motion is

$$v_y = v_{0y} - gt \qquad 0 = v_0 \sin \theta - gt \qquad t = \frac{v_0 \sin \theta}{g} \qquad (2)$$

The above value of $t$ is the time to reach the maximum height.
The equation for the $x$ motion is

$$x = v_{0x}t \qquad (3)$$

Using Eq. (2) in Eq. (3),

$$x = v_0 \cos \theta \, \frac{v_0 \sin \theta}{g}$$

Using $x = \frac{1}{2}x_{max} = 1.5/2$,

$$\frac{1.5}{2} = \frac{v_0^2 \sin \theta \cos \theta}{9.81} \qquad (4)$$

From Eq. (1),

$$v_0^2 = \frac{2(9.81)0.4}{\sin^2 \theta} \qquad (5)$$

$v_0^2$ is eliminated between Eqs. (4) and (5), to obtain

$$\tan \theta = \frac{(2)2(0.4)}{1.5} \qquad \theta = 46.8°$$

Using Eq. (1),

$$v_0 = 3.84 \text{ m/s}$$

(b) The initial velocity has the form

$$\mathbf{v}_0 = (v_0 \cos \theta)\mathbf{i} + (v_0 \sin \theta)\mathbf{j} = (3.84 \cos 46.8°)\mathbf{i} + (3.84 \sin 46.8°)\mathbf{j} = 2.63\mathbf{i} + 2.80\mathbf{j} \qquad \text{m/s}$$

**14.76** A person holds a garden hose, as shown in Fig. 14.76. The velocity of the water leaving the nozzle is 50 ft/s, and the stream of water may be imagined to be a succession of particles, each of which has the same initial conditions of velocity.

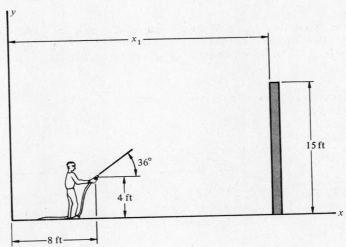

**Fig. 14.76**

(a) For what range of values of $x_1$ will the stream of water clear the vertical wall?

(b) Find the $y$ components of velocity corresponding to the extreme values of $x_1$ found in part (a).

▌ (a) From the figure,

$$x_0 = 8 \text{ ft} \qquad y_0 = 4 \text{ ft} \qquad v_{0x} = 50 \cos 36° = 40.5 \text{ ft/s} \qquad v_{0y} = 50 \sin 36° = 29.4 \text{ ft/s}$$

The time $t$ when the fluid stream is at the height of the wall will be found first. Using

$$y = y_0 + v_{0y}t - \tfrac{1}{2}gt^2 \qquad 15 = 4 + 29.4t - \tfrac{1}{2}(32.2)t^2 \qquad 16.1t^2 - 29.4t + 11 = 0$$

The solution to this quadratic equation is

$$t = \frac{29.4 \pm \sqrt{(-29.4)^2 - 4(16.1)(11)}}{2(16.1)} = 0.525 \text{ s}, \quad 1.30 \text{ s}$$

The distances $x$ which correspond to the above two times are

$$x = x_0 + v_{0x}t$$

$$x_{1,min} = 8 + 40.5(0.525) = 29.3 \text{ ft} \qquad x_{1,max} = 8 + 40.5(1.30) = 60.7 \text{ ft}$$

The stream will theoretically clear the wall if $29.3 \text{ ft} \le x_1 \le 60.7 \text{ ft}$.

(b)
$$v = v_{0y} - gt$$

At $x_{1,min}$, when $t = 0.525 \, s$,

$$v = 29.4 - 32.2(0.525) = 12.5 \text{ ft/s}$$

At $x_{1,max}$, when $t = 1.30 \, s$,

$$v = 29.4 - 32.2(1.30) = -12.5 \text{ ft/s}$$

It may be seen that at the minimum value of $x_1$, the stream has upward motion, and that at the maximum value of $x_1$, the stream has downward motion.

**14.77** (a) Find the value of $v_0$ for which the cannon projectile in Fig. 14.77a will just clear the top edge of the vertical wall.

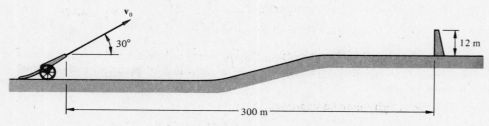

Fig. 14.77a

(b) Using the value of $v_0$ found in part (a), find the magnitude and direction of the velocity with which the projectile passes the wall. Express the results in scalar and vector forms.

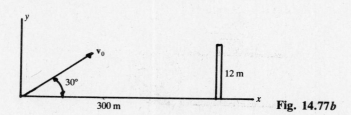

Fig. 14.77b

▌ (a) Figure 14.77b shows the coordinate system. The equation of motion in the y direction is

$$y = y_0 + v_{0y}t - \tfrac{1}{2}gt^2$$

Using the conditions $y = 12 \text{ m}$ and $y_0 = 0$, the above equation appears as

$$12 = (v_0 \sin 30°)t - \tfrac{1}{2}(9.81)t^2 \qquad (1)$$

The equation of motion in the x direction, with $x_0 = 0$, is

$$x = v_{0x}t \qquad 300 = (v_0 \cos 30°)t \qquad t = \frac{300}{v_0 \cos 30°} \qquad (2)$$

Using $t$ from Eq. (2) in Eq. (1) results in

$$12 = v_0 \sin 30°\left(\frac{300}{v_0 \cos 30°}\right) - \frac{1}{2}(9.81)\frac{300^2}{v_0^2 \cos 30°}$$

$$v_0^2 = \frac{\tfrac{1}{2}(9.81)300^2}{(300 \tan 30° - 12)\cos^2 30°}$$

$$v_0 = 60.4 \text{ m/s}$$

(*b*) Using Eq. (2),

$$t = \frac{300}{60.4 \cos 30°} = 5.74 \text{ s}$$

The velocity of the projectile in the *y* direction, at $t = 5.74$ s, is

$$v_y = v_{0y} - gt = 60.4 \sin 30° - 9.81(5.74) = -26.1 \text{ m/s}$$

The velocity in the *x* direction is

$$v_x = (60.4 \cos 30°) = 52.3 \text{ m/s}$$

The velocity with which the projectile passes the wall is

$$v = \sqrt{v_x^2 + v_y^2} = \sqrt{52.3^2 + (-26.1)^2} = 58.5 \text{ m/s}$$

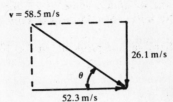

Fig. 14.77*c*

The direction $\theta$ of this velocity, using Fig. 14.77*c*, is found from

$$\tan \theta = \frac{26.1}{52.3} \qquad \theta = 26.5°$$

The velocity may be expressed in vector form as

$$\mathbf{v} = 52.3\mathbf{i} - 26.1\mathbf{j} \qquad \text{m/s}$$

**14.78** Figure 14.78*a* shows a conveyor belt system for moving metal castings. The castings are assumed to leave the upper conveyor belt at point *a* and strike the lower conveyor belt at point *b*. Find the maximum permissible value of the spacing dimension *x*.

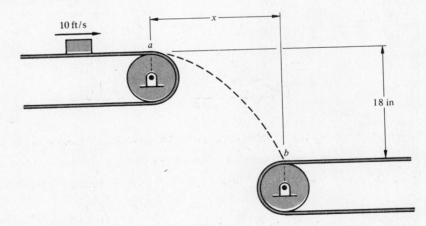

Fig. 14.78*a*

▮ Figure 14.78*b* shows the coordinate system. The initial conditions are

$$x_0 = y_0 = 0 \qquad v_{0x} = 10 \text{ ft/s} \qquad v_{0y} = 0$$

The time for the castings to fall from *a* to *b* is found from

$$y = y_0 + v_{0y}t - \tfrac{1}{2}gt^2 \qquad -\tfrac{18}{12} = -\tfrac{1}{2}(32.2)t^2 \qquad t = 0.305 \text{ s}$$

The value of *x* is then found from

$$x = v_{0x}t = 10(0.305) = 3.05 \text{ ft} = 36.6 \text{ in}$$

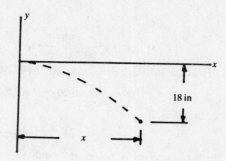

**Fig. 14.78b**

**14.79**   Figure 14.79a shows the vertical wall of a tank that contains water.   Fluid flows through two small orifices $A$ and $B$ in the wall.   If all friction losses are neglected, it can be shown that the velocity of the fluid jet is

$$v = \sqrt{2gh}$$

where $h$ is the vertical distance between the center of the orifice and the surface of the liquid.

(*a*)   Find the time for a particle of water leaving each orifice to reach the intersection point.

(*b*)   Find the locations $x$ and $y$ at which the centerlines of the two fluid streams intersect.

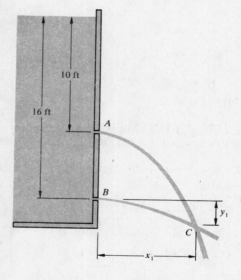

**Fig. 14.79a**

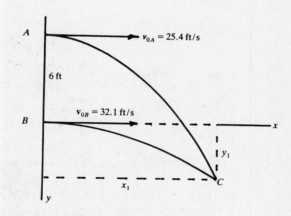

**Fig. 14.79b**

(*a*)   Figure 14.79b shows the trajectories of the two streams.   The velocities of the two streams at the orifices are found as

$$v_{0A} = \sqrt{2(32.2)10} = 25.4 \text{ ft/s} \qquad v_{0B} = \sqrt{2((32.2)16} = 32.1 \text{ ft/s}$$

$t_A$ is the time for a particle of water to travel from $A$ to $C$, and $t_B$ is the time for a particle of water to travel from $B$ to $C$.

For both cases, with   $x_0 = 0$,

$$x = v_{0x}t = 25.4t_A = 32.1t_B \qquad t_A = \frac{32.1}{25.4}\, t_B \tag{1}$$

$y_{AC}$ is the vertical distance between points $A$ and $C$, and $y_{BC}$ is the vertical distance between $B$ and $C$.   From Fig. 14.79b,

$$y_{AC} = 6 + y_{BC}$$

Using   $y = \frac{1}{2}gt^2$,

$$\tfrac{1}{2}(32.2)t_A^2 = 6 + \tfrac{1}{2}(32.2)t_B^2 \tag{2}$$

Equation (1) is used in Eq. (2), with the result

$$16.1\left(\frac{32.1}{25.4}\,t_B\right)^2 = 6 + 16.1t_B^2 \qquad t_B = 0.790\,\text{s}$$

From Eq. (1),

$$t_A = \frac{32.1}{25.4}\,(0.790) = 0.998\,\text{s}$$

(b) The position coordinates of the intersection point are

$$x_1 = v_{0A}t = 25.4(0.998) = 25.3\,\text{ft} \qquad y_1 = y_{BC} = \tfrac{1}{2}gt_B^2 = \tfrac{1}{2}(32.2)0.790^2 = 10.0\,\text{ft}$$

**14.80** Figure 14.80a shows two tanks on horizontal ground. Find the value of the distance $x$, if the shell fired from tank $A$ is to hit tank $B$.

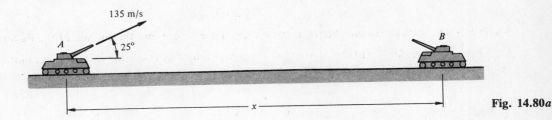

Fig. 14.80a

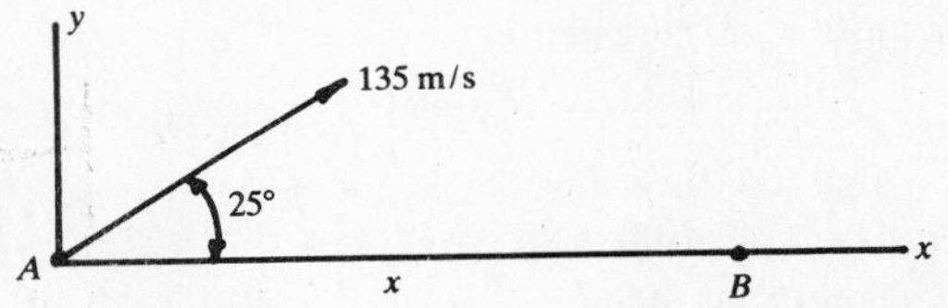

Fig. 14.80b

▌ Figure 14.80b shows the coordinate system. The $y$ equation of motion of the shell is

$$y = y_0 + v_{0y}t - \tfrac{1}{2}gt^2$$

When the shell hits the tank, $y = y_0 = 0$. Using the above equation,

$$0 = (135\sin 25°)t - \tfrac{1}{2}(9.81)t^2 = t[135\sin 25° - \tfrac{1}{2}(9.81)t] \qquad t=0\,\text{s}, \quad 11.6\,\text{s}$$

The time $t=0$ corresponds to the time of firing. Thus, the shell hits tank $B$ at $t=11.6\,\text{s}$. The $x$ displacement of the shell, with $x_0=0$, is then

$$x = v_{0x}t = (135\cos 25°)11.6 = 1.420\,\text{m}$$

**14.81** (a) Tank $A$ in Prob. 14.80 is stationary, and tank $B$ moves toward tank $A$ at a speed of 72 km/h. At $t=0$ the gunner in tank $A$ fires. At this instant, the separation distance of the two tanks is $x$. Find the value of $x$ if the shell is to hit tank $B$.

(b) Do the same as in part (a), if tank $B$ is stationary and tank $A$ advances toward tank $B$ at a speed of 72 km/h.

(c) Do the same as in part (a), if both tanks approach each other with speeds of 72 km/h.

▌ (a) The speed of tank $B$ is

$$v_B = 72\left(\frac{1,000}{3,600}\right) = 20\,\text{m/s}$$

Figure 14.81a shows a diagram of the events. From Prob. 14.80, the shell is in flight for 11.6 s, and lands 1,420 m to the right of $A$. During this flight time, tank $B$ moves through a distance

$$x_1 = v_B t = 20(11.6) = 232\,\text{m}$$

Thus, the required value of $x$ is given by

$$x = 1,420 + 232 = 1,650\,\text{m}$$

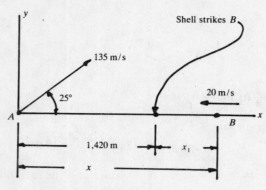

Fig. 14.81a

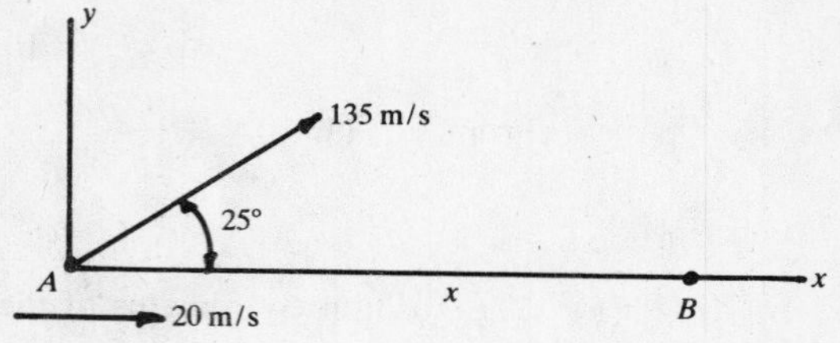

Fig. 14.81b

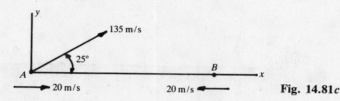

Fig. 14.81c

(b) A diagram of the events is shown in Fig. 14.81b. From Prob. 14.80, the shell is in flight for 11.6 s. The x motion of the shell is

$$x = v_{0x}t = (135 \cos 25° + 20)11.6 = 1,650 \text{ m}$$

It may be seen that this is the same result as in part (a), where tank A is stationary and tank B advances toward tank A at 20 m/s.

(c) Figure 14.81c shows the diagram of the events. From Prob. 14.80, the shell is in flight for 11.6 s. During this time, from part (b), the shell travels in the x direction through a distance 1,650 m. During this same time, from part (a), tank B moves through a distance 232 m. Thus, the total required separation distance between the two tanks is given by

$$x = 1,650 + 232 = 1,880 \text{ m}$$

**14.82** Figure 14.82a shows a gun emplacement and a stationary tank. The shell is fired from the gun and hits the tank.

(a) How long is the shell in flight?

(b) What is the value of the distance $x_1$?

(c) With what speed does the shell hit the tank?

(d) Give the position of the tank with respect to the gun emplacement, in both scalar and vector forms.

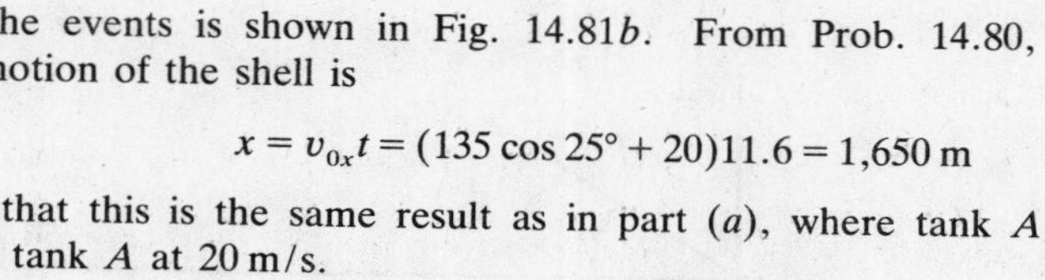

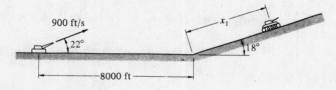

Fig. 14.82a

▌ (a) Figure 14.82b shows the coordinate system. The x motion of the shell, with $x_0 = 0$, is

$$x = v_{0x}t \qquad 8,000 + x_1 \cos 18° = (900 \cos 22°)t \qquad 8,000 + 0.951x_1 = 834t \qquad x_1 = \frac{834t - 8,000}{0.951} \quad (1)$$

The y motion of the shell, with $y_0 = 0$, is

$$y = v_{0y}t - \tfrac{1}{2}gt^2 \qquad x_1 \sin 18° = (900 \sin 22°)t - \tfrac{1}{2}(32.2)t^2 \qquad 0.309x_1 = 337t - 16.1t^2 \quad (2)$$

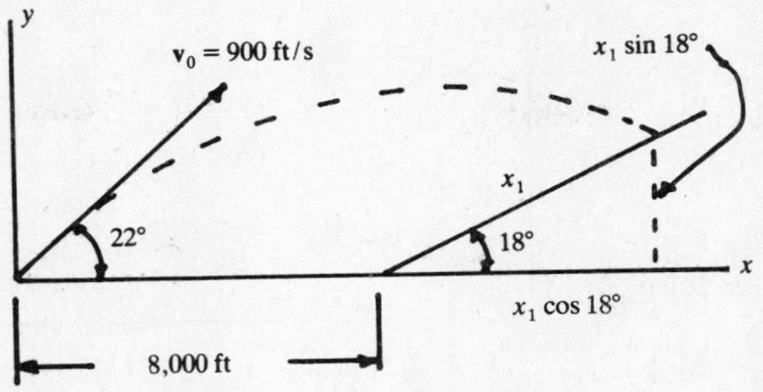

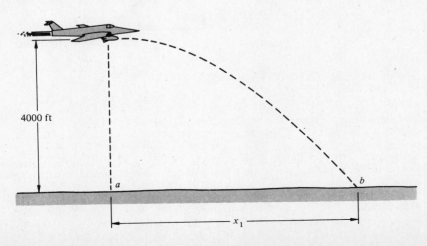

Fig. 14.82b

Using Eq. (1) in Eq. (2),

$$0.309\left[\frac{834t - 8,000}{0.951}\right] = 337t - 16.1t^2 \qquad 16.1t^2 - 66.0t - 2,600 = 0$$

$$t = \frac{66 \pm \sqrt{(-66)^2 - 4(16.1)(-2,600)}}{2(16.1)} = \frac{66 \pm 414}{2(16.1)} = 14.9 \text{ s}$$

(b) Using Eq. (1),

$$x_1 = \frac{834(14.9) - 8,000}{0.951} = 4,650 \text{ ft}$$

(c) The shell is in flight for 14.9 s. The components of velocity with which the shell strikes the tank are

$$v_x = v_{0x} = 900 \cos 22° = 834 \text{ ft/s} \qquad v_y = v_{0y} - gt = 900 \sin 22° - 32.2(14.9) = -143 \text{ ft/s}$$

The resultant impact speed is

$$v = \sqrt{v_x^2 + v_y^2} = \sqrt{834^2 + (-143)^2} = 846 \text{ ft/s}$$

(d) The position of the tank with respect to the origin of the coordinate system in Fig. 14.82b is

$$x = 8,000 + x_1 \cos 18° = 8,000 + 4,650 \cos 18° = 12,400 \text{ ft} \qquad y = x_1 \sin 18° = 4,650 \sin 18° = 1,440 \text{ ft}$$

In vector form,

$$\mathbf{s} = 12,400\mathbf{i} + 1,440\mathbf{j} \qquad \text{ft}$$

**14.83** A bomber flies in a horizontal direction with a speed of 400 mi/h, as shown in Fig. 14.83a. The bombardier wishes to hit a target at $b$.

(a) How many seconds before the plane is directly over the target should the bomb be released from the plane?

(b) Find the distance $x_1$ which corresponds to the solution to part (a).

(c) Find the magnitude and direction of the velocity with which the bomb strikes the target.

(d) Express the initial and final velocities of the bomb in formal vector notation.

4000 ft

Fig. 14.83a

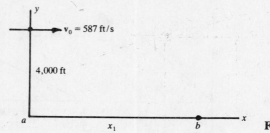

Fig. 14.83*b*

**❚** (*a*)  Figure 14.83*b* shows the coordinate system.  The initial speed of the plane is expressed in ft/s as

$$v_0 = v_{0x} = 400 \frac{\text{mi}}{\text{h}} \left( \frac{1\,\text{h}}{3600\,\text{s}} \right) \left( \frac{5{,}280\,\text{ft}}{1\,\text{mi}} \right) = 587\,\text{ft/s}$$

The equation of motion in the *y* direction is

$$y = y_0 + v_{0y}t - \tfrac{1}{2}gt^2$$

Using  $y = 0$,  $y_0 = 4{,}000\,\text{ft}$,  and  $v_{0y} = 0$,

$$0 = 4{,}000 - \tfrac{1}{2}(32.2)t^2 \qquad t = 15.8\,\text{s}$$

(*b*)  The equation of motion in the *x* direction, with  $x_0 = 0$,  is

$$x_1 = v_{0x}t = 587(15.8) = 9{,}270\,\text{ft} = 1.76\,\text{mi}$$

(*c*)  The *y* component of velocity is given by

$$v_y = v_{0y} - gt$$

With  $v_{0y} = 0$,  the above equation has the form

$$v_y = -32.2(15.8) = -509\,\text{ft/s}$$

From part (*a*),  $v_x = v_{0x} = 587\,\text{ft/s}$.  The velocity when the bomb hits the target is shown in Fig. 14.83*c*.  The magnitude of the velocity is given by

$$v = \sqrt{v_x^2 + v_y^2} = \sqrt{587^2 + 509^2} = 777\,\text{ft/s} = 530\,\text{mi/n}$$

The direction of the velocity is found from

$$\tan \theta = \frac{509}{587} \qquad \theta = 40.9°$$

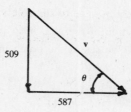

Fig. 14.83*c*

(*d*)  The initial velocity is

$$\mathbf{v}_0 = 587\mathbf{i} \qquad \text{ft/s}$$

The final velocity, when the bomb strikes the target, is

$$\mathbf{v} = 587\mathbf{i} - 509\mathbf{j} \qquad \text{ft/s}$$

**14.84**  Do the same as in Prob. 14.83 if the plane is in a glide of constant direction, with a speed of 400 mi/h, as shown in Fig. 14.84*a*.

**❚** (*a*)  The coordinate system is shown in Fig. 14.84*b*, together with the components of the initial velocity.  The equation of motion in the *y* direction is

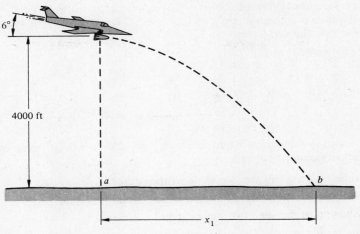

Fig. 14.84a

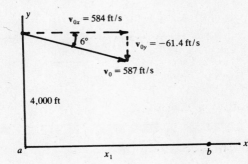

Fig. 14.84b

$$y = y_0 + v_{0y}t - \tfrac{1}{2}gt^2$$

Using $y = 0$, $y_0 = 4{,}000$ ft, and the value of $v_{0y}$ in Fig. 14.84b,

$$0 = 4{,}000 - 61.4t - \tfrac{1}{2}(32.2)t^2 \qquad 16.1t^2 + 61.4t - 4{,}000 = 0$$

$$t = \frac{-61.4 \pm \sqrt{61.4^2 - 4(16.1)(-4{,}000)}}{2(16.1)} = \frac{-61.4 \pm 511}{2(16.1)} = 14.0 \text{ s}$$

(b) The equation of motion in the $x$ direction, with $x_0 = 0$, is

$$x_1 = v_{0x}t = 584(14.0) = 8{,}180 \text{ ft} = 1.55 \text{ mi}$$

(c) The component of velocity in the $y$ direction is

$$v_y = v_{0y} - gt$$

Using the value of $v_{0y}$ from Fig. 14.84b,

$$v_y = -61.4 - 32.2(14.0) = -512 \text{ ft/s}$$

From Fig. 14.84b, $v_x = v_{0x} = 584$ ft/s, and the velocity has the form shown in Fig. 14.84c. The magnitude and direction of the velocity are found as

$$v = \sqrt{v_x^2 + v_y^2} = \sqrt{584^2 + 512^2} = 777 \text{ ft/s} = 530 \text{ mi/h} \qquad \tan \theta = \frac{512}{584} \qquad \theta = 41.2°$$

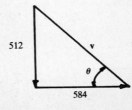

Fig. 14.84c

The results above may be compared with the solution to Prob. 14.84. When the plane is in a glide, both the time and the distance $x_1$ are reduced. The velocity with which the bomb strikes the target, however, has the same value for both cases.

(d) The initial velocity is

$$\mathbf{v}_0 = 584\mathbf{i} - 61.4\mathbf{j} \qquad \text{ft/s}$$

The final velocity is

$$\mathbf{v} = 584\mathbf{i} - 512\mathbf{j} \qquad \text{ft/s}$$

**14.85** Figure 14.85a shows an antiaircraft gun emplacement. At $t = 0$ a plane passes point $a$, traveling in a horizontal direction at constant speed.

(a) At what time of firing will the gun shell hit the plane?

(b) How far is the plane from the gun emplacement when it is hit?

(c) Find the velocity of the shell just before it hits the plane.

(d) Express the results in parts (b) and (c) in formal vector notation.

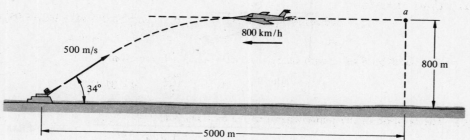

Fig. 14.85a

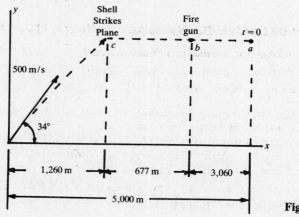

Fig. 14.85b

▮ (a) Figure 14.85b shows the coordinate system. Point $b$ in this figure is the position of the plane when the gun is fired, and point $c$ is the position of the plane when it is hit by the shell. The constant speed of the plane is

$$v_{\text{plane}} = 800 \text{ km/h} = 222 \text{ m/s}$$

The time for the shell to reach a height of 800 m, with the conditions $y_0 = 0$ and $y = 800$ m, is found from

$$y = y_0 + v_{0y}t - \tfrac{1}{2}gt^2$$
$$800 = (500 \sin 34°)t - \tfrac{1}{2}(9.81)t^2 \qquad 4.91t^2 - 280t + 800 = 0$$

$$t = \frac{280 \pm \sqrt{(-280)^2 - 4(4.91)800}}{2(4.91)} = \frac{280 \pm 250}{2(4.91)} = 3.05 \text{ s}, \quad 54.0 \text{ s}$$

For $t = 54.0$ s the travel of the plane in the $x$ direction is given by

$$x = v_{\text{plane}}t = (222)54 = 12,000 \text{ m} > 5,000 \text{ m}$$

Based on this result, the time of 3.05 s for the shell to reach the height of 800 m is used. The horizontal motion of the shell during this time is given by

$$x = v_{0x}t = (500 \cos 34°)3.05 = 1,260 \text{ m}$$

At $t = 0$, the plane is at point $a$ in Fig. 14.85$b$. 3.05 s after the shell is fired, the plane should be at point $c$. The distance traveled by the plane in 3.05 s is $222(3.05) = 677$ m. In order for the shell to hit the plane, the plane must be at location $b$ when the gun is fired. The time $t_1$ for the plane to travel from $a$ to $b$ is found from

$$v_{\text{plane}}t = s \qquad 222t_1 = 3,060 \qquad t_1 = 13.8 \text{ s}$$

The shell will hit the plane if the gun is fired at $t = 13.8$ s.

**(b)** From Fig. 14.85$b$, the plane is 1,260 m to the right, and 800 m above, the gun when it is hit. The distance $s_0$ from the gun to the plane is

$$s_0 = \sqrt{1,260^2 + 800^2} = 1,490 \text{ m}$$

**(c)** Using $t = 3.05$ s,

$$v_x = v_{0x} = 500 \cos 34° = 415 \text{ m/s} \qquad v_y = v_{0y} - gt = 500 \sin 34° - 9.81(3.05) = 250 \text{ m/s}$$

$$v = \sqrt{v_x^2 + v_y^2} = \sqrt{415^2 + 250^2} = 484 \text{ m/s}$$

The direction $\theta$ of $v$, with respect to the $x$ axis, is found from

$$\tan \theta = \frac{v_y}{v_x} = \frac{250}{415} \qquad \theta = 31.1°$$

**(d)** The position of the plane when the shell strikes it is

$$\mathbf{s} = 1,260\mathbf{i} + 800\mathbf{j} \qquad \text{m}$$

The velocity with which the shell strikes the plane is

$$\mathbf{v} = 415\mathbf{i} + 250\mathbf{j} \qquad \text{m/s}$$

## 14.4 ABSOLUTE AND RELATIVE DISPLACEMENT, VELOCITY, AND ACCELERATION

**14.86** **(a)** What is an *absolute*, or *inertial*, coordinate system?

**(b)** How may a particle, or point, be referenced to an absolute coordinate system.

**(c)** What is meant by the terms *absolute velocity* and *absolute acceleration*?

**❙ (a)** Figure 14.86 shows an *xyz* coordinate system imagined to be affixed to a distant star. A coordinate system such as this is referred to as an absolute, or inertial, coordinate reference system. An inertial coordinate system may also be described as a coordinate system within which Newton's laws of motion are valid. In the majority of engineering problems in dynamics, a set of axes that are imagined to be attached to the earth may be considered to be absolute coordinates.

**(b)** The particle or point $P$ may be referenced to the inertial coordinate system by the *absolute displacement, or position, vector* **s**, as shown in Fig. 14.86.

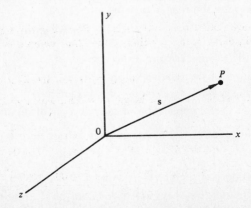

**Fig. 14.86**

(*c*) If velocity **v** and acceleration **a** of the particle are measured in the inertial coordinate system, then these quantities are referred to as absolute velocity and absolute acceleration.

**14.87** (*a*) Give the form for the *relative displacement* of two particles.

(*b*) Show a convention that may be used to distinguish between absolute and relative quantities.

▌ (*a*) A situation that occurs frequently in problems in dynamics is when one particle moves relative to a second particle that is also in motion.

Figure 14.87*a* shows two particles, *a* and *b*, which are constrained to move in the *xy* plane. These axes are assumed to be absolute coordinates. At the instant of observation shown in the figure, the two particles *a* and *b* have the displacements $s_a$ and $s_b$, and the absolute velocities $v_a$ and $v_b$.

The relative displacement $s_{ab}$ is defined to be

$$s_{ab} = s_a - s_b$$

The term $s_{ab}$ is read as "the displacement of *a* relative to *b*." Note that the above equation represents a *vector* subtraction. This operation was defined in Prob. 2.9. The equation may be written as

$$s_{ab} = s_a + (-s_b)$$

The construction is shown in Fig. 14.87*b*. It may be seen that, for the situation in Fig. 14.87*a*, particle *a* is located *relative* to particle *b* in a direction and sense that is generally down and to the left of particle *b*.

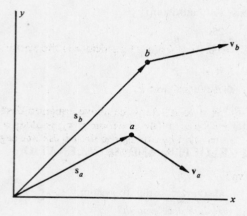

Fig. 14.87*a*

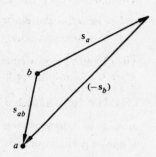

Fig. 14.87*b*

(*b*) A subscripting convention that may be used is that a single subscript on a displacement, velocity, or acceleration term represents an *absolute quantity*. Thus, $s_a$ and $s_b$ are absolute displacements. A double subscript on a term represents a *relative quantity*. The second subscript indicates the particle that is the reference point, while the first subscript identifies the particle whose relative motion quantity is being described. Thus, the term $s_{ab}$ is the relative displacement of particle *a* with respect to particle *b*, while the term $s_{ba}$ represents the relative displacement of particle *b* with respect to particle *a*. It is left as an exercise for the reader to show that

$$s_{ab} = -s_{ba}$$

**14.88** (*a*) Give the form for the *relative velocity* of two particles.

(*b*) Give the interpretation of the relative velocity.

(*c*) Give the form for the *relative acceleration* of two particles.

(*d*) What is an important useful application of the concepts of relative velocity and acceleration?

▌ (*a*) The relative velocity of article *a* with respect to particle *b* is defined to be

$$v_{ab} = v_a - v_b$$

The above equation is a vector subtraction.

(*b*) The concept of relative velocity may have either of the following two physical interpretations:

1. The relative velocity of $a$ with respect to $b$ is the velocity which an observer, who is imagined to travel with the motion of particle $b$, would perceive the velocity of particle $a$ to be.

2. If the velocity $-\mathbf{v}_b$ were added to both of the original velocities $\mathbf{v}_a$ and $\mathbf{v}_b$, then particle $b$ would have zero absolute velocity and the relative velocity $\mathbf{v}_{ab}$ would then be the absolute velocity of particle $a$. This effect is shown in Figs. 14.88$a$ through $c$.

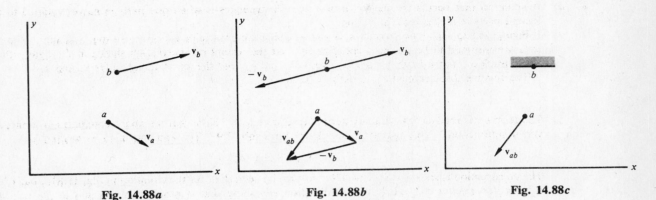

Fig. 14.88$a$          Fig. 14.88$b$          Fig. 14.88$c$

The equation in part ($a$) may be written as

$$\mathbf{v}_a = \mathbf{v}_b + \mathbf{v}_{ab}$$

The interpretation of this equation is that the *absolute velocity* of particle $a$ is the vector sum of

1. The *absolute velocity* of particle $b$
2. The *relative velocity* of particle $a$ with respect to particle $b$

The graphical construction used to find the relative velocity from the above equation is shown in Fig. 14.48$d$. $\mathbf{v}_a$ and $\mathbf{v}_b$ are first drawn from a common point. The direction of $\mathbf{v}_{ab}$ is along a line joining the tip ends of $\mathbf{v}_a$ and $\mathbf{v}_b$. The sense of $\mathbf{v}_{ab}$ is determined by placing the tip, on the line segment which represents $\mathbf{v}_{ab}$, in such a way that the vector equality in the equation

$$\mathbf{v}_a = \mathbf{v}_b + \mathbf{v}_{ab}$$

is satisfied.

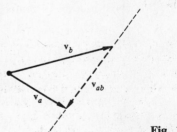

Fig. 14.88$d$

($c$)  The techniques presented above for the determination of the relative velocity may also be used to find the relative acceleration. If the absolute accelerations of the particles are $\mathbf{a}_a$ and $\mathbf{a}_b$, then the relative acceleration $\mathbf{a}_{ab}$ is defined to be

$$\mathbf{a}_{ab} = \mathbf{a}_a - \mathbf{a}_b$$

It is left as an exercise for the reader to show that

$$\mathbf{v}_{ab} = -\mathbf{v}_{ba} \qquad \mathbf{a}_{ab} = -\mathbf{a}_{ba}$$

($d$)  The concepts of relative displacement, velocity, and acceleration find widespread application in the motion analysis of linkages, cams, and gears. The study of these elements is included in the subject usually referred to as the kinematic analysis of mechanisms.

**14.89** Two particles move along parallel rectilinear paths, as shown in Fig. 14.89a. At time $t = 0$, both particles are in the positions shown in the figure. The motion of particle $a$ is given by

$$s_a = 12t^2 - 4t^3$$

where $s_a$ is in inches and $t$ is in seconds.

Particle $b$ moves with a constant velocity of

$$v_b = 12 \text{ in/s}$$

(*a*) Find the relative displacement of particle $a$ with respect to particle $b$ at $t = 1$ s. Express the result in both scalar and vector forms.

(*b*) Find the relative velocity of particle $a$ with respect to particle $b$ at times $t = 1, 2,$ and 3 s.

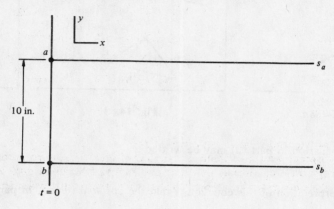

Fig. **14.89***a*

▎ (*a*) When $t = 1$ s,

$$s_a = (12t^2 - 4t^3)|_{t=1} = 12(1)^2 - 4(1)^3 = 8 \text{ in}$$

and

$$s_b = v_b t = 12(1) = 12 \text{ in}$$

The positions of the particles are shown in Fig. 14.89*b*, and point $O$ is arbitrarily chosen as the reference point for the displacement vectors $\mathbf{s}_a$ and $\mathbf{s}_b$. The construction for the relative displacement is shown in Fig. 14.89*c*. From the right triangle in this figure,

$$s_{ab} = \sqrt{10^2 + 4^2} = 10.8 \text{ in} \qquad \theta = \tan^{-1}\left(\frac{10}{4}\right) = 68.2°$$

The relative displacement of $a$ with respect to $b$ is seen to be "up and to the left" of particle $b$. This quantity may also be described in terms of components. Thus, from Fig. 14.89*c*,

$$x_{ab} = -4 \text{ in} \qquad y_{ab} = 10 \text{ in}$$

The physical interpretation of these results is that particle $a$ is 4 in to the left of, and 10 in above, particle $b$.

The above results may be written in vector form as

$$\mathbf{s}_{ab} = -4\mathbf{i} + 10\mathbf{j} \qquad \text{in}$$

The results obtained above are independent of the location of the arbitrary reference point $O$ in Fig. 14.89*b*, and the demonstration of this is left as an exercise for the reader.

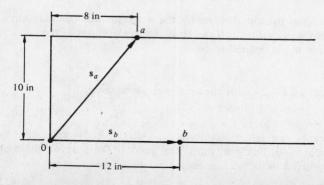

Fig. **14.89***b*

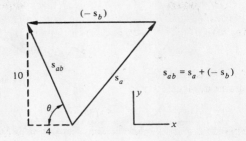

$$s_{ab} = s_a + (-s_b)$$

Fig. 14.89c

(b)  The velocity of particle $a$ is

$$v_a = \dot{s}_a = \frac{d}{dt}(12t^2 - 4t^3) = 24t - 12t^2$$

For $t = 1$ s:  $\qquad\qquad v_a = 24(1) - 12(1)^2 = 12 \text{ in/s}$

For $t = 2$ s:  $\qquad\qquad v_a = 24(2) - 12(2)^2 = 48 - 48 = 0$

For $t = 3$ s:  $\qquad\qquad v_a = 24(3) - 12(3)^2 = -36 \text{ in/s}$

For particle $b$:

$$v_b = 12 \text{ in/s} = \text{constant}$$

The absolute and relative velocities of the two particles at $t = 1, 2,$ and $3$ s are shown in Fig. 14.89d, together with a pictorial description of how an observer sitting on particle $b$ would see the motion of particle $a$. It may be observed that the separation distance between the paths of the two particles does *not* enter into the calculations for the relative velocities.

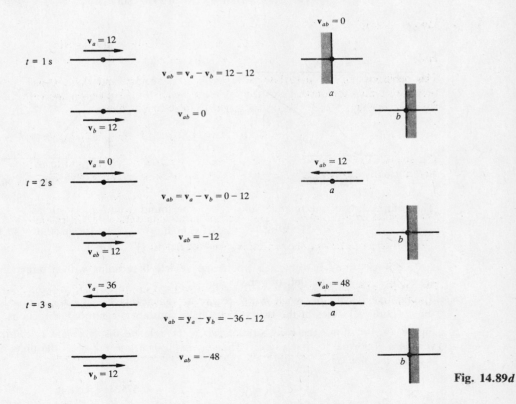

Fig. 14.89d

**14.90**  Two vehicles move with constant velocity along straight roadways, as shown in Fig. 14.90a. At $t = 0$, vehicle $a$ crosses roadway $b$. At this time, vehicle $b$ has the position shown in the figure. The two vehicles may be treated as particles, since there are no rotation effects of these elements in the plane of motion.

(a)  Find the relative velocity, at time $t = 0$, of vehicle $b$ with respect to vehicle $a$. Express the result in both scalar and vector forms.

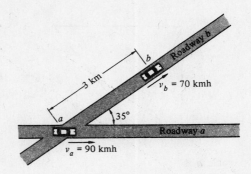

**Fig. 14.90a**

(**b**) Do the same as in part (*a*), for $t = 2$ s.

(**c**) Find the magnitude of the relative displacement of vehicle *b* with respect to vehicle *a*, when the direction of this quantity is normal to roadway *a*, and the corresponding time.

**▌** (**a**) The equation of relative velocity is

$$v_{ba} = v_b - v_a$$

The equation is shown in Fig. 14.90*b*. Using the law of cosines, we find that

$$v_{ba}^2 = 90^2 + 70^2 - 2(90)(70) \cos 35° \qquad v_{ba} = 51.8 \text{ km/h}$$

From the law of sines, we find that

$$\frac{v_{ba}}{\sin 35°} = \frac{v_b}{\sin \theta} \qquad \frac{51.8}{\sin 35°} = \frac{70}{\sin \theta} \qquad \theta = 50.8°$$

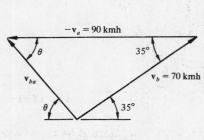

**Fig. 14.90b**

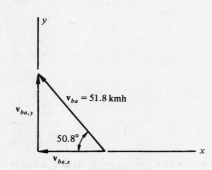

**Fig. 14.90c**

The components of the relative velocity may be found from Fig. 14.90*c* as

$$v_{ba,x} = -51.8 \cos 50.8° = -32.7 \text{ km/h} \qquad v_{ba,y} = 51.8 \sin 50.8° = 40.1 \text{ km/h}$$

$$\mathbf{v}_{ba} = -32.7\mathbf{i} + 40.1\mathbf{j} \qquad \text{km/h}$$

An observer sitting in vehicle *a* would see vehicle *b* receding with a direction and sense which is generally up and to the left of vehicle *a*.

(**b**) The relative velocity of vehicle *b* with respect to vehicle *a* is the same for *all* times in the problem, since the absolute velocities of the two vehicles have constant magnitudes, directions, and senses.

(**c**) Figure 14.90*d* shows the configuration when the relative displacement of vehicle *b* with respect to vehicle *a* is normal to roadway *a*. The corresponding time is *t*, with the units hours.
   From this figure,

$$\cos 35° = \frac{90t}{3 + 70t} \qquad t = 0.0752 \text{ h} = 4.51 \text{ min}$$

The displacements at this time are

$$s_a = 90(0.0752) = 6.77 \text{ km} \qquad s_b = 3 + 70(0.0752) = 8.26 \text{ km}$$

The magnitude of the relative displacement of vehicle *b* with respect to vehicle *a* is then

$$s_{ba} = \sqrt{s_b^2 - s_a^2} = \sqrt{8.26^2 - 6.77^2} = 4.73 \text{ km}$$

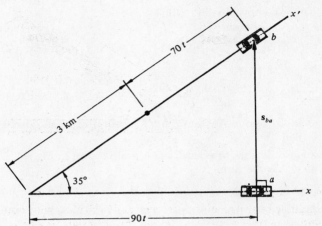

Fig. 14.90*d*

**14.91**  Figure 14.91*a* shows a Ferris wheel in an amusement park.  The wheel has eight equally spaced cars.  The absolute velocity of each car is 14 ft/s, with a direction which is tangent to the rim of the ferris wheel.

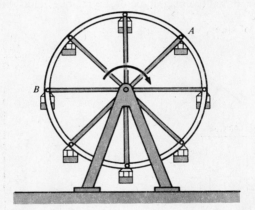

Fig. 14.91*a*

(*a*)  When the wheel is in the position shown in the figure, a passenger in car *A* looks out and observes a friend who is in car *B*.  Find the magnitude, direction, and sense of the velocity with which the friend in car *B* moves with respect to the passenger in car *A*.

(*b*)  Do the same as in part (*a*), when car *A* is in its lowest possible position.

(*c*)  Express the results in parts (*a*) and (*b*) in formal vector notation.

❚ (*a*)  Figure 14.91*b* shows the absolute velocities of points *A* and *B*.  The relative velocity $v_{BA}$ has the form

$$v_{BA} = v_B - v_A = v_B + (-v_A)$$

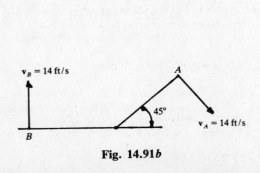

Fig. 14.91*b*

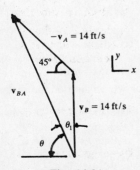

Fig. 14.91*c*

This equation is shown in Fig. 14.91c.

Using the law of cosines,

$$v_{BA}^2 = 14^2 + 14^2 - 2(14)^2 \cos 135° = 25.9 \text{ ft/s}$$

Using the law of sines,

$$\frac{25.9}{\sin 135°} = \frac{14}{\sin \theta_1} \qquad \theta_1 = 22.5° \qquad \theta = 90° - 22.5° = 67.5°$$

As observed from point $A$, point $B$ has a speed of 25.9 ft/s, up and to the left, at an angle of 67.5° with the horizontal direction.

(b) The absolute velocities of points $A$ and $B$ are shown in Fig. 14.91d. The relative velocity of $B$ with respect to $A$ has the form

$$v_{BA} = v_B - v_A = v_B + (-v_A)$$

Figure 14.91e shows this equation. The law of cosines is written as

$$v_{BA}^2 = 14^2 + 14^2 - 2(14)^2 \cos 135° = 25.9 \text{ ft/s}$$

Using the law of sines,

$$\frac{25.9}{\sin 135°} = \frac{14}{\sin \theta} \qquad \theta = 22.5°$$

It may be seen that the magnitude of the relative velocity is the same for part (a) as it is for part (b).

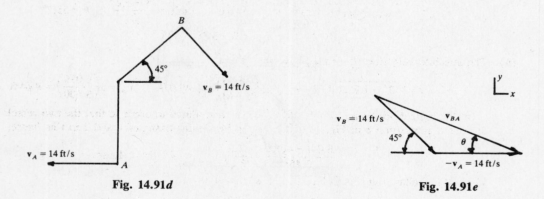

Fig. 14.91d                              Fig. 14.91e

(c) For the position in Fig. 14.91c,

$$\mathbf{v}_{ba} = (-25.9 \cos 67.5°)\mathbf{i} + (25.9 \sin 67.5°)\mathbf{j} = -9.91\mathbf{i} + 23.9\mathbf{j} \qquad \text{ft/s}$$

For the position in Fig. 14.91e,

$$\mathbf{v}_{ba} = (25.9 \cos 22.5°)\mathbf{i} - (25.9 \sin 22.5°)\mathbf{j} = 23.9\mathbf{i} - 9.91\mathbf{j} \qquad \text{ft/s}$$

From consideration of the results above for the relative velocities it may be seen that, for the two positions, the $x$ and $y$ components have interchanged their values.

14.92 A river, shown in Fig. 14.92a, flows at a rate of 1.2 knots. A freighter travels downstream at an absolute speed of 5 knots. A tugboat proceeds across the river, as shown in the figure, at a speed of 3 knots relative to the river. (A knot is a speed of one nautical mile per hour. One nautical mile is a length of 6,076 feet.)

(a) Find the relative velocity of the tugboat with respect to the freighter. Express the results in both scalar and vector forms.

(b) For what value of the distance $x$ would the two vessels collide?

▌ (a) The tugboat is designated $A$ and the freighter is designated $B$. Figure 14.92b shows the absolute velocities of $A$ and $B$. The absolute velocity of $A$ is the vector sum of the relative velocity of $A$ with respect to the stream, and the velocity of the stream. The relative velocity of the tugboat with respect to the freighter is

$$v_{AB} = v_A - v_B = v_A + (-v_B)$$

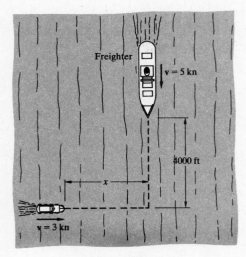

**Fig. 14.92a**

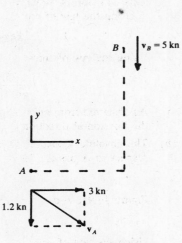

**Fig. 14.92b**

This equation, using the information in Fig. 14.92b, is shown in Fig. 14.92c. The magnitude and direction of the relative velocity are found as

$$v_{AB} = \sqrt{(5-1.2)^2 + 3^2} = 4.84 \text{ kn} \qquad \tan\theta = \frac{5-1.2}{3} \qquad \theta = 51.7°$$

$$\mathbf{v}_{ab} = 3\mathbf{i} + 3.8\mathbf{j} \qquad \text{kn}$$

(**b**) The absolute velocities $v_A$ and $v_B$ have the values

$$v_A = \sqrt{3^2 + 1.2^2} = 3.23 \text{ kn} = 3.23\left(\frac{6{,}076}{3{,}600}\right) = 5.45 \text{ ft/s} \qquad v_B = 5\left(\frac{6{,}076}{3{,}600}\right) = 8.44 \text{ ft/s}$$

In the original positions, $t = 0$. Motion is now assumed to occur so that the two vessels will collide at time $t$, at position $c$ in Fig. 14.92d. Using the similar triangles 1 and 2 in this figure,

$$\frac{8.44t - 4{,}000}{5.45t} = \frac{1.2}{3.23} \qquad t = 624 \text{ s} = 10.4 \text{ min}$$

Using the similar triangles in Fig. 14.92d a second time,

$$\frac{x}{3} = \frac{5.45t}{3.23} = \frac{5.45(624)}{3.23} \qquad x = 3{,}160 \text{ ft}$$

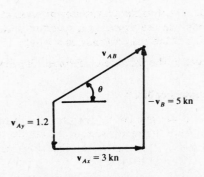

**Fig. 14.92c**

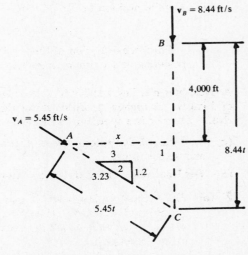

**Fig. 14.92d**

**14.93** Do the same as in Prob. 14.92, if the position of the tugboat with respect to the freighter is as shown in Fig. 14.93a.

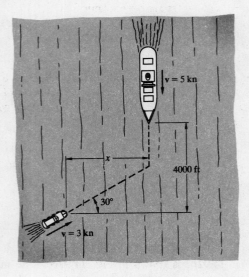

Fig. 14.93a

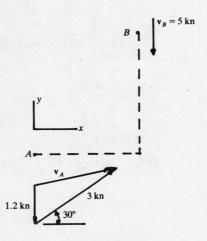

Fig. 14.93b

**▌(a)** Figure 14.93b shows the absolute velocities of A and B. The relative velocity of the tugboat with respect to the freighter is given by

$$v_{AB} = v_A - v_B = v_A + (-v_B)$$

The above equation is shown in Fig. 14.93c.

The components, magnitude, and direction of the relative velocity $v_{AB}$ are then found as

$$v_{AB,x} = 3\cos 30° = 2.60 \text{ kn} \qquad v_{AB,y} = -1.2 + 3\sin 30° + 5 = 5.30 \text{ kn}$$

$$v_{AB} = \sqrt{2.60^2 + 5.3^2} = 5.90 \text{ kn}$$

$$\tan\theta = \frac{5.3}{2.60} \qquad \theta = 63.9°$$

$$\mathbf{v}_{ab} = 2.60\mathbf{i} + 5.30\mathbf{j} \qquad \text{kn}$$

**(b)** The absolute velocities $v_A$ and $v_B$ have the values

$$v_{Ax} = 3\cos 30° = 2.60 \text{ kn} \qquad v_{Ay} = 3\sin 30° - 1.2 = 0.3 \text{ kn}$$

$$v_A = \sqrt{2.60^2 + 0.3^2} = 2.62 \text{ kn} = 2.62\left(\frac{6,076}{3,600}\right) = 4.42 \text{ ft/s}$$

$$v_B = 5\left(\frac{6,076}{3,600}\right) = 8.44 \text{ ft/s}$$

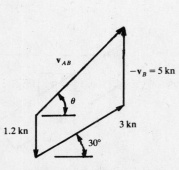

Fig. 14.93c

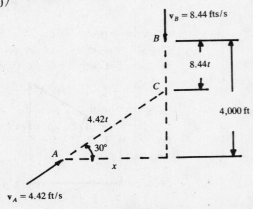

Fig. 14.93d

The two vessels are assumed to collide, at time $t_1$, at position $C$ in Fig. 14.93$d$. Using this figure,

$$\sin 30° = \frac{4,000 - 8.44t}{4.42t} \qquad t = 376 \text{ s} = 6.27 \text{ min}$$

$$\cos 30° = \frac{x}{4.42t} = \frac{x}{4.42(376)} \qquad x = 1,440 \text{ ft}$$

**14.94** Figure 14.94$a$ shows two concentric circular tracks that lie in a horizontal plane. Particle $a$ travels in a counterclockwise sense with a velocity of 10 m/s, and particle $b$ travels in a clockwise sense with a velocity of 26 m/s. Find the magnitude and direction of the relative displacement and velocity of particle $a$ with respect to particle $b$ for the following positions.

(a)  $\theta_a = \theta_b = 0$
(b)  $\theta_a = 0 \qquad \theta_b = 180°$
(c)  $\theta_a = 250° \qquad \theta_b = 130°$
(d)  Express the results in part ($c$) in formal vector notation.

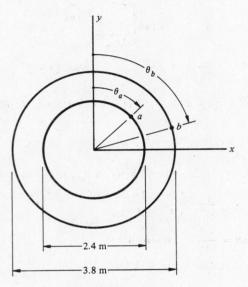

**Fig. 14.94$a$**

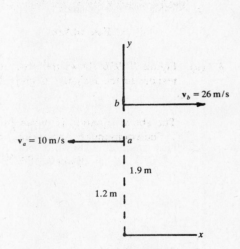

**Fig. 14.94$b$**

▌ (a)  The center of the circles is used as the reference for displacement. Figure 14.94$b$ shows the positions of the particles for $\theta_a = \theta_b = 0$. All displacements and velocities are considered to be positive if they act in the positive $xy$ coordinate senses. The relative displacement and velocity of particle $a$ with respect to particle $b$ are

$$s_{ab} = s_a - s_b = 1.2 - 1.9 = -0.7 \text{ m} \qquad v_{ab} = v_a - v_b = -10 - 26 = -36 \text{ m/s}$$

$s_{ab}$ acts downward in Fig. 14.94$b$ and $v_{ab}$ acts to the left.

(b)  Figure 14.94$c$ shows the positions of the particles for $\theta_a = 0$ and $\theta_b = 180°$. The relative displacement and velocity have the forms

$$s_{ab} = s_a - s_b = 1.2 - (-1.9) = 3.1 \text{ m} \qquad v_{ab} = v_a - v_b = -10 - (-26) = 16 \text{ m/s}$$

$s_{ab}$ acts upward in Fig. 14.94$c$ and $v_{ab}$ acts to the right.

(c)  The positions of the particles for $\theta_a = 250°$ and $\theta_b = 130°$ are seen in Fig. 14.94$d$. The relative displacement has the form

$$s_{ab} = s_a - s_b = s_a + (-s_b)$$

This result is shown in Fig. 14.94$e$.
Using the law of cosines,

$$s_{ab}^2 = 1.2^2 + 1.9^2 - 2(1.2)1.9 \cos 120° \qquad s_{ab} = 2.71 \text{ m}$$

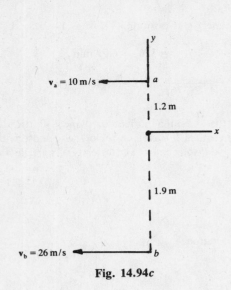

Fig. 14.94c

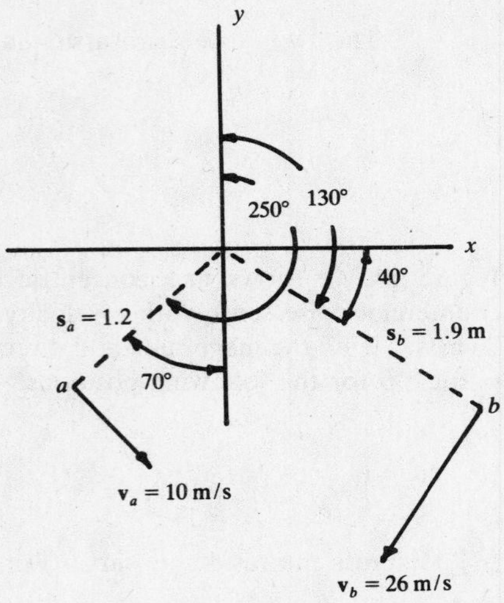

Fig. 14.94d

Using the law of sines,

$$\frac{1.9}{\sin \theta_1} = \frac{2.71}{\sin 120°} \qquad \theta_1 = 37.4°$$

The direction of $s_{ab}$ is given by

$$\theta = 70° + 37.4° - 90° = 17.4°$$

The relative velocity is written as

$$v_{ab} = v_a - v_b = v_a + (-v_b)$$

This equation is shown in Fig. 14.94f.

Using the law of cosines,

$$v_{ab}^2 = 26^2 + 10^2 - 2(26)10 \cos 60° \qquad v_{ab} = 22.7 \text{ m/s}$$

Using the law of sines,

$$\frac{10}{\sin \theta_1} = \frac{22.7}{\sin 60°} \qquad \theta_1 = 22.4°$$

The direction of $v_{ab}$ has the value

$$\theta = 50° - 22.4° = 27.6°$$

(d)      $s_{ab} = (-2.71 \cos 17.4°)\mathbf{i} + (2.71 \sin 17.4°)\mathbf{j} = -2.59\mathbf{i} + 0.810\mathbf{j}$     m

                   $v_{ab} = (22.7 \cos 27.6°)\mathbf{i} + (22.7 \sin 27.6°)\mathbf{j} = 20.1\mathbf{i} + 10.5\mathbf{j}$     m/s

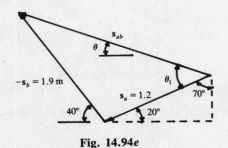

Fig. 14.94e

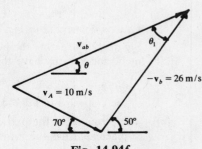

Fig. 14.94f

**14.95**    Two particles move along straight parallel paths, as shown in Fig. 14.95a. Particle $a$ moves with a velocity of 1.6 m/s. Particle $b$ accelerates at 0.4 m/s². At time zero the particles have the positions shown, and $v_b = 0.5$ m/s.

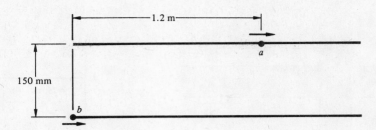

Fig. 14.95a

(a) At what time is the direction of the relative displacement of $b$ with respect to $a$ normal to the paths of motion?

(b) Find the corresponding values of the relative velocity and acceleration of particle $a$ with respect to particle $b$.

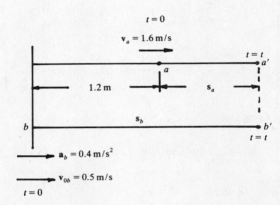

Fig. 14.95b

▌ (a) Figure 14.95b shows a diagram of the events. Time $t$ corresponds to the direction of the relative displacement of $b$ with respect to $a$, line $a'b'$ in Fig. 14.95b being normal to the paths of motion. From Fig. 14.95b,

$$s_b = 1.2 + s_a \qquad v_{0b}t + \tfrac{1}{2}a_B t^2 = 1.2 + v_a t$$
$$0.5t + \tfrac{1}{2}(0.4)t^2 = 1.2 + 1.6t \qquad 0.2t^2 - 1.1t - 1.2 = 0$$

$$t = \frac{1.1 \pm \sqrt{(-1.1)^2 - 4(0.2)(-1.2)}}{2(0.2)} = \frac{1.1 \pm 1.47}{2(0.2)} = 6.43 \text{ s}$$

(b) The absolute velocities $v_a$ and $v_b$ are

$$v_a = 1.6 \text{ m/s} \qquad v_b = v_{0b} + a_b t = 0.5 + 0.4(6.43) = 3.07 \text{ m/s}$$

The relative velocity and acceleration are then found as

$$v_{ab} = v_a - v_b = 1.6 - 3.07 = -1.47 \text{ m/s} \qquad a_{ab} = a_a - a_b = 0 - 0.4 = -0.4 \text{ m/s}^2$$

The senses of the relative velocity and acceleration of $a$ with respect to $b$ are to the left in Fig. 14.95b.

14.96   Figure 14.96a shows two vehicles on intersecting roads. At the instant shown, $v_A = 60$ km/h and $v_B = 45$ km/h. Vehicle $B$ accelerates at 2 m/s$^2$ and vehicle $A$ decelerates at 3.8 m/s$^2$. Find the relative displacement, velocity, and acceleration of vehicle $A$ with respect to vehicle $B$ at the instant shown.

▌ Figure 14.96b shows a diagram of the events. The relative displacement of $A$ with respect to $B$ is

$$s_{AB} = s_A - s_B = s_A + (-s_B)$$

The above vector equation is shown in Fig. 14.96c, and the magnitude of the relative displacement is

$$s_{AB} = \sqrt{1,250^2 + 750^2} = 1,460 \text{ m}$$

The direction of $s_{AB}$ is found from

$$\tan \theta = \frac{1,250}{750} \qquad \theta = 59.0°$$

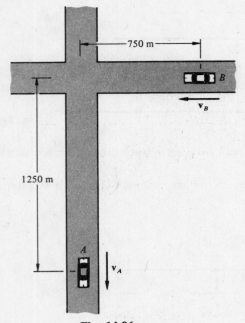

Fig. 14.96a

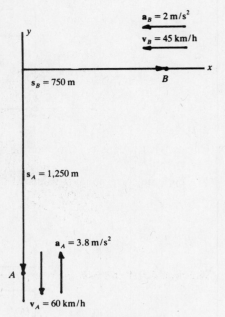

Fig. 14.96b

The relative velocity of $A$ with respect to $B$ is given by

$$v_{AB} = v_A - v_B = v_A + (-v_B)$$

This result is shown in Fig. 14.96d.
The magnitude and direction of $v_{AB}$ are found as

$$v_{AB} = \sqrt{60^2 + 45^2} = 75 \text{ km/h} \qquad \tan\theta = \frac{60}{45} \qquad \theta = 53.1°$$

The relative velocity of $A$ with respect to $B$ is given by

$$a_{AB} = a_A - a_B = a_A + (-a_B)$$

This equation is shown in Fig. 14.96e, and the magnitude and direction of $a_{AB}$ are

$$a_{AB} = \sqrt{3.8^2 + 2^2} = 4.29 \text{ m/s}^2 \qquad \tan\theta = \frac{3.8}{2} \qquad \theta = 62.2°$$

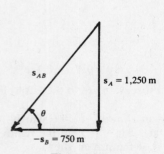

Fig. 14.96c

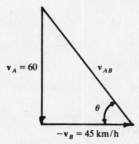

Fig. 14.96d

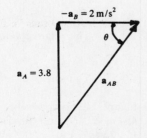

Fig. 14.96e

14.97    At $t = 0$ vehicle $B$ in Fig. 14.97a just passes beneath the overpass and vehicle $A$ is 1,000 ft from this structure. At this instant, $v_A = 35$ mi/h and vehicle $A$ starts to accelerate at 4 ft/s$^2$, while vehicle $B$ continues at a constant speed of 55 mi/h.

Find the relative displacement, velocity, and acceleration of vehicle $A$ with respect to vehicle $B$ at the time that vehicle $A$ crosses the overpass.

▌ Figure 14.97b shows a diagram of the events.   The speeds of $A$ and $B$ are

$$v_{0A} = 35 \text{ mi/h} = 51.3 \text{ ft/s} \qquad v_{0B} = v_B = 55 \text{ mi/h} = 80.7 \text{ ft/s} = \text{constant}$$

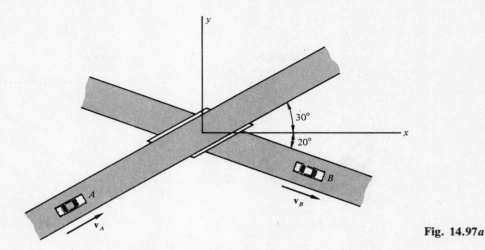

**Fig. 14.97a**

The time for $A$ to reach the overpass is found from

$$s = s_0 + v_0 t + \tfrac{1}{2} a t^2$$
$$1,000 = 51.3t + \tfrac{1}{2}(4)t^2 \qquad 2t^2 + 51.3t - 1,000 = 0$$

$$t = \frac{-51.3 \pm \sqrt{51.3^2 - 4(2)(-1,000)}}{2(2)} = \frac{-51.3 \pm 103}{2(2)} = 12.9 \text{ s}$$

The velocity of vehicle $A$ at the overpass is

$$v_A = v_{0A} + a_A t = 51.3 + 4(12.9) = 103 \text{ ft/s}$$

When $t = 12.9$ s, $A$ is at the overpass. The distance of $B$ from the overpass at this time is $s_B$, given by

$$s_B = v_{0B}t = 80.7(12.9) = 1,040 \text{ ft}$$

The absolute displacements of $A$ and $B$ are measured from the overpass, and this point is also the origin of the $xy$ coordinates. The absolute displacement of vehicle $A$ at the overpass is $s_A = 0$. The corresponding value of the relative displacement of $A$ with respect to $B$ is given by

$$s_{AB} = s_A - s_B = -s_B$$

$s_{AB} = 1,040$ ft, acting from $B$ to $A$. This result is shown in Fig. 14.97$c$.
The relative velocity of $A$ with respect to $B$ is

$$v_{AB} = v_A - v_B = v_A + (-v_B)$$

This result is shown in Fig. 14.97$d$.

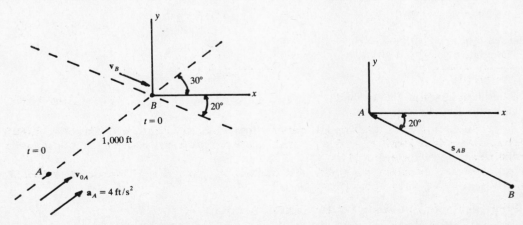

**Fig. 14.97b**                                           **Fig. 14.97c**

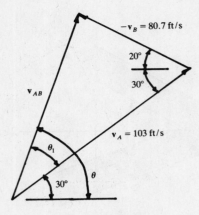

Fig. 14.97*d*

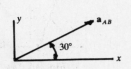

Fig. 14.97*e*

Using the law of cosines,

$$v_{AB}^2 = 103^2 + 80.7^2 - 2(103)80.7 \cos 50° \qquad v_{AB} = 80.2 \text{ ft/s}$$

Using the law of sines,

$$\frac{80.7}{\sin \theta_1} = \frac{80.2}{\sin 50°} \qquad \theta_1 = 50.4°$$

The direction $\theta$ of $v_{AB}$, shown in Fig. 14.97*d*, has the value

$$\theta = \theta_1 + 30° = 80.4°$$

The relative acceleration of $A$ with respect to $B$ is given by

$$a_{AB} = a_A - a_B \qquad a_B = 0$$
$$a_{AB} = a_A = 4 \text{ ft/s}^2 \qquad \theta = 30°$$

This result is shown in Fig. 14.97*e*.

**14.98** Do the same as in Prob. 14.97 if, at the instant that vehicle $B$ passes beneath the overpass, it starts to accelerate at 2.4 ft/s².

▎ As found in Prob. 14.97, vehicle $A$ reaches the overpass at $t = 12.9$ s, with $v_A = 103$ ft/s. The position of vehicle $B$ at $t = 13$ s is found from

$$s_B = s_0 + v_0 t + \tfrac{1}{2} a t^2 = 80.7(12.9) + \tfrac{1}{2}(2.4)12.9^2 = 1,240 \text{ ft}$$

The velocity $v_B$ at $t = 12.9$ s is given by

$$v_B = v_{0B} + a_B t = 80.7 + 2.4(12.9) = 112 \text{ ft/s}$$

The absolute displacements are measured from the overpass. When vehicle $A$ is at the overpass, $s_A = 0$ and

$$s_{AB} = s_A - s_B = -s_B = 1,250 \text{ ft}$$

The sense of $s_{AB}$ is from $B$ to $A$, and the direction of $s_{AB}$ is $\theta = 20°$.
   The relative velocity is given by

$$v_{AB} = v_A - v_B = v_A + (-v_B)$$

This result is shown in Fig. 14.98*a*.
   Using the law of cosines,

$$v_{AB}^2 = 103^2 + 112^2 - 2(103)112 \cos 50° \qquad v_{AB} = 91.2 \text{ ft/s}$$

Using the law of sines,

$$\frac{112}{\sin \theta_1} = \frac{91.2}{\sin 50°} \qquad \theta_1 = 70.2°$$

The direction of $v_{AB}$, shown in Fig. 14.98*a*, is

$$\theta = 180° - \theta_1 - 30° = 79.8°$$

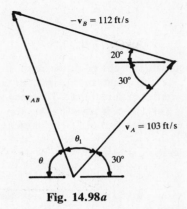

**Fig. 14.98a**

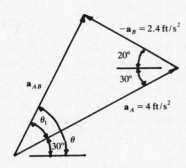

**Fig. 14.98b**

The relative acceleration $a_{AB}$ has the form

$$a_{AB} = a_A - a_B = a_A + (-a_B)$$

This result is shown in Fig. 14.98b.

Using the law of cosines and the law of sines,

$$a_{AB} = 4^2 + 2.4^2 - 2(4)2.4 \cos 50° = 3.07 \text{ ft/s}^2$$

$$\frac{2.4}{\sin \theta_1} = \frac{3.07}{\sin 50°} \qquad \theta_1 = 36.8° \qquad \theta = 30° + \theta_1 = 66.8°$$

**14.99**  Give a summary of the basic concepts of kinematics of particles presented in this chapter.

▌ A particle is a physical body whose dimensions are assumed to be vanishingly small. Thus, a particle cannot experience rotation effects.

Kinematics is the science that studies the motion of particles and bodies, without regard for the forces that produce these motions. The motion of a particle along a straight path is referred to as rectilinear translation. The particle motion is referred to as curvilinear translation if the motion is along a curved path.

The displacement of a particle is the position, or location, of the particle with respect to a given reference point. The velocity of a particle is given by the first time derivative of the displacement. The acceleration of a particle is given by the first time derivative of velocity, or by the second time derivative of displacement. Fundamentally, velocity is the time rate of change of displacement, and acceleration is the time rate of change of velocity. Displacement, velocity, and acceleration are vector quantities. The choice of the positive sense of a displacement coordinate automatically establishes the same positive sense for the velocity and acceleration. The magnitude of the velocity is called speed. Deceleration refers to an acceleration that acts in a sense to decrease velocity.

Graphical displays of the variation of displacement, velocity, and acceleration with respect to time are called motion diagrams. The magnitude of the velocity is equal to the slope of the displacement-time curve, and the magnitude of the acceleration is equal to the slope of the velocity-time curve. The area under the curve between two times on the velocity-time curve is equal to the change in displacement of the particle between these two times. The area under the acceleration curve between two times is equal to the change in the velocity between these two times.

The equations that relate displacement, velocity, acceleration, and time, of a particle which moves in rectilinear translation with constant acceleration $a$, are given by

$$v = v_0 + at \qquad s = s_0 + v_0 t + \tfrac{1}{2}at^2 \qquad v^2 = v_0^2 + 2a(s - s_0)$$

where $s_0$ and $v_0$ are the initial displacement and velocity, respectively, at $t = 0$, and $s$ and $v$ are the displacement and velocity at time $t$. For the case of a freely falling particle in the absence of air resistance, the above three equations have the forms

$$v = v_0 - gt \qquad s = s_0 + v_0 t - \tfrac{1}{2}gt^2 \qquad v^2 = v_0^2 - 2g(s - s_0)$$

where $g$ is the acceleration of the gravitational field and the positive sense of $s$ is upward from the surface of the earth.

The velocity of a particle that moves in plane curvilinear translation is always tangent to the path of motion. The acceleration of the particle has two distinct components. The magnitude of the normal component $a_n$ is expressed by

$$a_n = \frac{v^2}{\rho}$$

where $v$ is the speed of the particle and $\rho$ is the radius of curvature at the point where $a_n$ acts. The direction of $a_n$ is normal to the tangent line to the curved path at the point of interest, or along the radius of curvature. The sense of $a_n$ is always directed toward the center of curvature. The normal acceleration arises because of changes in the direction of the velocity of the particle as it travels along the curve. The normal acceleration is never zero for a particle that moves in curvilinear motion. The normal acceleration is also referred to as the centripetal acceleration.

The magnitude of the tangential component $a_t$ of acceleration is given by

$$a_t = \dot{v}$$

where $\dot{v}$ is the time rate of change of the speed along the curved path. The direction of the tangential acceleration is along a line tangent to the curve at the point of interest. The positive sense of $a_t$ is the same as the positive sense of $s$, the displacement along the curve.

The motion of a particle launched with initial velocity in a vertical plane may be expressed in terms of component motions. If the $x$ axis is horizontal and the $y$ axis is vertical, the component equations of motion are

$$v_x = v_{0x} = \text{constant} \qquad x = x_0 + v_{0x}t \qquad v_y = v_{0y} - gt$$
$$y = y_0 + v_{0y}t - \tfrac{1}{2}gt^2 \qquad v_y^2 = v_{0y}^2 - 2g(y - y_0) \qquad v_{0x} = v_0 \cos\theta \qquad v_{0y} = v_0 \sin\theta$$

where $v_0$ is the magnitude of the initial launch velocity, at the direction $\theta$ with the $x$ axis.

An absolute, or inertial, coordinate system is imagined to be fixed with respect to a distant star. Such a coordinate system may also be defined as a reference system in which Newton's laws of motion are valid. For most problems in engineering dynamics, the earth may be considered to be an inertial reference system. If the displacement, velocity, or acceleration of a particle is measured in an absolute coordinate system, these terms are referred to as absolute. An absolute quantity has a single subscript to indicate the particle or point under consideration. Thus, $s_a$, $v_a$, and $a_a$ are the absolute displacement, velocity, and acceleration, respectively, of particle $a$. A relative quantity is indicated by a double subscript, so that $s_{ab}$, $v_{ab}$, and $a_{ab}$ are the relative displacement, velocity, and acceleration, respectively, of particle $a$ with respect to particle $b$. A physical interpretation of a relative motion quantity is that it is that motion effect which a person would observe if the person moved with the motion of the reference particle or point. The relative motion terms are defined by

$$\mathbf{s}_{ab} = \mathbf{s}_a - \mathbf{s}_b \qquad \mathbf{v}_{ab} = \mathbf{v}_a - \mathbf{v}_b \qquad \mathbf{a}_{ab} = \mathbf{a}_a - \mathbf{a}_b$$

It should be noted that the above equations are vector operations in which the directions of the terms must be taken into account.

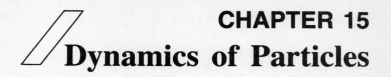

# CHAPTER 15
# Dynamics of Particles

## 15.1 NEWTON'S SECOND LAW, DYNAMICS OF PARTICLES IN RECTILINEAR TRANSLATION

**15.1**   What is studied in the subjects *kinetics* and *dynamics*?

▐   Kinetic studies the relationships between the forces which act on a particle, or a rigid body, and the motions of the particle or rigid body.   It is apparent that a necessary prerequisite for the study of kinetics is the study of kinematics.   The term kinematics was defined in Prob. 14.1, and Chap. 14 was devoted to the solution of problems in kinematics.   The subject dynamics includes both kinematics and kinetics.

**15.2**   What are the two general types of problems in dynamics?

▐   There are two general types of problems in dynamics.   In the first type, the motion of the particle or body is known and the forces or moments which produce this motion are to be found.   In the second general class of problems in dynamics, the forces and moments which act on a body, or the forces which act on a particle, are known and it is desired to find the motions which result from the application of these force effects.   This latter type of problem is the one usually encountered in the solution of engineering problems in dynamics.

**15.3**   (*a*)   Summarize the definition, and the possible motions, of a particle.

   (*b*)   Under what conditions can motion of a body be considered to be particle motion?

▐   (*a*)   In Prob. 14.1 a particle was defined to be a physical element, with mass, whose dimensions are all vanishingly small.   Consequently, no rotation effects need be considered when the motion of the particle is described.   It follows, then, that the motion of a particle is completely defined by describing its translational motion.

   The motion of a particle may be one of two general types.   The first kind occurs when the particle moves along a straight line.   This type of motion is described as rectilinear translation.   The second type occurs when the particle moves along a curved path.   This type of motion is called curvilinear translation.   This curved path may lie in either a two- or a three-dimensional space.   The former motion is described as plane curvilinear motion, while the latter is referred to as general curvilinear motion.

   (*b*)   In certain problems in this chapter, bodies will be considered that have finite dimensions.   From part (*a*) above, a particle is defined as a physical body with mass and vanishingly small dimensions.   Thus, the bodies used in the problems do not fit the strict definition of a particle.   It will be shown in Chap. 18, however, that the equations of particle dynamics accurately describe the translational motion of a body if *there are no rotation effects*.   The distinguishing characteristic of particle motion is the absence of all rotation effects.

**15.4**   (*a*)   Give the form of Newton's second law for the motion of a particle.

   (*b*)   What is the physical interpretation of Newton's second law for the motion of a particle?

▐   (*a*)   The fundamental basis of all engineering dynamics is Newton's second law, which states that a particle which is acted on by a resultant force will experience an acceleration which is directly proportional to this force and in the direction of the force.   In equation form, Newton's second law for a particle appears as

$$\mathbf{F} = m\mathbf{a}$$

where $m$ = mass of the particle
   $\mathbf{a}$ = acceleration of the particle
   $\mathbf{F}$ = resultant force acting on the particle
This resutant force is expressed by

$$\mathbf{F} = \sum_{i=1}^{i=n} \mathbf{F}_i$$

where $n$ is the number of forces $\mathbf{F}_i$ which act on the particle and the summation operation is understood to be a vector summation. The acceleration $\mathbf{a}$ in Newton's second law must be expressed in terms of an absolute inertial coordinate system. This type of coordinate system was defined in Prob. 14.86.

It may be seen from the above equation that $\mathbf{F}$ and $\mathbf{a}$ are two vector quantities related by the scalar multiplier $m$. It follows that $\mathbf{F}$ and $\mathbf{a}$ have the *same direction and sense* and differ only in *magnitude* by the positive factor $m$. The equation which expresses Newton's second law is an example of the multiplication of a vector quantity by a scalar quantity, and this operation was defined in Prob. 1.62. Throughout this text Newton's second law will be expressed in the typical forms

$$\sum F_x = ma \qquad \text{or} \qquad \sum F_s = ma$$

The summation sign is used to emphasize the fact that the left side of the equation is a *resultant force component*. The subscript is used to indicate both the direction of this component of the resultant force and the direction of the acceleration.

Newton's second law is often referred to as *the equation of motion*.

(*b*) The physical interpretation of Newton's second law is that if a *resultant* force $\mathbf{F}$ acts on a particle of mass $m$, then the particle will experience an acceleration which has the *direction* and *sense* of the resultant force. The magnitude of the acceleration is given by

$$a = \frac{1}{m} F$$

where $F$ and $a$ are the magnitudes of the resultant force and the acceleration.

The effect described above is shown in Fig. 15.4*a*, where a particle is acted on by the three forces $\mathbf{F}_1$, $\mathbf{F}_2$, and $\mathbf{F}_3$. The resultant force $\mathbf{F}$ is shown in Fig. 15.4*b*. From the definition of Newton's second law, it follows that the direction of the acceleration is along line $ab$, with a sense from $a$ to $b$.

*The reader is urged to master this fundamental interpretation of Newton's second law, because this relationship is used repeatedly throughout the study of engineering dynamics.*

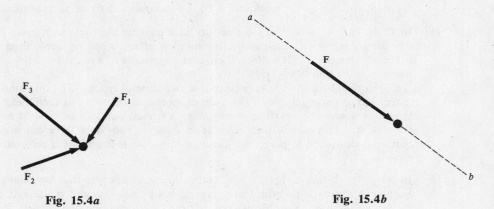

Fig. 15.4*a*                              Fig. 15.4*b*

**15.5** What type of motion may a particle have if the resultant force which acts on the particle is zero?

▌ If the resultant force which acts on a particle is zero, Newton's second law has the form

$$0 = m\mathbf{a} \qquad \mathbf{a} = 0$$

The components of the vector $\mathbf{a}$ along the three coordinate directions are $a_x$, $a_y$, and $a_z$. A necessary condition for the above equation to be true is

$$a_x = \ddot{x} = 0 \qquad a_y = \ddot{y} = 0 \qquad a_z = \ddot{z} = 0$$
$$v_x = \dot{x} = \text{constant} \qquad v_y = \dot{y} = \text{constant} \qquad v_z = \dot{z} = \text{constant}$$

If these equations are satisfied, it follows that the resultant force which acts on the particle is zero. It also follows that the particle is *at rest*, or *moving with constant velocity*. The above conditions of motion and resultant force are the necessary conditions for the *static equilibrium* of a particle, as may be seen in the definition of Newton's first law of motion in Prob. 4.26.

**15.6** (*a*) Give the USCS and SI units which are used with Newton's second law.

(*b*) Give the conversion factors between the USCS and SI units in part (*a*).

**(a)** The fundamental USCS units are length, time, and force. The length unit is the foot, the force unit is the pound, and the time unit is the second. The symbols used to represent these quantities are

Foot, ft

Pound, lb

Second, s

The value of the gravitational acceleration $g$ in USCS units is $32.17 \, \text{ft/s}^2$. Throughout the text, a rounded-off value of $32.2 \, \text{ft/s}^2$ is used. The mass unit is the slug. A *slug* is that mass which, when acted on by a force of one pound, will experience an acceleration of one foot per second squared. Using Newton's second law, we have

$$F = ma \qquad 1 \, \text{lb} = 1 \, \text{slug}\left(1 \, \frac{\text{ft}}{\text{s}^2}\right) \qquad 1 \, \text{slug} = 1 \, \frac{\text{lb} \cdot \text{s}^2}{\text{ft}}$$

The fundamental SI units are length, time, and mass. The length unit is the meter, the mass unit is the kilogram, and the time unit is the second. The symbols used to represent these quantities are

Meter, m

Kilogram, kg

Second, s

The value of the gravitational acceleration $g$ in SI units is $9.807 \, \text{m/s}^2$. Throughout the text, a rounded-off value of $9.81 \, \text{m/s}^2$ is used.

The derived unit of force is the *newton*, with the symbol N. A force of one newton will give a mass of one kilogram an acceleration of one meter per second squared. Using Newton's second law, we get

$$F = ma \qquad 1 \, \text{N} = 1 \, \text{kg}\left(1 \, \frac{\text{m}}{\text{s}^2}\right) = 1 \, \frac{\text{kg} \cdot \text{m}}{\text{s}^2}$$

**(b)** The factors for conversion between USCS and SI units are

$$1 \, \text{ft} = 0.3048 \, \text{m} \approx 0.305 \, \text{m} \qquad 1 \, \text{m} = 3.281 \, \text{ft} \approx 3.28 \, \text{ft}$$
$$1 \, \text{lb} = 4.448 \, \text{N} \approx 4.45 \, \text{N} \qquad 1 \, \text{N} = 0.2248 \, \text{lb} \approx 0.225 \, \text{lb}$$
$$1 \, \text{slug} = 14.59 \, \text{kg} \approx 14.6 \, \text{kg} \qquad 1 \, \text{kg} = 0.06854 \, \text{slug} \approx 0.0685 \, \text{slug}$$

As an estimate of the relative sizes of the above length and force units, it may be noted that 1 m is approximately 3 ft, while 1 N is approximately 0.25 lb.

**15.7**  Describe the problem where a particle is acted on by an applied force and moves in *rectilinear translation*.

Figure 15.7 shows the case where a particle is acted on by an applied force and moves in rectilinear translation. The $x$ axis is the axis of motion. A very important conclusion may be drawn from this figure. Since the particle is constrained to move in the $x$ direction *only*, the $y$ component of displacement and velocity of the particle must be zero. The consequence of this is that the $y$ component of the resultant of the applied and reaction forces which act on the particle must be zero. Thus, it is *only* the $x$ component of the resultant force which produces acceleration along the axis of translation. Moreover, *the particle must be in static equilibrium in the $y$ direction*.

The equation of motion for the particle in Fig. 15.7 is

$$\sum F_x = ma_x = ma$$

where $a = a_x$ is the acceleration of the particle along the $x$ axis.

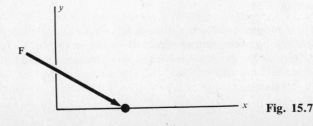

Fig. 15.7

**15.8** (*a*) Give a set of general rules to solve the problem of motion of a particle in rectilinear translation.

(*b*) Give a short summary of the rules in part (*a*).

▌ (*a*) A set of general rules for the solution of dynamics problems in rectilinear translation is as follows:

1. A coordinate is chosen to represent the displacement of the mass particle from some reference point. From the discussion in Probs. 14.4 and 14.5, *this definition of the positive sense of displacement automatically establishes the positive senses of the velocity and acceleration.* A variation of this step is to directly assume the sense of the acceleration.

2. A free-body diagram of the particle is drawn, and all the forces acting on the particle are shown. These forces will include both the known applied forces acting on the particle, and the unknown reaction forces exerted on the particle by the immediate supporting foundation which is imagined to be removed.

3. The components of the resultant force in the direction of the axis of motion, and normal to this direction, are computed. A force component is considered positive if it has the sense of a positive coordinate axis, or of the assumed sense of the acceleration.

4. The component of the resultant force which is normal to the direction of motion is set equal to zero, to satisfy the static equilibrium requirements in this direction.

5. The component of the resultant force in the direction of motion is then used in Newton's second law. Since the directions of the force and the acceleration are known, the equation will be in terms of the scalar magnitudes of the latter two quantities.

(*b*) The basic steps given in part (*a*) for solving for the acceleration of a mass particle in rectilinear translation are summarized below.

1. Choose the displacement coordinate, or sense of acceleration, of the particle.

2. Draw the free-body diagram of the particle.

3. Find the components of the resultant force normal to, and in the direction of, the motion.

4. Set the normal component of the resultant force equal to zero.

5. Use the component of the resultant force in the direction of the motion in Newton's second law, to find the acceleration of the particle.

**15.9** What equations are used to find the motion of a particle which is acted on by a *constant* force and moves in rectilinear translation?

▌ If the component of the resultant force in the direction of the motion has a constant value, it follows that the *acceleration of the article is constant*. The case of rectilinear translation with constant acceleration was presented in Prob. 14.31. The kinematic equations which describe this type of motion are

$$v = v_0 + at \qquad s = s_0 + v_0 t + \tfrac{1}{2}at^2 \qquad v^2 = v_0^2 + 2a(s - s_0)$$

where $s$ is a symbolic displacement coordinate. If the particle moves along the $x$ axis, for example, then $s$ is defined to be $x$. At $t = 0$, the initial displacement and velocity of the particle are $s_0$ and $v_0$, respectively. To solve a problem of particle dynamics in rectilinear translation, in which the force in the direction of motion is constant, first this force is used to find the *constant* value of the acceleration of the article. Then this acceleration is used with the above equations to obtain a complete description of the motion of the particle.

**15.10** A block of mass $m$ rests on a smooth, horizontal surface, as shown in Fig. 15.10a. A horizontal force of constant magnitude $P$ is applied to the block.

(*a*) State the equilibrium requirement in the direction normal to the surface, and find the acceleration of the block.

(*b*) Find the magnitude of the acceleration in part (*a*) if the mass of the block is 55 kg and the force $P$ has a magnitude of 275 N.

▌ (*a*) This is a problem where the force is known and the resulting motion is to be found.

The free-body diagram of the block is shown in Fig. 15.10b, together with an arbitrarily assumed positive sense for the displacement $x$. The known applied forces which act on the block are the weight force $mg$ and the applied force $P$. The unknown reaction force is designated $N$. For equilibrium in the $y$ direction,

$$\sum F_y = 0 \qquad -mg + N = 0 \qquad N = mg$$

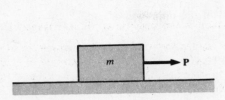

Fig. 15.10a

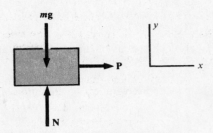

Fig. 15.10b

The equation of motion in the $x$ direction is

$$\sum F_x = ma \qquad P = ma \qquad a = \frac{P}{m} = \text{constant}$$

(b)   Using   $P = 275 \text{ N}$   and   $m = 55 \text{ kg}$,

$$a = \frac{P}{m} = \frac{275}{55} = 5 \text{ m/s}^2$$

15.11   A vehicle has a mass of 1,200 kg and moves in rectilinear translation.   The vehicle is initially at rest.   A resultant force of 535 N, in the direction of motion, acts on the vehicle.   Find the acceleration of the vehicle and the distance covered by the vehicle at the end of 15 s.

❙   The equation of motion is

$$\sum F_x = ma \qquad 535 = 1,200a \qquad a = 0.446 \text{ m/s}^2$$

Using   $s_0 = 0$   and   $v_0 = 0$,

$$s = s_0 + v_0 t + \tfrac{1}{2}at^2 = \tfrac{1}{2}(0.446)15^2 = 50.2 \text{ m}$$

15.12   A block of mass 4 kg slides along a straight frictionless track with a speed of 6.5 m/s, as shown in Fig. 15.12a.   At   $t = 0$   a force, shown as the dashed arrow in the figure, is applied to the particle and the resulting deceleration is 1.6 m/s².   Find the magnitude of the applied force.

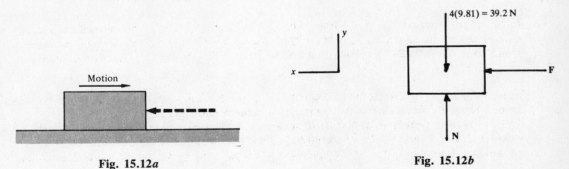

Fig. 15.12a

Fig. 15.12b

❙   This is a problem in which the motion is known and the force is to be found.
   Figure 15.12b shows the free-body diagram of the block, and the choice of $xy$ coordinates.   Using Newton's second law,

$$\sum F_x = ma \qquad F = 4(1.6) = 6.4 \text{ N}$$

For this case a summation of forces normal to the direction of motion is not required.   It may also be observed that the initial speed of 6.5 m/s does not enter into the problem.

15.13   An automobile moves along a straight roadway with a constant speed of 40 mi/h.   The automobile experiences a constant acceleration for 5 s and reaches a speed of 50 mi/h.   The weight of the automobile is 3,600 lb.

(a)   Find the required value of the resultant force acting on the automobile that will produce this acceleration.

(b)   Find the distance covered by the automobile during this period of time.

(c) Do the same as in parts (a) and (b) if the automobile accelerates for 5 s and changes the speed of the car from 50 to 60 mi/h.

▌ (a) The speeds of the automobile at the endpoints of the time interval are

$$40 \text{ mi/h} = 58.7 \text{ ft/s} \qquad 50 \text{ mi/h} = 73.3 \text{ ft/s}$$

The acceleration of the automobile is found from

$$v = v_0 + at \qquad 73.3 = 58.7 + a(5) \qquad a = 2.92 \text{ ft/s}^2$$

Newton's second law has the form

$$\sum F_x = ma \qquad F = \frac{3,600}{32.2}(2.92) = 326 \text{ lb}$$

(b) The distance covered by the automobile in the 5-s time interval is found from

$$v^2 = v_0^2 + 2a(s - s_0) \qquad 73.3^2 = 58.7^2 + 2(2.92)s \qquad s = 330 \text{ ft}$$

(c) Using $60 \text{ mi/h} = 88.0 \text{ ft/s}$ in the equation

$$v = v_0 + at$$

results in

$$88.0 = 73.3 + a(5) \qquad a = 2.94 \text{ ft/s}^2$$

Since $a = (v - v_0)/t$, it may be concluded that the required acceleration $2.94 \text{ ft/s}^2 \approx 2.92 \text{ ft/s}^2$ in parts (a) and (c) is the same.

The force is found from

$$\sum F_x = ma \qquad F = \frac{3,600}{32.2}(2.94) = 329 \text{ lb} \approx 326 \text{ lb}$$

The distance traveled in 5 s is found from

$$v^2 = v_0^2 + 2a(s - s_0) \qquad 88.0^2 = 73.3^2 + 2(2.94)s \qquad s = 403 \text{ ft}$$

15.14 (a) Do the same as in Prob. 15.13 if, at the time that the vehicle is traveling at 40 mi/h, the brakes are applied for a period of 3 s. The speed changes from 40 to 25 mi/h and the braking deceleration may be assumed to be constant.

(b) Do the same as in part (a) if the brakes are applied for a period of 6.5 s, and bring the vehicle to rest from a speed of 40 mi/h.

▌ (a) Using $25 \text{ mi/h} = 36.7 \text{ ft/s}$ in the equation

$$v = v_0 + at$$
$$36.7 = 58.7 + a(3) \qquad a = -7.33 \text{ ft/s}^2$$

Newton's second law is written as

$$\sum F_x = ma \qquad F = \frac{3,600}{32.2}(7.33) = 820 \text{ lb}$$

The distance covered during 3 s of braking is found from

$$v^2 = v_0^2 + 2a(s - s_0) \qquad 36.7^2 = 58.7^2 + 2(-7.33)s \qquad s = 143 \text{ ft}$$

(b) Using $40 \text{ mi/h} = 58.7 \text{ ft/s}$ in the equation

$$v = v_0 + at \qquad 0 = 58.7 + a(6.5) \qquad a = -9.03 \text{ ft/s}^2$$

the equation of motion is

$$\sum F_x = ma \qquad F = \frac{3,600}{32.2}(9.03) = 1,010 \text{ lb}$$
$$v^2 = v_0^2 + 2a(s - s_0) \qquad 0 = 58.7^2 + 2(-9.03)s \qquad s = 191 \text{ ft}$$

15.15 Figure 15.15a shows a simple model of a hoisting device to lift cargo out of the hold of a ship. The ship is anchored in calm water in a harbor. When the load is first raised, the hoisting motor accelerates the load at $1.5 \text{ ft/s}^2$. The total weight of the hoisting motor is 375 lb.

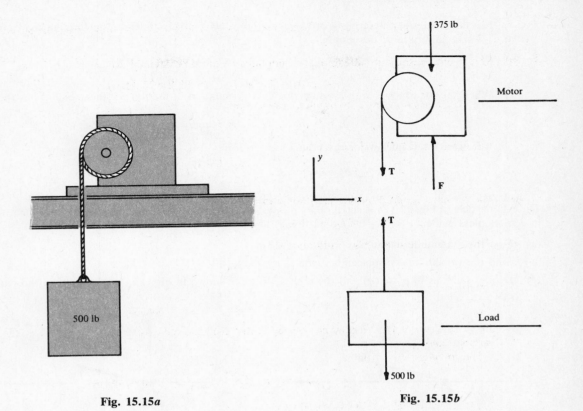

**Fig. 15.15a**                    **Fig. 15.15b**

(a)  Find the cable force, and the force exerted by the motor on the support beam, when the load is first raised.

(b)  Find the cable force, and the force exerted by the lifting motor on the support beam, at a later time when the load moves at a constant speed of 8.6 ft/s.

▌ (a)  Figure 15.15b shows the free-body diagrams of the motor and of the load.    $T$ is the cable tensile force and $F$ is the reaction force of the support beam on the hoisting motor.   Newton's second law for the load is written as

$$\sum F_y = ma \qquad T - 500 = \frac{500}{32.2}(1.5) \qquad T = 523\ \text{lb}$$

It may be seen that the requirement of 1.5 ft/s$^2$ acceleration of the load produces a   $523 - 500 = 23\ \text{lb}$   increase in force in the cable.
For equilibrium of the motor,

$$\sum F_y = 0 \qquad F - T - 375 = 0 \qquad F = 898\ \text{lb}$$

(b)  For equilibrium of the load, when this element moves with constant speed,

$$\sum F_y = 0$$
$$T - 500 = 0 \qquad T = 500\ \text{lb}$$

For equilibrium of the motor,

$$\sum F_y = 0 \qquad F - T - 375 = 0 \qquad F = 875\ \text{lb}$$

It may be seen that the speed of 8.6 ft/s does not enter the problem.

**15.16**  The ship of Prob. 15.15 is at sea in rough water.  Solve Prob. 15.15 if a wave causes a constant upward acceleration of the ship of 5 ft/s$^2$.

▌ (a)  When the load is first raised, using Fig. 15.15b,

$$\sum F_y = ma \qquad T - 500 = \frac{500}{32.2}(1.5 + 5) \qquad T = 601\ \text{lb}$$

For this case, the acceleration produces a $601 - 500 = 101$ lb increase in force in the cable. For equilibrium of the motor,

$$\sum F_y = ma \qquad F - T - 375 = \frac{375}{32.2}(5) \qquad F = 1,030 \text{ lb}$$

(b) When the load moves with constant speed, the equation of motion for this element has the form

$$\sum F_y = ma \qquad T - 500 = \frac{500}{32.2}(5) \qquad T = 578 \text{ lb}$$

Newton's second law for the motor has the form

$$\sum F_y = ma \qquad F - T - 375 = \frac{375}{32.2}(5) \qquad F = 1,010 \text{ lb}$$

**15.17** The 8-lb particle in Fig. 15.17a is acted on by the system of forces shown, and moves with rectilinear translation in a frictionless guide. At $t = 0$ the particle is at rest.

(a) Find the magnitude of the acceleration.

(b) Find the velocity and displacement when $t = 2$ s.

(c) Do the same as in parts (a) and (b) if the 3-lb force shown in Fig. 15.17b also acts on the particle.

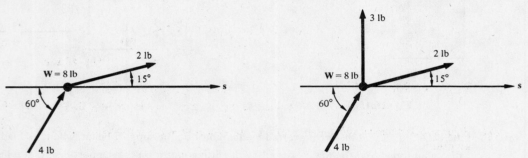

Fig. 15.17a          Fig. 15.17b

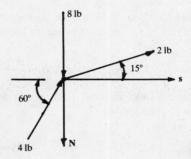

Fig. 15.17c

▌ (a) The free-body diagram of the particle is shown in Fig. 15.17c. Newton's second law has the form

$$\sum F_s = ma \qquad 2\cos 15° + 4\cos 60° = \frac{8}{32.2}a \qquad a = 15.8 \text{ ft/s}^2$$

(b) When $t = 2$ s,

$$v = v_0 + at = 15.8(2) = 31.6 \text{ ft/s} \qquad s = s_0 + v_0 t + \tfrac{1}{2}at^2$$
$$s = \tfrac{1}{2}(15.8)2^2 = 31.6 \text{ ft}$$

(c) Since the guide is frictionless, the results are the same as in parts (a) and (b).

**15.18** (a) Do the same as in Prob. 15.17, parts (a) and (b), if the direction of motion is that shown in Fig. 15.18a.

(b) Do the same as in part (a) if the applied forces are removed from the particle.

▌ (a) Figure 15.18b shows the free-body diagram of the particle. Newton's second law is written as

$$\sum F_s = ma \qquad 2\cos 15° + 4\cos 60° + 8\sin 30° = \frac{8}{32.2}a \qquad a = 31.9 \text{ ft/s}^2$$

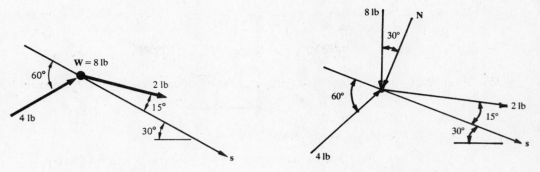

Fig. 15.18a

Fig. 15.18b

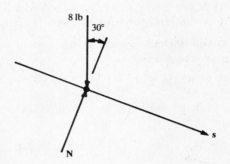

Fig. 15.18c

When $t = 2\,\text{s}$,

$$v = v_0 + at = 31.9(2) = 63.8\,\text{ft/s} \qquad s = s_0 + v_0 t + \tfrac{1}{2}at^2 = \tfrac{1}{2}(31.9)2^2 = 63.8\,\text{ft}$$

(b) Figure 15.18c shows the free-body diagram when the applied forces are removed. The equation of motion is

$$\sum F_s = ma \qquad 8 \sin 30° = \frac{8}{32.2}\,a \qquad a = 16.1\,\text{ft/s}^2$$

When $t = 2\,\text{s}$,

$$v = v_0 + at = 16.1(2) = 32.2\,\text{ft/s} \qquad s = s_0 + v_0 t + \tfrac{1}{2}at^2 = \tfrac{1}{2}(16.1)2^2 = 32.2\,\text{ft}$$

**15.19** (a) Do the same as in Prob. 15.17, parts (a) and (b), if the direction of motion is that shown in Fig. 15.19a.

(b) Do the same as in part (a) if the applied forces are removed from the particle.

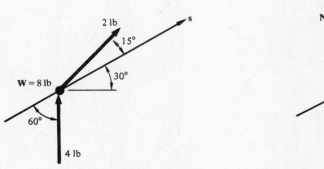

Fig. 15.19a

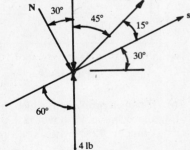

Fig. 15.19b

▌ (a) Figure 15.19b shows the free-body diagram of the particle, and Newton's second law has the form

$$\sum F_s = ma \qquad 2 \cos 15° - 8 \cos 60° + 4 \cos 60° = \frac{8}{32.2}\,a \qquad a = -0.274\,\text{ft/s}^2$$

Since the acceleration is negative, it may be concluded that the particle will move down and to the left in Fig. 15.19b.

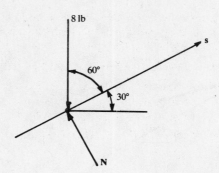

Fig. 15.19c

When $t = 2$ s,

$$v = v_0 + at = -0.274(2) = -0.548 \text{ ft/s} \qquad s = s_0 + v_0 t + \tfrac{1}{2}at^2 = \tfrac{1}{2}(-0.274)2^2 = -0.548 \text{ ft}$$

(b) The free-body diagram for the case where the applied forces are removed is shown in Fig. 15.19c. The acceleration is found from

$$\sum F_s = ma \qquad -8\cos 60° = \frac{8}{32.2}\, a \qquad a = -16.1 \text{ ft/s}^2$$

The motion is again down and to the left in Fig. 15.19c.
At $t = 2$ s,

$$v = v_0 + at = -16.1(2) = -32.2 \text{ ft/s} \qquad s = s_0 + v_0 t + \tfrac{1}{2}at^2 = \tfrac{1}{2}(-16.1)2^2 = -32.2 \text{ ft}$$

15.20 Body $A$ and cylinder $B$, shown in Fig. 15.20a, move in rectilinear translation with an acceleration of 2.7 m/s$^2$.

(a) Find the forces exerted on cylinder $B$ by body $A$.

(b) Express the results in part (a) in formal vector notation.

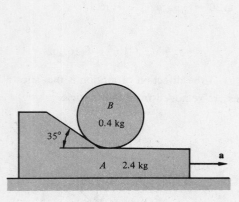

Fig. 15.20a

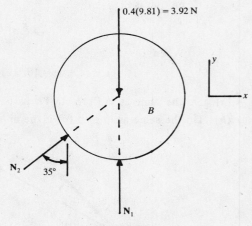

Fig. 15.20b

∎ (a) This is a problem in which the motion is given and the force is to be found. The free-body diagram of the cylinder is shown in Fig. 15.20b, and $N_1$ and $N_2$ are the reaction forces exerted by body $A$ on cylinder $B$. Newton's second law has the form

$$\sum F_x = ma \qquad N_2 \sin 35° = 0.4(2.7) \qquad N_2 = 1.88 \text{ N}$$

For equilibrium in the $y$ direction,

$$\sum F_y = 0 \qquad -3.92 + N_2 \cos 35° + N_1 = 0 \qquad N_1 = 2.38 \text{ N}$$

(b) The two forces acting on body $B$, using Fig. 15.20b, may be written in terms of unit vectors as

$$\mathbf{N}_1 = N_1 \mathbf{j} = 2.38\mathbf{j} \qquad \text{N}$$
$$\mathbf{N}_2 = (N_2 \sin 35°)\mathbf{i} + (N_2 \cos 35°)\mathbf{j} = (1.88 \sin 35°)\mathbf{i} + (1.88 \cos 35°)\mathbf{j} = 1.08\mathbf{i} + 1.54\mathbf{j} \qquad \text{N}$$

**15.21** Figure 15.21*a* shows an experimental device that may be used to measure the acceleration of a body moving in rectilinear translation. A small mass particle slides in a frictionless guide, in a vertical plane, with the form of a quarter circle. The device accelerates to the right in the figure with an acceleration of constant magnitude *a*.

(*a*) Find the relation between the observed angle $\theta$ and the acceleration *a* of the device.

(*b*) Find the magnitude of the normal force between the particle and the track, if the mass of the particle is 30 g and the acceleration of the device is $6 \, \text{m/s}^2$.

(*c*) Express the resultant force on the mass particle in formal vector notation.

(*d*) How could the device shown in Fig. 15.21*a* be used to measure constant deceleration of a body in rectilinear translation?

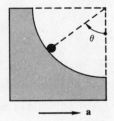
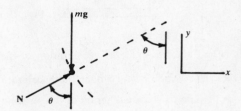

**Fig. 15.21*b***

▌ (*a*) The free-body diagram of the particle is shown in Fig. 15.21*b*. Newton's second law is written as

$$\sum F_x = ma \qquad N \sin \theta = ma \tag{1}$$

For equilibrium in the *y* direction,

$$\sum F_y = 0 \qquad N \cos \theta - mg = 0 \qquad N \cos \theta = mg \tag{2}$$

Dividing Eq. (1) by Eq. (2),

$$\tan \theta = \frac{a}{g} \qquad a = g \tan \theta \qquad \theta = \tan^{-1}\left(\frac{a}{g}\right) \tag{3}$$

(*b*)
$$m = 30 \, \text{g} = 0.030 \, \text{kg} \qquad a = 6 \, \text{m/s}^2$$

Using Eq. (3),

$$\theta = \tan^{-1}\left(\frac{a}{g}\right) = \tan^{-1}\left(\frac{6}{9.81}\right) = 31.5°$$

Using Eq. (2),

$$N = \frac{mg}{\cos \theta} = \frac{0.030(9.81)}{\cos 31.5°} = 0.345 \, \text{N}$$

(*c*) Using Fig. 15.21*b*, the resultant force **F** acting on the mass particle is

$$\mathbf{F} = (N \sin \theta)\mathbf{i} + (N \cos \theta - mg)\mathbf{j} = (0.345 \sin 31.5°)\mathbf{i} + [0.345 \cos 31.5° - 0.030(9.81)]\mathbf{j}$$
$$= 0.181\mathbf{i} + (0)\mathbf{j} = 0.181\mathbf{i} \qquad \text{N}$$

The zero value of the *y* component is an expected result, since $\sum F_y = 0$ is the equilibrium requirement which was used in part (*a*).

(*d*) To measure constant deceleration, one would operate the device in the sense opposite to that shown in Fig. 15.21*a*.

**15.22** A 180-lb person stands in an elevator, as shown in Fig. 15.22*a*.

(*a*) The elevator accelerates upward with a constant value of $8 \, \text{ft/s}^2$. What is the apparent weight of the person during this period of acceleration?

(*b*) Do the same as in part (*a*) for a constant downward acceleration of the elevator of $8 \, \text{ft/s}^2$.

(*c*) If the elevator cage weighs 1,450 lb, find the cable tensile force *T* which corresponds to the acceleration of part (*a*).

Fig. 15.22a

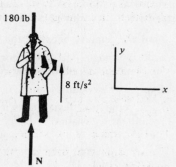

Fig. 15.22b

(d)  For what value of acceleration of the elevator would the person appear to be weightless?

▌  This is a problem where the acceleration is known and the value of the force is to be obtained.

(a)  The free-body diagram of the person is shown in Fig. 15.22b.  The equation of motion in the y direction is

$$\sum F_y = ma \qquad N - 180 = \frac{W}{g}\, a = \frac{180}{32.2}\,(8) \qquad N = 225\,\text{lb}$$

N is the force exerted by the elevator floor on the person.  From Newton's third law, the force exerted by the person on the elevator floor is *equal and opposite* to the force N shown in Fig. 15.22b.  This latter force is the "weight" of the person.  It may be seen that, for upward acceleration of the elevator, the person appears to weigh *more* than he or she actually does.

(b)  The free-body diagram of the person for downward acceleration of the elevator is shown in Fig. 15.22c.  The equation of motion is

$$\sum F_y = ma \qquad N - 180 = \frac{W}{g}\, a = \frac{180}{32.2}\,(-8) \qquad N = 135\,\text{lb}$$

For downward acceleration of the elevator, the person appears to weigh *less* than she or he actually does.

(c)  The free-body diagram of the elevator is shown in Fig. 15.22d, and T is the tensile force in the elevator cable.  The equation of motion is

$$\sum F_y = ma \qquad T - 1{,}630 = \frac{W}{g}\, a = \frac{1{,}630}{32.2}\,(8) \qquad T = 2{,}030\,\text{lb}$$

It may be observed that the normal force between the person and the elevator floor does *not* enter into this problem.  The reason is that this force is an *internal* force when the free-body diagram of the system (the elevator cage and the person) is considered.

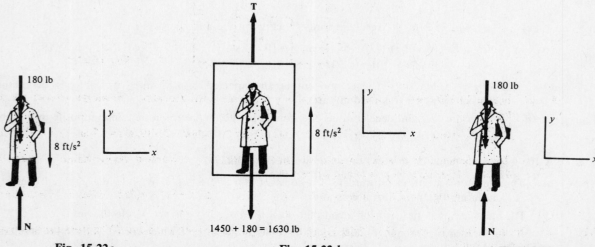

Fig. 15.22c

1450 + 180 = 1630 lb

Fig. 15.22d

Fig. 15.22e

(*d*)  The free-body diagram for general motion of the person is shown in Fig. 15.22*e*.  In order for the person to appear weightless, he or she must exert no force on the floor of the elevator, or  $N = 0$.  The equation of motion, in the negative *y*-coordinate sense, is then

$$\sum F_y = ma \qquad -180 = \frac{180}{g}\,a \qquad a = -g = -32.2 \text{ ft/s}^2$$

For this case the elevator is in free fall, as if the cable had been cut.

**15.23**   Figure 15.23*a* shows a person of mass 70 kg riding in a lightweight passenger elevator of mass 500 kg.  When the elevator moves upward with constant acceleration, the cable tensile force is 6,500 N.

(*a*)  Find the value of the acceleration of the elevator.

(*b*)  Do the same as in part (*a*) if the cable tensile force is 4,200 N as the elevator accelerates downward with a constant value.

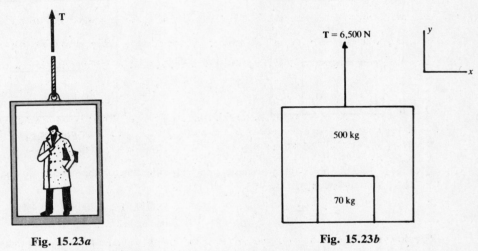

Fig. 15.23*a*                                   Fig. 15.23*b*

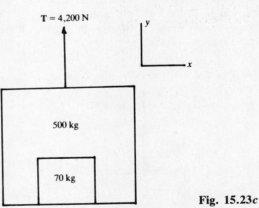

Fig. 15.23*c*

▌(*a*)  Figure 15.23*b* shows a schematic free-body diagram of the elevator.  Newton's second law has the form

$$\sum F_y = ma \qquad T - (500 + 70)9.81 = (500 + 70)a \qquad 6{,}500 - 570(9.81) = 570a \qquad a = 1.59 \text{ m/s}^2$$

(*b*)  The schematic free-body diagram of the elevator, for downward acceleration, is seen in Fig. 15.23*c*.  Newton's second law is written as

$$\sum F_y = ma \qquad (500 + 70)9.81 - T = (500 + 70)a \qquad 570(9.81) - 4{,}200 = 570a \qquad a = 2.44 \text{ m/s}^2$$

**15.24**   (*a*)  Do the same as in Prob. 15.23, part (*a*), if a second passenger with a mass of 50 kg joins the first passenger in the elevator.

(b)  Do the same as in Prob. 15.23, part (b), for the conditions of part (a) above.

❙ (a)  The schematic free-body diagram of the elevator is shown in Fig. 15.24a.  The equation of motion in the y direction is

$$\sum F_y = ma \qquad T - (500 + 70 + 50)9.81 = (500 + 70 + 50)a$$

$$6,500 - 620(9.81) = 620a \qquad a = 0.674 \text{ m/s}^2$$

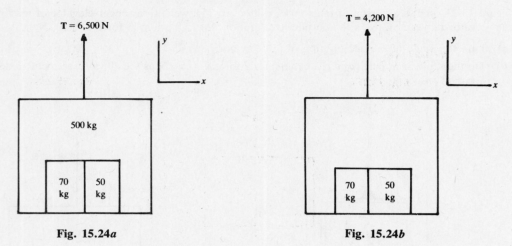

Fig. 15.24a                    Fig. 15.24b

(b)  Figure 15.24b shows the schematic free-body diagram for the case of downward acceleration.  Newton's second law is written as

$$\sum F_y = ma \qquad (500 + 70 + 50)9.81 - 4,200 = (500 + 70 + 50)a$$
$$620(9.81) - 4,200 = 620a \qquad a = 3.04 \text{ m/s}^2$$

15.25  The acceleration-time variation of a passenger elevator is shown in Fig. 15.25a.  The weight of the passenger is 165 lb and the weight of the elevator cage is 1,800 lb.  When the elevator accelerates upward, the passenger experiences an apparent weight gain of 18 lb.  When the elevator decelerates to zero velocity, the passenger experiences an apparent weight loss of 22 lb.

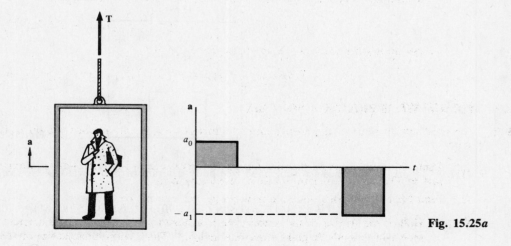

Fig. 15.25a

(a)  Find the magnitudes $a_0$ and $a_1$.

(b)  Find the corresponding values of the elevator cable tensile force $T$.

❙ (a)  The free-body diagram of the passenger for the case of upward acceleration is shown in Fig. 15.25b.  Newton's second law is written as

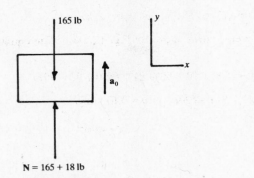

**Fig. 15.25b**

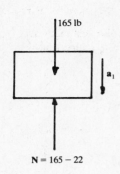

**Fig. 15.25c**

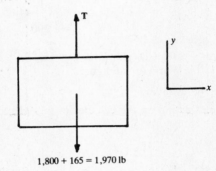

**Fig. 15.25d**

$$\sum F_y = ma \qquad (165 + 18) - 165 = \frac{165}{32.2}\, a_0 \qquad a_0 = 3.51\ \text{ft/s}^2$$

The free-body diagram of the passenger, for downward acceleration, is seen in Fig. 15.25c. Newton's second law for this case is

$$\sum F_y = ma \qquad -165 + (165 - 22) = \frac{165}{32.2}\, (-a_1) \qquad a_1 = 4.29\ \text{ft/s}^2$$

(b) Figure 15.25d shows the free-body diagram of the elevator for either upward or downward acceleration. The equation of motion for upward acceleration is

$$\sum F_y = ma \qquad T - 1{,}970 = \frac{1{,}970}{32.2}\, (3.51) \qquad T = 2{,}180\ \text{lb}$$

For downward acceleration, in the negative y sense,

$$\sum F_y = ma \qquad -1{,}970 + T = \frac{1{,}970}{32.2}\, (-4.29) \qquad T = 1{,}710\ \text{b}$$

## 15.2 MOTION WITH FRICTION FORCES

15.26 An approximate value of the coefficient of friction of a rubber automobile tire on an ice-covered pavement is 0.05.

(a) Using this data, estimate the minimum time required for a rear-wheel-drive automobile to accelerate from rest to 30 mi/h on a horizontal ice-covered pavement.

(b) Find the corresponding distance traveled by the automobile.

▌ (a) Half of the weight of the automobile is assumed to be supported by the rear wheels. Figure 15.26 shows a free-body diagram of the automobile. The friction forces acting on the rear wheels have their maximum possible values. Newton's second law has the form

$$\sum F_x - ma \qquad \mu N = ma \qquad 0.05\left(\frac{mg}{2}\right) = ma \qquad a = 0.05\left(\frac{32.2}{2}\right) = 0.805\ \text{ft/s}^2$$

Using $30\ \text{mi/h} = 44.0\ \text{ft/s}$ and $v_0 = 0$,

$$v = v_0 + at \qquad 44.0 = 0.805t \qquad t = 54.7\ \text{s}$$

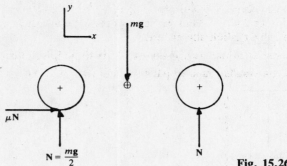

**Fig. 15.26**

(*b*) The distance traveled by the automobile is found from

$$v^2 = v_0^2 + 2a(s - s_0) \qquad 44.0^2 = 2(0.805)s$$
$$s = 1,200 \text{ ft}$$

**15.27** A cylinder of mass 6 kg rests in a trough, as shown in Fig. 15.27*a*. A force of 30 N is applied along the center axis of the cylinder, normal to the plane of the figure.

(*a*) Find the acceleration of the cylinder.

(*b*) Express the reaction forces acting on the cylinder in formal vector notation.

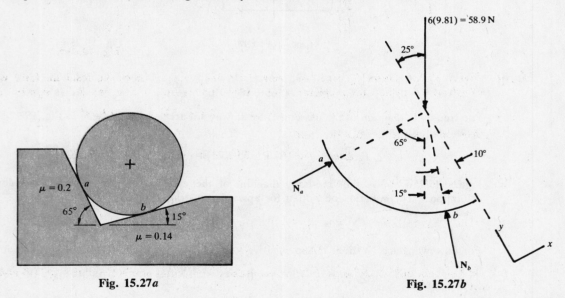

**Fig. 15.27*a***                      **Fig. 15.27*b***

❚ (*a*) Figure 15.27*b* shows the reaction forces which act on the cylinder. $N_a$ and $N_b$ are the normal forces exerted by the plane surfaces on the cylinder. The rotated $xy$ coordinates shown in the figure are used so that one of the two unknown reaction forces may be found directly. For equilibrium of the cylinder,

$$\sum F_y = 0 \qquad -58.9 \cos 25° + N_b \cos 10° = 0 \qquad N_b = 54.2 \text{ N}$$

$$\sum F_x = 0 \qquad N_a + N_b \sin 10° - 58.9 \sin 25° = 0 \qquad N_a = 15.5 \text{ N}$$

Motion in the $z$ direction occurs, so that the friction forces at $a$ and $b$ have their maximum possible values. Newton's second law in the $z$ direction then has the form

$$\sum F_z = ma \qquad 30 - 0.2N_a - 0.14N_b = 6a \qquad a = 3.22 \text{ m/s}^2$$

(*b*) The reaction forces acting on the cylinder, in terms of the unit vectors along the $xy$ axes shown in Fig. 15.27*b*, are

$$\mathbf{N}_a = N_a \mathbf{i} = 15.5\mathbf{i} \qquad \text{N}$$
$$\mathbf{N}_b = (N_b \sin 10°)\mathbf{i} + (N_b \cos 10°)\mathbf{j} = (54.2 \sin 10°)\mathbf{i} + (54.2 \cos 10°)\mathbf{j} = 9.41\mathbf{i} + 53.4\mathbf{j} \qquad \text{N}$$

**15.28** (*a*) Figure 15.28*a* shows a pile-driver weight. The total friction force exerted by the guides on the weight is estimated to be 350 N. Find the value of the constant cable tensile force $T$ that will raise the weight 8 m in 4 s. The weight is initially at rest.

(*b*) Do the same as in part (*a*), if the weight is to be raised through the distance 8 m in 2 s.

(*c*) Compare the results found in parts (*a*) and (*b*).

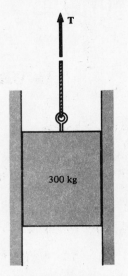

**Fig. 15.28*a***

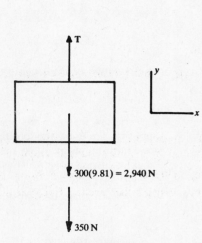

**Fig. 15.28*b***

▌ (*a*) Figure 15.28*b* shows the free-body diagram of the weight. Since the resultant force acting on the weight is a constant, the acceleration must also be constant. Using $s = 8$ m, $v_0 = 0$, and $t = 4$ s,

$$s = s_0 + v_0 t + \tfrac{1}{2}at^2 \qquad 8 = \tfrac{1}{2}(a)4^2 \qquad a = 1 \text{ m/s}^2$$

Newton's second law has the form

$$\sum F_y = ma \qquad T - 2{,}940 - 350 = 300(1) \qquad T = 3{,}590 \text{ N}$$

(*b*) With the conditions $s = 8$ m, $v_0 = 0$, and $t = 2$ s,

$$s = s_0 + v_0 t + \tfrac{1}{2}at^2 \qquad 8 = \tfrac{1}{2}(a)2^2 \qquad a = 4 \text{ m/s}^2$$

Using Fig. 15.28*b*,

$$\sum F_y = ma \qquad T - 2{,}940 - 350 = 300(4) \qquad T = 4{,}490 \text{ N}$$

(*c*) The percent difference between the two values of cable tensile force found in parts (*a*) and (*b*) is given by

$$\%\text{D} = \frac{4{,}490 - 3{,}590}{3{,}590}\,100 = 25\%$$

It may be seen that a 25 percent increase in the cable force increases the acceleration by a factor of 4.

**15.29** A 130-lb block slides along a smooth horizontal track with a velocity of 50 ft/s, as shown in Fig. 15.29*a*. At a certain point the smooth track joins a section of rough horizontal track. The coefficient of friction between the block and the rough track is 0.24.

(*a*) Find the deceleration of the block on the rough horizontal track.

(*b*) How will it take for the block to come to rest?

(*c*) How far along the rough track will the block slide before coming to rest?

▌ (*a*) The displacement coordinate $x$ is measured from the beginning of the rough track. The free-body diagram of the block, when sliding on the rough track, is shown in Fig. 15.29*b*.

For equilibrium in the $y$ direction,

$$\sum F_y = 0 \qquad -130 + N = 0 \qquad N = 130 \text{ lb}$$

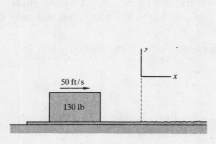

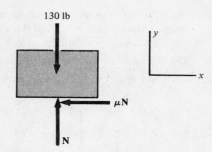

**Fig. 15.29a**    **Fig. 15.29b**

The equation of motion in the $x$ direction is

$$\sum F_x = ma \qquad -\mu N = ma \qquad -0.24(130) = \frac{130}{32.2}\,a \qquad a = -7.73 \text{ ft/s}^2$$

The friction force in the above equation is written with a minus sign, since this force acts in the opposite sense of the positive displacement coordinate $x$. The negative value obtained for the acceleration indicates that the sense of the acceleration is opposite to that of the velocity, so that the block is slowing down. Thus, the formal statement of the solution is that the deceleration of the block is 7.73 ft/s$^2$.

(**b**) The time to come to rest, with $v_0 = 50$ ft/s, is found from

$$v = v_0 + at \qquad 0 = 50 + (-7.73)t \qquad t = 6.47 \text{ s}$$

(**c**) The distance traveled by the block along the rough track before it comes to rest is given by

$$v^2 = v_0^2 + 2a(x - x_0)$$

With $x_0 = 0$ and $v_0 = 50$ ft/s,

$$0 = 50^2 + 2(-7.73)(x - 0) \qquad x = 162 \text{ ft}$$

This result may be checked by using the equation

$$x = x_0 + v_0 t + \tfrac{1}{2}at^2 = 0 + 50(6.47) + \tfrac{1}{2}(-7.73)6.47^2 = 162 \text{ ft}$$

The tacit assumption is made in this problem that the length of the roughened section of the track is at least 162 ft.

**15.30** The block shown in Fig. 13.30a slides along a horizontal plane with a constant value of deceleration. The speed of the block decreases from 35 in/s to zero over a length of 100 in.

(**a**) Find the value of the coefficient of friction.

(**b**) If a second block weighing 6 lb is attached to the top of the 14-lb block, find the distance through which the system of blocks will move before coming to rest. The initial velocity is 35 in/s.

❚ (**a**) The deceleration of the block is found first. With $v_0 = 35$ in/s and $s_0 = 0$,

$$v^2 = v_0^2 + 2a(s - s_0) \qquad 0 = 35^2 - 2a(100) \qquad a = -6.13 \text{ in/s}^2$$

Figure 15.30b shows the free-body diagram of the block. For equilibrium,

$$\sum F_y = 0 \qquad N - mg = 0 \qquad N = mg$$

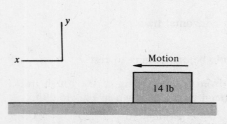

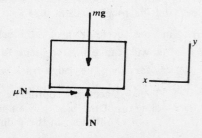

**Fig. 15.30a**    **Fig. 15.30b**

Newton's second law is written as

$$\sum F_x = ma \qquad -\mu N = ma \qquad -\mu mg = ma \qquad \mu = -\frac{a}{g} = \frac{-(-6.13)}{386} = 0.0159$$

(*b*)  Since the deceleration is independent of the mass of the block, the system of two blocks will come to rest in 100 in, as in the original problem statement.

**15.31**  A force of 100 lb, with variable direction $\theta$, acts on the block shown in Fig. 15.31*a*.

(*a*)  For what value of $\theta$ will the acceleration of the block be maximum?

(*b*)  Express this maximum acceleration in functional form.

(*c*)  Find the numerical value of the maximum acceleration.

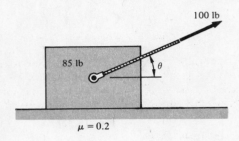

**Fig. 15.31*a***

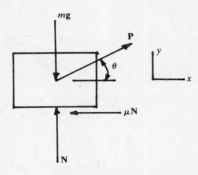

**Fig. 15.31*b***

(*a*)  Figure 15.31*b* shows the free-body diagram of the block.  The equations of equilibrium and motion have the forms

$$\sum F_y = 0 \qquad -mg + P \sin \theta + N = 0 \qquad N = mg - P \sin \theta \tag{1}$$

$$\sum F_x = ma \qquad P \cos \theta - \mu N = ma \tag{2}$$

Using Eq. (1) in Eq. (2),

$$P \cos \theta - \mu(mg - P \sin \theta) = ma \qquad a = \frac{P}{m}(\cos \theta + \mu \sin \theta) - \mu g \tag{3}$$

When $a$ is maximumn,  $da/d\theta = 0$.  Using Eq. (3),

$$\frac{da}{d\theta} = \frac{P}{m}(-\sin \theta + \mu \cos \theta) = 0 \qquad \sin \theta = \mu \cos \theta \qquad \tan \theta = \mu \qquad \theta = \tan^{-1} \mu$$

(*b*)  The relationship for $\theta$ found in part (*a*) is shown in Fig. 15.31*c*. From this figure,

$$\sin \theta = \frac{\mu}{\sqrt{1 + \mu^2}} \qquad \cos \theta = \frac{1}{\sqrt{1 + \mu^2}}$$

Using the above results in Eq. (3),

$$a_{max} = \frac{P}{m}\left[\frac{1}{\sqrt{1 + \mu^2}} + \mu\left(\frac{\mu}{\sqrt{1 + \mu^2}}\right)\right] - \mu g$$

$$= \frac{P}{m}\sqrt{1 + \mu^2} - \mu g$$

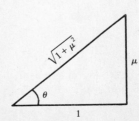

**Fig. 15.31*c***

(c) The numerical value of the maximum acceleration is

$$a_{max} = \frac{100}{85/32.2} \sqrt{1 + 0.2^2} - 0.2(32.2) = 32.2 \text{ ft/s}^2$$

**15.32** (a) Find the acceleration of the block shown in Fig. 15.32a, as this element slides down the inclined surface.

(b) If the block is released from rest, find the velocity at the end of 3 s.

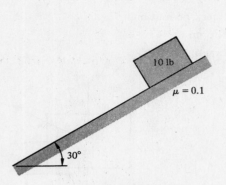

Fig. 15.32a

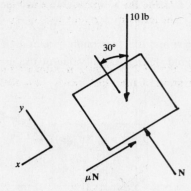

Fig. 15.32b

❚ (a) The free-body diagram of the block is seen in Fig. 15.32b. For equilibrium in the y direction,

$$\sum F_y = 0 \qquad -10 \cos 30° + N = 0 \qquad N = 10 \cos 30°$$

Newton's second law has the form

$$\sum F_x = ma \qquad 10 \sin 30° - \mu N = ma \qquad 10 \sin 30° - 0.1(10 \cos 30°) = \frac{10}{32.2} a \qquad a = 13.3 \text{ ft/s}^2$$

(b) Using $v_0 = 0$,

$$v = v_0 + at \qquad v = 13.3(3) = 39.9 \text{ ft/s}$$

**15.33** A block slides down an inclined plane with an acceleration of $1.62 \text{ m/s}^2$, as shown in Fig. 15.33a.

(a) Find the friction force that acts on the block.

(b) Find the value of the coefficient of friction between the block and the plane.

(c) Express the reaction force of the plane on the block in formal vector notation.

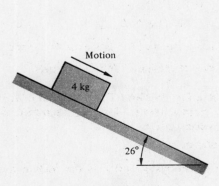

Fig. 15.33a

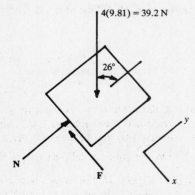

Fig. 15.33b

❚ (a) Figure 15.33b shows the free-body diagram of the block, and F is the friction force acting on this element. The equation of motion has the form

$$\sum F_x = ma \qquad 39.2 \sin 26° - F = 4(1.62) \qquad F = 10.7 \text{ N}$$

(*b*)  The normal force which acts on the block is found from

$$\sum F_y = 0 \qquad -39.2 \cos 26° + N = 0 \qquad N = 35.2 \text{ N}$$

Using the definition of the coefficient of friction,

$$\mu = \frac{F}{N} = \frac{10.7}{35.2} = 0.304$$

(*c*)  Using the *xy* coordinates in Fig. 15.33*b*, the reaction force **R** of the plane on the block may be written in terms of the unit vectors as

$$\mathbf{R} = -F\mathbf{i} + N\mathbf{j} = -10.7\mathbf{i} + 35.2\mathbf{j} \qquad \text{N}$$

**15.34**  When  $\theta = 27.2°$, sliding motion of the block shown in Fig. 15.34*a* is impending.  Find the value of the acceleration of the block when  $\theta = 40°$.

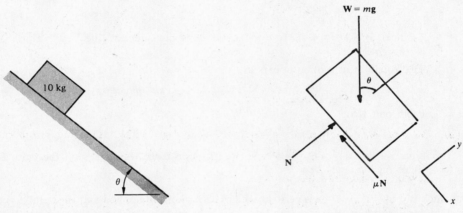

**Fig. 15.34*a***  \qquad\qquad  **Fig. 15.34*b***

❚  The free-body diagram of the block, for the case of impending motion, is seen in Fig. 15.34*b*.  For equilibrium of the block,

$$\sum F_y = 0 \qquad N - W \cos 27.2° = 0 \qquad N = W \cos 27.2° \qquad\qquad (1)$$

$$\sum F_x = 0 \qquad W \sin 27.2 - \mu N = 0 \qquad \mu N = W \sin 27.2° \qquad\qquad (2)$$

Dividing Eq. (2) by Eq. (1),

$$\mu = \tan 27.2° = 0.514$$

For the case of sliding motion, with  $\theta = 40°$,

$$\sum F_y = 0 \qquad N = mg \cos 40°$$

$$\sum F_x = ma \qquad mg \sin 40° - \mu N = ma \qquad mg \sin 40° - \mu(mg \cos 40°) = ma$$

$$a = g \sin 40° - \mu g \cos 40° = (\sin 40° - 0.514 \cos 40°)9.81 = 2.44 \text{ m/s}^2$$

**15.35**  (*a*)  A flat crate rests on the bed of a truck, as shown in Fig. 15.35*a*.  The coefficient of friction between the crate and the truck bed is 0.3, and the mass of the crate is 200 kg.  Find the magnitude of the friction force exerted by the bed on the crate when the truck decelerates at $1.2 \text{ m/s}^2$.

(*b*)  Do the same as in part (*a*), if the truck decelerates at $3.5 \text{ m/s}$.

❚  (*a*)  Figure 15.35*b* shows the free-body diagram of the crate, and *F* is the friction force acting on the crate.  Newton's second law has the form

$$\sum F_x = ma \qquad -F = 200(-1.2) = 240 \text{ N}$$

The maximum possible value of the friction force is next found, for comparison with the above value of 240 N.  For equilibrium in the *y* direction,

$$\sum F_y = 0 \qquad N - 1,960 = 0 \qquad N = 1,960 \text{ N}$$

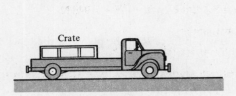

Fig. 15.35a

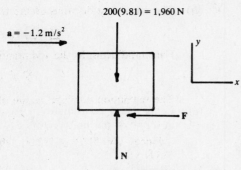

Fig. 15.35b

The maximum available friction force is

$$F_{max} = \mu N = 0.3(1,960) = 588 \text{ N}$$

Since $F = 240 \text{ N} < 588 \text{ N}$, the crate does not slide on the bed. Thus, the answer to part (a) is $F = 240 \text{ N}$.

(b) The equation of motion apears as

$$\sum F_x = ma \qquad -F = 200(-3.5) = 700 \text{ N}$$

From part (a),

$$F_{max} = 588 \text{ N}$$

Since $F = 700 \text{ N} > F_{max} = 588 \text{ N}$, the friction force which acts on the crate is 588 N, and the crate slides on the truck bed.

**15.36** The truck in Fig. 15.35a travels at 30 km/h. The brakes are applied and the vehicle is assumed to experience constant deceleration until it comes to rest. Find the minimum value of the stopping distance, if the crate is not to slide on the truck bed.

❚ From part (a) of Prob. 15.35,

$$F_{max} = 588 \text{ N}$$

Newton's second law has the form

$$\sum F_x = ma \qquad -588 = 200(a) \qquad a = -2.94 \text{ m/s}^2$$

Using $s_0 = 0$ and $v_0 = 30 \text{ km/h} = 8.33 \text{ m/s}$,

$$v^2 = v_0^2 + 2a(s - s_0) \qquad 0 = 8.33^2 + 2(-2.94)s$$
$$s = s_{min} = 11.8 \text{ m}$$

**15.37** (a) Do the same as in Prob. 15.35, part (a), if the truck travels on the downhill grade shown in Fig. 15.37a.

(b) Do the same as in Prob. 15.35, part (b), for the truck shown in Fig. 15.37a.

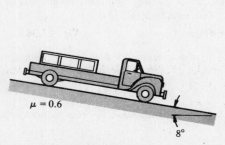

Fig. 15.37a

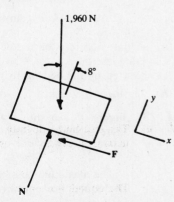

Fig. 15.37b

**(a)** The free-body diagram of the truck is shown in Fig. 15.37b. Newton's second law is written as

$$\sum F_x = ma \qquad -F + 1{,}960 \sin 8° = 200(-1.2) \qquad F = 513 \text{ N}$$

For equilibrium in the $y$ direction,

$$\sum F_y = 0 \qquad N - 1{,}960 \cos 8° = 0 \qquad N = 1{,}940 \text{ N}$$

The maximum available value of the friction force is

$$F_{max} = \mu N = 0.3(1{,}940) = 582 \text{ N}$$

Since $F = 513 \text{ N} < 582 \text{ N}$, the crate does not slide on the bed. The answer to part (a) is $F = 513 \text{ N}$.

**(b)** The equation of motion of the truck is

$$\sum F_x = ma \qquad -F + 1{,}960 \sin 8° = 200(-3.5) \qquad F = 973 \text{ N}$$

From part (a),

$$F_{max} = 582 \text{ N}$$

Since $F = 973 \text{ } N > F_{max} = 582 \text{ N}$, the friction force which acts on crate is equal to 582 N, and the crate slides on the truck bed.

**15.38** **(a)** For what value of acceleration $a$ will sliding motion of block $B$ with respect to block $A$, Fig. 15.38a, be impending?

**(b)** Find the required value of the horizontal force acting on block $A$ to produce the acceleration of part (a).

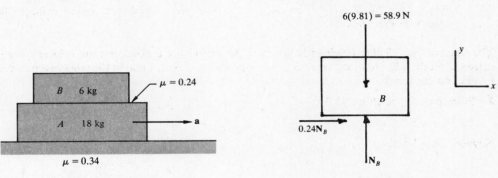

Fig. 15.38a       Fig. 15.38b

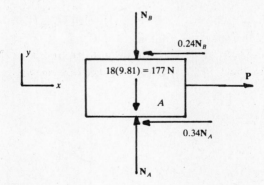

Fig. 15.38c

**(a)** The free-body diagram of block $B$ is shown in Fig. 15.38b. The friction force is assumed to have its maximum possible value. The normal force $N_b$ is found from

$$\sum F_y = 0 \qquad N_B - 58.9 = 0 \qquad N_B = 58.9 \text{ N}$$

The equation of motion of block $B$ is

$$\sum F_x = ma \qquad 0.24(58.9) = 6a \qquad a = 2.36 \text{ m/s}^2$$

(b) The free-body diagram of block $A$ is seen in Fig. 15.38c. Since motion of this block occurs, the friction force between the block and the plane surface has its maximum possible value. The normal force $N_a$ is found from

$$\sum F_y = 0 \qquad -N_B - 177 + N_A = 0 \qquad N_A = 236 \text{ N}$$

Newton's second law for block $A$ is then

$$\sum F_x = ma \qquad P - 0.24N_B - 0.34N_A = 18(2.36) \qquad P = 137 \text{ N}$$

**15.39** (a) Do the same as Prob. 15.38, if the blocks are arranged as shown in Fig. 15.39a.

(b) Express the reaction forces acting on block $A$ in formal vector notation.

(c) Compare the results found in part (a) above with those found in Prob. 15.38.

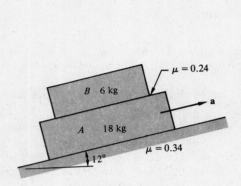

Fig. 15.39a

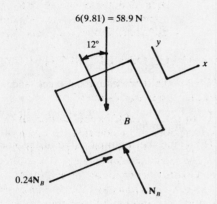

Fig. 15.39b

▌ (a) Using the free-body diagram in Fig. 15.39b, the equations of equilibrium and motion are

$$\sum F_y = 0 \qquad -58.9 \cos 12° + N_B = 0 \qquad N_B = 57.6 \text{ N}$$

$$\sum F_x = ma \qquad -58.9 \sin 12° + 0.24N_B = 6a \qquad a = 0.263 \text{ m/s}^2$$

Figure 15.39c shows the free-body diagram of block $A$. For equilibrium in the $y$ direction,

$$\sum F_y = 0 \qquad -N_B - 177 \cos 12° - P \sin 12° + N_A = 0 \qquad N_A = 231 + 0.208P \qquad (1)$$

The equation of motion is

$$\sum F_x = ma \qquad -0.24N_B - 177 \sin 12° + P \cos 12° - 0.34N_A = 18(0.263)$$

Using Eq. (1), the above equation may be written as

$$-0.24(57.6) - 177 \sin 12° + P \cos 12° - 0.34(231 + 0.208P) = 18(0.263)$$
$$P[\cos 12° - 0.34(0.208)] = 0.24(57.6) + 177 \sin 12° + 0.34(231) + 18(0.263) \qquad P = 148 \text{ N}$$

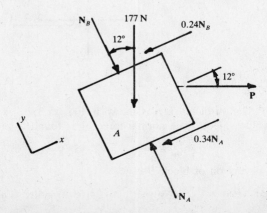

Fig. 15.39c

(*b*)   Using Fig. 15.39*c*,

$$\mathbf{N}_A = -0.34N_A\mathbf{i} + N_A\mathbf{j}$$

$N_A$ is found from

$$N_A = 231 + 0.208P = 231 + 0.208(148) = 262 \text{ N}$$

$N_A$ then has the form

$$\mathbf{N}_A = -0.34(262)\mathbf{i} + 262\mathbf{j} = -89.1\mathbf{i} + 262\mathbf{j} \qquad \text{N}$$

Again using Fig. 15.39*c*,

$$\mathbf{N}_B = -0.24N_B\mathbf{i} - N_B\mathbf{j} = -0.24(57.6)\mathbf{i} - 57.6\mathbf{j} = -13.8\mathbf{i} - 57.6\mathbf{j} \qquad \text{N}$$

(*c*)   From comparison of the acceleration value in part (*a*) above with that found in Prob. 15.38, it may be seen that these values differ by a factor of 9.   The forces required to produce the motions have approximately the same magnitudes.   It is left as an exercise for the reader to account for the large difference between the acceleration and force ratios.

## 15.3   DYNAMICS OF CONNECTED PARTICLES

**15.40**   What is the technique of solution for the motion of two bodies which are connected by a cable or link and move in rectilinear translation?

❚   In many problems the motion of two particles connected by cables or links is desired.   For this case, the assumptions are made that the cable or link is inextensible and that the mass of the cable or link is negligible compared with the masses of the particles.

The technique for solving for the motion of two connected particles is to draw individual free-body diagrams and write the equation of motion for each mass particle.   The cable, or link, force will appear in these free-body diagrams as a single unknown force.   Since the cable or link is assumed to be inextensible and the particle moves in rectilinear translation, it follows that the acceleration of both the connected bodies must be the same.

There are thus two equations of motion of the particles, in terms of the unknown cable force and acceleration.   These two equations may then be solved simultaneously to obtain the final solutions.

This method may be easily extended to the case where more than two bodies are connected to each other by a cable or link and move in rectilinear translation.

**15.41**   The system of two blocks in Fig. 15.41*a* is released from rest at   $t = 0$.   The pulley is massless, and the pulley bearings are frictionless.

(*a*)   Find the acceleration of the two blocks and the cable tensile force.

(*b*)   At what time will the two blocks pass each other?

(*c*)   What are the velocities of the blocks when they pass each other?

(*d*)   If block *B* were suddenly brought to rest 1.6 s after the onset of motion, describe the ensuing motion of block *A*.

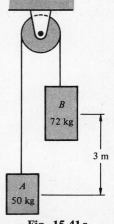

**Fig. 15.41*a***

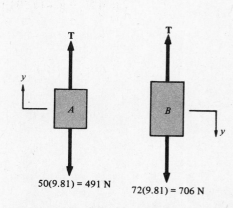

**Fig. 15.41*b***

▌ (a)  From physical considerations, it may be concluded that block $A$ moves upward and block $B$ moves downward. The free-body diagrams of the two blocks are shown in Fig. 15.41$b$, and $T$ is the magnitude of the unknown cable tensile force.

The equations of motion are

Block $A$: $\qquad\qquad\qquad\qquad \sum F_y = ma \qquad T - 491 = 50a$

Block $B$: $\qquad\qquad\qquad\qquad \sum F_y = ma \qquad 706 - T = 72a$

The above two equations are added, with the result

$$(50 + 72)a = 706 - 491 \qquad a = 1.76 \text{ m/s}^2$$

Using the first equation,

$$T = 491 + 50a = 579 \text{ N}$$

(b)  Both blocks start from rest with the *same* acceleration. The time $t$ for the two centerlines to be at the same horizontal level is found from

$$y = y_0 + v_0 t + \tfrac{1}{2}at^2 \qquad \tfrac{1}{2}(3) = 0 + 0(t) + \tfrac{1}{2}(1.76)t^2 \qquad t = 1.31 \text{ s}$$

(c)  The velocity of each block when they pass each other is obtained from

$$v = v_0 + at = 0 + 1.76(1.31) = 2.31 \text{ m/s}$$

(d)  The velocity of block $A$ at $t = 1.6$ s is

$$v = v_0 + at = 0 + 1.76(1.6) = 2.82 \text{ m/s}$$

At the instant that block $B$ comes to rest, the value of the cable force $T$ drops to zero. Thereafter, block $A$ behaves as a particle with an initial upward velocity in a gravitational field. The maximum *additional* height $y$ which it will attain is found from the equation

$$v^2 = v_0^2 + 2a(y - y_0)$$

Using $y_0 = 0$ and $a = -g = -9.81$ m/s$^2$ in the above equation results in

$$0 = 2.82^2 + 2(-9.81)y \qquad y = 0.400 \text{ m}$$

The time to reach this height is found from

$$v = v_0 + at \qquad 0 = 2.82 + (-9.81)t \qquad t = 0.287 \text{ s}$$

**15.42** (a)  When the system of two blocks shown in Fig. 15.42$a$ is released from rest, the force in the cable is 42 lb. Find the acceleration of the blocks, and the mass of block $B$.

(b)  Do the same as in part (a), if the force in the cable after release of the blocks is 78 lb.

(c)  Do the same as in part (a), if the force in the cable after release of the blocks is 60 lb.

▌ (a)  Figure 15.42$b$ shows the free-body diagrams of the two blocks. For block $a$,

$$\sum F_y = ma \qquad 60 - 42 = \frac{60}{32.2}\, a \qquad a = 9.66 \text{ ft/s}^2$$

For block $B$, using the above value of acceleration,

$$\sum F_y = ma$$

$$42 - W_B = \frac{W_B}{32.2}\,(9.66) \qquad W_B = 32.3 \text{ lb}$$

The mass of block $B$ is then found as

$$m_B = \frac{W_B}{g} = \frac{32.3}{32.2} = 1.00 \text{ slug}$$

(b)  The free-body diagrams of the blocks are shown in Fig. 15.42$c$. The equations of motion are

Block $A$: $\qquad\qquad\qquad\qquad \sum F_y = ma \qquad 78 - 60 = \frac{60}{32.2}\, a \qquad a = 9.66 \text{ ft/s}^2$

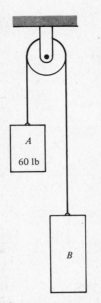

**Fig. 15.42a**

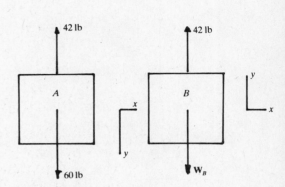

**Fig. 15.42b**

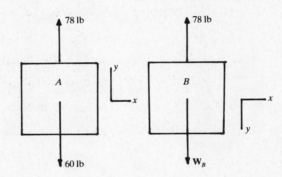

**Fig. 15.42c**

Block $B$:  $\sum F_y = ma$    $W_B - 78 = \dfrac{W_B}{32.2}(9.66)$    $W_B = 111$ lb

The mass of block $B$ is given by

$$m_B = \frac{W_B}{g} = \frac{111}{32.2} = 3.45 \text{ slugs}$$

It may be observed that the acceleration is the same in parts $(a)$ and $(b)$. This is because for both problems, the *net* force on block $A$ is 18 lb.

(c) Since the force in the cable is equal to the weight of block $A$, this element must be in static equilibrium. Thus, Block $B$ is also in static equilibrium, and $a = 0$. Using $W_B = 60$ lb,

$$m_B = \frac{W_B}{g} = \frac{60}{32.2} = 1.86 \text{ slugs}$$

**15.43**  At $t = 0$, block $A$ in Fig. 15.43a has an upward velocity of 1.85 m/s. Find the time at which block $B$ comes to rest.

❚ Figure 15.43b shows the free-body diagrams of the blocks. The equations of motion are as follows.

Block $A$:  $\sum F_y = ma$    $39.2 - T = 4a$          (1)

Block $B$:  $\sum F_y = ma$    $T - 27.5 = 2.8a$          (2)

Fig. 15.43a

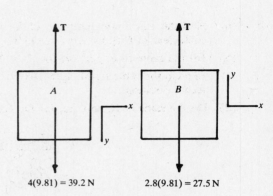

$4(9.81) = 39.2\,\text{N}$  $2.8(9.81) = 27.5\,\text{N}$

Fig. 15.43b

Equations (1) and (2) are added, with the result

$$11.7 = 6.8a \qquad a = 1.72\,\text{m/s}^2$$

The time at which block B comes to rest, using $v_0 = -1.85\,\text{m/s}$ is found from

$$v_B = v_0 + at \qquad 0 = -1.85 + 1.72t \qquad t = 1.08\,\text{s}$$

**15.44** (a) Do the same as in Prob. 15.43, if a mass of 1 kg is added to block B.

(b) Do the same as in Prob. 15.43, if a mass of 1 kg is added to block A.

▮ (a) Figures 15.43a and b are used, together with four significant figures to avoid loss of accuracy in the calculations. The weight of block B is given by

$$W_B = 3.8(9.81) = 37.28\,\text{N}$$

The weight of block A is

$$W_A = 4(9.81) = 39.24\,\text{N}$$

The equations of motion are as follows

Block A:  $\qquad \sum F_y = ma \qquad 39.24 - T = 4a$  (1)

Block B:  $\qquad \sum F_y = ma \qquad T - 37.28 = 3.8a$  (2)

Equations (1) and (2) are added to obtain

$$1.96 = 7.8a \qquad a = 0.251\,\text{m/s}^2$$

The time for block B to come to rest is found from

$$v_B = v_0 + at \qquad 0 = -1.85 + 0.251t \qquad t = 7.37\,\text{s}$$

(b) The weight of block A is found as

$$W_A = 5(9.81) = 49.1\,\text{N}$$

Using Fig. 15.43b, the equations of motion have the following forms:

Block A:  $\qquad \sum F_y = ma \qquad 49.1 - T = 5a$  (1)

Block B:  $\qquad \sum F_y = ma \qquad T - 27.5 = 2.8a$  (2)

Adding Eqs. (1) and (2),

$$21.6 = 7.8a \qquad a = 2.77 \text{ m/s}^2$$

The time for block $B$ to come to rest is found from

$$v_B = v_0 + at \qquad 0 = -1.85 + 2.77t$$
$$t = 0.668 \text{ s}$$

**15.45** (a) Find the magnitude and sense of the acceleration of the two blocks in Fig. 15.45a, if $W = 100$ lb. The coefficient of friction between the block and the plane is 0.3.

(b) Do the same as in part (a), if $W = 300$ lb.

(c) Express the cable tensile force and the acceleration of the blocks, for the condition of part (a), in formal vector notation.

(d) Do the same as in part (c), for the condition of part (b).

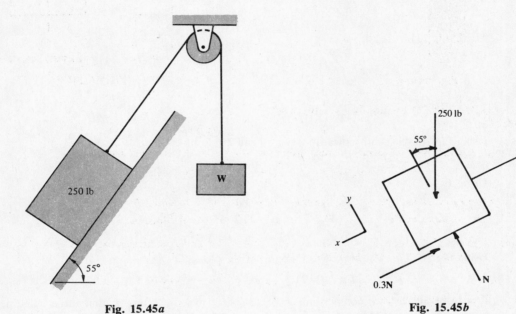

Fig. 15.45a                    Fig. 15.45b

(a) Figure 15.45b shows the free-body diagram of the 250-lb block, for assumed downward motion of this element. The equations of equilibrium and motion have the forms

$$\sum F_y = 0 \qquad -250 \cos 55° + N = 0 \qquad N = 143 \text{ lb}$$

$$\sum F_x = ma \qquad 250 \sin 55° - T - 0.3N = \frac{250}{32.2} a \qquad -T + 162 = 7.76a \qquad (1)$$

The free-body diagram of the 100-lb block is seen in Fig. 15.45c. Newton's second law has the form

$$\sum F_y = ma \qquad T - 100 = \frac{100}{32.2} a \qquad T - 100 = 3.11a \qquad (2)$$

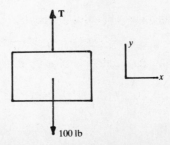

Fig. 15.45c

Adding Eqs. (1) and (2),

$$62 = 10.9a \qquad a = 5.69 \text{ ft/s}^2$$

(b) Upward motion of the 250-lb block is assumed. The free-body diagram in Fig. 15.45b, with opposite sense of the friction force shown, is used. The equation of motion for the 250-lb block is

$$\sum F_x = ma \qquad T - 0.3(143) - 250 \sin 25° = \frac{250}{32.2} a \qquad (3)$$

From Fig. 15.45c, using a weight force of 300 lb,

$$\sum F_y = ma \qquad 300 - T = \frac{300}{32.2} a \qquad (4)$$

Equations (3) and (4) are added, with the result

$$a = 8.87 \text{ ft/s}^2$$

(c) Using Fig. 15.45b,

$$\mathbf{T} = -T\mathbf{i}$$

Using Eq. (1),

$$-T + 162 = 7.76a \qquad T = 162 - 7.76(5.69) = 118 \text{ lb}$$
$$\mathbf{T} = -118\mathbf{i} \quad \text{lb} \qquad \mathbf{a} = a\mathbf{i} = 5.69\mathbf{i} \quad \text{ft/s}^2$$

Using Fig. 15.45c,

$$\mathbf{T} = T\mathbf{j} = 118\mathbf{j} \quad \text{lb} \qquad \mathbf{a} = a\mathbf{j} = 5.69\mathbf{j} \quad \text{ft/s}^2$$

(d) Using Fig. 15.45b, with upward motion of the 250-lb block,

$$\mathbf{T} = T\mathbf{i}$$

Using Eq. (4),

$$T = 300 - \frac{300}{32.2} a = 300 - \frac{300}{32.2}(8.87) = 217 \text{ lb} \qquad \mathbf{T} = 217\mathbf{i} \quad \text{lb}$$
$$\mathbf{a} = -a\mathbf{i} = -8.87\mathbf{i} \quad \text{ft/s}^2$$

Using Fig. 15.45c, with downward motion of the block,

$$\mathbf{T} = T\mathbf{j} = 217\mathbf{j} \quad \text{lb} \qquad \mathbf{a} = -a\mathbf{j} = -8.87\mathbf{j} \quad \text{ft/s}^2$$

It may be seen, from comparison of the above results for **T** and **a**, that these quantities must be interpreted in terms of the applicable set of xy coordinates, and unit vectors, used for their description.

**15.46** (a) Find the acceleration of the system of two blocks shown in Fig. 15.46a. The mass of block B is 160 kg.

(b) Find the required value of the mass of block B in Fig. 15.46a, if the acceleration of the system is to be 2.9 m/s².

▌ (a) Figure 15.46b shows the free-body diagram of block A. For equilibrium in the y direction,

$$\sum F_y = 0 \qquad -1{,}500 \cos 40° + N_A = 0 \qquad N_A = 1{,}150 \text{ N}$$

Newton's second law for the motion in the x direction is

$$\sum F_x = ma \qquad 1{,}500 \sin 40° - T \quad -0.1 N_A = \frac{1{,}500}{9.81} a \qquad (1)$$
$$-T + 849 = 153a \qquad (2)$$

The free-body diagram of block B is seen in Fig. 15.46c. The equations of equilibrium and motion are

$$\sum F_y = 0 \qquad -1{,}570 + N_B = 0 \qquad N_B = 1{,}570 \text{ N}$$
$$\sum F_x = ma \qquad T - 0.24 N_B = 160a \qquad (3)$$
$$T - 377 = 160a \qquad (4)$$

Adding Eqs. (2) and (4),

$$472 = 313a \qquad a = 1.5 \text{ m/s}^2$$

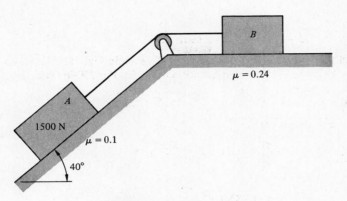

**Fig. 15.46a**

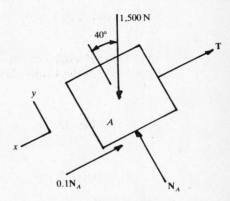

**Fig. 15.46b**

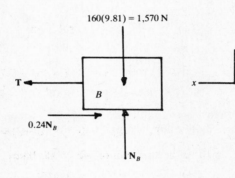

**Fig. 15.46c**

(*b*)  Using Eq. (1) from part (*a*), with  $a = 2.9 \text{ m/s}^2$,

$$\sum F_x = ma \qquad 1{,}500 \sin 40° - T - 0.1(1{,}150) = \frac{1{,}500}{9.81}(2.9) \qquad T = 406 \text{ N}$$

Using the above value of $T$ in Eq. (3) from part (*a*),

$$\sum F_x = ma \qquad 406 - 0.24 m_B (9.81) = m_B (2.9) \qquad m_B = 77.3 \text{ kg}$$

**15.47**  For what range of values of weight of block $A$ in Fig. 15.47$a$ will the *magnitude* of the acceleration of the blocks be less than, or equal to, $5 \text{ ft/s}^2$.

❚  Figure 15.47$b$ shows the free-body diagram of block $A$, for assumed downward motion of this element.  The equations of equilibrium and motion are

$$\sum F_y = 0 \qquad -W_A \cos 55° + N_A = 0 \qquad N_A = 0.574 W_A$$

$$\sum F_x = ma \qquad W_A \sin 55° - T - 0.2 N_A = \frac{W_A}{32.2}(5) \qquad T_A = 0.549 W_A \qquad (1)$$

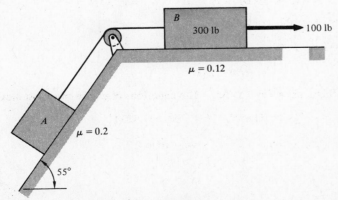

**Fig. 15.47a**

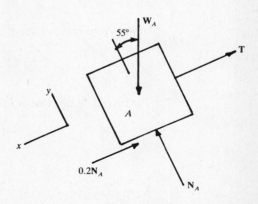

**Fig. 15.47b**

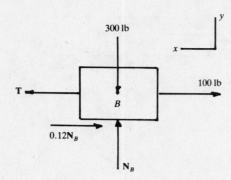

Fig. 15.47c

The free-body diagram of block $B$ is seen in Fig. 15.47c. The equations of equilibrium and motion are

$$\sum F_y = 0 \qquad N_B = 300 \text{ lb}$$

$$\sum F_x = ma \qquad T - 0.12N_B - 100 = \frac{300}{32.2} (5) \qquad T = 183 \text{ lb} \tag{2}$$

Using Eq. (2) in Eq. (1),

$$183 = 0.549W_A \qquad W_A = 333 \text{ lb}$$

For the case where block $A$ moves upward, the senses of both friction forces in Figs. 15.47b and c change. The equations of equilibrium and motion for this condition are as follows.

Block $A$: $\quad N_A = 0.574W_A \quad \sum F_x = ma \quad W_A \sin 55° - T + 0.2N_A = \dfrac{W_A}{32.2} (-5) \quad T = 1.09W_A \tag{3}$

Block $B$: $\quad N_B = 300 \text{ lb} \quad \sum F_x = ma \quad T + 0.12N_B - 100 = \dfrac{300}{32.2} (-5) \quad T = 17.4 \text{ lb} \tag{4}$

Using Eq. (4) in Eq. (3),

$$17.4 = 1.09W_A \qquad W_A = 16.0 \text{ lb}$$

The magnitude of the acceleration of the blocks will be less than, or equal to, $5 \text{ ft/s}^2$ if the weight $W_A$ of block $A$ is in the range

$$16.0 \text{ lb} \le W_A \le 333 \text{ lb}$$

**15.48** The system of blocks in Fig. 15.48a is released from rest at $t = 0$. The coefficient of friction on all sliding surfaces is 0.08. Find the magnitude, direction, and sense of the relative acceleration of block $A$ with respect to block $B$.

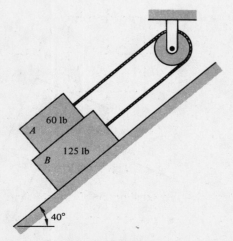

Fig. 15.48a

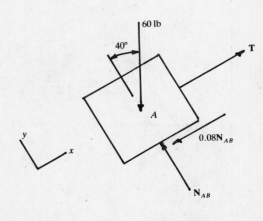

Fig. 15.48b

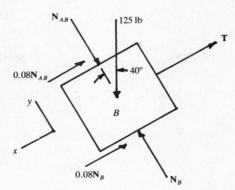

Fig. 15.48c

❚ Block A moves upward and block B moves downward.  The free-body diagrams of these two elements are shown in Figs. 15.48b and c.  The equations of equilibrium and motion for block A are

$$\sum F_y = 0 \qquad N_{AB} = 60 \cos 40° = 46.0 \text{ lb}$$

$$\sum F_x = ma \qquad T - 60 \sin 40° - 0.08 N_{AB} = \frac{60}{32.2} a \qquad T - 42.2 = 1.86a \tag{1}$$

The equations of equilibrium and motion for block B are

$$\sum F_y = 0 \qquad -N_{AB} - 125 \cos 40° + N_B = 0 \qquad N_B = N_{AB} + 125 \cos 40°$$

$$\sum F_x = ma \qquad 125 \sin 40° - 0.08 N_{AB} - 0.08 N_B - T = \frac{125}{32.2} a \qquad -T + 65.3 = 3.88a \tag{2}$$

Adding Eqs. (1) and (2),

$$5.74a = 23.1 \qquad a = 4.02 \text{ ft/s}^2$$

The relative acceleration of A with respect to B, defined in Prob. 14.88, has the form

$$a_{AB} = a_A - a_B = 4.02 - (-4.02) \qquad a_{AB} = 8.04 \text{ ft/s}^2$$

The direction and sense of $a_{AB}$ is up and to the right in Fig. 15.48a.

**15.49**  (a)  Block A in Fig. 15.49a is connected to the ground by the weightless link ab.  The masses of blocks A and B are 100 and 150 kg, respectively.  The coefficients of friction between the mating surfaces are $\mu_{AB} = 0.1$ and $\mu_{BC} = 0.14$.  Find the required magnitude of the horizontal force acting on block B that will cause this element to accelerate to the left at $2 \text{ m/s}^2$, and the corresponding value of the link force.

(b)  Do the same as in part (a), for rightward acceleration of the block of $2 \text{ m/s}^2$.

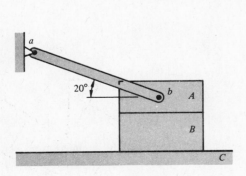

Fig. 15.49a

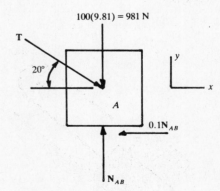

Fig. 15.49b

❚ (a)  Figure 15.49b shows the free-body diagram of block A.  For equilibrium of this element,

$$\sum F_x = 0 \qquad T \cos 20° - 0.1 N_{AB} = 0 \qquad T \cos 20° = 0.1 N_{AB} \tag{1}$$

$$\sum F_y = 0 \qquad -981 - T \sin 20° + N_{AB} = 0 \qquad T \sin 20° = N_{AB} - 981 \tag{2}$$

Equation (2) is divided by Eq. (3), to obtain

$$\tan 20° = \frac{N_{AB} - 981}{0.1 N_{AB}} \qquad N_{AB} = 1{,}020 \text{ N}$$

The link force is obtained from Eq. (1) as

$$T \cos 20° = 0.1(1{,}020)$$
$$T = 109 N$$

The free-body diagram of block $B$ is seen in Fig. 15.49$c$.   The equations of equilibrium and motion for this block have the forms

$$\sum F_y = 0 \qquad N_B - 1{,}470 - N_{AB} = 0 \qquad N_B = 2{,}490 \text{ N}$$

$$\sum F_x = ma \qquad P - 0.1 N_{AB} - 0.14 N_B = 150(2) \qquad P = 751 \text{ N}$$

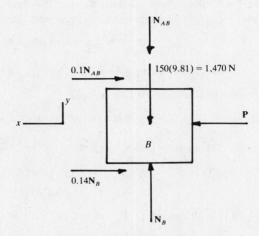

Fig. 15.49$c$

(**b**)   For rightward acceleration of block $B$, the free-body diagrams in Figs. 15.49$b$ and $c$ are used, with the senses of the link force, the applied force, and all friction forces changed.   For equilibrium of block $A$,

$$\sum F_x = 0 \qquad - T \cos 20° + 0.1 N_{AB} = 0 \qquad T \cos 20° = 0.1 N_{AB} \tag{3}$$

$$\sum F_y = 0 \qquad -981 + T \sin 20° + N_{AB} = 0 \qquad T \sin 20° = -N_{AB} + 981 \tag{4}$$

Equation (4) is divided by Eq. (3), to obtain

$$\tan 20° = \frac{-N_{AB} + 981}{0.1 N_{AB}} \qquad N_{AB} = 947 \text{ N}$$

Using Eq. (3),

$$T \cos 20° = 0.1(947) \qquad T = 101 \text{ N}$$

For block $B$,

$$\sum F_y = 0 \qquad N_B - 1{,}470 - N_{AB} = 0 \qquad N_B = 2{,}420 \text{ N}$$

$$\sum F_x = ma \qquad -P + 0.1 N_{AB} + 0.14 N_B = 150(-2) \qquad P = 734 \text{ N}$$

It may be seen that it requires a lower value of the applied force $P$ to accelerate block $B$ to the right.

**15.50**   Block $A$ in Fig. 15.50$a$ weighs 10 lb and block $B$ weighs 40 lb.   The link may be assumed to be weightless.   The coefficients of friction between blocks $A$ and $B$ and the plane are 0.12 and 0.18, respectively.

(**a**)   Find the acceleration of the system of blocks.

(**b**)   Find the force in the connecting link.

▮ (**a**)   The free-body diagrams of blocks $A$ and $B$ are shown in Figs. 15.50$b$ and $c$.   $P$ is the force in the link, and this force is assumed to be a tensile force.   For equilibrium and motion of block $A$,

$$\sum F_y = 0 \qquad -10 \cos 20° + N_A = 0 \qquad N_A = 9.40 \text{ lb}$$

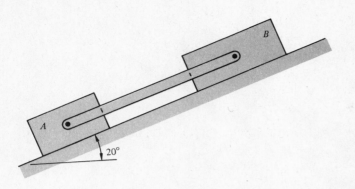

Fig. 15.50a

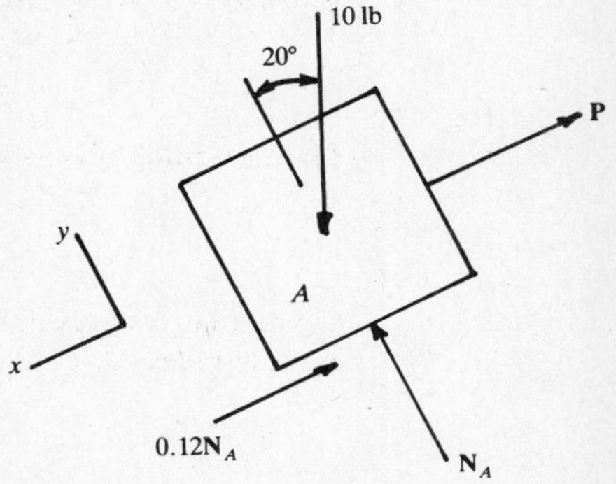

Fig. 15.50b

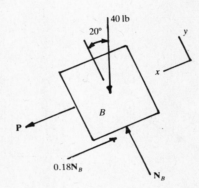

Fig. 15.50c

$$\sum F_x = ma \qquad 10\sin 20° - P - 0.12N_A = \frac{10}{32.2}\, a \qquad (1)$$

For equilibrium and motion of block $B$,

$$\sum F_y = 0 \qquad -40\cos 20° + N_B = 0 \qquad N_B = 37.6\,\text{lb}$$

$$\sum F_x = ma \qquad P + 40\sin 20° - 0.18N_B = \frac{40}{32.2}\, a \qquad (2)$$

Equations (1) and (2) are added, with the result

$$(10+40)\sin 20° - 0.12N_A - 0.18N_B = \frac{10+40}{32.2}\, a$$

Using the values of $N_A$ and $N_B$ above,

$$a = 5.93\,\text{ft/s}^2$$

(b) Using Eq. (1),

$$P = 10\sin 20° - 0.12N_A - \frac{10}{32.2}\, a = 0.45\,\text{lb}$$

Since the value for $P$ is positive, the assumption that $P$ is a tensile force is correct.

**15.51** The two blocks in Fig. 16.51a are connected by a massless link.

(a) Find the acceleration of the system and the force in the link.

(b) Express the forces acting on the ends of the link in formal vector notation.

❙ (a) The link may be recognized as being a two-force member since the forces acting on this element are applied at its ends only. Thus, the directions of these forces are collinear with the axis of the link. The free-body diagrams of the two blocks, with the components of the link and weight forces, are shown in Figs. 16.51c and d. $P$ is assumed to be a compressive force in the link.
The equilibrium and motion requirements of block $A$ are

$$\sum F_y = 0 \qquad -P\sin 10° - 10\cos 32° + N_A = 0 \qquad (1)$$

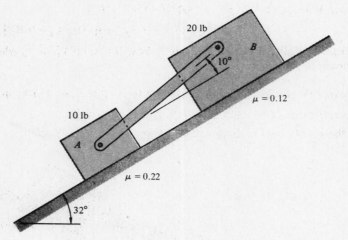

Fig. 15.51a

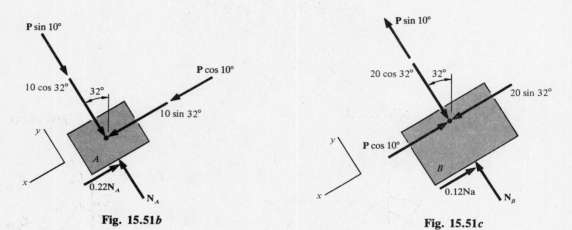

Fig. 15.51b          Fig. 15.51c

$$\sum F_x = ma \qquad P \cos 10° + 10 \sin 32° - 0.22N_A = \frac{10}{32.2} a \qquad (2)$$

The equilibrium and motion requirements of block $B$ are

$$\sum F_y = 0 \qquad P \sin 10° - 20 \cos 32° + N_B = 0 \qquad (3)$$

$$\sum F_x = ma \qquad -P \cos 10° + 20 \sin 32° - 0.12N_B = \frac{20}{32.2} a \qquad (4)$$

$N_A$ is eliminated between Eqs. (1) and (2), using four significant figures, to obtain

$$0.9466P + 3.433 = 0.3106a \qquad (5)$$

Elimination of $N_B$ between Eqs. (3) and (4) results in

$$-0.9640P + 8.563 = 0.6211a \qquad (6)$$

Equation (5) is multiplied by the fraction 0.9640/0.9466, with the result

$$0.9640P + 3.496 = 0.3163a \qquad (7)$$

Equations (6) and (7) are added, to obtain

$$12.06 = 0.9374a \qquad a = 12.87 \approx 12.9 \text{ ft/s}^2$$

Using Eq. (7),

$$-0.9640P + 8.563 = 0.6211(12.87) \qquad P = 0.591 \text{ lb}$$

(b)  The force $\mathbf{P}$ acting on the end of the link attached to block $A$ is

$$\mathbf{P} = (-P \cos 10°)\mathbf{i} + (P \sin 10°)\mathbf{j} = (-0.591 \cos 10°)\mathbf{i} + (0.591 \sin 10°)\mathbf{j} = -0.582\mathbf{i} + 0.103\mathbf{j} \quad \text{lb} \quad (8)$$

The force **P** acting on the end connected to block $B$ is

$$\mathbf{P} = (P \cos 10°)\mathbf{i} - (P \sin 10°)\mathbf{j} = 0.582\mathbf{i} - 0.103\mathbf{j} \qquad \text{lb} \tag{9}$$

The resultant of the two forces **P** given by Eqs. (8) and (9) is zero, which is the expected condition of equilibrium of the link.

**15.52**   Figure 15.52a shows the configuration when the positions of the two blocks in Prob. 15.51 are interchanged.

(*a*)   Find the acceleration of the system and the force in the link.

(*b*)   Compare the results found in part (*a*) with those found in Prob. 15.51.

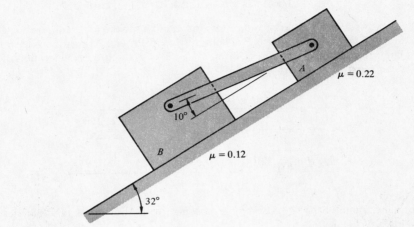

**Fig. 15.52a**

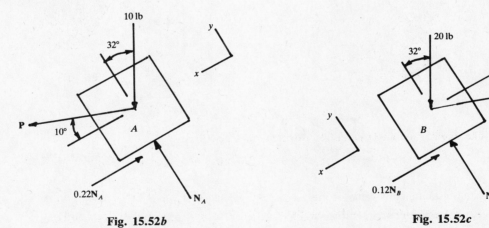

**Fig. 15.52b**                                   **Fig. 15.52c**

▌ (*a*)   Figures 15.52b and c show the free-body diagrams of blocks $A$ and $B$, and $P$ is the tensile force in the link.   The equilibrium and motion equations of block $A$ are

$$\sum F_y = 0 \qquad P \sin 10° - 10 \cos 32° + N_A = 0 \tag{1}$$

$$\sum F_x = ma \qquad P \cos 10° + 10 \sin 32° - 0.22N_A = \frac{10}{32.2} a \tag{2}$$

The equilibrium and motion requirements for block $B$ are

$$\sum F_y = 0 \qquad -P \sin 10° - 20 \cos 32° + N_B = 0 \tag{3}$$

$$\sum F_x = ma \qquad -P \cos 10° + 20 \sin 32° - 0.12N_B = \frac{20}{32.2} a \tag{4}$$

$N_A$ is eliminated between Eqs. (1) and (2), using four significant figures, to obtain

$$1.023P + 3.433 = 0.3106a \tag{5}$$

$N_B$ is eliminated between Eqs. (3) and (4), with the result

$$-1.006P + 8.563 = 0.6211a \qquad (6)$$

Equation (5) is multiplied by the factor 1.006/1.023, with the result

$$1.006P + 3.376 = 0.3054a \qquad (7)$$

Equations (6) and (7) are added, to obtain

$$11.94 = 0.9265a \qquad a = 12.89 \approx 12.9 \text{ ft/s}^2$$

Using Eq. (7),

$$1.006P + 3.376 = 0.3054(12.89) \qquad P = 0.557 \text{ lb}$$

(b) Table 15.1 shows a comparison of the results for $P$ and $a$ for the different positions of the link in Figs. 15.51a and 15.52a. The link force $P$ is compressive in Prob. 15.51 and tensile in Prob. 15.52. This force is larger for the configuration of Prob. 15.51, even though the acceleration of the system of two blocks is the same for both problems.

**TABLE 15.1**

|            | $P$, lb   | $a$, ft/s$^2$ |
|------------|-----------|---------------|
| Prob. 15.51 | 0.614(C) | 12.9          |
| Prob. 15.52 | 0.557(T) | 12.9          |

**15.53** The links in the plane frame in Fig. 15.53a may be assumed to be weightless, and all friction effects may be neglected. The mass of block $A$ is 4 kg.

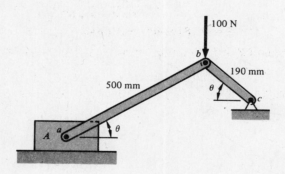

**Fig. 15.53a**

(a) Find the instantaneous value of the acceleration of block $A$ when $\theta = 40°$.

(b) Do the same as in part (a), if $\theta = 30°$.

(c) Do the same as in part (a), if $\theta = 20°$.

(d) Express the force exerted by link $ab$ on block $A$ in formal vector notation, for $\theta = 40°$.

▌ (a) Members $ab$ and $bc$ are two-force members. Figure 15.53b shows the free-body diagram of joint $b$. For equilibrium of this joint,

$$\sum F_x = 0 \qquad F \cos \theta - F_1 \cos \theta = 0 \qquad F_1 = F$$

$$\sum F_y = 0 \qquad 2F \sin \theta - 100 = 0 \qquad F = \frac{50}{\sin \theta} \qquad (1)$$

The free-body diagram of block $A$ is seen in Fig. 15.53c. Newton's second law has the form

$$\sum F_x = ma \qquad F \cos \theta = ma = 4a \qquad (2)$$

Using Eq. (1) in Eq. (2),

$$a = \frac{50}{4 \tan \theta}$$

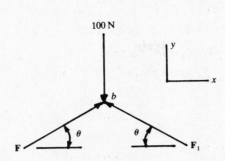

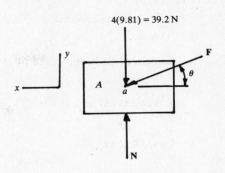

**Fig. 15.53***b*                                    **Fig. 15.53***c*

Using   $\theta = 40°$,

$$a = 14.9 \text{ m/s}^2$$

(*b*)   Using   $\theta = 30°$,

$$a = 21.7 \text{ m/s}^2$$

(*c*)   Using   $\theta = 20°$,

$$a = 34.3 \text{ m/s}^2$$

(*d*)   Using Fig. 15.53*c*,

$$\mathbf{F} = (F \cos \theta)\mathbf{i} - (F \sin \theta)\mathbf{j}$$

From Eq. (1),

$$F = \frac{50}{\sin \theta} = \frac{50}{\sin 40°} = 77.8 \text{ N}$$
$$\mathbf{F} = (77.8 \cos 40°)\mathbf{i} - (77.8 \sin 40°)\mathbf{j} = 59.6\mathbf{i} - 50.0\mathbf{j} \qquad \text{N}$$

The *y* component of **F** of 50 N, acting downward in Fig. 15.53*c*, represents half of the applied load of 100 N.

**15.54**   Do the same as in Prob. 15.53, if   $\mu = 0.1$   at the sliding surface between the block and the guide.

**▌** (*a*)   From Prob. 15.53,

$$F = \frac{50}{\sin \theta} \qquad\qquad (1)$$

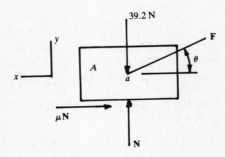

**Fig. 15.54**

Figure 15.54 shows the free-body diagram of block *A*.   Since motion is assumed to occur, the friction force has its maximum value $\mu N$.   For equilibrium of block *A* in the *y* direction,

$$\sum F_y = 0 \qquad -39.2 - F \sin \theta + N = 0 \qquad N = 39.2 + F \sin \theta \qquad (2)$$

Using Eq. (1) in Eq. (2),

$$N = 39.2 + \frac{50}{\sin \theta} \sin \theta = 89.2 \text{ N}$$

Newton's second law for block $A$ is written as

$$\sum F_x = ma \qquad F \cos \theta - \mu N = ma \qquad F \cos \theta - 0.1(89.2) = 4a \qquad (3)$$

Using Eq. (1) in Eq. (3),

$$\frac{50}{\sin \theta} \cos \theta - 8.92 = 4a \qquad a = \frac{1}{4} \left( \frac{50}{\tan \theta} - 8.92 \right)$$

Using $\theta = 40°$,

$$a = 12.7 \, \text{m/s}^2$$

(b)  Using $\theta = 30°$,

$$a = 19.4 \, \text{m/s}^2$$

(c)  Using $\theta = 20°$,

$$a = 32.1 \, \text{m/s}^2$$

It may be seen that the values of acceleration are smaller when friction is present, which is an expected result.

## 15.4   DYNAMICS OF PARTICLES IN PLANE CURVILINEAR TRANSLATION, NORMAL AND TANGENTIAL COMPONENT MOTIONS

**15.55**  (a)  Give the forms of Newton's second law for the *normal* and *tangential* component motion for a particle which moves in plane curvilinear translation.

(b)  Discuss the normal component of force which acts on the particle moving in plane curvilinear translation.

(c)  Under what conditions does a particle move with constant normal acceleration?

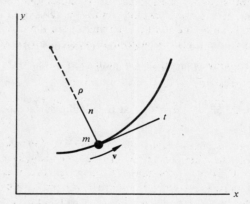

**Fig. 15.55a**

(a)  Figure 15.55a shows a mass particle which moves along a plane curve.   A set of *nt* axes are positioned on the curve, so that the *t* axis is tangent to the curve and the *n* axis is normal to this direction.   It was shown in Probs. 14.56 and 14.57 that a particle traveling on a plane curve experiences two distinct acceleration effects.   These two effects are related to the normal and tangential components of the *total* acceleration of the particle.

The first effect is called the normal, or centripetal, acceleration $a_n$.   This quantity is given by

$$a_n = \frac{v^2}{\rho}$$

where $v$ is the magnitude of the velocity, or the speed, along the curve and $\rho$ is the local radius of curvature.   The direction of $a_n$ is along the radius of curvature.   Its sense is *always* from the particle toward the center of curvature.

The second acceleration effect is referred to as the tangential acceleration $a_t$.   It is defined as

$$a_t = \dot{v} = \ddot{s}$$

where $s$ is the length coordinate along the curve.   The direction of $a_t$ is tangent to the curve, and $\dot{v}$ is the time rate of change of the magnitude of velocity, or of the speed, along the curve.   Figure 15.55b shows the positive senses of $a_n$ and $a_t$.

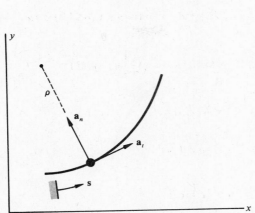

**Fig. 15.55b**

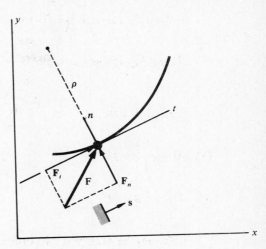

**Fig. 15.55c**

Newton's second law may be written for a particle which moves along a plane curvilinear path. Figure 15.55c shows the general appearance of a particle acted on by a resultant force **F**. This force may be resolved into normal and tangential components, as shown in the figure. Newton's second law in the component directions of the particle is then

$$F_n = ma_n = \frac{mv^2}{\rho} \qquad F_t = ma_t = m\dot{v} = m\ddot{s}$$

A set of general steps for the solution of dynamics problems in plane curvilinear motion, similar to the ones in Prob. 15.8 for the case of rectilinear translation, are presented below. In all cases, the *normal* direction is along the radius of curvature, with a positive sense which is from the path toward the center of curvature, and the *tangential* direction is along a straight line tangent to the path.

1. Draw a free-body diagram of the particle.
2. Find the normal and tangential components of the resultant force which acts on the particle.
3. Write the two component equations of motion given by

$$F_n = ma_n \qquad F_t = ma_t$$

(b) The two equations in part (a) reveal a very significant characteristic of plane curvilinear motion. From consideration of the first equation, it may be concluded that $F_n$ will be zero *only* if $v = 0$ or $\rho \to \infty$. The former condition is the trivial case of no motion of the particle, while the latter implies that the curvilinear path is a straight-line path. The above observations now lead to a very important conclusion: *If a mass particle moves with plane curvilinear motion, the normal component of the resultant force which acts on the particle is never zero.* This component of force is required to *continually* change the *direction* of the velocity vector of the particle, in order to have this quantity be tangent to the path. The results of the above case may be compared with the requirements for rectilinear translation of a particle. In this latter problem, the particle is always in static equilibrium in the direction normal to the motion.

(c) In certain problems, the normal acceleration $a_n$ may have a constant value. This motion, however, is *never* referred to as "motion with constant normal acceleration," since *the direction of the acceleration of the particle is continually changing.*

**15.56** A particle moves in plane curvilinear translation. Discuss the motion which would ensue if the normal force acting on the particle became equal to zero.

❚ The particle in Fig. 15.56 is assumed to rest on a frictionless horizontal plane and move along the curved path *abc*. The normal force $F_n$ which acts on the particle is the force required to make the particle follow the curved path. This normal force is now imagined to be suddenly removed when the particle reaches point *b*. At the instant that the force is removed, Newton's second law in the normal direction has the form

$$F_n = 0 = \frac{v^2}{\rho}$$

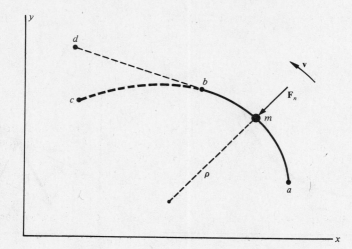

Fig. 15.56

When the particle reaches point $b$, it has a nonzero speed $v$. The only way that the above equation can be satisfied is for $\rho \to \infty$. Thus, the subsequent motion of the particle beyond point $b$ is along the straight line $bd$ shown in the figure. It also follows that the particle will move with *constant* speed $v$ along this straight line, since the resultant force that acts on the particle during this time is zero.

15.57 A particle of mass 0.8 kg moves along the curved track lying in a horizontal plane, which was shown in Fig. 14.59. The speed of the particle has a constant value of 0.15 m/s. Find the maximum value of normal force which the particle experiences as it moves from $a$ to $g$.

❚ The maximum value of the normal acceleration was found in Prob. 14.59 to occur along path $cd$, with the magnitude

$$a_n = 0.15 \text{ m/s}^2$$

Newton's second law for motion in the direction normal to the curve is then

$$F_n = ma_n = 0.8(0.15) = 0.12 \text{ N}$$

The above force is necessary to make the particle follow the given curvature of the path.

15.58 A fighter pilot is in a power dive, as shown in Fig. 15.58. The speed of the plane is 960 km/h, and the minimum radius of curvature of the flight path occurs at point $a$, as the pilot is pulling out of the dive.

(a) How many $g$'s of acceleration does the pilot experience as the plane passes point $a$?

(b) Find the force exerted by the pilot's seat on the pilot, at point $a$, if the mass of the pilot is 78 kg.

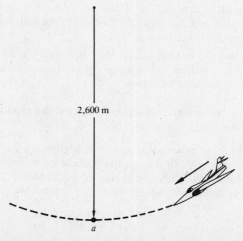

2,600 m

Fig. 15.58

I (a) The acceleration $a_n$ of the plane at point $a$ is

$$a_n = \frac{v^2}{\rho}$$

Using $v = 960(1,000/3,600) = 267\,\text{m/s}$ and $\rho = 2,600\,\text{m}$, we get

$$a_n = \frac{267^2}{2,600} = 27.4\,\text{m/s}^2$$

The quantity referred to as a $g$ is a measure of the magnitude of an acceleration. The magnitude of one $g$ is equal to the magnitude of the acceleration of the gravitational field. The number of $g$'s experienced by the pilot is thus $27.4/9.81 = 2.79$ $g$'s.

(b) Newton's second law in the normal direction at point $a$ has the form

$$F_n = ma_n$$

where $F_n$ is the force exerted by the seat on the pilot. For the numerical values of this problem,

$$F_n = 78(27.4) = 2,140\,\text{N}$$

An alternative way of finding the above result is as follows. The weight of the pilot is $78(9.81) = 765\,\text{N}$. From the definition of the $g$ of acceleration,

$$F_n = 2.79(765) = 2,130\,\text{N} \approx 2,140\,\text{N}$$

15.59 (a) A 300-g particle moves along the frictionless circular track in a vertical plane shown in Fig. 14.62a, with constant tangential acceleration $a_t = 0.28\,\text{m/s}^2$. The particle starts from rest at point $a$. Find the normal and tangential components of the force exerted on the article when it first reaches point $b$.

(b) Do the same as in part (a), for the case where the particle is at point $b$ the second time.

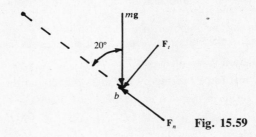

Fig. 15.59

I (a) Figure 15.59 shows the free-body diagram of the particle at point $b$. Newton's second law for the component motions has the forms

$$\sum F_n = ma_n \qquad F_n - mg \cos 20° = ma_n \qquad F_n = m(a_n + g \cos 20°) \tag{1}$$

$$F_t = ma_t$$

From Prob. 14.62, part (b),

$$a_n = 0.684\,\text{m/s}^2 \qquad \text{and} \qquad a_t = 0.28\,\text{m/s}^2$$

The values of $F_n$ and $F_t$ are then

$$F_n = 0.3(0.684 + 9.81 \cos 20°) = 2.97\,\text{N} \qquad F_t = 0.3(0.28) = 0.084\,\text{N}$$

(b) When the particle is at point $b$ the second time, from the solution to Prob. 14.62, part (c), $a_n = 4.20\,\text{m/s}^2$. Using Eq. (1) in part (a), the normal force which acts on the particle is found as

$$F_n = m(a_n + g \cos 20°) = 0.3(4.20 + 9.81 \cos 20°) = 4.03\,\text{N}$$

The tangential component of force which acts on the particle has the same value as that found in part (a), since $a_t$ is a constant in this problem.

15.60 A frictionless particle of weight 2 lb slides down a track in a vertical plane, as shown in Fig. 15.60a. At the lowest point of the track, the particle has an apparent weight which is 50 percent greater than its actual weight. Find the velocity of the particle at this point.

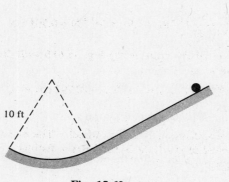

Fig. 15.60a

10 ft

Fig. 15.60b

▮ Figure 15.60b shows the free-body diagram of the particle at the lowest point of the track. Since the apparent weight of the particle is 50 percent greater than its actual weight, the normal force $N$ which acts on the particle is

$$N = 1.5mg$$

Newton's second law in the normal direction is then

$$\sum F_n = ma_n \qquad N - mg = 1.5mg - mg = m\frac{v^2}{10} \qquad v^2 = 0.5(10)g \qquad v = 12.7\text{ ft/s}$$

It may be observed that the above solution is independent of the mass of the particle.

**15.61** The particle of 65-mg mass shown in Fig. 15.61a travels in a horizontal circular path, at a speed of 2.2 m/s, around the surface of the stationary drum. All surfaces are assumed to be frictionless.

(a) Find the required value of $r$.

(b) Find the normal force exerted by the mass particle on the surface of the drum.

(c) Express the result in part (b) in formal vector notation.

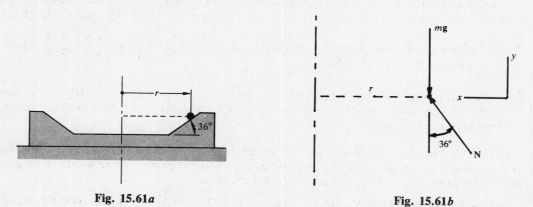

Fig. 15.61a          Fig. 15.61b

▮ (a) Figure 15.61b shows the free-body diagram of the particle. For equilibrium in the $y$ direction,

$$\sum F_y = 0 \qquad -mg + N\cos 36° = 0 \qquad N = 1.24mg \tag{1}$$

Newton's second law in the $x$ direction has the form

$$\sum F_x = ma_n \qquad N\sin 36° = m\frac{v^2}{r}$$

Using Eq. (1) in Eq. (2),

$$1.24mg \sin 36° = m\frac{v^2}{r}$$

$$r = \frac{v^2}{1.24g \sin 36°} = \frac{2.2^2}{1.24(9.81) \sin 36°} = 0.677 \text{ m} = 677 \text{ mm}$$

(*b*)  Using Eq. (1),

$$N = 1.24mg = 1.24\left[\frac{65}{1,000}\left(\frac{1}{1,000}\right)\right]9.81 = 7.91 \times 10^{-4} \text{ N} = 0.791 \text{ mN}$$

(*c*)  The normal force exerted by the mass particle on the drum, using Fig. 15.61*b* with the opposite sense of *N*, may be written in formal vector notation as

$$\mathbf{N} = (-N \sin 36°)\mathbf{i} - (N \cos 36°)\mathbf{j} = (-0.791 \sin 36°)\mathbf{i} - (0.791 \cos 36°)\mathbf{j}$$
$$= -0.465\mathbf{i} - 0.640\mathbf{j} \quad \text{mN}$$

**15.62**  Figure 15.62*a* shows an elementary model of one turbine blade on a turbine wheel.  The mean diameter of the circular path traveled by the blade is 28 in, and the speed of the blade is 1,260 ft/s.  Estimate the force exerted by the blade on the turbine wheel, if the blade weighs 0.38 lb.

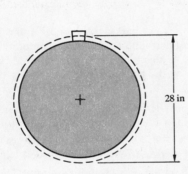

Fig. 15.62*a*

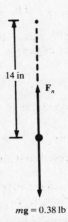

Fig. 15.62*b*

▌ The turbine blade is assumed to be a particle which travels on a 28-in-diameter circular path.  Figure 15.62*b* shows the free-body diagram of the turbine blade when this element is in its most downward position.  The force $F_n$ is the force exerted by the turbine wheel on the blade.  Newton's second law is written as

$$\sum F_n = ma_n \quad F_n - mg = ma_n \quad F_n = mg + ma_n = m(g + a_n) = m\left(g + \frac{v^2}{\rho}\right) = \frac{0.38}{32.2}\left(32.2 + \frac{1,260^2}{14/12}\right)$$
$$F = 16,100 \text{ lb} = 8.05 \text{ tons}$$

From Newton's third law, it may be concluded that the force exerted by the blade on the turbine wheel is 16,100 lb.  It may also be concluded that the static weight force of the blade is insignificant compared to the dynamic force due to rotation.

**15.63**  Figure 15.63*a* shows a drum that rotates in a horizontal plane with constant speed.  A small mass particle is held in position by the friction forces between the particle and the drum.  The weight of the particle is *W* and the coefficient of friction between the particle and the drum is $\mu$.

(*a*)  Find the general expression for the speed of the particle at which slipping motion is impending.

(*b*)  Find the numerical value of the speed in part (*a*) if  $r = 8$ in  and  $\mu = 0.2$.

▌ (*a*)  Figure 15.63*b* shows the free-body diagram of the particle, for impending sliding motion of this element in the *y* direction.  For equilibrium of the particle,

$$\sum F_y = 0 \quad -mg + \mu N = 0 \quad N = \frac{mg}{\mu}$$

Newton's second law has the form

$$\sum F_x = ma_n \quad N = m\frac{v^2}{r}$$

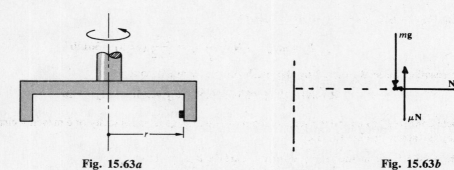

**Fig. 15.63a**　　　　　　　　　　　**Fig. 15.63b**

$N$ is eliminated between the above two equations, to obtain

$$v = \sqrt{\frac{rg}{\mu}}$$

(b)  Using  $r = 8$ in  and  $\mu = 0.2$,

$$v = \sqrt{\frac{\frac{8}{12}(32.2)}{0.2}} = 10.4 \text{ ft/s}$$

**15.64**  Figure 15.64a is a representation of a device seen in amusement parks.   A vertical drum in the form of a right circular cylinder is made of heavy steel mesh.   People stand around the inside circumference of the drum, and a motor brings the drum up to a constant angular speed.   At this time the floor is allowed to lower, as shown by the dashed outline, and the people are held in position by the friction forces exerted by the steel mesh of the drum on their bodies.

(a)  The coefficient of friction between the steel mesh and the person's clothing is estimated to be 0.5, and the speed of the person is 25.4 ft/s.   What is the limiting value of the drum diameter $d$ if the person is not to slip on the mesh?

(b)  What is the normal acceleration, for the condition of part (a), which the person experiences?

(c)  Express the reaction force exerted by the steel mesh on the person's body, in formal vector notation, if the weight of the person is 180 lb.

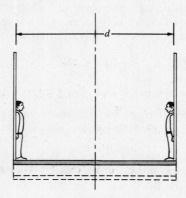

**Fig. 15.64a**

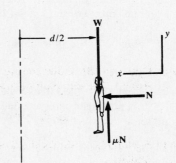

**Fig. 15.64b**

(a)  The free-body diagram of the person is shown in Fig. 15.64b.   Using the result from part (a) in Prob. 15.63,

$$v = \sqrt{\frac{rg}{\mu}} = \sqrt{\frac{dg}{2\mu}} \qquad d = \frac{2\mu v^2}{g} = \frac{2(0.5)25.4^2}{32.2} = 20.0 \text{ ft}$$

If $d$ is less than or equal to 20.0 ft, the person will not slip on the mesh of the drum.

(b)  The acceleration experienced by the person is

$$a_n = \frac{v^2}{\rho} = \frac{25.4^2}{10} = 64.5 \frac{\text{ft}}{\text{s}^2}$$

This acceleration may be expressed as  $64.5/32.2 = 2.00g$'s.

(c)  Using Fig. 15.64b, and the result in part (b),

$$\sum F_x = ma_n \qquad N = \frac{W}{g}\, a_n = \frac{180}{32.2}\,(64.5) = 361\ \text{lb}$$

The reaction force **R** exerted by the mesh on the person's body is

$$\mathbf{R} = N\mathbf{i} + \mu N\mathbf{j} = 361\mathbf{i} + 0.5(361)\mathbf{j} = 361\mathbf{i} + 181\mathbf{j} \qquad \text{lb}$$

**15.65**  A mass particle connected to an inextensible string moves with a constant velocity of 6 m/s in a circular path in a vertical plane, as shown in Fig. 15.65a.

(a)  Find the tensile force in the string when the particle is at point a.

(b)  Do the same as in part (a) when the particle is at point b.

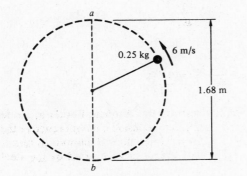

**Fig. 15.65a**

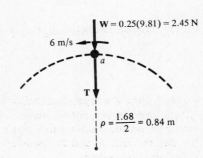

**Fig. 15.65b**

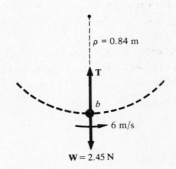

**Fig. 15.65c**

▌(a)  Figure 15.65b shows the free-body diagram of the particle at point a, and T is the tensile force in the string.  The equation of motion of the particle is

$$\sum F_n = ma_n \qquad W + T = \frac{mv^2}{\rho} \qquad 2.45 + T = \frac{0.25(6^2)}{0.84} \qquad T = 8.26\ \text{N}$$

(b)  Figure 15.65c shows the free-body diagram when the particle is at point b.  Newton's second law is written as

$$\sum F_n = ma_n \qquad T - W = \frac{mv^2}{\rho} \qquad T - 2.45 = \frac{0.25(6^2)}{0.84} \qquad T = 13.2\ \text{N}$$

The force T in both cases is the force exerted *by* the string *on* the mass.  By Newton's third law, the force exerted *on* the string *by* the mass is equal and opposite to T.  This latter force is also referred to as a *centrifugal force*.

**15.66**  A particle of 0.5-kg mass connected to an inextensible cable moves with constant speed in a circular path in a vertical plane, as shown in Fig. 15.66a.  When the particle is at point a, the tensile force in the string is 10 N.

(a)  Find the speed v of the particle.

(b)  Find the cable tensile force when the particle is at position b.

(c)  Find the minimum value of the speed of the particle in Fig. 15.66a if the cable is to always experience a tensile force.

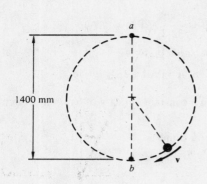

Fig. 15.66a

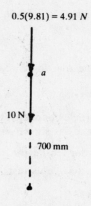

0.5(9.81) = 4.91 $N$

Fig. 15.66b

(a) Figure 15.66b shows the free-body diagram of the particle at point a. Newton's second law is written as

$$\sum F_n = ma_n \qquad 4.91 + 10 = 0.5\left(\frac{v^2}{0.7}\right) \qquad v = 4.57 \text{ m/s}$$

(b) The free-body diagram of the particle at point b is seen in Fig. 15.66c. The equation of motion has the form

$$\sum F_n = ma_n \qquad T - 4.91 = 0.5\left(\frac{4.57^2}{0.7}\right) \qquad T = 19.8 \text{ N}$$

Fig. 15.66c

Fig. 15.66d

(c) Figure 15.66d shows the free-body diagram for the general case of the particle at point a. Newton's second law has the form

$$\sum F_n = ma_n \qquad T + mg = m\frac{v^2}{\rho}$$

The minimum value of $T$ is $T = 0$. Using this result in the above equation,

$$g = \frac{v_{min}^2}{\rho} \qquad v_{min} = \sqrt{\rho g} = \sqrt{0.7(9.81)} = 2.62 \text{ m/s}$$

**15.67** The system shown in Fig. 15.67a is a conical pendulum. The mass particle, which is connected to an inextensible string, moves with constant velocity in a circular path in a horizontal plane. The weight of the particle is 4 N, and the maximum permissible value of the cable tensile force is 7 N.

(a) Find the maximum permissible value of the angle $\theta$ between the string and the vertical direction.

(b) Find the corresponding value of the speed of the particle.

(c) Express the resultant force on the mass particle in formal vector notation.

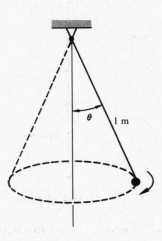

**Fig. 15.67a**

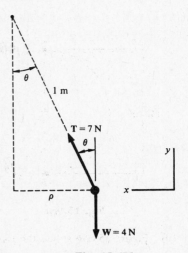

**Fig. 15.67b**

▌ (a) The free-body diagram of the particle is shown in Fig. 15.67b. For static equilibrium in the $y$ direction,

$$\sum F_y = 0 \qquad T \cos \theta - W = 7 \cos \theta - 4 = 0 \qquad \theta = 55.2°$$

For any value of $\theta$ greater than 55.2°, the maximum allowable cable tensile force of 7 N will be exceeded.

(b) Using the above value of $\theta$, we get

$$\rho = (1)(\sin \theta) = (1)(\sin 55.2°) = 0.821 \text{ m}$$

The equation of motion in the normal direction is

$$\sum F_x = ma_n \qquad T \sin \theta = m \frac{v^2}{\rho} \qquad 7 \sin 55.2° = \frac{4}{9.81}\left(\frac{v^2}{0.821}\right)$$

(c) Using Fig. 15.67b, the resultant force $\mathbf{F}$ on the mass particle is

$$\mathbf{F} = (T \sin \theta)\mathbf{i} + (T \cos \theta - W)\mathbf{j} = (7 \sin 55.2°)\mathbf{i} + (7 \cos 55.2° - 4)\mathbf{j} = 5.75\mathbf{i} \qquad \text{N}$$

The value of zero for the $y$ component of $\mathbf{F}$ is an expected result because of the use of the condition $\sum F_y = 0$ in part (a).

15.68 The mass particle in the conical pendulum in Fig. 15.68a moves with constant speed $v$, and $\theta = 20°$. Find the force in the inextensible string, if the particle has a weight of 0.6 lb, and find the speed $v$ of the particle.

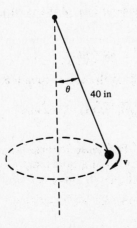

**Fig. 15.68a**

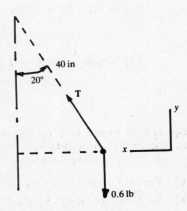

**Fig. 15.68b**

▌ Figure 15.68$b$ shows the free-body diagram of the particle, and $T$ is the string force. The equations of equilibrium and motion are

$$\sum F_y = 0 \qquad T \cos 20° - 0.6 = 0 \qquad T = 0.639 \text{ lb}$$

$$\sum F_x = ma_n$$

$$T \sin 20° = m \frac{v^2}{\rho} = \frac{0.6}{386} \left( \frac{v^2}{40 \sin 20°} \right) \qquad v = 43.9 \text{ in/s}$$

15.69 (a) The particle of 0.1-kg mass in Fig. 15.69$a$ travels at a speed of 3.2 m/s in a circular path. Find the forces in the two inextensible cables.

(b) Express the resultant force on the mass particle in formal vector notation.

(c) Find the limiting value of speed in part (a) at which a cable force just ceases to be tensile force.

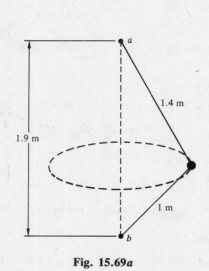

Fig. 15.69$a$

Fig. 15.69$b$

▌ (a) The free-body diagram of the particle is shown in Fig. 15.69$b$. Using the law of cosines,

$$1^2 = 1.9^2 + 1.4^2 - 2(1.9)1.4 \cos \theta_1 \qquad \theta_1 = 30.8°$$

Using the law of sines,

$$\frac{1}{\sin \theta_1} = \frac{1}{\sin 30.8°} = \frac{1.4}{\sin \theta_2} \qquad \theta_2 = 45.8°$$

The above values of $\theta_1$ and $\theta_2$ are used to find the directions, shown in Fig. 15.69$b$, of $T_1$ and $T_2$. For equilibrium in the $y$ direction,

$$\sum F_y = 0 \qquad T_1 \sin 59.2° - T_2 \sin 44.2° - 0.981 = 0 \qquad 0.859 T_1 - 0.697 T_2 - 0.981 = 0 \qquad (1)$$

The equation of motion has the form

$$\sum F_x = ma_n$$

$$T_1 \cos 59.2° + T_2 \cos 44.2° = 0.1 \frac{(3.2)^2}{0.717} \qquad 0.512 T_1 + 0.717 T_2 - 1.43 = 0 \qquad (2)$$

Equation (1) is multiplied by $-0.512/0.859$ and added to Eq. (2). The result is

$$1.13 T_2 - 0.845 = 0 \qquad T_2 = 0.748 \text{ N}$$

Using Eq. (1),

$$0.859 T_1 - 0.697(0.748) - 0.981 = 0 \qquad T_1 = 1.75 \text{ N}$$

(b)   The resultant force **F** acting on the mass particle, using Fig. 15.69b, is

$$\mathbf{F} = (T_1 \cos 59.2° + T_2 \cos 44.2°)\mathbf{i} + (T_1 \sin 59.2° - T_2 \sin 44.2° - 0.981)\mathbf{j}$$
$$= (1.75 \cos 59.2° + 0.748 \cos 44.2°)\mathbf{i} + (1.75 \sin 59.2° - 0.748 \sin 44.2° - 0.981)\mathbf{j} = 1.43\mathbf{i} \qquad \text{N}$$

(c)   From part (a),   $T_2 < T_1$.   The limiting condition of speed corresponds to   $T_2 = 0$.   The equations of equilibrium and motion, using Fig. 15.69b with   $T_2 = 0$,   are

$$\sum F_y = 0$$
$$T_1 \sin 59.2° - 0.981 = 0 \qquad T_1 = 1.14 \text{ N}$$

$$\sum F_x = ma_n$$

$$T_1 \cos 59.2° = m \frac{v^2}{\rho} \qquad 1.14 \cos 59.2° = 0.1\left(\frac{v^2}{0.717}\right) \qquad v = 2.05 \text{ m/s}$$

If   $v > 2.05$ m/s,   both cable force $T_1$ and $T_2$ in Fig. 15.69b will be tensile.

15.70   A 3,600-lb automobile is to travel along either of the two road surfaces shown in Fig. 15.70a.

Fig. 15.70a

(a)   How much would the vehicle appear to weigh if its speed were 60 mi/h at the position shown on road surface A?

(b)   Do the same as in part (a) for road surface B.

(c)   At what speed would the wheels of the automobile traveling on road A lose contact with the road surface?

▌ (a)   The free-body diagram of the automobile on road A is shown in Fig. 15.70b.   N is the total reaction force of the road on the four wheels of the vehicle.

The speed   $v = 60$ mi/h   may be expressed as

$$v = \left(60 \frac{\text{mi}}{\text{h}}\right)\left(\frac{5,280 \text{ ft}}{1 \text{ mi}}\right)\left(\frac{1 \text{ h}}{3,600 \text{ s}}\right) = 88.0 \frac{\text{ft}}{\text{s}}$$

The equation of motion is

$$\sum F_n = ma_n \qquad W - N = \frac{W}{g}\frac{v^2}{\rho} \qquad 3,600 - N = \frac{3,600}{32.2}\left(\frac{88.0^2}{800}\right) \qquad N = 2,520 \text{ lb}$$

The vehicle would appear to weigh 2,520 lb on road surface A, and the percent difference between this value and the actual weight of the car is

$$\%\text{D} = \frac{2,520 - 3,600}{3,600} \, 100 = -30\%$$

(b)   The free-body diagram of the automobile on road B is shown in Fig. 15.70c.   The equation of motion is

$$\sum F_n = ma_n \qquad N - W = \frac{W}{g}\frac{v^2}{\rho} \qquad N - 3,600 = \frac{3,600}{32.2}\left(\frac{88.0^2}{800}\right) \qquad N = 4,680 \text{ lb}$$

The percent difference between the above value and the actual weight of the car is

$$\%\text{D} = \frac{4,680 - 3,600}{3,600} \, 100 = +30\%$$

**Fig. 15.70b**

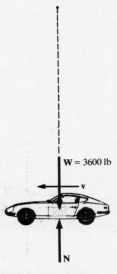

**Fig. 15.70c**

(*c*)  From consideration of Fig. 15.70*b*, the wheels of the automobile lose contact with road *A* when $N \to 0$.  Using the equation of motion in part (*a*),

$$\sum F_n = ma_n \qquad W - N = \frac{W}{g}\frac{v^2}{\rho} \qquad W - 0 = \frac{W}{g}\frac{v^2}{\rho} \qquad v = \sqrt{\rho g} = \sqrt{800(32.2)} = 160 \text{ ft/s} = 109 \text{ mi/h}$$

**15.71**  The brakes are suddenly applied and the wheels locked when the automobile is in either of the two positions shown in Fig. 15.70*a*.  The coefficient of friction between the tires and the road surface is assumed to be 0.7.

(*a*)  Find the maximum possible deceleration and the minimum stopping distance of the automobile, if it is assumed that this body moves in rectilinear translation.

(*b*)  Do the same as in part (*a*) for the case where the automobile moves along road *A*.  Assume that the vehicle moves in rectilinear translation with the constant maximum deceleration that it has at the highest position on road *A*.

(*c*)  Do the same as in part (*b*) for the case of the automobile moving along road *B*, with the maximum deceleration that it has at the lowest position on road *B*.

▌(*a*)  Figure 15.71*a* shows the free-body diagram of the car when the brakes are applied and the wheels are locked.  $\mu$ is the coefficient of friction between the tires and the road surface.
   The equation of motion for the sliding car is

$$\sum F_x = ma \qquad -\mu N = \frac{W}{g}a$$

For the case of rectilinear translation,  $N = W = $ constant,  and the above equation appears as

$$-\mu W = \frac{W}{g}a \qquad a = -\mu g = -0.7(32.2) = -22.5 \text{ ft/s}^2$$

The maximum possible deceleration of the automobile is 22.5 ft/s².  It may be seen that for the case of rectilinear translation of the automobile, the deceleration due to the frictional braking force is *independent* of the weight of the vehicle.
   The minimum stopping distance is found from

$$v^2 = v_0^2 + 2a(s - s_0) \qquad 0 = 88.0^2 + 2(-22.5)(s - 0) \qquad s = 172 \text{ ft}$$

(*b*)  For the case of the vehicle traveling on a curved roadway as shown in Fig. 15.70*a*, the frictional braking force is directly proportional to the *normal* force exerted by the road on the vehicle.  The mass of the car, by comparison, is the *actual* weight of the car divided by *g*.  The equation of motion for the case of the car sliding on road *A*, using the value of the normal force found in part (*a*) of Prob. 15.70, is

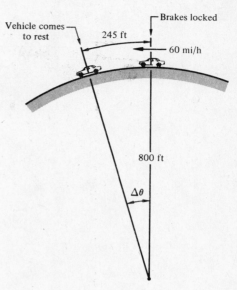

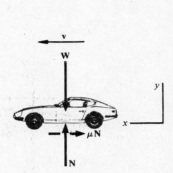

Fig. 15.71a

Fig. 15.71b

$$\sum F = ma \qquad -0.7(2,520) = \frac{3,600}{32.2}\, a \qquad a = -15.8\,\text{ft/s}^2$$

The maximum possible deceleration at the highest position on road $A$ is 15.8 ft/s$^2$.
The stopping distance, assuming rectilinear motion, is found from

$$v^2 = v_0^2 = 2a(s - s_0) \qquad 0 = 88.0^2 + 2(-15.8)(s - 0) \qquad s = 245\,\text{ft}$$

This distance is 42.4 percent greater than the required value of 172 ft for stopping with rectilinear translation. Figure 15.71b shows the stopping configuration for road $A$. The angle $\Delta\theta$ is found from

$$\Delta\theta = \frac{245}{800} = 0.306\,\text{rad} = 17.5°$$

It is left as an exercise for the reader to decide whether the result above supports the assumption that the *deceleration is constant* over the stopping distance.

(c) For road $B$, using the value of the normal force found in part (b) of Prob. 15.70,

$$\sum F = ma \qquad -0.7(4,680) = \frac{3,600}{32.2}\, a \qquad a = -29.3\,\text{ft/s}^2$$

The maximum possible deceleration at the lowest position of road $B$ is 29.3 ft/s$^2$. The minimum stopping distance is found from

$$v^2 = v_0^2 + 2a(s - s_0) \qquad 0 = 88.0^2 + 2(-29.3)(s - 0) \qquad s = 132\,\text{ft}$$

This value is 23.3 percent less than the required value of 172 ft for the rectilinear motion case in part (a).
The angle $\Delta\theta$ for road surface $B$, from a sketch similar to Fig. 15.71b, is

$$\Delta\theta = \frac{132}{800} = 0.165\,\text{rad} = 9.45°$$

This example illustrates the problems involved in braking on the crest of a hill where, in addition to the reduced visibility, the braking capacity may be significantly reduced.

15.72 An automobile drives around a horizontal, circular track at constant speed, as shown in Fig. 15.72a. The road surface is wet macadam, and the value of the coefficient of friction between the tires and the road is assumed to be 0.42. At what speed will sliding motion of the automobile on the road be impending?

▌ The free-body diagram of the automobile is shown in Fig. 15.72b, and this element is considered to be a mass particle in plane curvilinear transition. The direction and sense of the impending sliding motion are radially outward from the center of the track. When this sliding motion is impending, the friction force has its maximum possible value $\mu N$. For equilibrium in the $y$ direction,

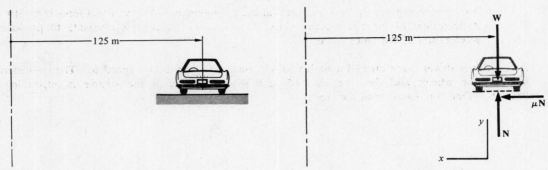

Fig. 15.72a          Fig. 15.72b

$$\sum F_y = 0 \qquad -W + N = 0 \qquad N = W$$

The equation of motion is

$$\sum F_x = ma_n \qquad \mu N = \frac{W}{g} a_n \qquad \mu W = \frac{W}{g} \frac{v^2}{\rho} \qquad v = \sqrt{\mu \rho g} = \sqrt{0.42(125)(9.81)} = 22.7 \, \text{m/s}$$

$$v = \left(22.7 \frac{\text{m}}{\text{s}}\right)\left(\frac{1 \, \text{km}}{1,000 \, \text{m}}\right)\left(\frac{3,600 \, \text{s}}{1 \, \text{h}}\right) = 81.7 \frac{\text{km}}{\text{h}}$$

It may be seen that the above result is *independent* of the weight of the car.

**15.73**    (*a*)   Do the same as in Prob. 15.72, if the road surface is banked as shown in Fig. 15.73*a*.

        (*b*)   Express the reaction force exerted by the track on the automobile in formal vector notation.

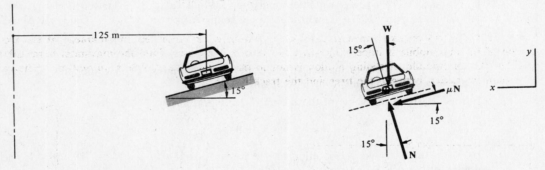

Fig. 15.73a          Fig. 15.73b

**❙**   (*a*)   Figure 15.73*b* shows the free-body diagram of the automobile for the case of the banked roadway. As before, the car is considered to be a mass particle in plane curvilinear translation. For static equilibrium in the *y* direction,

$$\sum F_y = 0 \qquad -W - \mu N \sin 15° + N \cos 15° = 0 \qquad N(\cos 15° - 0.42 \sin 15°) = W \qquad N = 1.17W$$

The equation of motion of the car is

$$\sum F_x = ma_n \qquad N \sin 15° + \mu N \cos 15° = \frac{W}{g} a_n = \frac{W}{g} \frac{v^2}{\rho} \qquad N(\sin 15° + \mu \cos 15°) = \frac{W}{g} \frac{v^2}{\rho}$$

$$1.17W(\sin 15° + 0.42 \cos 15°) = \frac{W}{9.81} \frac{v^2}{(125)} \qquad v = 30.9 \, \text{m/s} = 111 \, \text{km/h}$$

This above result is again independent of the weight of the vehicle.

     (*b*)   Using Fig. 15.73*b*, the reaction force **R** of the track on the automobile has the form

$$\mathbf{R} = (N \sin 15° + \mu N \cos 15°)\mathbf{i} + (N \cos 15° - \mu N \sin 15°)\mathbf{j} = [1.17W \sin 15° + 0.42(1.17W) \cos 15°]\mathbf{i}$$
$$+ [1.17W \cos 15° - 0.42(1.17W) \sin 15°]\mathbf{j} = 0.777W\mathbf{i} + W\mathbf{j} = (0.777\mathbf{i} + \mathbf{j})W$$

It may be seen from the above result that the $y$ component of the resultant force is exactly equal to the static weight of the automobile. The $x$ component is the force available to produce centripetal acceleration of the automobile.

**15.74** A test driver drives a car around a horizontal circular track at constant speed $v$. The coefficient of friction between the wheels and the track is 0.65, and sliding motion of the vehicle is impending when $v = 86$ mi/h. Find the diameter of the track.

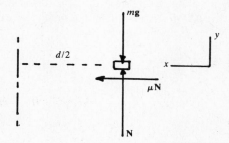

Fig. 15.74

▮ Figure 15.74 shows the free-body diagram of the car, and the diameter of the track is $d$. The equations of equilibrium and motion are

$$\sum F_y = 0 \qquad N - mg = 0 \qquad N = mg$$

$$\sum F_x = ma_n \qquad \mu N = ma_n = m\frac{v^2}{d/2} \qquad d = \frac{2v^2}{\mu g}$$

Using $86$ mi/h $= 126$ ft/s,

$$d = \frac{2(126)^2}{0.65(32.2)} = 1,520 \text{ ft}$$

**15.75** Figure 15.75a shows an experimental automobile testing track. For the embankment of the track, sliding motion of the automobile is impending when the vehicle is at rest. Find the maximum permissible constant speed of the automobile, if sliding motion is not to occur when the vehicle is in motion. The value of the coefficient of friction between the tires and the track is 0.58.

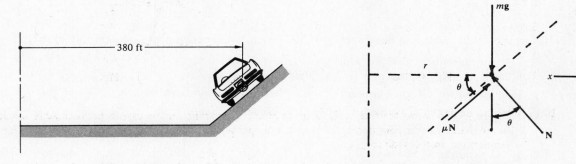

Fig. 15.75a                    Fig. 15.75b

▮ Figure 15.75b shows the free-body diagram of the vehicle, for the case of impending downward sliding motion when the vehicle is at rest. For equilibrium in the $x$ direction,

$$\sum F_x = 0 \qquad -\mu N \cos\theta + N \sin\theta = 0 \qquad \tan\theta = \mu = 0.58 \qquad \theta = 30.1°$$

The sense of the friction force is opposite to that shown in Fig. 15.75b when impending upward sliding motion occurs with the vehicle at its maximum speed. The equations of equilibrium and motion are

$$\sum F_y = 0 \qquad -mg - \mu N \sin\theta + N \cos\theta = 0 \qquad N = \frac{mg}{\cos\theta - \mu\sin\theta} \tag{1}$$

$$\sum F_x = ma_n \qquad N\sin\theta + \mu N \cos\theta = m\frac{v^2}{r} \qquad N(\sin\theta + \mu\cos\theta) = m\frac{v^2}{r} \tag{2}$$

$N$ is eliminated between Eqs. (1) and (2), with the result

$$v = \sqrt{gr\left(\frac{\sin\theta + \mu\cos\theta}{\cos\theta - \mu\sin\theta}\right)}$$

Using $r = 380$ ft and $\theta = 30.1°$, the final result is

$$v = 146 \text{ ft/s} = 99.5 \text{ mi/h}$$

**15.76**  A horizontal circular curve has the banking shown in Fig. 15.76a. The coefficient of friction between the tires and the road surface is 0.5. At what speed of the automobile will sliding motion be impending?

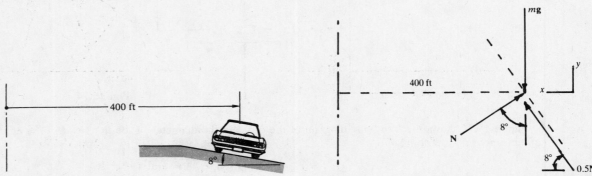

Fig. **15.76a**                    Fig. **15.76b**

❚  Figure 15.76b shows the forces which act on the automobile. The equations of equilibrium and motion, for the assumption of impending sliding motion, are

$$\sum F_y = 0 \qquad N\cos 8° + 0.5N\sin 8° - mg = 0 \qquad N(\cos 8° + 0.5\sin 8°) = m(32.2) \qquad (1)$$

$$\sum F_x = ma_n \qquad 0.5N\cos 8° - N\sin 8° = m\frac{v^2}{400} \qquad N(0.5\cos 8° - \sin 8°) = m\frac{v^2}{400} \qquad (2)$$

$N$ is eliminated between Eqs. (1) and (2) to obtain

$$v = \sqrt{400(32.2)\left(\frac{0.5\cos 8° - \sin 8°}{\cos 8° + 0.5\sin 8°}\right)} = 65.8 \text{ ft/s} = 44.9 \text{ mi/h}$$

**15.77**  A 3,200-lb vehicle drives from a straight road onto a curved road of circular shape, as shown in Fig. 15.77a. As the vehicle passes point $a$, it has a speed of 55 mi/h and is decelerating at a uniform rate of 6 ft/s².

(**a**)  Find the magnitude, direction, and sense of the acceleration of the vehicle when it passes point $b$. Point $b$ is 260 ft from point $a$, measured along the curved roadway.

(**b**)  Find the horizontal force at point $b$ exerted by the roadway on the vehicle.

(**c**)  Find the minimum value of the coefficient of friction at point $b$ if the vehicle is not to slide off the road.

(**d**)  Express the results in parts (a) and (b) in formal vector notation.

❚  (**a**)  The coordinate of length along the curve is designated $s$, and $s = 0$ corresponds to point $a$. The velocity at point $b$ is found from

$$v_b^2 = v_0^2 + 2as$$

Using $v_0 = 55$ mi/h $= 80.7$ ft/s, we get

$$v_b^2 = 80.7^2 + 2(-6)(260) \qquad v_b = 58.2 \text{ ft/s}$$

The centripetal acceleration at point $b$ is

$$a_n = \frac{v^2}{\rho} = \frac{58.2^2}{400} = 8.47 \frac{\text{ft}}{\text{s}^2}$$

The tangential acceleration has the constant value $a_t = -6$ ft/s² given in the problem statement. These two components of acceleration are shown in Fig. 15.77b. The magnitude of the total acceleration is

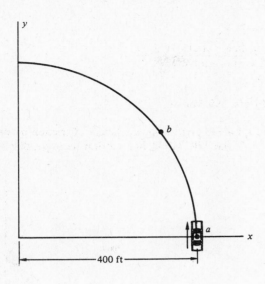

**Fig. 15.77a**                    **Fig. 15.77b**

$$a = \sqrt{a_n^2 + a_t^2} = \sqrt{8.47^2 + (-6)^2} = 10.4 \text{ ft/s}^2$$

The angles $\alpha$ and $\beta$ are found as

$$\beta = \frac{260}{400} = 0.650 \text{ rad} = 37.2° \qquad \tan \alpha = \frac{6}{8.47} \qquad \alpha = 35.3°$$

The angle $\zeta$ between the resultant acceleration vector and the $x$ axis is

$$\zeta = \beta + \alpha = 37.2° + 35.3° = 72.5°$$

(*b*)  The $x'$ axis is collinear with the direction of the resultant acceleration of the car, as shown in Fig. 15.77*b*.

By using Newton's second law, the force $F_{x'}$ exerted by the road on the car is

$$\sum F_{x'} = ma_{x'} \qquad F_{x'} = \frac{3,200}{32.2} (10.4) = 1,030 \text{ lb}$$

This force has the direction $\zeta$ shown in Fig. 15.77*b*.

(*c*)  When sliding motion of the vehicle is impending, the force $F_{x'}$ given above has its maximum value

$$F_{x'} = \mu N$$

The normal force is equal to the weight of the vehicle.  The minimum required value of the coefficient of friction, then, is given by

$$\mu_{min} = \frac{F_{x'}}{N} = \frac{1,030}{3,200} = 0.322$$

(*d*)  The acceleration **a** of the vehicle at point $b$, using Fig. 15.77*b*, may be expressed as

$$\mathbf{a} = (6 \sin 37.2° - 8.47 \cos 37.2°)\mathbf{i} - (6 \cos 37.2° + 8.47 \sin 37.2°)\mathbf{j} = -3.12\mathbf{i} - 9.90\mathbf{j} \qquad \text{ft/s}^2$$

As a check on the above result, using  $a = 10.4 \text{ ft/s}^2$  and  $\zeta = 72.5°$,

$$\mathbf{a} = (-10.4 \cos 72.5°)\mathbf{i} - (10.4 \sin 72.5°)\mathbf{j} = -3.13\mathbf{i} - 9.92\mathbf{j} \qquad \text{ft/s}^2$$

It may be noted that

$$-3.13 \approx -3.12 \qquad -9.92 \approx -9.90$$

The force **F** exerted by the road on the car, using Fig. 15.77*b*, is given by

$$\mathbf{F} = (-F_{x'} \cos \zeta)\mathbf{i} - (F_{x'} \sin \zeta)\mathbf{j} = (-1,030 \cos 72.5°)\mathbf{i} - (1,030 \sin 72.5°)\mathbf{j} = -310\mathbf{i} - 982\mathbf{j} \qquad \text{lb}$$

**15.78**  A racing car enters a circular track from a straight roadway, as shown in Fig. 15.78*a*.  When the car passes point $a$ it has a speed of 68 km/h and starts to accelerate with a constant value of $a_t$.  The car uses a

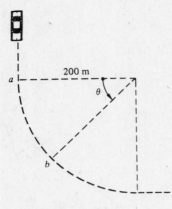

**Fig. 15.78a**

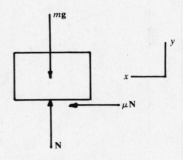

**Fig. 15.78b**

special-design low-profile radial tire, and it may be assumed that $\mu = 0.7$. When $\theta = 72°$, sliding motion of the car at point $b$ is impending.

(*a*)  Find $a_t$ when sliding motion is impending.

(*b*)  Find the time when sliding motion is impending, if $t = 0$ when the car is at point $a$.

**▮** (*a*)  The $x$ axis is along the direction of the total acceleration of the car. This direction lies in the plane of the track. Figure 15.78*b* shows the free-body diagram of the car when sliding motion is impending. The equations of equilibrium and motion are

$$\sum F_y = 0 \qquad N - mg = 0 \qquad N = mg$$

$$\sum F_x = ma \qquad \mu N = ma \qquad \mu mg = ma \qquad a = \mu g \qquad (1)$$

The car has acceleration components $a_n$ and $a_t$, and the resultant acceleration is

$$a = \sqrt{a_n^2 + a_t^2} \qquad a^2 = a_n^2 + a_t^2 \qquad (2)$$

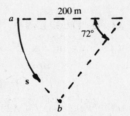

**Fig. 15.78c**

Figure 15.78*c* shows the distance coordinate $s$ measured along the track. $s$ is measured from point $a$. At the position $b$ where sliding motion motion is impending, at $\theta = 72°$,

$$s = 200(72°)\,\frac{\pi \text{ rad}}{180°} = 251 \text{ m}$$

The initial velocity of the car at point $a$ is

$$v_0 = 68 \text{ km/h} = 18.9 \text{ m/s}$$

The speed $v$ when the car reaches point $b$ is found from

$$v^2 = v_0^2 + 2a(s - s_0) = 18.9^2 + 2a_t(251)$$

From Eq. (1),

$$a = 0.7(9.81) \qquad (3)$$

Using Eq. (3), and the relationship $a_n = v^2/\rho$, in Eq. (2) results in

$$[0.7(9.81)]^2 = \left[\frac{18.9^2 + 2(251)a_t}{200}\right]^2 + a_t^2 \qquad 7.30a_t^2 + 8.96a_t - 44.0 = 0$$

$$a_t = \frac{-8.96 \pm \sqrt{8.96^2 - 4(7.3)(-44.0)}}{2(7.30)} = \frac{-8.96 \pm 36.9}{2(7.30)} = 1.91 \text{ m/s}^2$$

(b)  The speed of the car at point $b$ is found from

$$v^2 = v_0^2 + 2a(s - s_0) = 18.9^2 + 2a_t(251) = 18.9^2 + 2(1.91)251 \qquad v = 36.3 \text{ m/s}$$

The time to reach point $b$ is found from

$$v = v_0 + at \qquad 36.3 = 18.9 + 1.92t \qquad t = 9.11 \text{ s}$$

## 15.4  THE D'ALEMBERT, OR INERTIA, FORCE

**15.79**  (a)  Give an example of the *D'Alembert, or inertia, force*.

(b)  What are the characteristics of an inertia force?

(c)  Show how the use of the inertia force reduces the form of a dynamics problem to that of a statics problem.

▮ (a)  In this problem a technique is presented which converts the solution of a problem in dynamics to the form of a problem in statics.  This is accomplished by the introduction and use of a term referred to as the D'Alembert, or inertia, force.

The block on a smooth horizontal plane in Prob. 15.10 is shown in Fig. 15.79a, and the free-body diagram of the block is shown in Fig. 15.79b.  The equation of motion of the block was shown to be

$$\sum F_x = ma \qquad P = ma$$

This equation may be written as

$$P - ma = 0 \qquad P + (-ma) = 0 \qquad\qquad\qquad (1)$$

The quantity $-ma$ is referred to as the D'Alembert, or inertia, force which acts on the mass particle $m$.

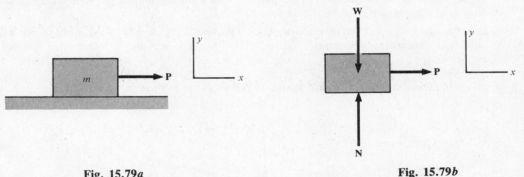

**Fig. 15.79a**          **Fig. 15.79b**

(b)  From consideration of Eq. (1) in part (a), and Fig. 15.79b, the following observations may be made.

1. The magnitude of the inertia force is the product of mass and acceleration.

2. *The inertia force exists only when the particle is accelerating.*  Thus, for the cases of a particle at rest or moving with constant velocity, the magnitude of the inertia force is identically zero.

3. Since the inertia force is related to the *mass* of the body, this force is a body force.

4. From the structure of Eq. (1), the sense of the inertia force $-ma$ is positive in the same sense as that of force $P$.  It follows then, from Fig. 15.79b, that the inertia force, as defined above, is positive in the sense of the positive sense of force $P$.

5. Because of the minus sign in the term $-ma$, and the fact that $m$ is always positive, the actual sense of the inertia force is always *opposite* to the *actual* sense of the acceleration.

(c)  If the inertia force were thought of as an *external, applied* force which acts on the mass, the free-body diagram of the block would have the form shown in Fig. 15.79c.  By including the inertia force in the free-body diagram, the dynamics problem is reduced to the form of a statics problem.  The requirement of *static equilibrium* is then

$$\sum F_x = 0 \qquad P + (-ma) = 0 \qquad P = ma$$

When using the inertia force, the particle is described as being in *dynamic equilibrium*.

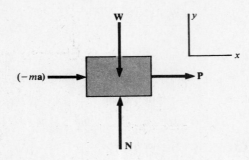

**Fig. 15.79c**

The use of the inertia force in the solution of problems in particle dynamics does not offer a significant advantage over the direct use of Newton's second law. The use of this technique, however, finds widespread application in the solution of problems in the dynamics of rigid bodies, and many examples of this will be seen in the problems in Chap. 18. This technique is used almost exclusively in the solution of problems in advanced dynamics.

**15.80** Give the general rules which are used when using the inertia force to solve problems in dynamics.

▐ There are three general rules to follow in using the inertia force in the solution of problems in particle dynamics.

1. Draw an arrow, on the free-body diagram of the mass particle, which has the direction of the acceleration of the particle and the same sense as the assumed, or actual, sense of this acceleration.

2. Designate the magnitude of the arrow drawn on the free-body diagram by $ma$, where $m$ is the mass of the particle and $a$ is the acceleration of the particle. In a given problem, the term $a$ may be known, or it may be the quantity which is to be solved for.

3. Insert a minus sign in front of the term $ma$. The quantity $-ma$ is now the inertia force which acts on the particle. This quantity is considered to be positive if it acts in the assumed, or actual, sense of the acceleration.

**15.81** Use the inertia force technique to find the acceleration of the particle in Fig. 15.17a.

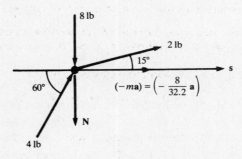

**Fig. 15.81**

▐ Figure 15.81 shows the free-body diagram of the particle, with the inertia force which acts on this mass element. For dynamic equilibrium of the particle,

$$\sum F_s = 0 \qquad 2\cos 15° + 4\cos 60° + \left(-\frac{8}{32.2}a\right) = 0 \qquad a = 15.8 \text{ ft/s}^2$$

**15.82** Use the inertia force technique to find the acceleration of the automobile in Prob. 15.26.

▐ The free-body diagram of the automobile is seen in Fig. 15.82. The automobile is assumed to accelerate to the right, and the inertia force has the form shown in the figure. For dynamic equilibrium of the automobile,

$$\sum F_x = 0 \qquad \mu N + (-ma) = 0 \qquad \mu N = ma \qquad 0.05\left(\frac{mg}{2}\right) = ma \qquad a = 0.805 \text{ ft/s}^2$$

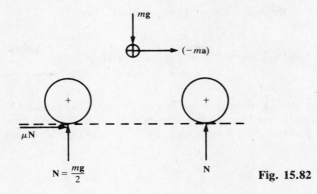

Fig. 15.82

**15.83** Use the inertia force technique to solve part (a) of Prob. 15.35.

▌ Figure 15,83 shows the free-body diagram of the crate. The actual sense of the acceleration is to the left, and the inertia force is drawn as shown in the figure.

For dynamic equilibrium of the crate,

$$\sum F_x = 0 \qquad -[-200(1.2)] - F = 0 \qquad F = 200(1.2) = 240 \text{ N}$$

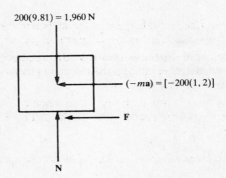

Fig. 15.83

**15.84** Use the inertia force technique to find the acceleration of the two blocks shown in Fig. 15.41a.

▌ Figure 15.84 shows the free-body diagrams of the two blocks and the positive senses of the two y-coordinate axes. The inertia forces are drawn on blocks A and B, in accordance with the general rules given in Prob. 15.80.

The requirement for dynamic equilibrium of block A is

$$\sum F_y = 0 \qquad T + (-m_A a) - 491 = 0 \qquad T - 491 = 50a \qquad (1)$$

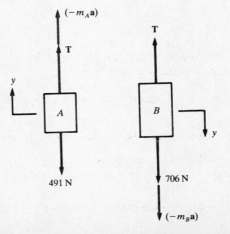

Fig. 15.84

The requirement for dynamic equilibrium of block $B$ is

$$\sum F_y = 0 \qquad 706 - T + (-m_B a) = 0 \qquad 706 - T = 72a \qquad (2)$$

Equations (1) and (2) are added, and $a$ is found as

$$a = 1.76 \text{ m/s}^2$$

**15.85** The system of Prob. 15.65 is repeated in Fig. 15.85a. The mass particle is connected to an inextensible string and moves with a constant velocity of 6 m/s in a circular path in a vertical plane. Use the inertia force technique to find the string tensile force when the particle is at points $a$ and $b$.

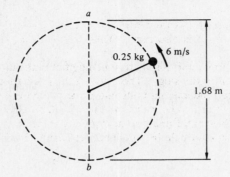

**Fig. 15.85a**

▌ Figure 15.85b shows the free-body diagram when the particle is at point $a$. $T$ is the cable tensile force. The positive sense of the acceleration $a_n$ is from the particle toward the center of the circular path. The inertia force which acts on the particle is shown in the figure.

For dynamic equilibrium of the particle,

$$\sum F_n = 0 \qquad -m\frac{v^2}{\rho} + W + T = -0.25\left(\frac{6^2}{0.84}\right) + 2.45 + T = 0 \qquad T = 8.26 \text{ N}$$

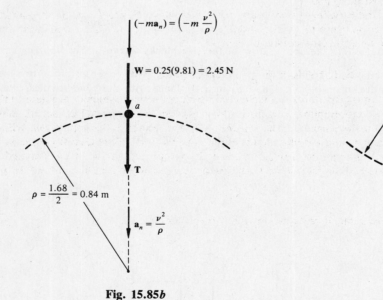

**Fig. 15.85b**                             **Fig. 15.85c**

When the particle is at point $b$, the free-body diagram has the appearance shown in Fig. 15.85c. The equilibrium requirement is

$$\sum F_n = 0 \qquad -m\frac{v^2}{\rho} - W + T = -0.25\left(\frac{6^2}{0.84}\right) - 2.45 + T = 0 \qquad T = 13.2 \text{ N}$$

**15.86** Give a summary of the basic concepts of particle dynamics.

The fundamental basis of all of engineering dynamics is Newton's second law, which states that a particle acted upon by a resultant force will experience an acceleration directly proportional to this force, and in the direction of the force. The equation form of Newton's second law for a particle is

$$\mathbf{F} = m\mathbf{a}$$

where $\mathbf{F}$ = resultant force
 $m$ = mass of particle
 $\mathbf{a}$ = acceleration

This acceleration has the same direction and sense as the resultant force. The above equation is often referred to as the equation of motion.

If a particle is at rest, or moving with constant velocity, the acceleration is zero. From the equation above, it follows that the resultant force is also zero. Thus, a particle at rest, or moving with constant velocity, is in static equilibrium.

In USCS units, length, time, and force are fundamental units and mass is a derived unit. The force unit is the pound and the mass unit is the slug. The value of the gravitational acceleration $g$ is 32.2 ft/s$^2$. In SI units, length, time, and mass are fundamental units and force is the derived unit. The force unit is the newton and the mass unit is the kilogram. The value of the gravitational acceleration $g$ is 9.81 m/s$^2$. One meter is approximately three feet in length; and one newton is a force of approximately one-quarter pound.

Newton's second law for a particle in rectilinear translation has the form

$$\sum F = ma$$

where $F$ is the component of the resultant force in the direction of motion. The boldface vector notation is not required for this case, since both the force and the acceleration have the known direction of the axis of motion. If the force in the direction of the rectilinear motion is a constant, it follows that the acceleration is a constant. The motion of a particle with constant acceleration is given by

$$v + v_0 + at \qquad s = s_0 + v_0 t + \tfrac{1}{2}at^2 \qquad v^2 = v_0^2 + 2a(s - s_0)$$

To solve the problem of motion of two connected particles in rectilinear translation, a free-body diagram of each particle is drawn. Newton's second law is written for each particle, considering the single cable or link force to be an unknown. The two equations obtained may then be solved simultaneously for the force and the common acceleration of the two particles.

When a particle moves in plane curvilinear translation, the normal force $F_n$ which acts on the particle is never zero, since this force is required to cause the particle to follow the curved path. The component equations of Newton's second law for a particle in plane curvilinear motion are

$$\sum F_n = ma_n = m\,\frac{v^2}{\rho} \qquad \sum F_t = ma_t = m\dot{v}$$

The use of the D'Alembert, or inertia, force is a technique which converts a dynamics problem to the form of a statics problem. An arrow, representing a vector, which acts in the assumed, or actual, sense of the acceleration, is first sketched on the particle. The magnitude of this vector is designated $ma$, where $a$ is the symbolic term for the acceleration, and a minus sign is affixed to this term. The quantity $-ma$ is the inertia force. This term is considered to be a force applied to the body, acting in the sense of the assumed, or actual, acceleration. With the inclusion of the inertia force, the dynamic equilibrium requirements of the particle are

$$\mathbf{F} = \sum_i \mathbf{F}_i = 0$$

Note that this equation has the form of an equation of static equilibrium.

# CHAPTER 16
# Kinematics of Plane Motion of a Rigid Body

## 16.1 RECTILINEAR AND CURVILINEAR TRANSLATION OF A RIGID BODY, ROTATION OF A RIGID BODY, ANGULAR DISPLACEMENT, VELOCITY, AND ACCELERATION

**16.1**     (*a*)   Define the term *rigid body*.

        (*b*)   What is the significant difference between *particle* motion and *rigid body* motion?

▌   (*a*)   In a rigid body, the distance between any two points is always the same, no matter what type of force system acts on the body. A rigid body thus may be viewed as having unchanging shape. Since the rigid body has nonzero length dimensions, it may experience rotation.

     (*b*)   It is the *absence* of any rotation effects which distinguishes the problem in particle kinematics from that in rigid-body kinematics.

**16.2**     Describe the motion of a rigid body in *rectilinear translation*.

▌   Figure 16.2 shows a rigid body. If the body moves so that *all* points in the body move along *straight parallel lines*, the motion is described as rectilinear translation. An alternative definition of this type of motion is that the body moves in such a way that any line, such as line *ab* in the figure, always has the same direction with respect to some fixed reference line. For the body in Fig. 16.2, the direction of line *ab* is given by the constant value of the angle $\beta$ between this line and a line parallel to the *y* axis.

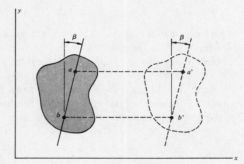

**Fig. 16.2**

**16.3**     Describe the motion of a rigid body in *curvilinear translation*.

▌   Figure 16.3 shows the situation when a rigid body moves with plane curvilinear motion. For this case, the two arbitrary points *a* and *b* on the body trace out *parallel* plane curvilinear paths. An alternative interpretation of this type of motion is that, as the body moves in plane curvilinear translation, any arbitrary line on the body is always parallel to its original position. It may be seen from the figure that lines *a'b'* and *a"b"* are always parallel to the original position *ab*. It follows from the discussion above that rectilinear translation is a special case of plane curvilinear translation, where the plane curved paths of all particles in the body have the limiting configuration of straight lines.

**16.4**     (*a*)   What is the relationship among the velocity and acceleration of arbitrary points on a rigid body which moves in rectilinear or curvilinear translation?

        (*b*)   What is the distinguishing characteristic of rectilinear or curvilinear translation?

▌   (*a*)   When a body moves with either rectilinear or curvilinear translation, the velocity and acceleration of *all* points on the body have the *same magnitude, direction, and sense*. This effect is shown in Fig. 16.4*a*, where a body moves with plane curvilinear translation with the known values $v_a$ and $a_a$ at point *a*. The velocity and acceleration of the two arbitrary points *b* and *c* are then as shown in Fig. 16.4*b*, and

$$v_a = v_b = v_c \qquad a_a = a_b = a_c$$

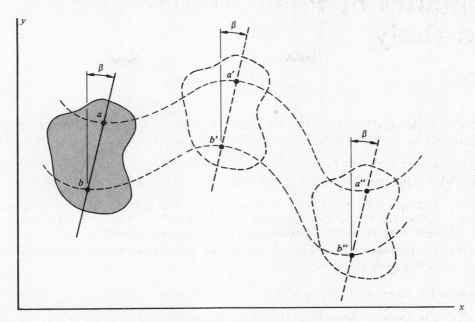

**Fig. 16.3**

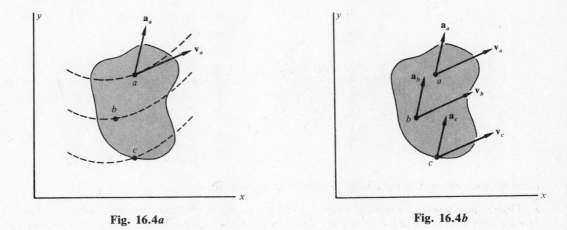

**Fig. 16.4a**                              **Fig. 16.4b**

    (*b*)  The distinguishing characteristic of rectilinear or curvilinear translational motion is the complete *absence* of any rotation effects of the rigid body.

**16.5**    (*a*)  Describe the possible motions of a link which is hinged to an absolute reference point, such as the earth.

    (*b*)  What is meant by the term *rotation*?

▌  (*a*)  Figure 16.5 shows a rigid link hinged to an absolute reference point such as the earth. The only possible motion of the link is rotation about the hinge. The axis of rotation is an axis which passes through the center of the of the hinge pin and is perpendicular to the plane of the figure. This axis is referred to as the *center of rotation*, or the *center of turning*.

    ·  (*b*)  The term rotation, in a general sense, is understood to mean motion in which a reference line on the rigid body experiences a *changing* angular position with respect to a fixed reference line in the plane of motion.

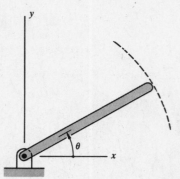

**Fig. 16.5**

**16.6** (*a*) Define the term *angular displacement*.

(*b*) Define the term *angular velocity*.

(*c*) Define the term *angular acceleration*.

❚ (*a*) The angular displacement of a link is given by angle $\theta$ shown in Fig. 16.5, and this displacement is a vector quantity. The direction of $\theta$ may be defined by the plane in which it is measured, or by the direction of a line normal to this plane. The basic magnitude of an angle is expressed in radians. The arrow representing $\theta$ in the figure indicates that the positive sense of $\theta$ is counterclockwise. The vector description of angular displacement is directly analogous to the vector description of a moment. The directions of all angular displacements are the same for the case of plane motion of a rigid body.

(*b*) The hinged link is redrawn in Fig. 16.6. The angular displacement $\theta$ corresponds to time $t$. At a later time $t + \Delta t$, the link has rotated to the new position shown by the dashed outline. The angular velocity $\omega$ of the link is defined to be the time rate of change of angular displacement, or

$$\omega = \lim_{\Delta t \to 0} \frac{\Delta \theta}{\Delta t} = \frac{d\theta}{dt} = \dot{\theta}$$

The basic units of angular velocity are radians per second, written as rad/s. These units are the same in both USCS and SI notation. Since $\Delta \theta$ is positive in the sense of increasing $\theta$ and $\Delta t$ is always positive, it follows that the angular velocity $\omega$ is always *positive in the same sense as $\theta$*.

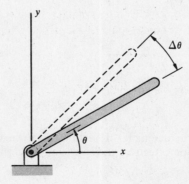

**Fig. 16.6**

(*c*) The *angular acceleration* $\alpha$ is defined to be the time rate of change of angular velocity. Thus,

$$\alpha = \frac{d\omega}{dt} = \frac{d}{dt}\,(\omega) = \frac{d}{dt}\left(\frac{d\theta}{dt}\right) = \frac{d^2\theta}{dt^2} = \ddot{\theta}$$

The basic units of angular acceleration are radians per second squared, written as (rad/s$^2$). These units are the same in both USCS and SI notation. Following a development similar to that in part (*b*), for angular velocity, it can be shown that *the angular acceleration $\alpha$ is always positive in the positive sense of $\theta$*.

**16.7**   The disk in Fig. 16.7*a* has an index line marked on it.   The angular displacement of the disk, and with it the reference line, is given by

$$\theta = 7.95t - 10.2t^3$$

where *t* is in seconds and $\theta$ is in radians.

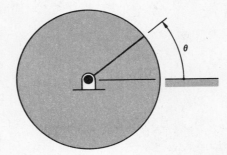

**Fig. 16.7*a***

(*a*)   Find the initial values of the angular displacement, velocity, and acceleration at   $t = 0$.

(*b*)   Find the time at which the disk reaches its maximum counterclockwise angular displacement, and the value of this maximum displacement.

(*c*)   At what time does the disk pass through the initial position again?

(*d*)   Find the angular velocity and angular acceleration corresponding to the time of part (*c*).

▌ (*a*)   The angular displacement of the disk is given by

$$\theta = 7.95t - 10.2t^3 \tag{1}$$

The angular velocity and acceleration are

$$\omega = \dot{\theta} = 7.95 - 3(10.2)t^2 = 7.95 - 30.6t^2 \text{ rad/s} \tag{2}$$
$$\alpha = \ddot{\theta} = -2(30.6t) = -61.2t \text{ rad/s}^2 \tag{3}$$

At   $t = 0$   the initial velocity $\omega_0$, from Eq. (2), is

$$\omega_0 = 7.95 - 30.6(0) = 7.95 \text{ rad/s}$$

The initial displacement, from Eq. (1), is zero and, from Eq. (3), the initial acceleration is zero.   The above results are shown in Fig. 16.7*b*.

(*b*)   It may be seen from Eq. (2) that the angular velocity is initially positive and at later times becomes negative.   Thus, when the angular velocity is zero, the disk reaches its position of maximum angular displacement, and

$$0 = 7.95 - 30.6t^2 \qquad t = 0.510 \text{ s}$$

The corresponding angular displacement is

$$\theta = 7.95t - 10.2t^3 = 7.95(0.510) - 10.2(0.510)^3 = 2.70 \text{ rad} = 155°$$

The angular acceleration at this position is

$$\alpha = -61.2t = -61.2(0.510) = -31.2 \text{ rad/s}^2$$

Figure 16.7*c* shows the above results.

(*c*)   When the disk passes through the initial position again,   $\theta = 0$.   Using Eq. (1), we get

$$0 = 7.95t - 10.2t^3 = t(7.95 - 10.2t^2) \qquad t = 0, \quad 0.883 \text{ s}$$

The time   $t = 0$   is the initial time of the problem.   At   $t = 0.883$ s   the disk again passes through the initial position.

(*d*)   The velocity and acceleration when the disk returns to the initial position are

$$\omega = 7.95 - 30.6t^2 = 7.95 - 30.6(0.833)^2 = -15.9 \text{ rad/s}$$
$$\alpha = -61.2t = -61.2(0.883) = -54.0 \text{ rad/s}^2$$

The results in parts (*c*) and (*d*) are displayed in Fig. 16.7*d*.

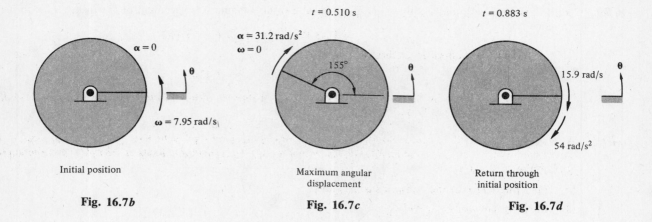

$t = 0.510$ s     $t = 0.883$ s

$\alpha = 0$

$\boldsymbol{\omega} = 7.95$ rad/s

Initial position

**Fig. 16.7b**

$\alpha = 31.2$ rad/s$^2$
$\omega = 0$

155°

Maximum angular
displacement

**Fig. 16.7c**

15.9 rad/s

54 rad/s$^2$

Return through
initial position

**Fig. 16.7d**

**16.8**    The disk shown in Fig. 16.8 rotates about a fixed point. The position of an index line on the disk is given by $\theta$. The angular displacement of the disk is given by $\theta = 2(1 + e^{-t})$, where $\theta$ is in radians and $t$ is in seconds.

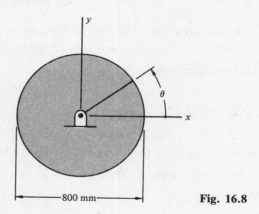

—800 mm—      **Fig. 16.8**

(*a*)   Find the initial values of the angular displacement, velocity, and acceleration when $t = 0$.
(*b*)   Find the angular displacement, velocity, and acceleration of the disk when $t = 1$ s and $t = 2$ s.
(*c*)   Find the time when the disk has completed one full revolution from its initial position.

**▌**  (*a*)   The angular displacement, velocity, and acceleration have the forms

$$\theta = 2(1 + e^{-t}) \text{ rad} \qquad \dot\theta = -2e^{-t} \text{ rad/s} \qquad \ddot\theta = 2e^{-t} \text{ rad/s}^2$$

At $t = 0$,

$$\theta = 2(1 + e^{-t}) = 2(1 + e^0) = 4 \text{ rad} \qquad \dot\theta = -2e^{-t} = -2e^0 = -2 \text{ rad/s}$$
$$\ddot\theta = 2e^{-t} = 2e^0 = 2 \text{ rad/s}^2$$

(*b*)   When $t = 1$ s,

$$\theta = 2(1 + e^{-t}) = 2(1 + e^{-1}) = 2.74 \text{ rad} \qquad \dot\theta = -2e^{-t} = -2e^{-1} = -0.736 \text{ rad/s}$$
$$\ddot\theta = 2e^{-t} = 2e^{-1} = 0.736 \text{ rad/s}^2$$

When $t = 2$ s,

$$\theta = 2(1 + e^{-t}) = 2(1 + e^{-2}) = 2.27 \text{ rad} \qquad \dot\theta = -2e^{-t} = -2e^{-2} = -0.271 \text{ rad/s}$$
$$\ddot\theta = 2e^{-t} = 2e^{-2} = 0.271 \text{ rad/s}^2$$

(*c*)   The initial value of $\theta$, found in part (*a*), is 4 rad. The angular displacement $\theta_1$, when the disk has completed one full revolution from its initial position, is given by

$$\theta_1 = 4 + 2\pi \text{ rad}$$

The time $t$ for this motion to occur is found from

$$4 + 2\pi = 2(1 + e^{-t}) \qquad e^{-t} = 4.14$$

Taking the natural logarithm of both sides,

$$-t = \ln 4.14 \qquad t = -1.42 \text{ s}$$

Since $t$ may not be negative, it may be concluded that the disk *will not* complete one full revolution from its initial position.

**16.9** A disk rotates about a fixed point, with the displacement $\theta = at^2 + bt + c$, where $a$, $b$, and $c$ are constants. When $t = 2$ s, the angular displacement is 5 rad, the angular velocity is $-8$ rad/s, and the angular acceleration is 3 rad/s$^2$.

(*a*) Find the values of the constants $a$, $b$, and $c$.

(*b*) Find the units of $a$, $b$, and $c$.

(*c*) Find $\theta$, $\omega$, and $\alpha$ when $t = 4$ s.

▌ (*a*) The angular displacement, velocity, and acceleration have the forms

$$\theta = at^2 + bt + c \tag{1}$$
$$\dot{\theta} = 2at + b \tag{2}$$
$$\ddot{\theta} = 2a \tag{3}$$

When $t = 2$ s, using Eq. (3),

$$\ddot{\theta} = 2a \qquad 3 = 2a \qquad a = 1.5$$

Using Eq. (2), with $t = 2$ s,

$$\dot{\theta} = 2at + b \qquad -8 = 2a(2) + b \qquad b = -14$$

Using Eq. (1), with $t = 2$ s,

$$\theta = at^2 + bt + c \qquad 5 = a(2)^2 + b(2) + c \qquad c = 27$$

The final numerical forms of the angular displacement, velocity, and acceleration are

$$\theta = 1.5t^2 - 14t + 27 \text{ rad} \qquad \dot{\theta} = 3t - 14 \text{ rad/s} \qquad \ddot{\theta} = 3 \text{ rad/s}^2$$

(*b*) From the form of Eq. (1),

$$at^2 \sim \theta \qquad \text{and} \qquad a \sim \frac{\theta}{t^2} = \text{rad/s}^2$$

$$bt \sim \theta \qquad \text{and} \qquad b \sim \frac{\theta}{t} = \text{rad/s}$$

$$c \sim \theta = \text{rad}$$

The constants $a$, $b$, and $c$ may then be written as

$$a = 1.5 \text{ rad/s}^2 \qquad b = -14 \text{ rad/s} \qquad c = 27 \text{ rad}$$

(*c*) When $t = 4$ s,

$$\theta = 1.5t^2 - 14t + 27 = 1.5(4)^2 - 14(4) + 27 = -5 \text{ rad} \qquad \omega = 3t - 14 = 3(4) - 14 = -2 \text{ rad/s}$$
$$\alpha = 3 \text{ rad/s}^2$$

**16.10** The angular acceleration of a body that rotates about a fixed axis is given by the function $\alpha = 3t - 5$. $\alpha$ is in radians per second squared and $t$ is in seconds. At $t = 0$ the angular displacement and velocity are zero.

(*a*) Find the angular velocity and angular displacement when $t = 1$ s and $t = 3$ s.

(*b*) Find the angular velocity and displacement when $\alpha = 0$.

(*c*) Sketch the variation of $\theta$, $\omega$, and $\alpha$ for $0 \le t \le 3$ s.

(*d*) Find the average values of angular velocity and angular acceleration in the interval $1 \le t \le 3$ s.

▌ (*a*) The angular velocity $\omega_1$, and angular displacement $\theta_1$, at $t = 1$ s are given by

$$\omega_1 = \int d\,dt = \int_0^1 (3t - 5)\,dt = \left(3\frac{t^2}{2} - 5t\right)\Big|_0^1 = -3.5 \text{ rad/s} \tag{1}$$

$$\theta_1 = \int \omega \, dt = \int_0^1 \left(3\frac{t^2}{2} - 5t\right) dt = \left[\frac{3}{2}\left(\frac{t^3}{3}\right) - 5\frac{t^2}{2}\right]\Big|_0^1 = -2 \text{ rad} \qquad (2)$$

The angular velocity $\omega_3$ at $t = 3$ s is found by using Eq. (1) with the limits $t = 0$ and $t = 3$ s. The result is

$$\omega_3 = \tfrac{3}{2}(3)^2 - 5(3) = -1.5 \text{ rad/s}$$

The angular displacement $\theta_3$ at $t = 3$ s is obtained by using Eq. (2) with the limits $t = 0$ and $t = 3$ s. This result has the form

$$\theta_3 = \frac{3}{2}\left(\frac{3^3}{3}\right) - 5\left(\frac{3^2}{2}\right) = -9 \text{ rad}$$

(b)    When $\alpha = 0$,

$$\alpha = 3t - 5 = 0 \qquad t = 1.67 \text{ s}$$

Using Eq. (1), with the limits $t = 0, 1.67$,

$$\omega = \tfrac{3}{2}(1.67)^2 - 5(1.67) = -4.17 \text{ rad/s}$$

Using Eq. (2), with the limits $t = 0, 1.67$,

$$\theta = \frac{3}{2}\left(\frac{1.67^3}{3}\right) - \frac{5}{2}(1.67)^2 = -4.64 \text{ rad}$$

(c)    Figure 16.10 shows the plots of $\theta$, $\omega$, and $\alpha$ versus time, for $0 \le t \le 3$ s.

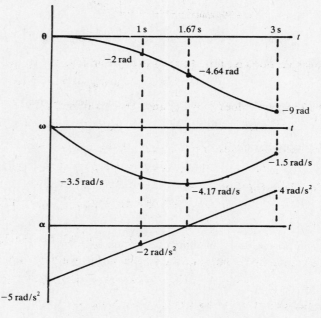

Fig. 16.10

(d)    The average values of $\omega$ and $\alpha$ over the time interval $t = 1$ s to $t = 3$ s have the forms

$$\omega_{avg} = \frac{\theta_3 - \theta_1}{3 - 1} = \frac{-9 - (-2)}{3 - 1} = -3.5 \text{ rad/s} \qquad \alpha_{avg} = \frac{\omega_3 - \omega_1}{3 - 1} = \frac{-1.5 - (-3.5)}{3 - 1} = 1 \text{ rad/s}^2$$

It may be seen that $\omega_{avg}$ is a fair representation of the actual values of $\omega$ over the time interval, while $\alpha_{avg}$ is a poor representation of the actual values.

16.11    A rigid body rotates about a fixed axis with the angular acceleration-time diagram shown in Fig. 16.11a. The initial angular displacement is zero and the initial angular velocity is $-110$ rad/s.

(a)    Find the angular velocity and angular displacement when $t = 4$ s.

(b)    Do the same as in part (a), for $t = 10$ s.

(c)    Sketch the curves of $\omega$ and $\theta$ for $0 \le t \le 10$ s and identify the extreme values of $\omega$ and $\theta$.

**(a)** The acceleration values in Fig. 16.11$a$ are expressed in radian units as

$$800 \frac{r}{min^2} \left( \frac{2\pi \, rad}{1 \, r} \right) \left( \frac{1 \, min}{60 \, s} \right)^2 = 1.40 \, rad/s^2 \qquad 600 \, r/min^2 = 1.05 \, rad/s^2$$

For $0 \le t \le 4\,s$, using Fig. 16.11$a$,

$$\alpha = \frac{1.40}{4} t = 0.35t$$

The angular velocity at $t = 4\,s$ is

$$\omega = \omega_0 + \int \alpha \, dt = -110 + \int_0^4 0.35t \, dt = -110 + 0.35 \left. \frac{t^2}{2} \right|_0^4 = -110 + \frac{0.35}{2} (4)^2 = -107 \, rad/s$$

The angular displacement at $t = 4\,s$ is

$$\theta = \int \omega \, dt = \int_0^4 \left( -110 + \frac{0.35}{2} t^2 \right) dt = \left[ -110t + \frac{0.35}{2} \left( \frac{t^3}{3} \right) \right]_0^4 = -110(4) + \frac{0.35}{2(3)} (4)^3 = -436 \, rad$$

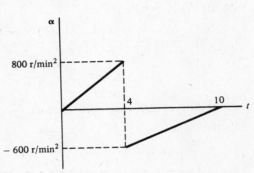

Fig. 16.11$a$

**(b)** The angular acceleration for $4\,s \le t \le 10\,s$, from Fig. 16.11$a$, has the form

$$\alpha = a_1 t + a_2 \tag{1}$$

Using the endpoint values $t = 4\,s$ and $\alpha = -1.05 \, rad/s^2$ in Eq. (1),

$$-1.05 = a_1(4) + a_2 \tag{2}$$

Using the endpoint values $t = 10\,s$ and $\alpha = 0$ in Eq. (1),

$$0 = a_1(10) + a_2 \qquad a_2 = -10a_1 \tag{3}$$

Using Eq. (3) in Eq. (2),

$$-1.05 = 4a_1 - 10a_1 \qquad a_1 = 0.175$$

Using Eq. (3),

$$a_2 = -10a_1 = -10(0.175) = -1.75$$

The acceleration in the interval $4\,s \le t \le 10\,s$ has the final form

$$\alpha = 0.175t - 1.75 \, rad/s^2$$

The angular velocity, for $4\,s \le t \le 10\,s$, is found as

$$\omega = \omega_0 + \int \alpha \, dt = \omega|_{t=4\,s} + \int_4^t (0.175t - 1.75) \, dt = -107 + \left. \left( 0.175 \frac{t^2}{2} - 1.75t \right) \right|_4^t$$

$$= -107 + \frac{0.175}{2} (t^2 - 4^2) - 1.75(t - 4) = 0.0875t^2 - 1.75t - 101$$

When $t = 10\,s$,

$$\omega = 0.0875(10)^2 - 1.75(10) - 101 = -110 \, rad/s$$

The angular displacement at $t = 10\,s$ is found as

$$\theta = \theta|_{t=4\,s} + \int_4^{10} \omega \, dt = -436 + \int_4^{10} (0.0875t^2 - 1.75t - 101) \, dt = -436 + \left. \left( 0.0875 \frac{t^3}{3} - 1.75 \frac{t^2}{2} - 101t \right) \right|_4^{10}$$

$$= -436 + \frac{0.0875}{3} (10^3 - 4^3) - \frac{1.75}{2} (10^2 - 4^2) - 101(10 - 4) = -1{,}090 \, rad$$

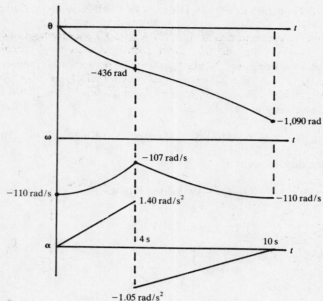

Fig. 16.11*b*

(*c*) Figure 16.11*b* shows the variation of $\theta$ and $\omega$ with time. The extreme values of these quantities are

$$\omega_{max} = -110 \text{ rad/s} \qquad \theta_{max} = -1{,}090 \text{ rad}$$

**16.12** What is the graphical interpretation of the definition of angular velocity as the first time derivative of the angular displacement?

❚ In Probs. 14.15 and 14.16 it was shown that the displacement, velocity, and acceleration of a particle in rectilinear translation have a very useful graphical interpretation. In this problem a similar interpretation will be made for the case of a rigid body rotating about a fixed axis. The similarity between this case and the case of rectilinear motion is that the displacement, velocity, and acceleration all have a common direction. The relationship between the angular velocity and displacement for the rigid body in rotation is

$$\omega = \frac{d\theta}{dt}$$

It may be seen from this equation that the angular velocity is equal to the slope of the angular displacement–time curve. The above equation may be written as

$$d\theta = \omega \, dt \qquad \int_{\theta_1}^{\theta_2} d\theta = \int_{t_1}^{t_2} \omega \, dt \qquad \theta_2 - \theta_1 = \int_{t_2}^{t_2} \omega \, dt$$

In these equations 1 and 2 designate the endpoints of a time interval of interest. The term on the right-hand side of the above equation is the area under the velocity diagram between times 1 and 2, and this area is equal to the change $\theta_2 - \theta_1$ in angular displacement during the time interval.

**16.13** What is the graphical interpretation of the definition of angular acceleration as the first time derivative of the angular velocity?

❚ The definition of angular acceleration for the rigid body in rotation is

$$\alpha = \frac{d\omega}{dt}$$

It follows from this result that the angular acceleration is equal to the slope of the angular velocity–time curve. This equation may be written in the form

$$d\omega = \alpha \, dt \qquad \int_{\omega_1}^{\omega_2} d\omega = \int_{t_1}^{t_2} \alpha \, dt \qquad \omega_2 - \omega_1 = \int_{t_1}^{t_2} \alpha \, dt$$

Thus, the change in the angular velocity is equal to the area under the angular acceleration–time curve.

The techniques presented in Prob. 14.17 for sketching positive or negative increasing or decreasing slopes may be used directly with the above equations.

**16.14**   Figure 16.14a shows a punch press. A heavy steel flywheel rotates at 20 r/min. During the punching operation a clutch engages the flywheel with the punching mechanism, and the corresponding acceleration diagram of the flywheel is shown in Fig. 16.14b. The metallic sheet stock to be punched consists of a thin, soft material on top and a thicker, hard material on the bottom, and the total punching operation takes place during the 1.7-s interval shown in the figure. After the punching operation is completed, the flywheel rotates through 0.5 r with constant angular velocity and engages with a motor which gives the flywheel a constant acceleration of 0.1 rad/s². When the flywheel reaches its original angular velocity, the motor is disengaged from the flywheel.

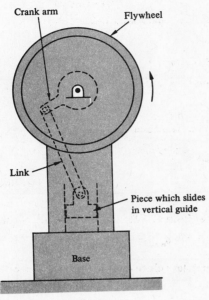

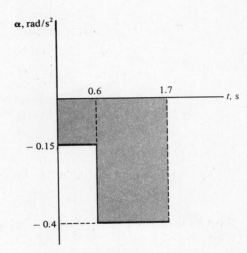

|  |  |
|---|---|
| **Fig. 16.14a** | **Fig. 16.14b** |

(a)   Find the time needed for the motor to return the flywheel to its original speed and the total time for one cycle of operation.

(b)   Show the variation of angular displacement and velocity for the entire cycle of operation of the punch press.

(c)   Find the total angular displacement of the flywheel for one cycle of operation.

▌ (a)   The initial angular velocity of the flywheel is

$$20\,\frac{\text{r}}{\text{min}}\left(\frac{2\pi\,\text{rad}}{1\,\text{r}}\right)\left(\frac{1\,\text{min}}{60\,\text{s}}\right) = 2.09\,\text{rad/s}$$

The angular velocity $\omega_2$ of the flywheel at $t = 1.7$ s, using Fig. 16.14b, is found from

$$\Delta\omega = \omega_2 - 2.09 = \int \alpha\,dt = -0.6(0.15) - (1.7 - 0.6)0.4 \qquad \omega_2 = 1.56\,\text{rad/s}$$

The time $\Delta t$ for the flywheel to complete one half revolution at the above angular velocity is found from

$$\theta = \omega_2\,\Delta t \qquad 0.5\,\text{r} = 1.56\,\frac{\text{rad}}{\text{s}}\left(\frac{1}{2\pi}\,\frac{\text{r}}{\text{rad}}\right)\Delta t \qquad \Delta t = 2.01\,\text{s}$$

The above results are shown in the acceleration–time diagram in Fig. 16.14c. The time $t_1$ required to return the flywheel to its original speed, using the area under the acceleration diagram, is found from

$$\Delta\omega = \int \alpha\,dt \qquad 2.09 - 1.56 = 0.1 t_1 \qquad t_1 = 5.3\,\text{s}$$

The total time $t$ for one cycle of operation is

$$t = 1.7 + 2.01 + 5.3 = 9.0\,\text{s}$$

(b)   The areas of the velocity diagram are computed, and these values are the changes in the angular displacement. The initial displacement of the flywheel at $t = 0$ is defined to be $\theta_0 = 0$. As the

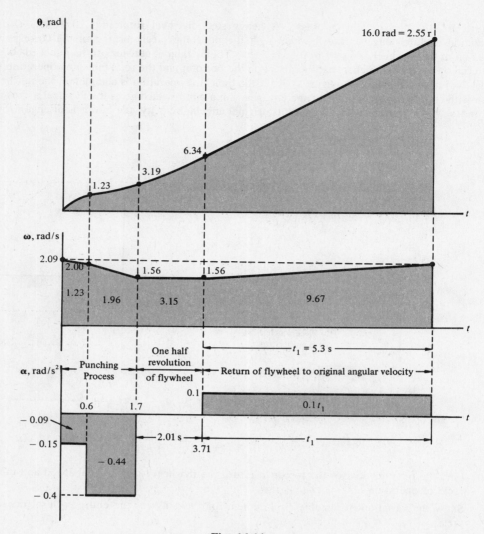

**Fig. 16.14c**

final step in the solution, the displacement diagram is sketched. These results are shown in Fig. 16.14c.

(c) The total angular displacement $\theta$ for one cycle of operation, from Fig. 16.14c, is

$$\theta = 2.55\,r$$

**16.15** The angular velocity–time diagram of a rigid body that rotates about a fixed axis is shown in Fig. 16.15a. All times are in seconds. At $t = 0$ the angular displacement is zero.

(a) Sketch the angular displacement and angular acceleration diagrams for the total time interval shown in the figure.

(b) Find the extreme value of the angular acceleration.

(c) Find the angular displacement of the body at the end of the time interval.

(d) Find the average values of the angular velocity and angular acceleration in the interval $0 \le t \le 4\,\text{s}$.

▌ (a) The plot of $\omega$ versus $t$ is redrawn in Fig. 16.15b. The slopes of the straight-line portions of this curve are equal to the magnitudes of the acceleration over the corresponding time interval. The areas under the $\omega - t$ diagram are next computed, and these values are shown in Fig. 16.15b. These values are the changes in the angular displacement over the corresponding time intervals. The final appearance of the $\theta - t$ curve is seen in Fig. 16.15b.

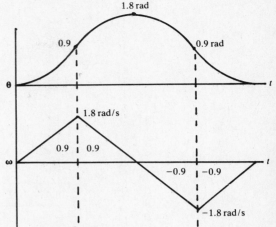

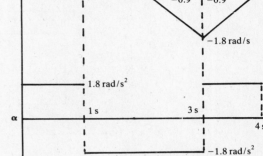

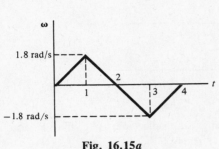

Fig. 16.15a                    Fig. 16.15b

(**b**)  The extreme values of the angular acceleration are

$$\alpha_{\max} = \pm 1.8 \, \text{rad/s}^2$$

(**c**)  At the end of the time interval, when  $t = 4 \, \text{s}$,

$$\theta = 0$$

(**d**)  The average values of the angular velocity and acceleration have the forms

$$\omega_{\text{avg}} = \frac{\theta_4 - \theta_0}{4 - 0} = \frac{0 - 0}{4 - 0} = 0 \qquad \alpha_{\text{avg}} = \frac{\omega_4 - \omega_0}{4 - 0} = \frac{0 - 0}{4 - 0} = 0$$

Both of the above average values are extremely poor representations of the actual values of $\omega$ and $\alpha$.

**16.16**  A rigid body rotates about a fixed axis with the angular acceleration–time diagram shown in Fig. 16.16a.  All times are in seconds, and the initial angular displacement is zero.

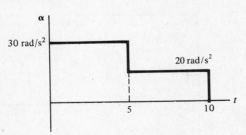

Fig. 16.16a

(**a**)  Sketch the angular velocity and angular displacement diagrams, and identify the extreme values of angular velocity and angular displacement, if the initial angular velocity is 80 rad/s.

(**b**)  Do the same as in part (**a**) if the initial velocity is −50 rad/s.

▌ (**a**)  Figure 16.16b shows the areas under the $\alpha - t$ diagram.  The values of these areas are used to find the changes in velocity over the time intervals.  The areas under the velocity diagrams are found as

$$A_1 = \frac{80 + 230}{2} (5) = 775 \qquad A_2 = \frac{230 + 330}{2} (10 - 5) = 1,400$$

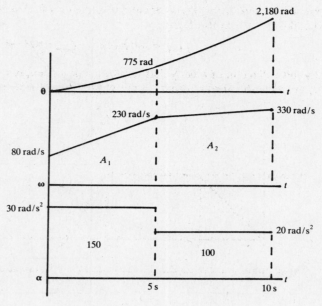

Fig. 16.16b

The values of these areas are used to find the changes in displacement over the time intervals, and the final results are shown in Fig. 16.16b. From this figure,

$$\omega_{\max} = 330 \text{ rad/s} \qquad \theta_{\max} = 2,180 \text{ rad}$$

(b) Figure 16.16c shows the $\omega - t$ curve for the initial velocity of $-50$ rad/s. Time $t_1$, when $\omega = 0$, is found by adding the area $30t_1$ in the $\alpha - t$ diagram to the initial value of $\omega$. The result is

$$30t_1 - 50 = 0 \qquad t_1 = 1.67 \text{ s}$$

The areas under the velocity diagram are found as

$$A_1 = \tfrac{1}{2}(1.67)(-50) = -41.8 \qquad A_2 = \tfrac{1}{2}(5 - 1.67)100 = 167 \qquad A_3 = \frac{100 + 200}{2}(10 - 5) = 750$$

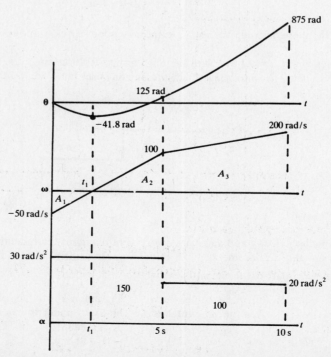

Fig. 16.16c

These areas are related to the displacement changes over the time intervals, and the final results are seen in Fig. 16.16c. From this figure,

$$\omega_{max} = 200 \text{ rad/s} \qquad \theta_{max} = 875 \text{ rad}$$

**16.17** The angular velocity–time diagram of a body that rotates about a fixed axis is shown in Fig. 16.17. Find the required value of $t_1$ at which the angular displacement of the body is 115° from the initial position.

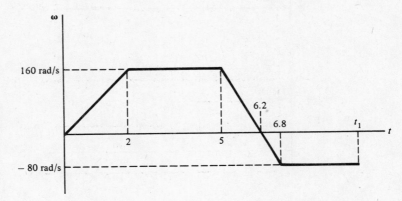

Fig. 16.17

▌ The total area under the $\omega - t$ diagram is equal to the change in the angular displacement over the corresponding time interval.

$$115° \left( \frac{\pi \text{ rad}}{180°} \right) = \tfrac{1}{2}(2)160 + 3(160) + \tfrac{1}{2}(1.2)160 - [\tfrac{1}{2}(0.6)80 + (t_1 - 6.8)80] \qquad t_1 = 15.7 \text{ s}$$

**16.18** A disk that rotates about a fixed axis has the angular acceleration diagram shown in Fig. 16.18a. The *magnitude* of the angular velocity may not exceed 4,000 r/min. Find the limiting values of the times $t_1$ and $t_2$, if $\omega = 0$ and $t = 0$.

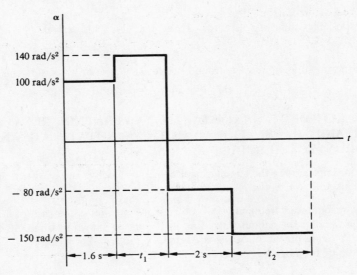

Fig. 16.18a

▌ The extreme value of angular velocity is

$$4,000 \text{ r/min} = 419 \text{ rad/s}$$

Using graphical methods, the $\omega - t$ curve is sketched, as shown in Fig. 16.18b. The maximum *positive* value of $\omega$ is given by

$$\omega = 140t_1 + 160$$

For $\omega_{max} = 419 \text{ rad/s}$,

$$140t_1 + 160 = 419 \qquad t_1 = 1.85 \text{ s}$$

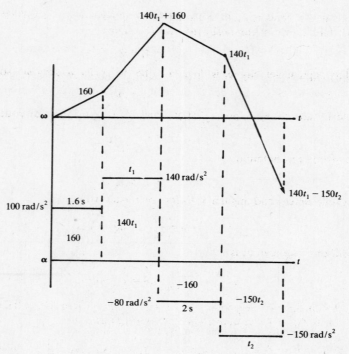

**Fig. 16.18*b***

The maximum *negative* value of $\omega$ is given by

$$\omega = 140t_1 - 150t_2$$

Using the extreme value $\omega = -419$ rad/s,

$$140t_1 - 150t_2 = -419 \qquad\qquad (1)$$

The minimum possible value of $t_1$ is zero. Using Eq. (1), with $t_1 = 0$,

$$-150t_2 = -419 \qquad t_2 = 2.79 \text{ s}$$

The maximum permissible value of $t_1$, found above, is $t_1 = 1.85$ s. Using this value in Eq. (1),

$$140(1.85) - 150t_2 = -419 \qquad t_2 = 4.52 \text{ s}$$

## 16.2 ANGULAR ROTATION WITH CONSTANT ACCELERATION, RELATIONSHIP BETWEEN ROTATIONAL AND TRANSLATIONAL MOTIONS, ROTATIONAL AND TRANSLATIONAL MOTION OF CONNECTED BODIES

16.19 Show the relationships among the angular displacement, velocity, and acceleration of a rigid body which rotates about a fixed axis with *constant* angular acceleration.

❚ A situation encountered frequently in problems in dynamics is the case of a rigid body which rotates, with *constant angular acceleration*, about a fixed axis. It will be shown in Chap. 18 that this case occurs when the *resultant moment* acting on the body has a *constant* value.

The magnitude of the angular acceleration is given by

$$\alpha = \frac{d\omega}{dt} = \text{constant} \qquad d\omega = \alpha\, dt$$

The integral form of this equation is

$$\int_{\omega_1}^{\omega_2} d\omega = \int_{t_1}^{t_2} \alpha\, dt$$

The numbers 1 and 2 define the endpoints of the time interval under consideration. Since $\alpha$ is a constant, it may be moved outside the integral sign. The result is

$$\int_{\omega_1}^{\omega_2} d\omega = \alpha \int_{t_1}^{t_2} dt \qquad \omega\Big|_{\omega_1}^{\omega_2} = \alpha t\Big|_{t_1}^{t_2} \qquad \omega_2 - \omega_1 = \alpha(t_2 - t_1)$$

The above equation is a general relationship among angular velocity, acceleration, and times $t_1$ and $t_2$. $t_1$ is defined to be the initial time of the problem, with the value

$$t_1 = 0$$

The corresponding angular velocity $\omega_1$ is defined to be the *initial velocity* $\omega_0$, so that

$$\omega_1 = \omega_0$$

The angular velocity and time, $\omega_2$ and $t_2$, at the end of the time interval under consideration are written as

$$\omega_2 = \omega \qquad t_2 = t$$

With the above changes in notation,

$$\omega = \omega_0 + \alpha t \tag{1}$$

The angular displacement and angular velocity are related by

$$\omega = \frac{d\theta}{dt} \qquad d\theta = \omega \, dt$$

The integral form of this equation is

$$\int_{\theta_1}^{\theta_2} d\theta = \int_{t_1}^{t_2} \omega \, dt$$

The angular displacement $\theta$ corresponding to $t_1 = 0$ is designated the initial displacement $\theta_0$. Using $\theta_2 = \theta$ and $t_2 = t$, and Eq. (1), the above equation appears as

$$\int_{\theta_0}^{\theta} d\theta = \int_0^t \omega \, dt = \int_0^t (\omega_0 + \alpha t) \, dt \qquad \theta \Big|_{\theta_0}^{\theta} = (\omega_0 t + \tfrac{1}{2}\alpha t^2) \Big|_0^t \qquad \theta - \theta_0 = \omega_0 t + \tfrac{1}{2}\alpha t^2$$

$$\theta = \theta_0 + \omega_0 t + \tfrac{1}{2}\alpha t^2 \tag{2}$$

If $t$ is eliminated between Eqs. (1) and (2), the result is

$$\omega^2 = \omega_0^2 + 2\alpha(\theta - \theta_0) \tag{3}$$

Equations (1), (2), and (3) are the three basic equations that describe the rotation of a rigid body with *constant angular acceleration* about a fixed axis. These equations are repeated here:

$$\omega = \omega_0 + \alpha t \qquad \theta = \theta_0 + \omega_0 t + \tfrac{1}{2}\alpha t^2 \qquad \omega^2 = \omega_0^2 + 2\alpha(\theta - \theta_0)$$

These above equations are identical in structure to the equations in part (c) of Prob. 14.31. These latter equations were used to describe the problem of rectilinear translation of a particle with constant acceleration.

**16.20** A machinist sharpens a chisel against a grinding wheel, as shown in Fig. 16.20. The chisel is held against the wheel for 5 s. During this time the speed of the wheel changes from 3,600 r/min to 3,450 r/min.

(a) Find the value of the angular deceleration of the grinding wheel, if this quantity is assumed to be constant.

(b) Find the angular displacement of the grinding wheel during the time interval.

(c) When the chisel is removed, the motor accelerates the grinding wheel at $65\ \text{rad/s}^2$. How long does it take for the wheel to return to its original angular velocity?

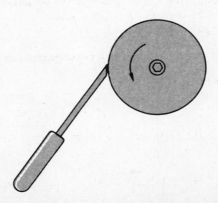

**Fig. 16.20**

▎ (a)  The initial and final magnitudes of the angular velocity are

$$3{,}600\ \text{r/min} = 377\ \text{rad/s} \qquad 3{,}450\ \text{r/min} = 361\ \text{rad/s}$$

The angular deceleration is then found from

$$\omega = \omega_0 + \alpha t \qquad 361 = 377 + \alpha(5) \qquad \alpha = -3.2\ \text{rad/s}^2$$

(b)  The total angular displacement during the 5-s interval is given by

$$\theta = \theta_0 + \omega_0 t + \tfrac{1}{2}\alpha t^2 = 377(5) + \tfrac{1}{2}(-3.2)5^2 = 1{,}850\ \text{rad} = 294\ \text{r}$$

(c)  The time for the wheel to return to its original speed is found from

$$\omega = \omega_0 + \alpha t \qquad 377 = 361 + 65t \qquad t = 0.246\ \text{s}$$

**16.21** (a)  How long after it is turned on does it take the grinding wheel in Prob. 16.20 to reach its rated speed of 3,600 r/min? Assume a constant value of acceleration of 65 rad/s².

(b)  Through how many revolutions does the grinding wheel turn in going from rest to its rated speed?

(c)  When the grinding wheel is turned off, it is observed that it takes 140 s for the wheel to come to rest from its rated speed of 3,600 r/min. Find the value of the deceleration of the wheel, assuming this quantity to be constant.

(d)  Through how many revolutions does the grinding wheel turn in coming to rest from its rated speed?

▎ (a)  Using $\alpha = 65\ \text{rad/s}^2$, with $\omega_0 = 0$,

$$\omega = \omega_0 + \alpha t \qquad 377 = 65t \qquad t = 5.8\ \text{s}$$

(b)  The number of revolutions it takes the grinding wheel to reach its rated speed, with $\theta_0 = 0$ and $\omega_0 = 0$, is given by

$$\theta = \theta_0 + \omega_0 t + \tfrac{1}{2}\alpha t^2 = \tfrac{1}{2}(65)5.8^2 = 1{,}090\ \text{rad} = 173\ \text{r}$$

(c)  The angular deceleration is found from

$$\omega = \omega_0 + \alpha t \qquad 0 = 377 + \alpha(140) \qquad \alpha = -2.69\ \text{rad/s}^2$$

(d)  The number of revolutions the grinding wheel turns before coming to rest is found from

$$\omega^2 = \omega_0^2 + 2\alpha(\theta - \theta_0)$$

With $\theta_0 = 0$,

$$0 = 377^2 + 2(-2.69)\theta \qquad \theta = 2.64 \times 10^4\ \text{rad} = 4{,}200\ \text{r}$$

**16.22** Figure 16.22 shows a radial-arm wood saw which has an operating speed of 1,500 r/min. When the saw is turned off, a magnetic brake brings the blade to rest in 1 min.

(a)  If the braking deceleration is assumed to be constant, through how many revolutions does the blade rotate before coming to rest?

(b)  In a certain operation the blade jams in a piece of material and comes to rest in $\tfrac{3}{4}$ r. Find the average values of the deceleration and the corresponding time needed for the blade to come to rest.

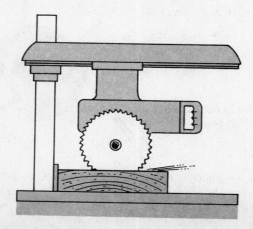

**Fig. 16.22**

▌ (a) The initial speed of the saw blade is

$$\omega_0 = 1{,}500 \ \frac{r}{min} \left( \frac{2\pi \ rad}{1 \ r} \right) \left( \frac{1 \ min}{60 \ s} \right) = 157 \ \frac{rad}{s}$$

The saw blade comes to rest in 1 min, so that

$$\omega = 0 = \omega_0 + \alpha t \qquad 0 = 157 + \alpha(60) \qquad \alpha = -2.62 \ rad/s^2$$

The total number of revolutions of the saw blade is found from

$$\omega^2 = \omega_0^2 + 2\alpha(\theta - \theta_0) \qquad 0 = 157^2 + 2(-2.62)(\theta - 0) \qquad \theta = 4{,}700 \ rad = 748 \ r$$

This result may be checked by using

$$\theta = \theta_0 + \omega_0 t + \tfrac{1}{2}\alpha t^2 = 0 + 157(60) + \tfrac{1}{2}(-2.62)(60^2) = 4{,}740 \ rad = 754 \ r \approx 748 \ r$$

(b) For the case where the saw blade jams, this element comes to rest in $0.75 \ r = 4.71 \ rad$, and the average value of the deceleration is found from

$$\omega^2 = \omega_0^2 + 2\alpha(\theta - \theta_0) \qquad 0 = 157^2 + 2\alpha(4.71 - 0) \qquad \alpha = -2620 \ rad/s^2$$

The corresponding time is found from

$$\omega = \omega_0 + \alpha t \qquad 0 = 157 + (-2620)t \qquad t = 0.0599 \ s = 59.9 \ ms$$

**16.23** In the manufacture of constant-angular-acceleration electric motors for a special application, a requirement is that the angular acceleration must not deviate more than 1 percent from a nominal design value of $\alpha = 80 \ rad/s^2$. Each motor coming off the assembly line is individually tested to meet this requirement.

(a) If each unit is started from rest and run for exactly 5 s, find the acceptable range of angular velocity, in revolutions per minute, that the motors should have at the end of this time interval.

(b) Find the corresponding range of total angular displacement during this time interval.

▌ (a) The maximum value of the angular acceleration is

$$\alpha_{max} = 80 + 1\%(80) = 80.8 \ rad/s^2$$

The maximum allowable value of the angular velocity at $t = 5 \ s$, with $\omega_0 = 0$, is then found from

$$\omega = \omega_0 + \alpha t \qquad \omega_{max} = 80.8(5) = 404 \ rad/s = 3{,}860 \ r/min$$

The minimum value of the angular acceleration is

$$\alpha_{min} = 80 - 1\%(80) = 79.2 \ rad/s^2$$

The minimum allowable value of the angular velocity at $t = 5 \ s$, with $\omega_0 = 0$, is given by

$$\omega = \omega_0 + \alpha t \qquad \omega_{min} = 79.2(5) = 396 \ rad/s = 3{,}780 \ r/min$$

The acceptable range of angular velocity for which the acceleration specification will be satisfied is

$$3{,}780 \ r/min \le \omega \le 3{,}860 \ r/min$$

(b) Using $\alpha_{max} = 80.8 \ rad/s^2$ and $\omega_{max} = 404 \ rad/s$,

$$\omega^2 = \omega_0^2 + 2\alpha(\theta - \theta_0) \qquad 404^2 = 2(80.8)\theta \qquad \theta = 1{,}010 \ rad = 161 \ r$$

Using $\alpha_{min} = 79.2 \ rad/s^2$ and $\omega_{min} = 396 \ rad/s$,

$$\omega^2 = \omega_0^2 + 2\alpha(\theta - \theta_0) \qquad 396^2 = 2(79.2)\theta \qquad \theta = 990 \ rad = 158 \ r$$

The acceptable range of angular displacement is

$$158 \ r \le \theta \le 161 \ r$$

**16.24** (a) What is the relationship between the angular velocity of a link which rotates about a fixed axis and the translational velocity of points on the link?

(b) What is the relationship between the angular velocity and angular acceleration, and the translational acceleration of points on the link of part (a)?

▌ (a) Figure 16.24a shows a link of length $r$ which rotates about a fixed axis. Point $a$ on the tip of the link may be considered to be a particle in curvilinear translation along the circular path. The relationships between the translational velocity of point $a$ and the angular velocity of the link will be developed now.

The velocity of point $a$ must be tangent to the path of motion. Thus, the direction of the velocity of point $a$ is normal to the link. The magnitude of this velocity is given by

$$v = \frac{ds}{dt} \qquad (1)$$

where $s$ is the coordinate of length along the curve. Point $a$ in Fig. 16.24a is assumed to move to position $a'$ as the link moves through the angle $\Delta\theta$ in time $\Delta t$. The arc length $\Delta s$ from $a$ to $a'$ is

$$\Delta s = r\,\Delta\theta$$

Both sides of this equation are divided by $\Delta t$, with the result

$$\frac{\Delta s}{\Delta t} = r\frac{\Delta\theta}{\Delta t}$$

$\Delta t$ is now allowed to approach zero and, in the limit,

$$\lim_{\Delta t \to 0}\frac{\Delta s}{\Delta t} = \lim_{\Delta t \to 0} r\frac{\Delta\theta}{\Delta t} \qquad \frac{ds}{dt} = r\frac{d\theta}{dt} = r\dot{\theta} = r\omega \qquad (2)$$

where $\omega$ is the angular velocity. Equations (1) and (2) are combined to obtain the final result

$$v = r\omega \qquad (3)$$

*Equation (3) is a fundamental relationship between the translational velocity of a point on a rigid link and the angular velocity of the link.* This equation may be plotted along the link, and the result is shown in Fig. 16.24b. This figure shows the linear variation of velocity along the length of the link.

(b) The normal component of acceleration of point $a$ has the form

$$a_n = \frac{v^2}{\rho}$$

The direction of $a_n$ is along the axis of the link, and the sense is from point $a$ toward the center of rotation. By using $v = r\omega$ and $\rho = r$, this equation appears as

$$a_n = \frac{(r\omega)^2}{r} = r\omega^2$$

The direction of the tangential component $a_t$ of the acceleration of point $a$ is along the path of motion. The magnitude of this term is equal to the first time derivative of the speed of the particle. Using Eq. (3), we get

$$a_t = \frac{dv}{dt} = r\frac{d\omega}{dt} = r\alpha$$

where $\alpha$ is the angular acceleration. The resultant acceleration of point $a$ is

$$a = \sqrt{a_n^2 + a_t^2}$$

It may be observed that the acceleration of point $a$ is a function of both the angular acceleration of the link *and* the angular velocity of the link.

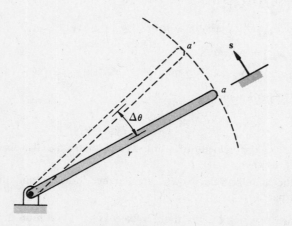

Fig. 16.24a

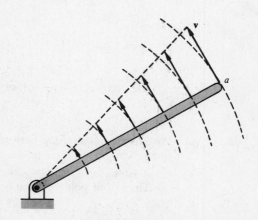

Fig. 16.24b

**16.25**  The saw blade of Prob. 16.22 is 10 in in diameter.  Point $a$ is located on the rim of the blade.  Find the magnitude of the velocity and acceleration of point $a$,

    (*a*)  When the blade rotates at 1,500 r/min

    (*b*)  At the instant that the magnetic braking starts

    (*c*)  At the instant that the blade jams, as described in part (*b*) of Prob. 16.22.

  **▮** (*a*)  When the blade rotates freely at 1,500 r/min,

$$v = r\omega = 5(157) = 785 \text{ in/s} = 65.4 \text{ ft/s}$$
$$a_n = r\omega^2 = 5(157^2) = 1.23 \times 10^5 \text{ in/s}^2 = 1.03 \times 10^4 \text{ ft/s} = 320g\text{'s} \qquad a_t = 0$$

    (*b*)  At the instant that the magnetic braking starts,

$$a_t = r\alpha = 5(-2.62) = -13.1 \text{ in/s}^2 = -1.09 \text{ ft/s}^2$$

$v$ and $a_n$ have the same values as in part (*a*).  It may be seen that $a_t \ll a_n$.

    (*c*)  When the saw blade jams,

$$a_t = r\alpha = 5(-2,620) = -13,100 \text{ in/s}^2 = 1,090 \text{ ft/s}^2 = 33.9g\text{'s}$$

$v$ and $a_n$ again have the same values as in part $a$.  The magnitude of the total acceleration is

$$a = \sqrt{a_n^2 + a_t^2} = \sqrt{(1.03 \times 10^4)^2 + 1,090^2} = 1.04 \times 10^4 \text{ ft/s}^2 = 323g\text{'s}$$

It may be seen that, even for the case of jamming of the saw blade, the normal acceleration is the major component of the resultant acceleration.

**16.26**  The link shown in Fig. 16.26*a* rotates about a fixed point with angular velocity $\omega = 6.5$ rad/s.

    (*a*)  For the position shown in the figure, find the magnitude, direction, and sense of the velocity and acceleration of points $a$ and $b$.

    (*b*)  Express the velocity and acceleration at point $a$ in formal vector notation.

  **▮** (*a*)  The velocities of points $a$ and $b$ are given by

$$v_a = r\omega = 0.540(6.5) = 3.51 \text{ m/s} \qquad v_b = r\omega = 0.225(6.5) = 1.46 \text{ m/s}$$

The centripetal, or normal, accelerations of points $a$ and $b$ are

$$a_a = a_{an} = \frac{v^2}{\rho} = \frac{v_a^2}{0.540} = \frac{3.51^2}{0.540} = 22.8 \text{ m/s}^2$$

$$a_b = a_{bn} = \frac{v^2}{\rho} = \frac{v_b^2}{0.225} = \frac{1.46^2}{0.225} = 9.47 \text{ m/s}^2$$

Figure 16.26*b* shows the directions and senses of the velocity and acceleration of points $a$ and $b$.

    (*b*)  Using Fig. 16.26*b*,

$$\mathbf{v}_a = (-v_a \cos 52°)\mathbf{i} + (v_a \sin 52°)\mathbf{j} = (-3.51 \cos 52°)\mathbf{i} + (3.51 \sin 52°)\mathbf{j} = -2.16\mathbf{i} + 2.77\mathbf{j} \qquad \text{m/s}$$
$$\mathbf{a}_a = (a_a \cos 38°)\mathbf{i} + (a_a \sin 38°)\mathbf{j} = (22.8 \cos 38°)\mathbf{i} + (22.8 \sin 38°)\mathbf{j} = 18.0\mathbf{i} + 14.0\mathbf{j} \qquad \text{m/s}^2$$

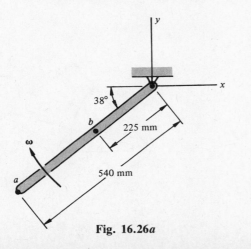

**Fig. 16.26*a***

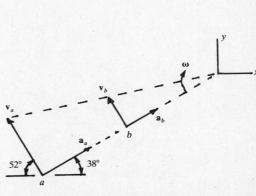

**Fig. 16.26*b***

**16.27** Do the same as in Prob. 16.26 if, at the position shown in Fig. 16.26a, the link has a counterclockwise acceleration of 10 rad/s².

▌ (a) All velocities and normal accelerations are the same as in Prob. 16.26, since these quantities are functions only of the magnitude of the angular velocity. Figure 16.27a shows the directions and senses of the acceleration components.

The tangential acceleration at $a$ is found as

$$a_{at} = r\alpha = 0.540(10) = 5.40 \text{ m/s}^2$$

The resultant acceleration at $a$ has the form

$$a_a = \sqrt{a_{an}^2 + a_{at}^2} = \sqrt{22.8^2 + 5.40^2} = 23.4 \text{ m/s}^2$$

This result is shown in Fig. 16.27b. The direction $\theta_2$ of $a_a$ is found from

$$\tan \theta_1 = \frac{5.40}{22.8} \qquad \theta_1 = 13.3° \qquad \theta_2 = 38° - \theta_1 = 24.7°$$

The tangential and resultant accelerations at $b$ are

$$a_{bt} = r\alpha = 0.225(10) = 2.25 \text{ m/s}^2 \qquad a_b = \sqrt{a_{bn}^2 + a_{bt}^2} = \sqrt{9.47^2 + 2.25^2} = 9.73 \text{ m/s}^2$$

The above result is seen in Fig. 16.27c. The direction $\theta_4$ of $a_b$ is found from

$$\tan \theta_3 = \frac{2.25}{9.47} \qquad \theta_3 = 13.4° \qquad \theta_4 = 38° - \theta_3 = 24.6°$$

(b) The velocity at point $a$ has the same form as in part (b) of Prob. 16.26. The acceleration at point $a$, using Fig. 16.27b, may be written in formal vector notation as

$$\mathbf{a}_a = (a_a \cos \theta_2)\mathbf{i} + (a_a \sin \theta_2)\mathbf{j} = (23.4 \cos 24.7°)\mathbf{i} + (23.4 \sin 24.7°)\mathbf{j} = 21.3\mathbf{i} + 9.78\mathbf{j} \qquad \text{m/s}^2$$

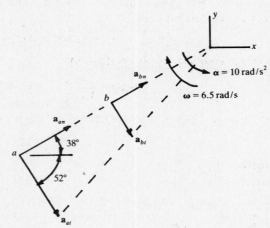

Fig. 16.27a

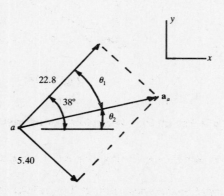

Fig. 16.27b

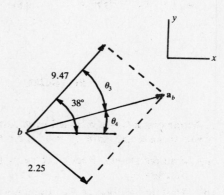

Fig. 16.27c

**16.28** At the instant shown in Fig. 16.28a, the magnitude and direction of the acceleration of end *a* of the rigid rod have the known values $a = 860 \text{ in/s}^2$ and $\theta = 67°$. The sense of the acceleration is as shown in the figure.

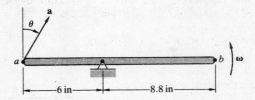

Fig. 16.28a

(*a*) Find the velocity of ends *a* and *b*, and the angular velocity, of the rod.

(*b*) Find the angular acceleration, and the acceleration of end *b*, of the rod.

(*c*) Is the rod speeding up or slowing down?

▌ (*a*) Using Fig. 16.28b,

$$a_{an} = 860 \sin 67° = \frac{v_a^2}{\rho} = \frac{va^2}{6} \qquad v_a = 68.9 \text{ in/s} \qquad v_a = r\omega \qquad 68.9 = 6\omega \qquad \omega = 11.5 \text{ rad/s}$$

$$v_b = r\omega = 8.8(11.5) = 101 \text{ in/s}$$

(*b*) From Fig. 16.28b,

$$a_{at} = 860 \cos 67° = r\alpha = 6\alpha \qquad \alpha = 56.0 \text{ rad/s}^2$$

The components of acceleration of end *b* are

$$a_{bn} = \frac{v_b^2}{\rho} = \frac{101^2}{8.8} = 1{,}160 \text{ in/s}^2 \qquad a_{bt} = r\alpha = 8.8(56.0) = 493 \text{ in/s}^2$$

The acceleration of end *b* is found as

$$a_b = \sqrt{a_{bn}^2 + a_{bt}^2} = \sqrt{1{,}160^2 + 493^2} = 1{,}260 \text{ in/s}^2$$

The direction of $a_b$, using Fig. 16.28c, is given by

$$\tan \theta = \frac{493}{1{,}160} \qquad \theta = 23.0°$$

(*c*) The rod is slowing down, since the sense of the angular acceleration is opposite the sense of the angular velocity.

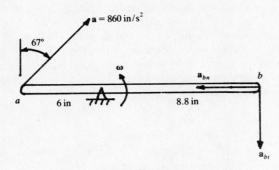

Fig. 16.28b

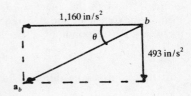

Fig. 16.28c

**16.29** A simple pendulum is released from rest at position *a* in Fig. 16.29. The velocity of the pendulum bob can be shown to have the form $v = \sqrt{2gl} \cos \theta$.

(*a*) Find the angular velocity of the pendulum and the normal acceleration of the bob when the pendulum is in position *b*.

(*b*) Find the position of the pendulum when the normal acceleration of the bob is half the value found in part *a*.

▌ (a)  At position $b$,  $\theta = 0$.  The velocity at point $b$ is

$$v = \sqrt{2gl \cos \theta} = \sqrt{2(9.81)1.3(1)} = 5.05 \text{ m/s}$$

The angular velocity of the pendulum at position $b$ is given by

$$\omega = \frac{v}{r} = \frac{5.05}{1.3} = 3.88 \text{ rad/s}$$

The normal acceleration of the bob at position $b$ is

$$a_n = \frac{v^2}{\rho} = \frac{5.05^2}{1.3} = 19.6 \text{ m/s}^2$$

(b)  Using  $v = \sqrt{2gl \cos \theta}$,

$$a_n = \frac{1}{2}(19.6) = \frac{v^2}{\rho} = \frac{2gl \cos \theta}{\rho} \qquad 9.8 = \frac{2(9.81)1.3 \cos \theta}{1.3} \qquad \theta = 60°$$

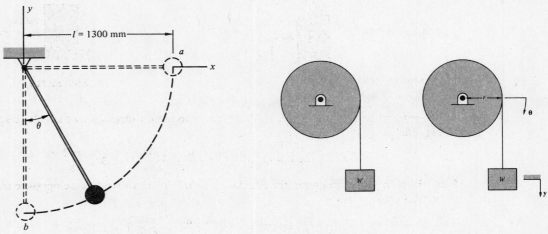

Fig. 16.29                     Fig. 16.30$a$                     Fig. 16.30$b$

**16.30**  What is the relationship between the translational and rotational motions of connected bodies?

▌ Figure 16.30$a$ shows a pulley hinged to the ground.  Around the pulley is wrapped an inextensible string which is connected to a weight.  It will now be shown that a unique relationship exists between the motions of the pulley and the weight.

   The positive senses of the displacement coordinates of the two bodies are shown in Fig. 16.30$b$.  The angular displacement of the pulley is $\theta$, and $y$ is the rectilinear displacement of the weight.  The pulley is now imagined to rotate through the angle $\theta$.  During this rotation, a length $r\theta$ of string is unwound from the pulley.  The string is inextensible, and the force in the string is assumed to be always a tensile force.  The displacement $y$ of the weight corresponding to the rotation of the pulley is then

$$y = r\theta$$

This equation may be differentiated twice with respect to time, with the results

$$\frac{dy}{dt} = r\frac{d\theta}{dt} \qquad v = r\omega \qquad \frac{d^2y}{dt^2} = r\frac{d^2\theta}{dt^2} \qquad a = r\alpha$$

$\omega$ and $\alpha$ are the angular velocity and acceleration, and $v$ and $a$ are the corresponding values of the translational velocity and acceleration.  The above two equations are very useful relationships among the displacement, velocity, and acceleration of bodies connected in the manner shown in Fig. 16.30$a$.

**16.31**  At a certain instant the pulley in Fig. 16.31$a$ has a counterclockwise angular velocity of 300 r/min and is decelerating at the rate of 8,600 r/min².

(a)  Express the angular velocity and acceleration in radian and second units.

(b)  Find the velocity and acceleration of the weights $A$ and $B$ in feet and second units.

(c)  Find the relative velocity and acceleration of weight $A$ with respect to weight $B$.

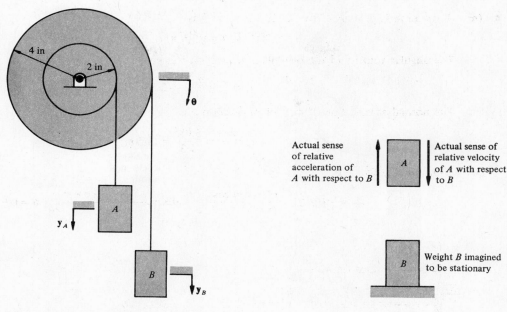

**Fig. 16.31*a***                                                    **Fig. 16.31*b***

▮ (*a*)   The angular velocity of the pulley, with reference to the positive sense of angular displacement shown in Fig. 16.31*a*, is

$$\omega = -300 \; \frac{r}{\min} \left( \frac{2\pi \, \text{rad}}{1 \, r} \right) \left( \frac{1 \, \min}{60 \, s} \right) = -31.4 \; \frac{\text{rad}}{s}$$

Since the pulley is decelerating, the angular acceleration must have a sense opposite that of the angular velocity, so that

$$\alpha = 8{,}600 \; \frac{r}{\min^2} \left( \frac{2\pi \, \text{rad}}{1 \, r} \right) \left( \frac{1 \, \min}{60 \, s} \right)^2 = 15.0 \; \frac{\text{rad}}{s^2}$$

(*b*)   The two weights move as particles in rectilinear translation. The velocity and acceleration for block *A* are

$$v_A = r_A \omega = -2 \, \text{in} \left( 31.4 \; \frac{\text{rad}}{s} \right) \left( \frac{1 \, \text{ft}}{12 \, \text{in}} \right) = -5.23 \; \frac{\text{ft}}{s}$$

$$a_A = r_A \alpha = 2 \, \text{in} \left( 15.0 \; \frac{\text{rad}}{s^2} \right) \left( \frac{1 \, \text{ft}}{12 \, \text{in}} \right) = 2.5 \; \frac{\text{ft}}{s^2}$$

For block *B*,

$$v_B = r_B \omega = -4(31.4)(\tfrac{1}{12}) = -10.5 \, \text{ft/s} \qquad a_B = r_B \alpha = 4(15.0)(\tfrac{1}{12}) = 5 \, \text{ft/s}^2$$

All of the above values are referenced to the positive senses of $y_A$ and $y_B$ shown in Fig. 16.31*a*.

(*c*)   The relative velocity and acceleration, $v_{AB}$ and $a_{AB}$, of weight *A* with respect to weight *B* are given by

$$v_{AB} = v_A - v_B = -5.23 - (-10.5) = 5.27 \, \text{ft/s} \qquad a_{AB} = a_A - a_B = 2.5 - 5 = -2.5 \, \text{ft/s}^2$$

$v_{AB}$ and $a_{AB}$ are both positive in the positive senses of $v_A$ and $a_A$. $v_A$ and $a_A$ are positive in the positive sense of the $y_A$ axis in Fig. 16.31*a*. To an observer positioned on weight *B*, weight *A* would appear to be slowing down as it moved toward her or him. These conclusions are shown in Fig. 16.31*b*.

**16.32**   Do the same as in parts (*b*) and (*c*) of Prob. 16.31, if weight *A* is supported from the left side of the inner pulley.

▮ The system is shown in Fig. 16.32*a*. The velocity and acceleration of weight *B* are the same as before, with the values

$$v_B = -10.5 \, \text{ft/s} \qquad a_B = 5 \, \text{ft/s}^2$$

For weight *A*,

$$v_A = r_A \omega = 2(31.4)(\tfrac{1}{12}) = 5.23 \, \text{ft/s} \qquad a_A = -(r_A \alpha) = -2(15)(\tfrac{1}{12}) = -2.5 \, \text{ft/s}^2$$

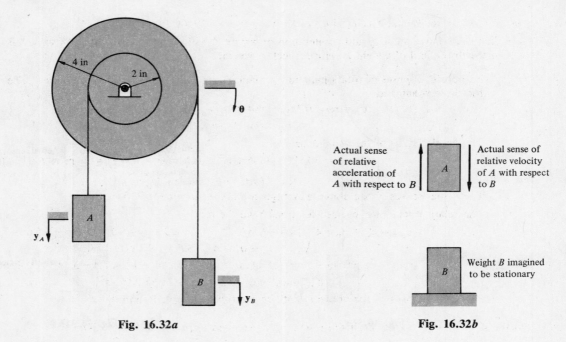

Fig. 16.32a

Fig. 16.32b

The relative velocity and acceleration of $A$ with respect to $B$ are

$$v_{AB} = v_A - v_B = 5.23 - (-10.5) = 15.7 \text{ ft/s} \qquad a_{AB} = a_A - a_B = -2.5 - 5 = -7.5 \text{ ft/s}^2$$

The senses of the relative velocity and acceleration, shown in Fig. 16.32b, are the same as in Prob. 16.31. The magnitudes of both these quantities, however, are greater than those in that problem.

**16.33** At the instant shown in Fig. 16.33a, the stepped pulley rotates clockwise at 25 rad/s, with an angular acceleration of 6 rad/s².

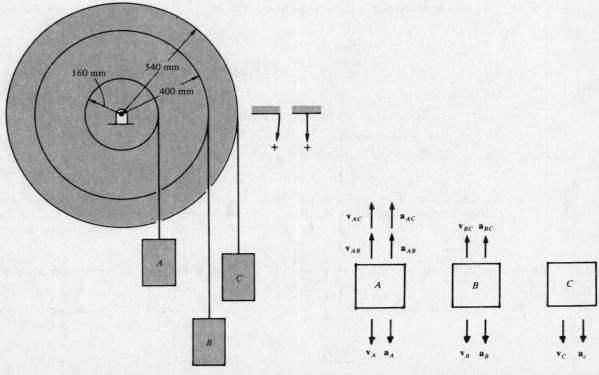

Fig. 16.33a

Fig. 16.33b

(*a*) Find the velocity and acceleration of the weights *A*, *B*, and *C*.

(*b*) Find the relative velocity and acceleration of weight *A* with respect to weight *B*, of weight *B* with respect to weight *C*, and of weight *A* with respect to weight *C*.

▮ (*a*) The positive senses of rotational and translational motion are shown in Fig. 16.33*a*. The velocities of the three weights are

$$v_A = r\omega = 0.16(25) = 4\ \text{m/s} \qquad v_B = r\omega = 0.4(25) = 10\ \text{m/s}$$
$$v_C = r\omega = 0.54(25) = 13.5\ \text{m/s}$$

The accelerations of the weights are

$$a_A = r\alpha = 0.16(6) = 0.96\ \text{m/s}^2 \qquad a_B = r\alpha = 0.4(6) = 2.4\ \text{m/s}^2$$
$$a_C = r\alpha = 0.54(6) = 3.24\ \text{m/s}^2$$

The actual senses of the above quantities are shown in Fig. 16.33*b*.

(*b*) The relative velocities and accelerations have the forms

$$v_{AB} = v_A - v_B = 4 - 10 = -6\ \text{m/s} \qquad v_{BC} = v_B - v_C = 10 - 13.5 = -3.5\ \text{m/s}$$
$$v_{AC} = v_A - v_C = 4 - 13.5 = -9.5\ \text{m/s}$$
$$a_{AB} = a_A - a_B = 0.96 - 2.4 = -1.44\ \text{m/s}^2 \qquad a_{BC} = a_B - a_C = 2.4 - 3.24 = -0.84\ \text{m/s}^2$$
$$a_{AC} = a_A - a_C = 0.96 - 3.24 = -2.28\ \text{m/s}^2$$

The actual senses of the above quantities are shown in Fig. 16.33*b*.

**16.34** Do the same as in Prob. 16.33, if the weights are arranged as shown in Fig. 16.34*a*.

▮ (*a*) The positive senses of rotational and translational motion are the same as those shown in Fig. 16.33*a*. The velocities and accelerations of the weights are

$$v_A = r\omega = -0.16(25) = -4\ \text{m/s} \qquad v_B = r\omega = 0.4(25) = 10\ \text{m/s}$$
$$v_C = r\omega = 0.54(25) = 13.5\ \text{m/s}$$
$$a_A = r\alpha = -0.16(6) = -0.96\ \text{m/s}^2 \qquad a_B = r\alpha = 0.4(6) = 2.4\ \text{m/s}^2$$
$$a_C = r\alpha = 0.54(6) = 3.24\ \text{m/s}^2$$

The actual senses of the above quantities are shown in Fig. 16.34*b*.

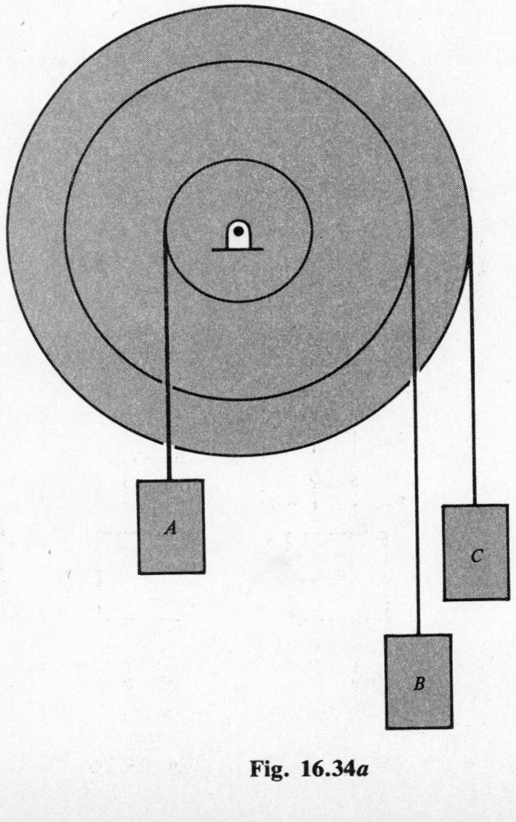


**Fig. 16.34*a***

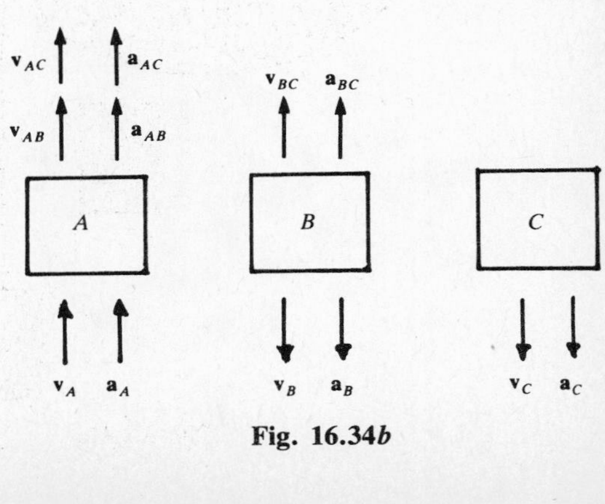


**Fig. 16.34*b***

(b) The relative velocities and accelerations are

$$v_{AB} = v_A - v_B = -4 - 10 = -14 \text{ m/s} \qquad v_{BC} = v_B - v_C = 10 - 13.5 = -3.5 \text{ m/s}$$
$$v_{AC} = v_A - v_C = -4 - 13.5 = -17.5 \text{ m/s}$$
$$a_{AB} = a_A - a_B = -0.96 - 2.4 = -3.36 \text{ m/s}^2 \qquad a_{BC} = a_B - a_C = 2.4 - 3.24 = -0.84 \text{ m/s}^2$$
$$a_{AC} = a_A - a_C = -0.96 - 3.24 = -4.2 \text{ m/s}^2$$

The actual senses of the relative velocities and accelerations are shown in Fig. 16.34b.

**16.35** A thin, inextensible string is wrapped around a cylinder, as shown in Fig. 16.35a. At $t = 2.8$ s the motion of the string is to the right, with a velocity of 2 m/s and a deceleration of 0.5 m/s². At this instant, point $a$ on the string is coincident with point $b$ on the cylinder.

(a) Find the velocity of points $a$ and $b$, when $t = 3.5$ s, and the position of point $b$.

(b) Find the acceleration of points $a$ and $b$ when $t = 3.5$ s.

▌ (a) Time $t = 0$ is defined to be the time when points $a$ and $b$ are coincident. The motion of point $a$ at this time is described by

$$v_{0a} = 2 \text{ m/s} \qquad a_a = -0.5 \text{ m/s}^2$$

At $t = 3.5 - 2.8 = 0.7$ s,

$$v_a = v_{0a} + a_a t = 2 + (-0.5)0.7 = 1.65 \text{ m/s}$$

The corresponding value of the speed $v_b$ of point $b$ is given by

$$v_b = v_a = 1.65 \text{ m/s}$$

The displacement of the string during the 0.7-s time interval is found next, and four significant figures are used to avoid loss of accuracy. With $s_0 = 0$,

$$s = s_0 + v_0 t + \tfrac{1}{2} a t^2 = 2(0.7) + \tfrac{1}{2}(-0.5)0.7^2 = 1.278 \text{ m}$$

The corresponding angle of rotation of the cylinder, with clockwise displacement considered to be positive, is found from

$$s = r\theta \qquad 1.278 = 0.080\theta \qquad \theta = 15.98 \text{ rad} = 916°$$

The *net* clockwise rotation of the wheel from the position shown in Fig. 16.35a is given by

$$\theta = 916° - 2(360°) = 196°$$

(b) The normal and tangential components of acceleration of point $b$, corresponding to the above angle of rotation, are

$$a_{bn} = \frac{v_b^2}{\rho} = \frac{1.65^2}{0.080} = 34.0 \text{ m/s}^2 \qquad a_{bt} = 0.5 \text{ m/s}^2$$

The two components of acceleration are shown in their actual senses in Fig. 16.35b. The resultant acceleration at point $b$ is

$$a_b = \sqrt{a_{bn}^2 + a_{bt}^2} = \sqrt{34.0^2 + 0.5^2} = 34.0 \text{ m/s}^2$$

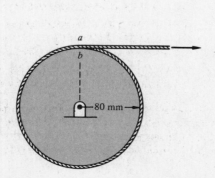

**Fig. 16.35a**

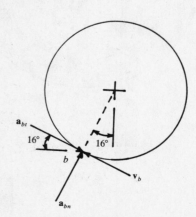

**Fig. 16.35b**

## 16.3 GENERAL PLANE MOTION OF A BODY, INSTANT CENTER OF ROTATION

**16.36** Show how the most general type of plane motion of a rigid body may be expressed as a combination of translational and rotational motions.

▮ The body in Fig. 16.36 is initially in position $A$. This position may be described by specifying the coordinates $x_A$ and $y_A$ of an arbitrary point $a$ on the body, as well as the angular displacement $\theta_A$ of the body with respect to an arbitrary reference line.

At a later time, the body is in position $B$. The total motion which is required for the body to move from the initial position $A$ to the final position $B$ may be envisioned to be the sum of the following two distinct motions:

1. *Translation* of the body until point $a$ has the position defined by $x = x_B$ and $y = y_B$
2. *Rotation* of the body, about the final position of point $a$, until the angle $\theta_B$ is obtained

The description of the initial position of the body is given by $x_A$, $y_A$, and $\theta_A$, and the final position is defined by $x_B$, $y_B$, and $\theta_B$. The *change* in the position may be found from the general equation which defines change, given by

$$\text{Change in condition} = (\text{final condition}) - (\text{initial condition})$$

The change in position of the body in Fig. 16.3 is found to be

$$\Delta x = x_B - x_A$$
$$\Delta y = y_B - y_A \qquad \Delta\theta = \theta_B - \theta_A$$

These results are for the change in displacement, or position, of the body in general plane motion. These results can be generalized readily to the case of the velocity or acceleration of the body. It can be shown that the velocity or acceleration of the rigid body in general plane motion is the sum of the translational velocity or acceleration of any arbitrary point on the body and the rotational velocity or acceleration with respect to this point.

The concept of viewing the general plane motion of a rigid body as a combination of translational and rotational motion finds very useful application in the solution of problems in rigid-body dynamics. This technique is also applied in describing a quantity which is referred to as the *kinetic energy* of the body. Here, one part of the total energy due to motion is related to the translational velocity of the body, while the remaining part of this energy is related to the angular velocity of the body with respect to a certain point on the body. This topic is studied in detail in the problems in Chap. 19.

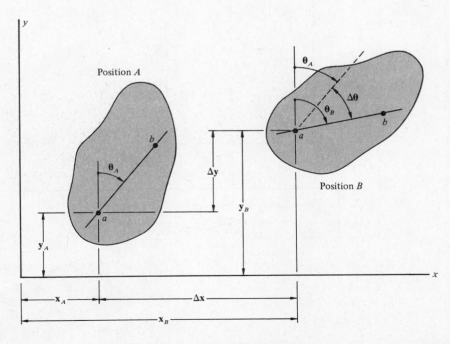

**Fig. 16.36**

**16.37** At $t = 0$ a flat, rigid plate has the position shown in Fig. 16.37a. The displacement of corner $a$ is given by

$$x = 3.94 + 4.82t \qquad y = 2.26 - 9.64t + 5t^2$$

where $t$ is in seconds and $x$ and $y$ are in inches. The angular position of the line $ab$ is expressed by

$$\theta = 1.98t^3 + 0.38$$

where $t$ is in seconds and $\theta$ is in radians.

(a) Sketch the position of the plate when $t = 1$ s.

(b) Describe the change in position of the plate during the time interval from 0 to 1 s.

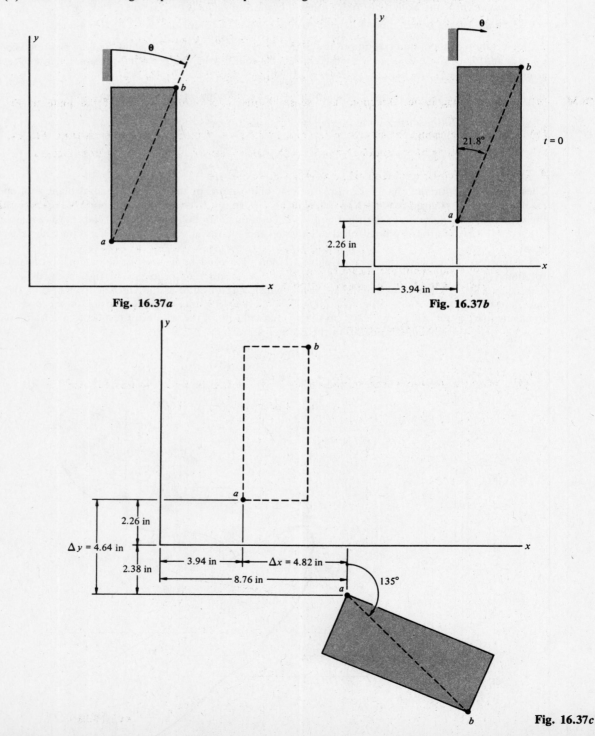

Fig. 16.37a

Fig. 16.37b

Fig. 16.37c

**▮** **(a)** The initial position of the plate at $t = 0$ is defined by

$$x = 3.94 \text{ in} \qquad y = 2.26 \text{ in} \qquad \theta = 0.38 \text{ rad} = 21.8°$$

The above values are shown in Fig. 16.37b.
   At the end of 1 s, the position of the plate is given by

$$x = 3.94 + 4.82t = 3.94 + 4.82(1) = 8.76 \text{ in}$$
$$y = 2.26 - 9.64t + 5t^2 = 2.26 - 9.64(1) + 5(1)^2 = -2.38 \text{ in}$$
$$\theta = 1.98t^3 + 0.38 = 1.98(1)^3 + 0.38 = 2.36 \text{ rad} = 135°$$

The position of the plate at the end of 1 s is shown in Fig. 16.37c.

**(b)** The change in position of the plate during the 1-s time interval is given by

$$\Delta x = x|_{t=1} - x|_{t=0} = 8.76 - 3.94 = 4.82 \text{ in}$$
$$\Delta y = y|_{t=1} - y|_{t=0} = -2.38 - 2.26 - 4.64 \text{ in}$$
$$\Delta \theta = \theta|_{t=1} - \theta|_{t=0} = 135° - 21.8° = 113°$$

**16.38** **(a)** Find the magnitude, direction, and sense of the velocity of point $a$, of the plate in Fig. 16.37a, when $t = 1$ s.

**(b)** Describe the change in velocity of corner $a$ of the plate during the time interval from 0 to 1 s.

**(c)** Find the change in the angular velocity of the plate during the time interval from 0 to 1 s.

**▮** **(a)** The velocity components of point $a$ in Fig. 16.37a are

$$v_x = \dot{x} = \frac{d}{dt}(3.94 + 4.82t) = 4.82 \text{ in/s} \qquad v_y = \frac{d}{dt}(2.26 - 9.64t + 5t^2) = -9.64 + 2(5)t \qquad \text{in/s}$$

At $t = 1$ s,

$$\dot{x}_1 = 4.82 \text{ in/s} \qquad \dot{y}_1 = -9.64 + 10(1) = 0.36 \text{ in/s}$$

These components are shown in Fig. 16.38.
   The magnitude of the velocity of point $a$ is

$$v_a = \sqrt{\dot{x}_1^2 + \dot{y}_1^2} = \sqrt{4.82^2 + 0.36^2} = 4.83 \text{ in/s}$$

The direction of this velocity is

$$\beta = \tan^{-1}\frac{0.36}{4.82} = 4.27°$$

**(b)** The velocity components of point $a$ at $t = 0$, using the results in part $(a)$, are

$$\dot{x}_0 = 4.82 \text{ in/s} \qquad \dot{y}_0 = -9.64 \text{ in/s}$$

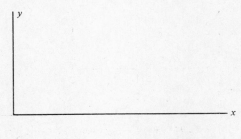

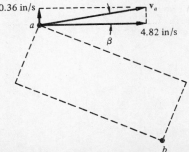

**Fig. 16.38**

The change in the velocity of point $a$ is given by

$$\Delta \dot{x} = \dot{x}_1 - \dot{x}_0 = 4.82 - 4.82 = 0 \qquad \Delta \dot{y} = \dot{y}_1 - \dot{y}_0 = 0.36 - (-9.64) = 10 \text{ m/s}$$

(c) The angular velocity of line $ab$, and of the plate, is

$$\dot{\theta} = \frac{d}{dt}(1.98t^3 + 0.38) = 3(1.98)t^2 = 5.94t^2 \qquad \text{rad/s}$$

The change in the angular velocity during the time interval is

$$\Delta \dot{\theta} = \dot{\theta}_1 - \dot{\theta}_0 = 5.94(1)^2 - 0 = 5.94 \text{ rad/s}$$

**16.39** The initial position at $t = 0$ of a flat plate moving with general plane motion is shown in Fig. 16.39a. The displacement of point $a$, and the angular displacement of line $ab$, are given by $x = -10(t^2 - 1)$, $y = 5t^2 + 8$, and $\theta = 15(2 - t^2)$, where $x$ and $y$ are in inches, $\theta$ is in degrees, and $t$ is in seconds.

(a) Find the position of the plate when $t = 2$ s.

(b) Find the magnitude, direction, and sense of the velocity of point $a$ when $t = 2$ s.

(c) Find the angular velocity of the plate, in radians per second, when $t = 2$ s.

❚ (a) At $t = 2$ s,

$$x = -10(t^2 - 1) = -10(2^2 - 1) = -30 \text{ in} \qquad y = 5t^2 + 8 = 5(2)^2 + 8 = 28 \text{ in}$$
$$\theta = 15(2 - t^2) = 15(2 - 2^2) = -30°$$

The position of the plate at $t = 2$ s is shown in Fig. 16.39b.

(b) The components of the velocity of point $a$ are

$$\dot{x} = -10(2t) = -20t \text{ in/s} \qquad \dot{y} = 10t \text{ in/s}$$

At $t = 2$ s,

$$\dot{x} = -20t = -20(2) = -40 \text{ in/s} \qquad \dot{y} = 10t = 10(2) = 20 \text{ in/s}$$

The two components of velocity at point $a$ are shown in Fig. 16.39c. The magnitude and direction of the velocity of point $a$ are

$$v_a = \sqrt{\dot{x}^2 + \dot{y}^2} = \sqrt{(-40)^2 + 20^2} = 44.7 \text{ in/s} \qquad \tan \theta_1 = \frac{20}{40} \qquad \theta_1 = 26.6°$$

(c) The angular velocity of the plate is

$$\dot{\theta} = 15(-2t) = -30t \qquad °/s$$

At $t = 2$ s,

$$\dot{\theta} = -30t = -30(2) = -60°/s \qquad \dot{\theta} = -60 \frac{°}{s}\left(\frac{2\pi \text{ rad}}{360°}\right) = -1.05 \text{ rad/s}$$

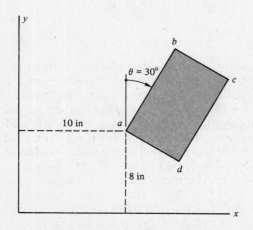

**Fig. 16.39a**

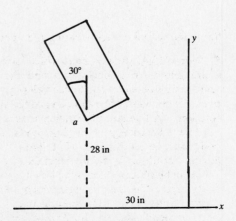

**Fig. 16.39b**

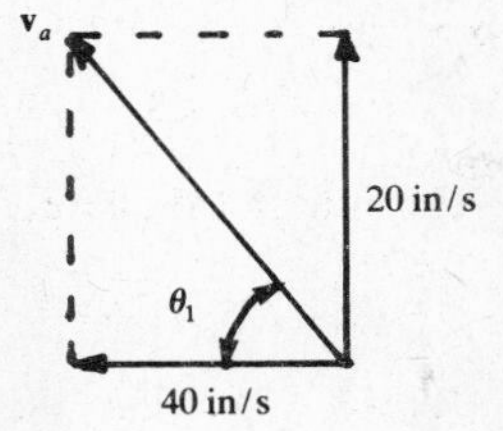

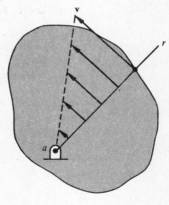

Fig. 16.39c

**16.40**   Give the definition of the *instant center of rotation* of a body.

▌ Figure 16.40 shows a plane rigid body which rotates about point *a*.   The radial coordinate *r*, with its origin at the center of rotation, is attached to the body at an arbitrary location.   The translational velocity *v* of points along the line *r* is related to the angular velocity *ω* by

$$v = r\omega$$

It follows from the above equation that *v* is a linear function of *r*, and this equation is plotted in the figure.   It may be seen that points closer to the center of rotation have smaller translational velocities than do points more distant from this reference axis.   At the center of rotation, defined by   $r = 0$,   the linear velocity is zero.   If this point is assumed to be physically part of the rigid body, then a very important conclusion may be drawn: *one point on the body, the center of rotation, has a velocity which is zero*.   This point of zero velocity on a rotating body is referred to as the *instant center of rotation*.   It is the point about which the body is rotating *at a given instant*.

In the above development, the instant center (IC), by implication, was seen to be located within the physical boundary of the rigid body.   It may be shown that *the instant center does not have to be located on the body*.

The concept of the instant center of rotation is used widely in the motion analysis of mechanical linkages, where the velocities of different points on the links are to be obtained.

Fig. 16.40

**16.41**   Show the technique for locating the instant center of rotation of a body.

▌ Figure 16.41 shows a body which moves with plane rotation.   The translational velocities of the points *a* and *b* are known; it is desired to find the location of the instant center of rotation.   From the fundamental relationship between the linear velocity of a point that is on a rotating body, and the angular velocity of the body, it follows that the direction of the velocity is *perpendicular* to the line that joins the point and the center of rotation.   A line *aa'*, perpendicular to $v_a$, is now drawn through point *a*, and it follows that the IC must lie on this line.   By using a similar line of reasoning, the line *bb'* normal to $v_b$ is drawn in Fig. 16.41, and the IC of the body must lie on this line.   It follows from the discussion above that the only possible location of the IC which lies on both lines is the intersection of the two lines.   This point is labeled IC in the figure.

It may be observed from the development above that the *magnitudes* of the translational velocities were never considered.   The only criteria which determine the location of the IC are the *directions* of the velocities at two points on the body.   The general rule, then, for finding the location of the instant center of rotation is to draw lines perpendicular to the *known* directions of the velocities of two points on the body.   The point of intersection of these two lines is then the instant center of rotation of the body.

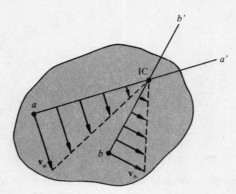

**Fig. 16.41**

A special case occurs if the known directions of the two velocities are parallel to each other. If the magnitudes of the two known velocities are not equal, then the location of the instant center may be found by using the triangular construction shown in Fig. 16.41. If the two magnitudes are equal, then the body is in translation, and the angular velocity is identically zero.

**16.42** Figure 16.42a shows a plane sliding mechanism. Slider block A moves leftward with a constant velocity of 4 m/s. For the position of the link shown in the figure,

(*a*) Find the angular velocity of the link.

(*b*) Find the velocity of slider block B.

(*c*) Find the magnitude, direction, and sense of the velocity of point *a*.

(*d*) Express the result in part (*c*) in formal vector notation.

▮ (*a*) The two slider blocks move as bodies in rectilinear translation, and the link has general plane motion. The system is redrawn in Fig. 16.42b. The directions of the two slider tracks define the directions of the slider block velocities. The IC of the link is found by drawing lines perpendicular to these tracks, as shown in the figure. This is an example of a case where the IC does *not* physically lie on the rigid body. In the position shown, link $AB$ and lines $0A$, $0B$, and $0a$ all instantaneously rotate about the IC with the same angular velocity $\omega$. The angular velocity of line $0A$ is found from

$$v_A = r\omega \qquad \omega = \frac{v_A}{r} = \frac{4}{376/1000} = 10.6 \text{ rad/s}$$

(*b*) Since all lines on the link rotate about the IC with the *same* angular velocity, the velocity of slider block B is

$$v_B = r\omega = \frac{706}{1000}(10.6) = 7.48 \text{ m/s}$$

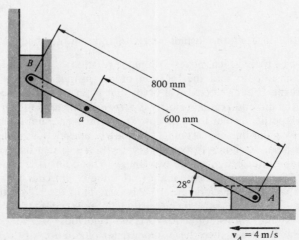

**Fig. 16.42a**

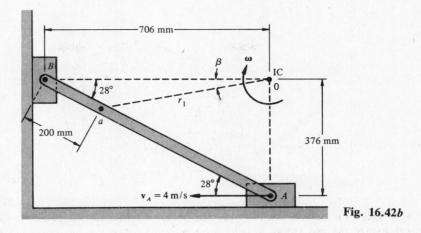

**Fig. 16.42b**

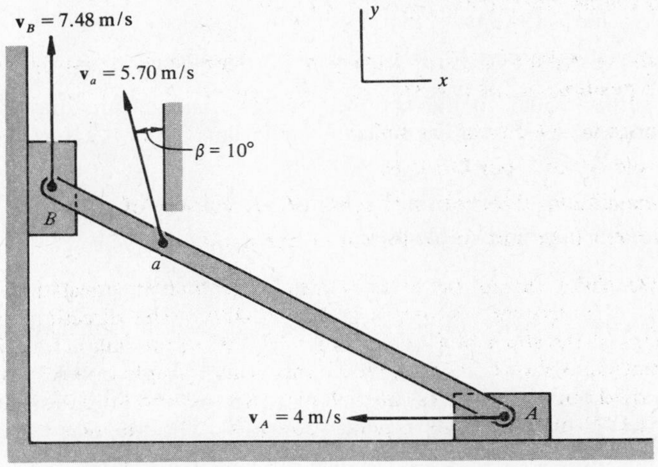

**Fig. 16.42c**

(*c*)  The length of line 0*a* is designated $r_1$.  From triangle *B*0*a*, using the law of cosines,

$$r_1^2 = 200^2 + 706^2 - 2(200)706 \cos 28° \qquad r_1 = 538 \text{ mm}$$

The angle $\beta$, using the law of sines, is found from

$$\frac{200}{\sin \beta} = \frac{r_1}{\sin 28°} \qquad \beta = 10°$$

The magnitude of $v_a$ is

$$v_a = r_1 \omega = \frac{538}{1000} (10.6) = 5.70 \text{ m/s}$$

The complete vector description of the velocities at the three points of interest is shown in Fig. 16.42*c*.

(*d*)  The velocity of point *a* may be written as

$$\mathbf{v}_a = (-5.70 \sin 10°)\mathbf{i} + (5.70 \cos 10°)\mathbf{j} = (-0.990\mathbf{i} + 5.61)\mathbf{j} \qquad \text{m/s}$$

**16.43**  The ladder in Fig. 16.43*a* is in a condition of impending sliding motion in the dashed position shown in the figure.  It starts to slide.  When it has rotated through 10°, it has an angular velocity of 4 rad/s.  Find the velocities of ends *a* and *b* of the ladder at this instant.

❚  This problem is kinematically equivalent to Prob. 16.42.  The construction used to find the location of the instant center, shown in Fig. 16.43*b*, is the same as that shown in Fig. 16.42*b*.  The velocities of ends *a* and *b* of the ladder are found to be

$$v_a = r\omega = 3.71(4) = 14.8 \text{ m/s} \qquad v_b = r\omega = 1.5(4) = 6 \text{ m/s}$$

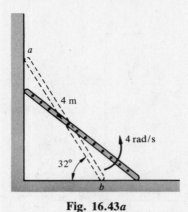

**Fig. 16.43a**

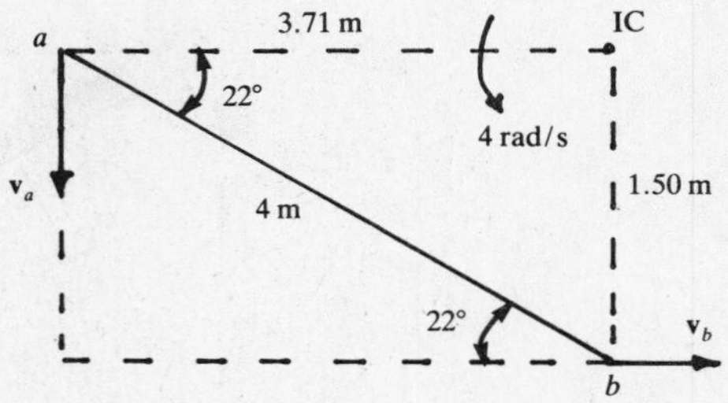

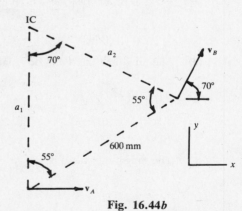

**Fig. 16.43b**

**16.44** For the position shown in Fig. 16.44a, the angular velocity of the link which connects the two blocks is 22 rad/s, in a counterclockwise sense.

(*a*) Find the velocities of the two blocks.

(*b*) Express the results in part (*a*) in formal vector notation.

▌ (*a*) Lines perpendicular to the two velocities $v_A$ and $v_B$ are drawn, as shown in Fig. 16.44b. The intersection of these two lines locates the instant center. Using the law of sines,

$$\frac{a_1}{\sin 55°} = \frac{600}{\sin 70°} = \frac{a_2}{\sin 55°} \qquad a_1 = a_2 = 523 \text{ mm}$$

The velocities of the two blocks are

$$v_A = v_B = r\omega = 0.523(22) = 11.5 \text{ m/s}$$

(*b*) The velocities of blocks $A$ and $B$ may be expressed in formal vector notation as

$$\mathbf{v}_A = v_A\mathbf{i} = 11.5\mathbf{i} \qquad \text{m/s}$$
$$\mathbf{v}_B = (v_B \cos 70°)\mathbf{i} + (v_B \sin 70°)\mathbf{j} = (11.5 \cos 70°)\mathbf{i} + (11.5 \sin 70°)\mathbf{j} = 3.93\mathbf{i} + 10.8\mathbf{j} \qquad \text{m/s}$$

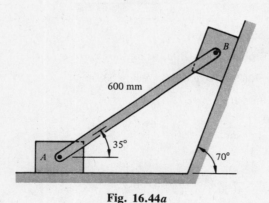

**Fig. 16.44a**

**Fig. 16.44b**

**16.45** The body in Fig. 16.45a moves in plane motion. The velocity of point *b* is known to be along the direction *ab*.

(*a*) Find the distance between the instant center and points *a* and *b*.

(*b*) Find the angular velocity of the body.

(*c*) Find the velocity of points *b* and *c*.

▌ (*a*) Figure 16.45b shows the construction used to find the location of the instant center. Line *ad* is perpendicular to the direction of $v_A$, and line *bd* is perpendicular to line *ab*. The lengths $a_1$ and $a_2$ are found as

$$a_1 = \frac{10}{\cos 30°} = 11.5 \text{ in} \qquad a_2 = 10 \tan 30° = 5.77 \text{ in}$$

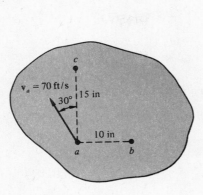

Fig. 16.45a

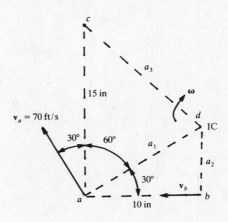

Fig. 16.45b

(b) Using the law of cosines,

$$a_3^2 = 15^2 + 11.5^2 - 2(15)11.5 \cos 60° \qquad a_3 = 13.6 \text{ in}$$

The angular velocity of the body is found from

$$v = r\omega \qquad v_a = a_1\omega \qquad 70(12) = 11.5\omega \qquad \omega = 73.0 \text{ rad/s}$$

(c) The velocities of points $b$ and $c$ are given by

$$v_b = r\omega = a_2\omega = 5.77(73.0) = 421 \text{ in/s} = 35.1 \text{ ft/s}$$
$$v_c = r\omega = a_3\omega = 13.6(73.0) = 993 \text{ in/s} = 82.8 \text{ ft/s}$$

**16.46** Do the same as in Prob. 16.45, for the body shown in Fig. 16.46a. The velocity of point $b$ is in a vertical direction.

▌ (a) The construction used to find the instant center is shown in Fig. 16.46b. Lines $ad$ and $bc$ are perpendicular to $v_a$ and $v_b$, respectively. The lengths $a_1$ and $a_2$ are given by

$$a_1 = \frac{280}{\cos 40°} = 366 \text{ mm} \qquad \tan 40° = \frac{a_3}{280} \qquad a_3 = 235 \text{ mm} \qquad a_2 = 340 - a_3 = 105 \text{ mm}$$

(b) The angular velocity of the body is given by

$$v_a = r\omega = a_1\omega \qquad 10 = \frac{366}{1,000}\omega \qquad \omega = 27.3 \text{ rad/s}$$

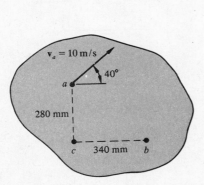

Fig. 16.46a

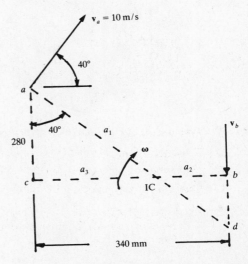

Fig. 16.46b

(c)   The velocities of points b and c are

$$v_b = r\omega = a_2\omega = \frac{105}{1,000}(27.3) = 2.87 \text{ m/s}$$

$$v_c = r\omega = a_3\omega = \frac{235}{1,000}(27.3) = 6.42 \text{ m/s}$$

**16.47**   The variation of velocity of points that lie on line ab on a rigid body rotating about a fixed axis is shown in Fig. 16.47a.

(a)   Find the angular velocity of the body.

(b)   Find the velocity of point c.

▮ (a)   The velocity distribution is shown in Fig. 16.47b, and point d is the instant center.   Using the similar triangles shown in Fig. 16.47b,

$$\frac{100}{a_1} = \frac{20}{14 - a_1} \qquad a_1 = 11.7 \text{ in}$$

The angular velocity of the body is found from

$$v_a = r\omega = a_1\omega \qquad 100 = 11.7\omega \qquad \omega = 8.55 \text{ rad/s}$$

(b)   From Fig. 16.47b,

$$a_2 = a_1 - 5 = 11.7 - 5 = 6.7 \text{ in} \qquad a_3 = \sqrt{4^2 + 6.7^2} = 7.80 \text{ in}$$

The magnitude of the velocity of point c is

$$v_c = r\omega = a_3\omega = 7.80(8.55) = 66.7 \text{ in/s}$$

Figure 16.47c shows the velocity of point c.   The direction of $v_c$ is given by

$$\tan\theta = \frac{6.7}{4} \qquad \theta = 59.2°$$

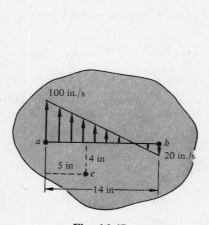

Fig. 16.47a

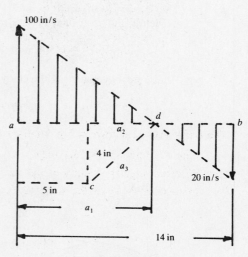

Fig. 16.47b

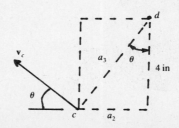

Fig. 16.47c

**16.48**    Do the same as in Prob. 16.47, for the body shown in Fig. 16.48a.

**▌** **(a)**    Figure 16.48b shows the velocity distribution.    The location of the instant center is found by using the construction shown in Fig. 16.41.   Using the similar triangles in Fig. 16.48b,

$$\frac{a_1}{5} = \frac{a_1 - 360}{3} \qquad a_1 = 900 \text{ mm}$$

The angular velocity is found from

$$v_a = r\omega = a_1\omega \qquad 5 = \frac{900}{1,000}\,\omega \qquad \omega = 5.56 \text{ rad/s}$$

**(b)**    From Fig. 16.48b,

$$a_2 = \sqrt{150^2 + 340^2} = 372 \text{ mm} \qquad a_3 = a_1 - 360 - 200 = 340 \text{ mm}$$

Figure 16.48c shows $v_c$.    The magnitude and direction of this velocity are found as

$$v_c = r\omega = a_2\omega = \frac{372}{1,000}\,(5.56) = 2.07 \text{ m/s} \qquad \tan\theta = \frac{150}{340} \qquad \theta = 23.8°$$

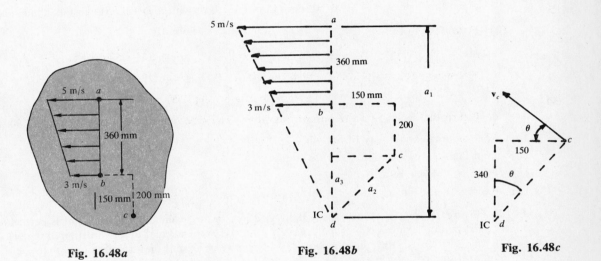

Fig. 16.48a                    Fig. 16.48b                    Fig. 16.48c

**16.49**    The crank arm ab shown in Fig. 16.49a rotates clockwise at 850 r/min.    For the position shown in the figure,

**(a)**    Find the direction of link bc.

**(b)**    Find the angular velocity of link bc.

**(c)**    Find the velocity of the slider block.

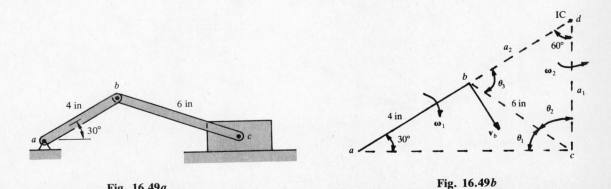

Fig. 16.49a                              Fig. 16.49b

▮ (a) The construction used to find the instant center is shown in Fig. 16.49b. Line $a_2$ is perpendicular to $v_b$. Line $a_1$ is perpendicular to the direction, given by line $ac$ in Fig. 16.49b, of the velocity $v_c$. Using the law of sines with triangle $abc$, the direction $\theta_1$ of link $bc$ is found as

$$\frac{4}{\sin\theta_1} = \frac{6}{\sin 30°} \qquad \theta_1 = 19.5°$$

(b) Angles $\theta_2$ and $\theta_3$ are found as

$$\theta_2 = 90 - \theta_1 = 70.5° \qquad \theta_3 = 180° - \theta_2 - 60° = 49.5°$$

Using the law of sines with triangle $bdc$,

$$\frac{6}{\sin 60°} = \frac{a_1}{\sin\theta_3} = \frac{a_2}{\sin\theta_2} \qquad a_1 = 5.27 \text{ in} \qquad a_2 = 6.53 \text{ in}$$

The angular velocity of the crank arm, and the velocity of point $b$, are found as

$$\omega_1 = 850 \text{ r/min} = 89.0 \text{ rad/s} \qquad v_b = r\omega_1 = 4(89.0) = 356 \text{ in/s}$$

The angular velocity of link $bc$ is found from

$$v_b = r\omega = a_2\omega_2 \qquad 356 = 6.53\omega_2 \qquad \omega_2 = 54.5 \text{ rad/s}$$

Since the angular velocity of all lines in triangle $bcd$, with respect to the instant center, are the same,

$$\omega_{bc} = \omega_2 = 54.5 \text{ rad/s}$$

(c) Using Fig. 16.49b,

$$v_c = r\omega = a_1\omega_2 = 5.27(54.5) = 287 \text{ in/s}$$

**16.50** Figure 16.50a shows an elementary model of the piston-cylinder-crank arrangement in an internal combustion engine. When the piston is in the position shown in the figure, it has a downward speed of 900 ft/min.

(a) Find the angular velocity of link $ab$.

(b) Find the velocity of point $b$ on the crank.

(c) Find the angular velocity of the crank $bc$.

(d) Express the result in part (b) in formal vector notation.

▮ (a) Figure 16.50b shows the construction used to find the instant center. Line $a_1$ is perpendicular to the direction of the velocity of the piston. Line $a_2$ is perpendicular to the direction of the velocity of point $b$. It follows, from the direction of the velocity of a point on a rotating link, that line $a_2$ is collinear with the axis of the crank $bc$. The intersection of lines $a_1$ and $a_2$ is the instant center of rotation of link $ab$.

Using the law of sines, with triangle $abc$ in Fig. 16.50b,

$$\frac{8.5}{\sin\theta_1} = \frac{2}{\sin 11°} \qquad \theta_1 = 54.2°, \quad 126°$$

From inspection of Fig. 16.50b, the value 126° is used. Angles $\theta_2$ and $\theta_3$, and the lengths $a_1$ and $a_2$, are then found as

$$\theta_2 = 180° - 11° - \theta_1 = 43° \qquad \theta_3 = 180° - \theta_1 = 54°$$

$$\tan\theta_2 = \frac{a_1}{8.5} \qquad a_1 = 7.93 \text{ in} \qquad \frac{a_2}{\sin 79°} = \frac{a_1}{\sin\theta_3} \qquad a_2 = 9.62 \text{ in}$$

The value of $v_a$ is given by

$$v_a = 900 \text{ ft/min} = 15 \text{ ft/s} = 180 \text{ in/s}$$

Using the instant center,

$$v_a = r\omega_1 = a_1\omega_1 \qquad 180 = 7.93\omega_1 \qquad \omega_1 = 22.7 \text{ rad/s}$$

$\omega_1$ is the angular velocity of link $ab$.

(b) The velocity of point $b$ on the crank is given by

$$v_b = r\omega = a_2\omega_1 = 9.62(22.7) = 218 \text{ in/s} = 18.2 \text{ ft/s}$$

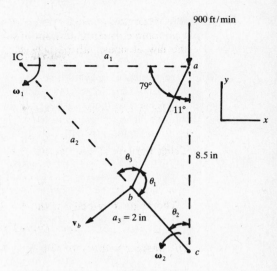

**Fig. 16.50a**                                    **Fig. 16.50b**

(c)  The angular velocity $\omega_2$ of the crank is found from

$$v_b = r\omega = a_3\omega_2 \qquad 218 = 2\omega_2 \qquad \omega_2 = 109 \text{ rad/s}$$

(d)  Using Fig. 16.50b,

$$\mathbf{v}_b = (-v_b \cos 43°)\mathbf{i} - (v_b \sin 43°)\mathbf{j} = (-18.2 \cos 43°)\mathbf{i} - (18.2 \sin 43°)\mathbf{j} = -13.3\mathbf{i} - 12.4\mathbf{j} \qquad \text{ft/s}$$

## 16.4  PURE ROLLING OF RIGID BODIES

**16.51**  Figure 16.51 shows a cylinder which contacts a plane track.

(a)  Describe the possible motions of the cylinder.

(b)  What is meant by the term *pure rolling*?

❚  (a)  In terms of the motion of the cylinder relative to the track, four states of motion are possible:

   1. The cylinder remains at rest.
   2. The cylinder does not rotate, and the surface of the cylinder slides along the track.
   3. The cylinder rolls along the track without slipping.
   4. The cylinder moves along the track with a combination of rolling and sliding motion.

Case 1 is a problem in statics.  In case 2, since the cylinder does not *rotate*, this element may be analyzed as a particle in translation.   The situation described in case 4 is an undefined problem, unless additional information is provided which describes the relationship between the rolling and sliding motions.   The following problems will be concerned with case 3, where the cylinder rolls without sliding.

(b)  The motion described in case 3 in part (a), when the cylinder rolls along the track *without sliding*, is referred to as pure rolling.

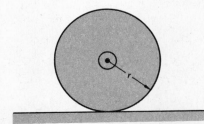

**Fig. 16.51**

**16.52** (*a*) What is the relationship between the angular velocity of a cylinder in a state of pure rolling and the translational velocity of the center of the cylinder?

(*b*) Give a physical interpretation of the effect of pure rolling of a cylinder.

❚ (*a*) Pure rolling of a cylindrical or spherical element is usually the desired operating condition for such a device. Typical examples include the wheel of a vehicle, a ball in the race of a ball bearing, and a roller used to paint a wall. From consideration of the motion of a cylinder which rolls without slipping, it may be concluded that this is a case of general plane motion. The center of the cylinder translates along a path parallel to the track, while the cylinder rotates about this center. The relationship between the angular motion of the cylinder and the translational motion of the center of the cylinder will now be obtained.

Figure 16.52 shows the senses of motion of the cylinder. The positive sense of the angular velocity $\omega$ is chosen to be clockwise, so that the center of the cylinder moves to the right with the velocity $v$. The direction of the velocity of the center of the cylinder is parallel to the track.

The contact point between the cylinder and the track is designated point *a*. This point may be considered to be simultaneously on the cylinder and the track. Since the problem of pure rolling implies no sliding motion between the cylinder and the track, point *a*, as a point on the cylinder which has zero velocity, must be *an instant center of rotation* of the cylinder. All lines on the cylinder, including line *r*, have the same angular velocity $\omega$. It may be concluded then, by considering the cylinder to instantaneously rotate about the contact point *a*, that

$$v = r\omega \qquad \omega = \frac{v}{r}$$

(*b*) The physical phenomenon of rolling of a cylinder may be envisioned to be a succession of instantaneous rotations of the cylinder about the contact point between the cylinder and the surface on which it rolls.

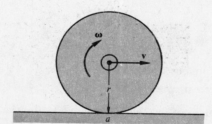

Fig. 16.52

**16.53** Figure 16.53*a* shows a yo-yo modeled as a right circular cylinder about which is wrapped a thin, inextensible string. Find the relationship between the angular velocity of the yo-yo and the translational velocity of its center.

❚ Figure 16.53*b* shows the positive senses of the angular velocity $\omega$ and the translational velocity $v$ of the center of the yo-yo. The point where the string contacts the yo-yo is an instant center of the yo-yo. The velocities $\omega$ and $v$ are related by

$$v = r\omega \qquad \omega = \frac{v}{r}$$

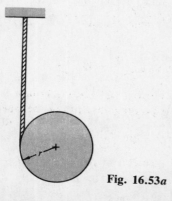

Fig. 16.53*a*

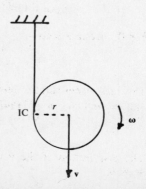

Fig. 16.53*b*

**16.54** Railroad wheels have flanges on them, as shown in Fig. 16.54a, so that the wheels do not slide sideways off the track. Show that, when the train is in motion, certain parts of the train are always moving in a sense *opposite* to the sense of the actual motion of the train.

▌ The train wheel is shown in Fig. 16.54b, and the center a of the wheel is assumed to move to the right. Pure rolling is assumed, and point b is the instant center of rotation of the wheel. The line abc lies on the wheel and rotates with the angular velocity ω of the wheel. The velocity distribution of points of the wheel on a vertical diameter is shown in Fig. 16.54c. It may be seen from this figure that the velocity relative to the track, and therefore to the ground, of all points on the flange of the wheel between b and c is *opposite* to the actual, rightward velocity of the train.

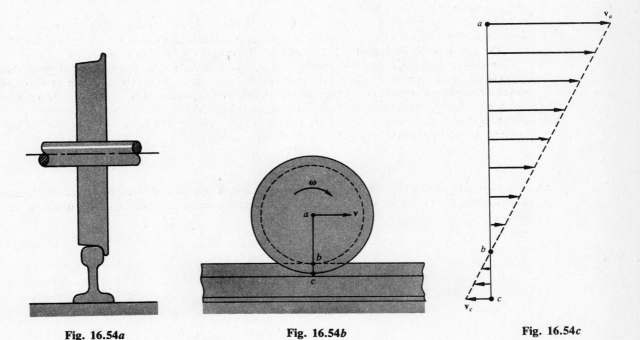

Fig. 16.54a         Fig. 16.54b         Fig. 16.54c

**16.55** The farm tractor shown in Fig. 16.55a moves with a constant speed of 10 mi/h.

(a) Find the angular velocity, in revolutions per minute, of the front and rear wheels.

(b) Find the velocities of points a and b on the tops of the wheels.

(c) Find the relative velocity of point a with respect to the axle of the rear wheel.

▌ (a) The translational velocity of both wheel centers is

$$v = 10 \frac{\text{mi}}{\text{h}} \left( \frac{5,280 \text{ ft}}{1 \text{ mi}} \right) \left( \frac{1 \text{ h}}{60 \text{ min}} \right) = 880 \frac{\text{ft}}{\text{min}}$$

For the rear wheels,

$$\omega = \frac{v}{r} = \frac{880 \text{ ft/min}}{3 \text{ ft}} = 293 \frac{\text{rad}}{\text{min}} \qquad \omega = 293 \frac{\text{rad}}{\text{min}} \left( \frac{1 \text{ r}}{2\pi \text{ rad}} \right) = 46.6 \frac{\text{r}}{\text{min}}$$

For the front wheels

$$\omega = \frac{v}{r} = \frac{880}{1} = 880 \frac{\text{rad}}{\text{min}} \qquad \omega = 880 \left( \frac{1}{2\pi} \right) = 140 \frac{\text{r}}{\text{min}}$$

(b) The rear wheel rotates about the contact point with the ground, which is the instant center, with the angular velocity of 293 rad/min. The velocity of point a on the upper rim of the wheel is then

$$v_a = r\omega = 6 \text{ ft} \left( 293 \frac{\text{rad}}{\text{min}} \right) = 1,760 \frac{\text{ft}}{\text{min}}$$

The distribution of the velocity of points on a vertical diameter of the rear wheel is shown in Fig. 16.55b.

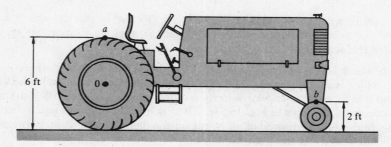

Fig. 16.55a

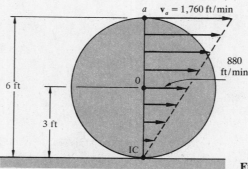

Fig. 16.55b

A similar analysis for the front wheel gives

$$v_b = r\omega = 2(880) = 1,760 \text{ ft/min}$$

It may be seen that $v_a = v_b$. It is left as an exercise for the reader to verify that this is an expected result, based on the fact that the velocity of the uppermost point on a cylinder in pure rolling is exactly twice the velocity of the center of the cylinder.

(c)  The relative velocity of point $a$ with respect to the wheel center is

$$v_{a0} = v_a - v_0 = 1,760 - 880 = 880 \text{ ft/min}$$

The sense of the above result is the same as the sense of $v_a$.

**16.56**  Figure 16.56a shows an arrangement of two concentric cylinders joined to each other. The unit rolls without slipping on the horizontal rail. At the instant shown in the figure, $t = 0$ and the counterclockwise angular velocity is 52 rad/s, with a constant deceleration of 10 rad/s². Find the magnitude, direction, and sense of the velocity of the uppermost points on the two cylindrical surfaces, when $t = 2$ s, by using (a) the definition of relative velocity and (b) the definition $v = r\omega$ for pure rolling.

**❚**  (a)  Figure 16.56b shows the rotational and translational velocities of the cylinder. When $t = 0$,

$$\theta_0 = 0 \qquad \omega_0 = 52 \text{ rad/s} \qquad \alpha = -10 \text{ rad/s}^2$$

When  $t = 2$ s,

$$\omega = \omega_0 + \alpha t = 52 + (-10)2 = 32 \text{ rad/s}$$

The velocity of the center of the unit is given by

$$v_c = r\omega = 5(32) = 160 \text{ in/s}$$

The velocity of point $a$ is given by

$$v_a = v_c + v_{ac} \qquad v_{ac} = r\omega = 10(32) = 320 \text{ in/s} \qquad v_a = 160 + 320 + 480 \text{ in/s}$$

The velocity of point $b$ is found from

$$v_b = v_c + v_{bc} \qquad v_{bc} = r\omega = 5(32) = 160 \text{ in/s} \qquad v_b = 160 + 160 = 320 \text{ in/s}$$

(b)  Using Fig. 16.56b,

$$v_a = r\omega = 15(32) = 480 \text{ in/s} \qquad v_b = r\omega = 10(32) = 320 \text{ in/s}$$

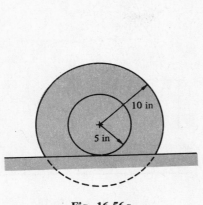

Fig. 16.56a

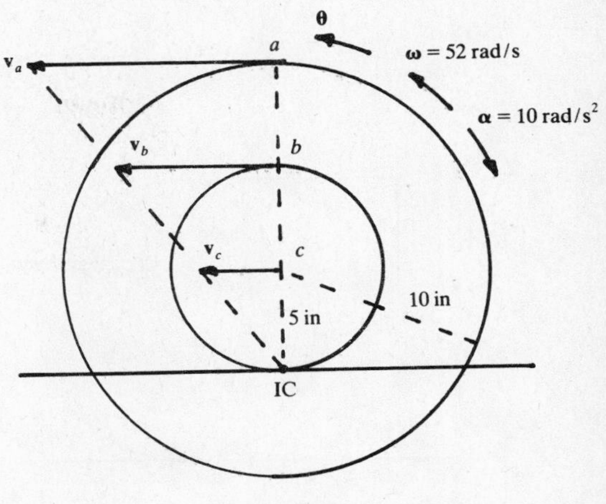

Fig. 16.56b

**16.57** A heavy crate is moved on rollers, as shown in Fig. 16.57a, and pure rolling is assumed. At the instant shown in the figure, $v = 4$ ft/min and $a = 2$ ft/min$^2$. Find the angular velocity of the rollers, and the velocity of the centers of these two elements.

▌ Figure 16.57b shows the velocity distribution along a vertical diameter of the rollers. The angular velocity of the rollers is found from

$$v_b = r\omega \qquad \omega = \frac{v_b}{r} = \frac{4}{1.2} = 3.33 \text{ rad/min}$$

The velocity of the centers of the rollers is given by

$$v_a = r\omega = 0.6(3.33) = 2 \text{ ft/min}$$

Fig. 16.57a

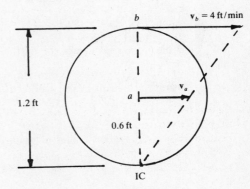

Fig. 16.57b

**16.58** A thin, inextensible string is wrapped around the inner diameter of the cylinder unit shown in Fig. 16.58. The absolute velocity of the string is $v = 8$ m/s, and the unit rolls without sliding.

(a) Find the velocity of the uppermost point on the larger cylinder.

(b) Find the relative velocity of the string with respect to the center of the cylinders.

(c) Find the rate, in m/s, at which the string unwinds from the cylinder.

▌ (a) Figure 16.58b shows the cylinder unit and the location of the instant center. The rotational and translational velocities are related by

$$v_a = r\omega \qquad 8 = \frac{275}{1,000}\,\omega \qquad \omega = 29.1 \text{ rad/s}$$

The velocity of point $b$ is given by

$$v_b = r\omega = \frac{400}{1,000}(29.1) = 11.6 \text{ m/s}$$

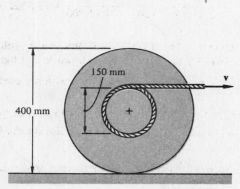

Fig. 16.58a

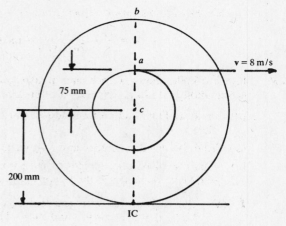

Fig. 16.58b

(b) The relative velocity of the string with respect to the center of the cylinders is $v_{ac}$. This quantity is found as

$$v_{ac} = r\omega = \frac{75}{1{,}000}\,(29.1) = 2.18\,\text{m/s}$$

(c) The rate of unwinding of the string is the same as the relative velocity of the string with respect to the center of the cylinders, given in part (b) as

$$v_{ac} = 2.18\,\text{m/s}$$

**16.59** A right circular cylinder rolls without slipping on a track in the form of a circular arc, as shown in Fig. 16.59a. The radius of the cylinder is $r$, and the radius of the track is $R$.

(a) Find the angular velocity of the cylinder, and of the radial line $0a$, if the velocity of the center of the roller is $v_a$.

(b) Find the numerical values for part (a), if $r = 20\,\text{mm}$, $R = 240\,\text{mm}$, and $v_a = 3\,\text{m/s}$.

▮ (a) Figure 16.59b shows the rotational and translational velocities. The angular velocity of the cylinder, using the instant center where the cylinder contacts the track, is found from

$$v_a = r\omega \qquad \omega = \frac{v_a}{r}$$

Using the instant center at point 0 of line $0a$,

$$v_a = r\omega = (R - r)\omega_1 \qquad \omega_1 = \frac{v_a}{R - r}$$

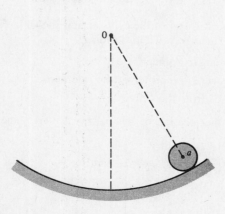

Fig. 16.59a

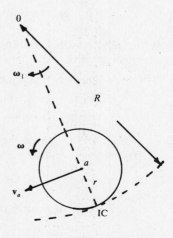

Fig. 16.59b

(*b*)  For the numerical values given,

$$\omega = \frac{v_a}{r} = \frac{3}{20/1,000} = 150 \, \text{rad/s} \qquad \omega_1 = \frac{v_a}{R-r} = \frac{3}{(240-20)/1,000} = 13.6 \, \text{rad/s}$$

**16.60**  Figure 16.60*a* shows a rack and pinion gear arrangement.  The rack moves downward with a constant acceleration of 15 ft/s².  At  *t* = 0,  the rack has zero velocity and the position shown in the figure.

(*a*)  Find the time, and corresponding velocity of the rack, when end *a* of the rack just loses contact with the pinion.

(*b*)  Find the angular velocity and angular acceleration of the pinion at the time found in part (*a*).

(*c*)  Find the magnitude, direction, and sense of the velocity and acceleration of a point on the left end of a horizontal diameter of the pinion, at the time found in part (*a*).

▌ (*a*)  The rack moves with constant acceleration.  Using  $s = s_0 + v_0 t + \frac{1}{2}at^2$,  with the conditions  $s = 5$ in,  $s_0 = 0$,  $v_0 = 0$,  and  $a = 15 \, \text{ft/s}^2$,

$$5 = \tfrac{1}{2}[15(12)]t^2 \qquad t = 0.236 \, \text{s}$$

The velocity $v_a$ of the rack, when this element loses contact with the pinion, is found from

$$v_a = v_0 + at = 15(12)0.236 = 42.5 \, \text{in/s}$$

(*b*)  The model of the rack and pinion gear assembly, as a cylindrical element which rolls without sliding on a straight element, is shown in Fig. 16.60*b*.  The angular velocity and acceleration of the pinion at  $t = 0.236$ s  are found as

$$v_a = r\omega \qquad \omega = \frac{v_a}{r} = \frac{42.5}{1} = 42.5 \, \text{rad/s} \qquad a_a = r\alpha \qquad \alpha = \frac{a_a}{r} = \frac{15(12)}{1} = 180 \, \text{rad/s}^2$$

(*c*)  Figure 16.60*c* shows point *b* on the left end of a horizontal diameter of the pinion.  The velocity of point *b* is found as

$$v_b = r\omega = 1(42.5) = 42.5 \, \text{in/s}$$

The direction and sense of this velocity are shown in Fig. 16.60*c*.
The acceleration of point *b* is found as

$$a_{bt} = r\alpha = 1(180) = 180 \, \text{in/s}^2$$

$$a_{bn} = \frac{v_b^2}{\rho} = \frac{42.5^2}{1} = 1,810 \, \text{in/s}^2 \qquad a_b = \sqrt{a_{bn}^2 + a_{bt}^2} = \sqrt{1,810^2 + 180^2} = 1,820 \, \text{in/s}^2$$

The direction $\theta$ of $a_b$, shown in Fig. 16.60*d*, is given by

$$\tan \theta = \frac{180}{1,810} \qquad \theta = 5.68°$$

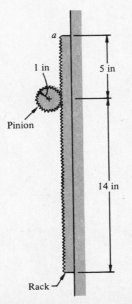

**Fig. 16.60*a***

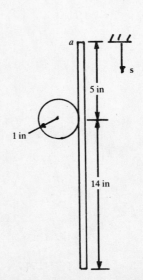

**Fig. 16.60*b***

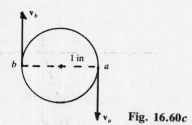

Fig. 16.60c

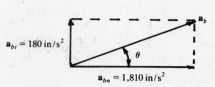

$a_{bt} = 180 \text{ in/s}^2$    $a_{bn} = 1{,}810 \text{ in/s}^2$    Fig. 16.60d

**16.61**    Figure 16.61a shows a pair of meshing gears that lie in a common plane. The gears are designated 1 and 2, with the corresponding number of teeth $N_1$ and $N_2$. The kinematic action of the set of gears is that of pure rolling of two cylinders with the pitch diameters $d_1$ and $d_2$. The equations that describe the motion and geometry relations between the gears are

$$\frac{\theta_1}{\theta_2} = \frac{\omega_1}{\omega_2} = \frac{d_2}{d_1} \qquad \frac{\omega_1}{\omega_2} = \frac{N_2}{N_1} \qquad \frac{N_1}{N_2} = \frac{d_1}{d_2}$$

At time $t = 0$, point $a$ on gear 1 and point $b$ on gear 2 are coincident points at the point of contact of the two pitch circles. $d_1 = 1 \text{ in}$ and $d_2 = 2.4 \text{ in}$, and gear 1 rotates at 1,000 r/min in a clockwise sense.

(**a**)   Find the magnitude, direction, and sense of the velocity and acceleration of point $b$ after gear 1 has completed one full revolution.

(**b**)   Express the results in part (**a**) in formal vector notation.

(**c**)   Do the same as in part (**a**), for $t = 1 \text{ s}$.

❙  (**a**)   The angular velocity of gear 1 is given by

$$\omega_1 = 1{,}000 \text{ r/min} = 105 \text{ rad/s}$$

Figure 16.61b shows the kinematic model of pure rolling of two cylinders. The angular displacement of gear 2, for one revolution of gear 1, is found from

$$\frac{\theta_1}{\theta_2} = \frac{d_2}{d_1} \qquad \frac{360°}{\theta_2} = \frac{2.4}{1} \qquad \theta_2 = 150°$$

The angular velocity of gear 2 is found from

$$\frac{\omega_1}{\omega_2} = \frac{d_2}{d_1} \qquad \frac{105}{\omega_2} = \frac{2.4}{1} \qquad \omega_2 = 43.8 \text{ rad/s}$$

The velocity and acceleration of point $b$ on gear 2 are given by

$$v_b = r\omega = 1.2(43.8) = 52.6 \text{ in/s} \qquad a_{bn} = r\omega^2 \qquad a_{bt} = 0 \qquad a_b = a_{bn} = 1.2(43.8)^2 = 2{,}300 \text{ in/s}^2$$

The above results are shown in Fig. 16.61c.

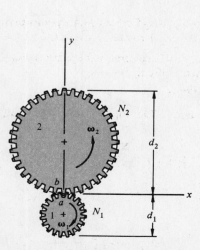

Fig. 16.61a

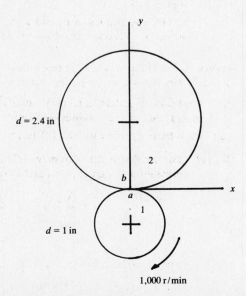

Fig. 16.61b

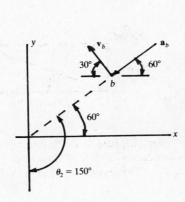

**Fig. 16.61c**

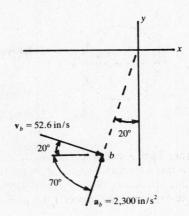

**Fig. 16.61d**

(*b*) The velocity and acceleration of point *b* may be written as

$$\mathbf{v}_b = (-v_b \cos 30°)\mathbf{i} + (v_b \sin 30°)\mathbf{j} = (-52.6 \cos 30°)\mathbf{i} + (52.6 \sin 30°)\mathbf{j} = -45.6\mathbf{i} + 26.3\mathbf{j} \text{ in/s}$$
$$\mathbf{a}_b = (-a_b \cos 60°)\mathbf{i} - (a_b \sin 60°)\mathbf{j} = (-2,300 \cos 60°)\mathbf{i} - (2,300 \sin 60°)\mathbf{j} = -1,150\mathbf{i} - 1,990\mathbf{j} \text{ in/s}^2$$

(*c*) Four significant figures are used to obtain accuracy in the calculation for the direction of the velocity and acceleration of point *b* at $t = 1$ s.

$$\omega_1 = 1,000 \text{ r/min} = 104.7 \text{ rad/s}$$

The position of gear 1 at $t = 1$ s, with $\theta_{01} = 0$ and $\omega_{01} = \omega_1$, is found from

$$\theta_1 = \theta_{01} + \omega_{01}t + \tfrac{1}{2}\alpha t^2 = \omega_{01}t = 104.7(1) = 104.7 \text{ rad}$$

The corresponding position of gear 2 is found from

$$\frac{\theta_1}{\theta_2} = \frac{d_2}{d_1} \qquad \frac{104.7}{\theta_2} = \frac{2.4}{1} \qquad \theta_2 = 43.63 \text{ rad}$$

Gear 2 completes six full revolutions, and a partial revolution, in 1 s. The displacement of this gear from its original position in Fig. 16.61*b* is

$$\theta_2 = 43.63 - 6(2\pi) = 5.931 \text{ rad} = 5.931\left(\frac{360°}{2\pi \text{ rad}}\right) = 340°$$

(If three significant figures had been used in the calculations, the result obtained would have been 350°.)

The magnitudes of $v_b$ and $a_b$ are the same as in part (*a*), since the gears rotate at constant angular velocity. Figure 16.61*d* shows the directions of $v_b$ and $a_b$ at $t = 1$ s.

**16.62** Figure 16.62*a* shows a set of three meshing gears. At $t = 0$, *a* and *b* are contact points between gears 1 and 2, and *c* and *d* are contact points between gears 2 and 3. Gear 1 rotates counterclockwise at 200 rad/s.

(*a*) Find the magnitude, direction, and sense of the velocity and acceleration of points *a* and *d* after gear 1 has completed one full revolution from the position shown in the figure.

(*b*) Express the results in part (*a*) in formal vector notation.

$\blacksquare$ (*a*) The results in the statement of Prob. 16.61 are used. The angular displacement of gear 3, for one revolution of gear 1, is found from

$$\frac{\theta_1}{\theta_2} = \frac{N_2}{N_1} \qquad \frac{\theta_2}{\theta_3} = \frac{N_3}{N_2} \qquad \frac{\theta_1}{\theta_3} = \frac{\theta_1}{\theta_2}\left(\frac{\theta_2}{\theta_3}\right) = \frac{N_2}{N_1}\left(\frac{N_3}{N_2}\right) = \frac{N_3}{N_1} = \frac{12}{18}$$

$$\frac{360°}{\theta_3} = \frac{12}{18} \qquad \theta_3 = 540°$$

The diameter of gear 3 is found from

$$\frac{d_2}{d_1} = \frac{N_2}{N_1} \qquad \frac{d_3}{d_2} = \frac{N_3}{N_2} \qquad \frac{d_3}{d_1} = \frac{d_3}{d_2}\left(\frac{d_2}{d_1}\right) = \frac{N_3}{N_2}\left(\frac{N_2}{N_1}\right) = \frac{N_3}{N_1} = \frac{12}{18}$$

202 □ CHAPTER 16

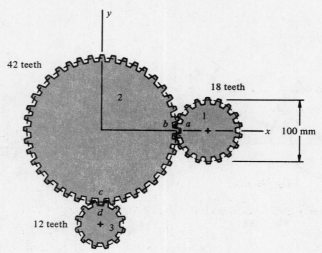

42 teeth

18 teeth

2

1

b a

x  100 mm

c

d

12 teeth  3

**Fig. 16.62a**

$$\frac{d_3}{100} = \frac{12}{18} \qquad d_3 = 66.7 \text{ mm}$$

The angular velocity of gear 3 is found from

$$\frac{\omega_1}{\omega_2} = \frac{N_2}{N_1} \qquad \frac{\omega_2}{\omega_3} = \frac{N_3}{N_2} \qquad \frac{\omega_1}{\omega_3} = \frac{\omega_1}{\omega_2}\left(\frac{\omega_2}{\omega_3}\right) = \frac{N_2}{N_1}\left(\frac{N_3}{N_2}\right) = \frac{N_3}{N_1} = \frac{12}{18}$$

$$\frac{200}{\omega_3} = \frac{12}{18} \qquad \omega_3 = 300 \text{ rad/s}$$

The model of the gears as rolling cylinders is shown in Fig. 16.62b.
The velocity and acceleration of point $a$ on gear 1 are

$$v_a = r\omega_1 = \frac{50}{1,000}(200) = 10.0 \text{ m/s}$$

$$a_{an} = r\omega_1^2 = \frac{50}{1,000}(200)^2 \qquad a_{at} = 0 \qquad a_a = a_{an} = 2,000 \text{ m/s}^2$$

Figure 16.62c shows the directions of $v_a$ and $a_a$.
The velocity and acceleration of point $d$ on gear 3 are

$$v_d = r\omega_3 = \frac{d_3}{2}\omega_3 = \frac{66.7}{2(1,000)}300 = 10.0 \text{ m/s}$$

$$a_{dn} = r\omega_3^2 = \frac{d_3}{2}\omega_3^2 = \frac{66.7}{2(1,000)}(300)^2 \qquad a_{dt} = 0 \qquad a_d = a_{dn} = 3,000 \text{ m/s}^2$$

The directions of $v_d$ and $a_d$ are seen in Fig. 16.62d.

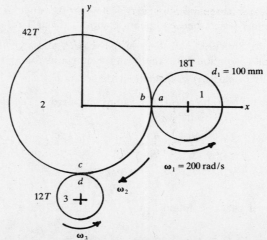

42T

18T  $d_1 = 100$ mm

2  b a 1  x

c

d  $\omega_2$

12T  3  $\omega_1 = 200$ rad/s

$\omega_3$

**Fig. 16.62b**

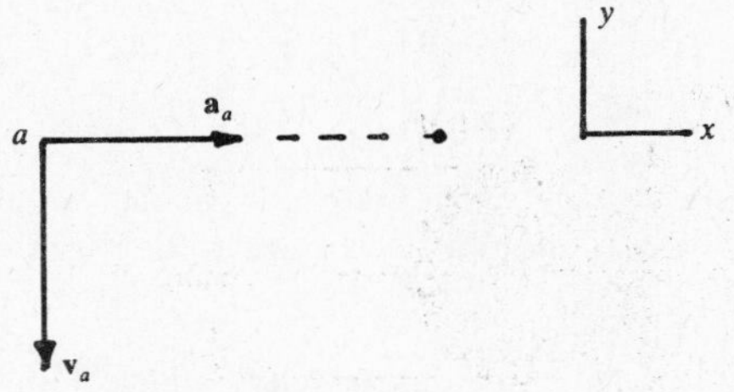

Fig. 16.62c

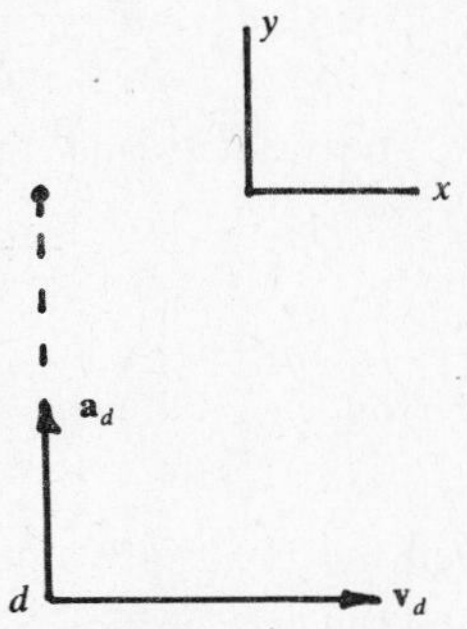

Fig. 16.62d

(b) The velocity and acceleration of points $a$ and $d$ may be written as

$$v_a = -10j \quad \text{m/s} \qquad a_a = 2{,}000i \quad \text{m/s}^2$$
$$v_d = 10i \quad \text{m/s} \qquad a_d = 3{,}000j \quad \text{m/s}^2$$

16.63   Do the same as in Prob. 16.62 if, at $t = 0$, gear 1 starts to decelerate at 65 rad/s².

▌ (a) Figure 16.62b is used, and $\alpha_1 = -65$ rad/s². The value of $\omega_1$ at the end of one revolution of gear 1, with $\theta_{01} = 0$ and $\omega_{01} = 200$ rad/s, is found from

$$\omega_1^2 = \omega_{01}^2 + 2\alpha_1(\theta_1 - \theta_{01}) = 200^2 + 2(-65)2\pi \qquad \omega_1 = 198 \text{ rad/s}$$

The angular velocity of gear 3 is given by

$$\omega_3 = \frac{18}{12}\,\omega_1 = \frac{18}{12}\,(198) = 297 \text{ rad/s}$$

The velocities of points $a$ and $d$ are found as

$$v_a = r\omega_1 = \frac{50}{1{,}000}\,(198) = 9.90 \text{ m/s} \qquad v_d = r\omega_3 = \frac{66.7}{2(1{,}000)}\,(297) = 9.90 \text{ ms/s}$$

The acceleration components of point $a$ are

$$a_{an} = r\omega_1^2 = \frac{50}{1{,}000}\,(198)^2 = 1{,}960 \text{ m/s}^2 \qquad a_{at} = r\alpha_1 = \frac{50}{1{,}000}\,(65) = 3.25 \text{ m/s}^2$$

Since $a_{at} \ll a_{an}$, $a_a \approx a_{an}$. The normal acceleration of point $d$ is given by

$$a_{dn} = r\omega_3^2 = \frac{66.7}{2(1{,}000)}\,(297)^2 = 2{,}940 \text{ m/s}^2$$

The angular acceleration of gear 3 is

$$\alpha_3 = \frac{18}{12}\,\alpha_1 = \frac{18}{12}\,(65) = 97.5 \text{ rad/s}^2$$

The tangential component of acceleration at point $d$ is then

$$a_{dt} = r\alpha_3 = \frac{66.7}{2(1{,}000)}\,(97.5) = 3.25 \text{ m/s}^2$$

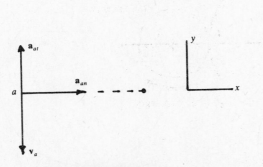

Fig. 16.63a

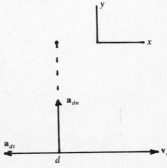

Fig. 16.63b

Since $a_{dt} \ll a_{dn}$, $a_d \approx a_{dn}$. The directions of the above components of velocity and acceleration are shown in Figs. 16.63a and b.

($b$) The velocity and acceleration of points $a$ and $d$ may be written as

$$\mathbf{v}_a = -9.90\mathbf{j} \quad \text{m/s} \qquad \mathbf{a}_a = 1{,}960\mathbf{i} + 3.25\mathbf{j} \quad \text{m/s}^2$$
$$\mathbf{v}_d = 9.90\mathbf{i} \quad \text{m/s} \qquad \mathbf{a}_d = -3.25\mathbf{i} + 2{,}940\mathbf{j} \quad \text{m/s}^2$$

**16.64** The friction drive turntable shown in Fig. 16.64a permits an infinitely variable angular velocity ratio $(\omega_2/\omega_1)$. $r_1 = 0.75$ in, $r_2$ may take on values between 1 in and 9 in, and pure rolling motion is assumed.

($a$) Find the range of values of $\omega_2/\omega_1$.

($b$) Find the acceleration of points on the rim of the disk if the angular velocity and acceleration of wheel 1 are 26 rad/s and 6 rad/s$^2$, respectively, and $r_2 = 7.6$ in.

❙ ($a$) Figure 16.64b shows the kinematic model of the two elements. For pure rolling motion,

$$v_a = v_b \qquad r_1\omega_1 = r_2\omega_2 \qquad \frac{\omega_2}{\omega_1} = \frac{r_1}{r_2}$$

The extreme values of $\omega_2/\omega_1$ are

$$\left.\frac{\omega_2}{\omega_1}\right|_{min} = \frac{0.75}{9} = 0.0833 \qquad \left.\frac{\omega_2}{\omega_1}\right|_{max} = \frac{0.75}{1} = 0.75$$

The range of the angular velocity ratio is given by

$$0.0833 \le \frac{\omega_2}{\omega_1} \le 0.75$$

($b$) Using $\omega_1 = 26$ rad/s and $r_2 = 7.6$ in, the angular velocity $\omega_2$ is found as

$$\frac{\omega_2}{\omega_1} = \frac{r_1}{r_2} \qquad \frac{\omega_2}{26} = \frac{0.75}{7.6} \qquad \omega_2 = 2.57 \text{ rad/s}$$

Using $\alpha_1 = 6$ rad/s$^2$ and $r_2 = 7.6$ in;

$$\frac{\alpha_2}{\alpha_1} = \frac{r_1}{r_2} \qquad \frac{\alpha_2}{6} = \frac{0.75}{7.6} \qquad \alpha_2 = 0.592 \text{ rad/s}^2$$

The components of acceleration of points on the rim of the disk, where $r = 10$ in, are

$$a_n = r\omega^2 = 10(2.57)^2 = 66.0 \text{ in/s}^2 \qquad a_t = r\alpha_2 = 10(0.592) = 5.92 \text{ in/s}^2$$

The resultant acceleration is

$$a = \sqrt{a_{an}^2 + a_{at}^2} = \sqrt{66.0^2 + 5.92^2} = 66.3 \text{ in/s}^2$$

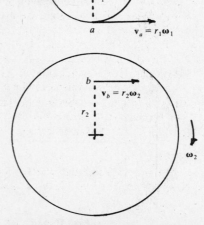

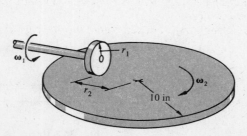

**Fig. 16.64a**

**Fig. 16.64b**

**16.65** Give a summary of the basic concepts of kinematics of plane motion of a rigid body.

❚ A rigid body is a body of unchanging dimensions. All points in a rigid body which moves with rectilinear, or curvilinear, translation move along parallel paths. The velocity and acceleration at all points in a body with this motion have the same magnitude, direction, and sense. If one point in a rigid body in plane motion is connected to an absolute reference system, such as the earth, the motion is referred to as rotation. Rotation, in its most general sense, is understood to mean motion in which a reference line on a rigid body experiences a changing angular position with respect to a fixed reference line in the plane of motion.

The angular displacement of the body with respect to the reference line is designated $\theta$. The angular velocity $\omega$ of the rigid body is the time rate of change of the angular displacement, or

$$\omega = \frac{d\theta}{dt} = \dot{\theta}$$

The basic units of $\omega$, in both USCS and SI units, are radians per second. The angular acceleration $\alpha$ of the rigid body is the time rate of change of the angular velocity, or

$$\alpha = \frac{d\omega}{dt} = \ddot{\theta}$$

The basic units of $\alpha$, in both USCS and SI units, are radians per second squared. The choice of a positive sense for $\theta$ automatically defines the same positive senses for $\omega$ and $\alpha$.

The plots of angular displacement, velocity, and acceleration vs. time are referred to as motion diagrams. The value of the angular velocity at any time is the slope of the angular displacement–time curve. The area under the angular velocity–time curve between two times is equal to the change in angular displacement during this time interval. The value of the angular acceleration at any time is the slope of the angular velocity–time curve. The area under the acceleration–time curve between two times is equal to the change in angular velocity during this time interval.

If a rigid body in plane motion has constant angular acceleration, the angular displacement, velocity, and acceleration are related by

$$\omega = \omega_0 + \alpha t \qquad \theta = \theta_0 + \omega_0 t + \tfrac{1}{2}\alpha t^2 \qquad \omega^2 = \omega_0^2 + 2\alpha(\theta - \theta_0)$$

where $\theta_0$ and $\omega_0$ are the initial angular displacement and velocity at initial time $t = 0$. These equations are the exact counterparts of the motion of a particle which moves in rectilinear translation with constant acceleration.

When a rigid body moves with plane rotation, the angular velocity $\omega$ is related to the translational velocity $v$ of an arbitrary point on the body by

$$v = r\omega$$

where $r$ is the length from the center of rotation to the point of interest. The direction of the velocity $v$ is normal to the direction of the line $r$. The components of acceleration of the point are related to the angular velocity and acceleration by

$$a_n = \frac{v^2}{r} = r\omega^2 \qquad a_t = r\alpha$$

where $n$ is the direction along the line $r$, and $t$ is a direction normal to the line $r$.

The most general type of plane motion of a rigid body is a combination of rotation and translation. The position of the body at any time may be described by specifying the two coordinates of an arbitrary point on the body, and the angular displacement of the body with respect to an arbitrary reference line.

All bodies which move with general plane motion may be considered to be instantaneously rotating about a fixed point in the plane. This point is called the instant center of rotation. The location of the instant center may be found if the directions of the absolute velocity of two points on the body are known. Two lines, which are normal to these directions, are drawn through the points. The intersection of these two lines locates the instant center. If the two known absolute velocities are parallel and unequal, the location of the instant center may be found from construction of the velocity distribution triangle. If the two known absolute velocities are parallel and equal, the body is in translation and does not rotate.

When a wheel experiences pure rolling along a surface, the translational velocity $v$ parallel to the track is related to the rotational velocity $\omega$ of the wheel by

$$v = r\omega$$

where $r$ is the radius of the wheel. Rolling of a wheel may be envisoned to be a succession of instantaneous rotations of the wheel about the contact between the wheel and the surface on which it rolls.

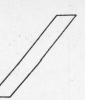

# CHAPTER 17
# Centroids, and Mass Moments and Products of Inertia, of Rigid Bodies

## 17.1 CENTROID OF A VOLUME, CENTER OF MASS, OR GRAVITY, OF A RIGID BODY, CENTROID OF A COMPOSITE RIGID BODY

**17.1**    Figure 17.1a shows a volume positioned with respect to a set of *xyz* coordinate axes.

    (*a*)    Find the coordinates of the *centroid* of the volume.

    (*b*)    How is the location of the centroid affected by the position of the *xyz* coordinates with respect to the volume?

**▌**  (*a*)    The volume is initially assumed to be a homogeneous solid of uniform density, so that it is a rigid body with an associated weight.    A rigid body is a body within which any two points always have the same separation distance between them, no matter what type of force system acts on the body.    As a consequence, the shape of a rigid body never changes.    Throughout this book the term body, or solid, will be understood to mean a rigid body.    A particular point in the body is defined to be the centroid of the volume.    Line *aa* in Fig. 17.1a is the top edge of an imagined knife edge, of vanishingly small lateral dimensions, which is parallel to the *zx* plane.    The gravitational acceleration vector **g** has the direction and sense of the negative *y* axis.    If the top edge of the knife is further imagined to pass through the centroid of the volume, then the body will be in perfect balance about this edge.

    Figure 17.1b shows a true view of the body, in a plane which is normal to the knife edge, and of a typical weight element *dW* of the body.    The *x* coordinate of the volume centroid is $x_c$.    The body is in equilibrium when resting on the imagined knife edge through the centroid.    Since the edge of the knife contacts the body only along a line through the centroid, it may be concluded that the reaction force of the knife edge on the body, and the weight force, are the only two forces acting on the body.    It follows, then, that these two forces constitute a collinear force system, and that the weight force must act through the centroid of the body.    Furthermore, the direction of these two forces must be the same as that of the gravitational acceleration vector.

    The moment of the weight force *W* about the *z* axis is given by

$$M_z = -Wx_c$$

This moment may also be expressed by summing all the moments of the elementary weight forces, or

$$M_z = -\int_v x \, dW$$

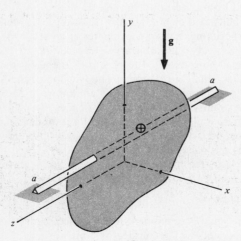

Fig. 17.1a

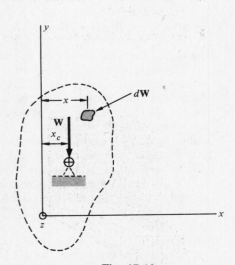

Fig. 17.1b

where $V$ is the volume of the body. It follows that

$$-Wx_c = -\int_v x\, dW \qquad x_c = \frac{\int_v x\, dW}{W}$$

Using $\gamma$, the specific weight of the material of the body, with the units of force per unit volume, the above equation may be written as

$$x_c = \frac{\int_v x\, dW}{\int_v dW} = \frac{\int_v x(\gamma\, dV)}{\int_v \gamma\, dV}$$

Since the body is assumed to be homogeneous, the specific weight is a constant, and this term may be moved outside the integral sign. The result is

$$x_c = \frac{\gamma \int_v x\, dV}{\gamma \int_v dV} = \frac{\int_v x\, dV}{V} \qquad (1)$$

By using an analysis similar to that shown above, the remaining two coordinates $y_c$ and $z_c$ of the centroid are

$$y_c = \frac{\int_v y\, dV}{V} \qquad (2)$$

$$z_c = \frac{\int_v z\, dV}{V} \qquad (3)$$

$x_c$, $y_c$, and $z_c$ are the centroidal coordinates of the volume $V$. It may be seen that these equations have the same forms as the equations which define the centroid of a plane area, with $A$ and $dA$ replaced by $V$ and $dV$, respectively. Equations (1), (2), and (3) are the formal definition of the coordinates of the *centroid of a volume*.

Table 17.13 at the end of this chapter gives the locations of the centroids of the volumes of several elementary homogeneous rigid bodies.

(b) The centroid of a volume is a point which has a *fixed* location with respect to the volume, and this location is *independent* of the position of the coordinate axes with respect to the volume.

17.2   (a) Show examples of volumes with one, two, and three planes of symmetry.

    (b) How is the determination of the coordinates of the centroid of a volume simplified if the volume has planes of symmetry?

❚ (a) Figures 17.2a, b, and c show examples of bodies with one, two, and three planes of symmetry.

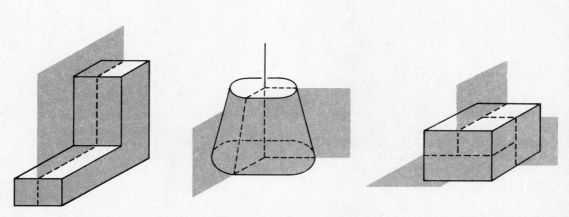

Fig. 17.2a           Fig. 17.2b           Fig. 17.2c

(*b*) If a volume has a plane of symmetry, Fig. 17.2*a*, then the centroid must lie in this plane. If there are two planes of symmetry, Fig. 17.2*b*, the centroid must lie in both planes. It follows that for this case the centroid must lie along the intersection line of the two planes. If the volume has three planes of symmetry, Fig. 17.2*c*, the centroid is located at the point of intersection of these three planes. For this last case, the common point of intersection of the three planes may be found by inspection.

**17.3** Define the *first moments* of a volume.

❚ Equations (1) through (3) in Prob. 17.1 may be written in the forms

$$x_c V = \int_V x \, dV \qquad y_c V = \int_V y \, dV \qquad z_c V = \int_V z \, dV$$

The terms $x_c V$, $y_c V$, and $z_c V$ are referred to as the first moments of the volume. In Problem 17.12 it will be shown how these terms are used in the computation of the centroidal coordinates of a composite volume.

**17.4** Find the centroidal coordinates of the volume of the right circular cone shown in Fig. 17.4*a*.

❚ From symmetry considerations, the centroid of the volume must lie on the *x* axis, so that

$$y_c = z_c = 0$$

A cross-sectional view of the cone is shown in Fig. 17.4*b*. The centroidal coordinate $x_c$ is defined by

$$x_c = \frac{\int_V x \, dV}{V} \tag{1}$$

The volume element $dV$ is chosen to be the disk-shaped element in Fig. 17.4*b* since, in the limit, all points in this element are at the same distance *x* from the *yz* plane. The differential volume is

$$dV = \pi y^2 \, dx \tag{2}$$

From the similar triangles in Fig. 17.4*b*,

$$\frac{y}{h-x} = \frac{d/2}{h}$$

$$y = \frac{d}{2}\left(1 - \frac{x}{h}\right) \tag{3}$$

Equation (3) is substituted into Eq. (2), to obtain:

$$dV = \pi \left(\frac{d}{2}\right)^2 \left(1 - \frac{x}{h}\right)^2 dx$$

The total volume of the right circular cone is

$$V = \frac{1}{3}\left(\frac{\pi d^2}{4}\right)h = \frac{1}{12}\pi d^2 h$$

Equation (1) now appears as

$$x_c = \frac{\int_V x \, dV}{V} = \frac{\int_0^h x\pi\left(\frac{d}{2}\right)^2\left(1-\frac{x}{h}\right)^2 dx}{\frac{1}{12}\pi d^2 h} = \frac{\frac{\pi d^2}{4}\int_0^h \left(x - \frac{2x^2}{h} + \frac{x^3}{h^2}\right)dx}{\frac{1}{12}\pi d^2 h} = \frac{3}{h}\left(\frac{x^2}{2} - \frac{2x^3}{3h} + \frac{x^4}{4h^2}\right)_0^h$$

$$= \frac{3}{h}\left(\frac{h^2}{2} - \frac{2h^3}{3h} + \frac{h^4}{4h^2}\right) = \frac{3}{h}\left(\frac{h^2}{12}\right) = \frac{1}{4}h$$

It was observed earlier that the centroid of an area or volume is a unique and unchanging location with respect to the boundaries of the area or volume. The values of the centroidal coordinates, however, are not unique, since these quantities are determined by the placement of the reference axes relative to the area or volume. If the axes were attached to the cone of the present problem as shown in Fig. 17.4*c*, the centroidal *x* coordinate would be

$$x_c = -\tfrac{3}{4}h$$

For placement of the axes as shown in Fig. 17.4*d*,

$$x_c = 0$$

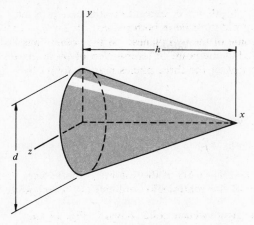

Fig. 17.4*a*

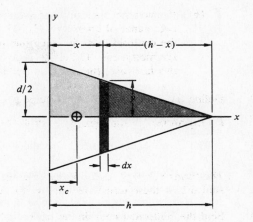

Fig. 17.4*b*

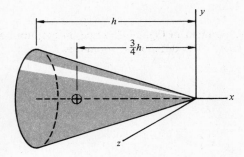

Fig. 17.4*c*

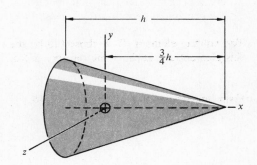

Fig. 17.4*d*

**17.5**  (*a*)  What is meant by the *center of mass* of a rigid body?

  (*b*)  What is an alternative term used to describe the center of mass of a rigid body?

  (*c*)  State two practical applications of the concept of the center of mass of a rigid body.

  (*d*)  How is the location of the center of mass of a body determined if the body is made of homogeneous material?

▎ (*a*)  The center of mass (CM) of a rigid body is a point in the body through which the total weight force acts.  It is also a point at which all the mass of the body may be imagined to be "concentrated."

  (*b*)  A term commonly used to describe the center of mass is *center of gravity*.

  (*c*)  An immediate, practical application of the concept of the center of mass is that the weight force of a body may be assumed to be a single resultant force, with magnitude equal to the total weight of the body, which acts through the center of mass.  It is evident that knowing the location of the center of mass of a rigid body is a prerequisite for showing the location of the weight force in a free-body diagram of the body.

   Another use of the center of mass, or gravity, occurs in many problems in dynamics, where it is convenient to describe the motion of the body as a combined effect of translation of the center of mass and rotation about this point.

  (*d*)  If the material of a body is homogeneous, the centroid of the volume of the body and the center of mass are coincident points.

**17.6**  How is the location of the center of mass of a body determined if the body is made of material which is not homogeneous?

▮ If the material of the body has a varying specific weight, the coordinates of the center of mass are given by

$$x_c = \frac{\int_v x\gamma \, dV}{\int_v \gamma \, dV} = \frac{\int_v x\gamma \, dV}{W} \qquad y_c = \frac{\int_v y\gamma \, dV}{\int_v \gamma \, dV} = \frac{\int_v y\gamma \, dV}{W} \qquad z_c = \frac{\int_v z\gamma \, dV}{\int_v \gamma \, dV} = \frac{\int_v z\gamma \, dV}{W}$$

where $\gamma$, the specific weight of the material of the body, is a known function of the coordinates and $W$ is the total weight of the body. $\gamma$ may *not* be moved outside the integral sign in the above equations, since it is a variable quantity.

The above equations are also valid if the specific weight term $\gamma$ is replaced by the mass density $\rho$.

**17.7** Figure 17.7a shows a container of rectangular cross section. A loose, soft material is packed, and compacted, into the container. The material at the bottom of the container is more tightly packed than the material at the top. An estimate of the variation of the specific weight with depth of packing is shown in Fig. 17.7b. Find the centroidal coordinate $y_c$ of the packed material in the container.

▮ The specific weight distribution curve is redrawn in Fig. 17.7c. $\gamma$ and $y$ are a pair of points on the curve. From the similar triangles in the figure,

$$\frac{960 - \gamma}{y} = \frac{960 - 780}{0.6} \qquad \gamma = 960 - 300y$$

where $\gamma$ is in newtons per cubic meter and $y$ is in meters.

The centroidal coordinate $y_c$, given in Prob. 17.1, has the form

$$y_c = \frac{\int_v y\gamma \, dV}{\int_v \gamma \, dV} \tag{1}$$

The volume element $dV$ is given by

$$dV = A \, dy$$

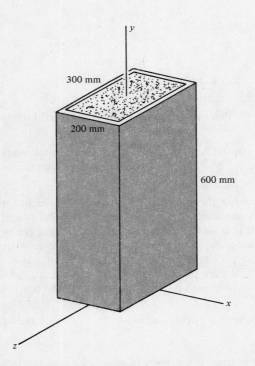

**Fig. 17.7a**

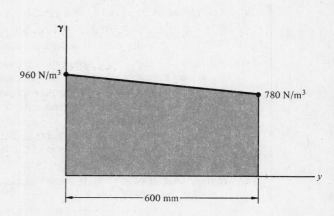

**Fig. 17.7b**

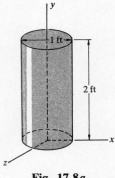

Fig. 17.7c

where $A = 0.2(0.3) \, \text{m}^2$ is the cross-sectional area of the container. Equation (1) may be written as

$$y_c = \frac{\int_0^{0.6} y(960 - 300y) \, A \, dy}{\int_0^{0.6} (960 - 300y) \, A \, dy} = \frac{[960(y^2/2) - 300(y^3/3)]_0^{0.6}}{[960y - 300(y^2/2)]_0^{0.6}} = 0.290 \, \text{m} = 290 \, \text{mm}$$

Had the material been homogeneous, the value of $y_c$, from inspection of the figure, would have been 300 mm.

**17.8**   The material of the cylinder shown in Fig. 17.8a has the variation in specific weight shown in Fig. 17.8b. The equation of this curve is $\gamma = 6.7y^2 - 3.4y + 80$, where $\gamma$ is in pounds per cubic feet and $y$ is in feet.

(**a**)   Find the coordinate $y_c$ of the center of mass of the material.

(**b**)   Find the percent difference between the result in part (*a*) and the centroidal coordinate $y_c$ of the volume of the cylinder.

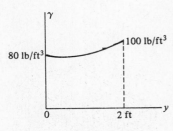

Fig. 17.8a                    Fig. 17.8b

▌ (a)   The functional forms for $\gamma$, and the differential volume element, are

$$\gamma = 6.7y^2 - 3.4y + 80 \qquad \text{lb/ft}^3 \qquad dV = \frac{\pi(1)^2}{4} \, dy \qquad \text{ft}^3$$

The centroidal coordinate $y_c$ is then found from

$$y_c = \frac{\int_V y\gamma \, dV}{\int_V \gamma \, dV} = \frac{\int_0^2 y(6.7y^2 - 3.4y + 80) \dfrac{\pi(1)^2}{4} \, dy}{\int_0^2 (6.7y^2 - 3.4y + 80) \dfrac{\pi(1)^2}{4} \, dy} = \frac{\int_0^2 (6.7y^3 - 3.4y^2 + 80y) \, dy}{\int_0^2 (6.7y^2 - 3.4y + 80) \, dy}$$

$$= \frac{\left[6.7 \dfrac{y^4}{4} - 3.4 \dfrac{y^3}{3} + 80 \dfrac{y^2}{2}\right]\Big|_0^2}{\left[6.7 \dfrac{y^3}{3} - 3.4 \dfrac{y^2}{2} + 80y\right]\Big|_0^2} = 1.04 \, \text{ft}$$

(b) The centroidal coordinate of the *volume* of the cylinder is $y_c = 1$ ft. The percent difference between the two centroidal coordinates is given by

$$\%D = \frac{1.04 - 1.00}{1.00} (100) = 4\%$$

**17.9** Figure 17.9a shows a container of rectangular cross section. Two experimental materials are to be cast in the container. Material 1, with the variation of specific weight shown in Fig. 17.9b, is cast in the container first. The specific weight variation of material 2 is shown in Fig. 17.9c, and this material is cast on top of material 1. The weight of the container may be neglected.

(a) Find the weight of the material in the container.

(b) Find the centroidal coordinate of the center of mass of the container.

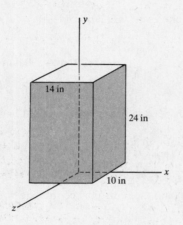

Fig. 17.9a

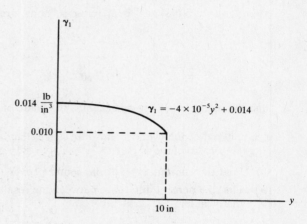

Fig. 17.9b

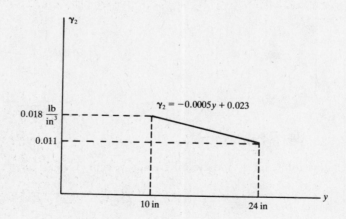

Fig. 17.9c

**▌** (a) The weight of the material in the container is given by

$$W = \int_V \gamma \, dV = \int_0^{10} (-4 \times 10^{-5} y^2 + 0.014)14(10) \, dy + \int_{10}^{24} (-0.0005y + 0.023)14(10) \, dy$$

$$= 140 \left[ -4 \times 10^{-5} \frac{y^3}{3} + 0.014y \right]_0^{10} + 140 \left[ -0.0005 \frac{y^2}{2} + 0.023y \right]_{10}^{24}$$

$$= 140 \left[ -4 \times 10^{-5} \left( \frac{10^3}{3} \right) + 0.014(10) - \frac{0.0005}{2} (24^2 - 10^2) + 0.023(24 - 10) \right]$$

$$= 46.2 \text{ lb}$$

(*b*) The centroidal coordinate $y_c$ of the material in the container is found from

$$y_c = \frac{\int_V y\gamma\, dV}{W} = \frac{\int_0^{10}(-4\times10^{-5}y^2+0.014)14(10)y\, dy + \int_{10}^{24}(-0.0005y+0.023)14(10)y\, dy}{46.2\text{ lb}}$$

$$= \frac{14(10)}{46.2}\left[-4\times10^{-5}\left(\frac{y^4}{4}\right)+0.014\left(\frac{y^2}{2}\right)\right]_0^{10} + \frac{14(10)}{46.2}\left[-0.0005\left(\frac{y^3}{3}\right)+0.023\left(\frac{y^2}{2}\right)\right]_{10}^{24}$$

$$= \frac{140}{46.2}\left[-4\times10^{-5}\left(\frac{10^4}{4}\right)+0.014\left(\frac{10^2}{2}\right)-\frac{0.0005}{3}(24^3-10^3)+\frac{0.023}{2}(24^2-10^2)\right] = 11.9\text{ in}$$

**17.10** The specific weight variation of the material from which the cone in Fig. 17.10*a* is fabricated is shown in Fig. 17.10*b*.

(*a*) Find the coordinate $y_c$ of the center of mass of the cone.

(*b*) Find the percent difference between the result in part (*a*) and the centroidal coordinate $y_c$ of the volume of the cone.

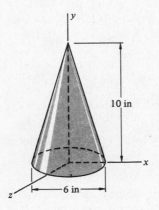

**Fig. 17.10*a***

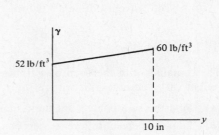

**Fig. 17.10*b***

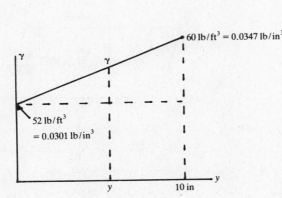

**Fig. 17.10*c***

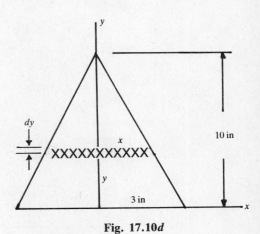

**Fig. 17.10*d***

▌ (*a*) The curve of $\gamma$ versus $y$ is redrawn in Fig. 17.10*c*. From the similar triangles in the figure

$$\frac{\gamma-0.0301}{y} = \frac{0.0347-0.0301}{10} \qquad \gamma = 0.00046y + 0.0301 \qquad \text{lb/in}^3$$

The cone is redrawn in Fig. 17.10*d*. From the similar tringles in this figure,

$$\frac{10-y}{x} = \frac{10}{3} \qquad x = 0.3(10-y)$$

The volume element has the form

$$dV = \pi x^2 \, dy = \pi [0.3(10 - y)]^2 \, dy$$

The expression for the coordinate of the center of mass of the cone has the form

$$y_c = \frac{\int_V y\gamma \, dV}{\int_V \gamma \, dV} = \frac{\cancel{\pi(0.3)^2} \int_0^{10} y(0.00046y + 0.0301)(100 - 20y + y^2) \, dy}{\cancel{\pi(0.3)^2} \int_0^{10} (0.00046y + 0.0301)(100 - 20y + y^2) \, dy}$$

$$= \frac{\int_0^{10} (0.00046y^4 + 0.0209y^3 - 0.556y^2 + 3.01y) \, dy}{\int_0^{10} (0.00046y^3 + 0.0209y^2 - 0.556y + 3.01) \, dy}$$

$$= \frac{\left[ 0.00046 \frac{y^5}{5} + 0.0209 \frac{y^4}{4} - 0.556 \frac{y^3}{3} + 3.01 \frac{y^2}{2} \right]\Big|_0^{10}}{\left[ 0.00046 \frac{y^4}{4} + 0.0209 \frac{y^3}{3} - 0.556 \frac{y^2}{2} + 3.01y \right]\Big|_0^{10}} = 2.56 \text{ in}$$

(**b**) The centroidal coordinate of the volume of the cone, from Case 6 of Table 17.13, is

$$y_c = \tfrac{1}{4}h = 2.50 \text{ in}$$

The percent difference between the two centroidal coordinates is given by

$$\%\text{D} = \frac{2.56 - 2.50}{2.50} \, (100) = 2.4\%$$

**17.11** The container with the form of a truncated cone, shown in Fig. 17.11a, is filled with material whose specific weight varies according to Fig. 17.7b.

(**a**) Find the coordinate $y_c$ of the center of mass of the material.

(**b**) Find the percent difference between the result in part (a) and the centroidal coordinate $y_c$ of the volume of the truncated cone.

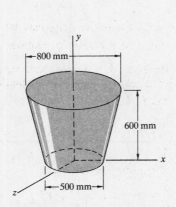

**Fig. 17.11a**

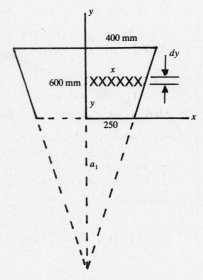

**Fig. 17.11b**

▌ (**a**) The cone is redrawn in Fig. 17.11b. From the similar triangles in this figure,

$$\frac{a_1 + 600}{400} = \frac{a_1}{250} \qquad a_1 = 1,000 \text{ mm} \qquad \frac{a_1 + 600}{400} = \frac{a_1 + y}{x} \qquad x = 250 + \tfrac{1}{4}y \qquad \text{mm}$$

The volume element has the form

$$dV = \frac{\pi x^2}{4}\, dy = \frac{\pi}{4}\,(250 + \tfrac{1}{4}y)^2\, dy$$

The variation of $\gamma$ with $y$, from Prob. 17.7, may be expressed as

$$\gamma = 960 - 0.3y \qquad N/m^3$$

where $y$ is in millimeters. The centroidal coordinate $y_c$ of the material in the truncated conical volume is found from

$$y_c = \frac{\int_V y\gamma\, dV}{\int_V \gamma\, dV} = \frac{\int_0^{600} y(960 - 0.3y)\frac{\pi}{4}(250 + \tfrac{1}{4}y)^2\, dy}{\int_0^{600} (960 - 0.3y)\frac{\pi}{4}(250 + \tfrac{1}{4}y)^2\, dy}$$

$$= \frac{\int_0^{600} (-0.0188y^4 + 22.5y^3 + 1.01 \times 10^5 y^2 + 6 \times 10^7 y)\, dy}{\int_0^{600} (-0.0188y^3 + 22.5y^2 + 1.01 \times 10^5 y + 6 \times 10^7)\, dy}$$

$$= \frac{\left[-0.0188\frac{y^5}{5} + 22.5\frac{y^4}{4} + 1.01 \times 10^5 \frac{y^3}{3} + 6 \times 10^7 \frac{y^2}{2}\right]\Big|_0^{600}}{\left[-0.0188\frac{y^4}{4} + 22.5\frac{y^3}{3} + 1.01 \times 10^5 \frac{y^2}{2} + 6 \times 10^7 y\right]\Big|_0^{600}} = 335 \text{ mm}$$

(b) The centroidal coordinate of the volume of the truncated cone is given by

$$y_c = \frac{\int_V y\, dV}{\int_V dV} = \frac{\int_0^{600} y\left(\frac{\pi}{4}\right)(250 + \tfrac{1}{4}y)^2\, dy}{\int_0^{600} \frac{\pi}{4}(250 + \tfrac{1}{4}y)^2\, dy} = \frac{\int_0^{600} (250^2 y + 125y^2 + \tfrac{1}{16}y^3)\, dy}{\int_0^{600} (250^2 + 125y + \tfrac{1}{16}y^2)\, dy}$$

$$= \frac{\left[6.25 \times 10^4 \frac{y^2}{2} + 125\frac{y^3}{3} + 0.0625\frac{y^4}{4}\right]\Big|_0^{600}}{\left[6.25 \times 10^4 y + 125\frac{y^2}{2} + 0.0625\frac{y^3}{3}\right]\Big|_0^{600}} = 345 \text{ mm}$$

The percent difference between the two centroidal coordinates is found as

$$\%D = \frac{345 - 335}{345}\,100 = 2.9\%$$

**17.12** (a) Show how the coordinates of the centroid of a homogeneous composite body may be found.

(b) How is the computation of the centroidal coordinates modified if the body contains holes, or cutouts?

❚ (a) In many problems the rigid body whose centroid is desired is not one of the simple shapes shown in Table 17.13 at the end of this chapter. For this case, it may be possible to subdivide the original body into elementary shapes whose centroidal coordinates are known. A centroidal coordinate of any body has the general form

$$x_c = \frac{\int_V dV}{V}$$

where $V$ is the total volume of the body. This equation may be written as

$$x_c = \frac{\int_V x\, dV}{V} = \frac{\int_{V_1} x\, dV_1 + \int_{V_2} x\, dV_2 + \cdots + \int_{V_n} x\, dV_n}{V_1 + V_2 + \cdots + V_n} \qquad (1)$$

where $n$ is the number of elementary volumes into which the original volume has been subdivided. The typical term $\int_{V_n} x\, dV_n$ in the numerator of the above equation, using the results in Prob. 17.3, may be written as

$$\int_{V_n} x\, dV_n = x_n V_n$$

where $x_n$ is the centroidal coordinate of the elementary volume $V_n$. The term $x_n V_n$ may be recognized as the first moment of volume of the volume element $V_n$ about the reference axis from which $x_n$ is measured. Using the above result, Eq. (1) may be written as

$$x_c = \frac{x_1 V_1 + x_2 V_2 + \cdots + x_n V_n}{V_1 + V_2 + \cdots + V_n} \qquad (2)$$

The remaining two centroidal coordinates have the forms

$$y_c = \frac{y_1 V_1 + y_2 V_2 + \cdots + y_n V_n}{V_1 + V_2 + \cdots + V_n} \qquad (3)$$

$$z_c = \frac{z_1 V_1 + z_2 V_2 + \cdots + z_n V_n}{V_1 + V_2 + \cdots + V_n} \qquad (4)$$

It may be seen that Eqs. (2) through (4) have the same forms as the equations used to find the centroid of a composite area, if the terms $V_n$ are replaced by $A_n$.

Since the composite body is homogeneous, the equations for the centroidal coordinates may be written in terms of the mass, or weight, elements of the elementary volumes, with the typical forms

$$x_c = \frac{x_1 m_1 + x_2 m_2 + \cdots + x_n m_n}{m_1 + m_2 + \cdots + m_n} = \frac{x_1 W_1 + x_2 W_2 + \cdots + x_n W_n}{W_1 + W_2 + \cdots + W_n}$$

(b) If the composite volume contains holes or cutout portions, the volumes, masses, or weights of these elements are considered to be *negative* quantities when used in the equations in part (a).

17.13 A tank of square cross section contains water and oil, as shown in Fig. 17.13a. The density of the water is $1,000 \text{ kg/m}^3$, and the density of the oil is $920 \text{ kg/m}^3$. Find the centroidal coordinate of the center of mass of the material in the tank.

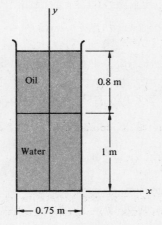

Fig. 17.13a

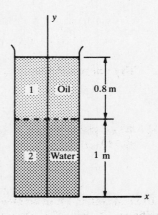

Fig. 17.13b

▌ The oil and water are designated masses 1 and 2, as shown in Fig. 17.13b. $A$ is the cross-sectional area of the tank, and the values of the two masses are

$$m_1 = 920 A (0.8) = 736 A \quad \text{kg} \qquad m_2 = 1,000 A (1) = 1,000 A \quad \text{kg}$$

The centroidal coordinate $y_c$ is then found from

$$y_c = \frac{y_1 m_1 + y_2 m_2}{m_1 + m_2} = \frac{1.4 (736 A) + 0.5 (1,000 A)}{736 A + 1,000 A} = 0.882 \text{ m}$$

It may be observed that the above result is independent of the value of the cross-sectional area $A$.

17.14 A cylindrical container exposed to freezing temperatures formed the ice layer shown in Fig. 17.14a. The specific weight of ice is $56 \text{ lb/ft}^3$, and the specific weight of water is $62.4 \text{ lb/ft}^3$. The mass of the container may be neglected.

(a) Find the $y$ coordinate of the center of mass of the container of water and ice.

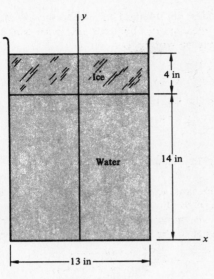

Fig. 17.14a

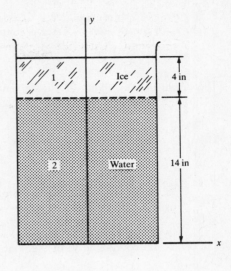

Fig. 17.14b

(**b**)   Find the centroidal $y$ coordinate at a later time when the ice has melted.   Neglect the small volume change due to the change in phase as the ice melts.

▮ (**a**)   Figure 17.14b shows the designation of the weights of ice and water as 1 and 2.   The values of these weights are

$$W_1 = \frac{56}{(12)^3} \, \frac{\pi(13)^2}{4} \, (4) = 17.2 \, \text{lb}$$

$$W_2 = \frac{62.4}{(12)^3} \, \frac{\pi(13)^2}{4} \, (14) = 67.1 \, \text{lb}$$

The centroidal coordinate $y_c$ of the container of ice and water is then found from

$$y_c = \frac{y_1 W_1 + y_2 W_2}{W_1 + W_2} = \frac{16(17.2) + 7(67.1)}{17.2 + 67.1} = 8.84 \, \text{in}$$

(**b**)   When the container is filled with water only, the centroidal coordinate $y_c$, from inspection of Fig. 17.14a, is given by

$$y_c = \frac{14 + 4}{2} = 9 \, \text{in}$$

**17.15**   A cylindrical tank is 1 ft in diameter and 4 ft high.   Into the tank is poured 12 gal of water and 10 gal of oil with a specific gravity of 0.89.   Find the location of the center of mass of the tank.   The mass of the tank may be neglected, and   7.48 gal = 1 ft³.

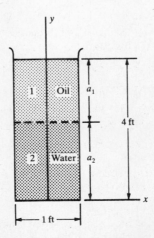

Fig. 17.15

❚ Figure 17.15 shows the tank. Since the oil is less dense than water, it floats on the top of the water. The height $a_2$ of the column of water is found from

$$12 \, \text{gal}\left(\frac{1 \, \text{ft}^3}{7.48 \, \text{gal}}\right) = \frac{\pi(12)^2}{4} a_2 \left(\frac{1 \, \text{ft}}{12 \, \text{in}}\right)^3 \qquad a_2 = 24.5 \, \text{in}$$

The height of the column of oil is given by

$$\frac{10}{7.48} = \frac{\pi(12)^2}{4} a_1 \left(\frac{1}{12}\right)^3 \qquad a_1 = 20.4 \, \text{in}$$

The total height of the two columns is

$$a_1 + a_2 = 44.9 \, \text{in} = 3.74 \, \text{ft}$$

Since 3.74 ft < 4 ft, the tank can hold the given quantities of oil and water. The weights of the oil and water are

$$W_1 = \frac{10}{7.48} (62.4)0.89 = 74.2 \, \text{lb} \qquad W_2 = \frac{12}{7.48} (62.4) = 100 \, \text{lb}$$

The centroidal coordinate $y_c$ of the container is then found from

$$y_c = \frac{y_1 W_1 + y_2 W_2}{W_1 + W_2} = \frac{\left(24.5 + \dfrac{20.4}{2}\right)74.2 + \dfrac{24.5}{2}(100)}{74.2 + 100} = 21.8 \, \text{in}$$

**17.16** (a) Find the $x$, $y$, and $z$ coordinates of the center of mass of the bracket shown in Fig. 17.16.

(b) Find the weight, and the mass in slugs, of the bracket.
The bracket is made of bronze, with a specific weight of 0.295 lb/in².

❚ (a) The bracket is divided into the three elementary volume shapes shown in Fig. 17.16. The volumes are

$$V_1 = 3.2(0.8)(2) = 5.12 \, \text{in}^3 \qquad V_2 = 1.4(1.8)(2) = 5.04 \, \text{in}^3 \qquad V_3 = \frac{\pi(1)^2}{4} 3 = 2.36 \, \text{in}^3$$

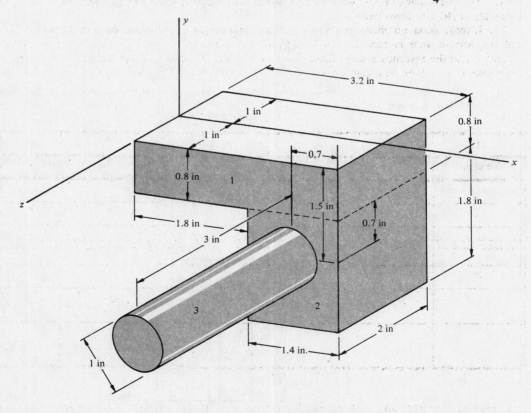

**Fig. 17.16**

The values of the nine required centroidal coordinates of the three elementary volumes are

$$x_1 = \frac{3.2}{2} = 1.6 \text{ in} \qquad y_1 = -\frac{0.8}{2} = -0.4 \text{ in} \qquad z_1 = 0$$

$$x_2 = 1.8 + \frac{1.4}{2} = 2.5 \text{ in} \qquad y_2 = -\left(0.8 + \frac{1.8}{2}\right) = -1.7 \text{ in} \qquad z_2 = 0$$

$$x_3 = 1.8 + \frac{1.4}{2} = 2.5 \text{ in} \qquad y_3 = -1.5 \text{ in} \qquad z_3 = 1 + \frac{3}{2} = 2.5 \text{ in}$$

The centroidal coordinates of the bracket are then

$$x_c = \frac{x_1 V_1 + x_2 V_2 + x_3 V_3}{V_1 + V_2 + V_3} = \frac{1.6(5.12) + 2.5(5.04) + 2.5(2.36)}{5.12 + 5.04 + 2.36} = 2.14 \text{ in} \qquad V = 12.5 \text{ in}^3$$

$$y_c = \frac{y_1 V_1 + y_2 V_2 + y_3 V_3}{V_1 + V_2 + V_3} = \frac{-0.4(5.12) - 1.7(5.04) - 1.5(2.36)}{12.5} = -1.13 \text{ in}$$

$$z_c = \frac{z_1 V_1 + z_2 V_2 + z_3 V_3}{V_1 + V_2 + V_3} = \frac{0 + 0 + 2.5(2.36)}{12.5} = 0.472 \text{ in}$$

(b) The weight of the bracket is

$$W = \gamma V = 0.295(12.5) = 3.69 \text{ lb}$$

The mass is

$$m = \frac{W}{g} = \frac{3.69}{32.2} = 0.115 \frac{\text{lb} \cdot \text{sec}^2}{\text{ft}} = 0.115 \text{ slug}$$

**17.17** Show a tabular format for organizing the computations for the centroidal coordinates for the case where the body is a complex shape which requires the description of several elementary shapes.

❚ A format for a tabular organization of the solutions for the centroidal coordinates is shown in Table 17.1. This particular example uses the volumes of the elementary shapes, although the masses or weights of the elements could also have been used.

The identification numbers of the elementary volumes are entered in column 1, and the values of these volumes are shown in column 2. The centroidal coordinates of the elementary volumes are listed in columns 3, 5, and 7, and the first moments of these volumes are shown in columns 4, 6, and 8. Columns 2, 4, 6, and 8 are then summed, and the final values for $x_c$, $y_c$, and $z_c$ are found by using the equations in the last row of the table.

**TABLE 17.1**

| (1) | (2) | (3) | (4) | (5) | (6) | (7) | (8) |
|---|---|---|---|---|---|---|---|
| Element | $V_i$ | $x_i$ | $x_i V_i$ | $y_i$ | $y_i V_i$ | $z_i$ | $z_i V_i$ |
| 1 | | | | | | | |
| 2 | | | | | | | |
| | | | | | | | |
| n | | | | | | | |
| | $V = \sum_i V_i$ | | $\sum_i x_i V_i$ | | $\sum_i y_i V_i$ | | $\sum_i z_i V_i$ |
| | | | $x_c = \dfrac{\sum_i x_i V_i}{V}$ | | $y_c = \dfrac{\sum_i y_i V_i}{V}$ | | $z_c = \dfrac{\sum_i z_i V_i}{V}$ |

**17.18** As part of a weight- and cost-reduction program, the bracket in Prob. 17.16 is modified by drilling a hole through it parallel to the y axis, as shown in Fig. 17.18.

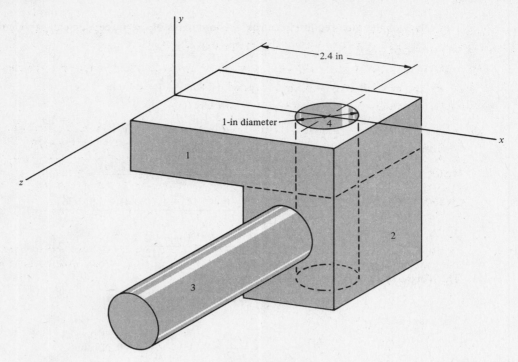

**Fig. 17.18**

(*a*)  Find the centroidal coordinates of the bracket after the hole has been drilled.

(*b*)  Find the weight of the modified bracket.

(*c*)  Find the percent reduction from the weight of the original bracket.

▌ (*a*)  The solution can be organized in a tabular format, as shown in Table 17.2.   The first three rows in the table record the results from Prob. 17.16.   The hole is designated element 4.   A minus sign is affixed to the volume of this element to show that it represents material *removed* from the bracket.   The new values of the centroidal coordinates are contained in the last row of the table.

(*b*)  The weight of the modifed bracket is

$$W = \gamma V = 0.295(10.5) = 3.10 \text{ lb}$$

(*c*)  The hole has the effect of reducing the original weight by   $[(3.69 - 3.10)/3.69]100 = 16.0$   percent.

| TABLE 17.2 | | | | | | | |
|---|---|---|---|---|---|---|---|
| Element | $V_i$ | $x_i$ | $x_i V_i$ | $y_i$ | $y_i V_i$ | $z_i$ | $z_i V_i$ |
| 1 | 5.12 | 1.6 | 8.19 | −0.4 | −2.05 | 0 | 0 |
| 2 | 5.04 | 2.5 | 12.6 | −1.7 | −8.57 | 0 | 0 |
| 3 | 2.36 | 2.5 | 5.9 | −1.5 | −3.54 | 2.5 | 5.90 |
| 4 | $\dfrac{-\pi(1)^2}{4}(2.6) = -2.04$ | 2.4 | −4.90 | −1.3 | 2.65 | 0 | 0 |
| | $V = \displaystyle\sum_i V_i$ <br> $V = 10.5$ | $\displaystyle\sum_i x_i V_i = 21.8$ | | $\displaystyle\sum_i y_i V_i = -11.5$ | | $\displaystyle\sum_i z_i V_i = 5.90$ | |
| | | $x_c = \dfrac{21.8}{10.5} = 2.08 \text{ in}$ | | $y_c = \dfrac{-11.5}{10.5} = -1.10 \text{ in}$ | | $z_c = \dfrac{5.90}{10.5} = 0.562 \text{ in}$ | |

**17.2 MOMENT OF INERTIA OF A RIGID BODY, RADIUS OF GYRATION, PARALLEL-AXIS, OR TRANSFER, THEOREM FOR MASS MOMENT OF INERTIA**

**17.19** **(a)** Give the formal definitions of the *mass moments of inertia* of a body about the *xyz* coordinate axes.

**(b)** What are the units of mass moment of inertia?

**(c)** Is the mass moment of inertia a scalar or vector quantity?

▎ **(a)** Figure 17.19a shows a homogeneous solid body. The typical mass element *dm* has the general position coordinates *x*, *y*, and *z*. The true view of the mass element in a plane parallel to the *xy* plane is shown in Fig. 17.19b. The *z* axis is normal to the plane of the figure, and $r_z$ is the perpendicular distance between the mass element and the *z* axis.

A quantity called the mass moment of inertia about the *z* axis, designated by the symbol $I_z$, is defined as

$$I_z = \int_V r_z^2 \, dm$$

In the basic formulation of the mass moment of inertia given above, each mass element of the body is multiplied by the square of its distance to the reference axis. These products are then summed over the volume of the body. Since

$$r_z^2 = x^2 + y^2$$

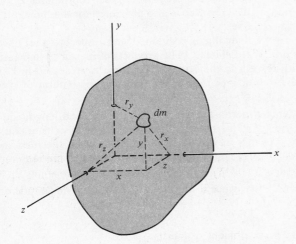

**Fig. 17.19a**

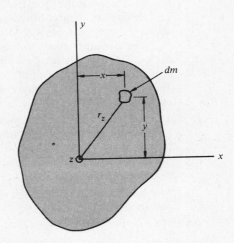

**Fig. 17.19b**

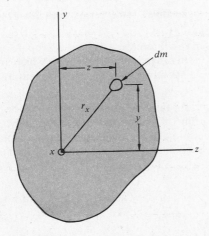

**Fig. 17.19c**

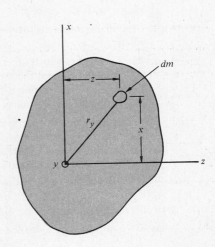

**Fig. 17.19d**

the mass moment of inertia about the $z$ axis may be written as

$$I_z = \int_V (x^2 + y^2)\, dm \qquad (1)$$

The true views of the mass element in planes parallel to the $yz$ and $zx$ planes are shown in Figs. 17.19$c$ and $d$. The mass moments of inertia about the $x$ and $y$ axes are defined as

$$I_x = \int_V r_x^2\, dm = \int_V (y^2 + z^2)\, dm \qquad (2)$$

$$I_y = \int_V r_y^2\, dm = \int_V (z^2 + x^2)\, dm \qquad (3)$$

Equations (1) through (3) are the formal definitions of the mass moments of inertia of the body about the three coordinate axes. The subscript on the symbol $I$ always designates the axis to which the mass moment of inertia is referenced.

The equations for $I_x$, $I_y$, and $I_z$ are usually not used directly to compute the moments of inertia, since integration over a volume is inherently a difficult triple-integration procedure. Rather, a technique is employed which combines known solutions for mass moments of inertia, and which uses a theorem which is referred to as the parallel-axis, or transfer, theorem. This latter relationship is developed in Prob. 17.28. Table 17.13 at the end of this chapter contains the equations for the centroidal coordinates and mass moments of inertia of rigid bodies of elementary geometry. These general results are presented for the moment of inertia with respect to a set of centroidal coordinates $x_0$, $y_0$, $z_0$ that pass through the center of mass of the body. The results are also presented for the moments of inertia about selected axes parallel to the centroidal axes.

There is another mass-related property of a body that is positioned with respect to a set of coordinate axes. This quantity is called the *mass product of inertia*. The products of inertia are a measure of the symmetry of placement of the axes with respect to the body. This topic is presented in Prob. 17.93, and used in Chap. 21, where the mass products of inertia are used in the solution of problems of dynamic unbalance.

(*b*)    The fundamental units of a mass moment of inertia are the product of mass and length squared. In USCS units, the typical units of mass moment of inertia are $lb \cdot sec^2 \cdot in$. The SI units of mass moment of inertia are $kg \cdot m^2$. On engineering drawings in which USCS units are used, the mass moment of inertia often is stated in $lb \cdot in^2$. This is a hybrid unit which may be converted to the required units by dividing by $g = 386\ in/s^2$.

(*c*)    The mass moment of inertia is a *scalar* quantity, since it requires a statement of magnitude only for its definition.

17.20    (*a*)    State three significant characteristics of the mass moment of inertia.

(*b*)    Give a physical interpretation of the mass moment of inertia.

▌ (*a*)    1. The mass moment of inertia of a body is *not* an inherent characteristic of the body. Rather, it is a property of the *combined* system of the body and its position with respect to a set of reference axes.

2. The mass moments of inertia are always positive, since terms such as $x^2$, $y^2$, $z^2$, and $dm$ are inherently positive.

3. Mass elements which are more distant from the reference axis have a proportionately greater effect on the magnitude of the mass moment of inertia than do mass elements which are closer to this axis, because of the squaring effect. Thus, the mass moment of inertia is a measure of the distribution, or placement, of the mass of the body.

(*b*)    The mass moment of inertia of a rigid body has a tangible physical interpretation. A reference axis, such as the $x$ axis, is imagined to be the centerline of a shaft attached to the body. The mass moment of inertia with respect to this axis is then a measure of the resistance of the body to angular acceleration about this axis. This effect will be seen in Chapter 18, when the rotational form of Newton's second law is introduced.

17.21    (*a*)    Give the formal definitions of the *radii of gyration* of a body about the $x$, $y$, and $z$ axes.

(*b*)    What is the physical significance of a radius of gyration?

▌ (*a*)    Each of the mass moments of inertia $I_x$, $I_y$, and $I_z$ about the $xyz$ coordinate axes has an associated term, the radius of gyration, defined by

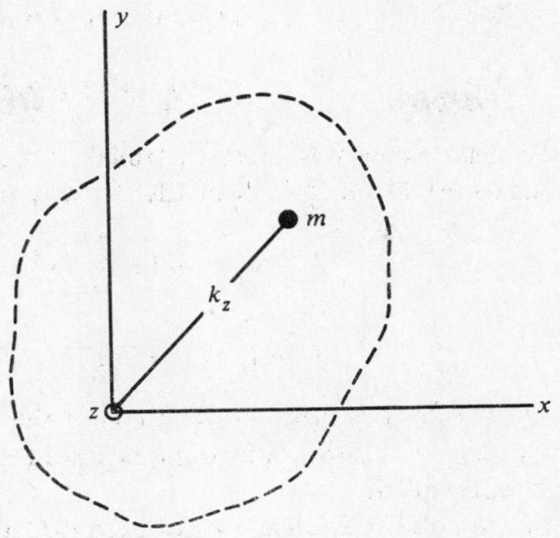

**Fig. 17.21**

$$I_x = k_x^2 m \qquad I_y = k_y^2 m \qquad I_z = k_z^2 m$$

$$k_x = \sqrt{\frac{I_x}{m}} \qquad k_y = \sqrt{\frac{I_y}{m}} \qquad k_z = \sqrt{\frac{I_z}{m}}$$

The radii of gyration have the unit of length, and $k_x$, $k_y$, and $k_z$ are normal to the $x$, $y$, and $z$ axes, respectively.

(**b**) A radius of gyration has no particular physical significance. If all the mass of the body were imagined to be concentrated at a single point, as shown in Fig. 17.21, then a product such as $k_z^2 m$ would be equal to the mass moment of inertia $I_z$ of the body about the $z$ axis.

**17.22** The thin, flat rectangular plate in Fig. 17.22*a* has mass $m$ and thickness $t$. The origin of the coordinate axes is located at the volume centroid of the plate.

(**a**) Find the mass moment of inertia and the radius of gyration with respect to the $x$ axis.

(**b**) Do the same as in part (*a*), but with respect to the $y$ axis.

(**c**) Compare the result in part (*a*) with that given in Case 10 in Table 17.13.

▮ (**a**) The mass element $dm$ shown in Fig. 17.22*b* is chosen. All particles in this element, in the limit, are assumed to be at the same distance $y$ from the reference $x$ axis. The magnitude of the differential mass is

$$dm = \rho(ta\,dy)$$

where $\rho$ is the mass density of the material. The differential moment of inertia is then found from

$$dI_x = y^2\,dm = y^2(\rho ta\,dy)$$

$$I_x = \int_{-b/2}^{b/2} y^2(\rho ta\,dy) = \rho ta \int_{-b/2}^{b/2} y^2\,dy = \rho ta\left(\frac{y^3}{3}\,\Big|_{-b/2}^{b/2}\right) = \frac{\rho ta}{3}\left[\left(\frac{b}{2}\right)^3 - \left(-\frac{b}{2}\right)^3\right] = \frac{1}{12}\,\rho tab^3$$

The mass $m$ of the plate is

$$m = \rho tab$$

and the final form for the mass moment of inertia is then

$$I_x = \tfrac{1}{12}mb^2$$

The radius of gyration is defined by

$$k_x = \sqrt{\frac{I_x}{m}} = \sqrt{\frac{mb^2}{12m}} = \frac{b}{\sqrt{12}} = 0.289b$$

This radius of gyration is drawn to scale in Fig. 17.22*c*.

(**b**) From the symmetry properties of the rectangular plate, with the dimensions $a$ and $b$ interchanged,

$$I_y = \tfrac{1}{12}ma^2 \qquad k_y = 0.289a$$

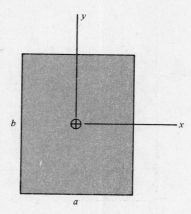

Fig. 17.22a

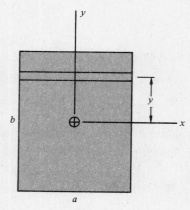

Fig. 17.22b

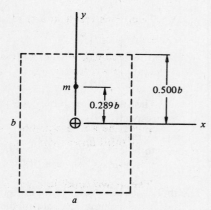

Fig. 17.22c

(c) From Case 10 in Table 17.13,

$$I_{0x} = \tfrac{1}{12}m(b^2 + c^2)$$ (1)

For the proportions of the plate in Fig. 17.22a,

$$t = c \ll a \text{ or } b \qquad I_{0x} = I_x$$

Thus,

$$c^2 <<< b^2$$

and Eq. (1) has the form

$$I_x \approx \tfrac{1}{12}mb^2$$

**17.23** A body in the form of a paraboloid is shown in Case 8 of Table 17.13.

(a) Verify the result $x_c = \tfrac{2}{3}h$.

(b) Verify the result $m = \tfrac{1}{8}\rho\pi d^2 h$.

(c) Verify the result $I_{0x} = I_x = \tfrac{1}{12}md^2$.

▌ (a) Figure 17.23 shows a parabola which has symmetry with respect to the $x$ axis. If this curve shape is rotated about the $x$ axis, the volume generated is called a paraboloid.

The general form of the plane curve in Fig. 17.23 is

$$x = ay^2 + by + c$$ (1)

A known point on the curve is

$$x = 0 \qquad y = 0$$

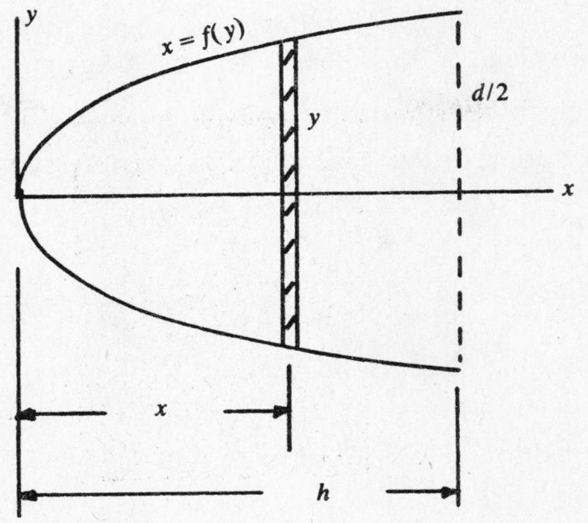

**Fig. 17.23**

Using this result in Eq. (1) gives

$$c = 0$$

A second known condition is that the parabola is tangent to the $y$ axis at $x = y = 0$. This condition may be written as

$$\left.\frac{dx}{dy}\right|_{\substack{x=0\\y=0}} = (2ay + b)\Big|_{y=0} = 0$$

From the above equation, $b = 0$.

A third known condition is

$$x = h \qquad y = \frac{d}{2}$$

The above conditions are now used with Eq. (1), with the final results

$$x = \frac{4h}{d^2}\, y^2 \qquad \text{or} \qquad y = \frac{d}{2}\sqrt{\frac{x}{h}}$$

The volume element shown in Fig. 17.23 has the form

$$dV = \pi y^2\, dx = \frac{\pi d^2}{4h}\, x\, dx$$

The centroidal coordinate $x_c$ is given by

$$x_c = \frac{\displaystyle\int_V x\, dv}{\displaystyle\int_V dV} = \frac{\displaystyle\int_0^h x\left(\frac{\pi d^2}{4h}\, x\, dx\right)}{\displaystyle\int_0^h \frac{\pi d^2}{4h}\, x\, dx} = \frac{\dfrac{\pi d^2}{4h}\dfrac{x^3}{3}\Big|_0^h}{\dfrac{\pi d^2}{4h}\dfrac{x^2}{2}\Big|_0^h} = \frac{h^3/3}{h^2/2} = \frac{2}{3}\, h$$

(**b**)  The mass of the body is given by

$$m = \int_V \rho\, dV = \int_0^h \rho\,\frac{\pi d^2}{4h}\, x\, dx = \rho\,\frac{\pi d^2}{4h}\frac{x^2}{2}\Big|_0^h = \rho\,\frac{\pi d^2}{4h}\frac{h^2}{2} = \frac{1}{8}\,\rho\pi d^2 h$$

(**c**)  The mass moment of inertia of a disk about its center axis, from Case 4 of Table 17.13, with $I_y \to I_x$, is

$$I_x = \tfrac{1}{8}md^2$$

The above value is used for the differential mass element shown in Fig. 17.23, with the result

$$dI_x = \tfrac{1}{8}(\pi y^2)\, dx\,\rho(2y)^2 = \tfrac{1}{2}\rho\pi y^4\, dx$$

Using the value of $y$ found in part (a),

$$dI_x = \tfrac{1}{2}\rho\pi\left(\frac{d}{2}\sqrt{\frac{x}{h}}\right)^4 = \frac{\rho\pi d^4}{32h^2}\, x^2\, dx$$

$$I_x = \int_0^h \frac{\rho\pi d^4}{32h^2}\, x^2\, dx = \frac{\rho\pi d^4}{32h^2}\frac{x^3}{3}\Big|_0^h = \frac{\rho\pi d^4 h^3}{3(32)h^2} = (\tfrac{1}{8}\rho\pi d^2 h)\frac{d^2}{12} = \frac{1}{12}\,md^2$$

Figure 17.24 shows a proposed design for the nose shape, in the form of a paraboloid, of an antiaircraft shell. The nose material is lead, with a density of 11,400 kg/m³.

(a) Find the mass moment of inertia of the nose about its axis of revolution.

(b) Find the radius of gyration of the nose about its axis of revolution.

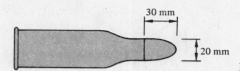

**Fig. 17.24**

∎ (a) Using the result in Prob. 17.23, or in Case 8 of Table 17.13, the mass moment of inertia of the nose may be written as

$$I_{0x} = \tfrac{1}{12} m d^2$$

where the mass $m$ is given by

$$m = \tfrac{1}{8} \rho \pi d^2 h$$

Using $d = 20$ mm and $h = 30$ mm,

$$m = \tfrac{1}{8}(11{,}400 \text{ kg/m}^3)\pi 20^2(30) \text{ mm}^3 \left(\frac{1 \text{ m}}{1{,}000 \text{ mm}}\right)^3 = 0.0537 \text{ kg}$$

$$I_{0x} = \tfrac{1}{12}(0.0537 \text{ kg})20^2 \text{ mm}^2 \left(\frac{1 \text{ m}}{1{,}000 \text{ mm}}\right)^2 = 1.79 \times 10^{-6} \text{ kg} \cdot \text{m}^2$$

(b) The radius of gyration is found as

$$k_{0x} = \sqrt{\frac{I_{0x}}{m}} = \sqrt{\frac{1.79 \times 10^{-6}}{0.0537}} = 0.00577 \text{ m} = 5.77 \text{ mm}$$

**17.25** The tolerances on the dimensions of a disk are shown in Fig. 17.25, and all dimensions are in inches. The disk is made of steel. The range of values of the specific weight of the steel used is 485 to 495 lb/ft³.

(a) Find the maximum and minimum computed values of the mass moment of inertia of the disk about its center axis. Use four significant figures in the calculations.

(b) Find the percent differences between the values found in part (a) and the nominal value of the mass moment of inertia.

∎ (a) From Case 5 in Table 17.13,

$$m = \tfrac{1}{4}\rho \pi h(d_0^2 - d_i^2) \qquad I_{0y} = I_y = \tfrac{1}{8}m(d_0^2 + d_i^2)$$

$m$ is eliminated between the above two equations, with the result

$$I_{0y} = \tfrac{1}{32}\rho \pi h(d_0^4 - d_i^4)$$

where $d_0$ is the outside diameter, $d_i$ is the inside diameter, and $h$ is the thickness of the disk.

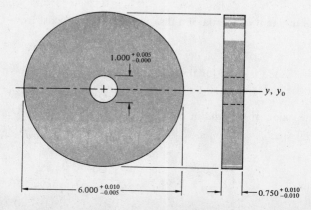

**Fig. 17.25**

The maximum value of $I_{0y}$ will occur with the values

$$d_{0,max} = 6.000 + 0.010 = 6.010 \text{ in} \qquad d_{i,min} = 1.000 - 0.000 = 1.000 \text{ in}$$
$$h_{max} = 0.750 + 0.010 = 0.760 \text{ in} \qquad \gamma_{max} = 495 \text{ lb/ft}^3$$

The maximum value of the mass moment of inertia is

$$I_{0y,max} = \tfrac{1}{32}\rho\pi h(d_0^4 - d_i^4) = \frac{1}{32}\left[\frac{495}{386(12)^3}\right]\pi(0.760)(6.010^4 - 1.000^4) = 0.07219 \text{ lb} \cdot \text{s}^2 \cdot \text{in}$$

The minimum value of $I_{0y}$ will occur with the values

$$d_{0,min} = 6.000 - 0.005 = 5.995 \text{ in} \qquad d_{i,max} = 1.000 + 0.005 = 1.005 \text{ in}$$
$$h_{min} = 0.750 - 0.010 = 0.740 \text{ in} \qquad \gamma_{min} = 485 \text{ lb/ft}^3$$

The minimum value of the mass moment of inertia is

$$I_{0y,min} = \tfrac{1}{32}\rho\pi h(d_0^4 - d_i^4) = \frac{1}{32}\left[\frac{485}{386(12)^3}\right]\pi(0.740)(5.995^4 - 1.005^4) = 0.06818 \text{ lb} \cdot \text{s}^2 \cdot \text{in}$$

(**b**) The nominal value of $I_{0y}$ is found with

$$d_0 = 6.00 \text{ in} \qquad d_i = 1.00 \text{ in} \qquad h = 0.750 \text{ in} \qquad \gamma = \gamma_{avg} = \frac{485 + 495}{2} = 490 \text{ lb/ft}^3$$

$$I_{0y} = \tfrac{1}{32}\rho\pi h(d_0^4 - d_i^4) = \frac{1}{32}\left[\frac{490}{386(12)^3}\right]\pi(0.750)(6^4 - 1^4) = 0.07005 \text{ lb} \cdot \text{s}^2 \cdot \text{in}$$

The percent difference between the maximum and nominal values of the mass moment of inertia is

$$\%\text{D} = \frac{0.07219 - 0.07005}{0.07005}(100) = 3.05\% \approx 3°$$

The percent difference between the minimum and nominal values of $I_{0y}$ is

$$\%\text{D} = \frac{0.06818 - 0.07005}{0.07005}(100) = -2.67\%$$

**17.26** Figure 17.26 shows a thin right circular cylinder. For what values of the length-to-diameter ratio $l/d$ will the approximate expression $I_{0x} = I_{0z} = \tfrac{1}{12}ml^2$ be no more than 5 percent in error?

❚ The exact form of the moments of inertia $I_{0x}$ and $I_{0z}$, from Case 4 in Table 17.13, is

$$I_{0x} = I_{0z} = \tfrac{1}{12}m(h^2 + \tfrac{3}{4}d^2)$$

where $d$ is the diameter and $h$ is the height of the cylinder. For the present problem, $h = l$. The percent difference between the exact and approximate forms of the mass moments of inertia is

$$\%\text{D} = \frac{I_{0x} - I_{0x,approx}}{I_{0x}} = \frac{\tfrac{1}{12}m(l^2 + \tfrac{3}{4}d^2) - \tfrac{1}{12}ml^2}{\tfrac{1}{12}m(l^2 + \tfrac{3}{4}d^2)} = \frac{1}{\tfrac{4}{3}(l/d)^2 + 1}$$

The limiting error of $5\% = 0.05$ is used, with the result

$$0.05 = \frac{1}{\tfrac{4}{3}(l/d)^2 + 1} \qquad \frac{l}{d} = 3.77$$

If $l/d \geqq 3.77$, the error between the exact and approximate forms will be less than or equal to 5 percent.

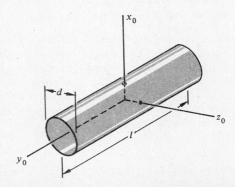

**Fig. 17.26**

**17.27** If the disk in Fig. 17.27 is considered to be thin, the mass moment of inertia about the $z_0$ axis has the approximate form $I_{0z} = \frac{1}{16}md^2$, where $m$ is the mass of the disk.

 (*a*) For what range of values of the ratio $a/d$ will the use of the approximate solution result in an error of no more than 5 percent in the value of the mass moment of inertia about the $z_0$ axis?

 (*b*) Do the same as in part (*a*), for an error of 10 percent.

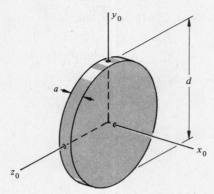

**Fig. 17.27**

 ▌ (*a*) The exact form of the mass moment of inertia $I_{0z}$, from Case 4 in Table 17.13, is

$$I_{0z} = \tfrac{1}{12}m(h^2 + \tfrac{3}{4}d^2)$$

where $d$ is the diameter and $h$ is the height of the cylinder. For this problem, $h = a$. The percent difference between the exact and approximate values of $I_{0z}$ is

$$\%\mathrm{D} = \frac{I_{0z} - I_{0z,\text{approx}}}{I_{0z}} = \frac{\tfrac{1}{12}m(a^2 + \tfrac{3}{4}d^2) - \tfrac{1}{16}md^2}{\tfrac{1}{12}m(a^2 + \tfrac{3}{4}d^2)} = \frac{(a/d)^2}{(a/d)^2 + \tfrac{3}{4}}$$

Using the limiting value $5\% = 0.05$,

$$0.05 = \frac{(a/d)^2}{(a/d)^2 + \tfrac{3}{4}} \qquad \frac{a}{d} = 0.199$$

If $a/d \le 0.199$, the error between the exact and approximate forms will be less than or equal to 5 percent.

 (*b*) For a maximum error of $10\% = 0.1$,

$$0.1 = \frac{(a/d)^2}{(a/d)^2 + \tfrac{3}{4}} \qquad \frac{a}{d} = 0.289$$

If $a/d \le 0.289$, the error between the exact and approximate forms will be less than or equal to 10 percent.

**17.28** (*a*) Show how the *parallel-axis, or transfer, theorem* is derived.

 (*b*) What is a common mistake made when using the parallel-axis theorem?

 (*c*) What is an inequality relationship between the moment of inertia of a body about a centroidal axis and the moment of inertia about an axis parallel to this centroidal axis?

 ▌ (*a*) A very useful relationship between the mass moments of inertia of a rigid body about a centroidal axis and about an axis parallel to this centroidal axis is now developed. This technique is the principal method used to find mass moments of inertia in problems in engineering dynamics.

 Figure 17.28 shows a plane section, normal to the $z$ axis, of the body. The $x_0$, $y_0$, $z_0$ axes, which pass through the center of mass of the body, are parallel to the $x$, $y$, $z$ axes. $x_c$, $y_c$, and $z_c$ are the position coordinates, measured in the $x$, $y$, $z$ system, of the center of mass of the body. $d_z$ is the normal separation distance between the $z$ and $z_0$ axes. The fundamental definition of the mass moment of inertia of the body about the $z$ axis is

$$I_z = \int_V (x^2 + y^2)\, dm \qquad (1)$$

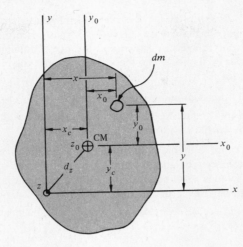

**Fig. 17.28**

From Fig. 17.28,

$$x = x_c + x_0 \qquad y = y_c + y_0$$

These values are substituted into Eq. ($1$), with the result

$$I_z = \int_V [(x_c + x_0)^2 + (y_c + y_0)^2]\, dm = \int_V [(x_c^2 + 2x_c x_0 + x_0^2) + (y_c^2 + 2y_c y_0 + y_0^2)]\, dm$$

$$= \int_V (x_c^2 + y_c^2)\, dm + 2x_c \int_V x_0\, dm + 2y_c \int_V y_0\, dm + \int_V (x_0^2 + y_0^2)\, dm \qquad (2)$$

The location of the center of mass CM of the body in the $x_0$, $y_0$ coordinate system is given by

$$x_{0c} = \frac{\int_V x_0\, dm}{m} \qquad y_{0c} = \frac{\int_V y_0\, dm}{m} \qquad (3)$$

Since the $x_0$ and $y$ axes both pass through the center of mass, the centroidal coordinates $x_{0c}$ and $y_{0c}$ must satisfy

$$x_{0c} = y_{0c} = 0$$

It follows from Eq. (3) that

$$\int_V x_0\, dm = 0 \qquad \int_V y_0\, dm = 0$$

and thus the middle two terms on the right side of Eq. (2) are identically zero. The last term on the right side of this equation is the mass moment of inertia of the body about the centroidal $z_0$ axis, defined by

$$I_{0z} = \int_V (x_0^2 + y_0^2)\, dm \qquad (4)$$

From Fig. 17.28,

$$x_c^2 + y_c^2 = d_z^2$$

The first term on the right side of Eq. (2) may then be written as

$$\int_V (x_c^2 + y_c^2)\, dm = \int_V d_z^2\, dm = d_z^2 \int_V dm = m d_z^2 \qquad (5)$$

where $m$ is the mass of the body. Equations (4) and (5) are substituted into Eq. (2), and the final result is

$$I_z = I_{0z} + m d_z^2 \qquad (6)$$

By using an analysis similar to that shown above, it can be shown that

$$I_x = I_{0x} + m d_x^2 \qquad (7)$$
$$I_y = I_{0y} + m d_y^2 \qquad (8)$$

where $d_x$ and $d_y$ are the distances between the center of mass and the $x$ and $y$ axes, respectively.

Equations (6), (7), and (8) are extremely useful results which are referred to as the parallel-axis, or transfer, theorems. In the usual application of these theorems, a mass moment of inertia, such as $I_{0x}$ about the centroidal $x_0$ axis, is known and the mass moment of inertia about an axis parallel to the centroidal axis is to be found. The theorem states that the moment of inertia about the parallel axis is equal to the sum of

1. The centroidal mass moment of inertia.
2. A correction term equal to the product of the mass of the body and the square of the separation distance between the two axes.

(*b*) When using the transfer theorem, the separation distance must be between the *centroidal* axis of the body and the reference axis about which the mass moment of inertia is desired. Failure to interpret this separation distance term correctly in a given problem is the usual reason for an incorrect result when using the parallel-axis, or transfer, theorem.

(*c*) Terms such as $md_z^2$ are always positive. It follows, from an equation of the form of Eq. (6), that

$$I_{0z} < I_z$$

This result shows that the mass moment of inertia about an axis through the *centroid* of a body is always less than the mass moment of inertia of the body about *any other axis* which is parallel to this centroidal axis.

**17.29** (*a*) Find the mass moment of inertia about the lower edge of the thin, flat rectangular plate in Prob. 17.22.

(*b*) Compare the results found for $I_x$ and $I_{x_1}$.

▮ (*a*) The $x_1$ axis is placed along the edge $a$, as shown in Fig. 17.29. Using the parallel-axis theorem,

$$I_{x_1} = I_x + md_x^2 = \tfrac{1}{12}mb^2 + m\left(\frac{b}{2}\right)^2 = \tfrac{1}{3}mb^2$$

(*b*) It may be seen that the mass moment of inertia about the edge of the plate is four times as great as the mass moment of inertia about the centroidal axis.

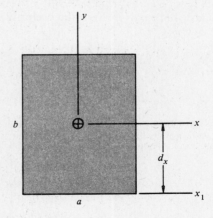

**Fig. 17.29**

**17.30** Use the parallel-axis theorem to verify the result for $I_{x_1}$ in Case 1 of Table 17.13.

▮ Using the parallel-axis theorem, $I_{x_1}$ may be written in the form

$$I_{x_1} = I_{0x} + md_{x_1}^2$$

where $d_{x_1}$ is the distance between the $x_0$ and $x_1$ axes. With $d_{x_1} = \tfrac{1}{2}d$, the result for $I_{x_1}$ is

$$I_{x_1} = \tfrac{1}{10}md^2 + m(\tfrac{1}{2}d)^2 = (\tfrac{1}{10} + \tfrac{1}{4})md^2 = \tfrac{7}{20}md^2$$

**17.31** Verify the results for $I_x$, $I_{x_1}$, and $I_{y_1}$, in Case 4 of Table 17.13, by using the parallel-axis theorem.

▮ Using the parallel-axis theorem, $I_{x_1}$ may be written as

$$I_{x_1} = I_{0x} + md_{x_1}^2$$

Using  $d_{x_1} = c/2$  in the above equation results in

$$I_{x_1} = \tfrac{1}{20}m(b^2 + c^2) + m\left(\frac{c}{2}\right)^2 = \tfrac{1}{20}mb^2 + (\tfrac{1}{20} + \tfrac{1}{4})mc^2 = \tfrac{1}{20}m(b^2 + 6c^2)$$

$I_{y_1}$  may be expressed as

$$I_{y_1} = I_{0y} + md_{y_1}^2$$

Since  $d_{y_1} = a/2$,  the above equation appears as

$$I_{y_1} = \tfrac{1}{20}m(a^2 + c^2) + m\left(\frac{a}{2}\right)^2 = \tfrac{1}{20}mc^2 + (\tfrac{1}{20} + \tfrac{1}{4})ma^2 = \tfrac{1}{20}m(c^2 + 6a^2)$$

$I_{z_1}$  has the form

$$I_{z_1} = I_{0z} + md_{z_1}^2$$

Using  $d_{z_1} = a/2$,

$$I_{z_1} = \tfrac{1}{20}m(a^2 + b^2) + m\left(\frac{a}{2}\right)^2 = \tfrac{1}{20}mb^2 + (\tfrac{1}{20} + \tfrac{1}{4})ma^2 = \tfrac{1}{20}m(b^2 + 6a^2)$$

**17.32**  Verify the results for $I_x$, $I_{x_1}$, and $I_{y_1}$, in Case 4 of Table 17.13, by using the parallel-axis theorem.

**▌** The form for $I_x$, using the parallel-axis theorem, is

$$I_x = I_{0x} + md_x^2$$

Using  $dx = \tfrac{1}{2}h$  in the above equation results in

$$I_x = \tfrac{1}{12}m(h^2 + \tfrac{3}{4}d^2) + m(\tfrac{1}{2}h)^2 = \tfrac{1}{3}mh^2 + \tfrac{1}{16}md^2 = \tfrac{1}{48}m(16h^2 + 3d^2)$$

The expression for $I_{x_1}$ is

$$I_{x_1} = I_{0x} + md_{x_1}^2 \tag{1}$$

$d_{x_1}$ is the separation distance between the $x_0$ and $x_1$ axes.  Using Fig. 17.32, this quantity may be expressed as

$$d_{x_1}^2 = (\tfrac{1}{2}d)^2 + (\tfrac{1}{2}h)^2 \tag{2}$$

$d_{x_1}$ is eliminated between Eqs. (1) and (2), and $I_{0x}$ from Case 4 is used, to obtain

$$I_{x_1} = \tfrac{1}{12}m(h^2 + \tfrac{3}{4}d^2) + m(\tfrac{1}{4}d^2 + \tfrac{1}{4}h^2) = \tfrac{1}{3}mh^2 + \tfrac{5}{16}md^2 = \tfrac{1}{48}m(16h^2 + 15d^2)$$

The form for $I_{y_1}$ is

$$I_{y_1} = I_{0y} + md_{y_1}^2$$

Using  $d_{y_1} = \tfrac{1}{2}d$  in the above equation, together with $I_{0y}$ from Case 4, results in

$$I_{y_1} = \tfrac{1}{8}md^2 + m(\tfrac{1}{2}d)^2 = (\tfrac{1}{8} + \tfrac{1}{4})md^2 = \tfrac{3}{8}md^2$$

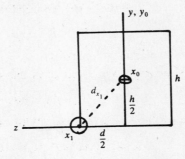

**Fig. 17.32**

**17.33**  Use the parallel-axis theorem to verify the results for $I_x$, $I_{x_1}$, $I_{x_2}$, and $I_{y_1}$ in Case 6 of Table 17.13.

**▌** Using the parallel-axis theorem, $I_x$ may be written as

$$I_x = I_{0x} + md_x^2$$

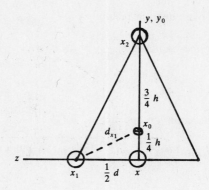

**Fig. 17.33**

$d_x = \frac{1}{4}h$, so that the above equation has the form

$$I_x = \tfrac{3}{80}m(d^2 + h^2) + m(\tfrac{1}{4}h)^2 = \tfrac{3}{80}md^2 + (\tfrac{3}{80} + \tfrac{1}{16})mh^2 = \tfrac{3}{80}md^2 + \tfrac{1}{10}mh^2 = \tfrac{1}{80}m(3d^2 + 8h^2)$$

The term $I_{x_1}$ may be expressed as

$$I_{x_1} = I_{0x} + md_{x_1}^2 \qquad (1)$$

$d_{x_1}$ is the distance between the $x_0$ and $x_1$ axes. Using Fig. 17.33, this quantity may be written in the form

$$d_{x_1}^2 = (\tfrac{1}{2}d)^2 + (\tfrac{1}{4}h)^2 \qquad (2)$$

$d_{x_1}$ is eliminated between Eqs. (1) and (2), to obtain

$$I_{x_1} = \tfrac{3}{80}m(d^2 + h^2) + m(\tfrac{1}{4}d^2 + \tfrac{1}{16}h^2) = (\tfrac{3}{80} + \tfrac{1}{4})md^2 + (\tfrac{3}{80} + \tfrac{1}{16})mh^2 = \tfrac{23}{80}md^2 + \tfrac{1}{10}mh^2 = \tfrac{1}{80}m(23d^2 + 8h^2)$$

$I_{x_2}$ is written as

$$I_{x_2} = I_{0x} + md_{x_2}^2$$

From Fig. 17.33, $d_{x_2} = \tfrac{3}{4}h$. This result is used in the above equation, to obtain

$$I_{x_2} = \tfrac{3}{80}m(d^2 + h^2) + m(\tfrac{3}{4}h)^2 = \tfrac{3}{80}m(d^2 + h^2) + \tfrac{9}{16}mh^2 = \tfrac{3}{80}md^2 + \tfrac{48}{80}mh^2 = \tfrac{3}{80}m(d^2 + 16h^2)$$

$I_{y_1}$ may be expressed as

$$I_{y_1} = I_{0y} + md_{y_1}^2$$

Since $d_{y_1} = \tfrac{1}{2}d$, the above equation has the form

$$I_{y_1} = \tfrac{3}{40}md^2 + m(\tfrac{1}{2}d)^2 = \tfrac{3}{40}md^2 + \tfrac{1}{4}md^2 = \tfrac{13}{40}md^2$$

**17.34**  Use the parallel-axis theorem to verify the results for $I_{x_1}$ in Case 2 of Table 17.13.

∎  Using the parallel-axis theorem, $I_{x_1}$ may be written in the form

$$I_{x_1} = I_{0x} + md_{x_1}^2$$

$d_{x_1} = \tfrac{1}{2}d_0$ is used in the above equation, to obtain

$$I_{x_1} = \tfrac{1}{10}m\left(\frac{d_0^5 - d_i^5}{d_0^3 - d_i^3}\right) + m(\tfrac{1}{2}d_0)^2 = \frac{\tfrac{1}{10}m(d_0^5 - d_i^5) + \tfrac{1}{4}md_0^2(d_0^3 - d_i^3)}{d_0^3 - d_i^3} = \frac{m[(\tfrac{1}{10} + \tfrac{1}{4})d_0^5 - \tfrac{1}{4}d_0^2d_i^3 - \tfrac{1}{10}d_i^5]}{d_0^3 - d_i^3}$$

$$= \frac{m(\tfrac{7}{20}d_0^5 - \tfrac{1}{4}d_0^2d_i^3 - \tfrac{1}{10}d_i^5)}{d_0^3 - d_i^3} = \tfrac{1}{20}m\,\frac{(7d_0^5 - 5d_0^2d_i^3 - 2d_i^5)}{d_0^3 - d_i^3}$$

**17.35**  Verify the results for $I_{x_1}$, $I_{x_2}$, and $I_{y_1}$, in Case 3 of Table 17.13, by using the parallel-axis theorem.

∎  The parallel-axis theorem may be used to express $I_{x_1}$ in the form

$$I_{x_1} = I_{0x} + md_{x_1}^2$$

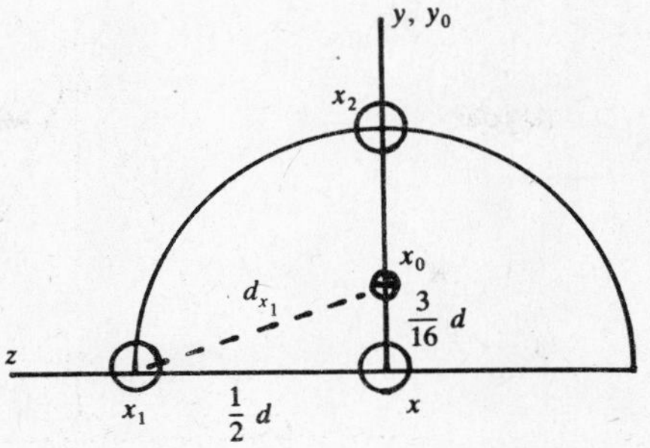

Fig. 17.35

From Fig. 17.35

$$d_{x_1}^2 = (\tfrac{1}{2}d)^2 + (\tfrac{3}{16}d)^2 = 0.285d^2$$

$d_{x_1}^2$ is eliminated between the above two equations, with the result

$$I_{x_1} = 0.0648md^2 + m(0.285d^2) = 0.350md^2$$

$I_{x_2}$ may be written as

$$I_{x_2} = I_{0x} + md_{x_2}^2$$

From Fig. 17.35,

$$d_{x_2} = \tfrac{1}{2}d - \tfrac{3}{16}d = \tfrac{5}{16}d$$

$d_{x_2}$ is eliminated between the above two equations, to obtain

$$I_{x_2} = 0.0648md^2 + m(\tfrac{5}{16}d)^2 = 0.162md^2$$

$I_{y_1}$ may be written in the form

$$I_{y_1} = I_{0y} + md_{y_1}^2$$

Using $d_{y_1} = \tfrac{1}{2}d$,

$$I_{y_1} = \tfrac{1}{10}md^2 + m(\tfrac{1}{2}d)^2 = \tfrac{7}{20}md^2$$

**17.36** Two spherical masses are attached to a thin, massless rod to form the dumbbell shown in Fig. 17.36a.

(a) Find the mass moment of inertia of the system about the $x$ axis.

(b) If the diameters $d$ of the spherical masses are small compared with the dimension $D$, the masses may be assumed to be point, or concentrated, masses located at $y = \pm D/2$. Based on this assumption, find the approximate value of the mass moment of inertia about the $x$ axis.

(c) Find the maximum permissible size of $d$ relative to $D$, if the approximate answer in part (b) is to be in error by no more than 5 percent.

(d) Do the same as in part (c), for a maximum error of 10 percent.

**▌** (a) The mass moment of inertia of a homogeneous sphere, of mass $m$ and diameter $d$, about a diameter is given by Case 1 in Table 17.13 as $I_0 = \tfrac{1}{10}md^2$. One of the spherical masses is shown in Fig. 17. 36b, and the centroidal $x_0$ axis passes through the CM. The mass moment of inertia $I_{0x}$ of the mass about its centroidal axis is

$$I_{0x} = \tfrac{1}{10}md^2$$

Using the parallel-axis theorem, the value of the mass moment of inertia of the dumbbell is

$$I_x = 2(I_{0x} + md_x^2) \qquad d_x = \frac{D}{2} \qquad I_x = 2\left[\tfrac{1}{10}md^2 + m\left(\frac{D}{2}\right)^2\right] = \tfrac{1}{10}m(5D^2 + 2d^2)$$

(b) The appearance of the system, for the assumption of point masses, is shown in Fig. 17.36c. The approximate value of the moment of inertia about the $x$ axis is

$$I_{x,\text{approx}} = 2\left[m\left(\frac{D}{2}\right)^2\right] = \frac{mD^2}{2}$$

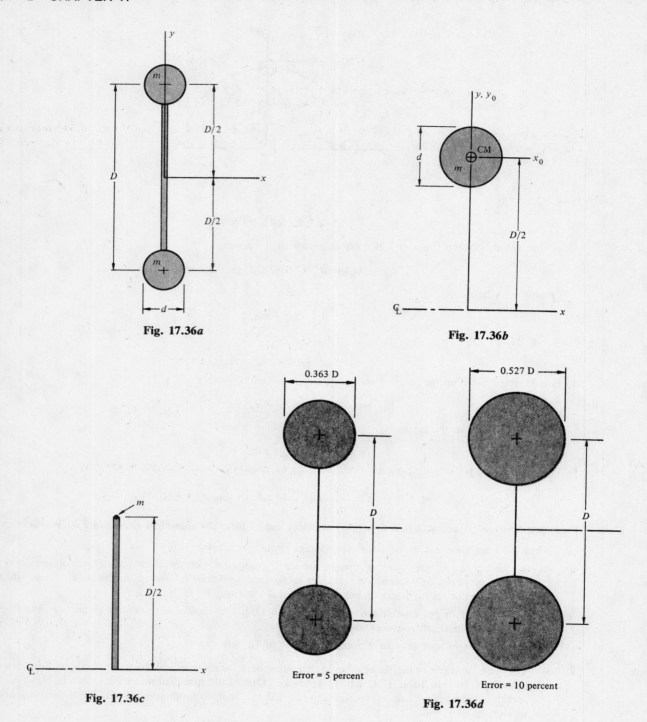

Fig. 17.36a

Fig. 17.36b

Fig. 17.36c

0.363 D

0.527 D

Error = 5 percent

Error = 10 percent

Fig. 17.36d

(c)  A percent difference between the exact and approximate values of $I_x$ is defined to be

$$\%\text{D} = \frac{I_x - I_{x,\text{approx}}}{I_x} = \frac{(m/10)(5D^2 + 2d^2) - mD^2/2}{(m/10)(5D^2 + 2d^2)} = \frac{1}{1 + 2.5(D/d)^2}$$

For an error of  5% = 0.05,

$$0.05 = \frac{1}{1 + 2.5(D/d)^2} \qquad d = 0.363D$$

If  $d \le 36.3$  percent of $D$, the error in the mass moment of inertia will be less than or equal to 5 percent.

(d) For an allowable error of $10\% = 0.10$,

$$0.10 = \frac{1}{1 + 2.5(D/d)^2} \qquad d = 0.527D$$

The error will be less than or equal to 10 percent if $d \le 52.7$ percent of $D$.
The results of parts (c) and (d) are drawn to scale in Fig. 17.36d.

**17.37** Find the error introduced in Prob. 17.36 by neglecting the mass moment of inertia of the rod. The masses and rod are a one-piece assembly of magnesium, with $\gamma = 108 \, \text{lb/ft}^3$, $d = 0.5 \, \text{in}$, and $D = 4 \, \text{in}$, and the diameter of the rod is 0.125 in.

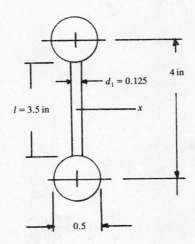

**Fig. 17.37**

❚ Figure 17.37 shows the dimensions of the rod. The mass $m_R$ of this element is given by

$$m_R = \rho\left(\frac{\pi d_1^2}{4}\right)l = \left[\frac{108}{386(12)^3}\right]\frac{\pi(0.125)^2}{4}(3.5) = 6.95 \times 10^{-6} \, \text{lb} \cdot \text{s}^2/\text{in}$$

The mass moment of inertia $I_R$ of the rod about the $x$ axis, using the approximate result in Prob. 17.26, is

$$I_R = \tfrac{1}{12}m_R l^2 = \tfrac{1}{12}(6.95 \times 10^{-6})3.5^2 = 7.09 \times 10^{-6} \, \text{lb} \cdot \text{s}^2 \cdot \text{in}$$

The mass $m$ of one of the spherical masses, from Case 1 in Table 17.13, is given by

$$m = \tfrac{1}{6}\rho\pi d^3 = \frac{1}{6}\left[\frac{108}{386(12)^3}\right]\pi(0.5)^3 = 1.06 \times 10^{-5} \, \text{lb} \cdot \text{s}^2/\text{in}$$

The mass moment of inertia of the dumbbell about the $x$ axis, for the assumption of a massless rod, was found in Prob. 17.36 to be

$$I_x = \tfrac{1}{10}m(5D^2 + 2d^2) = \tfrac{1}{10}(1.06 \times 10^{-5})[5(4)^2 + 2(0.5)^2] = 8.53 \times 10^{-5} \, \text{lb} \cdot \text{s}^2 \cdot \text{in}$$

The mass moment of inertia of the total dumbbell assembly about the $x$ axis is then

$$I_{\text{assembly}} = I_x + I_R = 8.53 \times 10^{-5} + 7.09 \times 10^{-6} = 9.24 \times 10^{-5} \, \text{lb} \cdot \text{s}^2 \cdot \text{in}$$

The percent error due to neglecting the mass of the rod is

$$\%D = \frac{I_R}{I_{\text{assembly}}}(100) = \frac{7.09 \times 10^{-6}}{9.24 \times 10^{-5}}(100) = 7.67\% \approx 8\%$$

**17.38** Two cube-shaped masses are attached to the end of a massless rod, as shown in Fig. 17.38. It is desired to find the mass moment of inertia of the assembly about the $x$ axis. As a first approximation, the masses may be assumed to be point masses at the ends of the rod of length $l$.

(a) For what range of values $a/l$ will the error in the above assumption not exceed 5 percent?

(b) Do the same as in part (a), for a maximum permissible error of 10 percent.

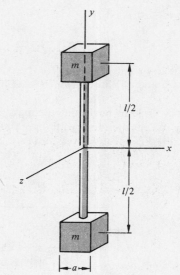

Fig. 17.38

▮ (a) If the masses are assumed to be point masses, the approximate value of the mass moment of inertia of the assembly about the $x$ axis is given by

$$I_{x,\text{approx}} = 2m\left(\frac{l}{2}\right)^2 = \frac{1}{2}\,ml^2$$

The mass moment of inertia of one cube-shaped mass about its centroidal axis, which is parallel to the $x$ axis, using Case 9 of Table 17.13, is

$$I_{0x} = \tfrac{1}{6}ma^2$$

The mass moment of inertia of the assembly about the $x$ axis, using the parallel-axis theorem, is then

$$I_x = 2\left[\frac{1}{6}ma^2 + m\left(\frac{l}{2}\right)^2\right] = \frac{1}{6}\,m(2a^2 + 3l^2)$$

The percent difference between the exact and approximate values of moment of inertia may next be written as

$$\%\text{D} = \frac{I_x - I_{x,\text{approx}}}{I_x} = \frac{\frac{1}{6}m(2a^2 + 3l^2) - \frac{1}{2}ml^2}{\frac{1}{6}m(2a^2 + 3l^2)} = \frac{2(a/l)^2}{2(a/l)^2 + 3}$$

Using $\%\text{D} = 5\% = 0.05$ in the above equation,

$$0.05 = \frac{2(a/l)^2}{2(a/l)^2 + 3} \qquad \frac{a}{l} = 0.281$$

If $a/l \le 0.281$, the error in the moment of inertia about the $x$ axis will be less than or equal to 5 percent.

(b) For $\%\text{D} = 10\% = 0.1$,

$$0.1 = \frac{2(a/l)^2}{2(a/l)^2 + 3} \qquad \frac{a}{l} = 0.408$$

If $a/l \le 0.408$, the error will be less than or equal to 10 percent.

## 17.3 COMPUTATION OF MASS MOMENT OF INERTIA USING THE TRANSFER THEOREM AND A SINGLE INTEGRATION, MASS MOMENT OF INERTIA OF A COMPOSITE BODY

17.39 (a) Find the mass moment of inertia $I_z$ of the body shown in Fig. 17.39 by using the transfer theorem, together with a single integration.

(b) Find the mass moment of inertia $I_{0z}$.

▮ (a) The method shown below may be used if the body has a cross section of known variation. This is the basic method by which all the results in Table 17.13 were obtained.

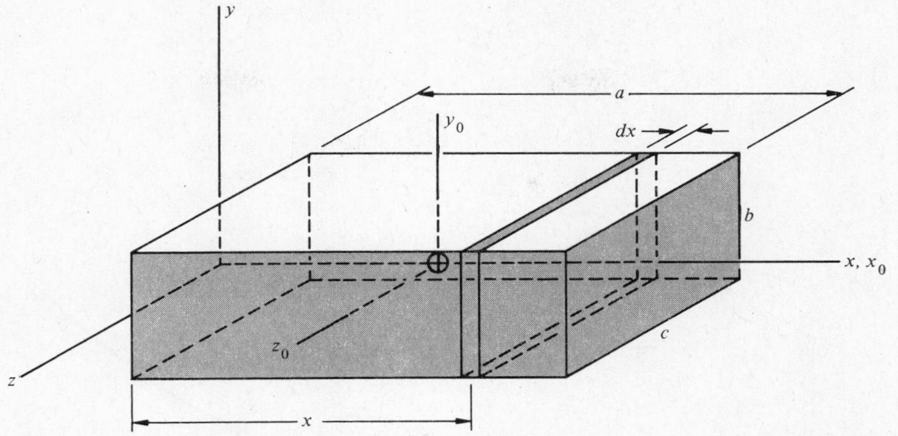

**Fig. 17.39**

The shaded mass element shown in Fig. 17.39 is a thin plane body. The mass moment of inertia $dI_{0z}$ of this element about an axis through its centroid, parallel to the $z$ axis, is now found. Using the result in Prob. 17.22, with $t \rightarrow dx$,

$$dI_{0z} = \frac{\rho \, dx \, cb^3}{12}$$

The moment of inertia of this element about the $z$ axis, using the parallel-axis theorem, is

$$dI_z = dI_{0z} + dm(x^2) = \frac{\rho cb^3}{12} \, dx + \rho bcx^2 \, dx$$

The above equation is integrated over the volume, with the result

$$I_z = \rho cb \int_0^a \left( \frac{b^2}{12} + x^2 \right) dx = \rho cb \left[ \frac{b^2}{12} x + \frac{x^3}{3} \right]_0^a = \rho cb \left( \frac{b^2 a}{12} + \frac{a^3}{3} \right) = \rho abc \left[ \frac{b^2}{12} + \frac{a^2}{3} \right]$$

The mass $m$ of the body is

$$m = \rho abc$$

The final form for the mass moment of inertia $I_z$ is then

$$I_z = \tfrac{1}{12} m(4a^2 + b^2)$$

(*b*) Using the parallel-axis theorem, we get

$$I_z = I_{0z} + md_z^2$$

Using $d_z = a/2$,

$$\tfrac{1}{12} m(4a^2 + b^2) = I_{0z} + m\left(\frac{a}{2}\right)^2 \qquad I_{0z} = \tfrac{1}{12} m(a^2 + b^2)$$

This equation is given in Case 10 in Table 17.13.

**17.40** The dimension $a$ of the rigid body in Prob. 17.39 is allowed to decrease until the shape of the body approaches a flat plate, as shown in Fig. 17.40a. The approximate mass moment of inertia about the $z_0$ axis, assuming the body to be a thin plane mass, was shown in Prob. 17.22 to have the form

$$I_{0z} = \tfrac{1}{12} mb^2$$

Compute and plot the percent error between the exact and approximate mass moments of inertia $I_{0z}$ for plate thicknesses up to one-half of the plate height $b$.

**▌** The exact solution, from Prob. 17.39, or from Case 10 in Table 17.13, is

$$I_{0z} = \tfrac{1}{12} m(a^2 + b^2)$$

The percent error is defined to be

$$\% \text{ Error} = \frac{\text{exact solution} - \text{approx. solution}}{\text{exact solution}} (100) = \frac{\tfrac{1}{12} m(a^2 + b^2) - \tfrac{1}{12} mb^2}{\tfrac{1}{12} m(a^2 + b^2)} (100) = \frac{1}{1 + (b/a)^2} (100) \qquad (1)$$

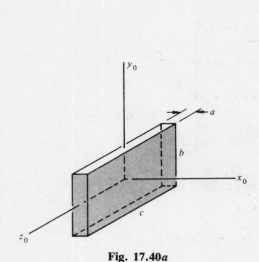

**Fig. 17.40a**

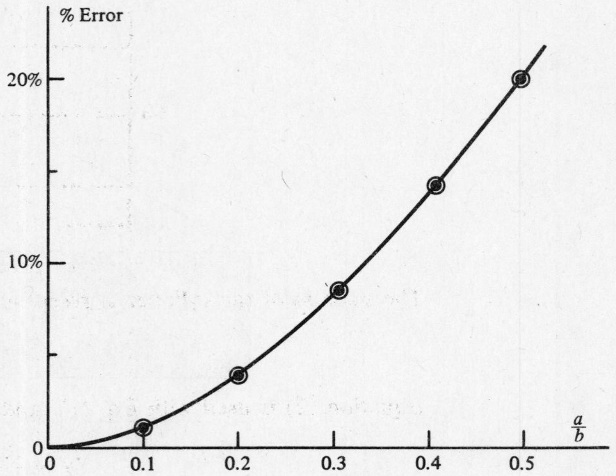

**Fig. 17.40b**

A plate thickness of one-half of the height of the plate corresponds to $a/b = \frac{1}{2}$. Equation (1) is solved for values of $a/b$ from 0 to 0.5. The results are listed in Table 17.3 and plotted in Fig. 17.40b. It may be seen that for small values of $a/b$, which characterize a thin plane body, the error is very small. It may also be observed that the results are completely independent of the plate width dimension $c$.

| TABLE 17.3 | |
|---|---|
| $\dfrac{a}{b}$ | Percent error |
| 0 | 0 |
| 0.1 | 1.0 |
| 0.2 | 3.8 |
| 0.3 | 8.3 |
| 0.4 | 13.8 |
| 0.5 | 20.0 |

**17.41**  (a)  Verify the results for the centroidal location, and the mass moment of inertia $I_x$, for the right circular cylinder in Case 4 of Table 17.13.

(b)  Using the results from part (a) and the parallel-axis theorem, verify the results for $I_{0x}$.

▌ (a)  From Prob. 17.27, for the case of a thin disk,

$$I_{0x} = \tfrac{1}{16} m d^2$$

For the differential mass element in Fig. 17.41,

$$dm = \rho\left(\frac{\pi d^2}{4}\right) dy \qquad dI_{0x} = \frac{1}{16}\,\rho\left(\frac{\pi d^2}{4}\right) d^2\,dy = \tfrac{1}{64}\rho\pi d^4\,dy$$

Using the parallel-axis theorem,

$$dI_x = dI_{0x} + (dm)y^2 = \frac{1}{64}\,\rho\pi d^4\,dy + \rho\left(\frac{\pi d^2}{4}\right)y^2\,dy$$

The mass moment of inertia of the cylinder about the $x$ axis then has the form

$$I_x = \int_0^h \left(\rho\,\frac{\pi d^4}{64} + \rho\,\frac{\pi d^2}{4}\,y^2\right) dy = \left(\rho\,\frac{\pi d^4}{64}\,y + \rho\,\frac{\pi d^2}{4}\,\frac{y^3}{3}\right)\Bigg|_0^h = \rho\,\frac{\pi d^4}{64}\,h + \rho\,\frac{\pi d^2}{4}\,\frac{h^3}{3} \qquad (1)$$

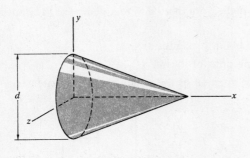

**Fig. 17.41**

The mass $m$ of the cylinder is given by

$$m = \rho \left( \frac{\pi d^2}{4} \right) h \tag{2}$$

Equation (2) is used with Eq. (1), and the result is

$$I_x = \frac{md^2}{16} + \frac{mh^2}{3} = \tfrac{1}{48} m (16h^2 + 3d^2)$$

The above equation is seen in Case 4 of Table 17.13.

(*b*) Using the parallel-axis theorem,

$$I_{0x} = I_x - md_x^2$$

With $d_x = h/2$,

$$I_{0x} = \tfrac{1}{48} m (16h^2 + 3d^2) - m \left( \frac{h}{2} \right)^2 = (\tfrac{1}{3} - \tfrac{1}{4})mh^2 + \tfrac{1}{16} md^2 = \tfrac{1}{12} mh^2 + \tfrac{1}{16} md^2 = \tfrac{1}{12} m (h^2 + \tfrac{3}{4} d^2)$$

The above equation for $I_{0x}$ is contained in Case 4 of Table 17.13.

**17.42**  Find the mass moment of inertia $I_z$ of the right circular cone of Prob. 17.4, shown in Fig. 17.42*a*.

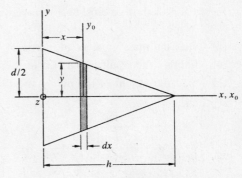

**Fig. 17.42*a***        **Fig. 17.42*b***

▌ A side view of the cone is shown in Fig. 17.42*b*.  The mass moment of inertia of a disk-shaped differential mass element about its centroidal $z_0$ axis was found in Prob. 17.41 to have the form

$$dI_{0z} = \rho \, \frac{\pi d^4}{64} \, dx$$

For the present problem, the variable diameter $d$ of the element is given by $d = y/2$, so that

$$dI_{0z} = \rho \, \frac{\pi y^4}{4} \, dx$$

Using the parallel-axis theorem, we find that the mass moment of inertia of the mass element about the $z$ axis is

$$dI_z = dI_{0z} + (dm)x^2 = \tfrac{1}{4} \rho \pi y^4 \, dx + (\rho \pi y^2 \, dx) x^2$$

$$I_z = \rho \pi \int_0^h (\tfrac{1}{4} y^4 + y^2 x^2) \, dx \tag{1}$$

The functional form of $y$ was given in Prob. 17.4 as

$$y = \frac{d}{2}\left(1 - \frac{x}{h}\right)$$

Using the above result in Eq. (1),

$$I_z = \rho\pi \int_0^h \left[\frac{1}{4}\left(\frac{d}{2}\right)^4\left(1 - \frac{4x}{h} + \frac{6x^2}{h^2} - \frac{4x^3}{h^3} + \frac{x^4}{h^4}\right) + \left(\frac{d}{2}\right)^2\left(1 - \frac{2x}{h} + \frac{x^2}{h^2}\right)x^2\right] dx$$

$$= \rho\pi\left[\frac{d^4}{64}\left(x - \frac{4x^2}{2h} + \frac{6x^3}{3h^2} - \frac{4x^4}{4h^3} + \frac{x^5}{5h^4}\right) + \frac{d^2}{4}\left(\frac{x^3}{3} - \frac{2x^4}{4h} + \frac{x^5}{5h^2}\right)\right]_0^h$$

$$= \rho\pi\left[\frac{d^4 h}{5(64)} + \frac{d^2 h^3}{4(30)}\right] \tag{2}$$

The mass of the cone is

$$m = \rho\left[\frac{1}{3}\left(\frac{\pi d^2}{4}\right)h\right] = \frac{1}{12}\rho\pi d^2 h \tag{3}$$

Equations (2) and (3) are combined to obtain the final value

$$I_z = \tfrac{1}{80}m(3d^2 + 8h^2)$$

This equation is given in Case 6 of Table 17.13.

**17.43** State the general technique of solution for the mass moment of inertia of a body which is not of simple geometric shape.

▮ If a body under consideration is not one of the simple shapes shown in Table 17.13, it may be possible to subdivide the body into several elementary shapes and sum the individual moments of inertia. In the process of this subdivision, elements are chosen whose centroidal mass moments of inertia about axes parallel to the desired axes are known. The parallel-axis theorem may then be used to find the mass moment of inertia of the element about the reference axis. If the body consists of many elements, it may be desirable to organize the solution in a tabular form. The technique for organization of such a tabular solution is shown in Prob. 17.57.

**17.44** Figure 17.44a shows a cross-sectional view of a steel pulley with three different diameters. The specific weight of steel is $489 \text{ lb/ft}^3$.

(a) Find the mass moment of inertia of the pulley about its center axis.

(b) Express the result in part (a) in the engineering units of $\text{lb} \cdot \text{in}^3$.

(c) Find the weight of the pulley and the radius of gyration of the pulley about its center axis.

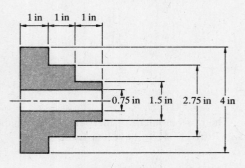

**Fig. 17.44a**

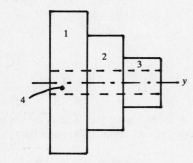

**Fig. 17.44b**

▮ (a) The pulley is divided into the four elements shown in Fig. 17.44b. Elements 1, 2, and 3 are solid cylinders, and element 4 is the cylinder of material which is absent. Using Case 4 in Table 17.13,

$$I_y = \tfrac{1}{8}md^2 = \tfrac{1}{8}(\tfrac{1}{4}\rho\pi d^2 h)d^2 = \tfrac{1}{32}\rho\pi h d^4 = \frac{\pi}{32}\left(\frac{489 \text{ lb/ft}^3}{386 \text{ in/s}^2}\right)\left(\frac{1 \text{ ft}}{12 \text{ in}}\right)^3 h d^4 = 7.20 \times 10^{-5} h d^4 \tag{1}$$

If $h$ and $d$ are in inches, the units of $I_y$ are $\text{lb} \cdot \text{s}^2 \cdot \text{in}$. The mass moment of inertia $I_y$ of the pulley is

$$I_y = I_1 + I_2 + I_3 - I_4$$

Using Eq. (1),

$$I_y = 7.20 \times 10^{-5}[1(4)^4 + 1(2.75)^4 + 1(1.5)^4 - 3(0.75)^4] = 0.0228 \text{ lb} \cdot \text{s}^2 \cdot \text{in}$$

(**b**)   The value for $I_y$ in part (*a*) may be expressed in terms of the engineering units lb · in² as

$$I_y = 0.0228 \text{ lb} \cdot \text{s}^2 \cdot \text{in}(386 \text{ in/s}^2) = 8.80 \text{ lb} \cdot \text{in}^2$$

(**c**)   The mass of a cylinder, from Case 4 in Table 17.13, is expressed as

$$m = \tfrac{1}{4}\rho\pi d^2 h = \frac{1}{4}\left(\frac{489 \text{ lb/ft}^3}{386 \text{ in/s}^2}\right)\left(\frac{1 \text{ ft}}{12 \text{ in}}\right)^3 \pi d^2 h = 5.76 \times 10^{-4} d^2 h$$

For the present problem,

$$m = m_1 + m_2 + m_3 - m_4 = 5.76 \times 10^{-4}[4^2(1) + 2.75^2(1) + 1.5^2(1) - 0.75^2(3)] = 0.0139 \text{ lb} \cdot \text{s}^2/\text{in}$$

The weight of the pulley is then

$$W = mg = 0.0139(386) = 5.37 \text{ lb}$$

The radius of gyration of the pulley about the *y* axis is given by

$$k_y = \sqrt{\frac{I_y}{m}} = \sqrt{\frac{0.0228}{0.0139}} = 1.28 \text{ in}$$

**17.45**   Do the same as in Prob. 17.44 for the steel pulley shown in Fig. 17.45*a*.   The density of steel is 7,830 kg/m³.

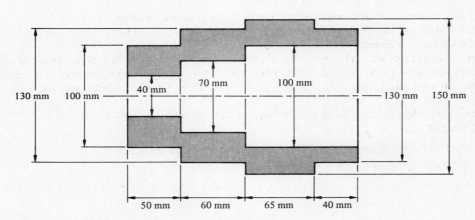

Fig. 17.45*a*

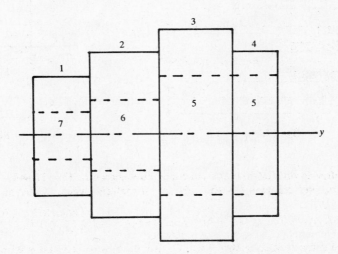

Fig. 17.45*b*

∎ (a)  The pulley is divided into the seven elements shown in Fig. 17.45b.  The mass moment of inertia of a cylinder about its center axis, from Prob. 17.44, and with $h$ and $d$ in millimeters, is

$$I_y = \tfrac{1}{32}\rho\pi hd^4 = \frac{\pi}{32}\left(7.830\,\frac{\text{kg}}{\text{m}^3}\right)hd^4\,\text{mm}^5\left(\frac{1\,\text{m}}{1,000\,\text{mm}}\right)^5 = 7.69\times10^{-13}hd^4$$

The units of $I_y$ in the above equation are $\text{kg}\cdot\text{m}^2$.  The mass moment of inertia of the pulley about its center axis may then be written as

$$I_y = I_1 + I_2 + I_3 + I_4 - I_5 - I_6 - I_7 = 7.69\times10^{-13}[50(100)^4 + 60(130)^4 + 65(150)^4$$
$$+ 40(130)^4 - 105(100)^4 - 60(70)^4 - 50(40)^4] = 0.0418\,\text{kg}\cdot\text{m}^2$$

(b)  $I_y$ is expressed in the units $\text{lb}\cdot\text{in}^2$ as

$$I_y = 0.0418\,\text{kg}\cdot\text{m}^2\left(9.81\,\frac{\text{m}}{\text{s}^2}\right)\left(\frac{1\,\text{lb}}{4.45\,\text{N}}\right)\left(\frac{3.28\,\text{ft}}{1\,\text{m}}\right)^2\left(\frac{12\,\text{in}}{1\,\text{ft}}\right)^2 = 143\,\text{lb}\cdot\text{in}^2$$

(c)  The mass of a cylinder, from Prob. 17.44, is

$$m = \tfrac{1}{4}\rho\pi d^2 h = \frac{1}{4}\left(7,830\,\frac{\text{kg}}{\text{m}^3}\right)\pi d^2 h\,\text{mm}^3\left(\frac{1\,\text{m}}{1,000\,\text{mm}}\right)^3 = 6.15\times10^{-6}\,d^2 h \qquad \text{kg}$$

The mass of the pulley is given by

$$m = m_1 + m_2 + m_3 + m_4 - m_5 - m_6 - m_7 = 6.15\times10^{-6}[100^2(50) + 130^2(60) + 150^2(65)$$
$$+ 130^2(40) - 100^2(105) - 70^2(60) - 40^2(50)] = 13.7\,\text{kg}$$

The weight of the pulley is given by

$$W = mg = 13.7(9.81) = 134\,\text{N}$$

The radius of gyration of the pulley about the $y$ axis is

$$k_y = \sqrt{\frac{I_y}{m}} = \sqrt{\frac{0.0418}{13.7}} = 0.0552\,\text{m} = 55.2\,\text{mm}$$

**17.46**  A hardened steel dowel pin has hemispherical ends, as shown in Fig. 17.46a.  The specific weight of steel is $489\,\text{lb/ft}^3$.

(a)  Find the mass moments of inertia $I_{0x}$ and $I_{0y}$.

(b)  Find the weight of the pin in ounces.

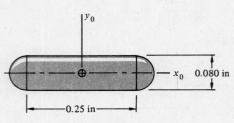

**Fig. 17.46a**

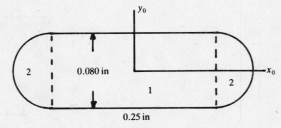

**Fig. 17.46b**

∎ (a)  The pin is divided into the simple volumes shown in Fig. 17.46b.  For element 1, using Case 4 in Table 17.13,

$$m_1 = \tfrac{1}{4}\rho\pi d^2 h = \frac{1}{4}\left[\frac{489}{386(12)^3}\right]\pi(0.080)^2 0.25 = 9.21\times10^{-7}\,\text{lb}\cdot\text{s}^2/\text{in}$$

$$I_{0x,1} = \tfrac{1}{8}m_1 d_1^2 = \tfrac{1}{8}(9.21\times10^{-7})0.080^2 = 7.37\times10^{-10}\,\text{lb}\cdot\text{s}^2\cdot\text{in} \qquad I_{0y,1} = \tfrac{1}{12}m_1(h^2 + \tfrac{3}{4}d^2)$$

Using $d = 0.080\,\text{in}$ and $h = 0.25\,\text{in}$,

$$I_{0y,1} = \tfrac{1}{12}(9.21\times10^{-7})[0.25^2 + \tfrac{3}{4}(0.080)^2] = 5.17\times10^{-9}\,\text{lb}\cdot\text{s}^2\cdot\text{in}$$

For elements 2, using Case 3 in Table 17.13,

$$m_2 = \tfrac{1}{12}\rho\pi d_2^3 = \frac{1}{12}\left[\frac{489}{386(12)^3}\right]\pi(0.080)^3 = 9.83\times10^{-8}\,\text{lb}\cdot\text{s}^2/\text{in}$$

$$y_c = \tfrac{3}{16}d = \tfrac{3}{16}(0.080) = 0.015 \text{ in}$$

$$I_{0x,2} = \tfrac{1}{10}m_2 d_2^2 = \tfrac{1}{10}(9.83 \times 10^{-8})0.080^2 = 6.29 \times 10^{-11} \text{ lb} \cdot \text{s}^2 \cdot \text{in}$$

$$I_{0y,2} = 0.0648 m_2 d_2^2 = 0.0648(9.83 \times 10^{-8})0.080^2 = 4.08 \times 10^{-11} \text{ lb} \cdot \text{s}^2 \cdot \text{in}$$

The mass moment of inertia about the centroidal $x_0$ axis is given by

$$I_{0x} = I_{0x,1} + 2I_{0x,2} = 7.37 \times 10^{-10} + 2(6.29 \times 10^{-11}) = 8.63 \times 10^{-10} \text{ lb} \cdot \text{s}^2 \cdot \text{in}$$

Using the parallel-axis theorem, the mass moment of inertia about the centroidal $y_0$ axis appears as

$$I_{0y} = I_{0y,1} + 2(I_{0y,2} + m_2 d_{0y}^2)$$

Using $d_{0y} = 0.25/2 + 0.015 = 0.140 \text{ in}$, the final result for $I_{0y}$ is

$$I_{0y} = 5.17 \times 10^{-9} + 2[4.08 \times 10^{-11} + 9.83 \times 10^{-8}(0.140)^2] = 9.10 \times 10^{-9} \text{ lb} \cdot \text{s}^2 \cdot \text{in}$$

($b$) The weight of the pin is found as

$$W = (m_1 + 2m_2)g = [9.21 \times 10^{-7} + 2(9.83 \times 10^{-8})]386 = 0.000431 \text{ lb}$$

$$W = 0.000431 \text{ lb}\left(\frac{16 \text{ oz}}{1 \text{ lb}}\right) = 0.00690 \text{ oz}$$

**17.47** Figure 17.47$a$ shows a steel pin. The density of steel is $7,830 \text{ kg/m}^3$.

($a$) Find the mass moments of inertia $I_x$ and $I_y$.

($b$) Find the radii of gyration $k_x$ and $k_y$.

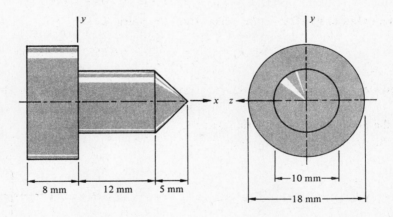

8 mm   12 mm   5 mm

10 mm

18 mm

**Fig. 17.47$a$**

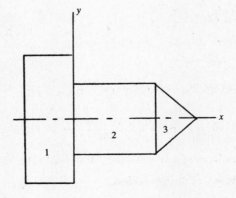

**Fig. 17.47$b$**

❚ ($a$) Figure 17.47$b$ shows the division of the pin into three elementary shapes. For the cylindrical shape elements, from Case 4 in Table 17.13,

$$m = \tfrac{1}{4}\rho\pi d^2 h \qquad I_x = \tfrac{1}{8}md^2 \qquad I_y = \tfrac{1}{48}m(16h^2 + 3d^2)$$

Using    $d_1 = 18\,\text{mm}$    and    $h_1 = 8\,\text{mm}$,

$$m_1 = \tfrac{1}{4}(7{,}830)\pi(18)^2 8\left(\frac{1}{1{,}000}\right)^3 = 0.0159\,\text{kg}$$

$$I_{x1} = \tfrac{1}{8}(0.0159)18^2\left(\frac{1}{1{,}000}\right)^2 = 6.44 \times 10^{-7}\,\text{kg}\cdot\text{m}^2$$

$$I_{y1} = \tfrac{1}{48}(0.0159)[16(8)^2 + 3(18)^2]\left(\frac{1}{1{,}000}\right)^2 = 6.61 \times 10^{-7}\,\text{kg}\cdot\text{m}^2$$

Using    $d_2 = 10\,\text{mm}$    and    $h_2 = 12\,\text{mm}$,

$$m_2 = \tfrac{1}{4}(7{,}830)\pi(10)^2 12\left(\frac{1}{1{,}000}\right)^3 = 0.00738\,\text{kg}$$

$$I_{x2} = \tfrac{1}{8}(0.00738)10^2\left(\frac{1}{1{,}000}\right)^2 = 9.23 \times 10^{-8}\,\text{kg}\cdot\text{m}^2$$

$$I_{y2} = \tfrac{1}{48}(0.00738)[16(12)^2 + 3(10^2)]\left(\frac{1}{1{,}000}\right)^2 = 4.00 \times 10^{-7}\,\text{kg}\cdot\text{m}^2$$

The mass properties of element 3 are found, from Case 6 in Table 17.13, as

$$m = \tfrac{1}{12}\rho\pi d^2 h \qquad I_x = \tfrac{3}{40}md^2 \qquad I_{0y} = \tfrac{3}{80}m(d^2 + h^2)$$

The centroidal coordinate of element 3 is

$$x_c = \tfrac{1}{4}h$$

Using    $d_3 = 10\,\text{mm}$    and    $h_3 = 5\,\text{mm}$,

$$m_3 = \tfrac{1}{12}(7{,}830)\pi(10)^2 5\left(\frac{1}{1{,}000}\right)^3 = 0.00102\,\text{kg}$$

$$I_{x3} = \tfrac{3}{40}(0.00102)10^2\left(\frac{1}{1{,}000}\right)^2 = 7.65 \times 10^{-9}\,\text{kg}\cdot\text{m}^2$$

$$I_{0y,3} = \tfrac{3}{80}(0.00102)(10^2 + 5^2)\left(\frac{1}{1{,}000}\right)^2 = 4.78 \times 10^{-9}\,\text{kg}\cdot\text{m}^2$$

Using the parallel-axis theorem, the mass moment of inertia of element 3 about the $y$ axis is written as

$$I_{y3} = I_{0y,3} + md_{y3}^2$$

Using

$$x_c = \tfrac{1}{4}h = \tfrac{1}{4}(5) = 1.25\,\text{mm} \qquad\text{and}\qquad d_{y3} = 12 + x_c = 12 + 1.25 = 13.3\,\text{mm}$$

the final value of $I_{y3}$ is

$$I_{y3} = 4.78 \times 10^{-9} + 0.00102(13.3)^2\left(\frac{1}{1{,}000}\right)^2 = 1.85 \times 10^{-7}\,\text{kg}\cdot\text{m}^2$$

The mass moment of inertia of the pin about the $x$ axis has the final form

$$I_x = I_{x1} + I_{x2} + I_{x3} = 6.44 \times 10^{-7} + 9.23 \times 10^{-8} + 7.65 \times 10^{-9} = 7.44 \times 10^{-7}\,\text{kg}\cdot\text{m}^2$$

The mass moment of inertia of the pin about the $y$ axis has the final form

$$I_y = I_{y1} + I_{y2} + I_{y3} = 6.61 \times 10^{-7} + 4.00 \times 10^{-7} + 1.85 \times 10^{-7} = 1.25 \times 10^{-6}\,\text{kg}\cdot\text{m}^2$$

(b)    The total mass $m$ of the pin is

$$m = m_1 + m_2 + m_3 = 0.0159 + 0.00738 + 0.00102 = 0.0243\,\text{kg}$$

The radii of gyration of the pin about the $x$ and $y$ axes are found as

$$k_x = \sqrt{\frac{I_x}{m}} = \sqrt{\frac{7.44 \times 10^{-7}}{0.0243}} = 0.00553\,\text{m} = 5.53\,\text{mm}$$

$$k_y = \sqrt{\frac{I_y}{m}} = \sqrt{\frac{1.25 \times 10^{-6}}{0.0243}} = 0.00717\,\text{m} = 7.17\,\text{mm}$$

**17.48** Figure 17.48a shows a cross-sectional view of an aluminium housing with a cone-cylinder form. Find the mass moment of inertia $I_x$ and the radius of gyration $k_x$. The density of aluminum is $2,770 \text{ kg/m}^3$.

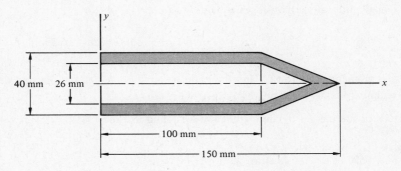

Fig. 17.48a

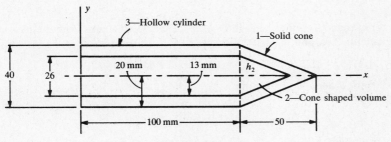

Fig. 17.48b                                      Fig. 17.48c

▌ The housing is divided into the elements shown in Fig. 17.48b, and a part of the cone-shaped element 2 is redrawn in Fig. 17.48c. From the similar triangles in this figure,

$$\frac{h_2}{13} = \frac{50}{20} \qquad h_2 = 32.5 \text{ mm}$$

From Case 6 in Table 17.13,

$$m = \tfrac{1}{12}\rho\pi d^2 h \qquad I_x = \tfrac{3}{40}md^2$$

From element 1, using $d_1 = 40 \text{ mm}$ and $h_1 = 50 \text{ mm}$,

$$m_1 = \tfrac{1}{12}(2,770)\pi(40)^2 50\left(\frac{1}{1,000}\right)^3 = 0.0580 \text{ kg}$$

$$I_{x1} = \tfrac{3}{40}(0.0580)40^2\left(\frac{1}{1,000}\right)^2 = 6.96 \times 10^{-6} \text{ kg} \cdot \text{m}^2$$

For element 2, using $d_2 = 26 \text{ mm}$ and $h_2 = 32.5 \text{ mm}$,

$$m_2 = \tfrac{1}{12}(2,770)\pi(26)^2 32.5\left(\frac{1}{1,000}\right)^3 = 0.0159 \text{ kg} \qquad I_{x2} = \frac{3}{40}(0.0159)26^2\left(\frac{1}{1,000}\right)^2 = 8.06 \times 10^{-7} \text{ kg} \cdot \text{m}^2$$

From Case 5 in Table 17.13,

$$m = \tfrac{1}{4}\rho\pi h(d_0^2 - d_i^2)$$
$$I_x = \tfrac{1}{8}m(d_0^2 + d_i^2)$$

For element 3, using $d_0 = 40 \text{ mm}$, $d_i = 26 \text{ mm}$, and $h = 100 \text{ mm}$,

$$m_3 = \tfrac{1}{4}(2,770)\pi(100)(40^2 - 26^2)\left(\frac{1}{1,000}\right)^3 = 0.201 \text{ kg}$$

$$I_{x3} = \tfrac{1}{8}(0.201)(40^2 + 26^2)\left(\frac{1}{1,000}\right)^2 = 5.72 \times 10^{-5} \text{ kg} \cdot \text{m}^2$$

The mass moment of inertia of the housing about the $x$ axis is then

$$I_x = I_{x1} - I_{x2} + I_{x3} = 6.96 \times 10^{-6} - 8.06 \times 10^{-7} + 5.72 \times 10^{-5} = 6.34 \times 10^{-5} \text{ kg} \cdot \text{m}^2$$

The mass of the housing is

$$m = m_1 - m_2 + m_3 = 0.0580 - 0.0159 + 0.201 = 0.243 \text{ kg}$$

The radius of gyration of the housing about the $x$ axis is

$$k_x = \sqrt{\frac{I_x}{m}} = \sqrt{\frac{6.34 \times 10^{-5}}{0.243}} = 0.0162 \text{ m} = 16.2 \text{ mm}$$

**17.49** Find the mass moment of inertia $I_y$, and the radius of gyration $k_y$, of the housing shown in Prob. 17.48. The density of aluminum is 2,770 kg/m³.

▮ The values of the masses of the elements were found in Prob. 17.48. From Case 6 in Table 17.13,

$$I_{0y} = \tfrac{3}{80}m(d^2 + h^2) \qquad x_c = \tfrac{1}{4}h$$

Using $m_1 = 0.0580$ kg, $d_1 = 40$ mm, and $h_1 = 50$ mm,

$$I_{0y,1} = \tfrac{3}{80}(0.0580)(40^2 + 50^2)\left(\frac{1}{1,000}\right)^2 = 8.92 \times 10^{-6} \text{ kg} \cdot \text{m}^2$$

$d_{y1}$ is the distance from the center of mass of element 1 to the $y$ axis. $x_{c1}$ is the distance from the base of cone 1 to the center of mass of this element. $x_{c1}$ and $d_{y1}$ have the values

$$x_{c1} = \tfrac{1}{4}(50) = 12.5 \text{ mm} \qquad d_{y1} = 100 + 12.5 = 113 \text{ mm}$$

Using $m_2 = 0.0159$ kg, $d_2 = 26$ mm, and $h_2 = 32.5$ mm,

$$I_{0y,2} = \tfrac{3}{80}(0.0159)(26^2 + 32.5^2)\left(\frac{1}{1,000}\right)^2 = 1.03 \times 10^{-6} \text{ kg} \cdot \text{m}^2$$

$d_{y2}$ is the distance from the center of mass of element 2 to the $y$ axis. $x_{c2}$ is the distance from the base of cone 2 to the center of mass of this element. $x_{c2}$ and $d_{y2}$ have the values

$$x_{c2} = \tfrac{1}{4}(32.5) = 8.13 \text{ mm} \qquad d_{y2} = 100 + 8.13 = 108 \text{ mm}$$

From Case 5 in Table 17.13,

$$I_{0y} = \tfrac{1}{48}m(3d_0^2 + 3d_i^2 + 4h^2)$$

Using $d_0 = 40$ mm, $d_i = 26$ mm, $h = 100$ mm, and $m_3 = 0.201$ kg,

$$I_{0y,3} = \tfrac{1}{48}(0.201)[3(40)^2 + 3(26)^2 + 4(100)^2]\left(\frac{1}{1,000}\right)^2 = 1.96 \times 10^{-4} \text{ kg} \cdot \text{m}^2$$

The distance from the center of mass of element 3 to the $y$ axis, $d_{y3}$, is

$$d_{y3} = \frac{100}{2} = 50 \text{ mm}$$

Using the parallel-axis theorem, the mass moment of inertia of the housing about the $y$ axis has the form

$$I_y = I_{y1} - I_{y2} + I_{y3} = \left[8.92 \times 10^{-6} + 0.0580(113)^2\left(\frac{1}{1,000}\right)^2\right] - \left[1.03 \times 10^{-6} + 0.0159(108)^2\left(\frac{1}{1,000}\right)^2\right]$$

$$+ \left[1.96 \times 10^{-4} + 0.201(50)^2\left(\frac{1}{1,000}\right)^2\right] = 0.00126 \text{ kg} \cdot \text{m}^2$$

From Prob. 17.48,

$$m = 0.243 \text{ kg}$$

The radius of gyration of the housing about the $y$ axis is then

$$k_y = \sqrt{\frac{I_y}{m}} = \sqrt{\frac{0.00126}{0.243}} = 0.0720 \text{ m} = 72 \text{ mm}$$

**17.50** The disk shown in Fig. 17.50 has two holes in it and is made of 0.5-in-thick aluminum plate. The specific weight of aluminum is 173 lb/ft³.

(a) Find the mass moment of inertia of the disk about its center axis.

(b) Find the radius of gyration of the disk about its center axis.

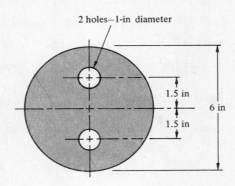

2 holes—1-in diameter

**Fig. 17.50**

▌ (*a*)  The solid disk and a single hole are designated elements 1 and 2, respectively.  The weights of these two elements, using Case 4 in Table 17.13, are

$$W_1 = \tfrac{1}{4}\gamma\pi d^2 h = \frac{1}{4}\left(\frac{173}{12^3}\right)\pi(6)^2 0.5 = 1.42\,\text{lb} \qquad W_2 = \frac{1}{4}\left(\frac{173}{12^3}\right)\pi(1)^2 0.5 = 0.0393\,\text{lb}$$

From Case 4,

$$I_{0y} = \tfrac{1}{8}md^2$$

The mass moment of inertia $I_0$ of the disk about its center axis, using the parallel-axis theorem, is then

$$I_0 = I_{01} - 2I_{02} = \frac{1}{8}\left(\frac{1.42}{386}\right)6^2 - 2\left[\frac{1}{8}\left(\frac{0.0393}{386}\right)1^2 + \frac{0.0393}{386}(1.5)^2\right] = 0.0161\,\text{lb}\cdot\text{s}^2\cdot\text{in}$$

(*b*)  The radius of gyration $k_0$ of the disk about its center axis is given by

$$k_0 = \sqrt{\frac{I_0}{m}} = \sqrt{\frac{0.0161}{[1.42 - 2(0.0393)](\tfrac{1}{386})}} = 2.15\,\text{in}$$

**17.51**  (*a*)  Find the mass moment of inertia about the center axis of the steel body shown in Fig. 17.51.  The specific weight of steel is 489 lb/ft³.  Neglect the effect of the center hole.

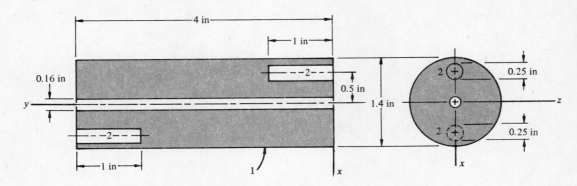

**Fig. 17.51**

(*b*)  What is the error introduced into the result in part (*a*) by neglecting the effect of the center hole?

▌ (*a*)  The solid cylinder is element 1, and each of the two end holes is element 2.  From Case 4 in Table 17.13,

$$m = \tfrac{1}{4}\rho\pi d^2 h$$

$$m_1 = \frac{1}{4}\left[\frac{489}{386(12)^3}\right]\pi(1.4)^2 4 = 0.00451\,\text{lb}\cdot\text{s}^2/\text{in}$$

$$m_2 = \frac{1}{4}\left[\frac{489}{386(12)^3}\right]\pi(0.25)^2 1 = 3.60\times10^{-5}\,\text{lb}\cdot\text{s}^2/\text{in}$$

The mass moment of inertia of the body about its center axis is

$$I_{0y} = I_{0y,1} - 2I_{0y,2}$$

Using Case 4 in Table 17.13,

$$I_{0y} = \tfrac{1}{8}md^2$$
$$I_{0y} = \tfrac{1}{8}(0.00451)1.4^2 - 2[\tfrac{1}{8}(3.60 \times 10^{-5})0.25^2 + 3.60 \times 10^{-5}(0.5)^2] = 0.00109 \text{ lb} \cdot \text{s}^2 \cdot \text{in}$$

(*b*)  The center hole is designated element 3.  The mass, and moment of inertia, of the material which would fill this hole are

$$m_3 = \frac{1}{4}\left[\frac{489}{386(12)^3}\right]\pi(0.16)^2 4 = 5.90 \times 10^{-5} \text{ lb} \cdot \text{s}^2/\text{in}$$

$$I_{0y,3} = \tfrac{1}{8}md^2 = \tfrac{1}{8}(5.90 \times 10^{-5})0.16^2 = 1.89 \times 10^{-7} \text{ lb} \cdot \text{s}^2 \cdot \text{in}$$

The mass moment of inertia of the hole material has the approximate value

$$I_{0y,3} \approx 0.0000002 \text{ lb} \cdot \text{s}^2 \cdot \text{in}$$

From part (*a*),

$$I_{0y} = 0.0010900 \text{ lb} \cdot \text{s}^2 \cdot \text{in}$$

The error introduced by neglecting the effect of the center hole is approximately 2 parts in 10,900 or

$$\frac{2}{10,900}(100) = 0.018\%$$

**17.52**  In a certain testing machine, a very precise value of mass moment of inertia is required.  This is accomplished by drilling pairs of holes on a diameter of a disk, as shown in Fig. 17.52.  Find the percent reduction in the mass moment of inertia about the center axis for each pair of holes drilled.  Use the mass moment of inertia of the undrilled disk with a center hole as the reference value, and use four significant figures in the calculations.  The disk material is aluminum, and  $\rho = 2{,}770 \text{ kg/m}^3$.

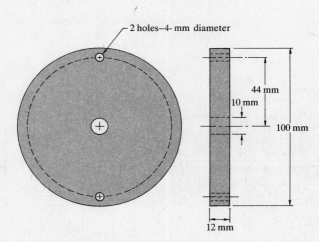

2 holes–4- mm diameter

44 mm

10 mm

100 mm

12 mm

**Fig. 17.52**

▮  For the undrilled disk with a center hole, using Case 5 of Table 17.13 and with $y_0$ as the center axis of the disk,

$$I_{\text{disk}} = I_{0y} = \tfrac{1}{8}m(d_0^2 + d_i^2) = \tfrac{1}{8}[\tfrac{1}{4}\rho\pi h(d_0^2 - d_i^2)](d_0^2 + d_i^2) = \tfrac{1}{32}\rho\pi h(d_0^4 - d_i^4)$$

Using  $d_0 = 100 \text{ mm}$  and  $d_i = 10 \text{ mm}$,

$$I_{\text{disk}} = \frac{1}{32}(2{,}770)\pi(12)(100^4 - 10^4)\left(\frac{1}{1{,}000}\right)^5 = 3.263 \times 10^{-4} \text{ kg} \cdot \text{m}^2$$

$I_{\text{holes}}$ is the mass moment of inertia about the $y_0$ axis of the material removed from a *pair* of holes.  Using the parallel-axis theorem,

$$I_{\text{holes}} = 2(\tfrac{1}{8}md^2 + md_{0y}^2)$$

Using $m = \frac{1}{4}\rho\pi d^2 h$, the above equation has the form

$$I_{\text{holes}} = 2m(\tfrac{1}{8}d^2 + d_{0y}^2) = 2(\tfrac{1}{4}\rho\pi d^2 h)(\tfrac{1}{8}d^2 + d_{0y}^2)$$

With $d_{0y}^2 = 44$ mm,

$$I_{\text{holes}} = 2(\tfrac{1}{4})2{,}770\pi(4)^2 12[\tfrac{1}{8}(4)^2 + 44^2]\left(\frac{1}{1{,}000}\right)^5 = 1.619 \times 10^{-6}\ \text{kg} \cdot \text{m}^2$$

The percent reduction in mass moment of inertia for each pair of holes drilled is given by

$$\%\text{D} = \frac{I_{\text{holes}}}{I_{\text{disk}}}(100) = \frac{1.619 \times 10^{-6}}{3.263 \times 10^{-4}}(100) = 0.496\% \approx 0.5\%$$

**17.53**  Figure 17.53a shows a flywheel made of die-cast zinc with a density of $7{,}050\ \text{kg/m}^3$.  The effect of the small center hole may be neglected.

(a) Find the mass, the polar mass moment of inertia, and the polar radius of gyration of the flywheel.

(b) In a proposed modification of this design the center section will have two holes, shown as the dashed circles in Fig. 17.53a.  Find the percent reduction in the mass, and in the polar mass moment of inertia, when the two holes are added to the original design.

(c) How would the results in part (a) change if the flywheel were fabricated of aluminum with a density of $2{,}700\ \text{kg/m}^3$?

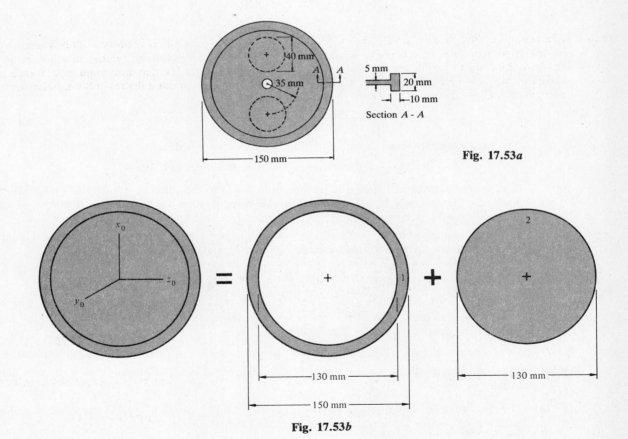

Fig. 17.53a

Fig. 17.53b

▌ (a)  The original design consists of a solid cylinder and a hollow cylinder, as shown in Fig. 17.53b.  The rim is designated element 1, and the plane center section is designated element 2.  The results in Cases 4 and 5 in Table 17.13 are used directly, and

$$m = m_1 + m_2 = \tfrac{1}{4}\rho\pi h_1(d_{01}^2 - d_{i1}^2) + \tfrac{1}{4}\rho\pi d_2^2 h_2$$

Using $d_{i1} = 130\,\text{mm}$, $d_{01} = 150\,\text{mm}$, $h_1 = 20\,\text{mm}$, $d_2 = 130\,\text{mm}$, and $h_2 = 5\,\text{mm}$, the mass of the flywheel has the value

$$m = \tfrac{1}{4}\left(7050\,\tfrac{\text{kg}}{\text{m}^3}\right)\pi(20\,\text{mm})(150^2 - 130^2)\,\text{mm}^2\left(\frac{1\,\text{m}}{1{,}000\,\text{mm}}\right)^3 + \tfrac{1}{4}(7050)\pi(130^2)5\left(\frac{1}{1{,}000}\right)^3$$

$$= 0.620 + 0.468 = 1.09\,\text{kg} \qquad\qquad (1)$$

The polar mass moment of inertia of the flywheel is

$$I_{0y} = I_{01} + I_{02} = \tfrac{1}{8}m_1(d_{01}^2 + d_{i1}^2) + \tfrac{1}{8}m_2 d_2^2 = \tfrac{1}{8}(0.620\,\text{kg})(150^2 + 130^2)\,\text{mm}^2\left(\frac{1\,\text{m}}{1{,}000\,\text{mm}}\right)^2$$

$$+ \tfrac{1}{8}(0.468)130^2\left(\frac{1}{1{,}000}\right)^2 = 0.00305 + 0.000989 = 0.00404\,\text{kg}\cdot\text{m}^2 \qquad\qquad (2)$$

It is interesting to observe that the contributions of the rim and center section from Eq. (2) are 76 and 24 percent, respectively, of the total mass moment of inertia. By comparison, from Eq. (1), the rim and center section contribute 57 and 43 percent, respectively, of the total mass of the assembly. The polar radius of gyration is found as

$$k_{0y} = \sqrt{\frac{I_{0y}}{m}} = \sqrt{\frac{0.00404}{1.09}} = 0.0609\,\text{m} = 60.9\,\text{mm}$$

(*b*) The mass of material which would occupy one hole, for $d = 40\,\text{mm}$, is

$$m = \tfrac{1}{4}\rho\pi d^2 h = \tfrac{1}{4}(7{,}050)\pi(40^2)5\left(\frac{1}{1{,}000}\right)^3 = 0.0443\,\text{kg}$$

The total mass of the proposed design is

$$m_{\text{net}} = 1.09 - 2(0.0443) = 1.00\,\text{kg}$$

The corresponding mass moment of inertia of the material of one hole about the center axis has the general form

$$I_y = I_{0y} + m d_y^2 = \tfrac{1}{8}m d^2 + m d_y^2 = \tfrac{1}{8}(0.0443)40^2\left(\frac{1}{1{,}000}\right)^2 + 0.0443\left(\frac{35}{1{,}000}\right)^2 = 6.31 \times 10^{-5}\,\text{kg}\cdot\text{m}^2$$

The mass moment of inertia of the flywheel with two holes is then

$$I_{\text{net}} = 0.00404 - 2(6.31 \times 10^{-5}) = 0.00391\,\text{kg}\cdot\text{m}^2$$

The percent reduction in mass with the proposed design modification is $[(1.09 - 1.00)/1.09]100 = 8.3$ percent. The corresponding reduction in the mass moment of inertia is $[(0.00404 - 0.00391)/0.00404]100 = 3.2$ percent.

(*c*) The mass moment of inertia is directly proportional to the density of the material. Thus, if aluminum were substituted for zinc, the new values of $m$ and $I_{0y}$ would be

$$m = \frac{2700}{7{,}050}\,(1.09) = 0.417\,\text{kg} \qquad I_{0y} = \frac{2{,}700}{7{,}050}\,(0.00404) = 0.00155\,\text{kg}\cdot\text{m}^2$$

**17.54** Figure 17.54 shows the cross section of a proposed lightweight flywheel design. The density of steel is $7{,}830\,\text{kg/m}^3$, and the density of aluminum is $2{,}770\,\text{kg/m}^3$.

(*a*) Find the weight and mass of the flywheel, and the mass moment of inertia about the center axis.

(*b*) Do the same as in part (*a*), if the flywheel is made entirely of steel. Compare the results with those in part (*a*).

▮ (*a*) The rim, web, and hub are designated elements 1, 2, and 3, respectively. All of these elements are hollow right circular cylinders. Using Case 5 in Table 17.13,

$$m = \tfrac{1}{4}\rho\pi h(d_0^2 - d_i^2) \qquad I_{oy} = \tfrac{1}{8}m(d_0^2 + d_i^2)$$

For the rim, element 1, with $d_i = 260\,\text{mm}$, $d_0 = 320\,\text{mm}$, and $h = 60\,\text{mm}$,

$$m_1 = \tfrac{1}{4}(7{,}830)\pi(60)(320^2 - 260^2)\left(\frac{1}{1{,}000}\right)^3 = 12.8\,\text{kg}$$

$$I_{01} = \tfrac{1}{8}(12.8)(320^2 + 260^2)\left(\frac{1}{1{,}000}\right)^2 = 0.272\,\text{kg}\cdot\text{m}^2$$

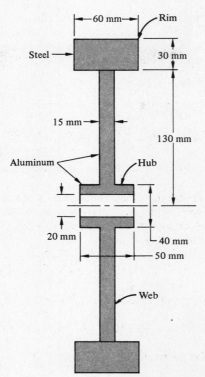

**Fig. 17·54**

For the web, element 2, with $d_i = 40\text{ mm}$, $d_0 = 260\text{ mm}$, and $h = 15\text{ mm}$,

$$m_2 = \tfrac{1}{4}(2{,}770)\pi(15)(260^2 - 40^2)\left(\frac{1}{1{,}000}\right)^3 = 2.15\text{ kg}$$

$$I_{02} = \tfrac{1}{8}(2.15)(260^2 + 40^2)\left(\frac{1}{1{,}000}\right)^2 = 0.0186\text{ kg}\cdot\text{m}^2$$

For the hub, element 3, with $d_i = 20\text{ mm}$, $d_0 = 40\text{ mm}$, and $h = 50\text{ mm}$,

$$m_3 = \tfrac{1}{4}(2{,}770)\pi(50)(40^2 - 20^2)\left(\frac{1}{1{,}000}\right)^3 = 0.131\text{ kg}$$

$$I_{03} = \tfrac{1}{8}(0.131)(40^2 + 20^2)\left(\frac{1}{1{,}000}\right)^2 = 3.28 \times 10^{-5}\text{ kg}\cdot\text{m}^2$$

The mass and weight of the flywheel are

$$m = m_1 + m_2 + m_3 = 12.8 + 2.15 + 0.131 = 15.1\text{ kg} \qquad W = mg = 15.1(9.81) = 148\text{ N}$$

The mass moment of inertia $I_0$ of the flywheel about the center axis is given by

$$I_0 = I_{01} + I_{02} + I_{03} = 0.272 + 0.0186 + 3.28 \times 10^{-5} = 0.291\text{ kg}\cdot\text{m}^2$$

(b) For the case where the flywheel is made entirely of steel, the mass properties of the web and hub must be increased by a factor $q$, the ratio of the densities of steel and aluminum. The value of $q$ is

$$q = \frac{7{,}830}{2{,}770}$$

The new mass properties are then

$$m_2 = q(2.15) = 6.08\text{ kg} \qquad I_{02} = q(0.0186) = 0.0526\text{ kg}\cdot\text{m}^2$$
$$m_3 = q(0.131) = 0.370\text{ kg} \qquad I_{03} = q(3.28 \times 10^{-5}) = 9.27 \times 10^{-5}\text{ kg}\cdot\text{m}^2$$

The mass, weight, and moment of inertia of the steel flywheel are then

$$m = m_1 + m_2 + m_3 = 12.8 + 6.08 + 0.370 = 19.3\text{ kg} \qquad W = mg = 19.3(9.81) = 189\text{ N}$$
$$I_0 = I_{01} + I_{02} + I_{03} = 0.272 + 0.0526 + 9.27 \times 10^{-5} = 0.325\text{ kg}\cdot\text{m}^2$$

The weights, and moments of inertia, of the original flywheel and the all-steel flywheel may now be compared. The mass properties of the original flywheel are designated $W$ and $I_0$, and those of the all-steel flywheel are $W_s$ and $I_{0s}$. The percent difference between the values, using the original design as the reference, are, for weight,

$$\%D = \frac{W_s - W}{W}(100) = \frac{189 - 148}{148}(100) = 27.7\%$$

For mass moment of inertia,

$$\%D = \frac{I_{0s} - I_0}{I_0}(100) = \frac{0.325 - 0.291}{0.291}(100) = 11.7\%$$

It may be seen that the percent increase in weight is more than twice the percent increase in mass moment of inertia.

**17.55**    Two cylindrical weights are attached to a circular disk, as shown in Fig. 17.55.

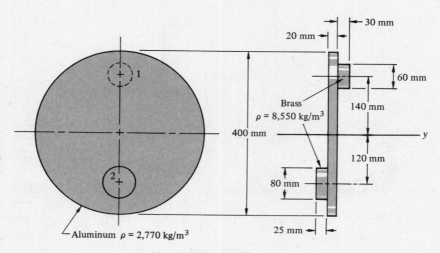

**Fig. 17.55**

(*a*)  Find the mass, and the mass moment of inertia, of the assembly about the center axis.

(*b*)  Find the radius of gyration of the assembly about the center axis.

▌ (*a*)  The two weights are identified as elements 1 and 2, as shown in Fig. 17.75.  The masses of the two weights are found from Case 4 of Table 17.13 as

$$m = \tfrac{1}{4}\rho\pi d^2 h \qquad m_1 = \frac{1}{4}\left(8{,}550\,\frac{\text{kg}}{\text{m}^3}\right)\pi(60)^2 30\text{ mm}^3\left(\frac{1\text{ m}}{1{,}000\text{ mm}}\right)^3 = 0.725\text{ kg}$$

$$m_2 = \tfrac{1}{4}(8{,}550)\pi(80)^2 25\left(\frac{1}{1{,}000}\right)^3 = 1.07\text{ kg}$$

The disk is element 3, with mass

$$m_3 = \tfrac{1}{4}(2{,}770)\pi(400)^2 20\left(\frac{1}{1{,}000}\right)^3 = 6.96\text{ kg}$$

The mass of the assembly is

$$m = m_1 + m_2 + m_3 = 0.725 + 1.07 + 6.96 = 8.76\text{ kg}$$

From Case 4 in Table 17.13,

$$I_{0y} = I_y = \tfrac{1}{8}md^2 = \tfrac{1}{8}m(\text{kg})d^2(\text{mm})^2\left(\frac{1\text{ m}}{1{,}000\text{ mm}}\right)^2$$

The mass moment of inertia of the assembly about the center axis is

$$I_y = I_{y1} + I_{y2} + I_{y3}$$

Using the parallel-axis theorem,

$$I_y = \{[\tfrac{1}{8}(0.725)60^2 + 0.725(140)^2] + [\tfrac{1}{8}(1.07)80^2 + 1.07(120)^2] + [\tfrac{1}{8}(6.96)400^2]\}\left(\frac{1}{1,000}\right)^2$$

$$= 0.170 \text{ kg} \cdot \text{m}^2$$

(*b*)  The radius of gyration of the assembly about the center axis is then found as

$$k_y = \sqrt{\frac{I_y}{m}} = \sqrt{\frac{0.170}{8.76}} = 0.139 \text{ m} = 13.9 \text{ mm}$$

**17.56**    Find the mass of the part shown in Fig. 17.56*a*, and the mass moment of inertia about the center axis.    The thickness of the disk is 40 mm and the material is bronze, with    $\rho = 8,800 \text{ kg/m}^3$.

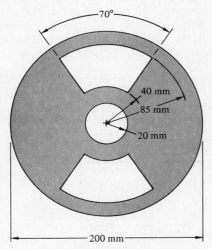

**Fig. 17.56*a***

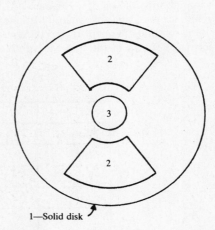

1—Solid disk

**Fig. 17.56*b***

▌ The mass of the solid disk is designated 1, and the cutout portions are designated 2 and 3, as shown in Fig. 17.56*b*.    From Case 4 in Table 17.13,

$$m = \tfrac{1}{4}\rho\pi d^2 h \qquad m_1 = \tfrac{1}{4}(8,800)\pi(200)^2 40\left(\frac{1}{1,000}\right)^3 = 11.1 \text{ kg} \qquad I_{0y} = \tfrac{1}{8}md^2$$

$$I_{0y,1} = \tfrac{1}{8}(11.1)200^2\left(\frac{1}{1,000}\right)^2 = 0.0555 \text{ kg} \cdot \text{m}^2 \qquad m_3 = \tfrac{1}{4}(8,800)\pi(40)^2 40\left(\frac{1}{1,000}\right)^3 = 0.442 \text{ kg}$$

$$I_{0y,3} = \tfrac{1}{8}(0.442)40^2\left(\frac{1}{1,000}\right)^2 = 8.84 \times 10^{-5} \text{ kg} \cdot \text{m}^2$$

Element 2 is treated as a portion of a hollow cylinder.    Its mass properties are a fraction of those of a full hollow cylinder, with the ratio $2(70°)/360°$.    From Case 5 of Table 17.13,

$$m_2 = \frac{2(70°)}{360°}\left[\tfrac{1}{4}\rho\pi h(d_0^2 - d_i^2)\right]$$

Using    $d_0 = 170$ mm    and    $d_i = 80$ mm,

$$m_2 = \frac{140°}{360°}\left[\frac{1}{4}(8,800)\pi(40)(170^2 - 80^2)\left(\frac{1}{1,000}\right)^3\right] = 2.42 \text{ kg}$$

The mass of the part is

$$m = m_1 - m_2 - m_3 = 11.1 - 2.42 - 0.442 = 8.24 \text{ kg}$$

The mass moment of inertia of element 2 is

$$I_{0y,2} = \frac{2(70°)}{360°}\left[\frac{1}{8}m(d_0^2 - d_i^2)\right] = \frac{140°}{360°}\left[\frac{1}{8}(2.42)(170^2 + 80^2)\left(\frac{1}{1,000}\right)^2\right] = 0.00415 \text{ kg} \cdot \text{m}^2$$

The mass moment of inertia of the part about the center axis is given by

$$I_{0y} = I_{0y,1} - I_{0y,2} - I_{0y,3} = 0.0555 - 0.00415 - 8.84 \times 10^{-5} = 0.0513 \, \text{kg} \cdot \text{m}^2$$

**17.57** Show a tabular format for organizing the computations for the mass moments of inertia for the case where a body is a complex shape which requires the description of several elementary shapes.

▌ A format for a tabular organization of the solution for the mass moment of inertia, using the $z$ axis as a typical reference axis, is shown in Table 17.4. The identification numbers of the mass elements are entered in column 1, and the values of these masses are shown in column 2. The reference axis is shown in the last row of the table. Two possible cases may now be identified. In the first, the form for the moment of inertia of a mass element about the reference axis is known. For this case, the equation for the moment of inertia is entered directly in column 6, and the numerical value is computed. In the second case the moment of inertia of the element about the reference axis is not known, and the parallel-axis theorem must be used to find this quantity. The centroidal moment of inertia of the element is entered in column 4, and the separation distance between the centroidal and reference axis is entered in column 3. The transfer term, with the form $md^2$, is computed in column 5. The sum of columns 4 and 5 is then the moment of inertia of the element about the reference axis, and this quantity is entered in column 6. As the final steps, columns 2 and 6 are summed, and the mass, mass moment of inertia, and radius of gyration may be found. An example of the use of a tabular format is shown in the following problem.

**TABLE 17.4**

| (1) | (2) | (3) | (4) | (5) | (6) |
|---|---|---|---|---|---|
| Element | $m_i$ | $d_{zi}$ | $I_{0z,i}$ | $m_i d_{zi}^2$ | $I_{zi}$ |
| 1 | $m_1$ | | $\cdots$ | $\cdots$ | $I_{z1}$ |
| 2 | $m_2$ | | | | $I_{z2}$ |
| 3 | $m_3$ | | | | $I_{z3}$ |
| | | | | | |
| $n$ | $m_n$ | | | | $I_{zn}$ |
| | $m = \sum\limits_{i=1}^{i=n} m_i$ | | | $I_z$ | |
| Reference axis: $z$ | | | | $k_z = \sqrt{\dfrac{I_z}{m}}$ | |

**17.58** The bracket of Prob. 17.16 is shown in Fig. 17.58a. The material is bronze, with a specific weight of $0.295 \, \text{lb/in}^3$. Find

(a) The mass moment of inertia $I_z$
(b) The centroidal mass moment of inertia $I_{0z}$
(c) The mass moment of inertia $I_{z_1}$
(d) The radius of gyration $k_{0z}$
(e) The radius of gyration $k_z$

Express all of the mass moments of inertia in the engineering units of $\text{lb} \cdot \text{in}^2$.

▌ (a) The bracket is divided into the same three elements used in Prob. 17.16 for the computation of the volume centroid. A set of centroidal coordinate axes is attached to elements 2 and 3, as shown in Fig. 17.58a.

Table 17.5 is next constructed. The solution for $I_{z_1}$ is found directly from Case 10 in Table 17.13 as the mass moment of inertia about an edge of a rectangular parallelepiped. The solutions for $I_{0z,2}$ and $I_{0z,3}$ are found from Cases 10 and 4, respectively, in the same table. The constructions used to find $d_{z2}$ and $d_{z3}$ are shown in Figs. 17.58b and c. These terms are

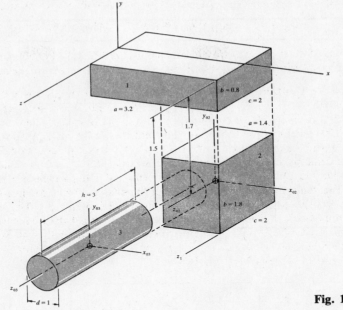

Fig. 17.58a

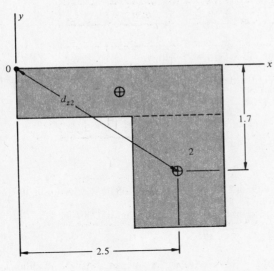

Fig. 17.58b

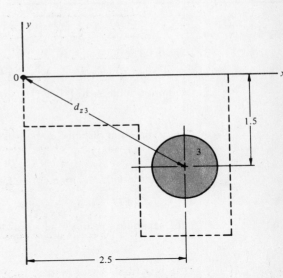

Fig. 17.58c

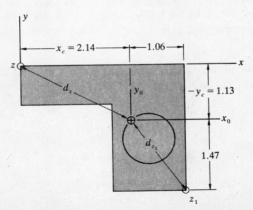

Fig. 17.58d

<table>
<tr><th colspan="6">TABLE 17.5</th></tr>
<tr><th>Element</th><th>$m_i$</th><th>$d_{zi}$</th><th>$I_{0z,i}$</th><th>$m_i d_{zi}^2$</th><th>$I_{zi}$</th></tr>
<tr>
<td>1</td>
<td>$m_1 = \rho abc$<br>$= \dfrac{0.295}{386}(3.2)(0.8(2)$<br>$= 0.00391 \text{ lb} \cdot \text{s}^2/\text{in}$</td>
<td>...</td>
<td>...</td>
<td>...</td>
<td>$I_{z1} = \frac{1}{3}m(a^2 + b^2)$<br>$= \frac{1}{3}(0.00391)(3.2^2 + 0.8^2)$<br>$= 0.0142 \text{ lb} \cdot \text{s}^2 \cdot \text{in}$</td>
</tr>
<tr>
<td>2</td>
<td>$m_2 = \rho abc$<br>$= \dfrac{0.295}{386}(1.4)1.8(2)$<br>$= 0.00385 \text{ lb} \cdot \text{s}^2/\text{in}$</td>
<td>$d_{z2} = 3.02$</td>
<td>$I_{0z,2} = \frac{1}{12}m(a^2 + b^2)$<br>$= \frac{1}{12}(0.00385)(1.4^2 + 1.8^2)$<br>$= 0.00167 \text{ lb} \cdot \text{s}^2 \cdot \text{in}$</td>
<td>$m_2 d_{z2}^2 = (0.00385)3.02^2$<br>$= 0.0351 \text{ lb} \cdot \text{s}^2 \cdot \text{in}$</td>
<td>$I_{z2} = I_{0z,2} + m_2 d_{z2}^2$<br>$= 0.00167 + 0.0351$<br>$= 0.0368 \text{ lb} \cdot \text{s}^2 \cdot \text{in}$</td>
</tr>
<tr>
<td>3</td>
<td>$m_3 = \frac{1}{4}\rho \pi d^2 h$<br>$= \frac{1}{4}\left(\dfrac{0.295}{386}\right)\pi(1^2)3$<br>$= 0.00180 \text{ lb} \cdot \text{s}^2/\text{in}$</td>
<td>$d_{z3} = 2.92$</td>
<td>$I_{0z,3} = \frac{1}{8}md^2$<br>$= \frac{1}{8}(0.00180)1^2$<br>$= 0.00023 \text{ lb} \cdot \text{s}^2 \cdot \text{in}$</td>
<td>$m_3 d_{z3}^2 = 0.0018(2.92^2)$<br>$= 0.0153 \text{ lb} \cdot \text{s}^2 \cdot \text{in}$</td>
<td>$I_{z3} = I_{0z,3} + m_3 d_{z3}^2$<br>$= 0.00023 + 0.0153$<br>$= 0.0155 \text{ lb} \cdot \text{s}^2 \cdot \text{in}$</td>
</tr>
<tr>
<td colspan="2">$m = \displaystyle\sum_{i=1}^{i=3} m_i$<br>$= 0.00956 \text{ lb} \cdot \text{s}^2/\text{in}$</td>
<td></td>
<td></td>
<td colspan="2">$I_z = 0.0665 \text{ lb} \cdot \text{s}^2 \cdot \text{in}$<br>$= 25.7 \text{ lb} \cdot \text{in}^2$</td>
</tr>
<tr>
<td colspan="3">Reference axis: $z$</td>
<td></td>
<td colspan="2">$k_z = \sqrt{\dfrac{I_z}{m}} = \sqrt{\dfrac{0.0665}{0.00956}} = 2.64 \text{ in}$</td>
</tr>
</table>

$$d_{z2} = \sqrt{2.5^2 + 1.7^2} = 3.02 \text{ in} \qquad d_{z3} = \sqrt{2.5^2 + 1.5^2} = 2.92 \text{ in}$$

The final result for $I_z$ is

$$I_z = 0.0665 \text{ lb} \cdot \text{s}^2 \cdot \text{in} = 25.7 \text{ lb} \cdot \text{in}^2$$

(b) From the solution to Prob. 17.16, the two centroidal coordinates required in the present calculation are

$$x_c = 2.14 \text{ in} \qquad y_c = -1.13 \text{ in}$$

The separation distance $d_z$ between the $z$ and $z_0$ axes, following Fig. 17.58d, is

$$d_z = \sqrt{2.14^2 + 1.13^2} = 2.42 \text{ in}$$

The centroidal mass moment of inertia $I_{0z}$ is then found from

$$I_z = I_{0z} + md_z^2 \qquad 0.0665 = I_{0z} + 0.00956(2.42)^2 \qquad I_{0z} = 0.0105 \text{ lb} \cdot \text{s}^2 \cdot \text{in} = 4.05 \text{ lb} \cdot \text{in}^2$$

(c) The separation distance between the $z_1$ and $z_0$ axes is found from Fig. 17.58d as

$$d_{z_1} = \sqrt{1.06^2 + 1.47^2} = 1.81 \text{ in}$$

The mass moment of inertia about the $z_1$ axis is then

$$I_{z_1} = I_{0z} + md_{z_1}^2 = 0.0105 + 0.00956(1.81)^2 = 0.0418 \text{ lb} \cdot \text{s}^2 \cdot \text{in}$$

A very common error, frequently made in the above type of calculation, is the attempt to find a moment of inertia, such as $I_{z_1}$, by using the transfer theorem directly from another axis, such as $z$, which is *not* a centroidal axis. In the present problem, this *incorrect* form would be written as

$$I_{z_1} \neq I_z + md_{z,z_1}^2$$

where $d_{z,z_1}$ is the separation distance between the $z$ and $z_1$ axes. The reader is urged to study the above equation carefully and understand why it is an *incorrect* statement of the parallel-axis theorem.

(d) The centroidal radius of gyration is found as

$$k_{0z} = \sqrt{\frac{I_{0z}}{m}} = \sqrt{\frac{0.0105}{0.00956}} = 1.05 \text{ in}$$

(e) The radius of gyration about the $z$ axis is

$$k_z = \sqrt{\frac{I_z}{m}} = \sqrt{\frac{0.0665}{0.00956}} = 2.64 \text{ in}$$

**17.59**    A copper spacer plate has the dimensions shown in Fig. 18.57a.   A square hole is cut through the plate thickness.   The density of the copper is 8,910 kg/m³.   Find the mass of the plate, and the mass moments of inertia $I_x$, $I_y$, and $I_z$.

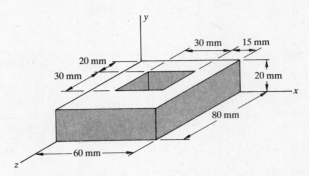

**Fig. 17.59a**

▌    The solid plate is element 1, and the mass that occupies the volume of the hole is element 2.   Using Case 10 in Table 17.13,

$$m = \rho abc$$

For element 1,

$$a = 60 \text{ mm} \qquad b = 20 \text{ mm} \qquad c = 80 \text{ mm}$$

For element 2,

$$a = 30 \text{ mm} \qquad b = 20 \text{ mm} \qquad c = 30 \text{ mm}$$

The values of the two masses are

$$m_1 = 8,910(60)20(80)\left(\frac{1}{1,000}\right)^3 = 0.855 \text{ kg} \qquad m_2 = 8,910(30)20(30)\left(\frac{1}{1,000}\right)^3 = 0.160 \text{ kg}$$

The mass of the spacer plate is given by

$$m = m_1 - m_2 = 0.855 - 0.160 = 0.695 \text{ kg}$$

The separation distances between the centroidal axes of element 2 and the $xyz$ axes are shown in Figs. 18.57b, c, and d.   From Case 10,

$$I_x = \tfrac{1}{3}m(b^2 + c^2) \qquad I_{0x} = \tfrac{1}{12}m(b^2 + c^2)$$

The moment of inertia of the spacer plate about the $x$ axis is given by

$$I_x = I_{x1} - I_{x2} = I_{x1} - (I_{0x,2} + m_2 d_{x2}^2) = \{\tfrac{1}{3}(0.855)(20^2 + 80^2) - [\tfrac{1}{12}(0.160)(20^2 + 30^2)$$
$$+ 0.160(10^2 + 35^2)]\}\left(\frac{1}{1,000}\right)^2 = 0.00171 \text{ kg} \cdot \text{m}^2$$

From Case 10,

$$I_y = \tfrac{1}{3}m(a^2 + c^2) \qquad I_{0y} = \tfrac{1}{12}m(a^2 + c^2)$$

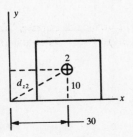

**Fig. 17.59b**

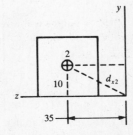

**Fig. 17.59c**

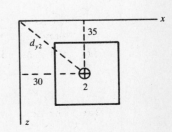

**Fig. 17.59d**

$I_y$ may be written as

$$I_y = I_{y1} - I_{y2} = I_{y1} - (I_{0y,2} + m_2 d_{y2}^2) = \{\tfrac{1}{3}(0.855)(60^2 + 80^2) - [\tfrac{1}{12}(0.160)(30^2 + 30^2)$$
$$+ 0.160(30^2 + 35^2)]\}\left(\frac{1}{1,000}\right)^2 = 0.00249 \text{ kg} \cdot \text{m}^2$$

From Case 10,

$$I_z = \tfrac{1}{3}m(a^2 + b^2) \qquad I_{0z} = \tfrac{1}{12}m(a^2 + b^2)$$

$I_z$ has the form

$$I_z = I_{z1} - I_{z2} = I_{z1} - (I_{0z,2} + m_2 d_{z2}^2) = \{\tfrac{1}{3}(0.855)(60^2 + 20^2) - [\tfrac{1}{12}(0.160)(30^2 + 20^2)$$
$$+ 0.160(30^2 + 10^2)]\}\left(\frac{1}{1,000}\right)^2 = 9.63 \times 10^{-4} \text{ kg} \cdot \text{m}^2$$

**17.60**   Find the centroidal coordinates $x_c$, $y_c$, and $z_c$, and the mass moments of inertia $I_{0x}$, $I_{0y}$, and $I_{0z}$ of the spacer plate in Prob. 17.59.

▌ The centroidal coordinates of elements 1 and 2 are

$$x_1 = 30 \text{ mm} \qquad x_2 = 30 \text{ mm} \qquad y_1 = 10 \text{ mm} \qquad y_2 = 10 \text{ mm} \qquad z_1 = 40 \text{ mm} \qquad z_2 = 35 \text{ mm}$$

From Prob. 17.59,

$$m_1 = 0.855 \text{ kg} \qquad m_2 = 0.160 \text{ kg} \qquad m = m_1 - m_2 = 0.695 \text{ kg}$$

The centroidal coordinates of the spacer plate are

$$x_c = \frac{x_1 m_1 - x_2 m_2}{m_1 - m_2} = \frac{30(0.855) - 30(0.160)}{0.695} = 30 \text{ mm}$$

$$y_c = \frac{y_1 m_1 - y_2 m_2}{m_1 - m_2} = \frac{10(0.855) - 10(0.160)}{0.695} = 10 \text{ mm}$$

$$z_c = \frac{z_1 m_1 - z_2 m_2}{m_1 - m_2} = \frac{40(0.855) - 35(0.160)}{0.695} = 41.2 \text{ mm}$$

The distances $d_x$, $d_y$, and $d_z$ between the centroidal axes of the plate and the $xyz$ axes are shown in Figs. 17.60$a$, $b$, and $c$. From Prob. 17.59,

$$I_x = 0.00171 \text{ kg} \cdot \text{m}^2 \qquad I_y = 0.00249 \text{ kg} \cdot \text{m}^2 \qquad I_z = 9.63 \times 10^{-4} \text{ kg} \cdot \text{m}^2$$

Using the parallel-axis theorem,

$$I_{0x} = I_x - md_x^2 = 0.00171 - 0.695(41.2^2 + 10^2)\left(\frac{1}{1,000}\right)^2 = 4.61 \times 10^{-4} \text{ kg} \cdot \text{m}^2$$

$$I_{0y} = I_y - md_y^2 = 0.00249 - 0.695(30^2 + 41.2^2)\left(\frac{1}{1,000}\right)^2 = 6.85 \times 10^{-4} \text{ kg} \cdot \text{m}^2$$

$$I_{0z} = I_z - md_z^2 = 9.63 \times 10^{-4} - 0.695(30^2 + 10^2)\left(\frac{1}{1,000}\right)^2 = 2.68 \times 10^{-4} \text{ kg} \cdot \text{m}^2$$

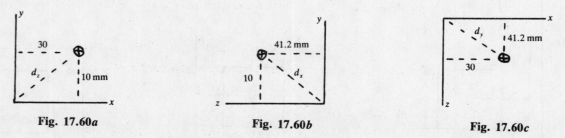

Fig. 17.60$a$          Fig. 17.60$b$          Fig. 17.60$c$

**17.61**   ($a$) A brass plate has two holes drilled through it, as shown in Fig. 17.61$a$. The specific weight of brass is 534 lb/ft$^3$, and the plate thickness is 1.75 in. Find the mass moment of inertia of the plate about an axis which is normal to the plane of the plate and which passes through the centroid of the plate.

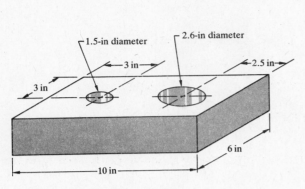

**Fig. 17.61a**

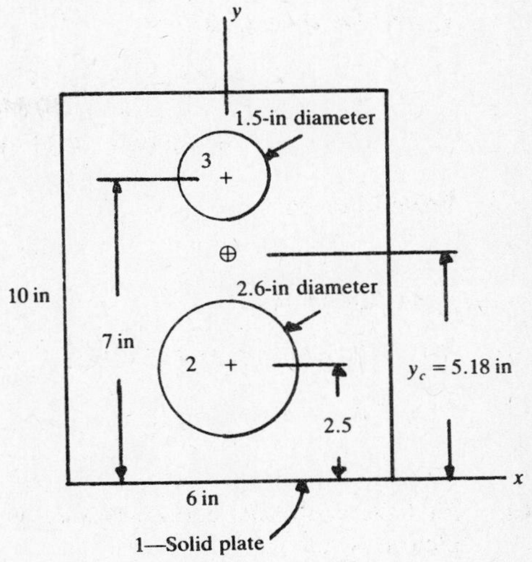

**Fig. 17.61b**

(*b*) Find the mass moment of inertia of the plate about the center axis of the smaller hole.

▌ (*a*) Figure 17.61*b* shows the identification of the elements of the plate. The masses are found as

$$m_1 = \left[\frac{534}{386(12)^3}\right]6(10)1.75 = 0.0841 \text{ lb} \cdot \text{s}^2/\text{in} \qquad m_2 = \left[\frac{534}{386(12)^3}\right]\frac{\pi(2.6)^2}{4}(1.75) = 0.00744 \text{ lb} \cdot \text{s}^2/\text{in}$$

$$m_3 = \left[\frac{534}{386(12)^3}\right]\frac{\pi(1.5)^2}{4}(1.75) = 0.00248 \text{ lb} \cdot \text{s}^2/\text{in}$$

The mass $m$ of the plate is given by

$$m = m_1 - m_2 - m_3 = 0.0742 \text{ lb} \cdot \text{s}^2/\text{in}$$

The centroidal coordinate $y_c$ of the plate is found from

$$y_c = \frac{y_1 m_1 - y_2 m_2 - y_3 m_3}{m_1 - m_2 - m_3} = \frac{5(0.0841) - 2.5(0.00744) - 7(0.00248)}{0.0742} = 5.18 \text{ in}$$

For element 1, from Case 10 in Table 17.13,

$$I_{0z} = \tfrac{1}{12}m(a^2 + b^2)$$

where $\qquad\qquad\qquad\qquad a = 6 \text{ in} \qquad \text{and} \qquad b = 10 \text{ in}$

For elements 2 and 3, from Case 4,

$$I_{0z} = \tfrac{1}{8}md^2$$

Using the parallel-axis theorem, the moment of inertia $I_0$ about the centroidal axis normal to the plate is

$$I_0 = I_{01} - I_{02} - I_{03} = [\tfrac{1}{12}(0.0841)(6^2 + 10^2) + 0.0841(5.18 - 5)^2] - [\tfrac{1}{8}(0.00744)2.6^2$$
$$+ 0.00744(5.18 - 2.5)^2] - [\tfrac{1}{8}(0.00248)1.5^2 + 0.00248(7 - 5.18)^2] = 0.887 \text{ lb} \cdot \text{s}^2 \cdot \text{in}$$

(*b*) From part (*a*),

$$I_0 = 0.887 \text{ lb} \cdot \text{s}^2 \cdot \text{in} \qquad \text{and} \qquad m = 0.0742 \text{ lb} \cdot \text{s}^2/\text{in}$$

The distance from the center of mass to the center of the smaller hole is

$$7 - 5.18 = 1.82 \text{ in}$$

The parallel-axis theorem has the form

$$I = I_0 + md^2 = 0.887 + 0.0742(1.82)^2 = 1.13 \text{ lb} \cdot \text{s}^2 \cdot \text{in}$$

**17.62** Find the mass moments of inertia $I_x$, $I_y$, and $I_z$ of the body shown in Fig. 17.62*a*. The material is steel, with $\rho = 7{,}830 \text{ kg/m}^3$.

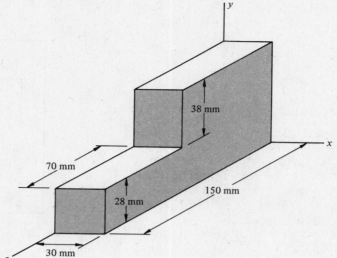

Fig. 17.62a

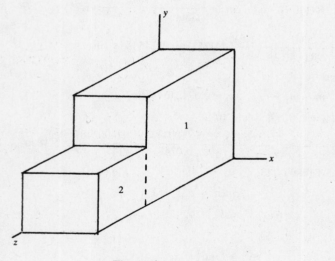

Fig. 17.62b

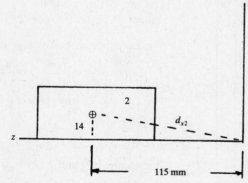

Fig. 17.62c

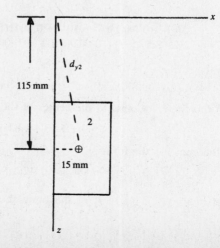

Fig. 17.62d

*▮* The body is divided into the two elements shown in Fig. 17.62b. From Case 10 of Table 17.13,

$$m = \rho abc$$

For element 1,

$$a = 30 \text{ mm} \qquad b = 66 \text{ mm} \qquad c = 80 \text{ mm}$$

For element 2,

$$a = 30 \text{ mm} \qquad b = 28 \text{ mm} \qquad c = 70 \text{ mm}$$

The masses of the two elements are then

$$m_1 = 7{,}830(30)66(80)\left(\frac{1}{1{,}000}\right)^3 = 1.24 \text{ kg} \qquad m_2 = 7{,}830(30)28(70)\left(\frac{1}{1{,}000}\right)^3 = 0.460 \text{ kg}$$

From Case 10,

$$I_x = \tfrac{1}{3}m(b^2 + c^2) \qquad I_{0x} = \tfrac{1}{12}m(b^2 + c^2)$$

The mass moment of inertia of the body about the $x$ axis has the form

$$I_x = I_{x1} + I_{x2} = [\tfrac{1}{3}m_1(b^2 + c^2)] + [\tfrac{1}{12}m_2(b^2 + c^2) + m_2 d_{x2}^2]$$

The separation distance $d_{x2}$ between the $x$ and $x_0$ axes is shown in Fig. 17.62c. $I_x$ is then found as

$$I_x = \{[\tfrac{1}{3}(1.24)(66^2 + 80^2)] + [\tfrac{1}{12}(0.460)(28^2 + 70^2) + 0.460(14^2 + 115^2)]\}\left(\frac{1}{1{,}000}\right)^2 = 0.0108 \text{ kg} \cdot \text{m}^2$$

The moment of inertia about the $y$ axis is

$$I_y = I_{y1} + I_{y2} = [\tfrac{1}{3}m_1(a^2 + c^2)] + [\tfrac{1}{12}m_2(a^2 + c^2) + m_2 d_{y2}^2]$$

The term $d_{y2}$ is shown in Fig. 17.62d, and

$$I_y = \{[\tfrac{1}{3}(1.24)(30^2 + 80^2)] + [\tfrac{1}{12}(0.460)(30^2 + 70^2) + 0.460(15^2 + 115^2)]\}\left(\frac{1}{1{,}000}\right)^2 = 0.00943 \text{ kg} \cdot \text{m}^2$$

The moment of inertia of the body about the $z$ axis is given by

$$I_z = I_{z1} + I_{z2}$$

Since the edges of both elements 1 and 2 lie on the $z$ axis, the parallel-axis theorem is not required. Using the form $I_z = \tfrac{1}{3}m(a^2 + b^2)$,

$$I_z = [\tfrac{1}{3}(1.24)(30^2 + 66^2) + \tfrac{1}{3}(0.460)(30^2 + 28^2)]\left(\frac{1}{1{,}000}\right)^2 = 0.00243 \text{ kg} \cdot \text{m}^2$$

**17.63** Find the centroidal coordinates $x_c$, $y_c$, and $z_c$, and the mass moments of inertia $I_{0x}$, $I_{0y}$, and $I_{0z}$ of the body in Prob. 17.62. The $x_0 y_0 z_0$ axes are parallel to the $xyz$ axes.

*▮* The centroidal coordinates of elements 1 and 2 are

$$x_1 = \tfrac{30}{2} = 15 \text{ mm} \qquad y_1 = \tfrac{66}{2} = 33 \text{ mm} \qquad z_1 = \tfrac{80}{2} = 40 \text{ mm}$$
$$x_2 = \tfrac{30}{2} = 15 \text{ mm} \qquad y_2 = \tfrac{28}{2} = 14 \text{ mm} \qquad z_2 = 150 - \tfrac{70}{2} = 115 \text{ mm}$$

From Prob. 17.62,

$$m_1 = 1.24 \text{ kg} \qquad m_2 = 0.460 \text{ kg} \qquad m = m_1 + m_2 = 1.24 + 0.460 = 1.70 \text{ kg}$$

The centroidal coordinates are found as

$$x_c = \frac{x_1 m_1 + x_2 m_2}{m_1 + m_2} = \frac{15(1.24) + 15(0.460)}{1.70} = 15.0 \text{ mm} \qquad y_c = \frac{y_1 m_1 + y_2 m_2}{m_1 + m_2} = \frac{33(1.24) + 14(0.460)}{1.70} = 27.9 \text{ mm}$$

$$z_c = \frac{z_1 m_1 + z_2 m_2}{m_1 + m_2} = \frac{40(1.24) + 115(0.460)}{1.70} = 60.3 \text{ mm}$$

The separation distances between the $x_0 y_0 z_0$ axes and the $xyz$ axes are shown in Figs. 17.63a, b, and c. Using the parallel-axis theorem,

$$I_{0x} = I_x - m d_x^2 = 0.0108 - 1.70(27.9^2 + 60.3^2)\left(\frac{1}{1{,}000}\right)^2 = 0.00330 \text{ kg} \cdot \text{m}^2$$

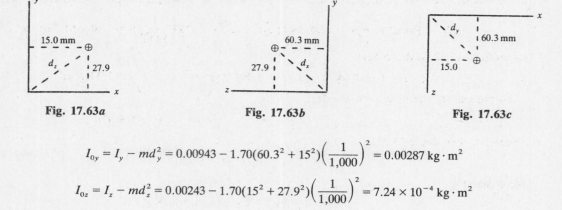

Fig. 17.63a      Fig. 17.63b      Fig. 17.63c

$$I_{0y} = I_y - md_y^2 = 0.00943 - 1.70(60.3^2 + 15^2)\left(\frac{1}{1,000}\right)^2 = 0.00287 \text{ kg} \cdot \text{m}^2$$

$$I_{0z} = I_z - md_z^2 = 0.00243 - 1.70(15^2 + 27.9^2)\left(\frac{1}{1,000}\right)^2 = 7.24 \times 10^{-4} \text{ kg} \cdot \text{m}^2$$

**17.64** Do the same as in Prob. 17.62 if a hole is drilled in the body, as shown in Fig. 17.64a.

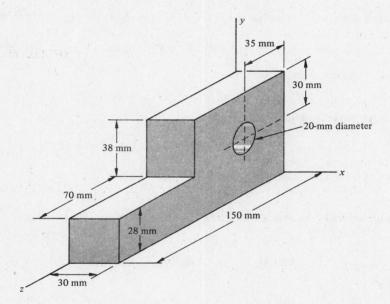

Fig. 17.64a

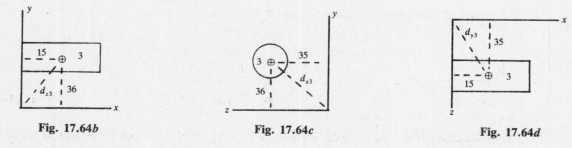

Fig. 17.64b      Fig. 17.64c      Fig. 17.64d

❚ The hole is element 3, and Figs. 17.64b, c, and d show the positions of the hole with respect to the x, y, and z axes. From Case 4 in Table 17.13,

$$m = \tfrac{1}{4}\rho\pi d^2 h$$

Using $d = 20$ mm and $h = 30$ mm,

$$m_3 = \tfrac{1}{4}(7,830)\pi(20)^2 30\left(\frac{1}{1,000}\right)^3 = 0.0738 \text{ kg}$$

The moment of inertia about the $x$ axis of the material which would fill the volume of the hole is given by

$$I_{x3} = \tfrac{1}{8}m_3 d^2 + m_3 d_{x3}^2 = 0.0738[\tfrac{1}{8}(20)^2 + (36^2 + 35^2)]\left(\frac{1}{1,000}\right)^2 = 1.90 \times 10^{-4}\ \text{kg} \cdot \text{m}^2$$

From Prob. 17.62, with no hole,

$$I_x = 0.0108\ \text{kg} \cdot \text{m}^2 \qquad I_y = 0.00943\ \text{kg} \cdot \text{m}^2 \qquad I_z = 0.00243\ \text{kg} \cdot \text{m}^2$$

For the case of the body with a hole,

$$I_{x,\text{net}} = I_x - I_{x3} = 0.0108 - 1.90 \times 10^{-4} = 0.0106\ \text{kg} \cdot \text{m}^2$$

From Case 4 of Table 17.13,

$$I_{0y} = \tfrac{1}{12}m(h^2 + \tfrac{3}{4}d^2)$$

For element 3,

$$d = 20\ \text{mm} \qquad \text{and} \qquad h = 30\ \text{mm}$$

Using the parallel-axis theorem,

$$I_{y3} = \tfrac{1}{12}m_3(h^2 + \tfrac{3}{4}d^2) + m_3 d_{y3}^2 = 0.0738\{\tfrac{1}{12}[30^2 + \tfrac{3}{4}(20)^2] + (15^2 + 35^2)\}\left(\frac{1}{1,000}\right)^2 = 1.14 \times 10^{-4}\ \text{kg} \cdot \text{m}^2$$

For the case of the body with a hole,

$$I_{y,\text{net}} = I_y - I_{y3} = 0.00943 - 1.14 \times 10^{-4} = 0.00932\ \text{kg} \cdot \text{m}^2$$

From Case 4,

$$I_{0z} = \tfrac{1}{12}m(h^2 + \tfrac{3}{4}d^2)$$

Using the parallel-axis theorem,

$$I_{z3} = \tfrac{1}{12}m_3(h^2 + \tfrac{3}{4}d^2) + m_3 d_{z3}^2$$

With $d = 20\ \text{mm}$ and $h = 30\ \text{mm}$,

$$I_{z3} = 0.0.0738\{[\tfrac{1}{12}[30^2 + \tfrac{3}{4}(20)^2] + (36^2 + 15^2)]\}\left(\frac{1}{1,000}\right)^2 = 1.20 \times 10^{-4}\ \text{kg} \cdot \text{m}^2$$

For the case of the body with a hole,

$$I_{z,\text{net}} = I_z - I_{z3} = 0.00243 - 1.20 \times 10^{-4} = 0.00231\ \text{kg} \cdot \text{m}^2$$

**17.65**  Find the centroidal coordinates $x_c$, $y_c$, and $z_c$, and the mass moments of inertia $I_{0x}$, $I_{0y}$, and $I_{0z}$, of the body shown in Prob. 17.64.

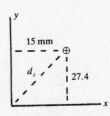

**Fig. 17.65a**

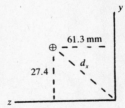

**Fig. 17.65b**

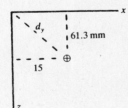

**Fig. 17.65c**

❚ The mass elements have the same designations as in Probs. 17.62 and 17.64.  The centroidal coordinates of elements 1 and 2, from Prob. 17.63, are

$$x_1 = 15\ \text{mm} \qquad y_1 = 33\ \text{mm} \qquad z_1 = 40\ \text{mm} \qquad x_2 = 15\ \text{mm} \qquad y_2 = 14\ \text{mm} \qquad z_2 = 115\ \text{mm}$$

The centroidal coordinates of the volume of the hole are

$$x_3 = \tfrac{30}{2} = 15\ \text{mm} \qquad y_3 = 36\ \text{mm} \qquad z_3 = 35\ \text{mm}$$

From Prob. 17.64,

$$m_3 = 0.0738\ \text{kg}$$

The mass of the body with a hole is

$$m = m_1 + m_2 - m_3 = 1.24 + 0.460 - 0.0738 = 1.63 \text{ kg}$$

The centroidal coordinates are then found as

$$x_c = \frac{x_1 m_1 + x_2 m_2 + x_3 m_3}{m} = \frac{15(1.24) + 15(0.460) - 15(0.0738)}{1.63} = 15.0 \text{ mm}$$

$$y_c = \frac{y_1 m_1 + y_2 m_2 + y_3 m_3}{m} = \frac{33(1.24) + 14(0.460) - 36(0.0738)}{1.63} = 27.4 \text{ mm}$$

$$z_c = \frac{z_1 m_1 + z_2 m_2 + z_3 m_3}{m} = \frac{40(1.24) + 115(0.460) - 35(0.0738)}{1.63} = 61.3 \text{ mm}$$

The centroidal $x_0 y_0 z_0$ axes are parallel to the $xyz$ axes. The separation distances between these sets of axes are shown in Figs. 17.65a, b, and c. Using the parallel-axis theorem,

$$I_{0x} = I_x - m d_x^2 = 0.0106 - 1.63(61.3^2 + 27.4^2)\left(\frac{1}{1,000}\right)^2 = 0.00325 \text{ kg} \cdot \text{m}^2$$

$$I_{0y} = I_y - m d_y^2 = 0.00932 - 1.63(15^2 + 61.3^2)\left(\frac{1}{1,000}\right)^2 = 0.00283 \text{ kg} \cdot \text{m}^2$$

$$I_{0z} = I_z - m d_z^2 = 0.00231 - 1.63(15^2 + 27.4^2)\left(\frac{1}{1,000}\right)^2 = 7.20 \times 10^{-4} \text{ kg} \cdot \text{m}^2$$

**17.66**  Find the mass moments of inertia $I_x$, $I_y$, and $I_z$ of the body shown in Fig. 17.66a. The body is made of die-cast zinc, with $\gamma = 410 \text{ lb/ft}^3$.

❚  The body is divided into the three elements shown in Fig. 17.66b. From Case 10 of Table 17.13,

$$m = \rho abc$$

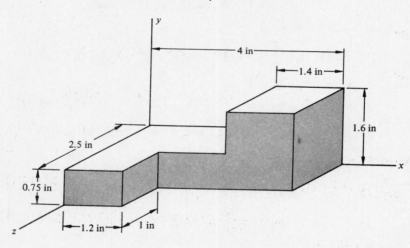

Fig. 17.66a

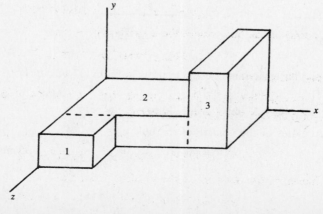

Fig. 17.66b

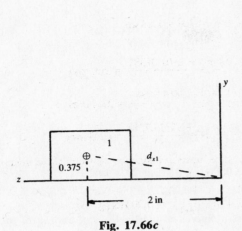

Fig. 17.66c

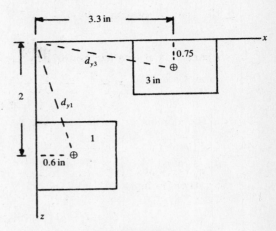

Fig. 17.66d

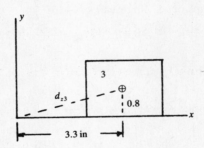

Fig. 17.66e

For element 1,

$$a = 1.2 \text{ in} \qquad b = 0.75 \text{ in} \qquad c = 1 \text{ in}$$

For element 2,

$$a = 2.6 \text{ in} \qquad b = 0.75 \text{ in} \qquad c = 1.5 \text{ in}$$

For element 3,

$$a = 1.4 \text{ in} \qquad b = 1.6 \text{ in} \qquad c = 1.5 \text{ in}$$

The masses of the three elements are found from

$$m_1 = \frac{410}{386(12)^3}(1.2)0.75(1) = 5.53 \times 10^{-4} \text{ lb} \cdot \text{s}^2/\text{in} \qquad m_2 = \frac{410}{386(12)^3}(2.6)0.75(1.5) = 0.00180 \text{ lb} \cdot \text{s}^2/\text{in}$$

$$m_3 = \frac{410}{386(12)^3}(1.4)1.6(1.5) = 0.00207 \text{ lb} \cdot \text{s}^2/\text{in}$$

From Case 10 of Table 17.13,

$$I_x = \tfrac{1}{3}m(b^2 + c^2) \qquad I_{0x} = \tfrac{1}{12}m(b^2 + c^2)$$

The moment of inertia of the body about the $x$ axis has the form

$$I_x = (I_{0x,1} + m_1 d_{x1}^2) + I_{x2} + I_{x3}$$

$d_{x1}$ is found from Fig. 17.66c, and

$$I_x = [\tfrac{1}{12}(5.53 \times 10^{-4})(0.75^2 + 1^2) + 5.53 \times 10^{-4}(2^2 + 0.375^2)] + [\tfrac{1}{3}(0.00180)(0.75^2 + 1.5^2)]$$
$$+ [\tfrac{1}{3}(0.00207)(1.6^2 + 1.5^2)] = 0.00737 \text{ lb} \cdot \text{s}^2 \cdot \text{in}$$

From Case 10,

$$I_y = \tfrac{1}{3}m(a^2 + c^2) \qquad I_{0y} = \tfrac{1}{12}m(a^2 + c^2)$$

The moment of inertia of the body about the $y$ axis is written as

$$I_y = (I_{0y,1} + m_1 d_{y1}^2) + I_{y2} + (I_{0y,3} + m_3 d_{y3}^2)$$

$d_{y1}$ and $d_{y3}$ are found from Fig. 17.66d, and

$$I_y = [\tfrac{1}{12}(5.53 \times 10^{-4})(1.2^2 + 1^2) + 5.53 \times 10^{-4}(2^2 + 0.6)^2] + [\tfrac{1}{3}(0.00180)(2.6^2 + 1.5^2)]$$
$$+ [\tfrac{1}{12}(0.00207)(1.4^2 + 1.5^2) + 0.00207(3.3^2 + 0.75^2)] = 0.0324 \text{ lb} \cdot \text{s}^2 \cdot \text{in}$$

From Case 10,

$$I_z = \tfrac{1}{3}m(a^2 + b^2) \qquad I_{0z} = \tfrac{1}{12}m(a^2 + b^2)$$

The moment of inertia about the $z$ axis is

$$I_z = I_{z1} + I_{z2} + (I_{0z,3} + m_3 d_{z3}^2)$$

$d_{z3}$ is found from Fig. 17.66e, and

$$I_z = [\tfrac{1}{3}(5.53 \times 10^{-4})(1.2^2 + 0.75^2)] + [\tfrac{1}{3}(0.00180)(2.6^2 + 0.75^2)] + [\tfrac{1}{12}(0.00207)(1.4^2 + 1.6^2)$$
$$+ 0.00207(3.3^2 + 0.8^2)] = 0.0294 \text{ lb} \cdot \text{s}^2 \cdot \text{in}$$

**17.67** Find the centroidal coordinates $x_c$, $y_c$, and $z_c$, and the mass moments of inertia $I_{0x}$, $I_{0y}$, and $I_{0x}$ of the body shown in Prob. 17.66.

**❚** The centroidal coordinates of the three elements are

$$x_1 = \frac{1.2}{2} = 0.6 \text{ in} \qquad y_1 = \frac{0.75}{2} = 0.375 \text{ in} \qquad z_1 = 2 \text{ in}$$

$$x_2 = \frac{2.6}{2} = 1.3 \text{ in} \qquad y_2 = \frac{0.75}{2} = 0.375 \text{ in} \qquad z_2 = \frac{1.5}{2} = 0.75 \text{ in}$$

$$x_3 = 4 - \frac{1.4}{2} = 3.3 \text{ in} \qquad y_3 = \frac{1.6}{2} = 0.8 \text{ in} \qquad z_3 = \frac{1.5}{2} = 0.75 \text{ in}$$

From Prob. 17.66,

$$m_1 = 5.53 \times 10^{-4} \text{ lb} \cdot \text{s}^2/\text{in} \qquad m_2 = 0.00180 \text{ lb} \cdot \text{s}^2/\text{in} \qquad m_3 = 0.00207 \text{ lb} \cdot \text{s}^2/\text{in}$$

The total mass of the body is

$$m = m_1 + m_2 + m_3 = 0.00442 \text{ lb} \cdot \text{s}^2/\text{in}$$

The centroidal coordinates of the body are found from

$$x_c = \frac{x_1 m_1 + x_2 m_2 + x_3 m_3}{m} = \left(\frac{1}{0.00442}\right)[0.6(5.53 \times 10^{-4}) + 1.3(0.00180) + 3.3(0.00207)] = 2.15 \text{ in}$$

$$y_c = \frac{y_1 m_1 + y_2 m_2 + y_3 m_3}{m} = \left(\frac{1}{0.00442}\right)[0.375(5.53 \times 10^{-4}) + 0.375(0.00180) + 0.8(0.00207)] = 0.574 \text{ in}$$

$$z_c = \frac{z_1 m_1 + z_2 m_2 + z_3 m_3}{m} = \left(\frac{1}{0.00442}\right)[2(5.53 \times 10^{-4}) + 0.75(0.00180) + 0.75(0.00207)] = 0.907 \text{ in}$$

The centroidal $x_0 y_0 z_0$ coordinates are parallel to the $xyz$ axes. The separation distances between these two sets of coordinates are shown in Figs. 17.67a, b, and c. From Prob. 17.66,

$$I_x = 0.00737 \text{ lb} \cdot \text{s}^2 \cdot \text{in} \qquad I_y = 0.0324 \text{ lb} \cdot \text{s}^2 \cdot \text{in} \qquad I_z = 0.0294 \text{ lb} \cdot \text{s}^2 \cdot \text{in}$$

Using the parallel-axis theorem,

$$I_{0x} = I_x - md_x^2 = 0.00737 - 0.00442(0.574^2 + 0.907^2) = 0.00228 \text{ lb} \cdot \text{s}^2 \cdot \text{in}$$

$$I_{0y} = I_y - md_y^2 = 0.0324 - 0.00442(2.15^2 + 0.907^2) = 0.00833 \text{ lb} \cdot \text{s}^2 \cdot \text{in}$$

$$I_{0z} = I_z - md_z^2 = 0.0294 - 0.00442(2.15^2 + 0.574^2) = 0.00751 \text{ lb} \cdot \text{s}^2 \cdot \text{in}$$

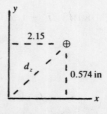

**Fig. 17.67a**

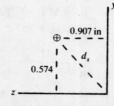

**Fig. 17.67b**

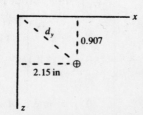

**Fig. 17.67c**

**17.68**  Do the same as in Prob. 17.66, if two holes are drilled in the body, as shown in Fig. 17.68*a*.

▌ Hole 4 is parallel to the *y* axis, and hole 5 is parallel to the *z* axis.  From Case 4 in Table 17.13,

$$m = \tfrac{1}{4}\rho\pi d^2 h$$

For hole 4,

$$d = 0.75 \text{ in} \qquad h = 0.75 \text{ in}$$

For hole 5,

$$d = 0.75 \text{ in} \qquad h = 1.5 \text{ in}$$

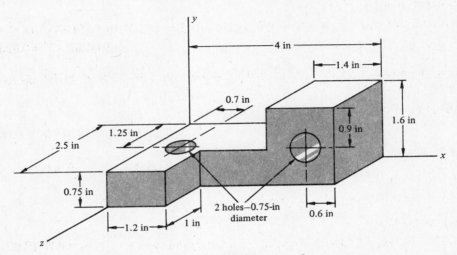

Fig. 17.68*a*

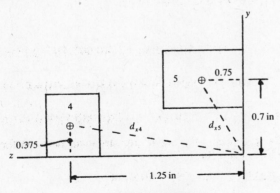

Fig. 17.68*b*

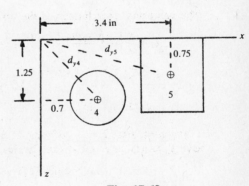

Fig. 17.68*c*

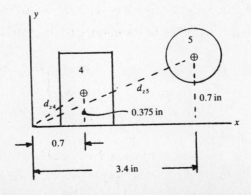

Fig. 17.68*d*

The masses of the volumes of material which would occupy the holes are

$$m_4 = \frac{1}{4}\left[\frac{410}{386(12)^3}\right]\pi(0.75)^2 0.75 = 2.04 \times 10^{-4} \text{ lb} \cdot \text{s}^2/\text{in}$$

$$m_5 = \frac{1}{4}\left[\frac{410}{386(12)^3}\right]\pi(0.75)^2 1.5 = 4.07 \times 10^{-4} \text{ lb} \cdot \text{s}^2/\text{in}$$

Using Case 4 in Table 17.13, the parallel-axis theorem, and the separation distances shown in Figs. 17.68*b*, *c*, and *d*,

$$I_{x4} = \tfrac{1}{12}m_4(h^2 + \tfrac{3}{4}d^2) + m_4 d_{x4}^2 = \tfrac{1}{12}(2.04 \times 10^{-4})[0.75^2 + \tfrac{3}{4}(0.75)^2]$$
$$+ 2.04 \times 10^{-4}(0.375^2 + 1.25^2) = 3.64 \times 10^{-4} \text{ lb} \cdot \text{s}^2 \cdot \text{in}$$

$$I_{x5} = \tfrac{1}{12}m_5(h^2 + \tfrac{3}{4}d^2) + m_5 d_{x5}^2 = \tfrac{1}{12}(4.07 \times 10^{-4})[1.5^2 + \tfrac{3}{4}(0.75)^2]$$
$$+ 4.07 \times 10^{-4}(0.75^2 + 0.7^2) = 5.19 \times 10^{-4} \text{ lb} \cdot \text{s}^2 \cdot \text{in}$$

$$I_{y4} = \tfrac{1}{8}m_4 d^2 + m_4 d_{y4}^2 = 2.04 \times 10^{-4}[\tfrac{1}{8}(0.75)^2 + (1.25^2 + 0.7^2)] = 4.33 \times 10^{-4} \text{ lb} \cdot \text{s}^2 \cdot \text{in}$$

$$I_{y5} = \tfrac{1}{12}m_5(h^2 + \tfrac{3}{4}d^2) + m_5 d_{y5}^2 = \tfrac{1}{12}(4.07 \times 10^{-4})[1.5^2 + \tfrac{3}{4}(0.75)^2]$$
$$+ 4.07 \times 10^{-4}(3.4^2 + 0.75^2) = 0.00502 \text{ lb} \cdot \text{s}^2 \cdot \text{in}$$

$$I_{z4} = \tfrac{1}{12}m_4(h^2 + \tfrac{3}{4}d^2) + m_4 d_{z4}^2 = \tfrac{1}{12}(2.04 \times 10^{-4})[0.75^2 + \tfrac{3}{4}(0.75)^2]$$
$$+ 2.04 \times 10^{-4}(0.375^2 + 0.7^2) = 1.45 \times 10^{-4} \text{ lb} \cdot \text{s}^2 \cdot \text{in}$$

$$I_{z5} = \tfrac{1}{8}m_5 d^2 + m_5 d_{z5}^2 = 4.07 \times 10^{-4}[\tfrac{1}{8}(0.75)^2 + (0.7^2 + 3.4^2)] = 0.00493 \text{ lb} \cdot \text{s}^2 \cdot \text{in}$$

From Prob. 17.66, for the body with no holes,

$$I_x = 0.00737 \text{ lb} \cdot \text{s}^2 \cdot \text{in} \qquad I_y = 0.0324 \text{ lb} \cdot \text{s}^2 \cdot \text{in} \qquad I_z = 0.0294 \text{ lb} \cdot \text{s}^2 \cdot \text{in}$$

For the body with holes,

$$I_{x,\text{net}} = I_x - I_{x4} - I_{x5} = 0.00737 - 3.64 \times 10^{-4} - 5.19 \times 10^{-4} = 0.00649 \text{ lb} \cdot \text{s}^2 \cdot \text{in}$$
$$I_{y,\text{net}} = I_y - I_{y4} - I_{y5} = 0.0324 - 4.33 \times 10^{-4} - 0.00502 = 0.0269 \text{ lb} \cdot \text{s}^2 \cdot \text{in}$$
$$I_{z,\text{net}} = I_z - I_{z4} - I_{z5} = 0.0294 - 1.45 \times 10^{-4} - 0.00493 = 0.0243 \text{ lb} \cdot \text{s}^2 \cdot \text{in}$$

**17.69** Find the centroidal coordinates $x_c$, $y_c$, and $z_c$, and the mass moments of inertia $I_{0x}$, $I_{0y}$, and $I_{0z}$, of the body shown in Prob. 17.68.

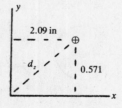

Fig. 17.69*a*

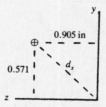

Fig. 17.69*b*

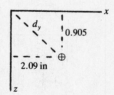

Fig. 17.69*c*

▮ The element identification is the same as that used in Probs. 17.66 and 17.68. The centroidal coordinates of the two hole volumes are

$$x_4 = 0.7 \text{ in} \qquad y_4 = 0.375 \text{ in} \qquad z_4 = 1.25 \text{ in} \qquad x_5 = 3.4 \text{ in} \qquad y_5 = 0.7 \text{ in} \qquad z_5 = 0.75 \text{ in}$$

From Prob. 17.67,

$$m_1 = 5.53 \times 10^{-4} \text{ lb} \cdot \text{s}^2/\text{in} \qquad m_2 = 0.00180 \text{ lb} \cdot \text{s}^2/\text{in} \qquad m_3 = 0.00207 \text{ lb} \cdot \text{s}^2/\text{in}$$

From Prob. 17.68,

$$m_4 = 2.04 \times 10^{-4} \text{ lb} \cdot \text{s}^2/\text{in} \qquad m_5 = 4.07 \times 10^{-4} \text{ lb} \cdot \text{s}^2/\text{in}$$

The mass of the body is given by

$$m = m_1 + m_2 + m_3 - m_4 - m_5 = 5.53 \times 10^{-4} + 0.00180 + 0.00207 - 2.04 \times 10^{-4} - 4.07 \times 10^{-4}$$
$$= 0.00381 \text{ lb} \cdot \text{s}^2/\text{in}$$

The centroidal coordinates of the body are then found as

$$x_c = \frac{x_1m_1 + x_2m_2 + x_3m_3 - x_4m_4 - x_5m_5}{m} = \left(\frac{1}{0.00381}\right)[0.6(5.53 \times 10^{-4}) + 1.3(0.00180) + 3.3(0.00207)$$
$$- 0.7(2.04 \times 10^{-4}) - 3.4(4.07 \times 10^{-4})] = 2.09 \text{ in}$$

$$y_c = \frac{y_1m_1 + y_2m_2 + y_3m_3 - y_4m_4 - y_5m_5}{m} = \left(\frac{1}{0.00381}\right)[0.375(5.53 \times 10^{-4}) + 0.375(0.00180) + 0.8(0.00207)$$
$$- 0.375(2.04 \times 10^{-4}) - 0.7(4.07 \times 10^{-4})] = 0.571 \text{ in}$$

$$z_c = \frac{z_1m_1 + z_2m_2 + z_3m_3 - z_4m_4 - z_5m_5}{m} = \left(\frac{1}{0.00381}\right)[2(5.53 \times 10^{-4}) + 0.75(0.00180) + 0.75(0.00207)$$
$$- 1.25(2.04 \times 10^{-4}) - 0.75(4.07 \times 10^{-4})] = 0.905 \text{ in}$$

The centroidal $x_0y_0z_0$ axes are parallel to the $xyz$ axes. The separation distances between these two axes are shown in Figs. 17.69$a$, $b$, and $c$. Using the parallel-axis theorem,

$$I_{0x} = I_x - md_x^2 = 0.00649 - 0.00381(0.571^2 + 0.905^2) = 0.00213 \text{ lb} \cdot \text{s}^2 \cdot \text{in}$$
$$I_{0y} = I_y - md_y^2 = 0.0269 - 0.00381(2.09^2 + 0.905^2) = 0.00714 \text{ lb} \cdot \text{s}^2 \cdot \text{in}$$
$$I_{0z} = I_z - md_z^2 = 0.0243 - 0.00381(0.571^2 + 2.09^2) = 0.00642 \text{ lb} \cdot \text{s}^2 \cdot \text{in}$$

**17.70** A machine part is fabricated by welding two cylindrical steel posts into a plate, as shown in Fig. 17.70$a$. Find the mass moments of inertia $I_x$, $I_{x_1}$, and $I_y$. The mass of the weld material may be neglected, and the density of steel is 7,830 kg/m³.

▌ Figure 17.70$b$ shows the division of the part into three elements. The masses of the cylindrical elements are

$$m = \rho \, \frac{\pi d^2}{4} \, h$$

$$m_1 = 7,830\left(\frac{\pi 20^2}{4}\right)20\left(\frac{1}{1,000}\right)^3 = 0.0492 \text{ kg} \qquad m_3 = 7,830\left(\frac{\pi 28^2}{4}\right)30\left(\frac{1}{1,000}\right)^3 = 0.145 \text{ kg}$$

The mass of element 2 is

$$m_2 = 7,830(28)40(140)\left(\frac{1}{1,000}\right)^3 = 1.23 \text{ kg}$$

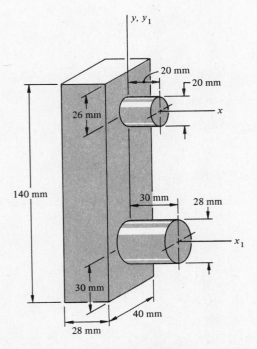

**Fig. 17.70$a$**

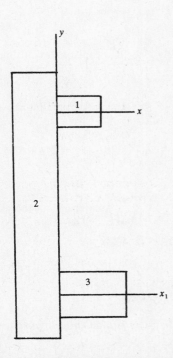

**Fig. 17.70$b$**

For elements 1 and 3, from Case 4 of Table 17.13,

$$I_{0x} = I_{x_1} = \tfrac{1}{8}md^2 \qquad I_y = \tfrac{1}{48}m(16h^2 + 3d^2)$$

For element 2, from Case 10,

$$I_{0x} = \tfrac{1}{12}m(b^2 + c^2) \qquad I_y = \tfrac{1}{12}m(4b^2 + c^2)$$

The moment of inertia of the part about the $x$ axis is

$$I_x = I_{0x,1} + [I_{0x,2} + m_2 d_{x2}^2 + [I_{0x,3} + m_3 d_{x3}^2]$$

Using $d_{x2} = 70 - 26 = 44$ mm and $d_{x3} = 140 - 26 - 30 = 84$ mm,

$$I_x = \{[\tfrac{1}{8}(0.0492)20^2] + [\tfrac{1}{12}(1.23)(40^2 + 140^2) + 1.23(44)^2] + [\tfrac{1}{8}(0.145)28^2 + 0.145(84)^2]\}\left(\frac{1}{1,000}\right)^2$$

$$= 0.00559 \text{ kg} \cdot \text{m}^2$$

The moment of inertia of the part about the $x_1$ axis is

$$I_{x_1} = (I_{0x,1} + m_1 d_{x_1,1}^2) + (I_{0x,2} + m_2 d_{x_1,2}^2) + I_{0x,3}$$

Using $d_{x_1,1} = 140 - 26 - 30 = 84$ mm and $d_{x_1,2} = 70 - 30 = 40$ mm

$$I_{x_1} = \{[\tfrac{1}{8}(0.0492)20^2 + 0.0492(84)^2] + [\tfrac{1}{12}(1.23)(40^2 + 140^2) + 1.23(40)^2] + [\tfrac{1}{8}(0.145)28^2]\}\left(\frac{1}{1,000}\right)^2$$

$$= 0.00450 \text{ kg} \cdot \text{m}^2$$

The moment of inertia of the part about the $y$ axis is

$$I_y = I_{y1} + I_{y2} + I_{y3} = (\{\tfrac{1}{48}(0.0492)[16(20)^2 + 3(20)^2]\} + \{\tfrac{1}{12}(1.23)[4(28)^2 + 40^2]\}$$

$$+ \{\tfrac{1}{48}(0.145)[16(30)^2 + 3(28)^2]\})\left(\frac{1}{1,000}\right)^2 = 5.44 \times 10^{-4} \text{ kg} \cdot \text{m}^2$$

**17.71** (a) Find the mass moment of inertia of the part in Prob. 17.70 about the centroidal $x_0$ axis. The $x_0$ axis is parallel to, and lies in the plane of, the $x$ and $x_1$ axes.

(b) Find the radius of gyration $k_{0x}$.

$\blacksquare$ (a) From Prob. 17.70,

$$m = m_1 + m_2 + m_3 = 1.42 \text{ kg} \qquad I_x = 0.00559 \text{ kg} \cdot \text{m}^2$$

The centroidal coordinate $y_c$, shown in Fig. 17.71, is found as

$$y_c = \frac{y_1 m_1 + y_2 m_2 + y_3 m_3}{m} = \frac{1}{1.42}[114(0.0492) + 70(1.23) + 30(0.145)] = 67.6 \text{ mm}$$

Using the parallel-axis theorem between the $x_0$ and $x$ axes,

$$I_{0x} = I_{x_1} - md_x^2$$

With $d_{x1} = 114 - 67.6 = 46.4$ mm,

$$I_{0x} = 0.00559 - 1.42(46.4^2)\left(\frac{1}{1,000}\right)^2 = 0.00253 \text{ kg} \cdot \text{m}^2$$

As a check on the above calculation, the parallel-axis theorem may be used between the $x_0$ and $x_1$ axes. The result is

$$I_{0x} = I_{x_1} - md_{x_1}^2$$

With $d_{x_1} = 67.6 - 30 = 37.6$ mm,

$$I_{0x} = 0.00450 - 1.42(37.6)^2\left(\frac{1}{1,000}\right)^2 = 0.00249 \text{ kg} \cdot \text{m}^2$$

The percent difference between the two values for $I_{0x}$ is

$$\%D = \frac{0.00253 - 0.00249}{0.00253}(100) = 1.6\%$$

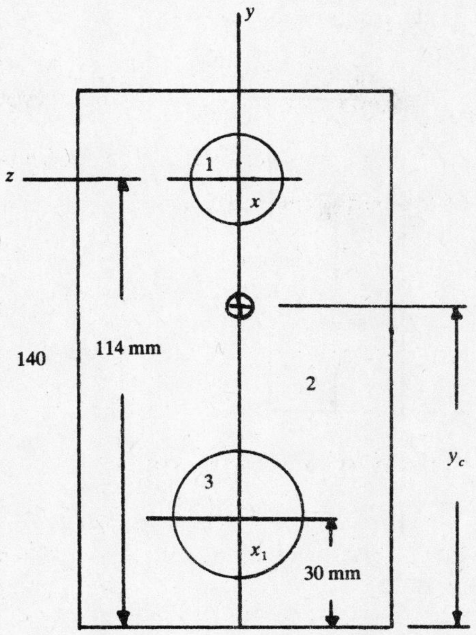

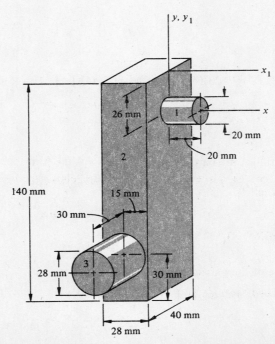

Fig. 17.71

This percent difference between the two results above is due to an accumulation of computational roundoff error.

(b)  The radius of gyration of the part about the $x_0$ axis is

$$k_{0x} = \sqrt{\frac{I_{0x}}{m}} = \sqrt{\frac{0.00253}{1.42}} = 0.0422 \text{ m} = 42.2 \text{ mm}$$

**17.72**  Do the same as in Prob. 17.70, for the welded machine part shown in Fig. 17.72a.

**┃**  From Prob. 17.70, using the same element numbering system,

$$m_1 = 0.0492 \text{ kg} \qquad m_2 = 1.23 \text{ kg} \qquad m_3 = 0.145 \text{ kg}$$

Fig. 17.72a

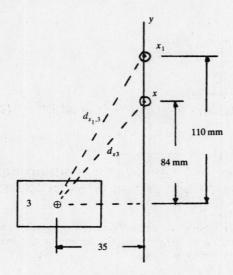

**Fig. 17.72b**

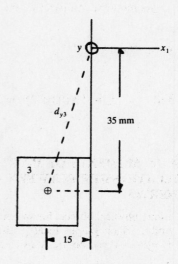

**Fig. 17.72c**

For element 1, using Case 4 of Table 17.13,

$$I_{0x} = I_{0x,1} = \tfrac{1}{8}md^2 \qquad I_y = \tfrac{1}{48}m(16h^2 + 3d^2)$$

For element 2, from Case 10,

$$I_{0x} = \tfrac{1}{12}m(b^2 + c^2) \qquad I_y = \tfrac{1}{12}m(4b^2 + c^2)$$

For element 3, from Case 4,

$$I_{0x} = \tfrac{1}{12}m(h^2 + \tfrac{3}{4}d^2)$$

The moment of inertia of the part about the $x$ axis is

$$I_x = I_{0x,1} + (I_{0x,2} + m_2 d_{x2}^2) + (I_{0x,3} + m_3 d_{x3}^2)$$

The term $d_{x2}$ has the value

$$d_{x2} = 70 - 26 = 44 \text{ mm}$$

From Fig. 17.72b,

$$d_{x3}^2 = 35^2 + 84^2$$

$I_x$ then has the value

$$I_x = \{[\tfrac{1}{8}(0.0492)20^2] + [\tfrac{1}{12}(1.23)(40^2 + 140^2) + 1.23(44)^2] + [\tfrac{1}{12}(0.145)[30^2 + \tfrac{3}{4}(28)^2] + 0.145(35^2 + 84^2)]\}$$

$$\times \left(\frac{1}{1,000}\right)^2$$

$$= 0.00578 \text{ kg} \cdot \text{m}^2$$

The moment of inertia of the part about the $x_1$ axis has the form

$$I_{x_1} = (I_{0x,1} + m_1 d_{x_1,1}^2) + (I_{0x,2} + m_2 d_{x_1,2}^2) + (I_{0x,3} + m_3 d_{x_1,3}^2)$$

where

$$d_{x_1,1} = 26 \text{ mm} \qquad d_{x_1,2} = 70 \text{ mm}$$

and, from Fig. 17.72b,

$$d_{x_1,3}^2 = 35^2 + 110^2$$

The moment of inertia of the part about the $x_1$ axis is then

$$I_{x_1} = \{[\tfrac{1}{8}(0.0492)20^2 + 0.0492(26)^2] + [\tfrac{1}{12}(1.23)(40^2 + 140^2) + 1.23(70)^2] + [\tfrac{1}{12}(0.145)[30^2 + \tfrac{3}{4}(28)^2]$$

$$+ 0.145(35^2 + 110^2)]\}\left(\frac{1}{1,000}\right)^2 = 0.0102 \text{ kg} \cdot \text{m}^2$$

The moment of inertia of the part about the $y$ axis is given by

$$I_y = I_{y1} + I_{y2} + (I_{0y,3} + m_3 d_{y3}^2)$$

From Fig. 17.72c,

$$d_{y3}^2 = 15^2 + 35^2$$

and
$$I_y = (\{\tfrac{1}{48}(0.0492)[16(20)^2 + 3(20)^2]\} + \{\tfrac{1}{12}(1.23)[4(28)^2 + 40^2]\} + \{\tfrac{1}{12}(0.145)[30^2 + \tfrac{3}{4}(28)^2]$$
$$+ 0.145(15^2 + 35^2)\})\left(\frac{1}{1,000}\right)^2 = 7.21 \times 10^{-4}\,\text{kg} \cdot \text{m}^2$$

## 17.4 MASS MOMENTS OF INERTIA OF HOMOGENEOUS, THIN PLANE RIGID BODIES, RELATIONSHIP BETWEEN AREA MOMENTS OF INERTIA AND MASS MOMENTS OF INERTIA

**17.73** Many physical objects for which a mass moment of inertia must be computed have the form of a thin, plane rigid body. Such shapes include disks and flat plates, and rods. Show how the forms for the mass moments of inertia of these bodies may be simplified.

▮ The descriptive term *thin* is understood to mean that the thickness dimension is much less than the dimensions measured in the plane of the body. The theory developed in this problem is approximate in nature.

Figure 17.73a shows a thin plane rigid body of thickness $t$, and it is desired to find the mass moment of inertia of the body about the $x$ axis. The formal definition of this quantity is

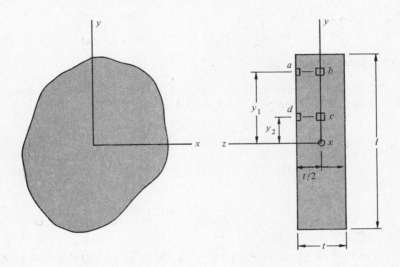

Fig. 17.73a

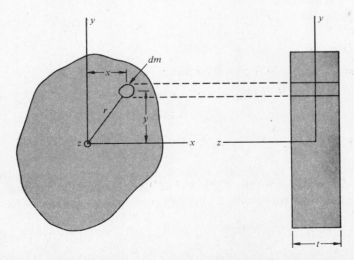

Fig. 17.73b

$$I_x = \int_V (y^2 + z^2)\, dm$$

The contributions of four typical mass elements, with the locations shown in the figure, will now be obtained. A typical position coordinate for mass elements $a$ and $b$ which are not near the $x$ axis is $y_1$. The differential mass element at $b$ lies on the $y$ axis and contributes

$$dI_{bx} = y_1^2\, dm$$

The mass element $a$ which forms part of the plane boundary surface of the body contributes the differential moment of inertia

$$dI_{ax} = \left[ y_1^2 + \left( \frac{t}{2} \right)^2 \right] dm$$

Since the plane body is assumed to be thin,

$$t \ll l \qquad \left( \frac{t}{2} \right)^2 <<< l^2$$

Since $y_1$ is the same order of magnitude as $l$,

$$\left( \frac{t}{2} \right)^2 <<< y_1^2 \qquad \text{and} \qquad y_1^2 + \left( \frac{t}{2} \right)^2 \approx y_1^2$$

The contribution of the mass element $a$ is then approximately

$$dI_{ax} \approx y_1^2\, dm \tag{1}$$

Mass element $c$ lies on the $y$ axis, where $z = 0$. The contribution of this element to $I_x$ is

$$dI_{cx} = y_2^2\, dm$$

The differential moment of inertia of element $d$ is

$$dI_{dx} = \left[ y_2^2 + \left( \frac{t}{2} \right)^2 \right] dm$$

Although $t/2$ is not small compared to $y_2$, it still obeys a relationship of the form of Eq. (1), so that

$$dI_{dx} \approx y_2^2\, dm$$

It may be observed that the contributions of the four typical mass elements to the moment of inertia of the body about the $x$ axis involve only the coordinate $y$ and are independent of the coordinate $z$. The above results may now be generalized, and the approximate form for the mass moment of inertia of the thin plane rigid body about the $x$ axis is

$$I_x \approx \int_V y^2\, dm$$

where the mass element $dm$ is taken to be a single mass element through the thickness of the plane body.

The thin plane body of Fig. 17.73$a$ is redrawn in Fig. 17.73$b$, and it is desired to find the mass moment of inertia of the body about the $z$ axis.

The formal definition of $I_z$ is

$$I_z = \int_V (x^2 + y^2)\, dm$$

From the figure,

$$x^2 + y^2 = r^2$$

and

$$I_z = \int_V r^2\, dm \tag{2}$$

Since all the mass elements across the thickness of the plane body, for a given value of $r$, are at the same distance from the reference $z$ axis, Eq. (2) is an *exact* solution for the mass moment of inertia about the $z$ axis. The quantity $I_z$, defined above for the thin plane rigid body, is also referred to as the *polar mass moment of inertia*.

An example of an approximate form for a mass moment of inertia, and the associated error, was seen earlier in Prob. 17.40.

**17.74** Show the relationship between the moments of inertia of a plane area and the mass moments of inertia of a homogeneous thin plane rigid body.

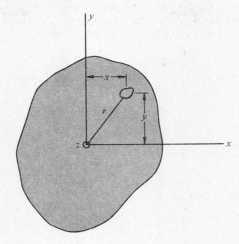

**Fig. 17.74**

❙ Figure 17.74 shows a thin plane body positioned with respect to a set of coordinate axes. The *area* moments of inertia $I_{xA}$ and $I_{yA}$ of the *plane surface area* of the body are

$$I_{xA} = \int_A y^2\, dA \qquad I_{yA} = \int_A x^2\, dA$$

The second subscript $A$ has been added to the two moments of inertia to emphasize that these are *area* moments of inertia.

It was shown in Prob. 17.73 that if the plane body is thin, the mass moments of inertia $I_{xM}$ and $I_{yM}$ are given by the approximate equations

$$I_{xM} \approx \int_V y^2\, dm \qquad I_{yM} \approx \int_V x^2\, dm \qquad (1)$$

where the second subscript $M$ has been added to indicate *mass* moment of inertia.

For a thin, homogeneous plane body, the surface area element $dA$ is related to the plane mass element $dm$ by the relationship

$$dm = \rho t\, dA$$

where $\rho$ is the mass density and $t$ is the thickness of the body.  By using the above result, Eqs. (1) may be written as

$$I_{xM} \approx \int_A y^2 \rho t\, dA = \rho t \int_A y^2\, dA \approx \rho t I_{xA} \qquad I_{yM} \approx \int_A x^2 \rho t\, dA = \rho t \int_A x^2\, dA \approx \rho t I_{yA}$$

The mass moment of inertia of the body in Fig. 17.74 about the $z$ axis is

$$I_{zM} = \int_V r^2\, dm \qquad (2)$$

The $\approx$ sign is *not* used in Eq. (2), since this equation is an exact relationship.  The polar moment of inertia about the $z$ axis of the plane bounding area of the thin body is expressed by

$$I_{zA} = \int_A r^2\, dA = I_{xA} + I_{yA} = J$$

where $J$ is a term referred to as the polar moment of inertia.  It follows that

$$I_{zM} = \int_V r^2 \rho t\, dA = \rho t \int_A r^2\, dA = \rho t I_{zA} = \rho t J$$

A very important conclusion may now be drawn: for a thin, homogeneous, rigid plane body, the mass moments of inertia of the body and the area moments of inertia of the plane bounding surface differ by only a constant factor.  The magnitude of this factor is the product of the thickness of the body and the mass density of the material.

Table 17.14, at the end of this chapter, contains the centroidal coordinates and moments of inertia of several elementary plane areas.

**17.75** The area moment of inertia of the surface area of the thin, rigid plate of mass $m$ shown in Fig. 17.75 is

$$I_{xA} = \tfrac{1}{12}ab^3$$

Find the mass moment of inertia $I_x$.

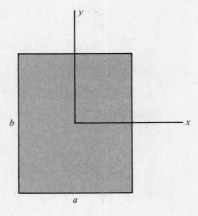

**Fig. 17.75**

▋ The area and mass moments of inertia, from Prob. 17.74, are related by

$$I_{xM} = \rho t I_{xA} \qquad I_{xM} = \rho t \left(\frac{ab^3}{12}\right)$$

The mass of the plate is

$$m = \rho tab$$

and the final form for the mass moment of inertia is

$$I_{xM} = \tfrac{1}{12}mb^2$$

The above equation is the same result which was obtained in Prob. 17.22.

**17.76** The steel disk shown in Fig. 17.76 is 0.5 in thick.

(a) Find the three mass moments of inertia $I_x$, $I_y$, and $I_z$, given that the area moments of inertia are

$$I_{xA} = I_{yA} = \frac{\pi d^4}{64} \qquad I_{zA} = J = \frac{\pi d^4}{32}$$

where $d$ is the diameter of the disk.

(b) Express the results in part (a) in the engineering units of lb · in², and in the units of slug · ft². The specific weight of steel is $\gamma = 490\,\text{lb/ft}^3$.

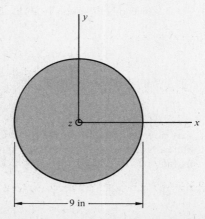

9 in

**Fig. 17.76**

▌ (a)  The area moments of inertia of the plane surface area are

$$I_{xA} = I_{yA} = \frac{\pi d^4}{64} = \frac{\pi(9)^4}{64} = 322 \text{ in}^4 \qquad J = 2I_{xA} = 644 \text{ in}^4$$

The mass density is found from the specific weight as

$$\rho = \frac{\gamma}{g} = \frac{(490 \text{ lb/ft}^3)(1 \text{ ft}^3/1{,}728 \text{ in}^3)}{386 \text{ in/s}^2} = 7.35 \times 10^{-4} \frac{\text{lb} \cdot \text{s}^2}{\text{in}^4}$$

The disk is assumed to be thin, and the mass moments of inertia are

$$I_{xM} = I_{yM} = \rho t I_{xA} = 7.35 \times 10^{-4} \frac{\text{lb} \cdot \text{s}^2}{\text{in}^4} (0.5 \text{ in})(322 \text{ in}^4) = 0.118 \text{ lb} \cdot \text{s}^2 \cdot \text{in}$$

$$I_{zM} = 2I_{xM} = 0.236 \text{ lb} \cdot \text{s}^2 \cdot \text{in}$$

(b)  The mass moments of inertia in terms of the engineering units of $\text{lb} \cdot \text{in}^2$ are

$$I_{xM} = I_{yM} = 0.118 \text{ lb} \cdot \text{s}^2 \cdot \text{in}\left(386 \frac{\text{in}}{\text{s}^2}\right) = 45.5 \text{ lb} \cdot \text{in}^2 \qquad I_{zM} = 2I_{xM} = 2(45.5) = 91 \text{ lb} \cdot \text{in}^2$$

$I_{xM}$ may be written in the form

$$I_{xM} = I_{yM} = 0.118 \text{ lb} \cdot \text{s}^2 \cdot \text{in}\left(\frac{1 \text{ ft}}{12 \text{ in}}\right)\left(\frac{1 \text{ ft}}{1 \text{ ft}}\right) = 0.00983 \frac{\text{lb} \cdot \text{s}^2}{\text{ft}} \cdot \text{ft}^2$$

Since 1 slug is equal to $1 \text{ lb} \cdot \text{s}^2/\text{ft}$, the final result is

$$I_{xM} = I_{yM} = 0.00983 \text{ slug} \cdot \text{ft}^2 \qquad I_{zM} = 2I_{xM} = 0.0197 \text{ slug} \cdot \text{ft}^2$$

**17.77**  Figure 17.77 shows a homogeneous right circular cylinder.  The differential mass element $dm$ has the form of a disk, with the general expression for the mass

$$m = \rho \frac{\pi d^2}{4} t$$

where $d$ is the diameter and $t$ is the thickness.  Find the mass moment of inertia $I_x$ and verify the result in Prob. 17.41.

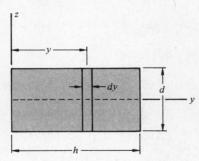

**Fig. 17.77**

▌ For  $t \to dy$,  the mass element is

$$dm = \rho \frac{\pi d^2}{4} dy$$

The moment of inertia of the differential mass element about a diameter is

$$I_{0x,M} = \rho t I_{0x,A}$$

The mass moment of inertia of the mass element about the $x$ axis, using the parallel-axis theorem, is

$$dI_x = \rho \, dy \frac{\pi d^4}{64} + \rho\left(\frac{\pi d^2}{4}\right) dy(y^2) = \rho\left(\frac{\pi d^4}{64} + \frac{\pi d^2 y^2}{4}\right) dy$$

$$I_x = \rho \int_0^h \left(\frac{\pi d^4}{64} + \frac{\pi d^2 y^2}{4}\right) dy = \rho\left[\frac{\pi d^4}{64} y + \frac{\pi d^2}{4}\left(\frac{y^3}{3}\right)\right]\Bigg|_0^h = \rho\left[\frac{\pi d^4}{64} h + \frac{\pi d^2}{4}\left(\frac{h^3}{3}\right)\right]$$

Using  $m = \rho(\pi d^2/4)h$  in the above equation results in

$$I_x = \left(\rho \frac{\pi d^2}{4} h\right)\frac{d^2}{16} + \left(\rho \frac{\pi d^2}{4} h\right)\frac{h^2}{3} = m\frac{d^2}{16} + m\frac{h^2}{3} = \frac{1}{48} m(16h^2 + 3d^2)$$

**17.78** Figure 17.78a shows a machine part made of thin, homogeneous material of constant thickness. The material of the part is bronze, with $\rho = 8,800\,\text{kg/m}^3$, and the thickness is 4 mm.

(a) Find the mass of the part.

(b) Find the mass moments of inertia $I_x$, $I_y$, and $I_z$.

(c) Find the centroidal mass moments of inertia $I_{0x}$, $I_{0y}$, and $I_{0z}$. The centroidal axes $x_0$, $y_0$, and $z_0$ are parallel to the $x$, $y$, and $z$ axes.

(d) Find the radii of gyration $k_x$, $k_y$, $k_z$, $k_{0x}$, $k_{0y}$, and $k_{0z}$.

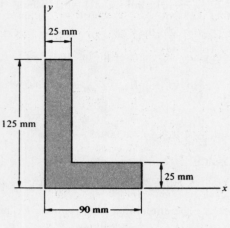

**Fig. 17.78a**

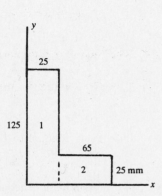

**Fig. 17.78b**

(a) The area is divided into the two elements shown in Fig. 17.78b. The total area of the plane surface is

$$A = A_1 + A_2 = 25(125) + 65(25) = 4,750\,\text{mm}^2$$

Using $\rho = 8,800\,\text{kg/m}^3$ and $t = 4\,\text{mm}$, the mass of the part is found to be

$$m = \rho t A = 8,800\,\frac{\text{kg}}{\text{m}^3}\,(4\,\text{mm})4,750\,\text{mm}^2\left(\frac{1\,\text{m}}{1,000\,\text{mm}}\right)^3 = 0.167\,\text{kg}$$

(b) The moments of inertia of the plane surface area of the part are

$$I_{xA} = \frac{25(125)^3}{3} + \frac{65(25)^3}{3} = 1.66 \times 10^7\,\text{mm}^4$$

$$I_{yA} = \frac{125(25)^3}{3} + \left[\frac{25(65)^3}{12} + 65(25)\left(25 + \frac{65}{2}\right)^2\right] = 6.60 \times 10^6\,\text{mm}^4$$

$$I_{zA} = I_{xA} + I_{yA} = 2.32 \times 10^7\,\text{mm}^4$$

The mass moments of inertia $I_M$ are related to the area moments of inertia $I_A$ by

$$I_M = \rho t I_A = \underbrace{8,800\,\frac{\text{kg}}{\text{m}^3}\,(4\,\text{mm})\left(\frac{1\,\text{m}}{1,000\,\text{mm}}\right)^5 I_A}_{q}\quad\text{mm}^4$$

The above equation may be written in the form

$$I_M = qI_A \qquad q = 3.52 \times 10^{-11}\,\text{kg}\cdot\text{m}^2/\text{mm}^4 \tag{1}$$

$I_x$, $I_y$, and $I_z$ are then found as

$$I_x = qI_{xA} = q(1.66 \times 10^7) = 5.84 \times 10^{-4}\,\text{kg}\cdot\text{m}^2$$
$$I_y = qI_{yA} = q(6.60 \times 10^6) = 2.32 \times 10^{-4}\,\text{kg}\cdot\text{m}^2$$
$$I_z = qI_{zA} = q(2.32 \times 10^7) = 8.17 \times 10^{-4}\,\text{kg}\cdot\text{m}^2$$

(c)  The centroidal coordinates of the plane surface area are

$$x_c = \frac{x_1 A_1 + x_2 A_2}{A_1 + A_2} = \frac{\frac{25}{2}(25)125 + [25 + \frac{1}{2}(65)]65(25)}{4,750} = 27.9 \text{ mm}$$

$$y_c = \frac{y_1 A_1 + y_2 A_2}{A_1 + A_2} = \frac{\frac{125}{2}(25)125 + \frac{25}{2}(65)25}{4,750} = 45.4 \text{ mm}$$

Using the parallel-axis theorem,

$$I_{0,A} = I - Ad^2 \qquad I_{0x,A} = 1.66 \times 10^7 - 4,750(45.4)^2 = 6.81 \times 10^6 \text{ mm}^4$$

$$I_{0y,A} = 6.60 \times 10^6 - 4,750(27.9)^2 = 2.90 \times 10^6 \text{ mm}^4 \qquad I_{0z,A} = I_{0x,A} + I_{0y,A} = 9.71 \times 10^6 \text{ mm}^4$$

The centroidal mass moments of inertia, using the above results with Eq. (1), are

$$I_{0x} = q(6.81 \times 10^6) = 2.40 \times 10^{-4} \text{ kg} \cdot \text{m}^2 \qquad I_{0y} = q(2.90 \times 10^6) = 1.02 \times 10^{-4} \text{ kg} \cdot \text{m}^2$$

$$I_{0z} = q(9.71 \times 10^6) = 3.42 \times 10^{-4} \text{ kg} \cdot \text{m}^2$$

(d)  The radii of gyration are given by

$$k_x = \sqrt{\frac{I_x}{A}} = \sqrt{\frac{1.66 \times 10^7}{4,750}} = 59.1 \text{ mm} \qquad k_y = \sqrt{\frac{I_y}{A}} = \sqrt{\frac{6.60 \times 10^6}{4,750}} = 37.3 \text{ mm}$$

$$k_z = \sqrt{\frac{I_z}{A}} = \sqrt{\frac{2.32 \times 10^7}{4,750}} = 69.9 \text{ mm}$$

$$k_{0x} = \sqrt{\frac{I_{0x}}{A}} = \sqrt{\frac{6.81 \times 10^6}{4,750}} = 37.9 \text{ mm} \qquad k_{0y} = \sqrt{\frac{I_{0y}}{A}} = \sqrt{\frac{2.90 \times 10^6}{4,750}} = 24.7 \text{ mm}$$

$$k_{0z} = \sqrt{\frac{I_{0z}}{A}} = \sqrt{\frac{9.71 \times 10^6}{4,750}} = 45.2 \text{ mm}$$

For the problem of a thin plate, the mass and area radii of gyration are the same.

**17.79**  Do the same as in Prob. 17.78, for the machine part shown in Fig. 17.79a. The material is bronze, with $\rho = 8,800 \text{ kg/m}^3$, and the thickness is 4 mm.

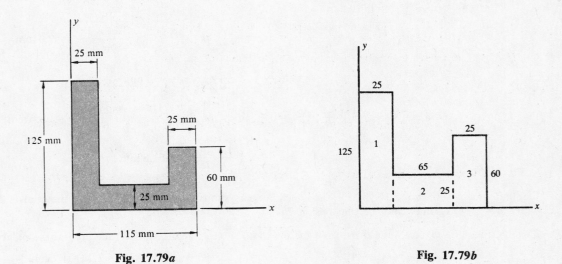

Fig. 17.79a                    Fig. 17.79b

**I** (a)  Figure 17.79b shows the area divided into three elements. The total area of the plane surface is

$$A = A_1 + A_2 + A_3 = 25(125) + 65(25) + 25(60) = 6,250 \text{ mm}^2$$

Using $\rho = 8,800 \text{ kg/m}^3$ and $t = 4 \text{ mm}$, the mass of the part is

$$m = \rho t A = 8,800(4)6,250\left(\frac{1}{1,000}\right)^3 = 0.220 \text{ kg}$$

(b) The area moments of inertia of the plane surface area of the part are

$$I_{xA} = \frac{25(125)^3}{3} + \frac{65(25)^3}{3} + \frac{25(60)^3}{3} = 1.84 \times 10^7 \text{ mm}^4$$

$$I_{yA} = \frac{125(25)^3}{3} + \left[ \frac{25(65)^3}{12} + 65(25)\left(25 + \frac{65}{2}\right)^2 \right] + \left[ \frac{60(25)^3}{12} + 25(60)\left(115 - \frac{25}{2}\right)^2 \right] = 2.24 \times 10^7 \text{ mm}^4$$

$$I_{zA} = I_{xA} + I_{yA} = 4.08 \times 10^7 \text{ mm}^4$$

From Prob. 17.78,

$$I_M = qI_A \qquad q = 3.52 \times 10^{-11} \text{ kg} \cdot \text{m}^2/\text{mm}^4$$

The mass moments of inertia have the values

$$I_x = qI_{xA} = q(1.84 \times 10^7) = 6.48 \times 10^{-4} \text{ kg} \cdot \text{m}^2$$
$$I_y = qI_{yA} = q(2.24 \times 10^7) = 7.88 \times 10^{-4} \text{ kg} \cdot \text{m}^2$$
$$I_z = qI_{zA} = q(4.08 \times 10^7) = 1.44 \times 10^{-3} \text{ kg} \cdot \text{m}^2$$

(c) The centroidal coordinates of the plane surface area are

$$x_c = \frac{x_1 A_1 + x_2 A_2 + x_3 A_3}{A_1 + A_2 + A_3} = \frac{\frac{25}{2}(25)125 + (25 + \frac{65}{2})65(25) + (25 + 65 + \frac{25}{2})25(60)}{6,250} = 45.8 \text{ mm}$$

$$y_c = \frac{\frac{125}{2}(25)125 + \frac{25}{2}(65)25 + \frac{60}{2}(25)60}{6,250} = 41.7 \text{ mm}$$

Using the parallel-axis theorem,

$$I_0 = I - Ad^2 \qquad I_{0x,A} = 1.84 \times 10^7 - 6,250(41.7)^2 = 7.53 \times 10^6 \text{ mm}^4$$
$$I_{0y,A} = 2.24 \times 10^7 - 6,250(45.8)^2 = 9.29 \times 10^6 \text{ mm}^4$$
$$I_{0z,A} = I_{0x,A} + I_{0y,A} = 1.68 \times 10^7 \text{ mm}^4$$

Using Eq. (1), the centroidal mass moments of inertia are found to be

$$I_{0x} = q(7.53 \times 10^6) = 2.65 \times 10^{-4} \text{ kg} \cdot \text{m}^2 \qquad I_{0y} = q(9.29 \times 10^6) = 3.27 \times 10^{-4} \text{ kg} \cdot \text{m}^2$$
$$I_{0z} = q(1.68 \times 10^7) = 5.91 \times 10^{-4} \text{ kg} \cdot \text{m}^2$$

(d) The radii of gyration are

$$k_x = \sqrt{\frac{I_x}{A}} = \sqrt{\frac{1.84 \times 10^7}{6,250}} = 54.3 \text{ mm} \qquad k_y = \sqrt{\frac{I_y}{A}} = \sqrt{\frac{2.24 \times 10^7}{6,250}} = 59.9 \text{ mm}$$

$$k_z = \sqrt{\frac{I_z}{A}} = \sqrt{\frac{4.08 \times 10^7}{6,250}} = 80.8 \text{ mm}$$

$$k_{0x} = \sqrt{\frac{I_{0x}}{A}} = \sqrt{\frac{7.53 \times 10^6}{6,250}} = 34.7 \text{ mm} \qquad k_{0y} = \sqrt{\frac{I_{0y}}{A}} = \sqrt{\frac{9.29 \times 10^6}{6,250}} = 38.6 \text{ mm}$$

$$k_{0z} = \sqrt{\frac{I_{0z}}{A}} = \sqrt{\frac{1.68 \times 10^7}{6,250}} = 51.8 \text{ mm}$$

**17.80** Do the same as in Prob. 17.78, for the machine part shown in Fig. 17.80a. The material is aluminum, with $\rho = 2,770 \text{ kg/m}^3$, and the thickness is 75 mm.

I (a) Fig. 17.80b shows the division of the area into three elements. The total area of the plane surface is

$$A = A_1 + A_2 + A_3 = \frac{1}{2}(713)850 + 928(850) + \frac{1}{2}(309)850 = 1.22 \times 10^6 \text{ mm}^2$$

Using $\rho = 2,770 \text{ kg/m}^3$ and $t = 75 \text{ mm}$, the mass of the part is

$$m = \rho t A = 2,770 \frac{\text{kg}}{\text{m}^3} (75 \text{ mm})1.22 \times 10^6 \text{ mm}^2 \left(\frac{1 \text{ m}}{1,000 \text{ mm}}\right)^3 = 253 \text{ kg}$$

(b) The area moments of inertia of the plane surface area of the part are

$$I_{xA} = \frac{713(850)^3}{12} + \frac{928(850)^3}{3} + \frac{309(850)^3}{12} = 2.42 \times 10^{11} \text{ mm}^4$$

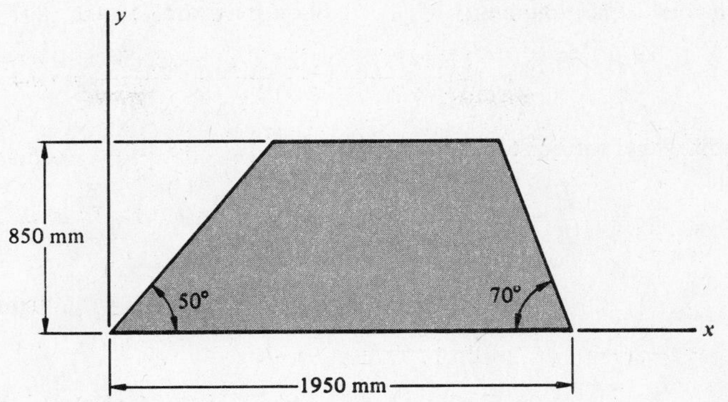

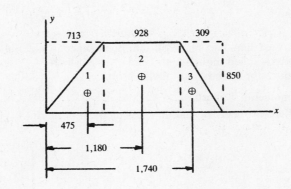

Fig. 17.80a

Fig. 17.80b

$$I_{yA} = \left[ \frac{850(713)^3}{36} + \frac{1}{2}(713)850(475)^2 \right] + \left[ \frac{850(928)^3}{12} + (928)850(1{,}180)^2 \right]$$

$$+ \left[ \frac{850(309)^3}{36} + \frac{1}{2}(309)850(1{,}740)^2 \right] = 1.63 \times 10^{12} \text{ mm}^4$$

$$I_{zA} = I_{xA} + I_{yA} = 1.87 \times 10^{12} \text{ mm}^4$$

The relationship between the mass moments of inertia and the area moments of inertia is given by

$$\underbrace{I_M = \rho t I_A = 2{,}770 \, \frac{\text{kg}}{\text{m}^3} \, (75 \text{ mm}) \left( \frac{1 \text{ m}}{1{,}000 \text{ mm}} \right)^5 I_A \qquad \text{mm}^4}_{q}$$

$$I_M = q I_A \qquad q = 2.08 \times 10^{-10} \text{ kg} \cdot \text{m}^2/\text{mm}^4$$

The mass moments of inertia $I_x$, $I_y$, and $I_z$ are then found as

$$I_x = q(2.42 \times 10^{11}) = 50.3 \text{ kg} \cdot \text{m}^2 \qquad I_y = q(1.63 \times 10^{12}) = 339 \text{ kg} \cdot \text{m}^2$$

$$I_z = q(1.87 \times 10^{12}) = 389 \text{ kg} \cdot \text{m}^2$$

(c)   The centroidal coordinates of the plane surface area are

$$x_c = \frac{x_1 A_1 + x_2 A_2 + x_3 A_3}{A} = \frac{\frac{2}{3}(713)\frac{1}{2}(713)850 + (713 + \frac{928}{2})928(850) + (713 + 928 + \frac{309}{3})\frac{1}{2}(309)850}{1.22 \times 10^6}$$

$$= 1{,}060 \text{ mm}$$

$$y_c = \frac{y_1 A_1 + y_2 A_2 + y_3 A_3}{A} = \frac{\frac{1}{3}(850)\frac{1}{2}(713)850 + \frac{850}{2}(928)850 + \frac{1}{3}(850)\frac{1}{2}(309)850}{1.22 \times 10^6} = 375 \text{ mm}$$

Using the parallel-axis theorem,

$$I_0 = I - Ad^2 \qquad I_{0x,A} = 2.42 \times 10^{11} - 1.22 \times 10^6 (375)^2 = 7.04 \times 10^{10} \text{ mm}^4$$

$$I_{0y,A} = 1.63 \times 10^{12} - 1.22 \times 10^6 (1{,}060)^2 = 2.59 \times 10^{11} \text{ mm}^4$$

$$I_{0z,A} = I_{0x,A} + I_{0y,A} = 3.29 \times 10^{11} \text{ mm}^4$$

The centroidal mass moments of inertia then have the forms

$$I_{0x} = q(7.04 \times 10^{10}) = 14.6\, \text{kg} \cdot \text{m}^2 \qquad I_{0y} = q(2.59 \times 10^{11}) = 53.9\, \text{kg} \cdot \text{m}^2$$

$$I_{0z} = q(3.29 \times 10^{11}) = 68.4\, \text{kg} \cdot \text{m}^2$$

(*d*)  The radii of gyration are

$$k_x = \sqrt{\frac{I_x}{A}} = \sqrt{\frac{2.42 \times 10^{11}}{1.22 \times 10^6}} = 445\, \text{mm} \qquad k_y = \sqrt{\frac{I_y}{A}} = \sqrt{\frac{1.63 \times 10^{12}}{1.22 \times 10^6}} = 1{,}160\, \text{mm}$$

$$k_z = \sqrt{\frac{I_z}{A}} = \sqrt{\frac{1.87 \times 10^{12}}{1.22 \times 10^6}} = 1{,}240\, \text{mm}$$

$$k_{0x} = \sqrt{\frac{I_{0x}}{A}} = \sqrt{\frac{7.04 \times 10^{10}}{1.22 \times 10^6}} = 240\, \text{mm} \qquad k_{0y} = \sqrt{\frac{I_{0y}}{A}} = \sqrt{\frac{2.59 \times 10^{11}}{1.22 \times 10^6}} = 461\, \text{mm}$$

$$k_{0z} = \sqrt{\frac{I_{0z}}{A}} = \sqrt{\frac{3.29 \times 10^{11}}{1.22 \times 10^6}} = 519\, \text{mm}$$

**17.81**  Figure 17.81*a* shows a machine part made of thin, homogeneous material of constant thickness. The material of the part is 0.3-in-thick rolled monel metal, and $\gamma = 555\, \text{lb/ft}^3$.
Organize the solutions for $I_x$ and $I_y$ in tabular form.

(*a*)  Find the mass moments of inertia $I_x$, $I_y$.

(*b*)  Using the parallel-axis theorem, find the mass moment of inertia $I_{0x}$.  The $x_0$ axis is parallel to the $x$ axis.

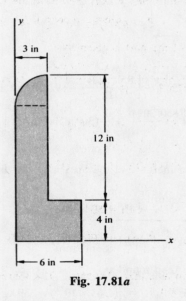

**Fig. 17.81*a***

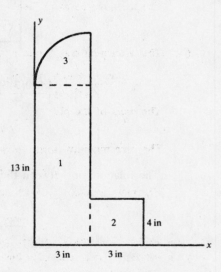

**Fig. 17.81*b***

▌ (*a*)  The area is divided into the three elements shown in Fig. 17.81*b*.  Using Case 6 in Table 17.14, for the case of a quarter circle,

$$x_c = y_c = \frac{4a}{3\pi} \qquad I_{0x} = I_{0y} = \frac{a^4}{2}\left(\frac{\pi}{8} - \frac{8}{9\pi}\right)$$

Table 17.6 contains the solutions for the area moments of inertia about the $x$ and $y$ axes.  The results are

$$I_{xA} = 3{,}710\, \text{in}^4 \qquad I_{yA} = 395\, \text{in}^4$$

The mass moments of inertia have the form

$$I_M = \rho t I_A = \underbrace{\frac{555\, \text{lb/ft}^3}{386\, \text{in/s}^2}\left(\frac{1\, \text{ft}}{12\, \text{in}}\right)^3 0.3\, \text{in}}_{q} I_A$$

**TABLE 17.6**

| Element | $I_{0i}$ | $d_i$ | $A_i$ | $A_i d_i^2$ | $I_i$ |
|---------|----------|-------|-------|-------------|-------|
| 1 | $\cdots$ | $\cdots$ | $\cdots$ | $\cdots$ | $3(13)^3/3 = 2{,}197$ |
| 2 | $\cdots$ | $\cdots$ | $\cdots$ | $\cdots$ | $3(4)^3/3 = 64$ |
| 3 | $\dfrac{3^4}{2}\left(\dfrac{\pi}{8} - \dfrac{8}{9\pi}\right) = 4.45$ | $13 + \dfrac{4(3)}{3\pi} = 14.3$ | $\dfrac{\pi(3)^2}{4} = 7.07$ | $1{,}450$ | $1{,}450$ |
| Axis: $x$ | | | | | $I_x = 3{,}710 \text{ in}^4$ |
| 1 | $\cdots$ | $\cdots$ | $\cdots$ | $\cdots$ | $13(3)^3/3 = 117$ |
| 2 | $\dfrac{4(3)^3}{12} = 9$ | $3 + 1.5 = 4.5$ | $3(4) = 12$ | $243$ | $252$ |
| 3 | $\dfrac{3^4}{2}\left(\dfrac{\pi}{8} - \dfrac{8}{9\pi}\right) = 4.45$ | $3 - \dfrac{4(3)}{3\pi} = 1.73$ | $\dfrac{\pi(3)^2}{4} = 7.07$ | $21.2$ | $25.7$ |
| Axis: $y$ | | | | | $I_y = 395 \text{ in}^4$ |

$$I_M = qI_A \qquad q = 2.50 \times 10^{-4} \text{ lb} \cdot \text{s}^2/\text{in}^3$$

The mass moments of inertia of the part about the $x$ and $y$ axes then have the values

$$I_x = q(3{,}710) = 0.928 \text{ lb} \cdot \text{s}^2 \cdot \text{in} \qquad I_y = q(395) = 0.0988 \text{ lb} \cdot \text{s}^2 \cdot \text{in}$$

(*b*) The centroidal coordinate $y_c$ of the plane area of the part is given by

$$y_c = \frac{y_1 A_1 + y_2 A_2 + y_3 A_3}{A_1 + A_2 + A_3} = \frac{\frac{13}{2}(3)13 + \frac{4}{2}(3)4 + [13 + 4(3)/3\pi]\frac{1}{4}\pi(3)^2}{3(13) + 3(4) + \frac{1}{4}\pi(3)^2} = 6.52 \text{ in} \qquad A = 58.1 \text{ in}^2$$

The mass of the plate is

$$m = \rho t A = qA = 2.50 \times 10^{-4}(58.1) = 0.0145 \text{ lb} \cdot \text{s}^2/\text{in}$$

The area moment of inertia about the centroidal $x_0$ axis is

$$I_0 = I - Ad^2 \qquad I_{0x} = I_x - Ay_c^2 = 3{,}710 - 58.1(6.52)^2 = 1{,}240 \text{ in}^4$$

The mass moment of inertia about the $x_0$ axis is

$$I_{0x,M} = qI_{0x,A} = 2.50 \times 10^{-4}(1{,}240) = 0.310 \text{ lb} \cdot \text{s}^2 \cdot \text{in}$$

**17.82** Do the same as in Prob. 17.81, for the machine part shown in Fig. 17.82*a*. The material is steel, with $\rho = 7{,}830 \text{ kg/m}^3$, and $t = 4 \text{ mm}$.

**❙** (*a*) Figure 17.82*b* shows the division of the plate into nine area elements. The solutions for $I_x$ and $I_y$ are contained in Tables 17.7 and 17.8. The results are

$$I_x = 5.50 \times 10^5 \text{ mm}^4 \qquad I_y = 5.21 \times 10^5 \text{ mm}^4$$

It may be observed in this table that the terms $A_i$ and $I_{0i}$, for elements 2 through 9, are negative quantities, since these terms represent areas which are absent. Using $\rho = 7{,}830 \text{ kg/m}^3$ and $t = 4 \text{ mm}$, the mass of the part is found as

$$m = \rho t A = 7{,}830 \, \frac{\text{kg}}{\text{m}^3} \, (4 \text{ mm})1{,}120 \text{ mm}^2\left(\frac{1 \text{ m}}{1{,}000 \text{ mm}}\right)^3 = 0.0351 \text{ kg}$$

The mass moments of inertia have the form

$$I_M = \rho t I_A = \underbrace{7{,}830 \, \frac{\text{kg}}{\text{m}^3} \, (4 \text{ mm})\left(\frac{1 \text{ m}}{1{,}000 \text{ mm}}\right)^5}_{q} I_A \qquad \text{mm}^4$$

$$I_M = qI_A \qquad q = 3.13 \times 10^{-11} \text{ kg} \cdot \text{m}^2/\text{mm}^4$$

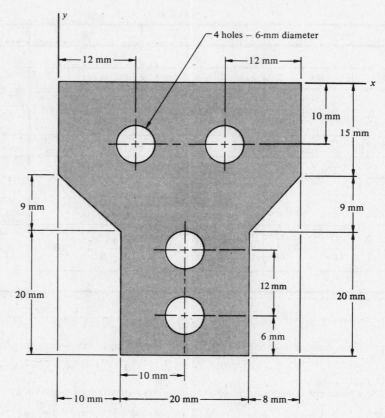

**Fig. 17.82a**

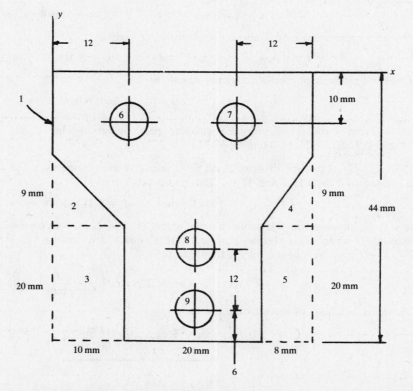

**Fig. 17.82b**

**TABLE 17.7**

| Element | $I_{0i}$ | $d_i$ | $A_i$ | $A_i d_i^2$ | $I_i$ |
|---|---|---|---|---|---|
| 1 | ... | ... | ... | ... | $38(44)^3/3 = 1.079 \times 10^6$ |
| 2 | $-10(9)^3/36 = -203$ | $15 + \frac{2}{3}(9) = 21$ | $-\frac{1}{2}(10)9 = -45$ | $-1.98 \times 10^4$ | $-2.0 \times 10^4$ |
| 3 | $-10(20)^3/12 = -6,670$ | $24 + \frac{20}{2} = 34$ | $-10(20) = -200$ | $-2.31 \times 10^5$ | $-2.38 \times 10^5$ |
| 4 | $-8(9)^3/36 = -162$ | $15 + \frac{2}{3}(9) = 21$ | $-\frac{1}{2}(8)9 = -36$ | $-1.59 \times 10^4$ | $-1.6 \times 10^4$ |
| 5 | $-8(20)^3/12 = -5,330$ | $24 + \frac{20}{2} = 34$ | $-8(20) = -160$ | $-1.85 \times 10^5$ | $-1.90 \times 10^5$ |
| 6 | $-\pi(6)^4/64 = -63.6$ | 10 | $-\pi(6)^2/4 = -28.3$ | $-2.8 \times 10^3$ | $-3 \times 10^3$ |
| 7 | $-63.6$ | 10 | $-28.3$ | $-2.8 \times 10^3$ | $3 \times 10^3$ |
| 8 | $-63.6$ | $44 - 18 = 26$ | $-28.3$ | $-1.91 \times 10^4$ | $-1.9 \times 10^4$ |
| 9 | $-63.6$ | $44 - 6 = 38$ | $-28.3$ | $-4.09 \times 10^4$ | $-4.0 \times 10^4$ |
| Axis: $x$ | | | | | $I_x = 5.50 \times 10^5 \text{ mm}^4$ |

**TABLE 17.8**

| Element | $I_{0i}$ | $d_i$ | $A_i$ | $A_i d_i^2$ | $I_i$ |
|---|---|---|---|---|---|
| 1 | ... | ... | ... | ... | $44(38)^3/3 = 8.048 \times 10^5$ |
| 2 | ... | ... | ... | ... | $-9(10)^3/12 = -8 \times 10^2$ |
| 3 | ... | ... | ... | ... | $-20(10)^3/3 = -6.7 \times 10^3$ |
| 4 | $-9(8)^3/36 = -128$ | $30 + \frac{2}{3}(8) = 35.3$ | $-\frac{1}{2}(8)9 = -36$ | $-4.49 \times 10^4$ | $-4.50 \times 10^4$ |
| 5 | $-20(8)^3/12 = -853$ | $30 + \frac{8}{2} = 34$ | $-8(20) = -160$ | $-1.85 \times 10^5$ | $-1.858 \times 10^5$ |
| 6 | $-\pi(6)^4/64 = -63.6$ | 12 | $-\pi(6)^2/4 = -28.3$ | $-4.07 \times 10^3$ | $-4.1 \times 10^3$ |
| 7 | $-63.6$ | 26 | $-28.3$ | $-1.91 \times 10^4$ | $-1.92 \times 10^4$ |
| 8 | $-63.6$ | 20 | $-28.3$ | $-1.13 \times 10^4$ | $-1.1 \times 10^4$ |
| 9 | $-63.6$ | 20 | $-28.3$ | $-1.13 \times 10^4$ | $-1.1 \times 10^4$ |
| Axis: $y$ | | | | | $I_y = 5.21 \times 10^5 \text{ in}^4$ |

**TABLE 17.9**

| Element | $A_i$ | $x_i$ | $x_i A_i$ | $y_i$ | $y_i A_i$ |
|---|---|---|---|---|---|
| 1 | $38(44) = 1,670$ | 19 | 31,730 | $-22$ | $-36,740$ |
| 2 | $-\frac{1}{2}(10)9 = -45$ | $\frac{10}{3} = 3.33$ | $-150$ | $-[15 + \frac{2}{3}(9)] = -21$ | 950 |
| 3 | $-10(20) = -200$ | 5 | $-1,000$ | $-34$ | 6,800 |
| 4 | $-\frac{1}{2}(8)9 = -36$ | $30 + \frac{2}{3}(8) = 35.3$ | $-1,270$ | $-[15 + \frac{2}{3}(9)] = -21$ | 760 |
| 5 | $-8(20) = -160$ | 34 | $-5,440$ | $-34$ | 5,440 |
| 6 | $-\pi 6^2/4 = -28$ | 12 | $-340$ | $-10$ | 280 |
| 7 | $-28$ | 26 | $-730$ | $-10$ | 280 |
| 8 | $-28$ | 20 | $-560$ | $-(44 - 18) = -26$ | 730 |
| 9 | $-28$ | 20 | $-560$ | $-(44 - 6) = -38$ | 1,060 |
| $A = \sum_i A_i = 1,120$ | | $\sum_i x_i A_i = 21,700$ | | $\sum_i y_i A_i = -20,400$ | |
| | | $x_c = 19.4 \text{ mm}$ | | $y_c = -18.3 \text{ mm}$ | |

The mass moments of inertia about the $x$ and $y$ axes have the values

$$I_x = q(5.50 \times 10^5) = 1.72 \times 10^{-5}\,\text{kg} \cdot \text{m}^2 \qquad I_y = q(5.21 \times 10^5) = 1.63 \times 10^{-5}\,\text{kg} \cdot \text{m}^2$$

(*b*) The solution for the centroidal coordinate $y_c$ is given in Table 17.9, and $y_c = -18.3$ mm. The area moment of inertia about the centroidal $x$ axis is then found as

$$I_0 = I - Ad^2 \qquad I_{0x,A} = 5.50 \times 10^5 - 1,120(18.3)^2 = 1.75 \times 10^5\,\text{mm}^4$$

The final result for the centroidal mass moment of inertia is

$$I_{0x} = q(1.75 \times 10^5) = 5.48 \times 10^{-6}\,\text{kg} \cdot \text{m}^2$$

**17.83** A plate with a rectangular cutout, shown in Fig. 17.83*a*, is made of brass strip of 0.125-in thickness. Find the mass moment of inertia of the plate about the $x$ axis. The specific weight of brass is 534 lb/ft³.

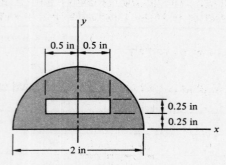

**Fig. 17.83*a***

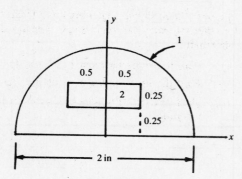

**Fig. 17.83*b***

▌ The plate area is divided into the two elements shown in Fig. 17.83*b*. Using Case 6 in Table 17.14, the area moment of inertia of element 1 about the $x$ axis is

$$I_{x1,A} = I_{0x} + Ay_c^2 = a^4\left(\frac{\pi}{8} - \frac{8}{9\pi}\right) + \frac{\pi a^2}{2}\left(\frac{4a}{3\pi}\right)^2 = a^4\left(\frac{\pi}{8} - \frac{8}{9\pi}\right) + \frac{16\pi a^4}{18\pi^2} = a^4\left(\frac{\pi}{8} - \frac{8}{9\pi} + \frac{8}{9\pi}\right) = \frac{\pi a^4}{8}$$

$$a = 1\,\text{in}$$

The area moment of inertia of the plate area about the $x$ axis has the form

$$I_{xA} = I_{x1,A} - I_{x2,A}$$

Using $a = 1$ in,

$$I_{xA} = \frac{\pi(1)^4}{8} - \left[\frac{1(0.25)^3}{12} + 1(0.25)\left(0.25 + \frac{0.25}{2}\right)^2\right] = 0.356\,\text{in}^4$$

The mass moment of inertia of the plate about the $x$ axis is given by

$$I_x = \rho t I_{xA} = \frac{534}{386}\left(\frac{1}{12}\right)^3 0.125(0.356) = 3.56 \times 10^{-5}\,\text{lb} \cdot \text{s}^2 \cdot \text{in}$$

**17.84** Do the same as in Prob. 17.83, if the design of the plate is changed to the elliptical shape with two rectangular cutouts shown in Fig. 17.84.

▌ The elliptical area is element 1, and the two hole areas are element 2. From Case 7 in Table 17.14,

$$I_{x1,A} = \frac{\pi a b^3}{64} \qquad a = 2\,\text{in} \qquad b = 1.6\,\text{in}$$

The moment of inertia of the plate area about the $x$ axis is

$$I_{xA} = I_{x1,A} - I_{x2,A} = \frac{\pi(2)1.6^3}{64} - 2\left[\frac{0.25(0.5)^3}{12}\right] = 0.397\,\text{in}^4$$

The mass moment of inertia of the plate about the $x$ axis is given by

$$I_x = \rho t I_{xA} = \frac{534}{386}\left(\frac{1}{12}\right)^3 0.125(0.397) = 3.97 \times 10^{-5}\,\text{lb} \cdot \text{s}^2 \cdot \text{in}$$

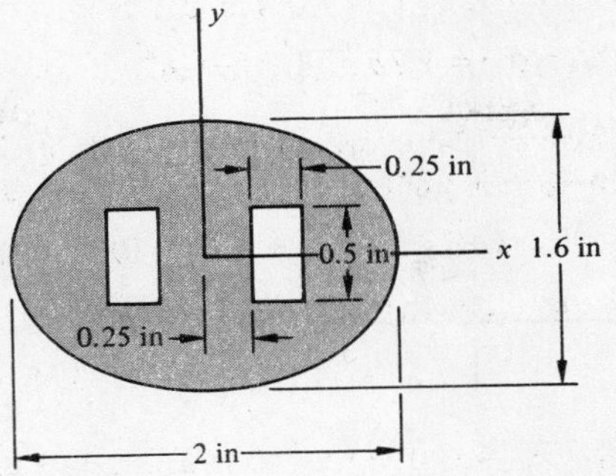

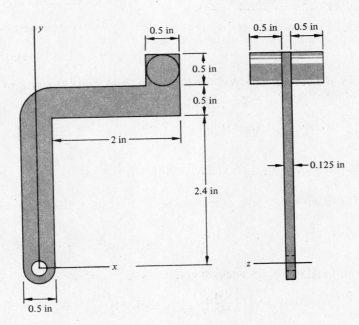

**Fig. 17.84**

**17.85** Figure 17.85a shows a latching lever formed from a steel strip, with attached cylindrical brass plugs. The effect of the hole may be neglected. The specific weight of brass is 534 lb/ft³, and the specific weight of steel is 489 lb/ft³.

(*a*) Find the mass moment of inertia $I_z$ of the lever and plug assembly.

(*b*) Find the mass and weight of the assembly.

(*c*) Show the contributions of the lever, and of the brass plugs, to the mass and mass moment of inertia of the assembly.

**Fig. 17.85a**

**▮** (*a*) Figure 17.85b shows the division of the area of the lever into five elements. For element 3, using Case 6 of Table 17.14,

$$b = y_c = \frac{4a}{3\pi} \qquad a = 0.5 \text{ in} \qquad b = \frac{4(0.5)}{3\pi} = 0.212 \text{ in}$$

The area moment of inertia of element 1 about the $x$ axis, using the result in Prob. 17.83, with $a = 0.25$ in, is

$$I_{x1} = \frac{\pi a^4}{8} = \frac{\pi (0.25)^4}{8} = 0.00153 \text{ in}^4$$

From Case 6 of Table 17.14,

$$I_{0x,3} = I_{0y,3} = \frac{a^4}{2}\left(\frac{\pi}{8} - \frac{8}{9\pi}\right) = \frac{0.5^4}{2}\left(\frac{\pi}{8} - \frac{8}{9\pi}\right) = 0.00343 \text{ in}^4$$

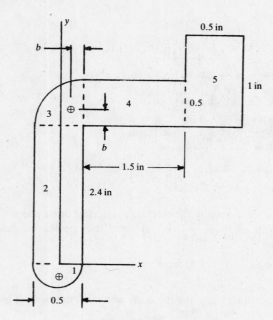

**Fig. 17.85b**

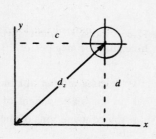

**Fig. 17.85c**

The area moments of inertia $I_{xA}$ and $I_{yA}$ for the lever area are next found, and the polar moment of inertia $I_z$ will subsequently be found as

$$I_{zA} = I_{xA} + I_{yA}$$

The moment of inertia of the lever area about the $x$ axis is

$$I_{xA} = I_{x1,A} + I_{x2,A} + I_{x3,A} + I_{x4,A} + I_{x5,A}$$

$$= 0.00153 + \left[\frac{0.5(2.4)^3}{3}\right] + \left[0.00343 + \frac{\pi(0.5)^2}{4}(2.4 + 0.212)^2\right]$$

$$+ \left[\frac{1.5(0.5)^3}{12} + 1.5(0.5)\left(2.4 + \frac{0.5}{2}\right)^2\right] + \left[\frac{0.5(1)^3}{12} + 0.5(1)(2.4 + 0.5)^2\right] = 13.2 \text{ in}^4$$

The moment of inertia of the lever area about the $y$ axis is found next. From Case 6 of Table 17.14, with $a = 0.25$ in,

$$I_{y1} = \frac{\pi a^4}{8} = \frac{\pi(0.25)^4}{8} = 0.00153 \text{ in}^4$$

$$I_{yA} = I_{y1,A} + I_{y2,A} + I_{y3,A} + I_{y4,A} + I_{y5,A}$$

$$= 0.00153 + \left[\frac{2.4(0.5)^3}{12}\right] + \left[0.00343 + \frac{\pi(0.5)^2}{4}(0.25 - 0.212)^2\right]$$

$$+ \left[\frac{0.5(1.5)^3}{12} + (1.5)0.5\left(0.25 + \frac{1.5}{2}\right)^2\right] + \left[\frac{1(0.5)^3}{12} + (0.5)1\left(0.25 + 1.5 + \frac{0.5}{2}\right)^2\right]$$

$$= 2.93 \text{ in}^4$$

The area moment of inertia about the $z$ axis is then

$$I_{zA} = I_{xA} + I_{yA} = 13.2 + 2.93 = 16.1 \text{ in}^4$$

The mass moment of inertia of the lever about the $z$ axis is

$$I_{zM} = \rho t I_{zA} = \left[\frac{489}{386(12)^3}\right]0.125(16.1) = 0.00148 \text{ lb} \cdot \text{s}^2 \cdot \text{in}$$

Figure 17.85c shows the position of the brass plugs. From this figure,

$$c = 0.25 + 1.5 + \frac{0.5}{2} = 2 \text{ in} \qquad d = 2.4 + 0.5 + \frac{0.5}{2} = 3.15 \text{ in}$$

From Case 4 in Table 17.13,

$$m = \frac{1}{4} \rho \pi d^2 h \qquad I_{0z} = \frac{1}{8} md^2 \qquad m = \frac{1}{4}\left[\frac{534}{386(12)^3}\right]\pi(0.5)^2 0.5 = 7.86 \times 10^{-5} \text{ lb} \cdot \text{s}^2/\text{in}$$

The mass moment of inertia of the plugs about the $z$ axis is given by

$$I_z = I_{0z} + md_z^2 = 2[\tfrac{1}{8}(7.86 \times 10^{-5})0.5^2 + (7.86 \times 10^{-5})(2^2 + 3.15^2)] = 0.00219 \text{ lb} \cdot \text{s}^2 \cdot \text{in}$$

The mass moment of inertia of the assembly about the $z$ axis is

$$I_{\text{assembly}} = 0.00148 + 0.00219 = 0.00367 \text{ lb} \cdot \text{s}^2 \cdot \text{in}$$

The area of the lever is found from

$$A = A_1 + A_2 + A_3 + A_4 + A_5 = \frac{\pi(0.25)^2}{2} + (0.5)2.4 + \frac{\pi(0.5)^2}{4} + 1.5(0.5) + (0.5)1 = 2.74 \text{ in}^2$$

The mass of the lever is

$$m = \rho t A = \left[\frac{489}{386(12)^3}\right]0.125(2.74) = 2.51 \times 10^{-4} \text{ lb} \cdot \text{s}^2/\text{in}$$

The total mass of the lever and plugs is

$$m = 2.51 \times 10^{-4} + 2(7.86 \times 10^{-5}) = 4.08 \times 10^{-4} \text{ lb} \cdot \text{s}^2/\text{in}$$

The weight of the lever assembly is

$$W = mg = 4.08 \times 10^{-4}(386) = 0.157 \text{ lb} = 2.51 \text{ oz}$$

(c)  The distribution of the masses of the lever and weights is shown in Table 17.10.  Table 17.11 shows the distribution of the mass moments of inertia.  It is interesting to note that the brass plugs contribute 60 percent of the mass moment of inertia, yet account for only 38 percent of the weight of the assembly.

<table>
<tr><td colspan="3">TABLE 17.10</td></tr>
<tr><td></td><td>$m$</td><td>%</td></tr>
<tr><td>Lever</td><td>$2.51 \times 10^{-4}$</td><td>62</td></tr>
<tr><td>Weights</td><td>$1.57 \times 10^{-4}$</td><td>38</td></tr>
<tr><td>Assembly</td><td>$4.08 \times 10^{-4}$</td><td>100</td></tr>
</table>

<table>
<tr><td colspan="3">TABLE 17.11</td></tr>
<tr><td></td><td>$I$</td><td>%</td></tr>
<tr><td>Lever</td><td>0.00148</td><td>40</td></tr>
<tr><td>Weights</td><td>0.00219</td><td>60</td></tr>
<tr><td>Assembly</td><td>0.00367</td><td>100</td></tr>
</table>

## 17.5  CENTER OF MASS, AND MASS MOMENT OF INERTIA, OF PLANE BODIES FORMED OF THIN ROD SHAPES, MASS PRODUCT OF INERTIA

17.86   Figure 17.86 shows a rigid body of plane form made of thin, homogeneous rod material with constant cross-sectional dimensions.  The body lies in the $xy$ plane, and the total length of the body, measured along the curvilinear axis, is $l$.

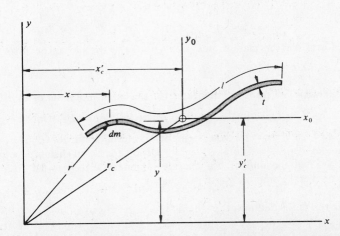

Fig. 17.86

(a) How is a determination made of whether the rod is thin?

(b) Find the coordinates of the center of mass of the plane rod shape.

❚ (a) The rod is considered to be thin if $t \ll l$.

(b) The centroidal coordinates of the center of mass of the body are defined by

$$x'_c = \frac{\int_l x\, dm}{\int_l dm} = \frac{\int_l x\, dm}{m} \qquad y'_c = \frac{\int_l y\, dm}{\int_l dm} = \frac{\int_l y\, dm}{m}$$

The mass density per unit length of the rod material is designated $\rho_0$, and the above equations may be written as

$$x'_c = \frac{\int_l x\rho_0\, dl}{\int_l \rho_0\, dl} \qquad y'_c = \frac{\int_l y\rho_0\, dl}{\int_l \rho_0\, dl}$$

Since the rod material is homogeneous,

$$x'_c = \frac{\rho_0 \int_l x\, dl}{\rho_0 \int_l dl} = \frac{\int_l x\, dl}{l} \qquad y'_c = \frac{\rho_0 \int_l y\, dl}{\rho_0 \int_l dl} = \frac{\int_l y\, dl}{l}$$

The above equations define the centroidal coordinates of a plane curve. Thus, the center of mass of the rigid plane body of thin rod shape is located at the centroid of the plane curve which defines its shape.

**17.87** (a) Find the mass moments of inertia about the $x$, $y$, and $z$ axes of the plane rod shape in Prob. 17.86.

(b) Give the form of the parallel-axis, or transfer, theorem for the plane rod shape.

❚ (a) The mass moments of inertia of the body are defined by

$$I_x = \int_l y^2\, dm \qquad I_y = \int_l x^2\, dm \qquad I_z = \int_l r^2\, dm = \int_l (x^2 + y^2)\, dm = I_x + I_y$$

Using $dm = \rho_0\, dl$, we may express the mass moments of inertia by

$$I_x = \int_l y^2\rho_0\, dl = \rho_0 \int_l y^2\, dl \qquad I_y = \int_l x^2\rho_0\, dl = \rho_0 \int_l x^2\, dl$$

$$I_z = \int_l (x^2 + y^2)\rho_0\, dl = \rho_0 \int_l (x^2 + y^2)\, dl$$

The integral terms on the right sides of the above equations may be recognized as the length moments of inertia of the plane curve about the $x$ and $y$ axes. With the notations for the moments of inertia of a plane curve, given by

$$I_{xl} = \int_l y^2\, dl \qquad I_{yl} = \int_l x^2\, dl \qquad I_{zl} = \int_l (x^2 + y^2)\, dl$$

the final forms for the mass moments of inertia of the plane rod form are

$$I_x = \rho_0 I_{xl} \qquad I_y = \rho_0 I_{yl} \qquad I_z = \rho_0 I_{zl}$$

(b) The parallel-axis, or transfer, theorems for the mass moments of inertia have the forms

$$I_x = I_{0x} + my'^2_c \qquad I_y = I_{0y} + mx'^2_c \qquad I_z = I_{0z} + m(x'^2_c + y'^2_c) = I_{0z} + mr^2_c$$

$I_{0x}$, $I_{0y}$, and $I_{0z}$ are the centroidal moments of inertia about the $x_0y_0z_0$ axes shown in Fig. 17.86, $x'_c$ and $y'_c$ are the centroidal coordinates, and $r_c$ is the separation distance between the $z$ and $z_0$ axes.

Table 17.15 at the end of this chapter contains the centroidal coordinates and mass moments of inertia for several common plane rod shapes.

**17.88** A 16-mm-diameter rod is formed into the stirring blade shape shown in Fig. 17.88a. When connected to a foot at the end of a shaft, the blade can be rotated about either the $y$ axis or the $y_1$ axis.

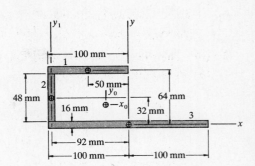

Fig. 17.88a                              Fig. 17.88b

(a)  Find the location of the centroid of the blade shape, and the mass of the blade.

(b)  Find the mass moment of inertia of the blade about the $y$ axis.

(c)  Find the mass moment of inertia of the blade about the $y_1$ axis.

The rod material is an alloy with a density of 6,760 kg/m³.

▌ (a)  The blade is divided into the three straight-line elements shown in Fig. 17.88b. This particular subdivision places the center of mass of element 3 on the reference $y$ axis, which will simplify the subsequent computations. The $x$ axis is placed along element 3. The centroidal coordinates are

$$x'_c = \frac{x_1 l_1 + x_2 l_2 + x_3 l_3}{l_1 + l_2 + l_3} = \frac{-50(100) - 92(48) + 0(200)}{100 + 48 + 200} = -27.1 \text{ mm} \qquad l = 348 \text{ mm}$$

$$y'_c = \frac{y_1 l_1 + y_2 l_2 + y_3 l_3}{l_1 + l_2 + l_3} = \frac{64(100) + 32(48) + 0(200)}{348} = 22.8 \text{ mm}$$

For a rod of constant cross-sectional area $A$ and length $l$, the mass density $\rho$ and mass density per unit length $\rho_0$ are related by

$$\rho A l = \rho_0 l \qquad \rho_0 = \rho A$$

For the present problem

$$\rho_0 = 6{,}760 \text{ kg/m}^3 \ \frac{\pi(16)^2}{4} \text{ mm}^2 \left(\frac{1 \text{ m}}{1{,}000 \text{ mm}}\right)^2 = 1.36 \text{ kg/m}$$

The masses of the three length elements are

$$m_1 = 1.36 \text{ kg/m} \left(\frac{100}{1{,}000}\right)\text{m} = 0.136 \text{ kg} \qquad m_2 = 1.36 \left(\frac{48}{1{,}000}\right) = 0.0653 \text{ kg}$$

$$m_3 = 1.36 \left(\frac{200}{1{,}000}\right) = 0.272 \text{ kg}$$

The total mass of the blade is

$$m = m_1 + m_2 + m_3 = 0.136 + 0.0653 + 0.272 = 0.473 \text{ kg}$$

(b)  The equations for the mass moment of inertia of a straight rod element are contained in Case 1 in Table 17.15. The mass moment of inertia about the $y$ axis is

$$I_y = I_{y1} + I_{y2} + I_{y3} = \tfrac{1}{3}m_1 l_1^2 + m_2 y_2^2 + \tfrac{1}{12}m_3 l_3^2$$

$$= \tfrac{1}{3}(0.136)\left(\frac{100}{1{,}000}\right)^2 + 0.0653\left(\frac{92}{1{,}000}\right)^2 + \tfrac{1}{12}(0.272)\left(\frac{200}{1{,}000}\right)^2 = 0.00191 \text{ kg} \cdot \text{m}^2$$

(c)  To find the mass moment of inertia about the $y_1$ axis, the centroidal mass moment of inertia $I_{0y}$ will first be found. Using the parallel-axis theorem,

$$I_y = I_{0y} + mx'^2_c \qquad I_{0y} = I_y - mx'^2_c = 0.00191 - 0.473\left(\frac{-27.1}{1{,}000}\right)^2 = 0.00156 \text{ kg} \cdot \text{m}^2$$

The separation distance between the centroidal $y_0$ axis and the $y_1$ axis is

$$d_{y_1} = 92 - 27.1 = 64.9 \text{ mm}$$

The mass moment of inertia about the $y_1$ axis, using the parallel-axis theorem, is

$$I_{y_1} = I_{0y} + md_{y_1}^2 = 0.00156 + 0.473\left(\frac{64.9}{1,000}\right)^2 = 0.00355 \text{ kg} \cdot \text{m}^2$$

A very common mistake made in problems of this type is to attempt to transfer the moment of inertia about axis $y$ directly to axis $y_1$. This *incorrect* application of the parallel-axis theorem would have the typical form

$$I_{y_1} \neq I_y + md_{y,y_1}^2 \qquad d_{y,y_1} = 92 \text{ mm}$$

The reader is urged to carefully study the above equation and to understand thoroughly why it is *not* a correct use of the parallel-axis theorem.

**17.89**  A stirring blade is formed of 0.75-in-diameter steel wire into the shape shown in Fig. 17.89a. The specific weight of steel is 489 lb/ft$^3$.

(*a*)  Find the mass, and the mass moment of inertia, about the $y$ axis.

(*b*)  Find the mass moment of inertia about the $y_1$ axis.

(*c*)  Find the mass moment of inertia about the vertical centroidal axis of the wire form.

Base all of the calculations on the length of the centerline of the blade shape.

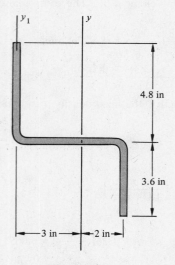

**Fig. 17.89a**

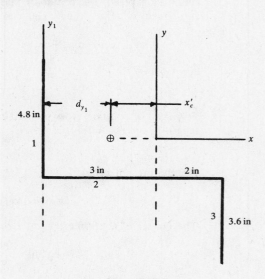

**Fig. 17.89b**

❚  (*a*)  Figure 17.89b shows the centerline of the blade shape, and this shape is divided into the three length elements shown. The mass per unit length is

$$\rho_0 = \rho A = \frac{489 \text{ lb/ft}^3}{386 \text{ in/s}^2}\left(\frac{1 \text{ ft}}{12 \text{ in}}\right)^3 \frac{\pi(0.75)^2}{4} \text{ in}^2 = 3.24 \times 10^{-4} \text{ lb} \cdot \text{s}^2/\text{in}^2$$

Using  $l_1 = 4.8 \text{ in}$,  $l_2 = 5 \text{ in}$,  and  $l_3 = 3.6 \text{ in}$,  the masses of the length elements are

$$m_1 = \rho_0(4.8) = 0.00156 \text{ lb} \cdot \text{s}^2/\text{in} \qquad m_2 = \rho_0(5) = 0.00162 \text{ lb} \cdot \text{s}^2/\text{in}$$
$$m_3 = \rho_0(3.6) = 0.00117 \text{ lb} \cdot \text{s}^2/\text{in}$$

The mass of the blade is

$$m = m_1 + m_2 + m_3 = 0.00435 \text{ lb} \cdot \text{s}^2/\text{in}$$

The mass moment of inertia of the blade about the $y$ axis has the form

$$I_y = I_{y1} + I_{y2} + I_{y3}$$

Using Case 1 of Table 17.15,

$$I_y = (0.00156)3^2 + [\tfrac{1}{12}(0.00162)5^2 + 0.00162(0.5)^2] + 0.00117(2)^2 = 0.0225 \text{ lb} \cdot \text{s}^2 \cdot \text{in}$$

(b) The moment of inertia of the blade about the $y_1$ axis has the form

$$I_{y_1} = I_{y1,1} + I_{y1,2} + I_{y1,3} = 0 + \tfrac{1}{3}(0.00162)5^2 + 0.00117(5)^2 = 0.0428 \text{ lb} \cdot \text{s}^2 \cdot \text{in}$$

(c) The centroidal coordinate $x'_c$ is given by

$$x'_c = \frac{x_1 l_1 + x_2 l_2 + x_3 l_3}{l_1 + l_2 + l_3} = \frac{-3(4.8) - 0.5(5) + 2(3.6)}{4.8 + 5 + 3.6} = -0.724 \text{ in}$$

The length $d_{y_1}$ in Fig. 17.89b is

$$d_{y_1} = 3 - 0.724 = 2.28 \text{ in}$$

Using the parallel-axis theorem,

$$I_{0y} = I_{y_1} - m d_{y_1}^2 = 0.0428 - 0.00435(2.28)^2 = 0.0202 \text{ lb} \cdot \text{s}^2 \cdot \text{in}$$

As a check on the above calculation,

$$I_{0y} = I_y - m x_c'^2 \overset{?}{=} 0.0225 - 0.00435(-0.724)^2 = 0.0202 \text{ lb} \cdot \text{s}^2 \cdot \text{in}$$

**17.90** Do the same as in Prob. 17.89, for the shape shown in Fig. 17.90a.

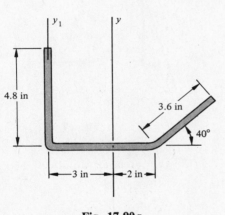

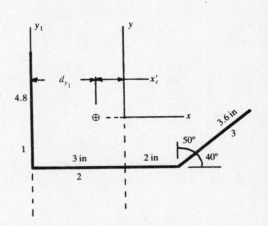

**Fig. 17.90a**          **Fig. 17.90b**

**❚** (a) Figure 17.90b shows the division of the blade into three elements. From Prob. 17.89,

$$m_1 = 0.00156 \text{ lb} \cdot \text{s}^2/\text{in} \qquad m_2 = 0.00162 \text{ lb} \cdot \text{s}^2/\text{in} \qquad m_3 = 0.00117 \text{ lb} \cdot \text{s}^2/\text{in}$$
$$m = 0.00435 \text{ lb} \cdot \text{s}^2/\text{in}$$

The mass moment of inertia of the blade about the $y$ axis is

$$I_y = I_{y1} + I_{y2} + I_{y3}$$

Using Case 2 of Table 17.15,

$$I_{y3} = m(\tfrac{1}{3} l^2 \sin^2 \theta + al \sin \theta + a^2) \qquad\qquad (1)$$

where

$$l = 3.6 \text{ in} \qquad a = 2 \text{ in} \qquad \theta = 50°$$

$I_y$ then is found from

$$I_y = 0.00156(3)^2 + [\tfrac{1}{12}(0.00162)5^2 + 0.00162(0.5)^2] + 0.00117[\tfrac{1}{3}(3.6)^2 \sin^2 50°$$
$$+ 2(3.6) \sin 50° + 2^2] = 0.0319 \text{ lb} \cdot \text{s}^2 \cdot \text{in}$$

(b) The moment of inertia about the $y_1$ axis may be expressed as

$$I_{y_1} = I_{y1,1} + I_{y1,2} + I_{y1,3}$$

Using Eq. (1) in Prob. 17.90 for $I_{y1,3}$, with $l = 3.6$ in, $a = 5$ in, and $\theta = 50°$,

$$I_{y1} = 0 + \tfrac{1}{3}(0.00162)5^2 + 0.00117[\tfrac{1}{3}(3.6)^2 \sin^2 50° + 5(3.6) \sin 50° + 5^2] = 0.0618 \text{ lb} \cdot \text{s}^2 \cdot \text{in}$$

(c) The centroidal coordinate $x_c'$ is given by

$$x_c' = \frac{x_1 l_1 + x_2 l_2 + x_3 l_3}{l_1 + l_2 + l_3} = \frac{-3(4.8) - 0.5(5) + (2 + \frac{3.6}{2} \cos 40°)3.6}{4.8 + 5 + 3.6} = -0.353 \text{ in}$$

From Fig. 17.90b,

$$d_{y1} = 3 - 0.353 = 2.65 \text{ in}$$

Using the parallel-axis theorem,

$$I_{0y} = I_{y1} - md_{y1}^2 = 0.0618 - 0.00435(2.65)^2 = 0.0313 \text{ lb} \cdot \text{s}^2 \cdot \text{in}$$

As a check on the result found above,

$$I_{0y} = I_y - mx_c'^2 \overset{?}{=} 0.0319 - 0.00435(-0.353)^2 = 0.0314 \approx 0.0313 \text{ lb} \cdot \text{s}^2 \cdot \text{in}$$

**17.91** Do the same as in Prob. 17.89, for the shape shown in Fig. 17.91a.

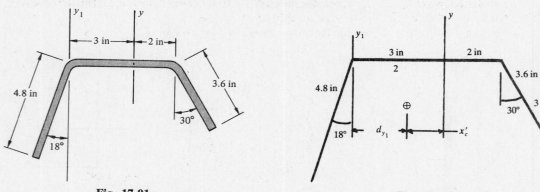

Fig. 17.91a          Fig. 17.91b

▌ (a) The three elements of the blade are shown in Fig. 17.91b. From Prob. 17.89,

$$m_1 = 0.00156 \text{ lb} \cdot \text{s}^2/\text{in} \quad m_2 = 0.00162 \text{ lb} \cdot \text{s}^2/\text{in} \quad m_3 = 0.00117 \text{ lb} \cdot \text{s}^2/\text{in} \quad m = 0.00435 \text{ lb} \cdot \text{s}^2/\text{in}$$

The mass moment of inertia $I_y$ is given by

$$I_y = I_{y1} + I_{y2} + I_{y3}$$

Using Case 2 of Table 17.15,

$$I_y = m(\tfrac{1}{3}l^2 \sin^2 \theta + al \sin \theta + a^2) \tag{1}$$

For element 1,

$$a = 3 \text{ in} \quad l = 4.8 \text{ in} \quad \theta = 18°$$

For element 3,

$$a = 2 \text{ in} \quad l = 3.6 \text{ in} \quad \theta = 30°$$

$I_y$ is then found as

$$I_y = 0.00156[\tfrac{1}{3}(4.8)^2 \sin^2 18° + 3(4.8) \sin 18° + 3^2] + [\tfrac{1}{12}(0.00162)5^2 + 0.00162(0.5)^2]$$
$$+ 0.00117[\tfrac{1}{3}(3.6)^2 \sin^2 30° + 2(3.6) \sin 30° + 2^2] = 0.0361 \text{ lb} \cdot \text{s}^2 \cdot \text{in}$$

(b) $I_{y1}$ is expressed as

$$I_{y1} = I_{y1,1} + I_{y1,2} + I_{y1,3}$$

Using Eq. (1), with

$$a = 0 \quad l = 4.8 \text{ in} \quad \theta = 18° \quad \text{(Element 1)}$$
$$a = 5 \text{ in} \quad l = 3.6 \text{ in} \quad \theta = 30° \quad \text{(Element 3)}$$

$I_{y_1}$ appears as

$$I_{y_1} = 0.00156[\tfrac{1}{3}(4.8)^2 \sin^2 18°] + \tfrac{1}{3}(0.00162)5^2 + 0.00117[\tfrac{1}{3}(3.6)^2 \sin^2 30°$$
$$+ 5(3.6)\sin 30° + 5^2] = 0.0557 \text{ lb} \cdot \text{s}^2 \cdot \text{in}$$

(c)    The centroidal coordinate $x_c'$ is found from

$$x_c' = \frac{x_1 l_1 + x_2 l_2 + x_3 l_3}{l_1 + l_2 + l_3} = \frac{-(3 + \tfrac{4.8}{2}\sin 18°)4.8 - 0.5(5) + (2 + \tfrac{3.6}{2}\sin 30°)3.6}{4.8 + 5 + 3.6} = -0.748 \text{ in}$$

Using Fig. 17.91b,

$$d_{y_1} = 3 - 0.748 = 2.25 \text{ in}$$

The moment of inertia about the centroidal y axis is then

$$I_{0y} = I_{y_1} - m d_{y_1}^2 = 0.0557 - 0.00435(2.25)^2 = 0.0337 \text{ lb} \cdot \text{s}^2 \cdot \text{in}$$

As a check on the above value,

$$I_{0y} = I_y - m x_c'^2 \overset{?}{=} 0.0361 - 0.00435(-0.748)^2 = 0.0337 \text{ lb} \cdot \text{s}^2 \cdot \text{in}$$

**17.92**    The circular hoop in Fig. 17.92 is formed of thin brass rod of 6.5 mm diameter.    The mean diameter of the rod is 100 mm, and the density of the brass is 8,550 kg/m$^3$.

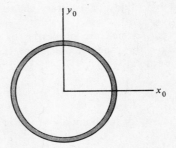

**Fig. 17.92**

(a)    Find the mass moment of inertia of the hoop about the $z_0$ axis.

(b)    Find the diameter of a homogeneous solid brass disk of 6.5 mm thickness which would have the same mass moment of inertia about the $z_0$ axis.

**I** (a)    From Case 4 of Table 17.15,

$$m = \rho_0 \pi d = \rho A \pi d$$

Using a rod diameter of 6.5 mm, with    $d = 100$ mm,

$$m = 8,550\left(\frac{\pi 6.5^2}{4}\right)\pi(100)\left(\frac{1}{1,000}\right)^3 = 0.0891 \text{ kg}$$

Using Case 4,

$$I_{0z} = \tfrac{1}{4}md^2 = \tfrac{1}{4}(0.0891)100^2\left(\frac{1}{1,000}\right)^2 = 2.23 \times 10^{-4} \text{ kg} \cdot \text{m}^2$$

(b)    Using Case 4 in Table 17.13,

$$m = \tfrac{1}{4}\rho \pi d^2 h$$

$$I_{0z} = \tfrac{1}{8}md^2 = \tfrac{1}{8}(\tfrac{1}{4}\rho\pi d^2 h)d^2 = \tfrac{1}{32}\rho\pi d^4 h = \tfrac{1}{32}(8,550)\pi d^4(6.5)\left(\frac{1}{1,000}\right)^5 = 2.23 \times 10^{-4}$$

$$d = 80.0 \text{ mm}$$

**17.93**    (a)    Give the expressions for the mass *products of inertia* of a rigid body.

(b)    Compare the mass moments, and mass products, of inertia.

(c)    What is meant by the term *principal axes*?

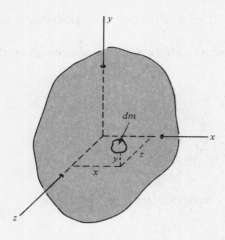

Fig. 17.93*a*

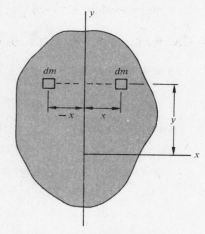

Fig. 17.93*b*

(*d*)  What is the physical interpretation of the mass products of inertia?

(*e*)  Give an important example of the use of mass products of inertia.

▌ (*a*)  Figure 17.93*a* shows a rigid body of arbitrary shape.  A set of *xyz* coordinates is imagined to be attached to the body.

Three mass-related terms which are properties of the mass distribution of the body and of the orientation of the body with respect to a set of coordinate axes are referred to as products of inertia.  These terms are defined by

$$I_{xy} = \int_V xy\, dm \qquad I_{yz} = \int_V yz\, dm \qquad I_{zx} = \int_V zx\, dm$$

(*b*)  The mass moments of inertia of the body about the *xyz* axes were presented earlier in Prob. 17.19 as

$$I_x = \int_V (y^2 + z^2)\, dm \qquad I_y = \int_V (z^2 + x^2)\, dm \qquad I_z = \int_V (x^2 + y^2)\, dm$$

where *V* is the volume of the body.

The three terms $I_x$, $I_y$, and $I_z$ were shown to be combined properties of the mass distribution of the rigid body and of the *position* of the *xyz* coordinate axes with respect to the body.  The additional physical interpretation was made that the magnitudes of the mass moments of inertia were measures of the resistance to angular acceleration of the body about the respective axes.  Finally, it was observed that the *mass moments of inertia are always positive qualities*.

The mass products of inertia are also combined properties of the mass distribution of the body, and of the position of the *xyz* coordinates with respect to the body.  It may be observed from the definitions of the products of inertia, however, that, unlike the mass moments of inertia, these terms may have *negative* as well as positive values.

(*c*)  By a judicious placement of the axes with respect to the body, one or more of the products of inertia may be made to vanish.  Figure 17.93*b* shows a body referenced to the coordinate axes in such a way that the *yz* plane is a plane of symmetry of the body.  For every mass element *dm* in the positive *x* region, there is a mirror-image mass element with the same *y* coordinate and an opposite-sense *x* coordinate.  When all these elements are summed over the volume of the body, in accordance with the expression in part (*a*) for $I_{xy}$, the result is

$$I_{xy} = 0$$

It is always possible to choose an orientation of the *xyz* axes with respect to the body in such a way that all three products of inertia are identically zero.  If the products of inertia with respect to a set of coordinate axes are zero, these axes are referred to as principal axes.

(*d*)  The physical interpretation of products of inertia is that *these quantities are a measure of the symmetry of placement of the xyz axes on the body*.

(*e*)  The mass products of inertia are used to find the dynamic forces produced when an unbalanced rigid body of arbitrary shape rotates about a fixed axis.  These types of problems are considered in Chapter 22.

**17.94** *(a)* Give the general forms of the parallel-axis, or transfer, theorems for mass products of inertia of a rigid body.

    *(b)* Give the forms of the parallel-axis theorems for mass products of inertia of a thin plane rigid body.

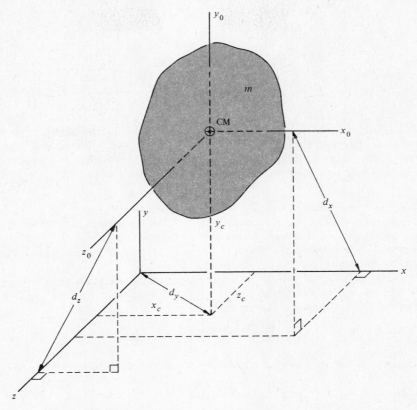

**Fig. 17.94a**

$\blacksquare$ *(a)* In Fig. 17.94a the rigid body is positioned with respect to two systems of *parallel* coordinate axes. The $x_0 y_0 z_0$ axes pass through the *center of mass* of the body, and $d_x$, $d_y$, $d_z$ are the perpendicular separation distances between the two sets of coordinate axes. The center of mass of the body is located by $x_c$, $y_c$, and $z_c$. The centroidal coordinates are defined *relative to the xyz coordinate system*. Thus, they may have negative as well as positive values.

    The products of inertia about the centroidal axes are assumed to be known, and it is desired to find the corresponding values of these quantities about the *xyz* axes. It was shown in Prob. 17.28 that the *parallel-axis*, or *transfer*, *theorems* for mass moments of inertia have the forms

$$I_x = I_{0x} + md_x^2 \qquad I_y = I_{0y} + md_y^2 \qquad I_z = I_{0z} + md_z^2$$

where $I_{0x}$, $I_{0y}$, and $I_{0z}$ are the mass moments of inertia about the centroidal axes, and $I_x$, $I_y$, and $I_z$ are the mass moments of inertia about the $x$, $y$, and $z$ axes. It can be shown that there exist similar transfer theorems for the products of inertia, with the forms

$$I_{xy} = I_{0,xy} + mx_c y_c \qquad I_{yz} = I_{0,yz} + my_c z_c \qquad I_{zx} = I_{0,zx} + mz_c x_c$$

$I_{0,xy}$, $I_{0,yz}$, and $I_{0,zx}$ are the products of inertia about the centroidal axes, and $I_{xy}$, $I_{yz}$, and $I_{zx}$ are the products of inertia about the parallel *xyz* axes.

    *(b)* The following problems will be limited to consideration of *slender*, or *thin*, *plane bodies*. The term slender, or thin, implies that the lateral, or thickness, dimension is much less than any other length dimensions of the body. The bodies will be assumed to lie in the *xy* plane, so that the *z* coordinate of any point on the thickness of the body satisfies the relationship

$$z \approx 0$$

With the above assumption the three products of inertia given in Prob. 17.93 have the forms

$$I_{xy} = \int_V xy \, dm \qquad I_{yz} = I_{zx} = 0$$

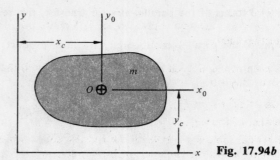

**Fig. 17.94*b***

The reference axes $xy$, which are *parallel* to the centroidal coordinates $x_0 y_0$, will be assumed to lie in the *same* plane as these latter coordinates. Thus, $z_c = 0$. The general appearance of the slender, or thin, plane body and its orientation with respect to the coordinate axes is shown in Fig. 17.94*b*.

The parallel-axis theorem for the product of inertia $I_{xy}$ of a thin plane body has the form

$$I_{xy} = I_{0,xy} + mx_c y_c \quad \text{and} \quad I_{yz} = I_{zx} = 0$$

**17.95** Figure 17.95 shows a shaft with two masses attached. The arms which support the masses may be assumed to be massless, and the two masses may be assumed to be point masses.

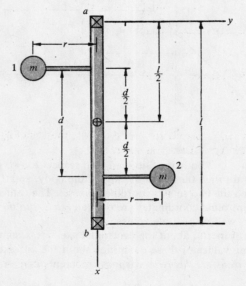

**Fig. 17.95**

(*a*) Find the product of inertia $I_{xy}$ of the shaft assembly.

(*b*) For what conditions would the result in part (*a*) be zero?

▌ (*a*) The product of inertia $I_{xy}$ of the shaft is zero, because of the symmetry of placement of the $xy$ axes on this element. Since the masses are assumed to be point masses,

$$I_{xy} = \int_V xy \, dm = x_1 y_1 m + x_2 y_2 m$$

Using

$$x_1 = \frac{l}{2} - \frac{d}{2} \qquad x_2 = \frac{l}{2} + \frac{d}{2} \qquad y_1 = -r \qquad y_2 = r$$

we find the product of inertia $I_{xy}$ to be

$$I_{xy} = \left(\frac{l}{2} - \frac{d}{2}\right)(-r)m + \left(\frac{l}{2} + \frac{d}{2}\right)rm = mrd$$

It may be observed that the above result is independent of the length $l$ of the shaft.

(b)  $I_{xy}$ will be zero if  $r = 0$  or  $d = 0$. If  $r = 0$,  the two masses lie on the centerline of the shaft. If  $d = 0$,  the two masses lie on a line, through the center of mass of the shaft, which is normal to the $x$ axis.

17.96    The slender rod shown in Fig. 17.96a lies in the $x_0 y_0$ plane, with the direction $\beta$.  The $x_0 y_0$ axes pass through the center of mass of the rod.

(a)    Find the product of inertia $I_{0,xy}$.

(b)    Find the product of inertia $I_{xy}$.

(c)    Find the numerical value for part (a) if the rod is made of white oak with a specific gravity of 0.77.  The rod is 1 m in length, with a diameter of 25 mm, and  $\beta = 20°$.

(d)    Do the same as in part (c), if  $\beta = 110°$.

(e)    For what values of $\beta$ will $I_{0,xy}$ be negative?

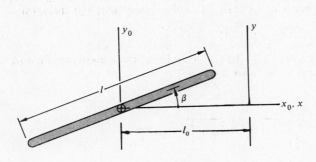

Fig. 17.96a

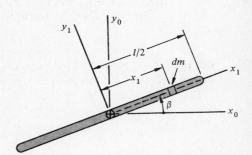

Fig. 17.96b

▌  (a)    A set of $x_1 y_1$ axes is placed on the rod, as shown in Fig. 17.96b.  Since the rod is slender, the coordinates of the typical mass element are

$$x_0 = x_1 \cos \beta \qquad y_0 = x_1 \sin \beta$$

The mass per unit length of the rod is $\rho_0$, so that

$$dm = \rho_0 \, dx_1$$

The product of inertia $I_{0,xy}$ of the rod is then

$$I_{0,xy} = \int_V x_0 y_0 \, dm = \int_{-l/2}^{l/2} (x_1 \cos \beta)(x_1 \sin \beta)\rho_0 \, dx_1 = \rho_0 \sin \beta \cos \beta \int_{-l/2}^{l/2} x_1^2 \, dx_1$$

$$= \frac{1}{2} \rho_0 \sin 2\beta \left( \frac{x_1^3}{3} \right)\Big|_{-l/2}^{l/2} = \frac{1}{2} \rho_0 \frac{\sin 2\beta}{3} \left[ \left( \frac{l}{2} \right)^3 - \left( -\frac{l}{2} \right)^3 \right]$$

$$= \frac{1}{24} \rho_0 l^3 \sin 2\beta$$

Using  $m = \rho_0 l$,

$$I_{0,xy} = \tfrac{1}{24} ml^2 \sin 2\beta$$

It may be observed that, for  $\theta = 0$ or 90°,  $I_{0,xy} = 0$.  This is consistent with the fact that, for these orientations, the $x_0$ and $y_0$ axes are axes of symmetry of the rod.

(b)    The product of inertia $I_{xy}$ is written as

$$I_{xy} = I_{0,xy} + m x_c y_c$$

Since  $y_c = 0$,  it follows that

$$I_{xy} = I_{0,xy} = \tfrac{1}{24} ml^2 \sin 2\beta$$

It may be observed that the dimension $l_0$ in Fig. 17.96a does not enter into the problem.

(c)   The density of the white oak rod, using $1,000 \text{ kg/m}^3$ for standard water as the reference, is

$$\rho = 0.77(1,000) = 770 \text{ kg/m}^3$$

The mass of the rod is

$$m = \rho \, \frac{\pi d^2}{4} \, l = 770\pi \, \frac{(25/1,000)^2}{4} \, (1) = 0.378 \text{ kg}$$

The centroidal product of inertia is then

$$I_{0,xy} = \tfrac{1}{24}ml^2 \sin 2\beta = \tfrac{1}{24}(0.378)(1)^2 \sin [2(20°)] = 0.0101 \text{ kg} \cdot \text{m}^2$$

(d)   For $\beta = 110°$,

$$I_{0,xy} = \tfrac{1}{24}ml^2 \sin 2\beta = \tfrac{1}{24}(0.378)(1)^2 \sin [2(110°)] = -0.0101 \text{ kg} \cdot \text{m}^2$$

The magnitudes of the results in parts (c) and (d) are the same, but the product of inertia, for the condition of part (d), is negative.

(e)   The product of inertia $I_{0,xy}$ has the form

$$I_{0,xy} = \tfrac{1}{24}ml^2 \sin 2\beta$$

$I_{0,xy}$ will be negative only when the term $\sin 2\beta$ is negative.   This occurs when

$$\pi < 2\beta < 2\pi \qquad \text{or} \qquad \frac{\pi}{2} < \beta < \pi$$

$$\pi = 180° < 2\beta < 2\pi = 360° \qquad \text{or} \qquad \frac{\pi}{2} = 90° < \beta < \pi = 180°$$

Figure 17.96c shows the typical position of the rod for the above range of $\beta$.

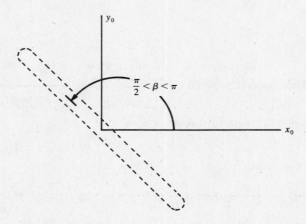

**Fig. 17.96c**

**17.97**   Figure 17.97a shows a stirring blade which is used in a paint manufacturing operation.   The blade is formed of $\tfrac{3}{4}$-in-diameter steel rod, and $\gamma = 0.283 \text{ lb/in}^3$.   Find the product of inertia $I_{xy}$ of the blade.

❚   The rod lengths which comprise the stirrer blade are assumed to be thin.   The required term $I_{xy}$ will be found by considering the stirrer blade to be a *composite plane curve*.   The blade is divided into the three straight elements shown in Fig. 17.97b.   Centroidal axes, which are parallel to the xy axes, are placed on each of the elements.   Since these axes are principal axes, all centroidal products of inertia are identically zero.   The term $I_{xy}$ then has the form

$$I_{xy} = m_1 x_1 y_1 + m_2 x_2 y_2 + m_3 x_3 y_3$$

The mass density per unit length of the material, using $\gamma = 0.283 \text{ lb/in}^3$, is

$$\rho_0 = \rho A = \frac{0.283}{386} \, \frac{\pi(0.75)^2}{4} = 3.24 \times 10^{-4} \, \frac{\text{lb} \cdot \text{s}^2}{\text{in}^2}$$

The computations are arranged in the tabular form shown in Table 17.12, and $I_{xy} = -0.175 \text{ lb} \cdot \text{s}^2 \cdot \text{in}$.

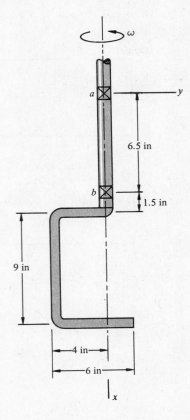

Fig. 17.97a

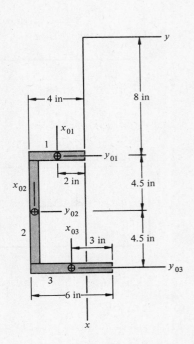

Fig. 17.97b

TABLE 17.12

| Element | $x_i$ | $y_i$ | $m_i = \rho_0 l_i$ | $m_i x_i y_i$ |
|---------|-------|-------|--------------------|---------------|
| 1 | 8 | $-2$ | $3.24 \times 10^{-4}(4) = 0.00130$ | $-0.0208$ |
| 2 | 12.5 | $-\left(4 - \dfrac{0.75}{2}\right) = -3.63$ | $3.24 \times 10^{-4}(8.25) = 0.00267$ | $-0.121$ |
| 3 | 17 | $-1$ | $3.24 \times 10^{-4}(6) = 0.00194$ | $-0.0330$ |
| Totals | | | $m = 0.00591 \, \dfrac{\text{lb} \cdot \text{s}^2}{\text{in}}$ | $I_{xy} = -0.175$ |

17.98 (a) Find the mass product of inertia $I_{xy}$ of the thin plate of mass 10 kg shown in Fig. 17.98a. Do the same as in part (a) for the position of the plate shown in

(b) Fig. 17.98b

(c) Fig. 17.98c

(d) Fig. 17.98d

▌ (a) Since $x_0$ and $y_0$ are axes of symmetry of the plate,

$$I_{0,xy} = 0$$

The mass product of inertia $I_{xy}$, with $x_c = 0.25 \, \text{m}$ and $y_c = 0.14 \, \text{m}$, is

$$I_{xy} = m x_c y_c = 10(0.25)(0.14) = 0.35 \, \text{kg} \cdot \text{m}^2$$

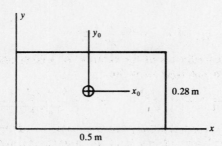

Fig. 17.98a

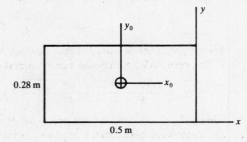

Fig. 17.98b

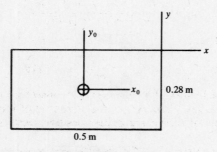

Fig. 17.98c

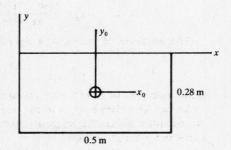

Fig. 17.98d

(b)   Using   $x_c = -0.25$ m   and   $y_c = 0.14$ m,

$$I_{xy} = mx_cy_c = 10(-0.25)(0.14) = -0.35 \text{ kg} \cdot \text{m}^2$$

(c)   Using   $x_c = -0.25$ m   and   $y_c = -0.14$ m,

$$I_{xy} = mx_cy_c = 10(-0.25)(-0.14) = 0.35 \text{ kg} \cdot \text{m}^2$$

(d)   Using   $x_c = 0.25$ m   and   $y_c = -0.14$ m,

$$I_{xy} = mx_cy_c = 10(0.25)(-0.14) = -0.35 \text{ kg} \cdot \text{m}^2$$

**17.99**   Give a summary of the basis concepts of centroids, and mass moments and products of inertia, of rigid bodies.

**▌** The centroid of a volume is a point whose location is a function only of the shape of the volume. The centroidal coordinates of a volume are

$$x_c = \frac{\int_v x\, dV}{V} \qquad y_c = \frac{\int_v y\, dV}{V} \qquad z_c = \frac{\int_v z\, dV}{V}$$

If a volume has a plane of symmetry, the centroid must lie in this plane. If there are two planes of symmetry, the centroid must lie along the intersection line of the planes. If there are three planes of symmetry, the centroid is the common intersection point of the three planes.

A rigid body is one within which any two points always have the same separation distance between them. Thus, the shape of a rigid body never changes. The center of mass (CM) of a rigid body is the point through which the weight force acts. If the material of the body is homogeneous, the center of mass and the centroid of the volume of the body are coincident points. If the material of the body is not homogeneous, the centroidal coordinates of the CM are given by

$$x_c = \frac{\int_v x\gamma\, dV}{\int_v \gamma\, dV} \qquad y_c = \frac{\int_v y\gamma\, dV}{\int_v \gamma\, dV} \qquad z_c = \frac{\int_v z\gamma\, dV}{\int_v \gamma\, dV}$$

where $\gamma$, the specific weight, is a known function of the coordinates. The above equations are also valid if $\rho$, the mass density, is substituted for $\gamma$.

The first moments of the volume are

$$x_c V = \int_V x \, dV \qquad y_c V = \int_V y \, dV \qquad z_c V = \int_V z \, dV$$

These quantities are used in the computation of the coordinates of the centroid of a composite volume.

The centroidal coordinates of a composite homogeneous rigid body are

$$x_c = \frac{x_1 V_1 + x_2 V_2 + \cdots + x_n V_n}{V_1 + V_2 + \cdots + V_n} \qquad y_c = \frac{y_1 V_1 + y_2 V_2 + \cdots + y_n V_n}{V_1 + V_2 + \cdots + V_n} \qquad z_c = \frac{z_1 V_1 + z_2 V_2 + \cdots + z_n V_n}{V_1 + V_2 + \cdots + V_n}$$

where $V_1, V_2, \ldots, V_n$ are the elementary volumes into which the body has been subdivided, and $x_n$, $y_n$, $z_n$ are the centroidal coordinates of these volumes. The volumes of any holes or cutouts are treated as negative quantities.

The above equations may also be written in terms of weight or mass, with the typical forms

$$x_c = \frac{x_1 m_1 + x_2 m_2 + \cdots + x_n m_n}{m_1 + m_2 + \cdots + m_n} \qquad x_c = \frac{x_1 W_1 + x_2 W_2 + \cdots + x_n W_n}{W_1 + W_2 + \cdots + W_n}$$

The moments of inertia of a rigid body are defined by

$$I_x = \int_V (y^2 + z^2) \, dm \qquad I_y = \int_V (z^2 + x^2) \, dm \qquad I_z = \int_V (x^2 + y^2) \, dm$$

Mass moments of inertia are a measure of both the distribution of mass in the body and of the placement of the reference axes with respect to the body. The terms are also a measure of the resistance of the body to angular acceleration about the reference axes. Mass moments of inertia are always positive, with the basic units of mass times length squared.

The parallel-axis, or transfer, theorems for mass moments of inertia have the typical form

$$I_x = I_{0x} + md_x^2$$

where $m$ is the mass of the body and $d_x$ is the separation distance between the centroidal $x_0$ axis and the parallel $x$ axis.

The mass radii of gyration are defined by

$$k_z = \sqrt{\frac{I_x}{m}} \qquad k_y = \sqrt{\frac{I_y}{m}} \qquad k_z = \sqrt{\frac{I_z}{m}}$$

The approximate solutions for the mass moments of inertia of a homogeneous, thin plane rigid body are

$$I_x = \int_V y^2 \, dm \qquad I_y = \int_V x^2 \, dm \qquad I_z = \int_V r^2 \, dm = I_x + I_y$$

where $r^2 = x^2 + y^2$, and the $xy$ axes lie in the plane surface area of the body. For this type of thin plane body, the area moment of inertia of the plane boundary area and the mass moment of inertia of the body are related by the approximate equations

$$I_{xM} = \rho t I_{xA} \qquad I_{yM} = \rho t I_{yA} \qquad I_{zM} = \rho t I_{zA} = \rho t (I_{xA} + I_{yA})$$

where the subscripts $A$ and $M$ refer to area and mass, respectively. The accuracy of these equations increases with decreasing thickness $t$ of the body.

The centroidal coordinates of a plane body formed of thin rod shapes are the same as the centroidal coordinates of the plane curve which defines the shape of the body, given by

$$x_c' = \frac{\int_l x \, dl}{l} \qquad y_c' = \frac{\int_l x \, dl}{l}$$

The mass moments of inertia are given by

$$I_x = \rho_0 I_{xl} \qquad I_y = \rho_0 I_{yl} \qquad I_z = \rho_0 I_{zl}$$

where $\rho_0$ is the mass density per unit length of the rod, given by

$$\rho_0 = \rho A$$

and $I_{xl}$, $I_{yl}$, and $I_{zl}$ are the length moments of inertia of the plane curve shape.

The products of inertia of a rigid body are defined by

$$I_{xy} = \int_V xy \, dm \qquad I_{yz} = \int_V yz \, dm \qquad I_{zy} = \int_V zx \, dm$$

Mass products of inertia are measures of the distribution of the mass in a body, and the placement of the reference axes with respect to the body. Unlike mass moments of inertia, the products of inertia may have negative values. The units of mass product of inertia are mass times length squared.

The parallel-axis, or transfer, theorems for mass products of inertia have the forms

$$I_{xy} = I_{0,xy} + mx_c y_c \qquad I_{yz} = I_{0,yz} + my_c z_c \qquad I_{zx} = I_{0,zx} + mz_c x_c$$

The mass products of inertia of a thin, plane rigid body, with the $xy$ coordinates in the plane boundary area of the body, are

$$I_{xy} = \int_V xy \, dm = I_{0,xy} + mx_c y_c \qquad I_{yz} = I_{zx} = 0$$

| TABLE 17.13   Mass Moments of Inertia of Elementary Homogeneous Rigid Bodies | | |
|---|---|---|
| **Body** | **Mass, and $x$, $y$, $z$ coordinates of CM** | **Mass moments of inertia** |
| 1  Solid sphere: | $m = \frac{1}{6}\rho\pi d^3$ | $I_{0x} = I_{0y} = I_{0z} = \frac{1}{10}md^2$ <br> $I_{x_1} = \frac{7}{20}md^2$ |
| 2  Hollow sphere: | $m = \frac{1}{6}\rho\pi(d_0^3 - d_i^3)$ | $I_{0x} = I_{0y} = I_{0z}$ <br> $I_{0x} = \frac{1}{10}m\left[\dfrac{d_0^5 - d_i^5}{d_0^3 - d_i^3}\right]$ <br> $I_{x_1} = \frac{1}{20}m\left[\dfrac{7d_0^5 - 5d_0^2 d_i^3 - 2d_i^5}{d_0^3 - d_i^3}\right]$ |
| 3  Hemisphere: | $m = \frac{1}{12}\rho\pi d^3$ <br> $x_c = z_c = 0$ <br> $y_c = \frac{3}{16}d$ | $I_{0x} = I_{0z} = 0.0648md^2$ <br> $I_x = I_y = I_z = I_{0y} = \frac{1}{10}md^2$ <br> $I_{x_1} = 0.350md^2$ <br> $I_{x_2} = 0.162md^2$ <br> $I_{y_1} = 0.350md^2$ |
| 4  Right circular cylinder: | $m = \frac{1}{4}\rho\pi d^2 h$ <br> $x_c = z_c = 0$ <br> $y_c = \frac{1}{2}h$ | $I_{0x} = I_{0z} = \frac{1}{12}m(h^2 + \frac{3}{4}d^2)$ <br> $I_{0y} = I_y = \frac{1}{8}md^2$ <br> $I_x = I_z = \frac{1}{48}m(16h^2 + 3d^2)$ <br> $I_{x_1} = \frac{1}{48}m(16h^2 + 15d^2)$ <br> $I_{y_1} = \frac{3}{8}md^2$ |

## TABLE 17.13 (cont.)

| Body | Mass, and $x$, $y$, $z$ coordinates of CM | Mass moments of inertia |
|---|---|---|
| 5  Hollow right circular cylinder: | $m = \frac{1}{4}\rho\pi h(d_0^2 - d_i^2)$ <br> $x_c = z_c = 0$ <br> $y_c = \frac{1}{2}h$ | $I_{0x} = I_{0z} = \frac{1}{48}m(3d_0^2 + 3d_i^2 + 4h^2)$ <br> $I_{0y} = I_y = \frac{1}{8}m(d_0^2 + d_i^2)$ <br> $I_{x_1} = \frac{1}{48}m(15d_0^2 + 3d_i^2 + 16h^2)$ <br> $I_{y_1} = \frac{1}{8}m(3d_0^2 + d_i^2)$ |
| 6  Right circular cone: | $m = \frac{1}{12}\rho\pi d^2 h$ <br> $x_c = z_c = 0$ <br> $y_c = \frac{1}{4}h$ | $I_{0x} = I_{0z} = \frac{3}{80}m(d^2 + h^2)$ <br> $I_{0y} = I_y = \frac{3}{40}md^2$ <br> $I_x = I_z = \frac{1}{80}m(3d^2 + 8h^2)$ <br> $I_{x_1} = \frac{1}{80}m(23d^2 + 8h^2)$ <br> $I_{x_2} = \frac{3}{80}m(d^2 + 16h^2)$ <br> $I_{y_1} = \frac{13}{40}md^2$ |
| 7  Ellipsoid: | $m = \frac{1}{6}\rho\pi abc$ | $I_{0x} = \frac{1}{20}m(b^2 + c^2)$ <br> $I_{0y} = \frac{1}{20}m(a^2 + c^2)$ <br> $I_{0z} = \frac{1}{20}m(a^2 + b^2)$ <br> $I_{x_1} = \frac{1}{20}m(b^2 + 6c^2)$ <br> $I_{y_1} = \frac{1}{20}m(c^2 + 6a^2)$ <br> $I_{z_1} = \frac{1}{20}m(b^2 + 6a^2)$ |

## TABLE 17.13 (cont.)

| Body | Mass, and $x$, $y$, $z$ coordinates of CM | Mass moments of inertia |
|---|---|---|
| 8 Paraboloid | $m = \frac{1}{8}\rho\pi d^2 h$ <br> $x_c = \frac{2}{3}h$ <br> $y_c = z_c = 0$ | $I_{0x} = I_x = \frac{1}{12}md^2$ <br> $I_{0y} = I_{0z} = \frac{1}{72}m(3d^2 + 4h^2)$ <br> $I_y = I_z = \frac{1}{24}m(d^2 + 12h^2)$ <br> $I_{y_1} = I_{z_1} = \frac{1}{24}m(d^2 + 4h^2)$ <br> $I_{x_1} = \frac{1}{3}md^2$ |
| 9 Cube: | $m = \rho a^3$ <br> $x_c = y_c = z_c = \frac{1}{2}a$ | $I_{0x} = I_{0y} = I_{0z} = \frac{1}{6}ma^2$ <br> $I_x = I_y = I_z = \frac{2}{3}ma^2$ <br> $I_{x_1} = \frac{5}{12}ma^2$ |
| 10 Rectangular parallelepipid: | $m = \rho abc$ <br> $x_c = \frac{1}{2}a$ <br> $y_c = \frac{1}{2}b$ <br> $z_c = \frac{1}{2}c$ | $I_{0x} = \frac{1}{12}m(b^2 + c^2)$ <br> $I_{0y} = \frac{1}{12}m(a^2 + c^2)$ <br> $I_{0z} = \frac{1}{12}m(a^2 + b^2)$ <br> $I_x = \frac{1}{3}m(b^2 + c^2)$ <br> $I_y = \frac{1}{3}m(a^2 + c^2)$ <br> $I_z = \frac{1}{3}m(a^2 + b^2)$ <br> $I_{x_1} = \frac{1}{12}m(4b^2 + c^2)$ |

**TABLE 17.14   Centroids, and Moments of Inertia, of Elementary Plane Areas**

| Case | Shape | $A$ | $x_c$ | $y_c$ | $I_{0x}$ | $I_{0y}$ | $I_{0z}$ |
|------|-------|-----|-------|-------|----------|----------|----------|
| 1 | | $a^2$ | $\dfrac{a}{2}$ | $\dfrac{a}{2}$ | $\dfrac{a^4}{12}$ | $\dfrac{a^4}{12}$ | $\dfrac{a^4}{6}$ |
| 2 | | $ab$ | $\dfrac{a}{2}$ | $\dfrac{b}{2}$ | $\dfrac{ab^3}{12}$ | $\dfrac{ba^3}{12}$ | $\dfrac{ab}{12}(a^2+b^2)$ |
| 3 | | $\dfrac{ab}{2}$ | $\dfrac{2a}{3}$ | $\dfrac{b}{3}$* | $\dfrac{ab^3}{36}$ | $\dfrac{ba^3}{36}$ | $\dfrac{ab}{36}(a^2+b^2)$ |
| 4 | | $\dfrac{ab}{2}$ | ... | $\dfrac{b}{3}$* | $\dfrac{ab^3}{36}$ | ... | ... |
| 5 | | $\pi a^2 = \dfrac{\pi d^2}{4}$ | $a=\dfrac{d}{2}$ | $a=\dfrac{d}{2}$ | $\dfrac{\pi a^4}{4}=\dfrac{\pi d^4}{64}$ | $\dfrac{\pi a^4}{4}=\dfrac{\pi d^4}{64}$ | $\dfrac{\pi a^4}{2}=\dfrac{\pi d^4}{32}$ |
| 6 | | $\dfrac{\pi a^2}{2}$ | $a$ | $\dfrac{4a}{3\pi}$ | $a^4\left(\dfrac{\pi}{8}-\dfrac{8}{9\pi}\right)$ | $\dfrac{\pi a^4}{8}$ | $a^4\left(\dfrac{\pi}{4}-\dfrac{8}{9\pi}\right)$ |
| 7 | | $\dfrac{\pi ab}{4}$ | $\dfrac{a}{2}$ | $\dfrac{b}{2}$ | $\dfrac{\pi ab^3}{64}$ | $\dfrac{\pi ba^3}{64}$ | $\dfrac{\pi ab}{64}(a^2+b^2)$ |

* The area centroid of any triangle is at the common intersection of the three angle bisectors, at a height above each base of $\frac{1}{3}$ of the altitude.

**TABLE 17.15   Mass Moments of Inertia of Plane Bodies of Thin Rod Shape\***

| Body shape | Mass, and $x$ and $y$ coordinates of CM | Mass moments of inertia |
|---|---|---|
| 1   Thin straight rod: | $m = \rho_0 l$ <br> $x'_c = \tfrac{1}{2}l$ <br> $y'_c = 0$ | $I_{0x} = I_x = 0$ <br> $I_{0y} = I_{0z} = \tfrac{1}{12}ml^2$ <br> $I_y = I_z = \tfrac{1}{3}ml^2$ |
| 2 | $m = \rho_0 l$ <br> $x'_c = a + \tfrac{1}{2}l \sin\theta$ <br> $y'_c = b + \tfrac{1}{2}l \cos\theta$ | $I_{0x} = \tfrac{1}{12}ml^2 \cos^2\theta$ <br> $I_{0y} = \tfrac{1}{12}ml^2 \sin^2\theta$ <br> $I_{0z} = \tfrac{1}{12}ml^2$ <br> $I_x = m(\tfrac{1}{3}l^2 \cos^2\theta + bl\cos\theta + b^2)$ <br> $I_y = m(\tfrac{1}{3}l^2 \sin^2\theta + al\sin\theta + a^2)$ <br> $I_z = m[\tfrac{1}{3}l^2 + l(a\sin\theta + b\cos\theta) + (a^2 + b^2)]$ |
| 3   Thin circular rod | $m = 2\rho_0 r\theta$ <br> $x'_c = \dfrac{r\sin\theta}{\theta}$ <br> $y'_c = 0$ | $I_{0x} = I_x = mr^2\left(\dfrac{\theta - \sin\theta\cos\theta}{2\theta}\right)$ <br> $I_{0y} = \tfrac{1}{2}mr^2\left(1 + \dfrac{\sin 2\theta}{2\theta} - \dfrac{2\sin^2\theta}{\theta^2}\right)$ <br> $I_{0z} = mr^2\left(1 - \dfrac{\sin^2\theta}{\theta^2}\right)$ <br> $I_y = mr^2\left(\dfrac{\theta + \sin\theta\cos\theta}{2\theta}\right)$ <br> $I_z = mr^2$ |
| 4   Thin ring: | $m = \rho_0 \pi d$ <br> $x'_c = y'_c = \dfrac{d}{2}$ | $I_{0x} = I_{0y} = \tfrac{1}{8}md^2$ <br> $I_{0z} = \tfrac{1}{4}md^2$ <br> $I_x = I_y = \tfrac{3}{8}md^2$ <br> $I_z = \tfrac{3}{4}md^2$ |

\*   $\rho_0$ = mass per unit length
   $m$ = total mass of body
$x_0, y_0, z_0$ = centroidal coordinates, with origin at CM of body

$I_{0x}, I_{0y}, I_{0z}$ = mass moments of inertia about centroidal axes
$x, y, z$ = axes which are parallel to centroidal axes $x_0, y_0, z_0$
$x'_c, y'_c$ = coordinates, in $xyz$ system, of CM of body.

# CHAPTER 18
# Dynamics of Rigid Bodies in Plane Motion

## 18.1 DYNAMIC MOTION OF A RIGID BODY ABOUT A FIXED POINT

**18.1** Give three general classifications of problems in plane motion of a rigid body.

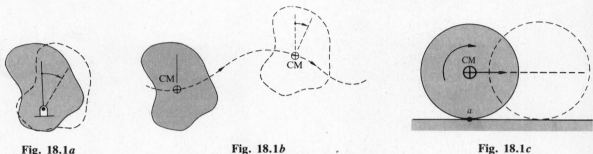

Fig. 18.1a          Fig. 18.1b          Fig. 18.1c

▎ There are three general classifications of problems in plane motion of a rigid body. In the first type, one point on the body is a fixed point which is attached to the ground. Figure 18.1a shows such a body. For this case, the motion of the body is fully described by defining the angular, or rotational, motion of the body about the fixed point. The case of rotation about a fixed point includes the very important case of one body rolling on another body. The point of contact between the two bodies may be considered to be a fixed point, or instant center, about which, at any given instant, one body rotates with respect to the other body.

In the second general type of plane motion, the total planar motion of the body is described by the translational motion of the center of mass (CM) and the rotation of the body about this point. This effect is illustrated in Fig. 18.1b, which shows the position of the body at two different times. The CM in this case may travel along either a straight or a curvilinear path.

The third general type of plane motion of a rigid body is similar to the case described above. The motion of the body is described by the translational motion of an arbitrary point on the body and the rotation about this point. The essential difference between the above two cases is that, for the last case, the reference point on the body is not the CM of the body. This last problem is of a more advanced nature than the first two types, and it is not considered further in this text.

Certain problems may be described as either rotation about a fixed point or motion of the CM. The homogeneous cylinder shown in Fig. 18.1c which rolls, without slipping, on a plane is such a problem. As the cylinder rolls along the plane, the CM translates along a straight line. The cylinder simultaneously rotates about the CM. The contact point $a$ between the cylinder and the plane is an instant center. The motion of the cylinder along the plane may thus be thought of as a succession of instantaneous rotations about the point of contact. Both of the above approaches will be used in the subsequent problems.

**18.2** Develop the equation for the plane motion of a rigid body about a fixed point.

▎ Figure 18.2a shows a body acted on by a system of external forces $\mathbf{F}_1, \mathbf{F}_2, \ldots, \mathbf{F}_n$. The body is attached to the ground at point $a$. The magnitude, direction, sense, and location of the resultant of this system of forces may be found by using the techniques of static force analysis. This resultant force is designated $\mathbf{F}$.

It is shown in statics that any force which acts on a body can be replaced by an equal, parallel force at another location, and a couple. The operation is now performed on the body shown in Fig. 18.2a and the result is seen in Fig. 18.2b. The term $M_a = Fd$ is the resultant external couple, or moment, about point $a$, which acts on the body.

It should be noted that Fig. 18.2b does *not* show a free-body diagram of the body. Rather, it portrays the operation of replacing a force by a force and a couple at another location. If a free-body diagram were drawn, the reaction force of the hinge pin in the body would have to be shown.

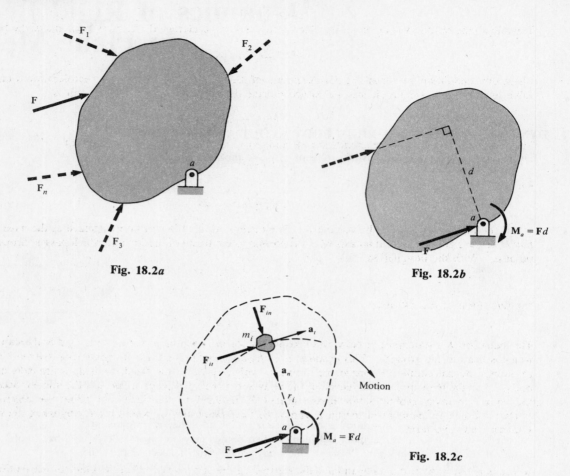

Fig. 18.2a

Fig. 18.2b

Fig. 18.2c

As the body rotates about the fixed point $a$, all the points on the body translate along circular paths which are concentric about the fixed point. A typical mass element $m_i$ of the body, located at distance $r_i$ from the fixed point, is shown in Fig. 18.2c. This element experiences motion as a particle in plane circular translation, with the normal and tangential acceleration components $a_n$ and $a_t$. The forces $F_{in}$ and $F_{it}$ are the rectangular components of the total force $\mathbf{F}_i$ which acts on the particular mass element $m_i$. This force may be either a reaction force exerted by adjacent mass elements on the element being considered, or an external force which acts directly on the element.

Newton's second law, for the mass element $m_i$ as a particle in curvilinear translation, is

$$F_{in} = m_i a_n \qquad F_{it} = m_i a_t$$

The particle $m_i$ moves along a circular path, so that

$$a_n = r_i \omega^2 \qquad a_t = r_i \alpha$$

where $r_i$ is the constant radial distance between the mass particle and the fixed point $a$, and $\omega$ and $\alpha$ are the angular velocity and acceleration, respectively, of the body about the fixed point. It may be noted that these latter two terms are not subscripted $i$, since they are the same for *all* mass elements of the body.

The two components of force which act on $m_i$ now have the forms

$$F_{in} = m_i r_i \omega^2 \qquad F_{it} = m_i r_i \alpha \qquad (1)$$

The relationship between the force components $F_{in}$ and $F_{it}$ on the mass element $m_i$, and the moment about the fixed point $a$ which is required to produce these forces, will now be developed. Since the line of action of $F_{in}$ passes through the fixed point, this force contributes no moment. The moment about the fixed point $a$, which is required to produce the force $F_{it}$, is

$$M_{ai} = F_{it} r_i \qquad (2)$$

$F_{it}$ is eliminated between Eqs. (1) and (2), with the result

$$M_{ai} = m_i r_i^2 \alpha$$

The magnitude of the total moment $M_a$ which is required to accelerate all the mass elements of the body is then

$$M_a = \sum_i M_{ai} = \sum_i m_i r_i^2 \alpha$$

where the summation is over all the mass elements of the body. Since the angular acceleration $\alpha$ has the same value for all mass elements, it may be moved outside the summation sign. The result is

$$M_a = \alpha \sum_i m_i r_i^2 \qquad (3)$$

The subdivision of the mass elements is now allowed to continue without limit, so that

$$m_i \rightarrow dm \qquad r_i \rightarrow r$$

Equation (3) now appears as

$$M_a = \alpha \int_V r^2 \, dm \qquad (4)$$

where the integration is over the volume $V$. The integral in Eq. (4) may be recognized as the *mass moment of inertia* of the rigid body about an axis which is normal to the plane of motion and which passes through the fixed point $a$. With the designation

$$I_a = \int_V r^2 \, dm$$

the final form of Eq. (4) is

$$M_a = I_a \alpha \qquad (5)$$

This equation is a statement of Newton's second law for the rotational motion of a rigid body about a point $a$ which is fixed to the ground. *This equation is of fundamental importance in engineering dynamics.* $M_a$ is the resultant external couple or torque, or moment, with respect to the fixed point $a$, which acts on the rigid body. $I_a$ is the mass moment of inertia of the body about the fixed point $a$, and $\alpha$ is the angular acceleration of the body. In many problems, the mass moment of inertia $I_0$ of the rigid body about the centroidal axis is known. The mass moment of inertia $I_a$ about the fixed point may be found readily by using the parallel-axis theorem, with the form

$$I_a = I_0 + md^2$$

where $m$ is the mass of the body and $d$ is the separation distance, measured in the plane of the motion, between the CM and the fixed point $a$. To emphasize the fact that $M_a$ in Eq. (5) is a *resultant* couple, or moment, this equation may be written in the form

$$\sum M_a = I_a \alpha \qquad (6)$$

Equations (5) and (6) are also referred to as *the equations of motion* of the body.

**18.3** How is the problem of plane motion of a rigid body about a fixed point simplified if the resultant couple or moment which acts on the body has a *constant* value?

❚ If $M_a$ in Prob. 18.2 has a constant magnitude, it follows that the angular acceleration $\alpha$ must be constant. For this case, all the equations presented in Prob. 16.19 for the case of rotational motion with constant acceleration are directly applicable. The forms of these equations are

$$\omega = \omega_0 + \alpha t \qquad \theta = \theta_0 + \omega_0 t + \tfrac{1}{2}\alpha t^2 \qquad \omega^2 = \omega_0^2 + 2\alpha(\theta - \theta_0)$$

where $\theta_0$ and $\omega_0$ are the initial angular displacement and velocity, respectively, at $t = 0$.

**18.4** (a) What is the usual first step when solving the problem of plane motion of a rigid body about a fixed point?

(b) Compare the resultant forces which act on bodies in static equilibrium with those which act on bodies in plane motion.

❚ (a) As in the analysis of problems in static equilibrium, an operation of fundamental importance is to draw a free-body diagram of the rigid body. On this diagram are shown both the applied forces or moments acting on the body, and the reaction forces exerted on the body by the ground which is imagined to be removed. If only the solution for the rotational motion is desired, the reaction forces may be omitted.

(b) In problems of static analysis, a necessary condition for equilibrium is that the resultants of all the forces and moments which act on a body must be identically zero. *In dynamics, the reverse of this situation must always be true,* unless the body moves with constant angular velocity. That is, a resultant force or moment on the body is *required* to impart to this element a rotational acceleration.

**18.5**   Which components of the force which act on a body in plane motion do not contribute to the plane motion?

▌ When a body moves in plane motion, all points in the body maintain the same respective distance, from some fixed reference plane, throughout the entire motion.  Not all the forces acting on the body must lie in the plane of motion of the body.  If any of these forces has a component normal to the plane of motion, then this component must be balanced out by a reaction force component, of the ground on the body, which is normal to the plane of motion.  *Only the components of force which lie in the plane, or are parallel to the plane, contribute to the motion in this plane.*

**18.6**   The disk in Fig. 18.6 is acted upon by a couple $M$ with constant magnitude $10\,\text{N}\cdot\text{m}$.  The mass of the disk is 5 kg.

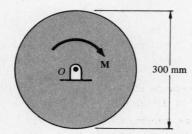

300 mm

**Fig. 18.6**

(*a*)   Find the angular acceleration of the disk.

(*b*)   Find the angular velocity after the disk has completed two revolutions, starting from rest.

▌ (*a*)   In this text, the subscript 0 is used on terms such as $M$ and $I$ if the fixed point is the *center of mass* of the body.
From Case 4 of Table 17.13,

$$I_0 = \tfrac{1}{8}md^2 = \tfrac{1}{8}(5)(300^2)\left(\frac{1}{1,000}\right)^2 = 0.0563\,\text{kg}\cdot\text{m}^2$$

Newton's second law has the form

$$M = I_0\alpha \qquad 10 = 0.0563\alpha \qquad \alpha = 178\,\text{rad/s}^2$$

For this case the term $M$ does *not* need the subscript 0, since a couple is a free vector which does *not* have a reference point about which it acts.

(*b*)   The angular velocity after 2 s is found from

$$\omega^2 = \omega_0^2 + 2\alpha(\theta - \theta_0) = 2(178)[2(2\pi)] \qquad \omega = 66.9\,\text{rad/s} = 639\,\text{r/min}$$

**18.7**   Figure 18.7 shows a slender, rigid rod of weight 10 lb.  Find the value of the moment $M_0$ that acts on the rod, if the magnitude of the angular acceleration is $26\,\text{rad/s}^2$.

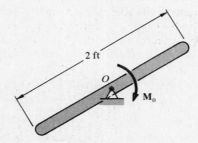

2 ft

$O$

$M_0$

**Fig. 18.7**

▌ Using Case 1 in Table 17.15,

$$I_0 = \tfrac{1}{12}ml^2 = \frac{1}{12}\left(\frac{10}{386}\right)24^2 = 1.24\,\text{lb}\cdot\text{s}^2\cdot\text{in}$$

Newton's second law appears as

$$M_0 = I_0\alpha = 1.24(26) = 32.2\,\text{in}\cdot\text{lb}$$

**18.8** A machinist sharpens a chisel against a grinding wheel, as shown in Fig. 18.8. The chisel is held against the wheel for 5 s. During this time the speed of the wheel changes from 3,600 r/min to 3,450 r/min. The mass moment of inertia of the wheel is 0.044 lb · s² · in.

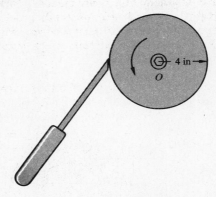

4 in

O

**Fig. 18.8**

**(a)** Find the tangential component of the force exerted by the chisel on the wheel. Assume that the chisel exerts a constant force on the wheel.

**(b)** The motor that drives the grinding wheel exerts a constant torque of 3 in · lb. How long does it take the wheel to return to its original speed after the chisel is removed?

▮ **(a)** The initial and final velocities are

$$\omega_0 = 3,600 \text{ r/min} = 377 \text{ rad/s} \qquad \omega = 3,450 \text{ r/min} = 361 \text{ rad/s}$$

The angular acceleration of the wheel is found from

$$\omega = \omega_0 + \alpha t \qquad 361 = 377 + \alpha(5) \qquad \alpha = -3.20 \text{ rad/s}^2$$

Using Newton's second law,

$$M_0 = I_0 \alpha = 0.044(3.20) = 0.141 \text{ in} \cdot \text{lb}$$

$F_t$ is the tangential force exerted by the chisel on the wheel. From the definition of a moment,

$$M_0 = F_t r \qquad 0.141 = F_t(4) \qquad F_t = 0.0353 \text{ lb}$$

**(b)** Newton's second law is written as

$$M_0 = I_0 \alpha \qquad 3 = 0.044(\alpha) \qquad \alpha = 68.2 \text{ rad/s}^2$$

From part (a),

$$\omega_0 = 361 \text{ rad/s} \qquad \omega = 377 \text{ rad/s}$$

The time for the wheel to return to its original speed is then found from

$$\omega = \omega_0 + \alpha t \qquad 377 = 361 + 68.2t \qquad t = 0.235 \text{ s}$$

**18.9** Figure 18.9 shows a heavy cast-iron flywheel that rotates at a constant speed of 150 r/min. A friction braking system is used to bring the flywheel to rest. When the flywheel is braked, it experiences a constant retarding moment of 5 N · m. The mass moment of inertia of the flywheel about its center axis is 6.25 kg · m².

O

**Fig. 18.9**

**(a)** How long does it take the flywheel to come to rest?

**(b)** Through how many revolutions does the flywheel rotate before coming to rest?

▌ (a) Using Newton's second law,

$$M_0 = I_0 \alpha \qquad -5 = 6.25\alpha \qquad \alpha = -0.8 \, \text{rad/s}^2$$

Using the above result, with the initial velocity $\omega_0 = 150 \, \text{r/min} = 15.7 \, \text{rad/s}$,

$$\omega = \omega_0 + \alpha t \qquad 0 = 15.7 + (-0.8)t \qquad t = 19.6 \, \text{s}$$

(b) Using $\omega_0 = 15.7 \, \text{rad/s}$, $\alpha = -0.8 \, \text{rad/s}^2$, and $\theta_0 = 0$,

$$\omega^2 = \omega_0^2 + 2\alpha(\theta - \theta_0) \qquad 0 = 15.7^2 + 2(-0.8)\theta \qquad \theta = 154 \, \text{rad} = 24.5r$$

**18.10** Figure 18.10 shows a cross-sectional view of a steel pulley with three different diameters. Find the value of the applied moment $M_0$, about the center axis, which will accelerate the pulley at $250 \, \text{rad/s}^2$. The specific weight of steel is $489 \, \text{lb/ft}^3$.

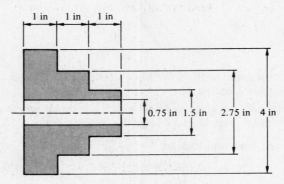

Fig. 18.10

▌ From Prob. 17.44,

$$I_0 = 0.0228 \, \text{lb} \cdot \text{s}^2 \cdot \text{in}$$

Newton's second law is written as

$$M_0 = I_0 \alpha = 0.0228(250) = 5.70 \, \text{in} \cdot \text{lb}$$

**18.11** Figure 18.11 shows the cross section of a proposed lightweight flywheel design. The density of steel is $7,830 \, \text{kg/m}^3$, and the density of aluminum is $2,770 \, \text{kg/m}^3$. It is to be used in a machine where it will be acted

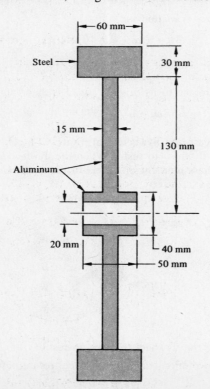

18.11

on by a moment, about the center axis of the flywheel, of $18\,\text{ft}\cdot\text{lb}$. Find the angular acceleration of the flywheel.

❚ From Prob. 17.54,

$$I_0 = 0.291\,\text{kg}\cdot\text{m}^2$$

Using Newton's second law,

$$M_0 = I_0\alpha = 18\,\text{ft}\cdot\text{lb}\left(\frac{1\,\text{m}}{3.28\,\text{ft}}\right)\left(\frac{4.48\,\text{N}}{1\,\text{lb}}\right) = (0.291\,\text{kg}\cdot\text{m}^2)\alpha \qquad \alpha = 84.5\,\text{rad/s}^2$$

**18.12** The disk shown in Fig. 18.12 is made of bronze, with a thickness of $40\,\text{mm}$, and $\rho = 8{,}800\,\text{kg/m}^3$. It is mounted on a shaft of negligible mass. When a constant moment $M_0$ is applied to the shaft, the disk reaches a speed of $1{,}500\,\text{r/min}$ in $10.8\,\text{s}$. Find the magnitude of the moment.

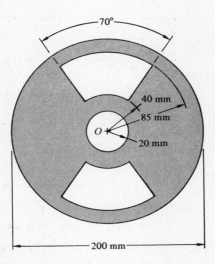

**Fig. 18.12**

❚ From Prob. 17.56,

$$I_0 = 0.0513\,\text{kg}\cdot\text{m}^2$$

Since the applied moment is constant, it follows that the angular acceleration of the disk must be constant. Using $\omega_0 = 0$ and $\omega = 1{,}500\,\text{r/min} = 157\,\text{rad/s}$,

$$\omega = \omega_0 + \alpha t \qquad 157 = \alpha(10.8) \qquad \alpha = 14.5\,\text{rad/s}^2$$

Using Newton's second law,

$$M_0 = I_0\alpha = 0.0513(14.5) = 0.744\,\text{N}\cdot\text{m}$$

**18.13** Figure 18.13 shows a latching lever formed from a steel strip, with attached cylindrical brass plugs. The specific weight of brass is $534\,\text{lb/ft}^3$, and the specific weight of steel is $489\,\text{lb/ft}^3$. Find the value of the constant moment $M_a$, about the $z$ axis, that will cause the latch to rotate from rest through $90°$ in $0.14\,\text{s}$.

❚ Since $M_a$ is a constant, it follows that $\alpha$ is a constant. The equation of motion of the lever assembly is

$$\theta = \theta_0 + \omega_0 t + \tfrac{1}{2}\alpha t^2$$

Using $\theta_0 = 0$ and $\omega_0 = 0$,

$$90°\left(\frac{\pi\,\text{rad}}{180°}\right) = \tfrac{1}{2}\alpha(0.14)^2 \qquad \alpha = 160\,\text{rad/s}^2$$

From Prob. 17.85,

$$I_{\text{assy}} = I_z = I_a = 0.00367\,\text{lb}\cdot\text{s}^2\cdot\text{in}$$

Newton's second law is written as

$$M_a = I_a\alpha = 0.00367(160) = 0.587\,\text{in}\cdot\text{lb}$$

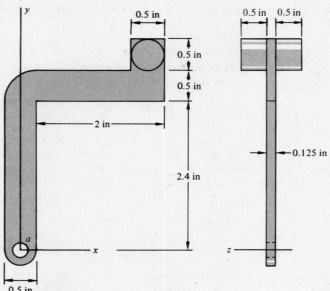

**Fig. 18.13**

**18.14**  Figure 18.14 shows a slender rod on which are positioned two masses.  These masses may be considered to be point masses, and the mass of the rod may be neglected.  A constant moment $M_0$, about the hinge, is applied to the system when the masses are in the solid positions shown, and the resulting acceleration is 68.4 rad/ $s^2$.  Find the acceleration if the same moment is applied to the system when the masses are in the dashed positions shown in the figure.

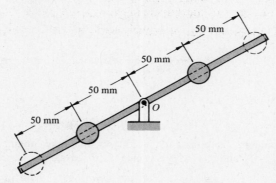

**Fig. 18.14**

▌  The mass moment of inertia $I_0$ of the assembly about the hinge, when the masses are in the inboard position, is

$$I_0 = 2(mr^2) = 2m(50)^2\left(\frac{1}{1,000}\right)^2$$

Newton's second law may be written as

$$M_0 = I_0\alpha = 2m(50)^2\left(\frac{1}{1,000}\right)^2 68.4 \qquad\qquad (1)$$

When the masses are in the outboard position,

$$I_0 = 2[m(2r)^2] = 2m(100)^2\left(\frac{1}{1,000}\right)^2$$

Newton's second law has the form

$$M_0 = I_0\alpha$$

Using $M_0$ from Eq. (1),

$$2\not{m}(50)^2\left(\frac{1}{1,000}\right)^2 68.4 = 2\not{m}(100)^2\left(\frac{1}{1,000}\right)^2\alpha \qquad \alpha = 17.1 \text{ rad/s}^2$$

It may be seen that the acceleration with the masses in the outboard position is one-quarter of the value of the acceleration with the masses in the inboard position.

**18.15** The angular velocity–time diagram of a rigid body that rotates about a fixed axis is shown in Fig. 18.15. If the mass moment of inertia of the body about the fixed axis is $0.125\ \text{lb} \cdot \text{s}^2 \cdot \text{in}$, find the value of the moment $M_0$ that acts on the body during the time interval $0 \le t \le 2\ \text{s}$.

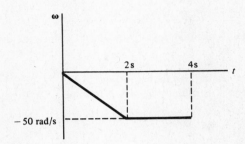

**Fig. 18.15**

▮ The angular acceleration of the body, given by the slope of the $\omega-t$ diagram, is

$$\alpha = \frac{d\omega}{dt} = \frac{-50 - 0}{2 - 0} = -25\ \text{rad/s}^2 = \text{constant}$$

Using Newton's second law,

$$M_a = I_a \alpha = 0.125(-25) = -3.13\ \text{in} \cdot \text{lb}$$

**18.16** (a) Figure 18.16a shows the angular acceleration–time diagram of a rigid body that rotates about a fixed axis. The maximum value of the magnitude of the moment, about the fixed axis, that acts on the body is $12\ \text{N} \cdot \text{m}$. Find the value of the mass moment of inertia of the body about the fixed axis.

(b) Do the same as in part (a), for the angular acceleration–time diagram shown in Fig. 18.16b.

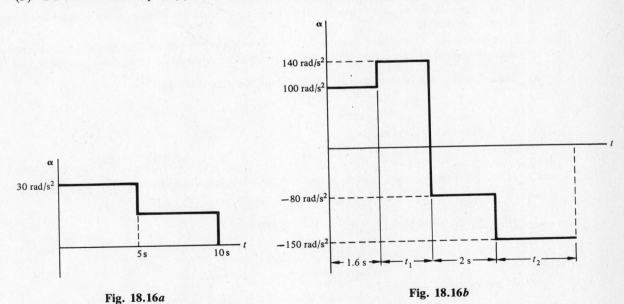

**Fig. 18.16a**

**Fig. 18.16b**

▮ (a) The maximum value of the moment $M_a$ corresponds to the maximum value of $\alpha$. From the problem statement,

$$M_{max} = 12\ \text{N} \cdot \text{m}$$

From Fig. 18.16a,

$$\alpha_{max} = 30\ \text{rad/s}^2$$

Newton's second law then has the form

$$M_a = I_a \alpha \qquad 12 = I_a(30) \qquad I_a = 0.4 \, \text{kg} \cdot \text{m}^2$$

(b)  From Fig. 18.16b,

$$\alpha_{max} = -150 \, \text{rad/s}^2$$

Using Newton's second law,

$$M_a = I_a \alpha \qquad -12 = I_a(-150) \qquad I_a = 0.08 \, \text{kg} \cdot \text{m}^2 \qquad (1)$$

$M_a$ must be used as a negative quantity in Eq. (1), since the mass moment of inertia is *always* a positive quantity.

18.17  A 20-lb disk is hinged at its center, as shown in Fig. 18.17.  A thin, inextensible string is wrapped around the outside of the disk.  The disk is initially at rest.  At time  $t = 0$,  a constant force of 10 lb is applied to the string.

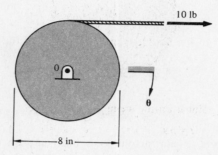

**Fig. 18.17**

(a)  Find the angular acceleration of the disk when the force is applied to the string.

(b)  Find the time, and the angular velocity of the disk in revolutions per minute, when 42 in of string has been unwound from the disk.

(c)  If the force is removed from the string at the time that the conditions of part (b) are reached, what constant tangential force on the rim of the disk would be required to bring this element to rest in 1 s?

▌(a)  The mass moment of inertia of the disk, from Case 4 in Table 17.13, is

$$I_0 = \tfrac{1}{8} md^2 = \frac{1}{8} \left( \frac{20}{386} \right)(8^2) = 0.415 \, \text{lb} \cdot \text{s}^2 \cdot \text{in}$$

Using Newton's second law,

$$M_0 = I_0 \alpha \qquad 10(4) = 0.415 \alpha \qquad \alpha = 96.4 \, \text{rad/s}^2 = \text{constant}$$

(b)  The angular displacement which corresponds to unwinding 42 in of string from the disk is found from

$$s = r\theta \qquad 42 = 4(\theta) \qquad \theta = 10.5 \, \text{rad}$$

The corresponding velocity $\omega$ and time $t$ are given by

$$\omega^2 = \omega_0^2 + 2\alpha(\theta - \theta_0) = 0 + 2(96.4)(10.5)$$

$$\omega = 45 \, \text{rad/s} = 45 \left( \frac{60}{2\pi} \right) = 430 \, \text{r/min}$$

$$\omega = \omega_0 + \alpha t \qquad 45 = 0 + 96.4t \qquad t = 0.467 \, \text{s}$$

(c)  At the instant that the force is removed from the string, the disk has an angular velocity of 45 rad/s.  The acceleration $\alpha_1$ required to bring the disk to rest in 1 s, with a constant retarding moment, is found from

$$\omega = \omega_0 + \alpha_1 t \qquad 0 = 45 + \alpha_1(1) \qquad \alpha_1 = -45 \, \text{rad/s}^2$$

The corresponding value of the moment is

$$M_0 = I_0 \alpha = 0.415(-45) = -18.7 \, \text{in} \cdot \text{lb}$$

The minus sign in the above result indicates that the sense of this moment is opposite to the sense of rotation at the beginning of the time interval under consideration.

The required value of the tangential rim force is

$$F = \frac{M_0}{r} = \frac{18.7}{4} = 4.68 \, \text{lb}$$

18.18    A thin, inextensible string is wrapped around a cylinder, as shown in Fig. 18.18.    At $t = 3\,\text{s}$ the motion of the string is to the right, with a velocity of 2 m/s and with constant acceleration.    At this instant, point $a$ on the string is coincident with point $b$ on the cylinder.    At $t = 5\,\text{s}$,    point $a$ has moved 5.4 m to the right of its original position.    If the string tensile force is 12 N, find the mass moment of inertia of the cylinder.

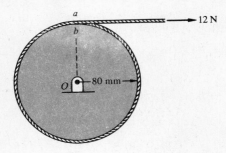

Fig. 18.18

**I**    The initial conditions of point $a$ on the string are

$$t = 3\,\text{s} \qquad s = 0 \qquad v = 2\,\text{m/s} \qquad a = \text{constant}$$

The final conditions are

$$t = 5\,\text{s} \qquad s = 5.4\,\text{m} \qquad a = \text{constant}$$

The times of the problem are redefined as follows:

$$t = 3\,\text{s} \quad \text{is} \quad t = 0$$
$$t = 5\,\text{s} \quad \text{is} \quad t = 2\,\text{s}$$

The motion of point $a$ on the string is given by

$$s = s_0 + v_0 t + \tfrac{1}{2}at^2 \qquad 5.4 = 2(2) + \tfrac{1}{2}a(2)^2 \qquad a = 0.7\,\text{m/s}^2$$

The angular acceleration of the string is found from

$$a = r\alpha \qquad \alpha = \frac{a}{r} = \frac{0.7}{80/1,000} = 8.75\,\text{rad/s}^2$$

Using Newton's second law for the motion of the cylinder,

$$M_0 = I_0\alpha \qquad 12(80)\left(\frac{1}{1,000}\right) = I_0(8.75) \qquad I_0 = 0.110\,\text{kg}\cdot\text{m}^2$$

18.19    The 4-kg disk shown in Fig. 18.19 is initially at rest at time $t = 0$.    A force of 50 N is applied to the thin, inextensible cable wrapped around the disk.

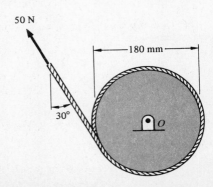

Fig. 18.19

(*a*) Find the angular acceleration of the disk.

(*b*) Find the angular velocity, and the angular displacement of the disk, 4 s after the force is applied.

(*c*) Find the angular velocity of the disk after this element has completed four revolutions.

(*d*) Do the same as in parts (*a*) through (*c*) if, at the time that the force is applied, the disk has a clockwise angular velocity of 230 rad/s.

(*e*) Do the same as in parts (*a*) through (*c*) if, in addition to the cable force, the disk is acted on by a counterclockwise couple of magnitude 2 N · m. The couple acts in the plane of the disk.

❙ (*a*) Using Case 4 in Table 17.13,

$$I_0 = \tfrac{1}{8}md^2 = \tfrac{1}{8}(4)(180)^2\left(\frac{1}{1,000}\right)^2 = 0.0162 \text{ kg} \cdot \text{m}^2$$

Newton's second law has the form

$$M_0 = I_0\alpha \qquad 50(90)\left(\frac{1}{1,000}\right) = 0.0162\alpha \qquad \alpha = 278 \text{ rad/s}^2$$

(*b*) At $t = 4$ s,

$$\omega = \omega_0 + \alpha t = 278(4) = 1,110 \text{ rad/s} \qquad \theta = \theta_0 + \omega_0 t + \tfrac{1}{2}\alpha t^2 = \tfrac{1}{2}(278)4^2 = 2,220 \text{ rad} = 354 \text{ r}$$

(*c*) When the disk has completed four revolutions,

$$\omega^2 = \omega_0^2 + 2\alpha(\theta - \theta_0) = 2(278)4(2\pi) \qquad \omega = 118 \text{ rad/s}$$

(*d*) From part (*a*),

$$\alpha = 278 \text{ rad/s}^2$$

Using $t = 4$ s and $\omega_0 = 230$ rad/s,

$$\omega = \omega_0 + \alpha t = 230 + 278(4) = 1,340 \text{ rad/s} \qquad \theta = \theta_0 + \omega_0 t + \tfrac{1}{2}\alpha t^2 = 230(4) + \tfrac{1}{2}(278)4^2 = 3,140 \text{ rad}$$

The angular velocity of the disk after four revolutions is found from

$$\omega^2 = \omega_0^2 + 2\alpha(\theta - \theta_0) = (230)^2 + 2(278)4(2\pi) \qquad \omega = 259 \text{ rad/s}$$

(*e*) From part (*a*),

$$I_0 = 0.0162 \text{ kg} \cdot \text{m}^2$$

Newton's second law is written as

$$\sum M_0 = I_0\alpha \qquad 50(90)\left(\frac{1}{1,000}\right) - 2 = 0.0162\alpha \qquad \alpha = 154 \text{ rad/s}^2$$

When $t = 4$ s,

$$\omega = \omega_0 + \alpha t = 154(4) = 616 \text{ rad/s} \qquad \theta = \theta_0 + \omega_0 t + \tfrac{1}{2}\alpha t^2 = \tfrac{1}{2}(154)4^2 = 1,230 \text{ rad}$$

After four revolutions of the disk,

$$\omega^2 = \omega_0^2 + 2\alpha(\theta - \theta_0) = 2(154)4(2\pi) \qquad \omega = 88.0 \text{ rad/s}$$

**18.20** Figure 18.20*a* shows a pulley about which passes a belt. At a certain instant, the belt forces have the constant values shown in the figure, and the mass moment of inertia of the pulley is 0.4 lb · s² · in. Find the angular acceleration of the pulley, if the belts are assumed to not slip.

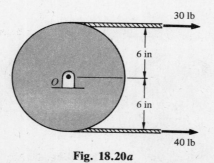

**Fig. 18.20*a***

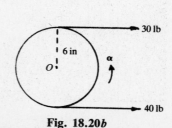

**Fig. 18.20*b***

❚ Figure 18.20*b* shows the sense of the angular acceleration $\alpha$. Newton's second law has the form

$$\sum M_0 = I_0\alpha \qquad (40-30)6 = 0.4\alpha \qquad \alpha = 150\,\text{rad/s}^2$$

**18.21** A belt drives a roller with a mass moment of inertia of $0.025\,\text{lb}\cdot\text{s}^2\cdot\text{in}$, as shown in Fig. 18.21*a*. At a certain time, the angular acceleration of the roller is $200\,\text{rad/s}^2$.

(*a*) Find the tangential force exerted by the belt on the roller at this time.

(*b*) Is it necessary in solving part (*a*) to assume that the belt does not slip with respect to the roller?

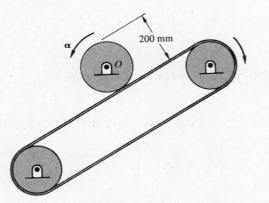

Fig. 18.21*a*

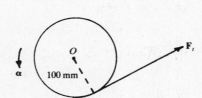

Fig. 18.21*b*

❚ (*a*) The tangential force $F_t$ exerted by the belt on the roller is shown in Fig. 18.21*b*. Newton's second law is

$$M_0 = I_0\alpha \qquad F_t(100\,\text{mm})\left(\frac{1\,\text{in}}{25.4\,\text{mm}}\right) = (0.025\,\text{lb}\cdot\text{s}^2\cdot\text{in})(200\,\text{rad/s}^2) = 1.27\,\text{lb}$$

(*b*) It is *not* necessary to assume no slipping of the belt on the roller. $F_t$ is the tangential force exerted by the belt on the roller. This force is not a function of whether slipping occurs, since, according to the problem statement, the roller has a known, constant angular acceleration. If slipping did ocur, the kinematic relationship between the roller velocity and the belt velocity would simply be indeterminate.

**18.22** A cylinder of mass 6 kg rests in a trough, as shown in Fig. 18.22*a*. A counterclockwise couple in the plane of the figure, of constant magnitude $3.8\,\text{N}\cdot\text{m}$, is applied to the cylinder.

(*a*) Find the resulting angular acceleration.

(*b*) Express the reaction forces acting on the cylinder in formal vector notation.

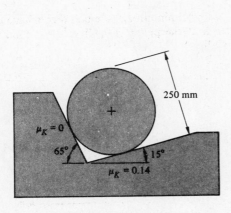

Fig. 18.22*a*

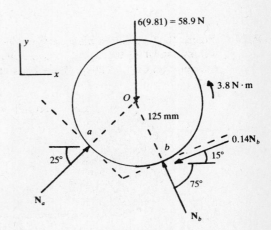

Fig. 18.22*b*

▌ (*a*) Figure 18.22*b* shows the free-body diagram of the cylinder. Since motion occurs, the friction force at point *b* has its maximum value. For static equilibrium of the cylinder,

$$\sum F_x = 0 \qquad N_a \cos 25° - N_b \cos 75° - 0.14N_b \cos 15° = 0 \qquad N_b = 2.30 N_a \qquad (1)$$

$$\sum F_y = 0 \qquad N_a \sin 25° - 58.9 - 0.14N_b \sin 15° + N_b \sin 75° = 0 \qquad (2)$$

Using Eq. (1) in Eq. (2),

$$N_a \sin 25° - 58.9 - 0.14(2.30 N_a) \sin 15° + (2.30 N_a) \sin 75° = 0 \qquad N_a = 23.0 \text{ N} \qquad (3)$$

Using Eq. (3) in Eq. (1),

$$N_b = 2.30(23.0) = 52.9 \text{ N}$$

From Case 4 in Table 17.13,

$$I_0 = \tfrac{1}{8}md^2 = \tfrac{1}{8}(6)\left(\frac{250}{1,000}\right)^2 = 0.0469 \text{ kg} \cdot \text{m}^2$$

Using Newton's second law,

$$\sum M_0 = I_0 \alpha \qquad 3.8 - 0.14N_b\left(\frac{125}{1,000}\right) = 0.0469\alpha \qquad 3.8 - 0.14(52.9)0.125 = 0.0469\alpha \qquad \alpha = 61.3 \text{ rad/s}^2$$

It may be observed in this problem that point *O*, the center of the cylinder, does not translate as the motion occurs. Thus, point *O* is a fixed point with respect to the ground.

(*b*) The reaction forces at *a* and *b* may be written as

$$\mathbf{N}_a = (N_a \cos 25°)\mathbf{i} + (N_a \sin 25°)\mathbf{j} = (23.0 \cos 25°)\mathbf{i} + (23.0 \sin 25°)\mathbf{j} = 20.8\mathbf{i} + 9.72\mathbf{j} \qquad \text{N}$$

$$\mathbf{N}_b = -(N_b \cos 75° + 0.14N_b \cos 15°)\mathbf{i} + (N_b \sin 75° - 0.14N_b \sin 15°)\mathbf{j}$$

$$= -[52.9 \cos 75° + 0.14(52.9) \cos 15°]\mathbf{i} + [52.9 \sin 75° - 0.14(52.9) \sin 15°]\mathbf{j}$$

$$= -20.8\mathbf{i} + 49.2\mathbf{j} \qquad \text{N}$$

As a check on the above results, considering the force equilibrium of the cylinder,

$$\mathbf{N}_a + \mathbf{N}_b + \mathbf{W} \overset{?}{=} 0 \qquad (20.8\mathbf{i} + 9.72\mathbf{j}) + (-20.8\mathbf{i} + 49.2\mathbf{j}) \overset{?}{=} -\mathbf{W} = -(-58.9\mathbf{i}) = 58.9\mathbf{i}$$

$$9.72 + 49.2 \overset{?}{=} 58.9 \qquad 58.9 = 58.9$$

**18.23** The cylinder arrangement shown in Fig. 18.23*a* rests in a trough. Find the value of the constant forces *P* applied to the thin, inextensible cables, if the unit experiences an angular acceleration of 40 rad/s². The mass moment of inertia of the cylinder arrangement is 0.164 kg·m², and the mass is 185 kg.

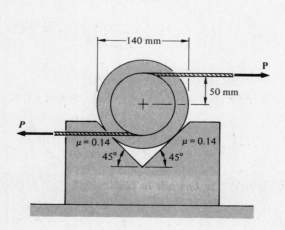

**Fig. 18.23*a***

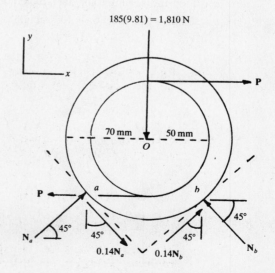

**Fig. 18.23*b***

▮ The free-body diagram of the cylinder arrangement is seen in Fig. 18.23$b$.  For equilibrium,

$$\sum F_x = 0 \qquad -P + N_a \cos 45° + 0.14 N_a \sin 45° + P + 0.14 N_b \sin 45° - N_b \cos 45° = 0$$

$$1.14 N_a - 0.86 N_b = 0$$

$$N_a = 0.754 N_b \qquad\qquad (1)$$

$$\sum F_y = 0 \qquad N_a \sin 45° - 0.14 N_a \cos 45° - 1{,}810 + 0.14 N_b \cos 45° + N_b \sin 45° = 0$$

$$0.608 N_a + 0.806 N_b = 1{,}810 \qquad\qquad (2)$$

Using Eq. (1) in Eq. (2),

$$0.608(0.754 N_b) + 0.806 N_b = 1{,}810 \qquad N_b = 1{,}430\ \text{N} \qquad\qquad (3)$$

Using Eq. (3) in Eq. (1),

$$N_a = 0.754(1{,}430) = 1{,}080\ \text{N} \qquad\qquad (4)$$

Point $O$ is a fixed point, and Newton's second law has the form

$$\sum M_0 = I_0 \alpha \qquad 2P\left(\frac{50}{1{,}000}\right) - 0.14 N_a\left(\frac{70}{1{,}000}\right) - 0.14 N_B\left(\frac{70}{1{,}000}\right) = 0.164(40) \qquad\qquad (5)$$

Using $N_a$ and $N_b$ from Eqs. (3) and (4) in Eq. (5),

$$P = 312\ \text{N}$$

**18.24**  Figure 18.24$a$ shows a slender rod of mass $m$ and length $l$.  The rod is released with zero initial velocity from the position $\theta = 0°$.

(*a*)  Find the maximum value of the angular acceleration of the rod, and the corresponding value of the tangential acceleration $a_t$ of the tip of the rod.

(*b*)  Find the numerical results for part (*a*), if $l = 1{,}200\ \text{mm}$  and  $m = 2.4\ \text{kg}$.

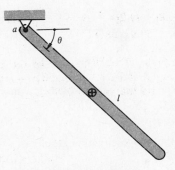

**Fig. 18.24$a$**

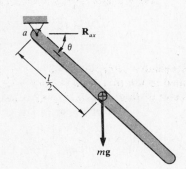

**Fig. 18.24$b$**

▮ (*a*)  Figure 18.24$b$ shows the weight force of the rod.  This force $mg$ produces a moment about point $a$.  The rotational form of Newton's second law for the rod is

$$M_a = I_a \alpha \qquad mg\left(\frac{l}{2}\cos\theta\right) = I_a \ddot\theta \qquad\qquad (1)$$

where $\ddot\theta$ is the angular acceleration of the rod.  The mass moment of inertia of a slender rod about its end, using Case 1 in Table 17.15, is

$$I_a = \tfrac{1}{3} m l^2$$

Equation (1) then appears as

$$mg\left(\frac{l}{2}\cos\theta\right) = \tfrac{1}{3} m l^2 \ddot\theta \qquad \ddot\theta = \frac{3g}{2l}\cos\theta \qquad\qquad (2)$$

The angular acceleration in this problem is seen to be a function of angle $\theta$, and *this acceleration is not constant*.  From Eq. (2), the angular acceleration is seen to decrease with increasing $\theta$.  It has its maximum value when $\theta = 0$, so that

$$\ddot{\theta}_{max} = \frac{3g}{2l}\cos 0° = \frac{3g}{2l}$$

The corresponding value of $a_t$ is

$$a_{t,max} = l\ddot{\theta}_{max} = l\left(\frac{3g}{2l}\right) = \frac{3}{2}g$$

The translational acceleration of the tip of the rod is seen to be 1.5 times the gravitational acceleration $g$. It may be observed from Eq. (2) that the acceleration is zero when $\theta = 90°$. This corresponds to the rod being in a vertical position.

(b) For $l = 1{,}200\ \text{mm}$,

$$\ddot{\theta}_{max} = \frac{3g}{2l} = \frac{3(9.81)}{2(1{,}200)/1{,}000} = 12.3\ \text{rad/s}^2 \qquad a_t = \tfrac{3}{2}g = \tfrac{3}{2}(9.81) = 14.7\ \text{m/s}^2$$

Since $\alpha$ is *not* a constant in this problem, the equations of motion given in Prob. 18.3 may *not* be used.

**18.25**   At the instant shown in Fig. 18.25a, the rod has an angular velocity $\omega$ of 26 r/min in a counterclockwise sense.

(a)   Find the angular acceleration of the rod at this instant.

(b)   State how the result in part (a) would change if the angular velocity of the rod, at the instant shown in the figure, was 50 r/min in a clockwise sense.

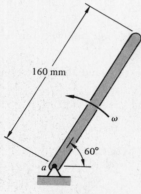

**Fig. 18.25a**

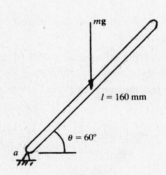

**Fig. 18.25b**

▌ (a)   Figure 18.25b shows the weight force $mg$ of the rod. This force produces a moment about point $a$. Newton's second law is written as

$$M_a = I_a\alpha$$

Using $I_a = \tfrac{1}{3}ml^2$, from Case 1 in Table 17.15, in the above equation results in

$$mg\left(\frac{l}{2}\cos\theta\right) = \tfrac{1}{3}ml^2\ddot{\theta} \qquad \ddot{\theta} = \frac{3g}{2l}\cos\theta = \frac{3(9.81)}{2(160)/1{,}000}\cos 60° = 46.0\ \text{rad/s}^2$$

(b)   There would be no change in the result in part (a), since the equation $M_a = I_a\alpha$ is *independent* of the angular velocity.

**18.26**   The rod of weight 6 lb in Fig. 18.26a is initially supported by a cable.

(a)   Find the angular acceleration of the rod at the instant that the cable is cut.

(b)   Find the angular velocity of the rod at the instant that the cable is cut.

▌ (a)   From Case 1 in Table 17.15,

$$I_a = \tfrac{1}{3}ml^2 = \frac{1}{3}\left(\frac{6}{386}\right)24^2 = 2.98\ \text{lb}\cdot\text{s}^2\cdot\text{in}$$

Figure 18.26b shows the weight force of the rod. At the instant that the cable is cut, this force produces a moment about point $a$. Using Newton's second law,

$$M_a = I_a\alpha \qquad 6(12\cos 40°) = 2.98\alpha \qquad \alpha = 18.5\ \text{rad/s}^2$$

(b)   At the instant that the string is cut,

$$\omega_0 = 0$$

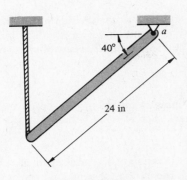

**Fig. 18.26a**

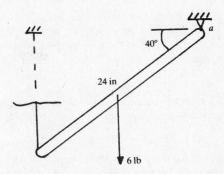

**Fig. 18.26b**

## 18.2 DYNAMIC MOTION DESCRIBED BY TRANSLATION OF THE CENTER OF MASS, AND ROTATION ABOUT THIS POINT

**18.27** The general plane body shown in Fig. 18.2a is redrawn in Fig. 18.27a. The body is no longer attached to the ground, but may have any general type of plane motion. Develop the equation which describes the *translational* motion of the body.

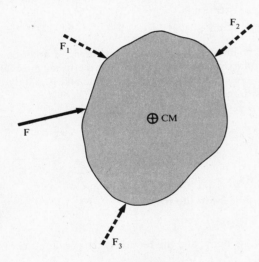

**Fig. 18.27a**

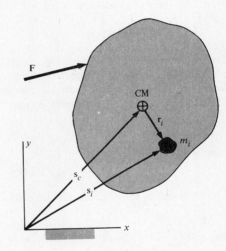

**Fig. 18.27b**

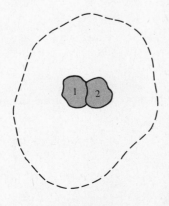

**Fig. 18.27c**

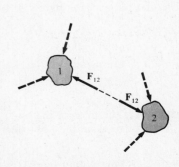

**Fig. 18.27d**

▮ Figure 18.27b shows the body positioned with respect to a set of coordinates which are attached to the ground. The CM is located with respect to the ground by the position vector $s_c$. The mass element $m_i$ is located relative to the CM by the displacement vector $r_i$. The *absolute displacement* of $m_i$ is given by the vector $s_i$ measured from the ground. From the figure,

$$s_i = s_c + r_i \qquad (1)$$

The above equation is a vector summation. Each of the mass elements $m_i$ may be envisioned to move in general curvilinear translation. Newton's second law for each such element is then

$$F_i = m_i \ddot{s}_i \qquad (2)$$

where $\ddot{s}_i$, the second time derivative of $s_i$, is the *absolute* acceleration of the mass element. $F_i$ may represent the force of adjacent mass elements on the element under consideration, or it may represent an external force which acts directly on the element. The resultant force which acts on the body is $F$, and, from the definition of a resultant force,

$$F = \sum_i F_i \qquad (3)$$

The summation on the right side of Eq. (3) is of *all* the forces acting on *all* the mass elements. In the process of performing this summation, all the internal forces in the body, of one mass element on another mass element, are identically self-canceling in pairs. This effect is illustrated in Fig. 18.27c, which shows two typical, adjacent mass elements. The force exerted by element 1 on element 2 is designated $F_{12}$, and $F_{21}$ is a similar designation for the force of element 2 on element 1. The free-body diagrams of these two particles are shown in Fig. 18.27d. The mass elements 1 and 2 physically contact each other. Thus, the forces $F_{12}$ and $F_{21}$, following Newton's third law, must be a pair of action-reaction forces, and

$$F_{12} = -F_{21}$$

The elements 1 and 2 in Fig. 18.27d may have other internal forces acting on them, in addition to external forces, if these elements form a portion of the boundary of the rigid body. All the other internal forces are self-canceling in pairs when taken together with the corresponding mating element, and it is only the *external forces* on the body which contribute to the resultant force.

Equations (2) and (3) are now combined, with the result

$$F = \sum_i F_i = \sum_i m_i \ddot{s}_i \qquad (4)$$

Equation (1) has the form

$$s_i = s_c + r$$

This equation is differentiated twice with respect to time, with the result

$$\ddot{s}_i = \ddot{s}_c + \ddot{r}_i \qquad (5)$$

The term $\ddot{s}_c$ is the absolute acceleration of the CM of the body, and the term $\ddot{r}_i$ is the acceleration of the typical mass particle $m_i$ with respect to the CM of the body.

Equation (5) is now substituted into Eq. (4), to obtain

$$F = \sum_i m_i (\ddot{s}_c + \ddot{r}_i) = \sum_i m_i \ddot{s}_c + \sum_i m_i \ddot{r}_i \qquad (6)$$

The term $\ddot{s}_c$, the absolute acceleration of the CM of the body, is written as

$$\ddot{s}_c = a_c$$

The first term on the right side of Eq. (6) now has the form

$$\sum_i m_i \ddot{s}_c = \sum_i m_i a_c$$

Since $a_c$ is the same for *all* mass elements of the body, it may be moved outside of the summation sign, so that

$$\sum_i m_i a_c = a_c \sum_i m_i = m a_c$$

where $m$ is the mass of the body.

The last term on the right side of Eq. (6) may be written as

$$\sum_i m_i \ddot{r}_i = \sum_i m_i \frac{d^2}{dt^2} r_i$$

Since the operations of time differentiation, and summation over the mass elements of the body, are independent of each other, the order of these operations may be interchanged, with the result

$$\sum_i m_i \ddot{\mathbf{r}}_i = \frac{d^2}{dt^2} \sum_i m_i \mathbf{r}_i \qquad (7)$$

It is shown in Prob. 18.28 that the quantity $\sum_i m_i \mathbf{r}_i$ is identically zero. Thus, the final form for the translational motion of the body in Fig. 18.27a is

$$\mathbf{F} = m\mathbf{a}_c \qquad (8)$$

*This equation is a very important fundamental relationship in engineering dynamics.* $\mathbf{F}$ is the resultant force which acts on the rigid body, and this quantity may be found by using the techniques of static analysis. The absolute *translational* acceleration of the CM is $\mathbf{a}_c$. If all the mass of the rigid body were imagined to be concentrated at the CM, then this mass would move as a particle in translation which is acted on by the force $\mathbf{F}$. All the previously developed equations in Chapter 15 for rectilinear or curvilinear translation of a particle may then be applied directly. It is emphasized that Eq. (8) is true for any direction of the force $\mathbf{F}$, and for any point of application of this force on the rigid body. It should also be noted that this equation yields *no* information whatsoever about the rotational motion of the body.

Equation (8) may be written in the form

$$\sum \mathbf{F} = m\mathbf{a}_c$$

to emphasize the fact that $\mathbf{F}$ is a resultant force.

**18.28** Show that the quantity $\sum_i m_i \mathbf{r}_i$ in Eq. (7) of Prob. 18.27 is identically zero.

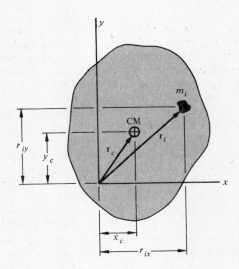

**Fig. 18.28**

▌ The quantity $\sum_i m_i \mathbf{r}_i$ has the form of a first moment of mass of the body. Figure 18.28 shows a set of $xy$ axes attached to the body. The center of mass of the body is located by $\mathbf{r}_c$. The position vector $\mathbf{r}_i$ is the displacement of the typical mass particle $m_i$. This vector has the components $r_{ix}$ and $r_{iy}$. The components of $\mathbf{r}_c$ are the centroidal coordinates $x_c$ and $y_c$, defined by

$$x_c = \frac{\sum_i r_{ix} m_i}{m} \qquad (1)$$

$$y_c = \frac{\sum_i r_{iy} m_i}{m} \qquad (2)$$

In Prob. 18.27 the quantity $\mathbf{r}_i$ is measured from the CM, as shown in Fig. 18.27b. Thus,

$$\mathbf{r}_c = 0$$

and, as a result,

$$x_c = y_c = 0$$

It follows from Eqs. (1) and (2) that

$$\sum_i r_{ix} m_i = 0 \qquad \sum_i r_{iy} m_i = 0$$

Since $r_{ix}$ and $r_{iy}$ are components of the vector $\mathbf{r}_i$, it may be concluded that

$$\sum_i m_i \mathbf{r}_i = 0$$

**18.29** Discuss the result found in Prob. 18.27 for the *translational* motion of a rigid body in plane motion.

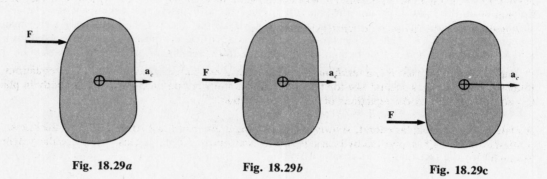

Fig. 18.29a          Fig. 18.29b          Fig. 18.29c

▌ Figures 18.29a, b, and c show a plane rigid body that has the *same* resultant force applied at three different locations on the body. In all three cases, the translational equation of motion would be written as

$$\mathbf{F} = m\mathbf{a}_c \qquad \mathbf{a}_c = \frac{\mathbf{F}}{m}$$

and the CM of the body would experience *the same magnitude, direction, and sense of acceleration* $\mathbf{a}_c$ at the instant shown. It may also be observed that *the line of action of the resultant force which acts on the body is generally not collinear with the line of action of the acceleration.* In all three cases shown in the figures, the line of action of the acceleration of the CM is the single horizontal line which passes through this point. The lines of action of the resultant forces, by comparison, are collinear with these forces and have the three different positions shown.

The reader may intuitively sense that these three cases are basically different problems, and indeed they are. In Fig. 18.29a the body would experience clockwise, angular acceleration. In Fig. 18.29b, the body would experience *no* angular acceleration. In Fig. 18.29c, the body would experience counterclockwise acceleration. The central observation in the above discussion is that *the translational motion of the center of mass CM is independent of the rotational motion of the body*, and is a function of *only* the magnitude, direction, and sense of the resultant force which acts on the body. This independence of the translation of the CM, and the rotation about this point, is the basis of the description of general plane motion as the sum of the translational motion of the center of mass, and rotational motion about this point.

**18.30** Develop the equation which describes the *rotational* motion of a body in general plane motion.

▌ The force components $F_{in}$ and $F_{it}$ which act on the mass elements of a rigid body were shown in Fig. 18.2c, and defined in Prob. 18.2 in terms of the acceleration components $a_n$ and $a_t$. These two terms, given by Eq. (1) in Prob. 18.2, are both functions of $r_i$, the distance from the reference point on the body to the mass element under consideration. It follows that both acceleration components are quantities which are *relative* to the reference point, since these terms vanish when $r_i \to 0$. Thus it may be concluded that the relationship between the resultant moment acting on the rigid body and the angular acceleration of the body holds true regardless of whether the reference point is stationary or moving. This reference point is now chosen to be the CM of the body, and the equation which describes the rotational motion about the CM, following Eq. (5) in Prob. 18.2, has the form

$$M_0 = I_0 \alpha$$

In this equation, $I_0$ is the mass moment of inertia of the rigid body about the CM, $M_0$ is the resultant external couple or torque, or moment, with respect to the CM, and $\alpha$ is the angular acceleration of the body.

The subscript 0 on the terms $M$ and $I$ will be used consistently in this text to indicate that the reference point for these quantities is the CM of the body.

**18.31** Summarize the equations which describe the plane motion of a rigid body.

$\blacksquare$ The complete description of the general plane motion of a rigid body is given by the two equations

$$\mathbf{F} = m\mathbf{a}_c \qquad (1)$$
$$M_0 = I_0\alpha \qquad (2)$$

The first equation describes the translational motion *of* the CM, while the second describes the rotational motion *about* the CM. It may be observed that Eq. (1) is a vector equation, and it can have at most two rectangular components in the plane of the motion. Equation (2) is written in scalar form since, if the rigid body moves in plane motion, the direction of the resultant moment $M_0$ on the body is always that of the plane of the motion.

Equations (1) and (2) may be written in the forms

$$\sum \mathbf{F} = m\mathbf{a}_c \qquad \sum M_0 = I_0\alpha$$

to emphasize the fact that $\mathbf{F}$ is a *resultant* force and $M_0$ is a *resultant* moment. The above equations, which are statements of Newton's second law for the translational and rotation motions of a rigid body in plane motion, are also referred to as the equations of motion of the body.

**18.32** A plate with a rectangular cutout, shown in Fig. 18.32a, is made of brass strip of 0.125 in thickness. The plate is attached to a shaft, supported by bearings, which is collinear with the $x$ axis. The specific weight of brass is $534 \text{ lb/ft}^3$.

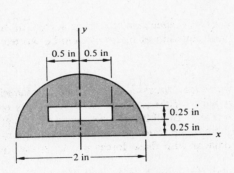

Fig. 18.32a

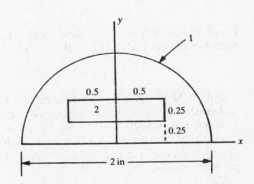

Fig. 18.32b

(a) Find the value of the moment $M_x$ acting on the plate, about the $x$ axis, that will produce an angular acceleration of $300 \text{ rad/s}^2$. All friction and drag effects may be neglected.

(b) Find the force exerted on the shaft when the plate rotates at $4,000 \text{ r/min}$.

$\blacksquare$ (a) From Prob. 17.83,

$$I_x = 3.56 \times 10^{-5} \text{ lb} \cdot \text{s}^2 \cdot \text{in}$$

Using Newton's second law,

$$M_x = I_x\alpha = 3.5 \times 10^{-5}(300) = 0.0107 \text{ in} \cdot \text{lb}$$

(b) The plate is divided into the two area elements shown in Fig. 18.32b. The plate area, and the centroidal coordinate $y_c$, are found as

$$A = \frac{\pi(1)^2}{2} - 1(0.25) = 1.32 \text{ in}^2$$

$$y_c = \frac{y_1A_1 - y_2A_2}{A} = \frac{\dfrac{4(1)}{3\pi}\left[\dfrac{\pi(1)^2}{2}\right] - \left(0.25 + \dfrac{0.25}{2}\right)(1)0.25}{1.32} = 0.434 \text{ in}$$

The mass of the plate is given by

$$m = \rho tA = \left[\frac{534}{386(12)^3}\right]0.125(1.32) = 1.32 \times 10^{-4} \text{ lb} \cdot \text{s}^2/\text{in}$$

The equation of translational motion of the plate is

$$F = ma_c$$

The acceleration of the CM of the plate is $a_c = y_c \omega^2$. Using 4,000 r/min = 419 rad/s,

$$F = my_c \omega^2 = 1.32 \times 10^{-4}(0.434)419^2 = 10.1 \text{ lb}$$

**18.33** Do the same as in Prob. 18.32, if the design of the plate is changed to the elliptical shape with two rectangular cutouts, shown in Fig. 18.33.

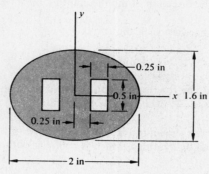

Fig. 18.33

**❚** (a) From Prob. 17.84,

$$I_x = 3.97 \times 10^{-5} \text{ lb} \cdot \text{s}^2 \cdot \text{in}$$

Newton's second law is written as

$$M_x = I_x \alpha = 3.97 \times 10^{-5}(300) = 0.0119 \text{ in} \cdot \text{lb}$$

(b) Since the center of mass lies on the axis of rotation,

$$F = 0$$

**18.34** The disk in Fig. 18.34a is attached to a shaft supported by two bearings equally spaced from the disk. When the disk rotates at a speed of 3,000 r/min, a force of 600 N is transmitted to the bearings. Find the eccentricity of the disk. (The eccentricity is the distance between the center of mass and the axis of rotation.) The thickness of the disk is 18 mm, and the density is 7,830 kg/m³.

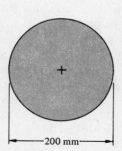

Fig. 18.34a

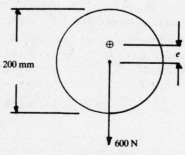

Fig. 18.34b

**❚** The mass of the disk is found from

$$m = 7,830 \left[ \frac{\pi(200)^2}{4} \right] 18 \left( \frac{1}{1,000} \right)^3 = 4.43 \text{ kg}$$

The equation of translational motion of the disk is

$$F = ma_c \qquad 600 = 4.43a_c \qquad a_c = 135 \text{ m/s}^2$$

$a_c$ is the centripetal acceleration, with the form

$$a_c = r\omega^2$$

Using $\omega = 3,000 \text{ r/min} = 314 \text{ rad/s}$, and $a_c = 135 \text{ m/s}^2$, in the above equation results in

$$135 = e(314)^2 \qquad e = 0.00137 \text{ m} = 1.37 \text{ mm}$$

**18.35** Two cylindrical weights are attached to a circular disk, shown in Fig. 18.35, which is attached to the midpoint of a shaft supported by bearings. Find the force exerted by the disk on the shaft when the disk rotates at 1,600 r/min.

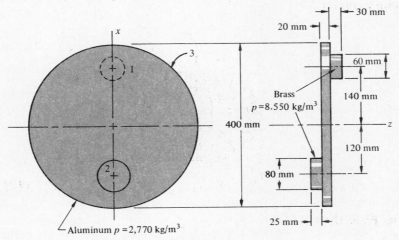

**Fig. 18.35**

▌ From Prob. 17.55,

$$m_1 = 0.725 \text{ kg} \qquad m_2 = 1.07 \text{ kg} \qquad m_3 = 6.96 \text{ kg}$$

The centroidal coordinate $x_c$ of the center of mass of the assembly is found from

$$x_c = \frac{x_1 m_1 + x_2 m_2 + x_3 m_3}{m_1 + m_2 + m_3} = \frac{140(0.725) - 120(1.07) + 0(6.96)}{0.725 + 1.07 + 6.96} = -3.07 \text{ mm} \qquad m = 8.76 \text{ kg}$$

The equation of motion is

$$F = ma_c = mx_c\omega^2$$

Using $160 \text{ r/min} = 168 \text{ rad/s}$,

$$F = 8.76\left(\frac{3.07}{1,000}\right)168^2 = 759 \text{ N}$$

**18.36** The motor shown in Fig. 18.36a is firmly mounted on the wall of a vertical chute through which fine granular material flows. The disk is mounted off center on the motor. When the unit is running, a pulsating force is transmitted through the motor base to the wall to prevent clogging of the material as it flows through the chute. The motor speed is 1,725 r/min and the density of the disk material is 7,830 kg/m³.

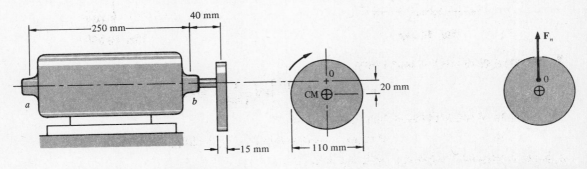

**Fig. 18.36a**                    **Fig. 18.36b**

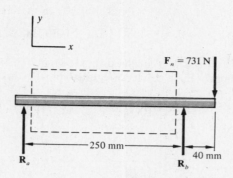

**Fig. 18.36c**

(a) Find the forces exerted on the bearings at $a$ and $b$ when the disk is in the position shown.

(b) Express the result in part (a) in formal vector notation.

❚ (a) The mass of the disk is

$$m = 7,830 \text{ kg/m}^3 \frac{\pi(110)^2}{4} 15 \text{ mm}^3 \left(\frac{1 \text{ m}}{1,000} \text{ mm}\right)^3 = 1.12 \text{ kg}$$

The disk is a body which rotates about a fixed point 0. Since the disk rotates with constant speed, the angular acceleration is zero. Thus, the external moment which acts on the disk is zero. The center of mass of the disk, however, travels in a circular path and experiences normal acceleration $a_n$. The free-body diagram of the disk is shown in Fig. 18.36b. $F_n$ is the force exerted on the disk by the shaft, with the known radial direction shown in the figure. Newton's second law for the motion of the disk is

$$F = ma_c \qquad F_n = ma_n = mr\omega^2 = 1.12 \text{ kg}\left(\frac{20}{1,000}\right)m\left[1,725\left(\frac{2\pi}{60}\right)\right]^2 \frac{\text{rad}^2}{\text{s}^2} = 731 \text{ N}$$

The free-body diagram of the shaft is shown in Fig. 18.36c. For equilibrium of this element,

$$\sum M_a = 0 \qquad R_b(250) - 731(290) = 0 \qquad R_b = 848 \text{ N}$$

$$\sum F_y = 0 \qquad R_a + R_b - 731 = 0 \qquad R_a = -117 \text{ N}$$

The forces which act on the bearings due to the rotating unbalance, for the position of the disk in Fig. 18.36a and using Newton's third law, are

$$R_a = 177 \text{ N} \qquad \text{(acting upward)}$$
$$R_b = 848 \text{ N} \qquad \text{(acting downward)}$$

(b) The forces acting on the bearings, using the $xy$ reference axes in Fig. 18.36c, may be written as

$$\mathbf{R}_a = 177\mathbf{j} \quad \text{N} \qquad \mathbf{R}_b = -848\mathbf{j} \quad \text{N}$$

**18.37** A simple pendulum is released from rest at position $a$ in Fig. 18.37a. The velocity of the pendulum bob can be shown to have the form $v = \sqrt{2gl \cos \theta}$. Find the force acting on the hinge pin when the pendulum is in position $b$. The pendulum rod may be assumed to be massless.

❚ Figure 18.37b shows the free-body diagram of the pendulum mass in position $b$. $F$ is the tensile force in the pendulum rod. Using

$$v = \sqrt{2gl \cos \theta}$$

the centripetal acceleration has the form

$$a_n = \frac{v^2}{\rho} = \frac{2gl \cos \theta}{l} = 2g \cos \theta$$

When the pendulum is in position $b$, $\theta = 0°$, and

$$a_n = 2g \cos 0° = 2g$$

The equation of translational motion of the mass is

$$\sum F = ma_c = ma_n = m(2g) \qquad F - 8.04 = 0.82(2)9.81 \qquad F = 24.1 \text{ N}$$

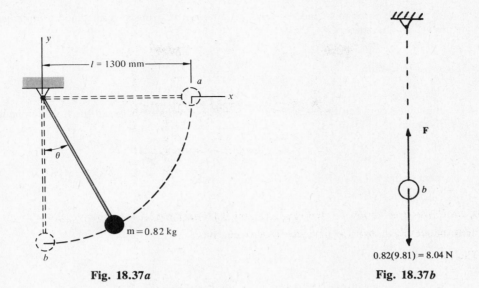

Fig. 18.37a          Fig. 18.37b

The tensile force $F$ is transmitted through the rod to the hinge pin.  Thus, the force acting on the hinge pin, when the pendulum is in position $b$ in Fig. 18.37a, is 24.1 N.

**18.38**   Figure 18.38a shows a yo-yo modeled as a homogeneous disk of mass $m$ connected to an inextensible string.

(*a*)   Find the translation and rotational accelerations of the yo-yo, and the string tensile force $P$, by writing the equations of motion of translation of the CM, and rotation about the CM.

(*b*)   Find the translational and rotational accelerations by using point $a$ as a fixed point about which the yo-yo instantaneously rotates.

(*c*)   If the yo-yo is released from rest, find the translational velocity, in ft/s, and the angular velocity, in revolutions per minute, after the center has moved through a distance of 30 in.   The radius of the yo-yo is 1.3 in.

(*d*)   If the string is cut at any arbitrary time, discuss the subsequent motion of the yo-yo.

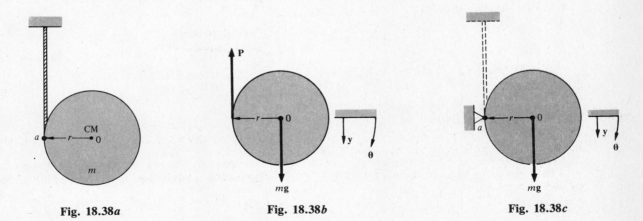

Fig. 18.38a          Fig. 18.38b          Fig. 18.38c

▌ (*a*)   The free-body diagram of the yo-yo is shown in Fig. 18.38b.   The vertical displacement of the CM is $y$, and $\theta$ is the angular motion about this point.   The string tensile force is designated $P$.   The resultant force on the body, acting vertically downward, is $mg - P$.   The equation of motion of the CM is

$$\sum F = ma_c \qquad mg - P = ma_y \tag{1}$$

The resultant moment about the CM is $Pr$.   The rotational equation of motion is

$$M_0 = I_0\alpha \qquad Pr = I_0\alpha \tag{2}$$

The string is assumed to be inextensible.  If a length of string $r\theta$ is imagined to be unwound, the CM will lower by the amount

$$y = r\theta$$

Since  $r = $ constant,

$$a_y = r\alpha \qquad (3)$$

$\alpha$ is eliminated between Eqs. (2) and (3), and $P$ is eliminated between the resulting equation, and Eq. (1), to obtain

$$a_y = \frac{g}{1 + I_0/(mr^2)} \qquad (4)$$

$a_y$ is eliminated between Eqs. (1) and (4), to obtain

$$P = \frac{mg}{1 + mr^2/I_0}$$

Equations (3) and (4) are combined, with the result

$$\alpha = \frac{g}{r[1 + I_0/(mr^2)]}$$

For the disk shown in Fig. 18.38a, using Case 4 in Table 17.13,

$$I_0 = \tfrac{1}{8}md^2 = \tfrac{1}{2}mr^2$$

The final results are then

$$a_y = \tfrac{2}{3}g \qquad \alpha = \frac{2g}{3r} \qquad P = \tfrac{1}{3}mg \qquad (5)$$

It may be seen that $a_y$, $\alpha$, and $P$ are all constants, and it is interesting to note that the string tensile force $P$ is only one-third of the static weight force of the disk.

(b) Since the string is inextensible, the distance from point $a$ to the attachment point of the string to the ground is a constant at any given instant of time.  Point $a$ thus behaves as an instantaneous fixed point, and it is an instant center of rotation of the disk with respect to the string.  This configuration is shown in Fig. 18.38c.

The equation of rotational motion of the body about the fixed point $a$ is given by

$$M_a = I_a\alpha$$

From Fig. 18.38c,

$$M_a = mgr \qquad I_a = I_0 + mr^2$$

so that

$$mgr = (I_0 + mr^2)\alpha \qquad \alpha = \frac{g}{r[1 + I_0/(mr^2)]}$$

The result above is the same as that found in part (a).

The relationship

$$a_y = r\alpha \qquad (4)$$

from part (a) is still true, and $a_y$ would have the form given by Eq. (4).  It may also be observed that the string force $P$ may not be obtained from the method of solution used in this part of the problem.

(c) From Eqs. (5) in part (a),

$$a_y = \tfrac{2}{3}g = \tfrac{2}{3}(32.2) = 21.5 \text{ ft/s}^2$$

The translational velocity when the CM has moved through 30 in is found from

$$v^2 = v_0^2 + 2a_ys \qquad v^2 = 2(21.5)\tfrac{30}{12} \qquad v = 10.4 \text{ ft/s}$$

The corresponding angular velocity is given by

$$v = r\omega \qquad \omega = \frac{v}{r} = \frac{10.4}{1.3/12} = 96 \text{ rad/s} \qquad \omega = 96\left(\frac{60}{2\pi}\right) = 917 \text{ r/min}$$

(d) If the string is cut, the string force $P$ becomes zero and thus vanishes from the problem.  Under these conditions, the method of solution in part (b) is no longer valid, since point $a$ is no longer a fixed point.   From consideration of the free-body diagram in Fig. 18.38$b$, the resultant force which acts on the disk is $mg$, the static weight of the disk.   The resultant moment about the CM is zero.   The equations of motion of the disk are then

$$F = ma_c \qquad mg = ma_y \qquad a_y = g$$
$$M_0 = I_0\alpha \qquad 0 = I_0\alpha \qquad \alpha = 0$$

Since the angular acceleration is zero, it follows that

$$\omega = \text{constant}$$

At the instant that the string is cut, the disk will have a downward acceleration which is equal to the gravitational acceleration of free fall.   In addition, it will have a *constant* angular velocity which is equal to the angular velocity it had at the instant the string was cut.

**18.39**   The disk which represents the yo-yo in Prob. 18.38 is assumed to have a symmetrical variable mass distribution.   The exterior shape of the yo-yo is assumed to be the right circular cylinder of radius $r$ and mass $m$ shown in Fig. 18.39$a$.   In one extreme configuration, all the mass is assumed to be located near the rim, as shown in Fig. 18.39$b$.   In the other extreme distribution of mass, all the mass is assumed to be located near the CM, as shown in Fig. 18.39$c$.   Find the range of values of the translational acceleration of the yo-yo for the extremes of mass distribution shown in the figures.

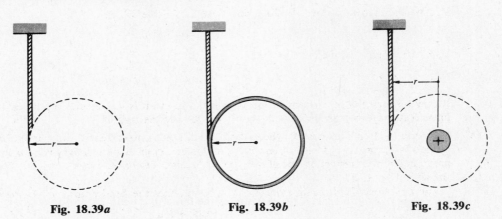

Fig. 18.39$a$                Fig. 18.39$b$                Fig. 18.39$c$

▌  $a_y$ was given by Eq. (4) in part ($a$) of Prob. 18.38 as

$$a_y = \frac{g}{1 + I_0/(mr^2)} \qquad (1)$$

$I_0$ is the mass moment of inertia of the disk about the CM, and this quantity may be expressed in terms of the radius of gyration $k_0$ by

$$I_0 = k_0^2 m$$

Equation (1) now appears as

$$a_y = \frac{g}{1 + mk_0^2/(mr^2)} = \frac{g}{1 + (k_0/r)^2}$$

For the case in Fig. 18.39$b$ of mass distribution near the rim,

$$I_0 \approx mr^2 = mk_0^2 \qquad k_0 = r$$
$$a_y = \frac{g}{1 + (r/r)^2} = \tfrac{1}{2}g$$

When the mass is distributed near the center,

$$I_0 \approx 0 \qquad k_0 \approx 0 \qquad a_y \approx g$$

It may be seen that the range of possible values of acceleration of the CM of the yo-yo is approximately 50 to 100 percent of the value of the gravitational acceleration.   For the case of mass distribution near the center, the motion of the CM is the same as the case where the string is cut.

It is left as an exercise for the reader to decide which of the two extreme constructions shown in Figs. 18.39*b* and *c* would result in a better yo-yo. This problem is considered further in Prob. 19.56.

**18.40** Figure 18.40*a* shows a yo-yo of weight 0.15 lb modeled as a cylindrical disk connected to an inextensible cable. At the instant shown in the figure, the yo-yo has a counterclockwise angular velocity of 30 rad/s.

(*a*) Find the maximum height that the yo-yo will attain.

(*b*) Find the time that corresponds to the motion in part (*a*).

(*c*) Do the same as in part (*a*) if, at a time 0.05 s later than that shown in Fig. 18.40*a*, the cable is cut.

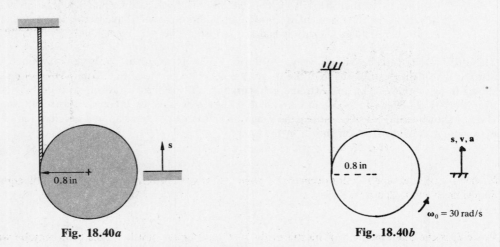

Fig. 18.40*a*          Fig. 18.40*b*

▌ (*a*) The initial condition of the disk is shown in Fig. 18.40*b*. Using Eq. (5) in part (*a*) of Prob. 18.38,

$$a_y = a = -\tfrac{2}{3}g = -\tfrac{2}{3}(386) = -257 \text{ in/s}^2$$

The initial translational velocity of the disk is given by

$$v_0 = r\omega_0 = 0.8(30) = 24 \text{ in/s}$$

*h* is the maximum height of the disk above the initial position, and this quantity is found from

$$v^2 = v_0^2 + 2a(s - s_0) \qquad 0 = 24^2 + 2(-257)h \qquad h = 1.12 \text{ in}$$

(*b*) The time *t* for the disk to reach the maximum height is found from

$$v = v_0 + at \qquad 0 = 24 + (-257)t \qquad t = 0.0934 \text{ s}$$

(*c*) The string is cut at $t = 0.05$ s. The translational equation of motion of the disk has the form

$$s = s_0 + v_0 t + \tfrac{1}{2}at^2$$

Using $a = -257 \text{ in/s}^2 = \text{constant}$ and $v_0 = 24 \text{ in/s}$, the height $h_1$ at $t = 0.05$ s is found as

$$s = h_1 = 24(0.05) + \tfrac{1}{2}(-257)0.05^2 = 0.879 \text{ in}$$

The velocity $v_1$ at time $t = 0.05$ s is given by

$$v_1 = v_0 + at = 24 - 257(0.05) = 11.2 \text{ in/s}$$

When the string is cut, the yo-yo experiences the gravitational acceleration $g = -386 \text{ in/s}^2$. The additional height increase $h_2$ is found from

$$v^2 = v_0^2 + 2a(s - s_0)$$
$$0 = 11.2^2 + 2(-386)h_2 \qquad h_2 = 0.162 \text{ in}$$

The maximum height that the yo-yo attains is

$$h = h_1 + h_2 = 1.04 \text{ in}$$

It may be seen that the height for the condition of part (*c*) is less than the height found in part (*a*). The difference between these two problems is as follows. In part (*a*), the disk has zero angular velocity at its maximum height. In part (*c*), when the yo-yo reaches its maximum height, it still has the angular velocity it possessed at the time that the string was cut.

## 18.3 PURE ROLLING OF A CYLINDRICAL BODY

**18.41** Figure 18.41a shows a cylindrical body which is acted on by a horizontal force $P$, and which rolls along a straight, horizontal track. Find the condition for the cylinder to roll without slipping.

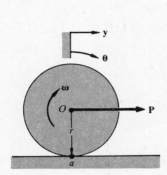

**Fig. 18.41a**

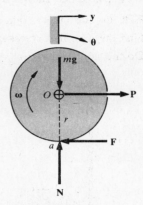

**Fig. 18.41b**

❙ If slipping is assumed to not occur, then point $a$ is an instant center of rotation. The equation of motion of the cylinder about this point is

$$M_a = I_a \alpha \qquad Pr = I_a \alpha \qquad (1)$$

where $I_a$ is the mass moment of inertia of the cylinder about point $a$. This term may be written as

$$I_a = I_0 + mr^2$$

where $I_0$ is the centroidal mass moment of inertia. Equation (1) now appears as

$$Pr = (I_0 + mr^2)\alpha \qquad \alpha = \frac{Pr}{I_0 + mr^2} \qquad (2)$$

This equation gives the relationship between the angular acceleration and the applied force $P$, based on the assumption of no slipping.

The free-body diagram of the cylinder is shown in Fig. 18.41b. The force $F$ is the friction force exerted by the track on the cylinder. It is this force, *and this force alone*, which causes angular acceleration of the cylinder. Since the actual sense of the angular acceleration is known to be clockwise, the force $F$ must act to the left, as shown in Fig. 18.41b. The equation of motion of the cylinder about its CM is

$$M_0 = I_0 \alpha \qquad Fr = I_0 \alpha \qquad (3)$$

$\alpha$ is eliminated between Eqs. (2) and (3), with the result

$$F = \frac{P}{1 + mr^2/I_0} \qquad (4)$$

It may be seen that the applied force $P$ and the friction force $F$ are directly proportional to each other. If the applied force $P$ increases, the friction force $F$ will increase until it reaches its maximum possible value. This limiting value is given by

$$F_{max} = \mu N \qquad (5)$$

where $\mu$ is the coefficient of friction and $N$ is the normal force.

From consideration of Fig. 18.41b, we find that

$$N = mg$$

and Eq. (5) then has the form

$$F_{max} = \mu mg \qquad (6)$$

The maximum permissible value $P_{max}$ of the applied force $P$, for no slipping, is found by combining Eqs. (4) and (6):

$$P_{max} = \mu mg\left(1 + \frac{mr^2}{I_0}\right) \qquad (7)$$

The above equation is valid *only* when slipping motion of the cylinder on the track is impending.

If the body is a homogeneous right circular cylinder,

$$I_0 = \tfrac{1}{2}mr^2$$

and the maximum force which may be applied without causing slipping motion is

$$P_{max} = \mu mg\left(1 + \frac{mr^2}{\tfrac{1}{2}mr^2}\right) = 3\mu mg$$

The quantity $\mu mg$ is the maximum friction force which could exist between the track and the cylinder if nonrolling, sliding motion of the cylinder were impending. It may be seen that the applied force $P$ may have a magnitude which is up to 3 times this value before slipping of the cylinder occurs.

The maximum acceleration of the cylinder, using Eq. (2), is given by

$$\alpha_{max} = \frac{P_{max}r}{I_0 + mr^2} \tag{8}$$

Using Eq. (7) in Eq. (8),

$$\alpha_{max} = \frac{\mu mg(I_0 + mr^2/I_0)r}{I_0 + mr^2} = \frac{\mu mgr}{I_0} \tag{9}$$

**18.42** Figure 18.42a shows an arrangement of two concentric cylinders joined to each other and moving in the plane of the figure. The mass is 6 kg and the centroidal mass moment of inertia is $0.1\ \text{kg} \cdot \text{m}^2$.

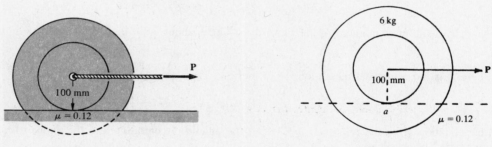

**Fig. 18.42a**                    **Fig. 18.42b**

(*a*) Find the maximum permissible value of the cable tensile force $P$ if the unit is to roll without slipping.

(*b*) Find the maximum value of the angular acceleration of the cylinder, if the unit is to roll without slipping.

❚ (*a*) Using Eq. (7) in Prob. 18.41,

$$P_{max} = \mu mg\left(1 + \frac{mr^2}{I_0}\right) = 0.12(6)9.81\left[1 + \frac{6(100/1,000)^2}{0.1}\right] = 11.3\ \text{N}$$

(*b*) Using Eq. (9) in Prob. 18.41,

$$\alpha_{max} = \frac{\mu mgr}{I_0} = \frac{0.12(6)9.81(0.1)}{0.1} = 7.06\ \text{rad/s}^2$$

**18.43** The cable attached to the cylinder unit in Fig. 18.42a is now inclined, as shown in Fig. 18.43a.

(*a*) Derive general expressions for the maximum value of force $P$, and the maximum angular acceleration $\alpha$, for which the cylinder unit may roll without slipping. Express the answers in terms of $\mu$, $r$, $m$, $\theta$ and $I_0$, the centroidal mass moment of inertia.

(*b*) Find the numerical values of the results in part (*a*), and compare them with the solution to Prob. 18.42. $\mu = 0.12$, $r = 100$ mm, $m = 6$ kg, and $I_0 = 0.1\ \text{kg} \cdot \text{m}^2$. Use values of $\theta$ of 15°, 30°, and 45°.

(*c*) For what angle $\theta$ will the value of $P$ found in part (*a*) have a minimum value?

❚ (*a*) Figure 18.43b shows the instant center at point *a*. For motion about this point,

$$M_a = I_a\alpha \qquad (P\cos\theta)r = (I_0 + mr^2)\alpha \qquad \alpha = \frac{Pr\cos\theta}{I_0 + mr^2} \tag{1}$$

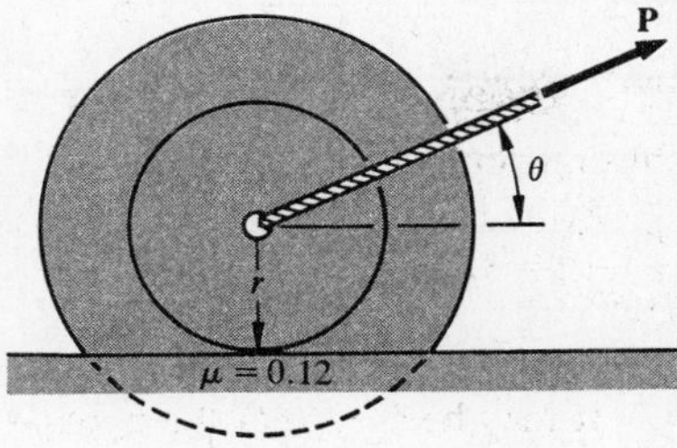

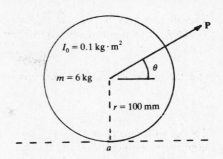

Fig. 18.43a

Fig. 18.43b

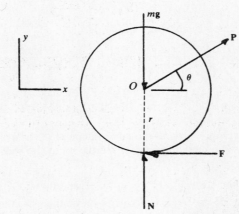

Fig. 18.43c

For motion about the center of mass,

$$M_0 = I_0 \alpha$$

From the free-body diagram in Fig. 18.43c

$$Fr = I_0 \alpha \tag{2}$$

$\alpha$ is eliminated between Eqs. (1) and (2), to obtain

$$F = \frac{P \cos \theta}{1 + mr^2/I_0} \tag{3}$$

For equilibrium in the $y$ direction,

$$\sum F_y = 0 \qquad N = mg - P \sin \theta \tag{4}$$

The maximum available value $F_{max}$ of the friction force is given by

$$F_{max} = \mu N \tag{5}$$

Using Eq. (4) in Eq. (5) results in

$$F_{max} = \mu(mg - P \sin \theta) \tag{6}$$

$F = F_{max}$ is eliminated between Eqs. (3) and (6), and the final result is

$$\mu(mg - P_{max} \sin \theta) = \frac{P_{max} \cos \theta}{1 + mr^2/I_0} \qquad P_{max} = \frac{\mu mg}{(\cos \theta)/(1 + mr^2/I_0) + \mu \sin \theta} \tag{7}$$

Equation (7) is used in Eq. (1), with the result

$$\alpha_{max} = \frac{\mu mgr \cos \theta}{I_0[\cos \theta + \mu(1 + mr^2/I_0) \sin \theta]} \tag{8}$$

(b)   The results for $P_{max}$ and $\alpha_{max}$, using   $\mu = 0.12$,   $r = 100$ mm,   $m = 6$ kg,   and   $I_0 = 0.1$ kg·m², are shown in Table 18.1.

TABLE 18.1

| $\theta$ | $P_{max}$, N | $\alpha_{max}$, rad/s$^2$ |
|---|---|---|
| 0° | 11.3 | 7.06 |
| 15° | 11.1 | 6.72 |
| 30° | 11.7 | 6.36 |
| 45° | 13.4 | 5.90 |

(c)  The minimum value of $P_{max}$, Eq. (7), occurs when

$$\frac{dP_{max}}{d\theta} = 0$$

The condition, from the structure of Eq. (7), is

$$\frac{d}{d\theta}\left(\frac{\cos\theta}{1 + mr^2/I_0} + \mu\sin\theta\right) = \frac{-\sin\theta}{1 + mr^2/I_0} + \mu\cos\theta = 0$$

$$\tan\theta = \mu\left(1 + \frac{mr^2}{I_0}\right) = 0.12\left[1 + \frac{6(0.1)^2}{0.1}\right] = 0.192 \qquad \theta = 10.9°$$

18.44  A thin, inextensible string is wrapped around the inner diameter of the cylinder unit shown in Fig. 18.44a.  The mass of the unit is 8.2 kg, and the mass moment of inertia about the center axis is 0.15 kg·m$^2$.  The unit is to roll without slipping.

(a)  Find the maximum permissible value of the cable tensile force.

(b)  Find the maximum value of the angular acceleration of the cylinder unit.

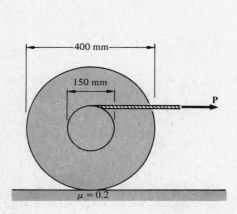

Fig. 18.44a

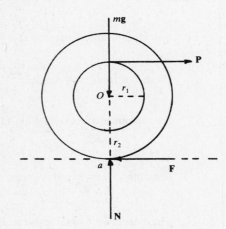

Fig. 18.44b

▮ (a)  The free-body diagram of the cylinder unit is shown in Fig. 18.44b.  For motion about point $a$,

$$M_a = I_a\alpha \qquad P(r_1 + r_2) = I_a\alpha \qquad P(r_1 + r_2) = (I_0 + mr_2^2)\alpha$$

$$\alpha = \frac{P(r_1 + r_2)}{I_0 + mr_2^2} \tag{1}$$

For motion about the center of mass,

$$M_0 = I_0\alpha \qquad Pr_1 + Fr_2 = I_0\alpha \qquad \alpha = \frac{Pr_1 + Fr_2}{I_0} \tag{2}$$

$\alpha$ is eliminated between Eqs. (1) and (2), to obtain

$$\frac{P(r_1 + r_2)}{I_0 + mr_2^2} = \frac{Pr_1 + Fr_2}{I_0} \qquad F = \frac{I_0}{r_2}\left[\frac{P(r_1 + r_2)}{I_0 + mr_2^2}\right] - P\frac{r_1}{r_2} \tag{3}$$

The maximum available value of the friction force is given by

$$F_{max} = \mu N = \mu mg \tag{4}$$

$F = F_{max}$ is eliminated between Eqs. (3) and (4), to obtain

$$P_{max} = \mu mg\left(\frac{1 + mr_2^2/I_0}{1 - mr_1r_2/I_0}\right) \tag{5}$$

Using $m = 8.2$ kg, $I_0 = 0.15$ kg·m², $\mu = 0.2$, $r_1 = 75$ mm, and $r_2 = 200$ mm,

$$P_{max} = 0.2(8.2)9.81\left[\frac{1 + 8.2(0.2)^2/0.15}{1 - 8.2(0.075)(0.2)/0.15}\right] = 285 \text{ N}$$

(b)  Using $P_{max}$ from Eq. (5) in Eq. (1) results in

$$\alpha_{max} = \frac{\mu mg(r_1 + r_2)}{I_0 - r_1 r_2 m} = \frac{0.2(8.2)9.81(0.075 + 0.2)}{0.15 - 0.075(0.2)8.2} = 164 \text{ rad/s}^2$$

**18.45**  (a)  Do the same as in Prob. 18.44, if the string is arranged as shown in Fig. 18.45a.

(b)  Show how the solutions to part (a) can be obtained, in one step, from the solutions to Prob. 18.44.

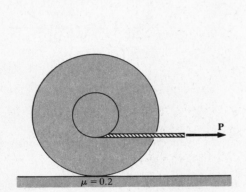

Fig. 18.45a

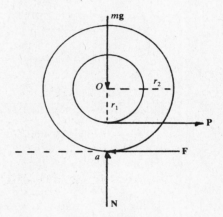

Fig. 18.45b

▌ (a)  Figure 18.45b shows the free-body diagram of the cylinder unit.  The steps in the solution are the same as those used in Prob. 18.44.

$$M_a = I_a\alpha \qquad P(r_2 - r_1) = I_a\alpha \qquad P(r_2 - r_1) = (I_0 + mr_2^2)\alpha$$

$$\alpha = \frac{P(r_2 - r_1)}{I_0 + mr_2^2} \tag{1}$$

$$M_0 = I_0\alpha \qquad Fr_2 - Pr_1 = I_0\alpha \qquad \alpha = \frac{Fr_2 - Pr_1}{I_0} \tag{2}$$

$\alpha$ is eliminated between Eqs. (1) and (2), with the result

$$\frac{P(r_2 - r_1)}{I_0 + mr_2^2} = \frac{Fr_2 - Pr_1}{I_0} \qquad F = \frac{I_0}{r_2}\left[\frac{P(r_2 - r_1)}{I_0 + mr_2^2}\right] + P\frac{r_1}{r_2} \tag{3}$$

$$F_{max} = \mu N = \mu mg \tag{4}$$

$F = F_{max}$ is eliminated between Eqs. (3) and (4), with the result

$$P_{max} = \mu mg\left(\frac{1 + mr_2^2/I_0}{1 + mr_1r_2/I_0}\right) \tag{5}$$

Using $m = 8.2$ kg, $I_0 = 0.15$ kg·m², $\mu = 0.2$, $r_1 = 75$ mm, and $r_2 = 200$ mm,

$$P_{max} = 0.2(8.2)9.81\left[\frac{1 + 8.2(0.2)^2/0.15}{1 + 8.2(0.075)0.2/0.15}\right] = 28.2 \text{ N}$$

Using $P_{max}$ from Eq. (5) in Eq. (1) gives

$$\alpha_{max} = \frac{\mu mg(r_2 - r_1)}{I_0 + r_1 r_2 m} = \frac{0.2(8.2)9.81(200 - 75)(1/1,000)}{0.15 + 75(200)(1/1,000)^2 8.2} = 7.37 \, \text{rad/s}^2$$

It may be seen that the value of $\alpha$ for the cylinder unit arrangement in Fig. 18.45a is much lower than the value of $\alpha$ for the arrangement shown in Fig. 18.44a.

(b) The solutions to this problem can be obtained directly from the solutions to Prob. 18.44, if the term $r_1$ is replaced by the term $-r_1$ in all of the equations in Prob. 18.44.

**18.46** Do the same as in Prob. 18.44, if the cable which acts on the cylinder unit is inclined, as shown in Fig. 18.46a.

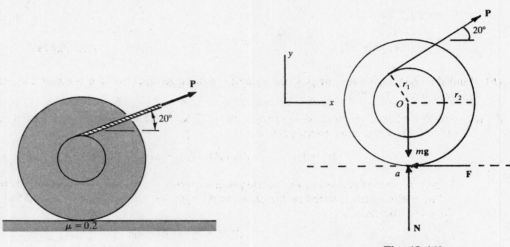

**Fig. 18.46a**          **Fig. 18.46b**

▮ (a) Figure 18.46b shows the free-body diagram of the cylinder unit. For motion about point $a$,

$$M_a = I_a \alpha \qquad (P \cos 20°)(r_1 \cos 20° + r_2) + (P \sin 20°)r_1 \sin 20° = I_a \alpha = (I_0 + mr_2^2)\alpha \qquad (1)$$

Using $r_1 = 75 \, \text{mm}$, $r_2 = 200 \, \text{mm}$, $I_0 = 0.15 \, \text{kg} \cdot \text{m}^2$, $m = 8.2 \, \text{kg}$, and $\mu = 0.2$, the numerical form of Eq. (1) is

$$\alpha = 0.550P \qquad (2)$$

For equilibrium in the $y$ direction,

$$\sum F_y = 0 \qquad P \sin 20° - mg + N = 0 \qquad N = mg - P \sin 20°$$

The maximum available value of the friction force is

$$F = \mu N = \mu(mg - P \sin 20°)$$

For motion about the center of mass,

$$M_0 = I_0 \alpha \qquad Pr_1 + Fr_2 = I_0 \alpha \qquad Pr_1 + \mu(mg - P \sin 20°)r_2 = I_0 \alpha \qquad (3)$$

The numerical form of Eq. (3) is

$$\alpha = 0.409P + 21.5 \qquad (4)$$

$\alpha$ is eliminated betwen Eqs. (2) and (4), with the result

$$0.550P = 0.409P + 21.5 \qquad P_{max} = 152 \, \text{N}$$

(b) Using Eq. (2),

$$\alpha_{max} = 0.550P_{max} = 0.550(152) = 83.6 \, \text{rad/s}^2$$

**18.47** The cylindrical body shown in Fig. 18.47a rolls, without slipping, down the inclined plane with direction $\beta$.

(a) Derive an expression for the minimum required value of the coefficient of friction $\mu$ if slipping is not to occur.

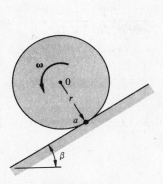

**Fig. 18.47a**

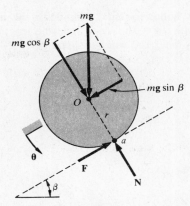

**Fig. 18.47b**

(**b**) Find the numerical value of $\mu$ if the cylinder is homogeneous, with a mass of 2 kg and a diameter of 150 mm, and $\beta = 30°$.

▌ (**a**) The free-body diagram of the cylinder is shown in Fig. 18.47b. Point $a$ is an instant center. Newton's second law for motion about point $a$ is

$$M_a = I_a \alpha \qquad (mg \sin \beta)r = (I_0 + mr^2)\alpha \qquad \alpha = \frac{mgr \sin \beta}{I_0 + mr^2} \tag{1}$$

It may be seen that the angular acceleration increases if the slope of the plane increases.

The friction force exerted by the plane on the cylinder is $F$. Newton's second law for motion about the center of mass is

$$M_0 = I_0 \alpha \qquad Fr = I_0 \alpha \tag{2}$$

If $\alpha$, given by Eq. (1), increases with increasing values of $\beta$, then Eq. (2) shows that the required friction force must also increase with increasing $\beta$.

The limiting value of $F$, when sliding motion is impending, is given by

$$F_{max} = \mu N \tag{3}$$

where

$$N = mg \cos \beta \tag{4}$$

Equations (1) through (4) are now combined, to obtain

$$\mu = \frac{\tan \beta}{1 + mr^2/I_0} \tag{5}$$

This equation gives the *minimum required value of* $\mu$ if the cylinder is to roll, without slipping, down the plane of inclination $\beta$.

For a homogeneous cylinder,

$$I_0 = \tfrac{1}{2}mr^2$$

and Eq. (5) appears as

$$\mu = \frac{\tan \beta}{1 + mr^2/(\tfrac{1}{2}mr^2)} = \frac{1}{3} \tan \beta \tag{6}$$

It is interesting to compare this result with the case where the cylinder is a nonrolling body. The value of $\mu$ at which sliding motion for this case is impending is

$$\mu = \tan \beta$$

It may be seen that the *tangents of the angles* of the inclines differ by a factor of 3, and that the minimum value of the coefficient of friction for pure rolling is less than the corresponding value for sliding motion.

(**b**) It follows from Eq. (6) that, for a homogeneous cylinder, $\mu$ *is independent of both the mass and the dimensions of the cylinder*. This quantity is a function *only* of the inclination angle of the plane on which the cylinder rolls. When $\beta = 30°$, the minimum required value of $\mu$ is

$$\mu = \tfrac{1}{3} \tan \beta = \tfrac{1}{3} \tan 30° = 0.192 \approx 0.2$$

**18.48**  A hollow cylinder rolls down the incline shown in Fig. 18.48.  The material of the cylinder is aluminium, with  $\gamma = 0.1 \, \text{lb/in}^3$,  and the length is 3.6 in.  For what range of values of the coefficient of friction will the cylinder roll without slipping?

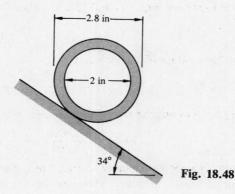

**Fig. 18.48**

**▮**  Using Eq. (5) in Prob. 18.47,

$$\mu_{min} = \frac{\tan \beta}{1 + mr^2/I_0}$$

From Case 5 in Table 17.13,

$$I_{0y} = \tfrac{1}{8}m(d_0^2 + d_i^2)$$

Using  $d_0 = 2.8 \, \text{in}$  and  $d_i = 2 \, \text{in}$,

$$\mu_{min} = \frac{\tan 34°}{1 + \not{m}(1.4)^2/\tfrac{1}{8}\not{m}(2.8^2 + 2^2)}$$

$$\mu_{min} = 0.290$$

If  $\mu \geq 0.290$,  the cylinder will roll without sliding.

**18.49**  A cylinder is released from rest in the position shown in Fig. 18.49.

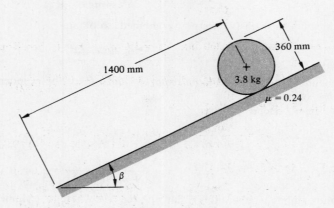

**Fig. 18.49**

(*a*)  For what value of $\beta$ will the cylinder roll down the plane in the minimum time?

(*b*)  Find the time for the motion in part (*a*).

**▮**  (*a*)  Motion down the plane in minimum time corresponds to maximum friction force, which produces maximum angular acceleration.  Using Eq. (6) in Prob. 18.47,

$$\mu = \tfrac{1}{3} \tan \beta = 0.24 \qquad \beta = 35.8°$$

(*b*)  For the cylinder,

$$I_0 = \tfrac{1}{2}mr^2 \qquad\qquad (1)$$

Using Eq. (1) in Prob. 18.47,

$$\alpha = \frac{mgr \sin \beta}{I_0 + mr^2} \qquad (2)$$

$I_0$ is eliminated between Eqs. (1) and (2), to obtain

$$\alpha = \frac{2g}{3r} \sin \beta$$

For the numerical values of this problem,

$$\alpha = \frac{2(9.81) \sin 35.8°}{3(0.180)} = 21.3 \text{ rad/s}^2$$

The translational and angular motions are related by

$$s = r\theta \qquad \frac{1,400}{1,000} = \frac{180}{1,000}\theta \qquad \theta = 7.78 \text{ rad}$$

The minimum time for the cylinder to translate a distance of 1,400 mm, starting from rest, is found from

$$\theta = \theta_0 + \omega_0 t + \tfrac{1}{2}\alpha t^2 \qquad 7.78 = \tfrac{1}{2}(21.3)t^2 \qquad t = 0.855 \text{ s}$$

**18.50** Do the same as in Prob. 18.49 if the cylinder is hollow, as shown in Fig. 18.50.

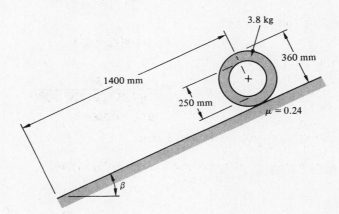

**Fig. 18.50**

▌ (a) Using $m = 3.8$ kg, $d_0 = 360$ mm, $d_i = 250$ mm, $r = d_0/2 = 180$ mm, and Case 5 in Table 17.13,

$$I_0 = \tfrac{1}{8}m(d_0^2 + d_i^2) = \tfrac{1}{8}(3.8)(360^2 + 250^2)\left(\frac{1}{1,000}\right)^2 = 0.0912 \text{ kg} \cdot \text{m}^2$$

Using Eq. (5) in Prob. 18.47,

$$\mu = \frac{\tan \beta}{1 + mr^2/I_0} \qquad 0.24 = \frac{\tan \beta}{1 + 3.8(180/1,000)^2/0.0912} \qquad \beta = 29.4°$$

(b) Using Eq. (1) in Prob. 18.47

$$\alpha = \frac{mgr \sin \beta}{I_0 + mr^2} = \frac{3.8(9.81)180/1,000 \sin 29.4°}{0.0912 + 3.8(180/1,000)^2} = 15.4 \text{ rad/s}^2$$

The displacement and rotation of the cylinder are related by

$$s = r\theta \qquad \frac{1,400}{1,000} = \frac{180}{1,000}\theta \qquad \theta = 7.78 \text{ rad}$$

The minimum time for the hollow cylinder to move through a distance of 1,400 mm, starting from rest, is found from

$$\theta = \theta_0 + \omega_0 t + \tfrac{1}{2}\alpha t^2 \qquad 7.78 = \tfrac{1}{2}(15.4)t^2 \qquad t = 1.01 \text{ s}$$

The time for the hollow cylinder to roll down the plane is greater than that for the solid cylinder in Prob. 18.49.

**18.51** (*a*) Find the maximum value of $\beta$ for which the solid sphere shown in Fig. 18.51 can roll down the plane without slipping.

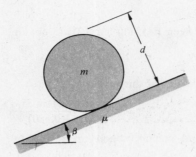

**Fig. 18.51**

(*b*) Find the numerical value of $\beta$ if the coefficient of friction is 0.23.

(*c*) Do the same as in parts (*a*) and (*b*), if the sphere is hollow, with an outside diameter 40 percent greater than the inside diameter.

▮ (*a*) Using Eq. (5) in Prob. 18.47,

$$\mu = \frac{\tan \beta}{1 + mr^2/I_0} \tag{1}$$

From Case 1 in Table 17.13,

$$I_0 = \tfrac{1}{10}md^2 = \tfrac{4}{10}mr^2$$

The above two equations are combined, to obtain

$$\mu = \frac{\tan \beta}{1 + mr^2/\tfrac{4}{10}mr^2} = \frac{\tan \beta}{1 + \tfrac{5}{2}} \qquad \tan \beta = 3.5\mu$$

(*b*) Using $\mu = 0.23$,

$$\tan \beta = 3.5(0.23) \qquad \beta_{\text{max}} = 38.8°$$

(*c*) Using Case 2 in Table 17.13,

$$I_0 = \frac{1}{10} m \left[ \frac{d_0^5 - d_i^5}{d_0^3 - d_i^3} \right]$$

The outer and inner diameters are related by $d_0 = 1.40d_i$, so that

$$I_0 = \frac{1}{10} m \left[ \frac{d_0^5 - (d_0/1.40)^5}{d_0^3 - (d_0/1.40)^3} \right] = 0.128md_0^2$$

Equation (1) in part (*a*) is

$$\mu = \frac{\tan \beta}{1 + mr^2/I_0}$$

Using $r = d_0/2$ in the above equation results in

$$\mu = \frac{\tan \beta}{1 + m(d_0/2)^2/0.128md_0^2)} = \frac{\tan \beta}{2.95} \qquad \tan \beta = 2.95\mu$$

For $\mu = 0.23$,

$$\tan \beta = 2.95(0.23) \qquad \beta = 34.2°$$

It may be seen that the hollow sphere will start to slide at a lower value of angle $\beta$ than that for impending sliding motion of the solid sphere.

**18.52** A homogeneous cylinder of mass 6.75 kg is released from rest at position *a* in Fig. 18.52*a*. It rolls without sliding until it reaches position *b*. Length *bc* of the inclined plane is contaminated with lubricant, and, for the purpose of this problem, the coefficient of friction on this surface may be assumed to be zero.

(*a*) Find the initial angular acceleration of the cylinder.

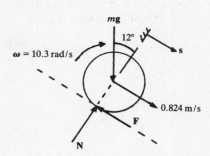

**Fig. 18.52a**                                         **Fig. 18.52b**

(**b**)   Find the value of the friction force that acts on the cylinder in the regime *ab*.

(**c**)   Find the angular velocity and the velocity of the center of the cylinder when the cylinder reaches position *b*.

(**d**)   Do the same as in part (*c*), when the cylinder reaches position *c*.

(**a**)   The minimum value of $\mu$ for which the cylinder will roll without sliding is found from Eq. (6) in Prob. 18.47 as

$$\mu = \tfrac{1}{3}\tan\beta = \tfrac{1}{3}\tan 12° = 0.0709$$

For the present problem,

$$\mu = 0.12 > 0.0709$$

Thus, the friction force which acts on the cylinder *is not the maximum available friction force.*   Using Eq. (1) in Prob. 18.47, with $I_0 = \tfrac{1}{2}mr^2$,

$$\alpha = \frac{mgr\sin\beta}{I_0 + mr^2} = \frac{mgr\sin\beta}{\tfrac{1}{2}mr^2 + mr^2} = \frac{2g}{3r}\sin\beta = \frac{2(9.81)\sin 12°}{3(80/1{,}000)} = 17.0\ \text{rad/s}^2$$

(**b**)   Using Eq. (2) in Prob. 18.47,

$$F = \frac{I_0\alpha}{r} = \frac{\tfrac{1}{2}mr^2\alpha}{r} = \frac{mr\alpha}{2} = \frac{6.75(80/1{,}000)17.0}{2} = 4.59\ \text{N}$$

(**c**)   As the cylinder rolls from *a* to *b*,

$$s = r\theta \qquad \frac{250}{1{,}000} = \frac{80}{1{,}000}\theta \qquad \theta = 3.13\ \text{rad}$$

The cylinder starts from rest at *a*.   The angular velocity at *b* is found from

$$\omega^2 = \omega_0^2 + 2\alpha(\theta - \theta_0) = 2(17.0)3.13 \qquad \omega = 10.3\ \text{rad/s}$$

The velocity of the center of the cylinder at *b* is given by

$$v = r\omega = \frac{80}{1{,}000}(10.3) = 0.824\ \text{m/s}$$

The free-body diagram of the cylinder at location *b*, and the translational and rotational velocities, are shown in Fig. 18.52*b*.

(**d**)   Length *bc* is frictionless.   Thus, the angular velocity does *not* change as the cylinder translates from *b* to *c*.
   Newton's second law for translation of the cylinder has the form

$$F = ma_c \qquad mg\sin 12° = ma = 9.81\sin 12° \qquad a = g\sin 12° = 2.04\ \text{m/s}^2$$

Using the conditions

$$v_0 = 0.824\ \text{m/s} \qquad \text{and} \qquad s = 430\ \text{mm}$$

the velocity at *c* is found from

$$v^2 = v_0^2 + 2a(s - s_0) = 0.824^2 + 2(2.04)\frac{430}{1{,}000} \qquad v = 1.56\ \text{m/s}$$

When the cylinder reaches location *c* it has a translational velocity of 1.56 m/s and a clockwise angular velocity of 10.3 rad/s.

**18.53** Figure 18.53a shows two cylinders at rest on an inclined plane. Both cylinders have identical *dimensions* and identical *weights*. The surfaces of both cylinders, and of the plane, are slightly roughened, and angle β is sufficiently small so that the cylinders roll without sliding. Both cylinders are released from rest at the same instant. A short time later, cylinder A has advanced down the plane an observably greater distance than cylinder B.

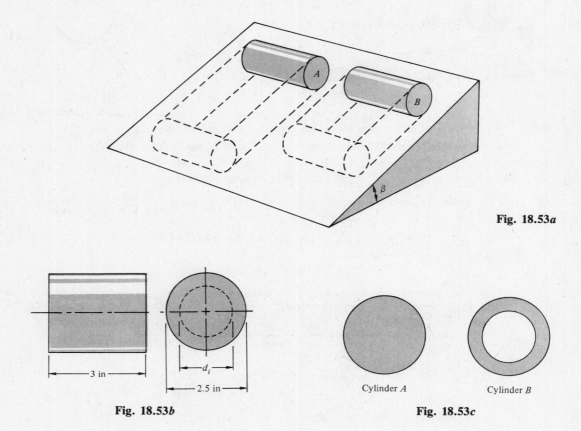

**Fig. 18.53a**

Fig. 18.53b

Fig. 18.53c

Cylinder A          Cylinder B

3 in          2.5 in          $d_i$

(*a*) Explain how this phenomenon is possible.

(*b*) The dimensions of the cylinders are chosen as shown in Fig. 18.53b. Cylinder A is made of aluminum, with γ = 173 lb/ft³, and cylinder B is made of steel, with γ = 489 lb/ft³. The inclination of the plane is 10°. Find the distance between the two cylinders 1 s after they are released from rest.

▌ (*a*) The equation for the angular acceleration which governs the motion of a right circular cylinder which rolls without sliding down an inclined plane was given by Eq. (1) in Prob. 18.47, which is repeated here:

$$\alpha = \frac{mgr \sin \beta}{I_0 + mr^2} \tag{1}$$

Since the cylinders have the same weight, $m$ is the same for both. The dimensions are the same, so that $r$ is a constant. β is a constant which defines the inclination of the plane. The only term in the above equation which may vary, and thus explain the behavior of the cylinders, is $I_0$, the mass moment of inertia about the center axis of the cylinder. For cylinders of the same weight and outside dimensions, the term $I_0$ will vary *if the cylinders are made of different materials*. Figure 18.53c shows the cross sections of the two cylinders. Cylinder A is a homogeneous solid cylinder. Cylinder B is a homogeneous *hollow* cylinder made of material with a density greater than that of cylinder A. Thus,

$$I_{0B} > I_{0A}$$

and, from Eq. (1),

$$\alpha_B < \alpha_A$$

Thus, in a given time, cylinder A will have traveled farther than cylinder B.

(*b*)  For cylinder *A*,

$$W_A = \frac{\pi(2.5)^2}{4}\,(3)\,\text{in}^3\left(\frac{1\,\text{ft}^3}{1{,}728\,\text{in}^3}\right)\!\left(173\,\frac{\text{lb}}{\text{ft}^3}\right) = 1.47\,\text{lb}$$

$$I_{0A} = \frac{1}{8}\,m_A d_A^2 = \frac{1}{8}\left(\frac{1.47}{386}\right)(2.5)^2 = 0.00298\,\text{lb}\cdot\text{s}^2\cdot\text{in}$$

The inside diameter $d_i$ of cylinder *B* is found from

$$\frac{\pi}{4}\,(2.5^2 - d_i^2)(3)\left(\frac{489}{1{,}728}\right) = 1.47\,\text{lb} \qquad d_i = 2.01\,\text{in}$$

The mass moment of inertia of cylinder *B* may be found from

$$I_{0B} = \rho t J$$

where *J* is the area moment of inertia of the *annular cross-sectional area* about the center axis, and $\rho$ is the mass density.

  For a hollow cylinder,

$$J = \frac{\pi}{32}\,(d_0^4 - d_i^4)$$

where $d_0$ and $d_i$ are the outer and inner diameters.

$$I_{0B} = 489\,\frac{\text{lb}}{\text{ft}^3}\left(\frac{1\,\text{ft}^3}{1{,}728\,\text{in}^3}\right)\!\left(\frac{1}{386\,\text{in/s}^2}\right)(3\,\text{in})\left(\frac{\pi}{32}\right)[(2.5^4 - 2.01^4)]\,\text{in}^4 = 0.00491\,\text{lb}\cdot\text{s}^2\cdot\text{in}$$

The angular accelerations of the cylinders, from Eq. (1), are

$$\alpha = \frac{mgr\sin\beta}{I_0 + mr^2}$$

$$\alpha_A = \frac{1.47(2.5/2)\sin 10°}{0.00298 + (1.47/386)(2.5/2)^2} = 35.7\,\text{rad/s}^2 \qquad \alpha_B = \frac{1.47(2.5/2)\sin 10°}{0.00491 + (1.47/386)(2.5/2)^2} = 29.4\,\text{rad/s}^2$$

If the cylinders start from rest, the angular displacements are

$$\theta = \tfrac{1}{2}\alpha t^2$$

The translational displacements *s* of the rolling cylinders are given by   $s = r\theta$.   At the end of 1 s,

$$\theta_A = \tfrac{1}{2}(35.7)(1^2) = 17.9\,\text{rad} \qquad s_A = \frac{2.5}{2}\,(17.9) = 22.4\,\text{in}$$

$$\theta_B = \tfrac{1}{2}(29.4)(1^2) = 14.7\,\text{rad} \qquad s_B = \frac{2.5}{2}\,(14.7) = 18.4\,\text{in}$$

The separation distance $\Delta s$ between the two cylinders at this time is

$$\Delta s = s_A - s_B = 22.4 - 18.4 = 4\,\text{in}$$

Figure 18.54 shows the two cylinders of Prob. 18.53.   Both cylinders start from rest and move down the inclined plane, and   $\beta = 10°$.   Find the position of cylinder *B*, given by the distance *l*, if both cylinders reach the 18-in displacement at the same time.   The cylinders may be assumed to roll without sliding.

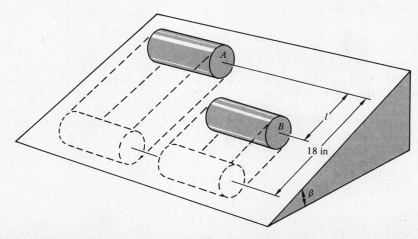

**Fig. 18.54**

▌ From Prob. 18.53,

$$\alpha_A = 35.7 \, \text{rad/s}^2 \qquad \alpha_B = 29.4 \, \text{rad/s}^2$$

The displacements of the cylinders are given by

$$s = r\theta$$

For cylinder $A$,

$$18 = \frac{2.5}{2} \theta_A$$

The times for the motions are found from

$$\theta = \theta_0 + \omega_0 t + \tfrac{1}{2}\alpha t^2$$

The time for cylinder $A$ to move through 18 in is given by

$$\theta_A = \frac{2}{2.5}(18) = \tfrac{1}{2}\alpha_A t^2 = \tfrac{1}{2}(35.7)t^2 \qquad t^2 = 0.807 \qquad (1)$$

The motion of cylinder $B$ is expressed as

$$\theta_B = \frac{s_B}{r} = \frac{2}{2.5}(18 - l) = \tfrac{1}{2}\alpha_B t^2$$

Using Eq. (1) with the above result,

$$\frac{2}{2.5}(18 - l) = \tfrac{1}{2}(29.4)0.807 \qquad l = 3.17 \, \text{in}$$

## 18.4 DYNAMIC MOTION OF CONNECTED RIGID BODIES

18.55   The system of Prob. 15.41 is repeated in Fig. 18.55a. In this earlier problem the pulley was considered to be massless. The translational acceleration of the blocks was found to be

$$a_y = 1.76 \, \text{m/s}^2$$

and the cable tensile force $T$ had the value

$$T = 579 \, \text{N}$$

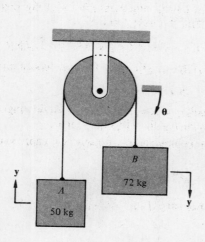

Fig. 18.55a

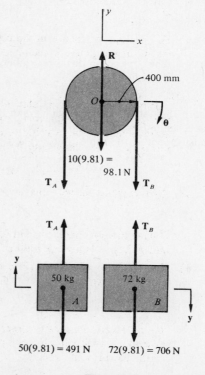

Fig. 18.55b

The pulley has a mass of 10 kg, is 800 mm in diameter, and has the shape of a disk. The hinge is assumed to be frictionless.

(*a*) If the cable is assumed to not slip on the rim of the pulley, find the force in each cable, the acceleration of the masses, and the hinge pin force acting on the pulley.

(*b*) What is the minimum required value of the coefficient of friction between the cable and the rim of the pulley if slipping of the cable on the pulley is not to occur?

▌ (*a*) In Prob. 15.40 the motion of connected particles was considered. The technique of solution consisted of drawing free-body diagrams, and writing the equations of motion, for *each* mass particle. The forces in elements such as cables or links which connect the particles are unknown quantities which may be found from solution of the several equations of motion of the entire system. This method of solution may be readily extended to the case of connected rigid bodies with translational and rotational motion.

The free-body diagrams of the three mass elements of the system are shown in Fig. 18.55b. $R$ is the force exerted by the hinge pin on the pulley, and $T_A$ and $T_B$ are the cable tensile forces. The mass moment of inertia $I_0$ of the pulley is

$$I_0 = \tfrac{1}{8}md^2 = \tfrac{1}{8}(10)\left(\frac{800}{1,000}\right)^2 = 0.8 \text{ kg} \cdot \text{m}^2$$

The equation of motion of the pulley is

$$\sum M_0 = I_0\alpha \qquad (T_B - T_A)\left(\frac{400}{1,000}\right) = 0.8\alpha \qquad (1)$$

The equations of motion of the masses are

$$\sum F = m_A a_y \qquad T_A - 491 = 50a_y \qquad (2)$$

$$\sum F = m_B a_y \qquad 706 - T_B = 72a_y \qquad (3)$$

The kinematic relationship between $a_y$ and $\alpha$ is

$$a_y = r\alpha = \frac{400}{1,000} = 0.4\alpha \qquad (4)$$

Equations (1) through (4) are solved simultaneously, and the results are

$$a_y = 1.69 \text{ m/s}^2 \qquad T_A = 576 \text{ N} \qquad T_B = 584 \text{ N}$$

The percent difference in the computed values of the acceleration of the masses, when the moment of inertia of the pulley is included, is

$$\%\text{D} = \frac{1.69 - 1.77}{1.77}\,100 = -5\%$$

It may be seen that a relatively small error is introduced by neglecting the mass moment of inertia of the pulley. In engineering calculations, unless the dimensions of the pulley are large, with a correspondingly large value of the mass moment of inertia, the mass effects of the pulley are usually negligible. The hinge pin force $R$, from Fig. 18.55*b*, is found from

$$\sum F_y = 0 \qquad R - T_A - T_B - 98.1 = 0 \qquad R = 575 + 584 + 98.1 = 1,260 \text{ N}$$

As a comparison with the above value, the total static weight $W$ of the two blocks and the pulley is

$$W = 491 + 706 + 98.1 = 1,300 \text{ N}$$

During motion, the hinge pin force acting on the pulley is less than the total static weight of the blocks and pulley. Thus, the motion has the effect of *reducing* the hinge pin force.

(*b*) The equation which describes the condition of impending sliding motion of a cable with respect to a curved surface is

$$\frac{T_1}{T_2} = e^{\mu\beta} \qquad (1)$$

$T_1$ and $T_2$ are the tensile forces in the cable on either side of the pulley, and $T_1 > T_2$. $\beta$ is the angle of contact, and $\mu$ is the coefficient of friction.

For the present example

$$T_1 = T_B = 584 \text{ N} \qquad T_2 = T_A = 576 \text{ N} \qquad \beta = 180° = \pi \text{ rad}$$

Equation (1) then appears as

$$\frac{T_1}{T_2} = \frac{584}{575} = e^{\mu\beta} = e^{\mu\pi} \qquad 1.014 = e^{\mu\pi} \qquad \ln 1.014 = \mu\pi \qquad \mu = 0.0044$$

If $\mu \geq 0.0044$, the cable will not slip on the pulley.

**18.56** Because of poor lubrication, the pulley in Prob. 18.55 must overcome a constant friction moment of $12\,\text{N}\cdot\text{m}$ as it rotates. The cable is assumed to not slip on the pulley.

(a) Find the cable tensile forces, the acceleration of the masses, and the hinge pin force acting on the pulley.

(b) Find the velocity of the blocks after block $B$, starting from rest, has moved downward 1 m.

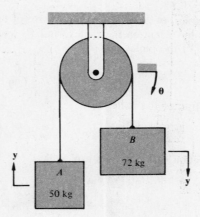

**Fig. 18.56**

❚ (a) Using Fig. 18.55b,

$$\sum M_0 = I_0\alpha \qquad (T_B - T_A)\left(\frac{400}{1,000}\right) - 12 = 0.8\alpha \tag{1}$$

$$\sum F = m_A a_y \qquad T_A - 491 = 50a_y \tag{2}$$

$$\sum F = m_B a_y \qquad 706 - T_B = 72a_y \tag{3}$$

$$a_y = r\alpha = \left(\frac{400}{1,000}\right)\alpha = 0.4\alpha \tag{4}$$

The solutions to the above equations are

$$a_y = 1.46\,\text{m/s}^2 \qquad T_A = 564\,\text{N} \qquad T_B = 601\,\text{N}$$

The hinge pin force acting on the pulley is given by

$$\sum F_y = 0 \qquad R - T_A - T_B - 98.1 = 0 \qquad R = 1{,}260\,\text{N}$$

(b) The velocity of the blocks, after a displacement of 1 m, is found from

$$v^2 = v_0^2 + 2a(s - s_0) = 2(1.46)1 \qquad v = 1.71\,\text{m/s}$$

**18.57** At the instant shown in Fig. 18.57a, the stepped pulley rotates clockwise at 25 rad/s, with a clockwise angular acceleration of $6\,\text{rad/s}^2$. The masses of the three blocks are $m_A = 2\,\text{kg}$, $m_B = 1.5\,\text{kg}$, and $m_C = 1.2\,\text{kg}$, and the cables are inextensible. Find the mass moment of inertia of the pulley.

❚ Figure 18.57b shows the free-body diagrams of the pulley and the blocks. The accelerations of the blocks are

$$a_A = r_A\alpha = \frac{160}{1,000}\,(6) = 0.96\,\text{m/s}^2 \qquad a_B = r_B\alpha = \frac{400}{1,000}\,(6) = 2.4\,\text{m/s}^2$$

$$a_C = r_C\alpha = \frac{540}{1,000}\,(6) = 3.24\,\text{m/s}^2$$

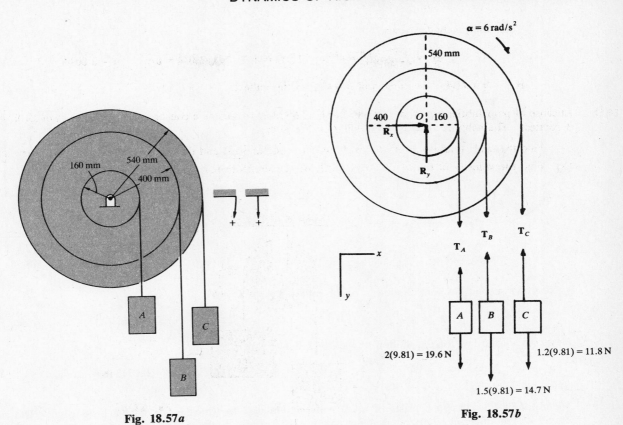

Fig. 18.57a

Fig. 18.57b

Newton's second law for the blocks appears as

$$\sum F_A = m_A a_A \qquad 19.6 - T_A = 2(0.96) \qquad T_A = 17.7 \text{ N}$$

$$\sum F_B = m_B a_B \qquad 14.7 - T_B = 1.5(2.4) \qquad T_B = 11.1 \text{ N}$$

$$\sum F_C = m_C a_C \qquad 11.8 - T_C = 1.2(3.24) \qquad T_C = 7.91 \text{ N}$$

Newton's second law for the motion of the pulley is

$$\sum M_0 = I_0 \alpha \qquad T_A\left(\frac{160}{1,000}\right) + T_B\left(\frac{400}{1,000}\right) + T_C\left(\frac{540}{1,000}\right) = I_0(6)$$

Using the values of $T_A$, $T_B$, and $T_C$ found above,

$$I_0 = 1.92 \text{ kg} \cdot \text{m}^2$$

**18.58** The disk shown in Fig. 18.58a has two holes in it and is made of 0.5-in-thick aluminum plate. The specific weight of aluminum is 173 lb/ft³. A thin, inextensible cable is wrapped around the rim of the disk and attached to a block, as shown in Fig. 18.58b. When the system is released from rest, the weight moves downward through a distance of 26 in in 1.2 s. Find the weight, in ounces, of the block (16 oz = 1 lb).

▌ From Prob. 17.50,

$$I_0 = 0.0161 \text{ lb} \cdot \text{s}^2 \cdot \text{in}$$

Using $s = 26$ in and $t = 12$ s,

$$s = s_0 + v_0 t + \tfrac{1}{2} a t^2 \qquad 26 = \tfrac{1}{2} a (1.2)^2 \qquad a = 36.1 \text{ in/s}^2$$

The angular acceleration of the disk is given by

$$\alpha = \frac{a}{r} = \frac{36.1}{3} \text{ rad/s}^2$$

The free-body diagrams of the disk and the block are shown in Fig. 18.58c.

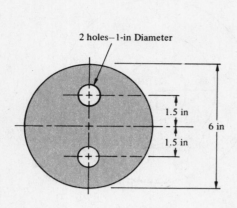

Fig. 18.58a

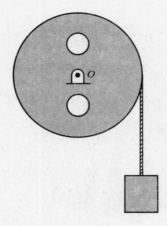

Fig. 18.58b

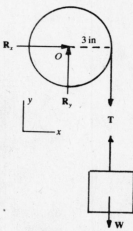

Fig. 18.58c

Newton's second law for the motion of the pulley has the form

$$M_0 = I_0 \alpha \qquad T(3) = 0.0161 \alpha = 0.0161 \left( \frac{0.361}{3} \right) \qquad T = 0.0646 \, \text{lb}$$

Newton's second law for the motion of the block has the form

$$\sum F = ma \qquad W - T = \frac{W}{g} a \qquad W - 0.0646 = \frac{W}{386} (36.1) \qquad W = 0.0713 \, \text{lb} = 1.14 \, \text{oz}$$

**18.59** The rotor in Fig. 18.59a consists of a solid aluminium disk to which is attached a brass plug. The rotor assembly is mounted in a horizontal plane, and a set of $xyz$ axes is attached to the disk. An inextensible string is wrapped around the rim of the disk and is connected to a 600-g mass. The density of brass is 8,550 kg/m$^3$ and the density of aluminum is 2,770 kg/m$^3$.

(a) Find the mass, and the location of the center of mass, of the rotor assembly, and the mass moment of inertia of this assembly about the axis of rotation.

(b) Find the acceleration of the 600-g mass and of the rotor assembly, and the cable tensile force.

(c) Find the horizontal components, and the magnitude, of the force of the hinge pin on the rotor assembly hinge pin 2 s after the mass is released from rest. Assume that the initial angular position of the rotor assembly was such that, at $t = 2 \, \text{s}$, the string force is parallel to the $y$ axis, and acting in the negative sense of this axis.

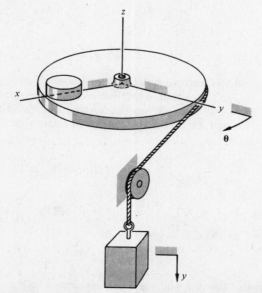

Fig. 18.59a

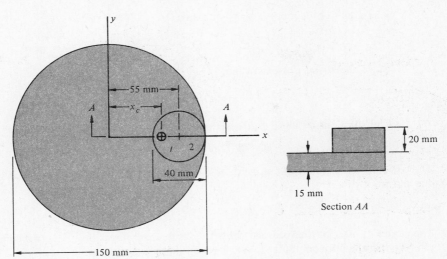

**Fig. 18.59b**

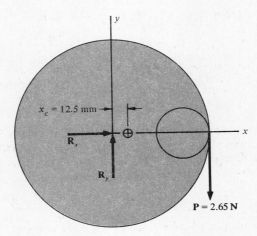

**Fig. 18.59c**

(*d*) Express the forces acting on the rotor assembly, for the position of part (*c*), in formal vector notation.

▮ (*a*) Figure 18.59*b* shows the rotor assembly. The aluminum disk is element 1 and the brass plug is element 2. The masses of these two elements are

$$m_1 = \frac{\pi}{4}(150)^2 15 \text{ mm}^3(2770 \text{ kg/m}^3)\left(\frac{1 \text{ m}}{1,000 \text{ mm}}\right)^3 = 0.734 \text{ kg}$$

$$m_2 = \frac{\pi}{4}(40)^2 20(8,550)\left(\frac{1}{1,000}\right)^3 = 0.215 \text{ kg} \qquad m = m_1 + m_2 = 0.734 + 0.215 = 0.949 \text{ kg}$$

The centroidal coordinate $x_c$ is found from

$$x_c = \frac{x_1 m_1 + x_2 m_2}{m_1 + m_2} = \frac{0 + 55(0.215)}{0.949} = 12.5 \text{ mm}$$

The mass moment of inertia of the rotor assembly about the axis of rotation is

$$I_z = I_{01} + I_2 = I_{01} + I_{02} + md^2$$

Using Case 4 in Table 17.13,

$$I_z = \tfrac{1}{8}m_1 d_1^2 + \tfrac{1}{8}m_2 d_2^2 + m_2 d^2 = \tfrac{1}{8}(0.734)\left(\frac{150}{1,000}\right)^2 + \tfrac{1}{8}(0.215)\left(\frac{40}{1,000}\right)^2 + 0.215\left(\frac{55}{1,000}\right)^2 = 2.76 \times 10^{-3} \text{ kg} \cdot \text{m}^2$$

(*b*) The tensile force in the string is designated *P*. The equation of motion of the rotor is

$$M_z = I_z \alpha \qquad P\left(\frac{75}{1,000}\right) \text{N} \cdot \text{m} = (2.76 \times 10^{-3} \text{ kg} \cdot \text{m}^2)\alpha$$

The equation of motion of the mass is

$$\sum F = ma_z \qquad \frac{600}{1,000}(9.81) - P = \frac{600}{1,000}a_z$$

Since the string is inextensible,

$$a_z = r\alpha \qquad a_z = \frac{75}{1,000}\alpha$$

The solutions to these equations are

$$a = 72.0 \text{ rad/s}^2 \qquad P = 2.65 \text{ N} \qquad a_y = 5.40 \text{ in/s}^2$$

(c) The angular acceleration of the rotor is constant, and the angular velocity of this element 2 s after starting from rest is

$$\omega = \omega_0 + \alpha t = 72.0(2) = 144 \text{ rad/s}$$

The center of mass of the rotor assembly travels in a circular path. The two components of the hinge pin force, in the horizontal plane, which act on the disk are shown in Fig. 18.59c. The components of the acceleration of the center of mass are

$$a_n = x_c\omega^2 = \frac{12.5}{1,000}(144)^2 = 259 \text{ m/s}^2 \qquad a_t = x_c\alpha = \frac{12.5}{1,000}(72.0) = 0.9 \text{ m/s}^2$$

The equations for the translational motion of the CM are

$$\mathbf{F} = m\mathbf{a}_c \qquad \sum F_x = ma_x \qquad -R_x = ma_n = 0.949(259) = 246 \text{ N} \qquad R_x = -246 \text{ N}$$
$$\sum F_y = ma_y \qquad -R_y + P = ma_t \qquad -R_y + 2.65 = 0.949(0.9) \qquad R_y = 1.8 \text{ N}$$

It may be seen that the normal force component $R_x$ is the predominant force effect.

(d) The weight force $\mathbf{W}$ is given by

$$\mathbf{W} = -mg\mathbf{k} = -0.949(9.81)\mathbf{k} = -9.31\mathbf{k} \qquad \text{N}$$

Using Fig. 18.59c,

$$\mathbf{P} = -2.65\mathbf{j} \qquad \text{N} \qquad \mathbf{R} = R_x\mathbf{i} + R_y\mathbf{j} = -246\mathbf{i} + 1.8\mathbf{j} \qquad \text{N}$$

**18.60** Figure 18.60a shows a system of weights and pulleys; the cable is assumed to not slip on the pulleys. The masses and mass moments of inertia are $m_A = 20 \text{ kg}$, $m_B = 14 \text{ kg}$, $I_{0C} = 0.005 \text{ kg} \cdot \text{m}^2$, $I_{0D} = 0.010 \text{ kg} \cdot \text{m}^2$.

(a) Neglect the mass moments of inertia of the pulleys, and find the acceleration of the weights, the tensile force in the cable, and the hinge pin forces.

(b) Using the acceleration in part (a), find the velocities of the weights and pulleys when weight B, starting from rest, has moved upward 900 mm.

(c) Do the same as in part (a), but include the mass moments of inertia of the pulleys.

(d) Do the same as in part (b), for the conditions of part (c).

▌ (a) Using Case 4 in Table 17.13,

$$I_0 = \tfrac{1}{8}md^2 \qquad I_{0C} = 0.005 = \tfrac{1}{8}m_C\left(\frac{180}{1,000}\right)^2 \qquad m_C = 1.23 \text{ kg}$$

$$I_{0D} = 0.010 = \tfrac{1}{8}m_D\left(\frac{260}{1,000}\right)^2 \qquad m_D = 1.18 \text{ kg}$$

Figures 18.60b and c show the free-body diagrams of the weights and pulleys. The cable tensile force has the constant value $T$. The equation of motion for weight $A$ is

$$\sum F_y = ma_y \qquad 196 - T = 20a_y \tag{1}$$

The equation of motion for weight $B$ is

$$\sum F_y = ma_y \qquad T - 137 = 14a_y \tag{2}$$

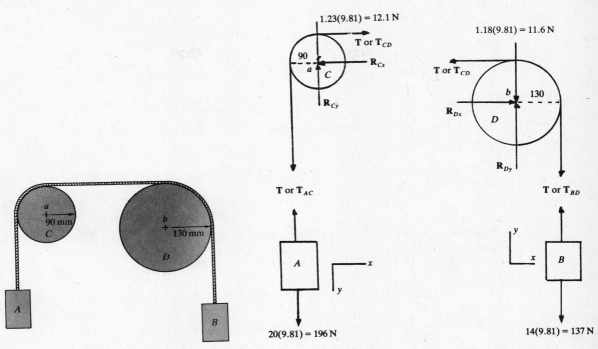

Fig. 18.60a

Fig. 18.60b

Fig. 18.60c

Adding Eqs. (1) and (2),

$$59 = 34a_y \qquad a_y = 1.74 \, \text{m/s}^2$$

Using Eq. (1),

$$196 - T = 20(1.74) \qquad T = 161 \, \text{N}$$

For equilibrium of pulley $C$,

$$\sum F_y = 0 \qquad -T + R_{Cy} - 12.1 = 0 \qquad R_{Cy} = 173 \, \text{N} \qquad R_C = \sqrt{R_{Cx}^2 + R_{Cy}^2} = \sqrt{161^2 + 173^2} = 236 \, \text{N}$$

For equilibrium of pulley $D$,

$$\sum F_x = 0 \qquad R_{DX} - T = 0 \qquad R_{Dx} = 161 \, \text{N}$$

$$\sum F_y = 0 \qquad R_{Dy} - 11.6 - T = 0 \qquad R_{Dy} = 173 \, \text{N} \qquad R_D = \sqrt{R_{Dx}^2 + R_{Dy}^2} = \sqrt{161^2 + 173^2} = 236 \, \text{N}$$

(b)  Using  $s = 900$ mm,

$$v^2 = v_0^2 + 2a(s - s_0) = 2(1.74)\left(\frac{900}{1,000}\right) \qquad v = 1.77 \, \text{m/s}$$

The pulley angular velocity $\omega$ is related to the weight velocity $v$ by

$$v = r\omega$$

For pulley $C$, $\qquad\qquad 1.77 = \dfrac{90}{1,000}\omega_C \qquad \omega_C = 19.7 \, \text{rad/s}$

For pulley $D$, $\qquad\qquad 1.77 = \dfrac{130}{1,000}\omega_D \qquad \omega_D = 13.6 \, \text{rad/s}$

(c)  When the mass moments of inertia are included, the cable tensile forces acting on pulley $C$ are $T_{AC}$ and $T_{CD}$, and the cable forces on pulley $D$ are $T_{BD}$ and $T_{CD}$. The equations of motion of the weights and pulleys are

Weight A: $\qquad\qquad \sum F_y = ma_y \qquad 196 - T_{AC} = 20a_y$ $\qquad\qquad$ (3)

Pulley C:
$$\sum M_0 = I_0\alpha \qquad (T_{AC} - T_{CD})\frac{90}{1,000} = 0.005\alpha_C \qquad (4)$$

Pulley D:
$$\sum M_0 = I_0\alpha \qquad (T_{CD} - T_{BD})\frac{130}{1,000} = 0.010\alpha_D \qquad (5)$$

Weight B:
$$\sum F_y = ma_y \qquad T_{BD} - 137 = 14a_y \qquad (6)$$

The translational and rotational motions are related by

$$a = r\alpha$$

$$a_y = \frac{90}{1,000}\alpha_C \qquad (7)$$

$$a_y = \frac{130}{1,000}\alpha_D \qquad (8)$$

Equations (7) and (8) are used to eliminate $\alpha_C$ and $\alpha_D$ in Eqs. (4) and (5). The equilibrium equations then have the forms

$$196 - T_{AC} = 20ay \qquad (9)$$

$$T_{AC} - T_{CD} = \frac{0.005}{0.090}\left(\frac{a_y}{0.090}\right) \qquad (10)$$

$$T_{CD} - T_{BD} = \frac{0.010}{0.130}\left(\frac{a_y}{0.130}\right) \qquad (11)$$

$$T_{BD} - 137 = 14a_y \qquad (12)$$

Equations (9) through (12) are added, to obtain

$$59 = 35.2a_y \qquad a_y = 1.68 \, \text{m/s}^2$$

The percent decrease in acceleration, from the result in part (a), is

$$\%\text{D} = \frac{1.68 - 1.74}{1.73}(100) = -3.4\%$$

Using Eq. (9),

$$T_{AC} = 162 \, \text{N}$$

Using Eq. (10),

$$T_{CD} = 161 \, \text{N}$$

Using Eq. (11)

$$T_{BD} = 160 \, \text{N}$$

For equilibrium of pulley C,

$$\sum F_y = 0 \qquad -T_{AC} + R_{Cy} - 12.1 = 0 \qquad R_{Cy} = 174 \, \text{N}$$

$$\sum F_x = 0 \qquad -R_{Cx} + T_{CD} = 0 \qquad R_{Cx} = 161 \, \text{N} \qquad R_C = \sqrt{R_{Cx}^2 + R_{Cy}^2} = \sqrt{161^2 + 174^2} = 237 \, \text{N}$$

For equilibrium of pulley D,

$$\sum F_x = 0 \qquad R_{Dx} - T_{CD} = 0 \qquad R_{Dx} = 161 \, \text{N}$$

$$\sum F_y = 0 \qquad R_{Dy} - 11.6 - T_{BD} = 0 \qquad R_{Dy} = 172 \, \text{N}$$

$$R_D = \sqrt{R_{Dx}^2 + R_{Dy}^2} = \sqrt{161^2 + 172^2} = 236 \, \text{N}$$

(d) Using $s = 900 \, \text{mm}$,

$$v^2 = v_0^2 + 2a(s - s_0) = 2(1.68)\left(\frac{900}{1,000}\right) \qquad v = 1.74 \, \text{m/s}$$

$$v = r\omega \qquad 1.74 = \frac{90}{1,000}\omega_C \qquad \omega_C = 19.3 \, \text{rad/s}$$

$$1.74 = \frac{130}{1,000}\omega_D \qquad \omega_D = 13.4 \, \text{rad/s}$$

It may be seen that the inclusion of the mass moments of inertia of the pulleys in this problem has a very small effect on the cable forces and on the acceleration of the blocks.

**18.61** **(a)** Do the same as in Prob. 18.60, parts (a) through (d), if the system is modified to that shown in Fig. 18.61a. Pulley E has the same dimensions, and mass properties, as pulley D.

**(e)** Express the resultant cable force on pulley D in formal vector notation.

**▮ (a)** If the mass moments of inertia of the pulleys are neglected, the acceleration of the weights, and the cable tensile force, have the same values as in part (a) of Prob. 18.60, given by

$$a_y = 1.74 \text{ m/s}^2 \qquad T_{AC} = T_{CD} = T_{DE} = T_{BE} = 161 \text{ N}$$

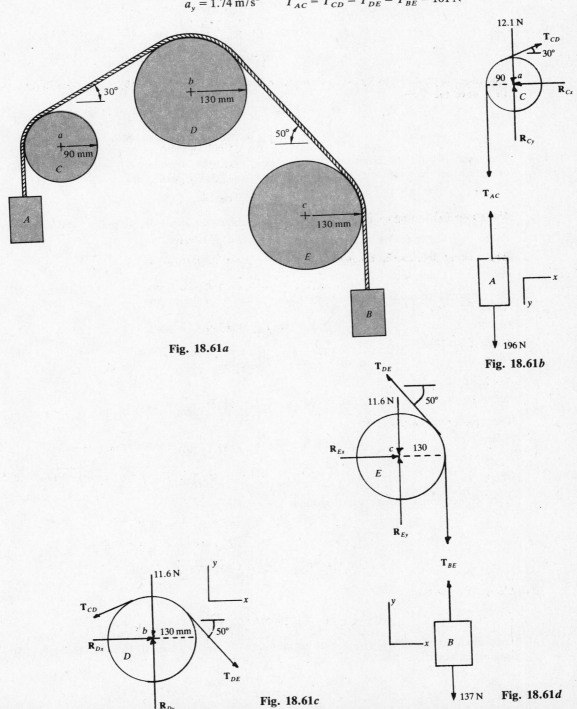

Fig. 18.61a

Fig. 18.61b

Fig. 18.61c

Fig. 18.61d

The free-body diagrams of the weights and pulleys are shown in Figs. 18.61$b$ through $d$. For equilibrium of the pulleys,

Pulley C:
$$\sum F_x = 0 \quad -R_{Cx} + T_{CD}\cos 30° = 0 \quad R_{Cx} = 139\,\text{N} \tag{1}$$

$$\sum F_y = 0 \quad -T_{AC} + R_{Cy} - 12.1 + T_{CD}\sin 30° = 0 \quad R_{Cy} = 92.6\,\text{N} \tag{2}$$

$$R_C = \sqrt{R_{Cx}^2 + R_{Cy}^2} = \sqrt{139^2 + 92.6^2} = 167\,\text{N}$$

Pulley D:
$$\sum F_x = 0 \quad -T_{CD}\cos 30° + R_{Dx} + T_{DE}\cos 50° = 0 \tag{3}$$

$$R_{Dx} = 35.9\,\text{N}$$

$$\sum F_y = 0 \quad -T_{CD}\sin 30° - 11.6 + R_{Dy} - T_{DE}\sin 50° = 0 \quad R_{Dy} = 215\,\text{N} \tag{4}$$

$$R_D = \sqrt{R_{Dx}^2 + R_{Dy}^2} = \sqrt{35.9^2 + 215^2} = 218\,\text{N}$$

Pulley E:
$$\sum F_x = 0 \quad R_{Ex} - T_{DE}\cos 50° = 0 \quad R_{Ex} = 103\,\text{N} \tag{5}$$

$$\sum F_y = 0 \quad R_{Ey} - 11.6 + T_{DE}\sin 50° - T_{BE} = 0 \quad R_{Ey} = 49.3\,\text{N} \tag{6}$$

$$R_E = \sqrt{R_{Ex}^2 + R_{Ey}^2} = \sqrt{103^2 + 49.3^2} = 114\,\text{N}$$

($b$) The results for the velocities are the same as in part ($b$) of Prob. 18.60, since, from part ($a$) of this problem, the acceleration of the blocks is the same. These results are

$$v = 1.77\,\text{ms/s} \quad \omega_C = 19.7\,\text{rad/s} \quad \omega_D = 13.6\,\text{rad/s}$$

($c$) When the mass moments of inertia are included, the cable forces on the pulleys are as shown below.

Pulley C: Cable forces $T_{AC}$ and $T_{CD}$
Pulley D: Cable forces $T_{CD}$ and $T_{DE}$
Pulley E: Cable forces $T_{BE}$ and $T_{DE}$

The equations of motion of the weights and pulleys have the forms

Weight A:
$$\sum F_y = ma_y \quad 196 - T_{AC} = 20a_y \tag{7}$$

Pulley C:
$$\sum M_0 = I_0\alpha \quad (T_{AC} - T_{CD})\frac{90}{1,000} = 0.005\alpha_C \tag{8}$$

Pulley D:
$$\sum M_0 = I_0\alpha \quad (T_{CD} - T_{DE})\frac{130}{1,000} = 0.010\alpha_D \tag{9}$$

Pulley E:
$$\sum M_0 = I_0\alpha \quad (T_{DE} - T_{BE})\frac{130}{1,000} = 0.010\alpha_E \tag{10}$$

Weight B:
$$\sum F_y = ma_y \quad T_{BE} - 137 = 14a_y \tag{11}$$

The translational and rotational motions are related by

$$a = r\alpha$$

$$a_y = \frac{90}{1,000}\alpha_C \tag{12}$$

$$a_y = \frac{130}{1,000}\alpha_D \tag{13}$$

$$a_y = \frac{130}{1,000}\alpha_E \tag{14}$$

Equations (12) through (14) are used to eliminate $\alpha_C$, $\alpha_D$, and $\alpha_E$ in Eqs. (8) through (10). The equilibrium equations then appear as

$$196 - T_{AC} = 20a_y \tag{15}$$

$$T_{AC} - T_{CD} = \frac{0.005}{0.090}\left(\frac{a_y}{0.090}\right) \tag{16}$$

$$T_{CD} - T_{DE} = \frac{0.010}{0.130}\left(\frac{a_y}{0.130}\right) \qquad (17)$$

$$T_{DE} - T_{BE} = \frac{0.010}{0.130}\left(\frac{a_y}{0.130}\right) \qquad (18)$$

$$T_{BE} - 137 = 14a_y \qquad (19)$$

Equations (15) through (19) are added, to obtain

$$59 = 36.0a_y \qquad a_y = 1.64 \text{ m/s}^2$$

Using Eq. (15),

$$T_{AC} = 163 \text{ N}$$

Using Eq. (16),

$$T_{CD} = 162 \text{ N}$$

Using Eq. (17),

$$T_{DE} = 161 \text{ N}$$

Using Eq. (18)

$$T_{BE} = 160 \text{ N}$$

As a check, using Eq. (19),

$$T_{BE} \overset{?}{=} 137 + 14a_y \qquad 160 \overset{?}{=} 137 + 14(1.64) \qquad 160 = 160$$

Using Eqs. (1) and (2),

$$-R_{Cx} + 162 \cos 30° = 0 \qquad R_{Cx} = 140 \text{ N}$$

$$-163 + R_{Cy} - 12.1 + 162 \sin 30° = 0 \qquad R_{Cy} = 94.1 \text{ N} \qquad R_C = \sqrt{R_{Cx}^2 + R_{Cy}^2} = \sqrt{140^2 + 94.1^2} = 169 \text{ N}$$

Using Eqs. (3) and (4),

$$-162 \cos 30° + R_{Dx} + 161 \cos 50° = 0 \qquad R_{Dx} = 36.8 \text{ N}$$

$$-162 \sin 30° - 11.6 + R_{Dy} - 161 \sin 50° = 0 \qquad R_{Dy} = 216 \text{ N}$$

$$R_D = \sqrt{R_{Dx}^2 + R_{Dy}^2} = \sqrt{36.8^2 + 216^2} = 219 \text{ N}$$

Using Eqs. (5) and (6),

$$R_{Ex} - 161 \cos 50° = 0 \qquad R_{Ex} = 103 \text{ N}$$

$$R_{Ey} - 11.6 + 161 \sin 50° - 160 = 0 \qquad R_{Ey} = 48.3 \text{ N}$$

$$R_E = \sqrt{R_{Ex}^2 + R_{Ey}^2} = \sqrt{103^2 + 48.3^2} = 114 \text{ N}$$

(d) Using $s = 900$ mm,

$$v^2 = v_0^2 + 2a(s - s_0) = 2(1.64)\left(\frac{900}{1,000}\right) \qquad v = 1.72 \text{ m/s}$$

$$v = r\omega$$

$$1.72 = \frac{90}{1,000}\omega_C \qquad \omega_C = 19.1 \text{ rad/s} \qquad 1.72 = \frac{130}{1,000}\omega_D \qquad \omega_D = 13.2 \text{ rad/s}$$

$$\omega_E = \omega_D = 13.2 \text{ rad/s}^2$$

As was seen in Prob. 18.60, the mass moments of inertia of the pulleys have an insignificant effect on the cable forces and on the motions of the weights and pulleys.

(e) The resultant cable force $\mathbf{T}$ on pulley $D$, from Fig. 18.61c, is

$$\mathbf{T} = \mathbf{T}_{CD} + \mathbf{T}_{DE} = -(T_{CD} \cos 30°)\mathbf{i} - (T_{CD} \sin 30°)\mathbf{j} + (T_{DE} \cos 50°)\mathbf{i} - (T_{DE} \sin 50°)\mathbf{j}$$

Using $T_{CD} = 162$ N and $T_{DE} = 161$ N,

$$\mathbf{T} = -(162 \cos 30°)\mathbf{i} - (162 \sin 30°)\mathbf{j} + (161 \cos 50°)\mathbf{i} - (161 \sin 50°)\mathbf{j}$$

$$= -36.8\mathbf{i} - 204\mathbf{j} \quad \text{N}$$

$$T = \sqrt{T_x^2 + T_y^2} = \sqrt{(-36.8)^2 + (-20.4)^2} = 42.1 \text{ N}$$

As a check on the above calculations, for equilibrium of pulley $D$,

$$\mathbf{R}_D + \mathbf{T}_{CD} + \mathbf{T}_{DE} + \mathbf{W} \overset{?}{=} 0 \qquad 36.8\mathbf{i} + 216\mathbf{j} - 36.8\mathbf{i} - 204\mathbf{j} - 11.6\mathbf{j} \overset{?}{=} 0$$

$$216 \overset{?}{=} 204 + 11.6 \qquad 216 \approx 212$$

**18.62** A body with the form of a rectangular parallelepiped moves without slipping on two cylindrical rollers, as shown in Fig. 18.62a. The mass of each roller is 1,200 kg, and the mass of the body is 2,200 kg. Find the acceleration of the body, and the translational and angular acceleration of the rollers.

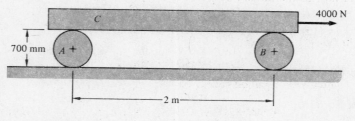

Fig. 18.62a

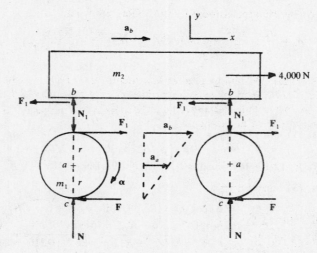

Fig. 18.62b

❙ Figure 18.62b shows the free-body diagrams of the body and the rollers. $F$ is the friction force between the rollers and the ground, and $F_1$ is the friction force between the rollers and the body. The rotational equation of motion of the rollers is

$$\sum M_0 = I_0\alpha \qquad F_1 r + Fr = I_0\alpha \tag{1}$$

From Case 4 in Table 17.13,

$$I_0 = \tfrac{1}{8}md^2 = \tfrac{1}{2}m_1 r^2$$

Using the above result, and $\alpha = a_a/r$, in Eq. (1) yields

$$F_1 + F = \left(\frac{\tfrac{1}{2}m_1 r^2}{r}\right)\frac{a_a}{r} = \tfrac{1}{2}m_1 a_a \tag{2}$$

The translational equation of motion of a roller is

$$\sum F_x = ma \qquad F_1 - F = m_1 a_a \tag{3}$$

Equations (2) and (3) are added, with the result

$$2F_1 = \tfrac{3}{2}m_1 a_a$$
$$F_1 = \tfrac{3}{4}m_1 a_a \tag{4}$$

An alternative method of finding $F_1$ is to sum moments about the instant center, point $c$. The result is

$$M_c = I_c\alpha \qquad F_1(2r) = (I_0 + m_1 r^2)\alpha \tag{5}$$

Using

$$I_0 = \tfrac{1}{2}m_1 r^2 \qquad \text{and} \qquad \alpha = \frac{a_a}{r}$$

in Eq. (5) yields

$$F_1(2r) = (\tfrac{1}{2}m_1 r^2 + m_1 r^2)\frac{a_a}{r} \qquad F_1 = \tfrac{3}{4}m_1 a_a$$

The acceleration of point $b$ is

$$a_b = (2r)\alpha = (2r)\left(\frac{a_a}{r}\right) = 2a_a \tag{6}$$

Newton's second law for the motion of the body is

$$\sum F_x = ma \qquad 4{,}000 - 2F_1 = m_2 a_b \tag{7}$$

Using Eq. (4) in Eq. (7) results in

$$4{,}000 - 2(\tfrac{3}{4} m_1 a_a) = m_2 a_b$$

Using Eq. (6) in the above equation results in

$$4{,}000 - 2\left[\frac{3}{4}(1{,}200)\frac{a_b}{2}\right] = 2{,}200 a_b \qquad a_b = 1.29 \text{ m/s}^2$$

The translational and angular accelerations of the rollers are found to be

$$a_a = \frac{a_b}{2} = \frac{1.29}{2} = 0.645 \text{ m/s}^2 \qquad \alpha = \frac{a_a}{r} = \frac{0.645}{350/1{,}000} = 1.84 \text{ rad/s}^2$$

**18.63**  Figure 18.63$a$ shows a rack-and-pinion gear arrangement.  The rack may be approximated as a steel rod of 0.5-in by 0.5-in square cross section, and the pinion may be approximated as a 0.5-in-thick steel disk.  All frictional effects may be neglected.  The specific weight of steel is 489 lb/ft$^3$.  At  $t = 0$  the system is released from rest in the position shown in the figure.

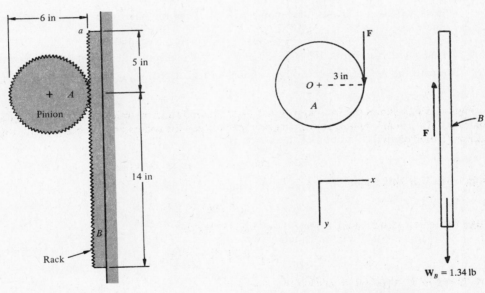

**Fig. 18.63$a$**                         **Fig. 18.63$b$**

(a)  Find the acceleration of the rack.

(b)  Find the time at which the rack loses contact with the pinion.

(c)  Find the speed of the rack at the position where the rack loses contact with the pinion.

(d)  How would the results in parts (a) through (c) change if the pinion had zero mass?

▌ (a)  Figure 18.63$b$ shows the tangential force $F$ transmitted between the rack and the pinion.  The weights of these two elements are

$$W_A = \frac{\pi(6)^2}{4}(0.5)\frac{489}{(12)^3} = 4.00 \text{ lb} \qquad W_B = 0.5(0.5)19\left(\frac{489}{12^3}\right) = 1.34 \text{ lb}$$

Newton's second law for the motion of the pinion is

$$M_0 = I_0 \alpha$$

Using $I_0 = \frac{1}{8}md^2$ in the above equation yields

$$F(3) = \left[\frac{1}{8}\left(\frac{4.00}{386}\right)6^2\right]\alpha$$

Using $a = r\alpha$, or $a = 3\alpha$, in the above equation results in

$$3F = \frac{1}{8}\left(\frac{4.00}{386}\right)6^2\left(\frac{a}{3}\right) \qquad F = 0.00518a \tag{1}$$

The equation of motion of the rack is

$$\sum F_y = ma \qquad 1.34 - F = \frac{1.34}{386}a \qquad F = 1.34 - 0.00347a \tag{2}$$

$F$ is eliminated between Eqs. (1) and (2), to obtain

$$0.00518a = 1.34 - 0.00347a \qquad a = 155 \text{ in/s}^2$$

(b)  With $v_0 = 0$, $s_0 = 0$, and $s = 5$ in,

$$s = s_0 + v_0t + \frac{1}{2}at^2 \qquad 5 = \frac{1}{2}(155)t^2 \qquad t = 0.254 \text{ s}$$

(c)  The speed of the rack when it loses contact with the pinion is found from

$$v = v_0 + at = 155(0.254) = 39.4 \text{ in/s}$$

(d)  If the pinion has zero mass,

$$a_{\text{rack}} = g = 386 \text{ in/s}^2 \qquad s = s_0 + v_0t + \frac{1}{2}at^2 \qquad 5 = \frac{1}{2}(386)t^2 \qquad t = 0.161 \text{ s}$$
$$v = v_0 + at = 386(0.161) = 62.1 \text{ in/s}$$

It may be seen that the mass moment of inertia of the pinion has a significant effect on the motion of the rack.

18.64  (a)  Do the same as in Prob. 18.63, if the pinion must overcome a constant friction moment of 1.2 in · lb.
(b)  Compare the results in part (a) with those in Prob. 18.63.

▎(a)  From Prob. 18.63,

$$W_A = 4.00 \text{ lb} \qquad W_B = 1.34 \text{ lb}$$

The equation of motion of the pinion is

$$\sum M_0 = I_0\alpha$$

Using $I_0 = \frac{1}{8}md^2$ in the above equation results in

$$F(3) - 1.2 = \left[\frac{1}{8}\left(\frac{4.00}{386}\right)6^2\right]\alpha$$

Using $\alpha = a/3$ in the equation above gives

$$3F - 1.2 = \frac{1}{8}\left(\frac{4.00}{386}\right)6^2\left(\frac{a}{3}\right) \qquad F = 0.4 + 0.00518a \tag{1}$$

The equation of motion of the rack is

$$\sum F_y = ma \qquad 1.34 - F = \frac{1.34}{386}a \qquad F = 1.34 - 0.00347a \tag{2}$$

Equations (1) and (2) are equated, to obtain

$$0.4 + 0.00518a = 1.34 - 0.00347a \qquad a = 109 \text{ in/s}^2$$

Using $v_0 = 0$ and $s_0 = 0$,

$$s = s_0 + v_0t + \frac{1}{2}at^2 \qquad 5 = \frac{1}{2}(109)t^2 \qquad t = 0.303 \text{ s}$$

The speed of the rack when it loses contact with the pinion is

$$v = v_0 + at = 109(0.303) = 33.0 \text{ in/s}$$

The equation of motion of the pinion, if this element has zero mass, is

$$\sum M_0 = 0 \qquad F(3) - 1.2 = 0 \qquad F = 0.4 \text{ lb}$$

The equation of motion of the rack is

$$\sum F_y = ma \qquad 1.34 - F = 1.34 - 0.4 = \frac{1.34}{386} a \qquad a = 271 \text{ in/s}^2$$

$$s = s_0 + v_0 t + \tfrac{1}{2} at^2 \qquad 5 = \tfrac{1}{2}(271)t^2 \qquad t = 0.192 \text{ s} \qquad v = v_0 + at = 271(0.192) = 52.0 \text{ in/s}$$

(b)  Table 18.2 shows a comparison of the results of Prob. 18.63 with the results of part (a) above, giving the times at which the rack loses contact with the pinion and the speeds of the rack at those times.

**TABLE 18.2**

|  | t, s | v, in/s |
|---|---|---|
| Pinion mass, with friction | 0.303 | 33.0 |
| Pinion mass, with no friction | 0.254 | 39.4 |
| No pinion mass, with friction | 0.192 | 52.0 |
| No pinion mass, with no friction | 0.161 | 62.1 |

18.65  Figure 18.65a shows a pair of meshing gears that lie in a common plane. The gears are designated 1 and 2, with the corresponding number of teeth $N_1$ and $N_2$. The kinematic action of the set of gears is that of pure rolling of two cylinders, with the pitch diameters $d_1$ and $d_2$. The equations which describe the motion and geometry relationships between the gears are

$$\frac{\theta_1}{\theta_2} = \frac{\omega_1}{\omega_2} = \frac{\alpha_1}{\alpha_2} = \frac{d_2}{d_1} = \frac{N_2}{N_1}$$

The mass moment of inertia of a gear may be assumed to be the same as that of a disk whose diameter is equal to that of the pitch diameter of the gear, and with the thickness of the gear. The gears are steel, with $\gamma = 489 \text{ lb/ft}^3$, $d_1 = 4 \text{ in}$, and $d_2 = 10 \text{ in}$. The thickness of the gears is 1 in.

(a)  Find the accelerations of gears 1 and 2, if a clockwise moment of magnitude 15 in · lb, in the plane of the gear, is applied to gear 1.

(b)  Do the same as in part (a), if the moment is applied to gear 2 instead of gear 1.

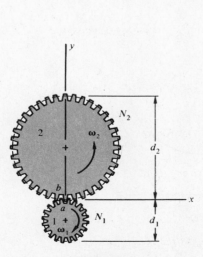

Fig. 18.65a

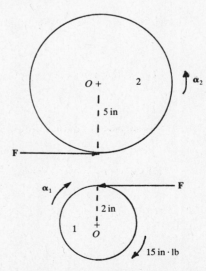

Fig. 18.65b

▌ (a) Figure 18.65b shows the tangential force $F$ transmitted between the gears. Using Case 4 in Table 17.13,

$$I_0 = \tfrac{1}{8}md^2 \qquad I_{01} = \frac{1}{8}\left[\frac{\pi(4)^2}{4}(1)\frac{489}{386(12)^3}\right]4^2 = 0.0184\,\text{lb}\cdot\text{s}^2\cdot\text{in}$$

$$I_{02} = \frac{1}{8}\left[\frac{\pi(10)^2}{4}(1)\frac{489}{386(12)^3}\right]10^2 = 0.720\,\text{lb}\cdot\text{s}^2\cdot\text{in}$$

The equation of motion of gear 1 is

$$\sum M_0 = I_0\alpha \qquad 15 - F(2) = 0.0184\alpha_1 \qquad\qquad (1)$$

The equation of motion of gear 2 is

$$M_0 = I_0\alpha \qquad F(5) = 0.720\alpha_2$$

Using $\alpha_2 = \tfrac{4}{10}\alpha_1$, the above equation apears as

$$5F = 0.720(\tfrac{4}{10}\alpha_1) \qquad\qquad (2)$$

$F$ is eliminated between Eqs. (1) and (2), to obtain

$$15 - 2\left[\frac{0.720}{5}\left(\frac{4}{10}\alpha_1\right)\right] = 0.0184\alpha_1 \qquad \alpha_1 = 112\,\text{rad/s}^2$$

$\alpha_2$ is then found as

$$\alpha_2 = \tfrac{4}{10}\alpha_1 = 44.8\,\text{rad/s}^2$$

(b) For the case where the clockwise moment of 15 in·lb is applied to gear 2,

$$M_0 = I_0\alpha \qquad F(2) = 0.0184\alpha_1 \qquad\qquad (3)$$

$$\sum M_0 = I_0\alpha \qquad 15 - F(5) = 0.720\alpha_2 \qquad\qquad (4)$$

$$\alpha_2 = \tfrac{4}{10}\alpha_1 \qquad\qquad (5)$$

$F$ is eliminated between Eqs. (3) and (4), using Eq. (5), with the results

$$15 - 5\left(\frac{0.0184\alpha_1}{2}\right) = 0.720(\tfrac{4}{10}\alpha_1) \qquad \alpha_1 = 44.9\,\text{rad/s} \qquad \alpha_2 = \tfrac{4}{10}\alpha_1 = 18.0\,\text{rad/s}$$

When the moment is applied to gear 1, the angular acceleration of the gears is $112/44.9 = 2.49$ times greater than the angular acceleration for the case where the moment is applied to gear 2. It may thus be seen that the magnitudes of the angular accelerations of the gears are highly dependent on the gear to which the moment is applied.

**18.66** Do the same as in Prob. 18.65, if the same clockwise moment of 15 in·lb is applied to both gears.

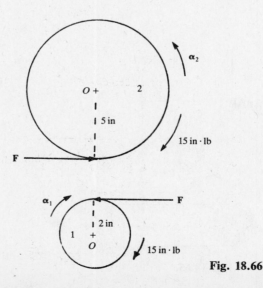

**Fig. 18.66**

❚ Figure 18.66 shows the tangential force $F$ transmitted between the gears. The equations of motion of these elements are

Gear 1:  $\sum M_0 = I_0\alpha$     $15 - F(2) = 0.0184\alpha_1$                                    (1)

Gear 2:  $\sum M_0 = I_0\alpha$     $F(5) - 15 = 0.720\alpha_2$

Using  $\alpha_2 = \frac{4}{10}\alpha_1$  in the above equation results in

$$5F - 15 = 0.720(\tfrac{4}{10}\alpha_1)$$                                    (2)

$F$ is eliminated between Eqs. (1) and (2), with the final results

$$\frac{15 - 0.0184\alpha_1}{2} = \frac{15 + 0.720(\tfrac{4}{10}\alpha_1)}{5}   \qquad \alpha_1 = 67.4 \text{ rad/s}^2 \qquad \alpha_2 = \tfrac{4}{10}\alpha_1 = 27.0 \text{ rad/s}^2$$

It may be seen that, when the same moment is applied to both gears, the angular accelerations lie between the limiting values found in Prob. 18.66.

18.67   A counterclockwise moment of magnitude $1.2\,\text{N}\cdot\text{m}$ is applied to gear 1 in Fig. 18.67a. Find the angular accelerations of the three gears. The gears are made of 20-mm-thick gray cast iron, with  $\rho = 7,080 \text{ kg/m}^3$.

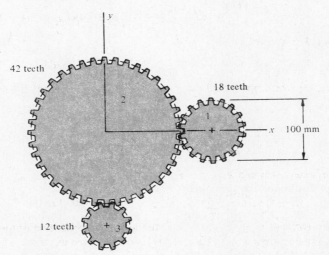

Fig. 18.67a

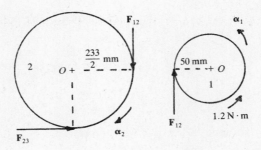

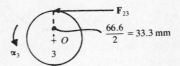

Fig. 18.67b

❚ The diameters of gears 2 and 3 are found as

$$\frac{d_2}{d_1} = \frac{N_2}{N_1} \qquad \frac{d_2}{100} = \frac{42}{18} \qquad d_2 = 233 \text{ mm}$$

$$\frac{d_2}{d_3} = \frac{N_2}{N_3} \qquad \frac{233}{d_3} = \frac{42}{12} \qquad d_3 = 66.6 \text{ mm}$$

Figure 18.67 shows the tangential forces transmitted between each pair of gears. The mass moments of inertia of the gears, using Case 4 in Table 17.13, are

$$m = \tfrac{1}{4}\rho\pi d^2 h \qquad I_0 = \tfrac{1}{8}md^2 = \tfrac{1}{32}\rho\pi d^4 h = \underbrace{\tfrac{1}{32}(7{,}080)\pi(20)\left(\frac{1}{1{,}000}\right)^5 d^4}_{q}$$

$$q = 1.39 \times 10^{-11} \text{ kg}\cdot\text{m}^2/\text{mm}^4 \qquad I_0 = qd^4$$

For gear 1,

$$I_{01} = q(100)^4 = 0.00139 \text{ kg}\cdot\text{m}^2$$

The mass moment of inertia of gear 2 is given by

$$I_{02} = q(233)^4 = 0.0410 \text{ kg}\cdot\text{m}^2$$

The mass moment of inertia of gear 3 is found as

$$I_{03} = q(66.6)^4 = 2.73 \times 10^{-4} \text{ kg}\cdot\text{m}^2$$

Using

$$\frac{\alpha_1}{\alpha_2} = \frac{N_2}{N_1} = \frac{42}{18}$$

$$\alpha_2 = \frac{18}{42}\alpha_1$$

Using

$$\frac{\alpha_2}{\alpha_3} = \frac{N_3}{N_2} = \frac{12}{42}$$

$$\alpha_3 = \tfrac{42}{12}\alpha_2 = \tfrac{42}{12}\left(\tfrac{18}{42}\alpha_1\right) = \tfrac{18}{12}\alpha_1$$

The equations of motion of the three gears have the forms

Gear 1: $\qquad \sum M_0 = I_0\alpha \qquad 1.2 - F_{12}\left(\frac{50}{1{,}000}\right) = 0.00139\alpha_1$ $\qquad\qquad$ (1)

Gear 2: $\qquad \sum M_0 = I_0\alpha \qquad (F_{12} - F_{23})\left[\frac{233}{2(1{,}000)}\right] = 0.0410\alpha_2 = 0.0410(\tfrac{18}{42}\alpha_1)$ $\qquad$ (2)

Gear 3: $\qquad M_0 = I_0\alpha \qquad F_{23}\left(\frac{33.3}{1{,}000}\right) = 2.73 \times 10^{-4}\alpha_3 = 2.73 \times 10^{-4}(\tfrac{18}{12}\alpha_1)$ $\qquad$ (3)

From Eq. (1),

$$F_{12} = 24 - 0.0278\alpha_1 \qquad\qquad (4)$$

From Eq. (3),

$$F_{23} = 0.0123\alpha_1 \qquad\qquad (5)$$

Equations (4) and (5) are used in Eq. (2), with the results

$$\frac{233}{2(1{,}000)}(24 - 0.0278\alpha_1 - 0.0123\alpha_1) = 0.0410(\tfrac{18}{42}\alpha_1)$$

$$\alpha_1 = 126 \text{ rad/s}^2 \qquad \alpha_2 = \tfrac{18}{42}\alpha_1 = 54.0 \text{ rad/s}^2 \qquad \alpha_3 = \tfrac{18}{12}\alpha_1 = 189 \text{ rad/s}^2$$

**18.68** Do the same as in Prob. 18.67, if the moment is applied to gear 2.

❚ Figure 18.68 shows the tangential forces transmitted between each pair of gears. The values of $I_0$, $d_2$, and $d_3$, and the relationships among $\alpha_1$, $\alpha_2$, and $\alpha_3$, found in Prob. 18.67, are used.
The equations of motion of the gears have the forms

Gear 1: $\qquad M_0 = I_0\alpha \qquad F_{12}\left(\frac{50}{1{,}000}\right) = 0.00139\alpha_1$ $\qquad\qquad$ (1)

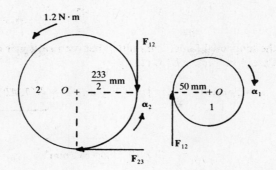

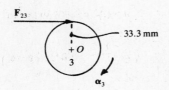

Fig. 18.68

*Gear 2:* $\qquad \sum M_0 = I_0\alpha \qquad 1.2 - (F_{12} + F_{23})\left[\dfrac{233}{2(1,000)}\right] = 0.0410\alpha_2 = 0.0410(\tfrac{18}{42}\alpha_1)$ $\qquad\qquad$ (2)

*Gear 3:* $\qquad M_0 = I_0\alpha \qquad F_{23}\left[\dfrac{33.3}{(1,000)}\right] = 2.73 \times 10^{-4}\alpha_3 = 2.73 \times 10^{-4}(\tfrac{18}{12}\alpha_1)$ $\qquad\qquad$ (3)

From Eq. (1),

$$F_{12} = 0.0278\alpha_1 \qquad\qquad (4)$$

From Eq. (3),

$$F_{23} = 0.0124\alpha_1 \qquad\qquad (5)$$

Equations (4) and (5) are used in Eq. (2), with the results

$$1.2 - (0.0278\alpha_1 + 0.0124\alpha_1)\left(\frac{233}{2(1,000)}\right) = 0.0410(\tfrac{18}{42}\alpha_1)$$

$$\alpha_1 = 53.9\,\text{rad/s}^2 \qquad \alpha_2 = \tfrac{18}{42}\alpha_1 = 23.1\,\text{rad/s}^2 \qquad \alpha_3 = \tfrac{18}{12}\alpha_1 = 80.9\,\text{rad/s}^2$$

**18.69** (*a*)  Do the same as in Prob. 18.67, if the moment is applied to gear 3.

$\qquad$ (*b*)  Compare the results obtained in part (*a*) with those in Probs. 18.67 and 18.68.

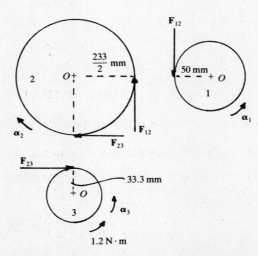

Fig. 18.69

**┃** (a) Figure 18.69 shows the tangential forces transmitted between each pair of gears. The values of $I_0$, $d_2$, and $d_3$, and the relationships among $\alpha_1$, $\alpha_2$, and $\alpha_3$, found in Prob. 18.67, are used.

*Gear 1:* 
$$M_0 = I_0 \alpha \qquad F_{12}\left(\frac{50}{1,000}\right) = 0.00139 \alpha_1 \tag{1}$$

*Gear 2:* 
$$\sum M_0 = I_0 \alpha \qquad (F_{23} - F_{12})\left[\frac{233}{2(1,000)}\right] = 0.0410 \alpha_2 = 0.0410(\tfrac{18}{42}\alpha_1) \tag{2}$$

*Gear 3:* 
$$\sum M_0 = I_0 \alpha \qquad 1.2 - F_{23}\left[\frac{33.3}{(1,000)}\right] = 2.73 \times 10^{-4}\alpha_3 = 2.73 \times 10^{-4}(\tfrac{18}{12}\alpha_1) \tag{3}$$

From Eq. (1),

$$F_{12} = 0.0278 \alpha_1 \tag{4}$$

From Eq. (3),

$$F_{23} = 36.0 - 0.0123 \alpha_1 \tag{5}$$

Equations (4) and (5) are used in Eq. (2), to obtain

$$[(36.0 - 0.0124\alpha_1) - 0.0278\alpha_1]\left[\frac{233}{2(1,000)}\right] = 0.0410(\tfrac{18}{42}\alpha_1)$$

$$\alpha_1 = 189 \text{ rad/s}^2 \qquad \alpha_2 = \tfrac{18}{42}\alpha_1 = 81.0 \text{ rad/s}^2 \qquad \alpha_3 = \tfrac{18}{12}\alpha_1 = 284 \text{ rad/s}^2$$

(b) The values of the angular accelerations of the three gears, for application of the moment to either gear 1, gear 2, or gear 3, are shown in Table 18.3. A general conclusion to be drawn from these problems is that application of the moment to the *smallest* gear produces the *largest* angular acceleration of the system of gears.

**TABLE 18.3**

| moment applied to | $d$, mm | $\alpha_1$, rad/s$^2$ | $\alpha_2$, rad/s$^2$ | $\alpha_3$, rad/s$^2$ |
|---|---|---|---|---|
| Gear 3 | 66.6 | 189 | 81.0 | 284 |
| Gear 1 | 100 | 126 | 54.0 | 189 |
| Gear 2 | 233 | 53.9 | 23.1 | 80.9 |

**18.70** The gears of Prob. 18.65 are shown in Fig. 18.70a. An inextensible cable is wrapped around a cylinder, of negligible mass, attached to gear 1. An 8-lb weight is attached to the cable.

(a) Find the acceleration of the weight and the force in the cable.

(b) Find the speed of the weight after this element, starting from rest, has lowered through a distance of 20 in.

**┃** (a) Figure 18.70b shows the tangential force $F_{12}$ transmitted between gears 1 and 2, and the cable tensile force $T$. From Prob. 18.65,

$$I_{01} = 0.0184 \text{ lb} \cdot \text{s}^2 \cdot \text{in}$$
$$I_{02} = 0.720 \text{ lb} \cdot \text{s}^2 \cdot \text{in}$$

Newton's second law for the motions of the weight and gears have the forms

*Weight:* 
$$\sum F_y = ma \qquad 8 - T = \tfrac{8}{386}a \tag{1}$$

*Gear 1:* 
$$\sum M_0 = I_0 \alpha \qquad T(1) - F_{12}(2) = 0.0184\alpha_1 \qquad r\alpha = a \qquad 1(\alpha_1) = a \qquad \alpha_1 = \frac{a}{1}$$

$$T - 2F_{12} = 0.0184a \tag{2}$$

*Gear 2:* 
$$M_0 = I_0 \alpha \qquad F_{12}(5) = 0.720\alpha_2 \qquad \alpha_2 = \tfrac{4}{10}\alpha_1 = \tfrac{4}{10}a \qquad 5F_{12} = 0.720(\tfrac{4}{10}a) \tag{3}$$

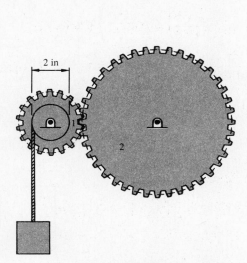

Fig. 18.70a                                          Fig. 18.70b

$T$ is eliminated between Eqs. (1) and (2), with the result

$$8 - \tfrac{8}{386}a = 2F_{12} + 0.0184a \qquad F_{12} = -0.0196a + 4 \qquad (4)$$

Using Eq. (4) in Eq. (3),

$$5(-0.0196a + 4) = 0.720(\tfrac{4}{10}a) \qquad a = 51.8 \text{ in/s}^2$$

Using Eq. (1),

$$T = 8 - \tfrac{8}{386}(51.8) = 6.93 \text{ lb}$$

(b)   The speed of the weight, after moving through 20 in, is found from

$$v^2 = v_0^2 + 2a(s - s_0) = 2(51.8)20 \qquad v = 45.5 \text{ in/s}$$

**18.71**   (a)   Do the same as in Prob. 18.70, if the gears and weight are arranged as shown in Fig. 18.71a.

(b)   Compare the results in part (a) with those in Prob. 18.70.

▌ (a)   The tangential $F_{12}$ transmitted between the gears, and the cable tensile force $T$, are shown in Fig. 18.71b.   From Prob. 18.65,

$$I_{01} = 0.0184 \text{ lb} \cdot \text{s}^2 \cdot \text{in} \qquad I_{02} = 0.720 \text{ lb} \cdot \text{s}^2 \cdot \text{in}$$

The equations of motion of the weight and gears are

*Weight:*           $\sum F_y = ma \qquad 8 - T = \tfrac{8}{386}a$                         (1)

*Gear 2:*           $\sum M_0 = I_0\alpha \qquad T(1) - F_{12}(5) = 0.720\alpha_2$

$$r\alpha = a \qquad (1)\alpha_2 = a \qquad \alpha_2 = \frac{a}{1} \qquad T - 5F_{12} = 0.720a \qquad (2)$$

*Gear 1:*           $M_0 = I_0\alpha \qquad F_{12}(2) = 0.0184\alpha_1 \qquad \alpha_1 = \tfrac{10}{4}\alpha_2 = \tfrac{10}{4}a$

$$2F_{12} = 0.0184(\tfrac{10}{4}a) \qquad (3)$$

$T$ is eliminated between Eqs. (1) and (2), to obtain

$$8 - \tfrac{8}{386}a = 5F_{12} + 0.720a \qquad F_{12} = 1.6 - 0.148a \qquad (4)$$

Using Eq. (4) in Eq. (3),

$$2(1.6 - 0.148a) = 0.0184(\tfrac{10}{4}a) \qquad a = 9.36 \text{ in/s}^2$$

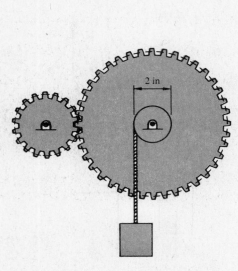

Fig. 18.71*a*

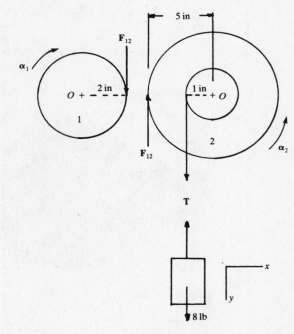

Fig. 18.71*b*

Using Eq. (1),

$$T = 8 - \tfrac{8}{386}(9.36) = 7.81 \text{ lb}$$

The speed of the weight, after moving through 20 in, is found from

$$v^2 = v_0^2 + 2a(s - s_0) = 2(9.36)20 \qquad v = 19.3 \text{ in/s}$$

(*b*)  The arrangement in Fig. 18.71*a* results in a lower acceleration of the weight, with a percent difference given by

$$\%D = \frac{9.36 - 51.8}{51.8}\,100 = -82\%$$

The speed of the weight is also lower for the system in Fig. 18.71*a*, with a percent difference of

$$\%D = \frac{19.3 - 45.5}{45.5}\,100 = -58\%$$

## 18.5  SOLUTIONS USING THE D'ALEMBERT, OR INERTIA, FORCES AND MOMENTS, CRITERIA FOR SLIDING OR TIPPING, CENTER OF PERCUSSION

18.72  Show how the D'Alembert, or inertia, moment may be used to solve problems in plane motion of a rigid body.

❚  In Prob. 15.79 the concept of the D'Alembert, or inertia, force acting on a mass particle was introduced.   It was seen that the addition of another force, namely the inertia force, to the particle had the effect of transforming a problem in dynamics to a problem in static equilibrium.

The concept of an inertia force acting on a particle may be extended readily to the case of a rigid body in plane motion.   Figure 18.72 shows a body which is hinged to the ground.   The equation of rotational motion is

$$M_a = I_a \alpha$$

This equation may be written in the form

$$M_\alpha - I_a \alpha = 0 \qquad M_a + (-I_a \alpha) = 0$$

The quantity $(-I_a \alpha)$ is called the D'Alembert, or inertia, moment.   If this quantity is imagined to be an applied moment which acts on the body, then the equation of motion would have the form

$$\sum M = 0$$

which is the basic form of moment equilibrium in statics.

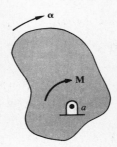

**Fig. 18.72**

The general rules which govern the formulation of the inertia force or moment acting on a rigid body are as follows.

1. An arrow is drawn on the free-body diagram of the body in the assumed, or actual, sense of the acceleration. In the case of a rigid body, this arrow would be drawn through the CM, for the case of translation, or about a known reference point, for the case of rotation.

2. The magnitude of the force or moment which this arrow represents is indicated by the product of mass and acceleration, for the case of translation, or by the product of mass moment of inertia and angular acceleration, for the case of angular motion.

3. *A minus sign is inserted in front of the quantity described in step 2.* The quantity formulated above, *including* the minus sign, is then the inertia force or moment which acts on the element under consideration. This quantity is considered to be positive if it acts in the assumed, or actual, sense of the acceleration, and it may then be treated as simply another force or moment which acts in the system. The equations of static equilibrium may then be used to solve for the unknowns in the problem.

In the following problems, the concept of an inertia force will be used to characterize certain types of problems in general plane motion. For these problems, the inertia force technique will be found to be a particularly convenient way of formulating the problem. In Prob. 18.86, the solutions will be obtained by using both the inertia force and the direct application of Newton's second law, and the two methods of solution may be compared.

**18.73** A crate slides down an incline, as shown in Fig. 18.73*a*.

(*a*) Find the normal and friction forces exerted on edges *a* and *b*.

(*b*) Express the reaction forces on edges *a* and *b* in formal vector notation.

▌ (*a*) Figure 18.73*b* shows the free-body diagram of the crate, with the inertia force acting through the center of mass. For force equilibrium of the crate,

$$\sum F_x = 0 \qquad (-60a_x) - 0.18(N_a + N_b) + 589 \sin 30° = 0 \qquad (1)$$

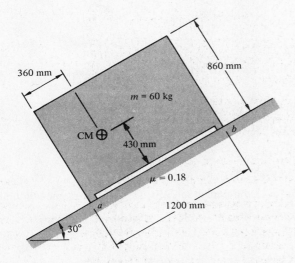

**Fig. 18.73*a***

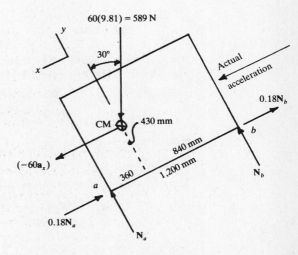

**Fig. 18.73*b***

$$\sum F_y = 0 \qquad N_a + N_b - 589 \cos 30° = 0 \qquad N_a + N_b = 510\,\text{N} \tag{2}$$

Using Eq. (2) in Eq. (1),

$$a_x = 3.38\,\text{m/s}^2$$

For moment equilibrium,

$$\sum M_a = 0 \qquad (-60a_x)430 + (589 \sin 30°)430 - (589 \cos 30°)360 + N_b(1,200) = 0$$

$$N_b = 120\,\text{N}$$

Using Eq. (2),

$$N_a = 390\,\text{N}$$

The friction forces acting on edges $a$ and $b$ are then found as

$$F_a = \mu N_a = 0.18(390) = 70.2\,\text{N} \qquad F_b = \mu N_b = 0.18(120) = 21.6\,\text{N}$$

(b) The reaction forces at $a$ and $b$ may be written as

$$\mathbf{R}_a = -0.18N_a\mathbf{i} + N_a\mathbf{j} = -0.18(390)\mathbf{i} + 390\mathbf{j} = -70.2\mathbf{i} + 390\mathbf{j} \qquad \text{N}$$
$$\mathbf{R}_b = -0.18N_b\mathbf{i} + N_b\mathbf{j} = -0.18(120)\mathbf{i} + 120\mathbf{j} = -21.6\mathbf{i} + 120\mathbf{j} \qquad \text{N}$$

**18.74** (a) Do the same as in Prob. 18.73, part (a), if the crate slides down the incline in the position shown in Fig. 18.74a.

(b) Compare the results in part (a) with those in Prob. 18.73.

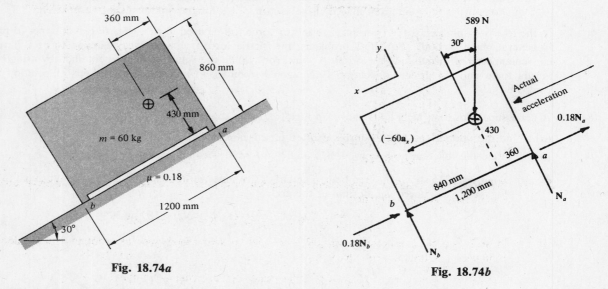

Fig. 18.74a    Fig. 18.74b

▮ (a) The free-body diagram of the crate, with the inertia force acting through the center of mass, is shown in Fig. 18.74b. For force equilibrium of the crate,

$$\sum F_x = 0 \qquad (-60a_x) - 0.18(N_a + N_b) + 589 \sin 30° = 0 \tag{1}$$
$$\sum F_y = 0 \qquad N_a + N_b - 589 \cos 30° = 0 \qquad N_a + N_b = 510 \tag{2}$$

Equation (2) is used in Eq. (1), with the result

$$a_x = 3.38\,\text{m/s}^2$$

The above value of acceleration is the same as that obtained in Prob. 18.73. For moment equilibrium,

$$\sum M_b = 0 \qquad (-60a_x)430 + (589 \sin 30°)430 - (589 \cos 30°)840 + N_a(1,200) = 0$$

$$N_a = 324\,\text{N}$$

Using Eq. (2),

$$N_b = 186\,\text{N}$$

The friction forces acting on edges $a$ and $b$ are given by

$$F_a = \mu N_a = 0.18(324) = 58.3 \text{ N} \qquad F_b = \mu N_b = 0.18(186) = 33.5 \text{ N}$$

(**b**) The maximum value of normal force is $N_a = 390 \text{ N}$, for the position shown in Fig. 18.73a, with a friction force $F_a = 70.2 \text{ N}$. The maximum value of normal force is $N_a = 324 \text{ N}$, for the position shown in Fig. 18.74a, with a friction force $F_a = 58.3 \text{ N}$. The acceleration of the crate is the same for both positions of the crate.

**18.75** A cabinet on casters is pulled to the right along a horizontal floor by a constant force $P$, as shown in Fig. 18.75a. The mass of the cabinet is 160 kg.

(**a**) Find the value of the acceleration of the cabinet that will just cause the rear wheels to lose contact with the floor.

(**b**) Find the value of the force $P$ that corresponds to part (a).

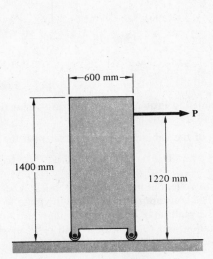

Fig. 18.75a

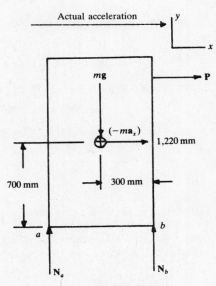

Fig. 18.75b

▮ (**a**) The free-body diagram of the cabinet is shown in Fig. 18.75b, together with the inertia force. For equilibrium in the $x$ direction,

$$\sum F_x = 0 \qquad (-ma_x) + P = 0 \qquad P = ma_x \qquad (1)$$

A limiting condition occurs when $N_a \rightarrow 0$ and the rear wheels just lose contact with the floor. For moment equilibrium, with $N_a = 0$,

$$\sum M_b = 0 \qquad mg(300) - (-ma_x)700 - P(1,220) = 0 \qquad (2)$$

Using $P$ from Eq. (1) in Eq. (2),

$$mg(300) + ma_x(700) - (ma_x)1,220 = 0 \qquad a_x = 0.577g = 0.577(9.81) = 5.66 \text{ m/s}^2$$

It may be seen that the above result is independent of the mass of the cabinet.

(**b**) The force $P$ that corresponds to the acceleration found in part (a) is given by

$$P = ma_x = 160(5.66) = 906 \text{ N}$$

**18.76** Do the same as in Prob. 18.75, if the direction of the force is that shown in Fig. 18.76a.

▮ (**a**) Figure 18.76b shows the free-body diagram of the cabinet. For force equilibrium in the $x$ direction,

$$\sum F_x = 0 \qquad (-ma_x) + P \cos 40° = 0 \qquad P \cos 40° = ma_x \qquad (1)$$

For moment equilibrium, with $N_a = 0$,

$$\sum M_b = 0 \qquad mg(300) - (-ma_x)700 - (P \cos 40°)1,220 = 0 \qquad (2)$$

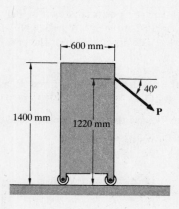

Fig. 18.76a

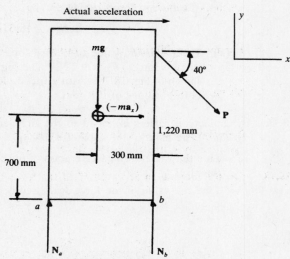

Fig. 18.76b

Using $P$ from Eq. (1) in Eq. (2),

$$mg(300) + ma_x(700) - ma_x(1,220) = 0 \qquad a_x = 0.577g = 0.577(9.81) = 5.66\ \text{m/s}^2$$

The above result for $a_x$ is the same as that obtained in Prob. 18.75. This result is again independent of the mass of the cabinet.

(b) The force $P$ required to cause the rear wheels of the cabinet to lose contact with the floor is given by

$$P \cos 40° = ma_x = 160(5.66) \qquad P = 1,180\ \text{N}$$

18.77 Figure 18.77a shows a heavy metal plate that moves on rollers along a conveyor rail. The plate is made of 0.5-in steel plate, with a specific weight of 489 lb/ft³. The constant force $P$ causes the plate to accelerate to the right. Find the maximum value of acceleration that the plate may have, if the rollers are not to lose contact with the rail, by direct use of Newton's laws for translation and rotation of a rigid body in plane motion.

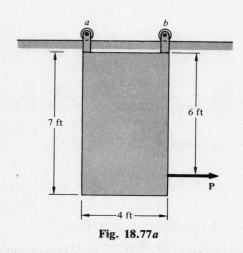

Fig. 18.77a

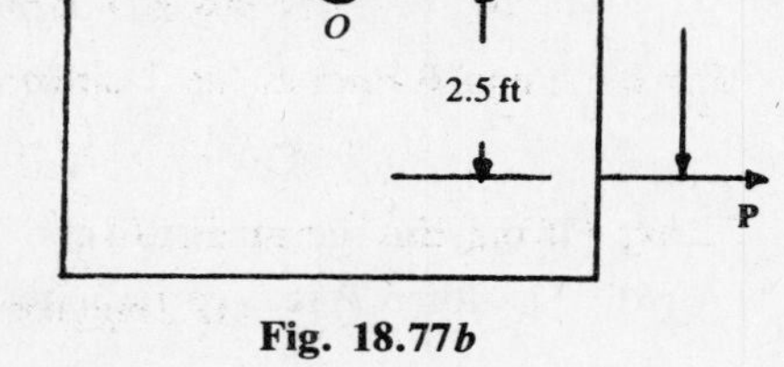

Fig. 18.77b

❙ Newton's second law for motion about the center of mass is

$$M_0 = I_0 \alpha = 0$$

The limiting condition occurs when $N_b = 0$. Using this result in the above equation,

$$M_0 = P(2.5) - N_a(2) = 0 \qquad (1)$$

For force equilibrium in the $y$ direction,

$$\sum F_y = 0 \qquad N_a - mg = 0 \qquad N_a = mg \qquad (2)$$

Using $N_a$ from Eq. (2) in Eq. (1),

$$2.5P - 2mg = 0 \qquad P = 0.8\,mg \tag{3}$$

For force equilibrium in the $x$ direction,

$$F_x = ma_x \qquad P = ma_x \tag{4}$$

Using $P$ from Eq. (3) in Eq. (4),

$$0.8mg = ma_x \qquad a_x = 0.8g = 0.8(32.2) = 25.8 \text{ ft/s}^2$$

The above result is seen to be independent of the mass. It is left as an exercise for the reader to solve this problem by using the inertia force method.

**18.78** Do the same as in Prob. 18.77, if the force has the direction shown in Fig. 18.78a.

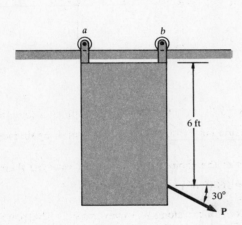

Fig. 18.78a

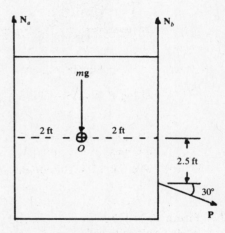

Fig. 18.78b

▎ The equation of motion about the center of mass is

$$M_0 = I_0 \alpha = 0$$

Using $N_b = 0$ in the above equation,

$$(P \cos 30°)2.5 - (P \sin 30°)2 - N_a(2) = 0 \tag{1}$$

For force equilibrium in the $y$ direction,

$$\sum F_y = 0 \qquad N_a - mg - P \sin 30° = 0 \qquad N_a = mg + P \sin 30° \tag{2}$$

Using $N_a$ from Eq. (2) in Eq. (1),

$$(P \cos 30°)2.5 - (P \sin 30°)2 - (mg + P \sin 30°)2 = 0 \qquad P = 12.1\,mg \tag{3}$$

For force equilibrium in the $x$ direction,

$$F_x = ma_x \qquad P \cos 30° = ma_x \tag{4}$$

Using $P$ from Eq. (3) in Eq. (4),

$$(12.1mg) \cos 30° = ma_x \qquad a_x = 10.5g = 10.5(32.2) = 338 \text{ ft/s}^2$$

It may be seen, from comparison of the above result with that in Prob. 18.77, that the maximum permissible acceleration of the plate is much larger when the force has the direction shown in Fig. 18.78a.

**18.79** Do the same as in Prob. 18.77, if the plate is modified by cutting a hole through it, as shown in Fig. 18.79a.

▎ Figure 18.79b shows the centroidal coordinates $x_c$ and $y_c$ of the plate. Element 1 is the plate area with no cutout, and element 2 is the area of the cutout. The centroidal coordinates are found to be

$$x_c = \frac{x_1 A_1 - x_2 A_2}{A_1 - A_2} = \frac{\frac{4}{2}(4)7 - (0.5 + \frac{1.5}{2})1.5(4)}{4(7) - 1.5(4)} = 2.20 \text{ ft} \qquad A = 22 \text{ ft}^2$$

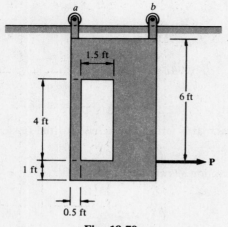

Fig. 18.79a

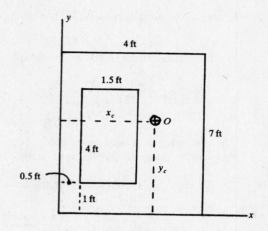

Fig. 18.79b

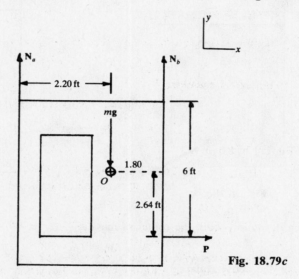

Fig. 18.79c

$$y_c = \frac{y_1 A_1 - y_2 A_2}{A} = \frac{\frac{7}{2}(4)7 - (1 + \frac{4}{2})1.5(4)}{22} = 3.64 \text{ ft}$$

Figure 18.79c shows the free-body diagram of the plate. The equation of motion about the center of mass is

$$M_0 = I_0 \alpha = 0$$

Using $N_b = 0$ in the above equation results in

$$P(2.64) - N_a(2.20) = 0 \qquad (1)$$

For force equilibrium in the y direction,

$$\sum F_y = 0 \qquad N_a - mg = 0 \qquad N_a = mg \qquad (2)$$

Using $N_a$ from Eq. (2) in Eq. (1),

$$2.64P - 2.20mg = 0 \qquad P = 0.833mg \qquad (3)$$

For force equilibrium in the x direction,

$$F_x = ma_x \qquad P = ma_x \qquad (4)$$

Using $P$ from Eq. (3) in Eq. (4),

$$0.833mg = ma_x$$

$$a_x = 0.833g = 0.833(32.2) = 26.8 \text{ ft/s}^2$$

**18.80**   Do the same as in Prob. 18.77 if the modified plate of Fig. 18.79a is mounted as shown in Fig. 18.80a.

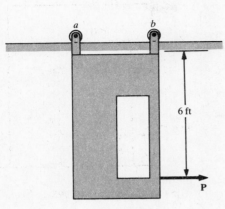

Fig. 18.80a

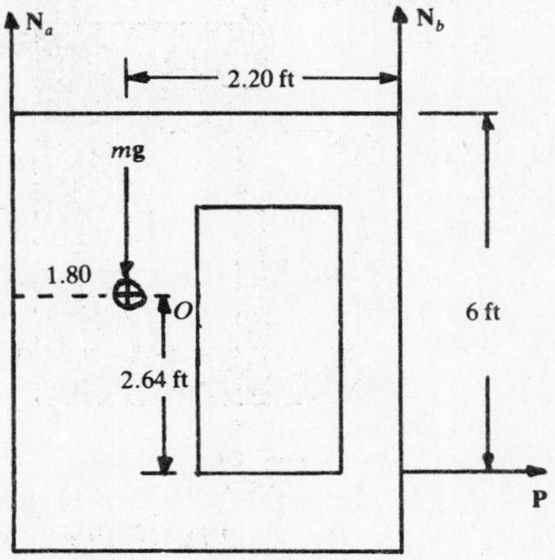

Fig. 18.80b

❚ The free-body diagram of the plate is shown in Fig. 18.80b.  The solution follows that given in Prob. 18.79.

$$M_0 = I_0\alpha = 0$$

Using   $N_b = 0$   in the above equation,

$$P(2.64) - N_a(1.80) = 0 \qquad\qquad (1)$$

$$\sum F_y = 0 \qquad N_a - mg = 0 \qquad N_a = mg \qquad\qquad (2)$$

Using $N_a$ from Eq. (2) in Eq. (1),

$$2.64P - 1.80mg = 0 \qquad P = 0.682mg \qquad\qquad (3)$$

$$F_x = ma_x \qquad P = ma_x \qquad\qquad (4)$$

Using $P$ from Eq. (3) in Eq. (4),

$$0.682mg = ma_x \qquad a_x = 0.682g = 0.682(32.2) = 22.0 \text{ ft/s}^2$$

**18.81**   Figure 18.81a shows a conveyor system used to transport heavy steel plates.  The two carriages have shoes which slide along an inclined steel rail, with clamping mechanisms to attach to the plate.  The maximum value of the coefficient of friction is assumed to be 0.15.  The plates are released with zero velocity from the top of the incline.

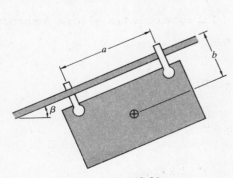

Fig. 18.81a

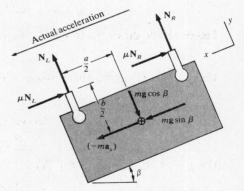

Fig. 18.81b

(a)   Find the acceleration of the plate.

(b)   Find the normal contact forces between the shoes and the rail.

(c)   Find the numerical results for parts (a) and (b), if   $a = 6$ ft,   $b = 3$ ft,   $\beta = 28°$,   and the weight of the plate is 750 lb.

(d)   Are there any conditions which would cause a shoe to lose contact with the rail?

**▮** (*a*) The free-body diagram of the carriage-plate assembly is shown in Fig. 18.81*b*, together with the inertia force. For dynamic equilibrium of the plate,

$$\Sigma F_x = 0 \qquad -\mu N_L + (-ma_x) - \mu N_R + mg \sin \beta = 0 \tag{1}$$

$$\Sigma F_y = 0 \qquad N_L - mg \cos \beta + N_R = 0 \tag{2}$$

$$\Sigma M_{N_L} = 0 \qquad -(mg \sin \beta - ma_x)\frac{b}{2} - (mg \cos \beta)\frac{a}{2} + N_R a = 0 \tag{3}$$

From Eq. (2),

$$N_R = -N_L + mg \cos \beta \tag{4}$$

Using the above result in Eq. (1),

$$-\mu N_L - ma_x - \mu(-N_L + mg \cos \beta) + mg \sin \beta = 0$$

The terms $\mu N_L$ cancel in the above equation, and the final result for $a_x$ is

$$a_x = g(\sin \beta - \mu \cos \beta) \tag{5}$$

(*b*) Using Eqs. (4) and (5) in Eq. (3) results in

$$-[mg \sin \beta - mg(\sin \beta - \mu \cos \beta)]\frac{b}{2} - (mg \cos \beta)\frac{a}{2} + (-N_L + mg \cos \beta)a = 0$$

$$N_L = \left(\frac{a - \mu b}{2a}\right)mg \cos \beta$$

Using Eq. (4),

$$N_R = \left(\frac{a + \mu b}{2a}\right)mg \cos \beta$$

It may be seen, from the above results that, for $\mu \neq 0$,

$$N_R > N_L$$

(*c*) The numerical results are

$$a_x = g(\sin \beta - \mu \cos \beta) = 32.2(\sin 28° - 0.15 \cos 28°) = 10.9 \text{ ft/s}^2$$

$$N_L = \left(\frac{a - \mu b}{2a}\right)mg \cos \beta = \left[\frac{6 - 0.15(3)}{2(6)}\right]750 \cos 28° = 306 \text{ lb}$$

$$N_R = \left(\frac{a + \mu b}{2a}\right)mg \cos \beta = \left[\frac{6 + 0.15(3)}{2(6)}\right]750 \cos 28° = 356 \text{ lb}$$

It can be shown that, if the rail is frictionless, $N_L = N_R = 331$ lb. Thus, the friction forces have the effect of increasing the right-side normal reaction force $N_R$ and decreasing the left-side normal reaction force $N_L$.

(*d*) The left shoe will lose contact with the rail for computed values of $N_L \leq 0$. The required condition for this shoe to remain in contact with the rail, from Eq. (6), is

$$\mu b < a \qquad \mu < \frac{a}{b} = \frac{6}{3} = 2$$

For the present problem, $\mu = 0.15$. Thus, the shoes remain in contact with the rail.

**18.82** (*a*) Find the value of $\theta$ that will result in the maximum acceleration of the crate in Fig. 18.82*a*.

(*b*) Do the same as in part (*a*), if the 16-N·m couple is removed.

**▮** (*a*) Figure 18.82*b* shows the free-body diagram of the crate, together with the inertia force. It is assumed that $N_a$ is positive, subject to later verification. For dynamic equilibrium of the crate,

$$\Sigma F_y = 0 \qquad N_a - 196 + N_b + 100 \sin \theta = 0 \qquad N_a + N_b = 196 - 100 \sin \theta \tag{1}$$

$$\Sigma F_x = 0 \qquad -0.18N_a + (-20a_x) + 100 \cos \theta - 0.18N_b = 0$$

$$-0.18(N_a + N_b) - 20a_x + 100 \cos \theta = 0 \tag{2}$$

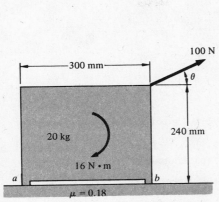

**Fig. 18.82a**

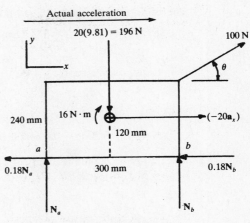

**Fig. 18.82b**

$(N_a + N_b)$ from Eq. (1) is substituted into Eq. (2), with the result

$$-0.18(196 - 100 \sin \theta) - 20a_x + 100 \cos \theta = 0 \qquad (3)$$

$$a_x = 5 \cos \theta + 0.9 \sin \theta - 1.76 \qquad (4)$$

The first derivative of the above equation is set equal to zero to find the extreme value of $a_x$. The result is

$$\frac{da_x}{d\theta} = -5 \sin \theta + 0.9 \cos \theta = 0 \qquad \tan \theta = \frac{0.9}{5} \qquad \theta = 10.2° \qquad (5)$$

The determination of whether $a_x$ is a maximum or minimum value is made from

$$\left.\frac{d^2 a_x}{d\theta^2}\right|_{\theta = 10.2°} = (-5 \cos \theta - 0.9 \sin \theta)\Big|_{\theta=10.2°} = -5.08$$

Since the second derivative is negative, $a_x$ is a maximum value for $\theta = 10.2°$.

The maximum value of $a_x$, using Eq. (4), is

$$a_{x,\max} = (5 \cos \theta + 0.9 \sin \theta - 1.76)|_{\theta=10.2°} = 5 \cos 10.2° + 0.9 \sin 10.2° - 1.76 = 3.32 \text{ m/s}^2 \qquad (6)$$

As the final step in the solution, the value of $N_a$ must be found to confirm the initial assumption that $N_a$ is positive. For moment equilibrium about point $b$,

$$\sum M_b = 0 \qquad -(100 \cos 10.2°) \frac{240}{1,000} - 16 + 196\left(\frac{150}{1,000}\right) - [-20(3.32)]\frac{120}{1,000}$$

$$-N_a\left(\frac{300}{1,000}\right) = 0 \qquad N_a = -7.51 \text{ N} \qquad (7)$$

$N_a$ was assumed to be a compressive reaction force, with the sense shown in Fig. 18.82b. Since $N_a$ is negative for $\theta = 10.2°$, *the preceding solution is not valid.*

The limiting condition will occur when tipping is impending about edge $b$. For this case, $N_a = 0$. The corresponding value of $\theta$ is now found.

$$\sum M_b = 0 \qquad -(100 \cos \theta) \frac{240}{1,000} - 16 + 196\left(\frac{150}{1,000}\right) - (-20a_x)\frac{120}{1,000} = 0 \qquad (8)$$

Equation (3) is still valid, and this equation has the form

$$-0.18(196 - 100 \sin \theta) - 20a_x + 100 \cos \theta = 0 \qquad (9)$$

The term $(-20a_x)$ is eliminated between Eqs. (8) and (9), to obtain

$$(100 \cos \theta) \frac{240}{1,000} + 16 + [0.18(196 - 100 \sin \theta) - 100 \cos \theta]\frac{120}{1,000} - 196\left(\frac{150}{1,000}\right) = 0$$

$$12 \cos \theta - 2.16 \sin \theta - 9.17 = 0 \qquad 12 \cos \theta = 2.16 \sin \theta + 9.17$$

Both sides of the above equation are squared, with the result

$$12^2 \cos^2 \theta = 12^2 (1 - \sin^2 \theta) = 2.16^2 \sin^2 \theta + 2(2.16)9.17 \sin \theta + 9.17^2$$

$$149 \sin^2 \theta + 39.6 \sin \theta - 59.9 = 0$$

The solution to the above quadratic equation is

$$\sin \theta = \frac{-39.6 \pm \sqrt{39.6^2 - 4(149)(-59.9)}}{2(149)} = \frac{-39.6 \pm 193}{2(149)}$$

Using the positive root,

$$\theta = 31.0°$$

Using Eq. (9),

$$20a_x = 100 \cos 31.0° - 0.18(196 - 100 \sin 31.0°) \qquad a_{x,\text{max}} = 2.99 \text{ m/s}^2$$

(b) If the 16-N · m couple is removed, the solution is given by Eq. (5) in part (a), subject to a check of the sign of $N_a$. Solving Eq. (7), with the term (−16) deleted,

$$N_a = 45.8 \text{ N} > 0$$

Since $N_a$ is positive, the maximum value of the acceleration of the crate, with $\theta = 10.2°$, is

$$a_{x,\text{max}} = 3.32 \text{ m/s}^2$$

**18.83** Figure 18.83a shows a motorcycle driving around a horizontal circular track at constant speed. The total weight of the motorcycle and rider is 450 lb.

(a) Find the value of angle $\beta$ when the motorcycle travels at 40 mi/h.

(b) Find the maximum value of the speed of the motorcycle, if sliding motion is not to occur.

(c) Find the angle $\beta$ that corresponds to the motion in part (b).

(d) Express the forces acting on the motorcycle and rider in formal vector notation.

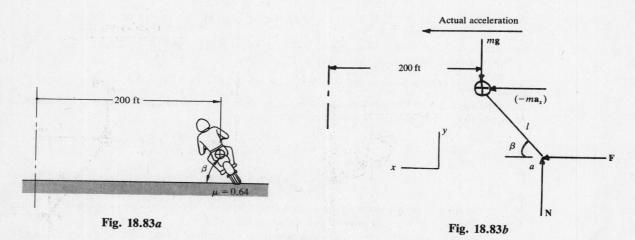

Fig. 18.83a            Fig. 18.83b

▌ (a) The free-body diagram of the motorcycle is shown in Fig. 18.83b, together with the inertia force which acts through the center of mass of the combined system of the motorcycle and rider. For moment equilibrium about point $a$,

$$\sum M_a = 0 \qquad mgl \cos \beta + (-ma_x)l \sin \beta = 0$$

Using $a_x = v^2/\rho$ in the above equation, where $\rho$ is the radius of the track, results in

$$mgl \cos \beta = m \frac{v^2}{\rho} l \sin \beta \qquad \tan \beta = \frac{\rho g}{v^2} \qquad\qquad (1)$$

The speed of the motorcycle is

$$v = 40 \text{ mi/h} = 58.7 \text{ ft/s}$$

so that

$$\tan \beta = \frac{200(32.2)}{58.7^2} \qquad \beta = 61.9°$$

(*b*) Sliding motion is impending when

$$F = \mu N$$

For force equilibrium of the motorcycle,

$$\sum F_y = 0 \qquad N = mg$$

$$\sum F_x = 0 \qquad \mu N + (-ma_x) = 0 \qquad \mu mg = m\frac{v^2}{\rho} \qquad v = \sqrt{\mu \rho g}$$

$$v_{max} = \sqrt{0.64(200)32.2} = 64.2 \text{ ft/s} = 43.8 \text{ mi/h}$$

(*c*) Using Eq. (1) in part (*a*),

$$\tan \beta = \frac{\rho g}{v^2} = \frac{200(32.2)}{64.2^2} \qquad \beta = 57.4°$$

(*d*) The forces which act on the motorcycle and rider are

$$\mathbf{W} = m\mathbf{g} = -450\mathbf{j} \qquad \text{lb}$$
$$\mathbf{R}_a = \mathbf{F} + \mathbf{N} = -\mu N\mathbf{i} + N\mathbf{j} = -\mu mg\mathbf{i} + mg\mathbf{j} = 0.64(450)\mathbf{i} + 450\mathbf{j} = 288\mathbf{i} + 450\mathbf{j} \qquad \text{lb}$$

**18.84** A stunt driver in a circus drives a motorcycle around the inside wall of a cylindrical drum, as shown in Fig. 18.84*a*. The coefficient of friction between the tires and the wall is 0.6.

(*a*) Find the minimum value of speed required if the motorcycle is not to slide down the wall.

(*b*) Find the maximum value of speed of the rider if, for physiological reasons, the rider's body must not experience a total acceleration greater than six times the gravitational acceleration.

(*c*) Find the angle $\beta$ that corresponds to the motion in part (*b*).

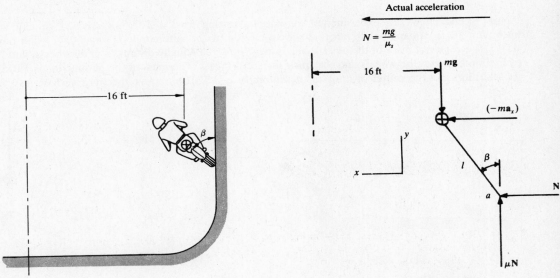

Fig. 18.84*a*                                      Fig. 18.84*b*

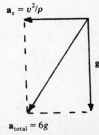

Fig. 18.84*c*

**▌ (a)** The free-body diagram, together with the inertia force, is shown in Fig. 18.84b. The friction force is assumed to have its maximum value $\mu N$. For force equilibrium,

$$\sum F_y = 0 \qquad \mu N - mg = 0 \tag{1}$$

$$\sum F_x = 0 \qquad N + (-ma_x) = 0 \tag{2}$$

The acceleration $a_x$ is given by

$$a_x = \frac{v^2}{\rho}$$

Equating the values of $N$ from Eqs. (1) and (2),

$$N = \frac{mg}{\mu} = \frac{mv^2}{\rho} \qquad v_{min} = \sqrt{\frac{\rho g}{\mu}} = \sqrt{\frac{16(32.2)}{0.6}} = 29.3 \text{ ft/s} = 20.0 \text{ mi/h}$$

**(b)** The resultant acceleration $a_{total}$ of the rider is given by

$$a_{total} = \sqrt{\left(\frac{v^2}{\rho}\right)^2 + g^2}$$

Using the limiting value $a_{total} = 6g$ in the above equation, and squaring both sides, gives

$$\frac{v^4}{\rho^2} + g^2 = (6g)^2 \qquad v^4 = 35g^2\rho^2 = 35(32.2)^2 16^2 \qquad v = 55.2 \text{ ft/s} = 37.6 \text{ mi/h}$$

**(c)** For moment equilibrium about point $a$,

$$\sum M_a = 0 \qquad (-ma_x)l \cos \beta + mgl \sin \beta = 0 \qquad \frac{v^2}{\rho} \cos \beta = g \sin \beta$$

$$\tan \beta = \frac{v^2}{\rho g} = \frac{55.2^2}{16(32.2)} \qquad \beta = 80.4°$$

**18.85** Figure 18.85a shows an experimental automobile testing track. The track has a circular shape, with a maximum angle of banking $\beta$ such that sliding motion of the automobile is impending when the vehicle is at rest. The center of mass of the automobile is assumed to be equally distant from all wheels, and the coefficient of friction between the wheels and the track surface is 0.81. The mass of the automobile is 1,075 kg, the tread is 1,480 mm, and the center of mass of the automobile is 520 mm above the track surface.

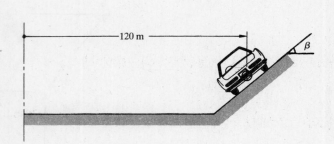

Fig. 18.85a

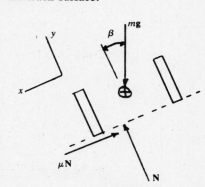

Fig. 18.85b

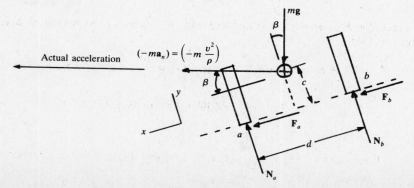

Fig. 18.85c

(a) Find the angle of banking of the track.

(b) Find the total friction force acting on the wheels when the automobile travels at 160 km/h.

(c) Find the normal forces on the inner and outer wheels, for the condition of part (b).

(d) Find the range of values of the friction forces exerted on the inner and outer wheels, for the condition of part (b).

▌ (a) The free-body diagram, for impending sliding motion down the track when the vehicle is at rest, is shown in Fig. 18.85b. For equilibrium,

$$\sum F_y = 0 \qquad -mg \cos \beta + N = 0 \qquad N = mg \cos \beta \tag{1}$$

$$\sum F_x = 0 \qquad -mg \sin \beta + \mu N = 0 \qquad N = \frac{mg \sin \beta}{\mu} \tag{2}$$

Equations (1) and (2) are equated, to obtain

$$mg \cos \beta = \frac{mg \sin \beta}{\mu} \qquad \tan \beta = \mu = 0.81 \qquad \beta = 39.0°$$

(b) When the vehicle is in motion, the free-body diagram has the form shown in Fig. 18.85c. $N_a$ and $N_b$ are the normal forces on the inner and outer wheels, with the associated friction forces $F_a$ and $F_b$. For equilibrium of the automobile,

$$\sum M_a = 0 \qquad (-ma_n)c \cos \beta + (-ma_n) \frac{d}{2} \sin \beta - mg \frac{d}{2} \cos \beta + mgc \sin \beta + N_b d = 0$$

$$-ma_n\left(c \cos \beta + \frac{d}{2} \sin \beta\right) + mg\left(c \sin \beta - \frac{d}{2} \cos \beta\right) + N_b d = 0 \tag{3}$$

$$\sum F_x = 0 \qquad (-ma_n) \cos \beta + mg \sin \beta + F_a + F_b = 0 \tag{4}$$

$$\sum F_y = 0 \qquad (-ma_n) \sin \beta - mg \cos \beta + N_a + N_b = 0 \tag{5}$$

Equations (3) through (5) are a set of three equations in the four unknowns $N_a$, $N_b$, $F_a$, and $F_b$, and thus may not be solved for those unknowns. These equations may, however, be solved by using the definition of the *resultant* friction force $F$, given by

$$F = F_a + F_b$$

The centripetal acceleration $a_n$ is given by

$$a_n = \frac{v^2}{\rho}$$

Using Eq. (4),

$$F = F_a + F_b = ma_n \cos \beta - mg \sin \beta = m\left(\frac{v^2}{\rho} \cos \beta - g \sin \beta\right) \tag{6}$$

Using

$$v = 160 \, \frac{\text{km}}{\text{h}} \left(\frac{1,000 \, \text{m}}{1 \, \text{km}}\right)\left(\frac{1 \, \text{h}}{3,600 \, \text{s}}\right) = 44.4 \, \text{m/s} \qquad m = 1,075 \, \text{kg} \qquad \beta = 39° \qquad \rho = 120 \, \text{m}$$

in Eq. (6) results in

$$F = F_a + F_b = 1,075\left(\frac{44.4^2}{120} \cos 39° - 9.81 \sin 39°\right) = 7,090 \, \text{N} \tag{7}$$

The above result for $F$ must now be compared with the value of the maximum available friction force. From Eq. (5),

$$N_a + N_b = ma_n \sin \beta + mg \cos \beta = m\left(\frac{v^2}{\rho} \sin \beta + g \cos \beta\right)$$

$$= 1,075\left(\frac{44.4^2}{120} \sin 39° + 9.81 \cos 39°\right) = 19,300 \, \text{N} \tag{8}$$

The maximum available friction force is given by

$$F_{\text{max}} = \mu N \qquad (F_a + F_b)_{\text{max}} = \mu(N_a + N_b) = 0.81(19,300) = 15,600 \, \text{N}$$

The result obtained in Eq. (7) is less than the maximum possible value, so that the result

$$F_a + F_b = 7,090 \text{ N}$$

may be accepted as being the correct value.

(c) Using

$$m = 1,075 \text{ kg} \qquad d = 1,480 \text{ mm} \qquad c = 520 \text{ mm} \qquad a_n = \frac{v^2}{\rho} = \frac{44.4^2}{120} = 16.4 \text{ m/s}^2$$

the numerical form of Eq. (3) is

$$\sum M_a = 0$$

$$-1,075(16.4)\left(520 \cos 39° + \frac{1,480}{2} \sin 39°\right) + 1,075(9.81)\left(520 \sin 39° - \frac{1,480}{2} \cos 39°\right) + N_b(1,480) = 0$$

$$N_b = 12,100 \text{ N}$$

Using Eq. (8),

$$N_a = 19,300 - N_b = 19,300 - 12,100 = 7,200 \text{ N}$$

It may be seen that $N_b > N_a$.

(d) From Eq. (7),

$$F_a + F_b = 7.090 \text{ N} \qquad\qquad (9)$$

The two normal forces were found in part (c) as

$$N_a = 7,200 \text{ N} \qquad N_b = 12,100 \text{ N}$$

It is now assumed that $F_a$ has its maximum possible value, given by

$$F_{a,\max} = \mu Na = 0.81(7,200) = 5,830 \text{ N}$$

From Eq. (9), the corresponding value of $F_b$ is given by

$$F_b = 7,090 - F_a = 7,090 - 5,830 = 1,260 \text{ N}$$

It is next assumed that $F_b$ has its maximum possible value, given by

$$F_{b,\max} = \mu N_b = 0.81(12,100) = 9,800 \text{ N}$$

The corresponding value of $F_a$, from Eq. (9), is

$$F_a = 7,090 - F_b = 7,090 - 9,800 = -2,710 \text{ N}$$

$F_a$ may not be negative, and the minimum value that this force may have is zero. Using $F_a = 0$ in Eq. (9),

$$F_{b,\max} = 7,090 \text{ N}$$

The range of values of the friction forces is then

$$0 \le F_a \le 5,830 \text{ N} \qquad 1,260 \text{ N} \le F_b \le 7,090 \text{ N}$$

18.86 Figure 18.86a shows a side view of an automobile of mass $m$.

(a) Find the maximum possible acceleration of the automobile, and the reaction forces of the roadway on the wheels, for the case of a rear-wheel drive only. Use both the inertia force method and the direct application of Newton's second law.

(b) Do the same as in part (a), for the case of front-wheel drive only, but use only the inertia force method.

(c) Do the same as in part (a), for the case of four-wheel drive, but use only the direct application of Newton's second law.

▌ (a) *Solution for rear-wheel drive only, using the inertia force.* Since rear-wheel drive is assumed, the friction forces of the ground on the tires act at the rear wheels only, and $\mu$ is the coefficient of friction. The free-body diagram for the case of rear-wheel drive only is shown in Fig. 18.86b, and the inertia force is shown acting through the center of mass.

The equations of equilibrium are

$$\sum M_d = 0 \qquad -mg(a) + (-ma_x)c + N_R l = 0$$

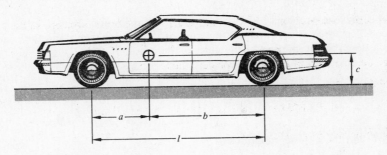

Fig. 18.86a

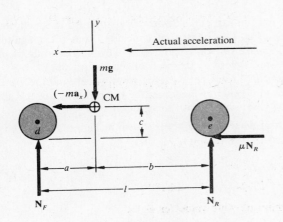

Fig. 18.86b

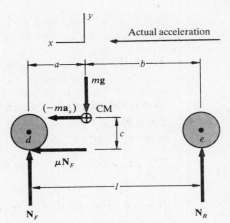

Fig. 18.86c

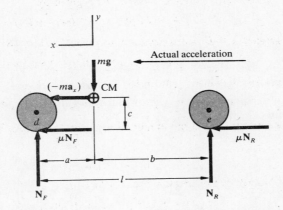

Fig. 18.86d

$$\sum F_x = 0 \qquad \mu N_R + (-ma_x) = 0$$

$$\sum F_y = 0 \qquad N_F - mg + N_R = 0$$

The solutions to the above equations are:

$$a_x = \frac{\mu a g}{l - \mu c} \tag{1}$$

$$N_F = mg \frac{(b - \mu c)}{l - \mu c} \tag{2}$$

$$N_R = \frac{mga}{l - \mu c} \tag{3}$$

*Solutions for rear-wheel drive only, by direct use of Newton's second law.* The automobile is a body in general plane motion. Thus, the general equations of motion are

$$\sum M_0 = I_0 \alpha \qquad \sum F = ma_c = ma_x$$

In the present problem $\alpha = 0$, so that the summation of moments about the center of mass is zero. The force $F$ is expressed in terms of its components along the $x$ and $y$ axes. The equations which describe the motion of the automobile are then

$$\sum F_x = ma_x \qquad \mu N_R = ma_x$$

$$\sum F_y = ma_y = 0 \qquad N_F - mg + N_R = 0$$

$$\sum M_0 = 0 \qquad -N_F a + N_R b - \mu N_R c = 0$$

$a_x$ is found from these equations as

$$a_x = \frac{\mu ag}{l - \mu c} \qquad \text{(rear-wheel drive only)} \qquad (4)$$

Equation (4) is the result obtained previously in Eq. (1). Several very interesting conclusions may be drawn from the above equations. From Eq. (4), it may be concluded that the magnitude of the maximum possible acceleration $a_x$ will increase if $a$ or $c$ increases. This corresponds to placing the center of mass as far from the ground, and as close to the rear wheels, as possible.

From Eq. (2), it may be seen that the normal force between the front wheels and the ground will be negative if $\mu c > b$. Since $N_F$ is defined as a *compressive* force, the interpretation of the above result is that, when $\mu c = b$, the front wheels lose contact with the ground.

(*b*) The free-body diagram for the case of front-wheel drive only is shown in Fig. 18.86*c*. The equations of equilibrium are

$$\sum M_e = 0 \qquad -N_F l + (-ma_x)c + mgb = 0$$

$$\sum F_x = 0 \qquad \mu N_F + (-ma_x) = 0$$

$$\sum F_y = 0 \qquad N_F - mg + N_R = 0$$

The solution to this system of equations is

$$a_x = \frac{\mu bg}{l + \mu c} \qquad \text{(front-wheel drive only)}$$

$$N_F = \frac{bmg}{l + \mu c} \qquad N_R = \frac{(a + \mu c)mg}{l + \mu c}$$

It may be observed that $N_F$ and $N_R$ are always positive. Thus, there is no condition of motion with front-wheel drive only which would permit the wheels to lose contact with the ground.

(*c*) The free-body diagram, for the case of four-wheel drive, is shown in Fig. 18.86*d*. Using Newton's second law directly, we get

$$\sum F_x = ma_x \qquad \mu N_R + \mu N_F = ma_x$$

$$\sum F_y = 0 \qquad N_F - mg + N_R = 0$$

$$\sum M_0 = 0 \qquad -N_F a + N_R b - \mu N_F c - \mu N_R c = 0$$

The solutions to these equations are

$$a_x = \mu g \qquad N_F = \frac{(b - \mu c)mg}{l} \qquad N_R = \frac{(a + \mu c)mg}{l} \qquad \text{(four-wheel drive)}$$

It may be seen from these equations that, as in the case of rear-wheel drive only, the front wheels lose contact with the ground when $\mu c \geq b$.

It is interesting to note that all of the results obtained in parts (*a*), (*b*), and (*c*) for the acceleration are *independent* of the mass of the vehicle.

**18.87** The vehicle described in Prob. 18.86 has the following dimensions and weight:

$$a = 45 \text{ in} \qquad b = 65 \text{ in} \qquad c = 21 \text{ in} \qquad l = a + b = 110 \text{ in} \qquad W = mg = 3{,}000 \text{ lb}$$

The coefficient of friction between the tires and the pavement is assumed to have the value 0.68. It is assumed that the vehicle has the required power to satisfy the conditions below.

(*a*) Find the minimum time to accelerate from rest to 60 mi/h, and the normal forces $N_F$ and $N_R$ which correspond to maximum acceleration of the automobile, for rear-wheel drive only.

(b)  Do the same as in part (a), for front-wheel drive only.

(c)  Do the same as in part (a), for four-wheel drive.

(d)  Compare the values in parts (a) through (c) with the corresponding values of the normal forces when the vehicle is at rest.

(e)  Express the reaction forces of the roadway on the wheels, in formal vector notation, for rear-, front-, and four-wheel drives.

▌ (a)  For the case of rear-wheel drive only,

$$a_x = \frac{\mu a g}{l - \mu c} = \frac{0.68(45)(32.2)}{110 - 0.68(21)} = 10.3 \text{ ft/s}^2$$

The final velocity is

$$v = \left(60 \frac{\text{mi}}{\text{h}}\right)\left(\frac{5,280}{1 \text{ mi}}\right)\left(\frac{1 \text{ h}}{3,600 \text{ s}}\right) = 88 \text{ ft/s}$$

The required time to reach 60 mi/h is then found from

$$v = v_0 + at \qquad 88 = 0 + 10.3t \qquad t = 8.54 \text{ s}$$

$$N_F = mg \frac{(b - \mu c)}{l - \mu c} = 3,000 \frac{[65 - 0.68(21)]}{110 - 0.68(21)} = 1,590 \text{ lb} \qquad N_R = \frac{amg}{l - \mu c} = \frac{45(3,000)}{110 - 0.68(21)} = 1,410 \text{ lb}$$

(b)  When only the front wheels drive the vehicle,

$$a_x = \frac{\mu b g}{l + \mu c} = \frac{0.68(65)(32.2)}{110 + 0.68(21)} = 11.5 \text{ ft/s}^2 \qquad v = v_0 + at$$

$$88 = 0 + 11.5t \qquad t = 7.65 \text{ s}$$

$$N_F = \frac{bmg}{l + \mu c} = \frac{65(3,000)}{110 + 0.68(21)} = 1,570 \text{ lb}$$

$$N_R = \frac{(a + \mu c)mg}{l + \mu c} = \frac{[45 + 0.68(21)]}{110 + 0.68(21)} 3,000 = 1,430 \text{ lb}$$

(c)  When all four wheels drive the vehicle,

$$a_x = \mu g = 0.68(32.2) = 21.9 \text{ ft/s}^2$$
$$v = v_0 + at \qquad 88 = 0 + 21.9t \qquad t = 4.02 \text{ s}$$

$$N_F = \frac{(b - \mu c)}{l} mg = \frac{[65 - 0.68(21)]}{110} 3,000 = 1,380 \text{ lb}$$

$$N_R = \frac{(a + \mu c)}{l} mg = \frac{[45 + 0.68(21)]}{110} 3,000 = 1,620 \text{ lb}$$

(d)  When the vehicle is at rest, using Fig. 18.86d,

$$\sum M_d = 0 \qquad -(mg)a + N_R l = 0 \qquad N_R = \frac{a}{l} mg = \frac{45}{110}(3,000) = 1,230 \text{ lb}$$

$$\sum M_e = 0 \qquad -N_F l + (mg)b = 0 \qquad N_F = \frac{b}{l}(mg) = \frac{65}{110}(3,000) = 1,770 \text{ lb}$$

The values of $N_F$ and $N_R$, for front-, rear-, and four-wheel drives, are summarized in Table 18.4. It may be seen that the effect of the acceleration of the vehicle, for all three types of drive, is to "shift the weight" from the front wheels to the rear wheels. The effect is most pronounced in the case of four-wheel drive.

It follows from the solutions above that front-wheel drive is more effective traction than rear-wheel drive. Although the four-wheel-drive vehicle has the maximum traction, it is unlikely that such a vehicle would be equipped with an engine of sufficient power output to satisfy the conditions of this problem.

This vehicle problem is considered further in Probs. 18.88 through 18.90. Here, the effectiveness of front-wheel and rear-wheel braking is compared, in addition to the cases of accelerating or braking in traveling up or down an inclined surface.

(e)  For rear-wheel drive,

$$\mathbf{R}_F = 1,590\mathbf{j} \quad \text{lb} \qquad \mathbf{R}_R = 0.68(1,410)\mathbf{i} + 1,410\mathbf{j} = 959\mathbf{i} + 1,410\mathbf{j} \quad \text{lb}$$

**TABLE 18.4**

|  | $a_x$, ft/s$^2$ | $N_F$, lb | %D[†] | $N_R$, lb | %D[†] |
|---|---|---|---|---|---|
| Rear-wheel drive | 10.3 | 1,590 | −10 | 1,410 | 15 |
| Front-wheel drive | 11.5 | 1,570 | −11 | 1,430 | 16 |
| Four-wheel drive | 21.9 | 1,380 | −22 | 1,620 | 32 |
| Automobile stationary | — | 1,770 | | 1,230 | |

[†] Reference is normal force when automobile is stationary.

For front-wheel drive,

$$\mathbf{R}_F = 0.68(1,570)\mathbf{i} + 1,570\mathbf{j} = 1,070\mathbf{i} + 1,570\mathbf{j} \quad \text{lb} \qquad \mathbf{R}_R = 1,430\mathbf{j} \quad \text{lb}$$

For four-wheel drive,

$$\mathbf{R}_F = 0.68(1,380)\mathbf{i} + 1,380\mathbf{j} = 938\mathbf{i} + 1,380\mathbf{j} \quad \text{lb}$$
$$\mathbf{R}_R = 0.68(1,620)\mathbf{i} + 1,620\mathbf{j} = 1,100\mathbf{i} + 1,620\mathbf{j} \quad \text{lb}$$

**18.88** The automobile of Prob. 18.86, shown in Fig. 18.88a, moves along a horizontal roadway. The coefficient of friction between the tires and the roadway is 0.68, the dimensions of the vehicle are $a = 45$ in, $b = 65$ in, and $c = 21$ in, and the weight of the vehicle is 3,000 lb.

**(a)** The automobile is assumed to have brakes on the rear wheels only. Find the maximum possible braking deceleration of the automobile, and the reaction forces of the roadway on the wheels, by using the inertia force technique.

**(b)** Do the same as in part (a), if the automobile has brakes on the front wheels only.

**(c)** Do the same as in part (a), if the automobile has brakes on all four wheels.

**(d)** Find the reaction forces of the roadway on the wheels for the case where the automobile is stationary, and compare these values with those of the reaction forces when the vehicle experiences maximum deceleration.

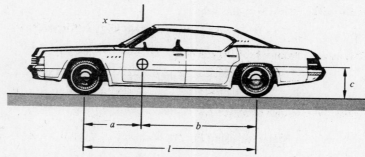

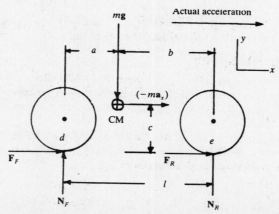

Fig. 18.88b

**(a)** A general solution is developed for parts (a) through (d).   Figure 18.88b shows the free-body diagram of the automobile.   The braking acceleration is to the right in this figure, so that the inertia force $(-ma_x)$ must act to the right.   $N_F$ and $N_R$ are the normal forces of the roadway on the front and rear wheels, and $F_F$ and $F_R$ are the corresponding values of the friction forces.   All of these forces are shown in Fig. 18.88b in their actual senses.

For the general case of equilibrium of the automobile

$$\sum F_x = 0 \qquad F_F + F_R + (-ma_x) = 0 \tag{1}$$

$$\sum F_y = 0 \qquad N_F - mg + N_R = 0 \tag{2}$$

$$\sum M_d = 0 \qquad -mga - (-ma_x)c + N_R l = 0 \tag{3}$$

For the case of rear-wheel brakes only,

$$F_F = 0 \tag{4}$$
$$F_R = \mu N_R \tag{5}$$

From Eq. (3),

$$N_R = \frac{mga - mca_x}{l} \tag{6}$$

Using Eqs. (4) through (6) in Eq. (1),

$$\mu\left(\frac{mga - mca_x}{l}\right) + (-ma_x) = 0 \qquad a_x = \frac{\mu ag}{l + \mu c} \tag{7}$$

Using   $a = 45$ in,   $b = 65$ in,   $c = 21$ in,   $l = 110$ in,   and   $\mu = 0.68$,

$$a_x = \frac{0.68(45)32.2}{110 + 0.68(21)} = 7.93 \text{ ft/s}^2 \qquad \text{(rear-wheel brakes only)}$$

Using Eq. (7) in Eq. (6),

$$N_R = \frac{mga}{l + \mu c} = \frac{3,000(45)}{110 + 0.68(21)} = 1,090 \text{ lb}$$

From Eq. (2),

$$N_F = mg - N_R = 3,000 - 1,090 = 1,910 \text{ lb}$$

**(b)**   For the case of front-wheel brakes only,

$$F_F = \mu N_F \tag{8}$$
$$F_R = 0 \tag{9}$$

For moment equilibrium about $e$,

$$\sum M_e = 0 \qquad -N_F l + mgb - (-ma_x)c = 0 \qquad N_F = \frac{mgb + mca_x}{l} \tag{10}$$

Using Eqs. (8) through (10) in Eq. (1),

$$a_x = \frac{\mu bg}{l - \mu c} \tag{11}$$

$$a_x = \frac{0.68(65)32.2}{110 - 0.68(21)} = 14.9 \text{ ft/s}^2 \qquad \text{(front-wheel brakes only)}$$

Using Eq. (11) in Eq. (10),

$$N_F = \frac{mgb}{l - \mu c} = \frac{3,000(65)}{110 - 0.68(21)} = 2,040 \text{ lb}$$

From Eq. (2),

$$N_R = mg - N_F = 3,000 - 2,040 = 960 \text{ lb}$$

**(c)**   For the case of brakes on all four wheels,

$$F_F = \mu N_F \tag{12}$$
$$F_R = \mu N_R \tag{13}$$

Using Eqs. (12) and (13) in Eqs. (1) and (2),

$$N_F + N_R = \frac{ma_x}{\mu} \qquad (14)$$

$$N_F + N_R = mg \qquad (15)$$

The right sides of Eqs. (14) and (15) are equated, with the result

$$\frac{ma_x}{\mu} = mg \qquad a_x = \mu g \qquad (16)$$

$$a_x = 0.68(32.2) = 21.9 \text{ ft/s}^2 \qquad \text{(four-wheel brakes)}$$

Using Eq. (16) in Eq. (3),

$$N_R = mg\left(\frac{a - \mu c}{l}\right) = 3,000\,\frac{[45 - 0.68(21)]}{110} = 838 \text{ lb}$$

From Eq. (2),

$$N_F = mg - N_R = 3,000 - 838 = 2,160 \text{ lb}$$

(d) For the case where the automobile is stationary, using Eq. (3) with $a_x = 0$,

$$N_R = \frac{mga}{l} = \frac{3,000(45)}{110} = 1,230 \text{ lb}$$

From Eq. (2),

$$N_F = mg - N_R = 3,000 - 1,230 = 1,770 \text{ lb}$$

A summary of the normal forces for the cases of rear-wheel, front-wheel, and four-wheel braking is shown in Table 18.5.

TABLE 18.5

| | $a_x$, ft/s$^2$ | $N_F$ | | $N_R$ | |
|---|---|---|---|---|---|
| | | lb | %D[†] | lb | %D[†] |
| Rear-wheel brakes | 7.93 | 1,910 | 8 | 1,090 | −11 |
| Front-wheel brakes | 14.9 | 2,040 | 15 | 960 | −22 |
| Four-wheel brakes | 21.9 | 2,160 | 22 | 838 | −32 |
| Automobile stationary | ⋯ | 1,770 | ⋯ | 1,230 | |

[†] Reference is normal force when automobile is stationary.

**18.89** The automobile in Prob. 18.88 ascends the straight inclined roadway shown in Fig. 18.89a. $\beta = 15°$.

(a) The automobile is assumed to have rear-wheel drive only. Find the maximum possible acceleration of the automobile, and the reaction forces of the roadway on the wheels, using the inertia force technique.

(b) Do the same as in part (a), if the automobile has front-wheel drive only.

(c) Do the same as in part (a), if the automobile has four-wheel drive.

(d) Find the reaction forces of the roadway on the wheels for the case where the automobile is stationary, and compare these values with those of the reaction forces when the vehicle experiences maximum acceleration.

(e) Find the maximum possible braking deceleration of the automobile, and the reaction forces of the roadway on the wheels, for the cases of rear-, front-, and four-wheel brakes.

▮ (a) A general solution is developed for parts (a) through (d). Figure 18.89b shows the free-body diagram of the automobile. $N_F$ and $N_R$ are the normal forces of the roadway on the front and rear wheels, and $F_F$ and $F_R$ are the corresponding values of the friction forces. All of these forces are shown in Fig. 18.89b in their actual senses.

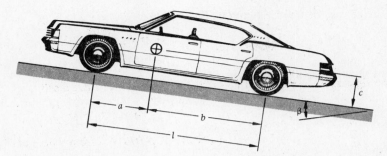

**Fig. 18.89a**

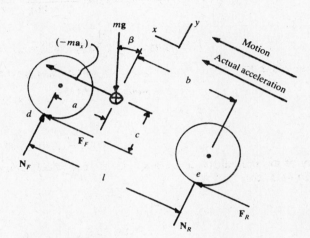

**Fig. 18.89b**

For the general case of equilibrium of the automobile,

$$\sum F_x = 0 \qquad F_F - mg \sin \beta + F_R + (-ma_x) = 0 \tag{1}$$

$$\sum F_y = 0 \qquad N_F - mg \cos \beta + N_R = 0 \tag{2}$$

$$\sum M_d = 0 \qquad -(mg \cos \beta)a - (mg \sin \beta)c + (-ma_x)c + N_R l = 0 \tag{3}$$

For the case of rear-wheel drive only,

$$F_F = 0 \tag{4}$$
$$F_R = \mu N_R \tag{5}$$

From Eq. (3),

$$N_R = \frac{mga \cos \beta + mgc \sin \beta + mca_x}{l} \tag{6}$$

Using Eqs. (4) through (6) in Eq. (1),

$$-mg \sin \beta + \mu \left( \frac{mga \cos \beta + mgc \sin \beta + mca_x}{l} \right) - ma_x = 0 \qquad a_x = \left( \frac{\mu a \cos \beta}{l - \mu c} - \sin \beta \right) g \tag{7}$$

Using $a = 45$ in, $b = 65$ in, $c = 21$ in, $l = 110$ in, $\mu = 0.68$, and $\beta = 15°$,

$$a_x = \left[ \frac{0.68(45) \cos 15°}{110 - 0.68(21)} - \sin 15° \right] 32.2 = 1.61 \text{ ft/s}^2 \qquad \text{(rear-wheel drive only)}$$

Using Eq. (7) in Eq. (6),

$$N_R = \frac{mga \cos \beta}{l - \mu c} = \frac{3,000(45) \cos 15°}{110 - 0.68(21)} = 1,360 \text{ lb}$$

From Eq. (2),

$$N_F = mg \cos \beta - N_R = 3,000 \cos 15° - 1,360 = 1,540 \text{ lb}$$

(*b*)  For the case of front-wheel drive only,

$$F_F = \mu N_F \tag{8}$$
$$F_R = 0 \tag{9}$$

For moment equilibrium about *e*,

$$\sum M_e = 0 \qquad -N_F l + (mg \cos \beta)b - (mg \sin \beta)c + (-ma_x)c = 0$$

$$N_F = \frac{mg(b \cos \beta - c \sin \beta) - mca_x}{l} \tag{10}$$

Using Eqs. (8) through (10) in Eq. (1),

$$a_x = \left( \frac{\mu b \cos \beta}{l + \mu c} - \sin \beta \right) g \tag{11}$$

$$a_x = \left( \frac{0.68(65) \cos 15°}{110 + 0.68(21)} - \sin 15° \right) 32.2 = 2.73 \text{ ft/s}^2 \qquad \text{(front-wheel drive only)}$$

Using Eq. (11) in Eq. (10),

$$N_F = \frac{mgb \cos \beta}{l + \mu c} = \frac{3{,}000(65) \cos 15°}{110 + 0.68(21)} = 1{,}520 \text{ lb}$$

From Eq. (2),

$$N_R = mg \cos \beta - N_F = 3{,}000 \cos 15° - 1{,}520 = 1{,}380 \text{ lb}$$

(*c*)  For the case of four-wheel drive,

$$F_F = \mu N_F \tag{12}$$
$$F_R = \mu N_R \tag{13}$$

Using Eqs. (12) and (13) in Eqs. (1) and (2),

$$N_F + N_R = \frac{mg \sin \beta + ma_x}{\mu} \tag{14}$$

$$N_F + N_R = mg \cos \beta \tag{15}$$

The right sides of Eqs. (14) and (15) are equated, with the result

$$a_x = (\mu \cos \beta - \sin \beta)g \tag{16}$$
$$a_x = (0.68 \cos 15° - \sin 15°)32.2 = 12.8 \text{ ft/s}^2 \qquad \text{(four-wheel drive)}$$

Using Eq. (16) in Eq. (3),

$$N_R = \frac{(a + \mu c)mg \cos \beta}{l} = \frac{[45 + 0.68(21)]3{,}000 \cos 15°}{110} = 1{,}560 \text{ lb}$$

From Eq. (2),

$$N_F = mg \cos \beta - N_R = 3{,}000 \cos 15° - 1{,}560 = 1{,}340 \text{ lb}$$

(*d*)  For the case where the automobile is stationary, using Eq. (3) with $a_x = 0$,

$$N_R = \left( \frac{a \cos \beta + c \sin \beta}{l} \right) mg = \left( \frac{45 \cos 15° + 21 \sin 15°}{110} \right) 3{,}000 = 1{,}330 \text{ lb}$$

From Eq. (2),

$$N_F = mg \cos \beta - N_R = 3{,}000 \cos 15° - 1{,}330 = 1{,}570 \text{ lb}$$

(*e*)  It is left as an exercise for the reader to verify the following results for the maximum braking deceleration, and reaction forces of the roadway on the wheels, as the automobile ascends the straight roadway shown in Fig. 18.89*a*.

*Rear-wheel brakes only:*   $a_x = 16 \text{ ft/s}^2$    $N_F = 1{,}850 \text{ lb}$    $N_R = 1{,}050 \text{ lb}$
*Front-wheel brakes only:*   $a_x = 22.7 \text{ ft/s}^2$    $N_F = 1{,}970 \text{ lb}$    $N_R = 928 \text{ lb}$
*Four-wheel brakes:*    $a_x = 29.5 \text{ ft/s}^2$    $F_F = 2{,}090 \text{ lb}$    $N_R = 810 \text{ lb}$

**18.90**  The automobile in Prob. 18.88 descends the straight inclined roadway shown in Fig. 18.90*a*.  $\beta = 15°$.

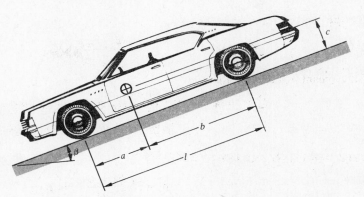

**Fig. 18.90a**

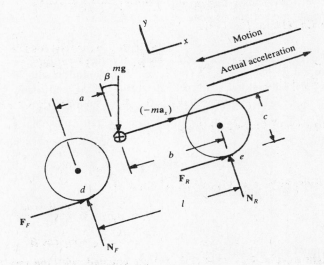

**Fig. 18.90b**

(a) The automobile is assumed to have rear-wheel brakes only. Find the maximum possible braking deceleration of the automobile, and the reaction forces of the roadway on the wheels, by using the inertia force technique.

(b) Do the same as in part (a), if the automobile has front-wheel brakes only.

(c) Do the same as in part (a), if the automobile has brakes on all four wheels.

(d) Find the reaction forces of the roadway on the wheels for the case where the automobile is stationary, and compare these values with those of the reaction forces when the vehicle experiences maximum deceleration.

(e) Find the maximum possible acceleration of the automobile, and the reaction forces of the roadway on the wheels, for the cases of rear-, front-, and four-wheel drives.

▌ (a) A general solution is developed for parts (a) through (d). Figure 18.90b shows the free-body diagram of the automobile. $N_F$ and $N_R$ are the normal forces of the roadway on the front and rear wheels, and $F_F$ and $F_R$ are the corresponding values of the friction forces. All of these forces are shown in Fig. 18.90b in their actual senses. For the general case of equilibrium of the automobile,

$$\sum F_x = 0 \qquad F_F - mg \sin \beta + F_R + (-ma_x) = 0 \tag{1}$$

$$\sum F_y = 0 \qquad N_F - mg \cos \beta + N_R = 0 \tag{2}$$

$$\sum M_d = 0 \qquad -(mg \cos \beta)a + (mg \sin \beta)c - (-ma_x)c + N_R l = 0 \tag{3}$$

For the case of rear-wheel brakes only,

$$F_F = 0 \tag{4}$$

$$F_R = \mu N_R \tag{5}$$

From Eq. (3),

$$N_R = \frac{mga \cos \beta - mgc \sin \beta - mca_x}{l} \qquad (6)$$

Using Eqs. (4) through (6) in Eq. (1),

$$-mg \sin \beta + \mu\left(\frac{mga \cos \beta - mgc \sin \beta - mca_x}{l}\right) - ma_x = 0 \qquad a_x = \left(\frac{\mu a \cos \beta}{l + \mu c} - \sin \beta\right)g \qquad (7)$$

Using $a = 45$ in, $b = 65$ in, $c = 21$ in, $l = 110$ in, $\mu = 0.68$, and $\beta = 15°$,

$$a_x = \left[\frac{0.68(45) \cos 15°}{110 + 0.68(21)} - \sin 15°\right]32.2 = -0.676 \text{ ft/s}^2 \qquad \text{(rear-wheel brakes only)}$$

The result obtained above for $a_x$ has a significant implication. If braking, with reduction of velocity, is to occur then $a_x$, as used in this problem, must be a *positive* quantity. The fact that a negative value is obtained for $a_x$ indicates that *the vehicle with rear-wheel brakes only does not have adequate braking effect, when descending the inclined roadway, to bring the vehicle to rest.*

In the absence of any braking effect, the acceleration of the automobile down the incline is given by

$$a_x = -g \sin \beta = -32.2 \sin 15° = -8.33 \text{ ft/s}^2$$

The use of maximum rear-wheel braking force has the effect of changing the above value of acceleration to $-0.676$ ft/s². Thus, even with the full braking action, the automobile with rear-wheel brakes only will continue to accelerate down the inclined roadway. Using Eq. (7) in Eq. (6),

$$N_R = \frac{mga \cos \beta}{l + \mu c} = \frac{3,000(45) \cos 15°}{110 + 0.68(21)} = 1,050 \text{ lb}$$

From Eq. (2),

$$N_F = mg \cos \beta - N_R = 3,000 \cos 15° - 1,050 = 1,850 \text{ lb}$$

(b) For the case of front-wheel brakes only,

$$F_F = \mu N_F \qquad (8)$$
$$F_R = 0 \qquad (9)$$

For moment equilibrium about $e$,

$$\sum M_e = 0 \qquad -N_F l + (mg \cos \beta)b + (mg \sin \beta)c - (-ma_x)c = 0$$
$$N_F = \frac{mg(b \cos \beta + c \sin \beta) + mca_x}{l} \qquad (10)$$

Using Eqs. (8) through (10) in Eq. (1),

$$a_x = \left(\frac{\mu b \cos \beta}{l - \mu c} - \sin \beta\right)g \qquad (11)$$

$$a_x = \left[\frac{0.68(65) \cos 15°}{110 - 0.68(21)} - \sin 15°\right]32.2 = 6.03 \text{ ft/s}^2 \qquad \text{(front-wheel brakes only)}$$

Using Eq. (11) in Eq. (10),

$$N_F = \frac{mgb \cos \beta}{l - \mu c} = \frac{3,000(65) \cos 15°}{110 - 0.68(21)} = 1,970 \text{ lb}$$

From Eq. (2),

$$N_R = mg \cos \beta - N_F = 3,000 \cos 15° - 1,970 = 928 \text{ lb}$$

(c) For the case of four-wheel brakes,

$$F_F = \mu N_F \qquad (12)$$
$$F_R = \mu N_R \qquad (13)$$

Using Eqs. (12) and (13) in Eqs. (1) and (2),

$$N_F + N_R = \frac{mg \sin \beta + ma_x}{\mu} \qquad (14)$$
$$N_F + N_R = mg \cos \beta \qquad (15)$$

The right sides of Eqs. (14) and (15) are equated, with the result

$$a_x = (\mu \cos \beta - \sin \beta)g \qquad (16)$$
$$a_x = (0.68 \cos 15° - \sin 15°)32.2 = 12.8 \text{ ft/s}^2 \qquad \text{(four-wheel brakes)}$$

Using Eq. (16) in Eq. (3),

$$N_R = \frac{(a - \mu c)mg \cos \beta}{l} = \frac{[45 - 0.68(21)]3,000 \cos 15°}{110} = 809 \text{ lb}$$

From Eq. (2),

$$N_F = mg \cos \beta - N_R = 3,000 \cos 15° - 809 = 2,090 \text{ lb}$$

(*d*)  For the case where the automobile is stationary, using Eq. (3) with  $a_x = 0$,

$$N_R = \left( \frac{a \cos \beta - c \sin \beta}{l} \right) mg = \left( \frac{45 \cos 15° + 21 \sin 15°}{110} \right) 3,000 = 1,330 \text{ lb}$$

From Eq. (2),

$$N_F = mg \cos \beta - N_R = 3,000 \cos 15° - 1,330 = 1,570 \text{ lb}$$

(*e*)  It is left as an exercise for the reader to verify the following results for the maximum acceleration, and reaction forces of the roadway on the wheels, as the automobile descends the straight roadway shown in Fig. 18.90*a*.

*Rear-wheel drive only:*  $a_x = 18.3 \text{ ft/s}^2$  $N_F = 1,540 \text{ lb}$  $N_R = 1,360 \text{ lb}$
*Front-wheel drive only:*  $a_x = 19.4 \text{ ft/s}^2$  $N_F = 1,620 \text{ lb}$  $N_R = 1,380 \text{ lb}$
*Four-wheel drive:*  $a_x = 29.5 \text{ ft/s}^2$  $N_F = 1,340 \text{ lb}$  $N_R = 1,560 \text{ lb}$

**18.91**  Give a summary of the maximum values of the acceleration and deceleration, and the corresponding values of the normal forces of the roadway on the wheels, for the automobile motion in Probs. 18.86 through 18.90.

**❚**  The maximum values of the acceleration, deceleration, and normal forces are shown in Table 18.6, page 398.  For the dimensions of the automobile considered, the following effects may be observed.

**1.**  For all cases, the maximum values of both acceleration and deceleration increase as the design evolves from rear-wheel to front-wheel to four-wheel drive or braking.

**2.**  For all cases there is a "shift of weight" from the front wheels to the rear wheels during acceleration.

**3.**  For all cases there is a "shift of weight" from the rear wheels to the front wheels during deceleration.

**4.**  The front- and rear-wheel normal forces for the stationary automobile are the same for both the upward-inclined roadway and the downward-inclined roadway.

**5.**  For one case—braking with rear-wheel brakes only on the downward-inclined roadway—the braking force is inadequate to stop the automobile.

## 18.6  CRITERIA FOR SLIDING OR TIPPING, CENTER OF PERCUSSION

**18.92**  Show how the inertia force may be used to determine whether an object will slide or tip.

**❚**  A typical example is shown in Fig. 18.92*a*, which shows a crate resting on a flatbed truck.  With no loss of generality, the crate may be assumed to be made of a homogeneous material, so that the CM is located at the centroid of the volume.  As the driver applies the brakes, the crate may decelerate with the truck, slide relative to the flatbed, or tip over.  Each case will be considered now.

When the truck is at rest or moving with constant velocity, the free-body diagram of the crate is as shown in Fig. 18.92*b*.  The weight force *mg* is equal to the normal force *N* of the bed of the truck on the crate.

Figure 18.92*c* shows the free-body diagram when the driver brakes.  The positive *x* coordinate axis shown in the figure has the sense of the *actual* acceleration.  The truck is decelerating, so that $a_x$ is positive.  *The actual sense of the inertia force, interpreted as an external force applied to the crate, is to the right.*  Since this force tends to rotate the crate clockwise, the line of action of the reaction force of the bed of the truck on the crate moves to the right of its original position.  The force *F* is the friction force exerted by the truck bed on the crate.

The magnitude of the deceleration is now imagined to increase.  If the crate is temporarily assumed not to tip, sliding motion of this object will be impending when the friction force attains its maximum value of

$$F_{\max} = \mu N$$

**TABLE 18.6**

| condition | maximum acceleration | | | maximum deceleration | | |
|---|---|---|---|---|---|---|
| | $a_x$, ft/s² | $N_F$, lb | $N_R$, lb | $a_x$, ft/s² | $N_F$, lb | $N_R$, lb |
| **level roadway** | | | | | | |
| Rear-wheel drive, or brakes | 10.3 | 1,590 | 1,410 | 7.93 | 1,910 | 1,090 |
| Front-wheel drive, or brakes | 11.5 | 1,570 | 1,430 | 14.9 | 2,040 | 960 |
| Four-wheel drive, or four-wheel brakes | 21.9 | 1,380 | 1,620 | 21.9 | 2,160 | 838 |
| Automobile stationary | · · · | 1,770 | 1,230 | · · · | 1,770 | 1,230 |
| **upward-inclined roadway** | | | | | | |
| Rear-wheel drive, or brakes | 1.61 | 1,540 | 1,360 | 16 | 1,850 | 1,050 |
| Front-wheel drive, or brakes | 2.73 | 1,520 | 1,380 | 22.7 | 1,970 | 928 |
| Four-wheel drive, or four-wheel brakes | 12.8 | 1,340 | 1,560 | 29.5 | 2,090 | 810 |
| Automobile stationary | · · · | 1,570 | 1,330 | · · · | 1,570 | 1,330 |
| **downward-inclined roadway** | | | | | | |
| Rear-wheel drive, or brakes | 18.3 | 1,540 | 1,360 | −0.676[†] | 1,850 | 1,050 |
| Front-wheel drive, or brakes | 19.4 | 1,620 | 1,380 | 6.03 | 1,970 | 928 |
| Four-wheel drive, or four-wheel brakes | 29.5 | 1,340 | 1,560 | 12.8 | 2,090 | 809 |
| Automobile stationary | · · · | 1,570 | 1,330 | · · · | 1,570 | 1,330 |

[†] No braking effect available.

where $\mu$ is the coefficient of friction. The equations of motion for this case are

$$\sum F_y = 0 \qquad -mg + N = 0 \qquad N = mg$$

$$\sum F_x = 0 \qquad (-ma_x) + \mu N = 0$$

$$ma_x = \mu N = \mu mg \qquad a_x = \mu g$$

If the magnitude of the deceleration of the truck exceeds $\mu g$, the crate will slide.

As the magnitude of the deceleration of the truck increases, the line of action of the reaction force of the crate continues to move rightward. A limiting condition is reached when this force acts at the forward edge $c$ of the crate, as shown in Fig. 18.92d. If the friction force $F$ at this condition is less than the maximum possible value $\mu N$, the crate will be in a condition of impending tipping motion about the forward edge $c$. The equilibrium requirements are

$$\sum F_y = 0 \qquad -mg + N = 0 \qquad N = mg$$

$$\sum M_c = 0 \qquad mg\,\frac{b}{2} + (-ma_x)\left(\frac{a}{2}\right) = 0 \qquad a_x = \frac{b}{a}\,g$$

It may be observed that $b/a$ defines the shape of the crate. A short, squat crate would have a large $b/a$ ratio, while a tall, slender crate would have a small value of $b/a$.

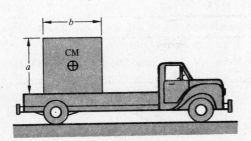

**Fig. 18.92a**

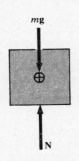

**Fig. 18.92b**

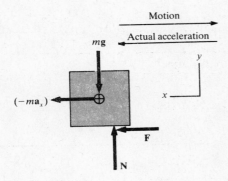

**Fig. 18.92c**

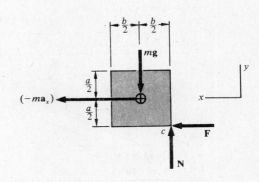

**Fig. 18.92d**

The following possible motions of the crate on the bed of the truck may now be identified.

1. If $a_x \le \mu g$ and $a_x \le (b/a)g$, the crate will neither slide nor tip.
2. If $a_x > \mu g$ and $\mu < b/a$, the crate will slide without tipping.
3. If $a_x > (b/a)g$ and $\mu > b/a$, the crate will tip without sliding.

A special case occurs if $\mu = b/a$. For this case, sliding and tipping are equally likely and these effects would occur simultaneously.

**18.93** A crate rests on the back of a truck, as shown in Fig. 18.93. The truck bed is made of rough wood, and the coefficient of friction between the bed and the crate is estimated to be 0.6. Find the magnitude of the maximum deceleration which the truck may experience if the crate is not to move relative to the bed.

▌ Sliding motion of the crate is impending when

$$a_x = \mu g = 0.6(9.81) = 5.89 \text{ m/s}^2$$

Tipping will occur if

$$a_x = \frac{b}{a} g = \frac{1.2}{1.84} 9.81 = 6.40 \text{ m/s}^2$$

Sliding motion will occur first, and the maximum permissible deceleration is $5.89 \text{ m/s}^2$.

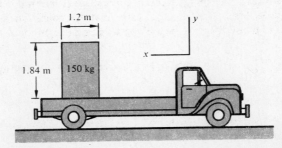

**Fig. 18.93**

**18.94** A truck ascends the straight inclined roadway shown in Fig. 18.94a.

(a) Develop the criteria for sliding or tipping motion of the crate as the truck accelerates.

(b) Do the same as in part (a), for the case when the truck decelerates.

(c) Find the numerical results for parts (a) and (b), using $\beta = 15°$ and the numerical data from Prob. 18.93.

(d) Do the same as in part (c) but with a coefficient of friction between the crate and the bed of the truck of 0.7.

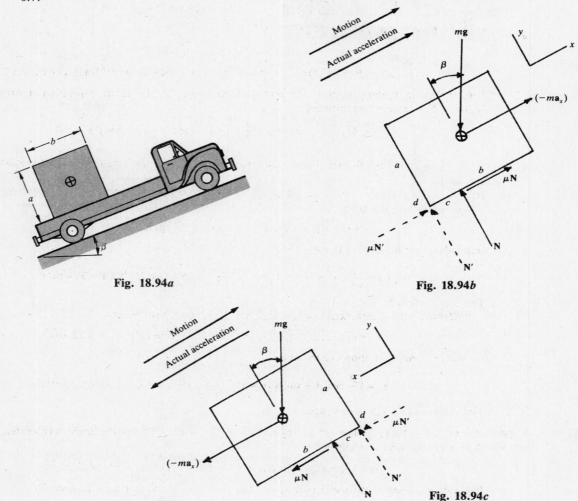

Fig. 18.94a

Fig. 18.94b

Fig. 18.94c

▮ (a) When the truck accelerates, the normal force $N$ is at some intermediate location $c$, as shown in the free-body diagram in Fig. 18.94b. The friction force is assumed to have its maximum value $\mu N$. For equilibrium of the crate, when sliding motion is impending,

$$\sum F_y = 0 \qquad -mg \cos \beta + N = 0 \qquad N = mg \cos \beta \tag{1}$$

$$\sum F_x = 0 \qquad -mg \sin \beta + (-ma_x) + \mu N = 0 \tag{2}$$

Using $N$ from Eq. (1) in Eq. (2),

$$-mg \sin \beta - ma_x + \mu mg \cos \beta = 0$$

$$a_x = (\mu \cos \beta - \sin \beta)g \qquad \text{(impending sliding motion when truck accelerates)}$$

For impending tipping motion, the normal force $N'$ is at the limiting position $d$ shown in Fig. 18.94b. For moment equilibrium,

$$\sum M_d = 0 \qquad -(mg \cos \beta)\frac{b}{2} + (mg \sin \beta)\frac{a}{2} - (-ma_x)\frac{a}{2} = 0$$

$$a_x = \left(\frac{b}{a}\cos\beta - \sin\beta\right)g \qquad \text{(impending tipping motion when truck accelerates)}$$

(b) When the truck decelerates, the free-body diagram is as shown in Fig. 18.94c, and the normal force $N$ is at some intermediate location $c$. For equilibrium of the crate, when sliding motion is impending,

$$\sum F_y = 0 \qquad -mg\cos\beta + N = 0 \qquad N = mg\cos\beta \qquad (3)$$

$$\sum F_x = 0 \qquad (-ma_x) + mg\sin\beta + \mu N = 0 \qquad (4)$$

Using $N$ from Eq. (3) in Eq. (4),

$$-ma_x + mg\sin\beta + \mu mg\cos\beta = 0$$

$$a_x = (\mu\cos\beta + \sin\beta)g \qquad \text{(impending sliding motion when truck decelerates)}$$

For impending tipping motion, the normal force $N'$ is at the limiting position $d$ shown in Fig. 18.94c. For moment equilibrium,

$$\sum M_d = 0 \qquad (-ma_x)\frac{a}{2} + (mg\sin\beta)\frac{a}{2} + (mg\cos\beta)\frac{b}{2} = 0$$

$$a_x = \left(\frac{b}{a}\cos\beta + \sin\beta\right)g \qquad \text{(impending tipping motion when truck decelerates)}$$

(c) $\beta = 15°$, $a = 1.84\,\text{m}$, $b = 1.2\,\text{m}$, and $\mu = 0.6$ are used. When the truck accelerates, impending sliding motion occurs when

$$a_x = (\mu\cos\beta - \sin\beta)g = (0.6\cos 15° - \sin 15°)9.81 = 3.15\,\text{m/s}^2$$

Impending tipping motion occurs when

$$a_x = \left(\frac{b}{a}\cos\beta - \sin\beta\right)g = \left(\frac{1.2}{1.84}\cos 15° - \sin 15°\right)9.81 = 3.64\,\text{m/s}^2$$

The crate will slide before it tips.
When the truck decelerates, impending sliding motion occurs when

$$a_x = (\mu\cos\beta + \sin\beta)g = (0.6\cos 15° + \sin 15°)9.81 = 8.22\,\text{m/s}^2$$

Impending tipping motion occurs when

$$a_x = \left(\frac{b}{a}\cos\beta + \sin\beta\right)g = \left(\frac{1.2}{1.84}\cos 15° + \sin 15°\right)9.81 = 8.72\,\text{m/s}^2$$

The crate will slide before it tips.

(d) $\beta = 15°$, $a = 1.84\,\text{m}$, $b = 1.2\,\text{m}$, and $\mu = 0.7$ are used. When the truck accelerates, impending sliding motion occurs when

$$a_x = (\mu\cos\beta - \sin\beta)g = (0.7\cos 15° - \sin 15°)9.81 = 4.09\,\text{m/s}^2$$

Impending tipping motion occurs when $a_x$ has the value found in part (c), given by

$$a_x = 3.64\,\text{m/s}^2$$

The crate will tip before it slides.
When the truck decelerates, impending sliding motion occurs when

$$a_x = (\mu\cos\beta + \sin\beta)g = (0.7\cos 15° + \sin 15°)9.81 = 9.17\,\text{m/s}^2$$

Impending tipping motion occurs when $a_x$ has the value found in part (c), given by

$$a_x = 8.72\,\text{m/s}^2$$

The crate will tip before it slides.

18.95 For what range of values of $h$ will the crate in Fig. 18.95a slide without tipping?

❙ Figure 18.95b shows the free-body diagram of the crate with the inertia force acting to the right, in the sense of the actual acceleration. When tipping is impending, the normal reaction force is at the forward edge $a$ in Fig. 18.95b. Since sliding motion of the crate occurs, the friction force has its maximum value $\mu N = 0.18N$. For equilibrium of the crate,

$$\sum F_y = 0 \qquad -196 + N = 0 \qquad N = 196\,\text{N}$$

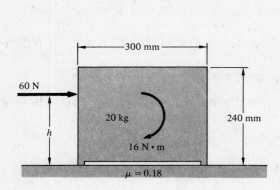

Fig. 18.95a

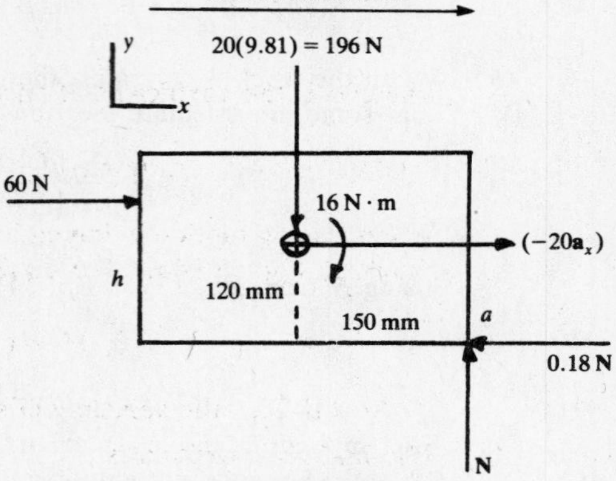

Fig. 18.95b

$$\sum F_x = 0 \qquad 60 + (-20a_x) - 0.18(196) = 0 \qquad a_x = 1.24 \text{ m/s}^2$$

$$\sum M_a = 0 \qquad -60h - 16 - [-20(a_x)]\frac{120}{1,000} + 196\left(\frac{150}{1,000}\right) = 0 \qquad h = 0.273 \text{ m} = 273 \text{ mm}$$

If $h \le 273$ mm, the crate will slide without tipping. Since the height of the crate is 240 mm, the crate will slide for *any* position of the 60-N force.

**18.96** A vehicle travels with constant speed around a horizontal circular track, as shown in Fig. 18.96a.

(*a*) Develop the criteria for sliding or tipping of the vehicle.

(*b*) If $c = 21$ in, $d = 57$ in, $\rho = 300$ ft, and $\mu = 0.81$, find the limiting values of the speed at which sliding or tipping will occur.

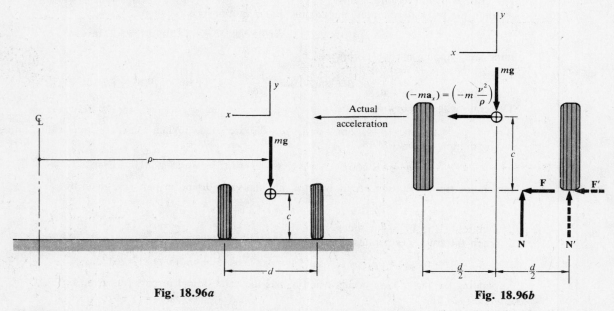

Fig. 18.96a

Fig. 18.96b

**▌** (*a*) The center of mass of the vehicle travels with constant speed in a plane circular path which is parallel to the track. The plane motion which is considered in this problem is in a radial plane which is normal to the track, as shown in Fig. 18.96a. The free-body diagram of the vehicle is shown in Fig. 18.96b. The centripetal acceleration has the magnitude $v^2/\rho$ and is directed toward the center of the track, in the sense of positive $x$. For the case where tipping is *not* impending, the normal force $N$ has the general position shown in the figure. The equilibrium requirements are

$$\sum F_y = 0 \qquad -mg + N = 0 \qquad N = mg$$

$$\sum F_x = 0 \qquad \left(-m\frac{v^2}{\rho}\right) + F = 0 \qquad F = \frac{mv^2}{\rho} \tag{1}$$

When sliding motion is impending,

$$F = F_{max} = \mu N = \mu mg \tag{2}$$

where $\mu$ is the coefficient of friction. $F$ is eliminated between Eqs. (1) and (2), with the result

$$\mu mg = m\frac{v_{max}^2}{\rho} \qquad v_{max} = \sqrt{\mu\rho g} \qquad \text{(impending sliding motion)}$$

When tipping motion is impending, the normal reaction force component $N'$ has the limiting position shown in Fig. 18.96b. For equilibrium,

$$\sum M_e = 0 \qquad \left(-\frac{mv^2}{\rho}\right)c + mg\,\frac{d}{2} = 0 \qquad v_{max} = \sqrt{\frac{d}{2c}\rho g}$$

If $\mu < d/2c$, the vehicle will slide before it tips.

**(b)** The limiting condition is

$$\mu \overset{?}{<} \frac{d}{2c} = \frac{57}{2(21)} = 1.36$$

Since $\mu = 0.81 < 1.36$, the vehicle will slide before it tips. The value of the speed when sliding motion is impending is

$$v_{max} = \sqrt{\mu\rho g} = \sqrt{0.81(300)32.2} = 88.5\,\text{ft/s} = 60.3\,\text{mi/h}$$

**18.97** **(a)** The vehicle of Prob. 18.96 travels with constant speed around the horizontal circular track shown in Fig. 18.97a. Find the speeds at which sliding or tipping of the vehicle occurs.

**(b)** Do the same as in part (a), if the track is banked as shown in Fig. 18.97b.

**(c)** How can the results in part (b) be obtained directly from the solution to part (a)?

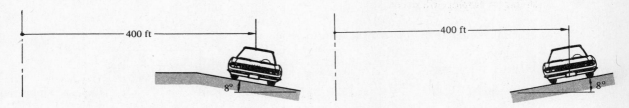

Fig. 18.97a        Fig. 18.97b

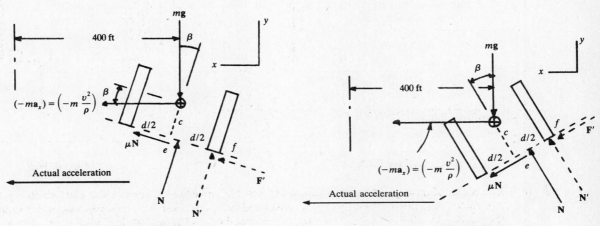

Fig. 18.97c        Fig. 18.97d

**❚ (a)** For impending sliding motion the normal force $N$ is at some intermediate location $e$, as shown in Fig. 18.97c. For equilibrium of the vehicle,

$$\sum F_y = 0 \qquad -mg + N\cos\beta + \mu N\sin\beta = 0 \qquad N = \frac{mg}{\cos\beta + \mu\sin\beta} \tag{1}$$

$$\sum F_x = 0 \qquad \left(-m\,\frac{v^2}{\rho}\right) - N \sin\beta + \mu N \cos\beta = 0 \qquad N = \frac{m(v^2/\rho)}{\mu\,\cos\beta - \sin\beta} \tag{2}$$

Equations (1) and (2) are equated, with the result

$$v = \sqrt{\left(\frac{\mu\,\cos\beta - \sin\beta}{\cos\beta + \mu\,\sin\beta}\right)\rho g} = \sqrt{\left(\frac{\mu - \tan\beta}{1 + \mu\,\tan\beta}\right)\rho g}$$

Using $c = 21\,\text{in}$, $d = 57\,\text{in}$, $\mu = 0.81$, $\rho = 400\,\text{ft}$, and $\beta = 8°$ in the above equation,

$$v = \sqrt{\left(\frac{0.81 - \tan 8°}{1 + 0.81\,\tan 8°}\right)400(32.2)} = 88.0\,\text{ft/s} = 60\,\text{mi/n}$$

For impending tipping motion the normal force $N'$ is at the limiting position $f$ in Fig. 18.97c. For equilibrium,

$$\sum M_f = 0 \qquad -\left(-m\,\frac{v^2}{\rho}\right)c\,\cos\beta - \left(-m\,\frac{v^2}{\rho}\right)\frac{d}{2}\,\sin\beta - (mg\,\cos\beta)\frac{d}{2} + (mg\,\sin\beta)c = 0$$

$$v = \sqrt{\left(\frac{(d/2)\cos\beta - c\,\sin\beta}{c\,\cos\beta + (d/2)\sin\beta}\right)\rho g} = \sqrt{\left(\frac{(d/2) - c\,\tan\beta}{c + (d/2)\tan\beta}\right)\rho g} = \sqrt{\left(\frac{(57/2) + 21\,\tan 8°}{21 - (57/2)\tan 8°}\right)400(32.2)}$$

$$= 115\,\text{ft/s} = 78.2\,\text{mi/h}$$

From the above results, it may be concluded that the vehicle would slide before it would tip.

(b) For impending sliding motion the normal force $N$ is at some intermediate location $e$, as shown in Fig. 18.97d. For equilibrium,

$$\sum F_y = 0 \qquad -mg - \mu N \sin\beta + N \cos\beta = 0 \qquad N = \frac{mg}{\cos\beta - \mu\,\sin\beta} \tag{3}$$

$$\sum F_x = 0 \qquad \left(-m\,\frac{v^2}{\rho}\right) + \mu N \cos\beta + N \sin\beta = 0 \qquad N = \frac{m(v^2/\rho)}{\mu\,\cos\beta + \sin\beta} \tag{4}$$

Equations (3) and (4) are equated, to obtain

$$v = \sqrt{\left(\frac{\mu\,\cos\beta + \sin\beta}{\cos\beta - \mu\,\sin\beta}\right)\rho g} = \sqrt{\left(\frac{\mu + \tan\beta}{1 - \mu\,\tan\beta}\right)\rho g}$$

Using $c = 21\,\text{in}$, $d = 57\,\text{in}$, $\mu = 0.81$, $\rho = 400\,\text{ft}$, and $\beta = 8°$ in the above equation results in

$$v = \sqrt{\left(\frac{0.81 + \tan 8°}{1 - 0.81\,\tan 8°}\right)400(32.2)} = 118\,\text{ft/s} = 80.5\,\text{mi/h}$$

For impending tipping motion, the normal force $N'$ is at the limiting position $f$ shown in Fig. 18.97d. For equilibrium,

$$\sum M_f = 0 \qquad \left(-m\,\frac{v^2}{\rho}\right)c\,\cos\beta - \left(-m\,\frac{v^2}{\rho}\right)\frac{d}{2}\,\sin\beta + (mg\,\cos\beta)\frac{d}{2} + (mg\,\sin\beta)c = 0$$

$$v = \sqrt{\left(\frac{(d/2)\cos\beta + c\,\sin\beta}{c\,\cos\beta - (d/2)\sin\beta}\right)\rho g} = \sqrt{\left(\frac{d/2 + c\,\tan\beta}{c - d/2\,\tan\beta}\right)\rho g} = \sqrt{\left[\frac{57/2 + 21\,\tan 8°}{21 - (57/2)\tan 8°}\right]400(32.2)}$$

$$= 154\,\text{ft/s} = 105\,\text{mi/h}$$

The vehicle would slide before it would tip.

(c) The results in part (b) can be obtained directly from the solution in part (a) by replacing $\beta$ by $-\beta$ and $\tan(-\beta)$ by $-\tan\beta$.

**18.98** (a) What is meant by the *center of percussion*?

(b) How is the location of the center of percussion determined?

▮ (a) When a rigid body experiences *angular acceleration about a fixed point* there is a point on the body, which is usually not coincident with the center of mass (CM), that has a unique property. This property is that the sum about this point of all the moments which act on the body is always zero. This point is defined to be the center of percussion. A familiar example is when a baseball is struck by a bat at the "wrong" location along the bat and an unpleasant stinging sensation is felt in the hands.

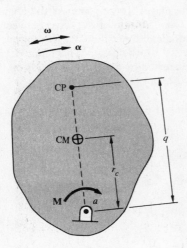

**Fig. 18.98a**

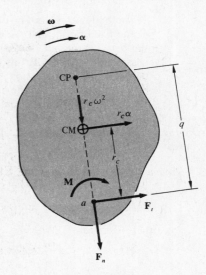

**Fig. 18.98b**

Figure 18.98a shows a rigid body which moves in plane motion about a fixed point $a$. The point CP is defined to be the center of percussion. This point lies on a straight line through the CM and the fixed point, at the distance $q$ from the fixed point. The body is acted upon by the resultant couple $M$.

The free-body diagram is shown in Fig. 18.98b, and on this diagram are shown the acceleration components of the CM. The forces $F_n$ and $F_t$, with the directions shown in the figure, are the two components of the total force exerted by the hinge pin on the body. These two force components act in the actual senses of the two acceleration components. The weight force of the body is neglected.

The equations of motion of the CM of the body are

$$\mathbf{F} = m\mathbf{a}_c \qquad F_n = mr_c\omega^2 \qquad F_t = mr_c\alpha \tag{1}$$

In any given problem, these equations may be used directly to find the hinge pin force.

The equation of motion about the fixed point is

$$M = I_a\alpha = (I_0 + mr_c^2)\alpha \tag{2}$$

where $I_0$ is the mass moment of inertia of the body about the CM.

The CP is the point about which the sum of all the moments acting on the body is identically zero. This condition may be written as

$$\sum M_{CP} = 0 \qquad -M + F_tq = 0 \tag{3}$$

$F_t$ and $M$ from Eqs. (1) and (2) are substituted into Eq. (3), with the result

$$-(I_0 + mr_c^2)(\alpha) + mr_c\alpha q = 0$$

$$q = \frac{I_0 + mr_c^2}{mr_c} = r_c + \frac{I_0}{mr_c} \tag{4}$$

The term $I_0$ may be written as

$$I_0 = k_0^2 m$$

where $k_0$ is the centroidal radius of gyration. Equation (4) now appears as

$$q = r_c + \frac{k_0^2 m}{mr_c} = r_c + \frac{k_0^2}{r_c} \tag{5}$$

The separation distance $\delta$ between the points CP and CM is

$$\delta = q - r_c = r_c + \frac{k_0^2}{r_c} - r_c \qquad \delta = \frac{k_0^2}{r_c}$$

It may be seen from the above equation that the only condition for which $\delta = 0$ would be for $k_0 = 0$. This corresponds to the condition where the body is a point mass, and this case is considered in Prob. 18.100.

CHAPTER 18

The body shown in Fig. 18.98*b* is now imagined to represent a baseball bat. The player's hands are at the fixed point, and the couple applied by the player's hands imparts the angular acceleration $\alpha$ to the bat. If the bat strikes the ball so that the line of action of the force between these two elements passes through the CP, this force will cause no moment about the CP. If, however, the bat strikes the ball at a different location, a moment about the CP is produced. This moment must be counteracted by a sudden change in the value of $F_t$, the force exerted by the player's hands on the bat handle. The case of a ball striking a bat is treated in Prob. 20.70.

**18.99** Figure 18.99 shows a slender rod of mass $m$ which is hinged at one end. Find the location of the center of percussion.

▮ For a slender rod, from Case 1 in Table 17.15,

$$I_0 = \tfrac{1}{12}ml^2 \quad \text{and} \quad k_0 = \sqrt{\frac{I_0}{m}} = \sqrt{\frac{\tfrac{1}{12}ml^2}{m}} = 0.289l$$

Using Eq. (5) in Prob. 18.98,

$$q = r_c + \frac{k_0^2}{r_c} = 0.5l + \frac{(0.289l)^2}{0.5l} = 0.667l$$

The distance between the CP and the hinge is two-thirds the length of the rod.

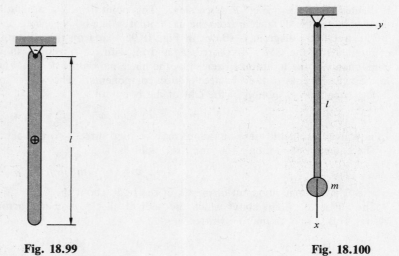

**Fig. 18.99**                    **Fig. 18.100**

**18.100** Find the $x$ coordinate of the center of percussion of the plane body shown in Fig. 18.100. The mass of the rod may be neglected, and the mass of the body at the end of the rod may be considered to be a point mass.

▮ For this limiting case,

$$k_0 = 0 \quad \text{and} \quad q = r_c = l$$

**18.101** The disk in Fig. 18.101*a* is hinged to the ground. Find the location of the CP.

▮ Figure 18.101*b* shows the CM and CP of the disk. From Case 4 in Table 17.13,

$$I_{0z} = \tfrac{1}{8}md^2$$

The radius of gyration is defined by

$$k_0^2 = \frac{I_{0z}}{m} = \tfrac{1}{8}d^2$$

The location of the CP is given by Eq. (5) in Prob. 18.98 as

$$q = r_c + \frac{k_0^2}{r_c} = \frac{d}{2} + \frac{\tfrac{1}{8}d^2}{d/2} = \tfrac{3}{4}d$$

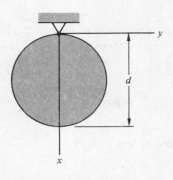

**Fig. 18.101a**

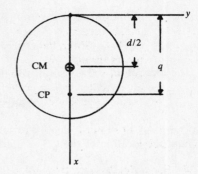

**Fig. 18.101b**

**18.102** The plane body in Fig. 18.102a is hinged to the ground. Find the location of the CP of the body.

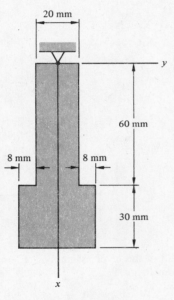

**Fig. 18.102a**

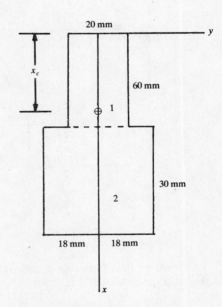

**Fig. 18.102b**

▌ The plane area of the body is divided into the two elements shown in Fig. 18.102b. The coordinate $x_c$ of the center of mass is given by

$$x_c = \frac{x_1 A_1 + x_2 A_2}{A_1 + A_2} = \frac{30(20)60 + 75(36)30}{20(60) + 36(30)} = 51.3 \text{ mm} \qquad A = 2{,}280 \text{ mm}^2$$

The area moment of inertia about the $x$ axis is

$$I_x = \frac{60(20)^3}{12} + \frac{30(36)^3}{12} = 1.57 \times 10^5 \text{ mm}^4$$

The area moment of inertia about the $y$ axis is given by

$$I_y = \frac{20(60)^3}{3} + \left[ \frac{36(30)^3}{12} + (36)30(75^2) \right] = 7.60 \times 10^6 \text{ mm}^4$$

The polar moment of inertia $I_z$ has the form

$$I_z = I_x + I_y = 7.76 \times 10^6 \text{ mm}^4$$

The centroidal polar moment of inertia $I_{0z}$ is given by

$$I_{0z} = I_z - A x_c^2 = 7.76 \times 10^6 - 2{,}280(51.3)^2 = 1.76 \times 10^6 \text{ mm}^4$$

The centroidal radius of gyration is found from

$$k_0^2 = \frac{I_{0z}}{A} = \frac{1.76 \times 10^6}{2,280} = 772 \text{ mm}^2$$

The location of the CP, using Eq. (5) in Prob. 18.98, is

$$q = r_c + \frac{k_0^2}{r_c} = 51.3 + \frac{772}{51.3} = 66.3 \text{ mm}$$

**18.103** Find the location of the CP of the latching lever assembly in Prob. 18.13.

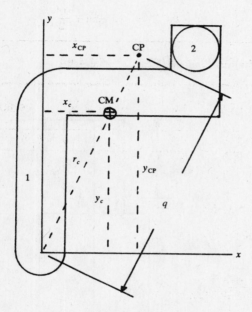

Fig. 18.103a

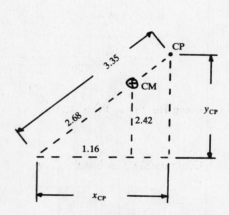

Fig. 18.103b

▌ Figure 18.103a shows the locations of the CM and the CP of the lever assembly. Element 1 is the mass of the strip, and element 2 is the mass of the two plugs. From Prob. 17.85,

$$m_1 = 2.51 \times 10^4 \text{ lb} \cdot \text{s}^2/\text{in} \qquad m_2 = 2(7.86 \times 10^{-5}) = 1.57 \times 10^{-4} \text{ lb} \cdot \text{s}^2/\text{in} \qquad I_a = I_z = 0.00367 \text{ lb} \cdot \text{s}^2 \cdot \text{in}$$

The total mass of the lever assembly is

$$m = m_1 + m_2 = 4.08 \times 10^{-4} \text{ lb} \cdot \text{s}^2/\text{in}$$

As a first step, the centroidal coordinates of the area of the lever are found. This area is divided into the five elements shown in Fig. 17.85b. The computations for the centroidal coordinates $x_c$ and $y_c$ are shown in Table 18.7. The results are

$$x_c = 0.640 \text{ in} \qquad y_c = 1.96 \text{ in}$$

The centroidal coordinates of the CM of the lever assembly are given by

$$x_c = \frac{x_1 m_1 + x_2 m_2}{m_1 + m_2} = \frac{0.640(2.52 \times 10^{-4}) + 2(1.57 \times 10^{-4})}{4.09 \times 10^{-4}} = 1.16 \text{ in}$$

$$y_c = \frac{y_1 m_1 + y_2 m_2}{m_1 + m_2} = \frac{1.96(2.52 \times 10^{-4}) + 3.15(1.57 \times 10^{-4})}{4.09 \times 10^{-4}} = 2.42 \text{ in}$$

The polar centroidal coordinate $r_c$, seen in Fig. 18.103a, is

$$r_c = \sqrt{x_c^2 + y_c^2} = \sqrt{1.16^2 + 2.42^2} = 2.68 \text{ in}$$

The term $M$ is eliminated between Eqs. (2) and (3) in Prob. 18.98, to obtain

$$I_a \alpha = F_t q$$

**TABLE 18.7**

| element | $A_i$ | $x_i$ | $x_iA_i$ | $y_i$ | $y_iA_i$ |
|---------|-------|-------|----------|-------|----------|
| 1 | $\dfrac{1}{2}\left[\dfrac{\pi(0.5)^2}{4}\right]$ | 0 | 0 | $\dfrac{-4(0.25)}{3\pi}$ | $-0.010$ |
| 2 | $0.5(2.4)$ | 0 | 0 | $\dfrac{2.4}{2}$ | $1.44$ |
| 3 | $\frac{1}{4}\pi(0.5)^2$ | $0.25-0.212$ | $0.0075$ | $2.4+0.212$ | $0.513$ |
| 4 | $1.5(0.5)$ | $0.25+\dfrac{1.5}{2}$ | $0.750$ | $2.4+\dfrac{0.5}{2}$ | $1.988$ |
| 5 | $0.5(1)$ | $0.25+1.5+\dfrac{0.5}{2}$ | $1.000$ | $2.4+\frac{1}{2}$ | $1.450$ |

|  |  |  |  |
|---|---|---|---|
| $A=2.745$ | $\sum x_iA_i=1.758$ | | $\sum y_iA_i=5.381$ |
| | $x_c=0.640\text{ in}$ | | $y_c=1.96\text{ in}$ |

Using Eq. (1) in Prob. 18.98,

$$F_t q = mr_c\alpha q = I_a\alpha \qquad q = \frac{I_a}{mr_c} \tag{1}$$

Equation (1) now has the form

$$q = \frac{I_a}{mr_c} = \frac{0.00367}{4.08\times10^{-4}(2.68)} = 3.36\text{ in}$$

From the similar triangles in Fig. 18.103b,

$$\frac{x_{\text{CP}}}{1.16} = \frac{y_{\text{CP}}}{2.42} = \frac{3.35}{2.68} \qquad x_{\text{CP}} = 1.45\text{ in} \qquad y_{\text{CP}} = 3.03\text{ in}$$

**18.104** Give a summary of the basic concepts of dynamics of rigid bodies in plane motion.

▐ When a rigid body moves in plane motion, all points in the body maintain the same respective distance, from some fixed reference plane, throughout the entire motion. Only the components of force which lie in the plane of motion contribute to the motion of the body.

A preliminary, necessary step in the solution of any problem in plane motion of a rigid body is to draw a free-body diagram. If the body rotates about a fixed point $a$, the rotational equation of motion is

$$M_a = I_a\alpha$$

where $M_a$ = resultant external couple or torque, or moment which acts on the body
    $I_a$ = mass moment of inertia about the fixed point $a$
    $\alpha$ = angular acceleration of the body
The force **F** exerted by the ground on the body at the fixed point is found from

$$\mathbf{F} = m\mathbf{a}_c$$

where $\mathbf{a}_c$ is the translational acceleration of the center of mass and $m$ is the mass of the body. The force **F** can have at most two rectangular components which lie in the plane of the motion. If the resultant external moment $M$ has a constant magnitude, the angular acceleration is constant. For this case, the angular velocity and displacement satisfy the equations

$$\omega = \omega_0 + \alpha t \qquad \theta = \theta_0 + \omega_0 t + \tfrac{1}{2}\alpha t^2 \qquad \omega^2 = \omega_0^2 + 2\alpha(\theta - \theta_0)$$

The case of general plane motion of a rigid body may be described by the sum of the translational motion of the center of mass, and the rotational motion about this point. The two governing equations of motion are

$$\mathbf{F} = m\mathbf{a}_c \qquad M_0 = I_0\alpha$$

where $I_0$ is the mass moment of inertia about the center of mass.

For pure rolling of a rigid cylindrical body on a surface,

$$v_c = r\omega \qquad a_c = r\alpha$$

where $v_c$ and $a_c$ = translational velocity and acceleration, respectively, of the center of the cylindrical body
      $\omega$ and $\alpha$ = angular velocity and acceleration, respectively, of the body
          $r$ = radius of the body

In order to have pure rolling motion, the tangential friction force $F$ exerted by the surface on the body must satisfy the equations

$$M_0 = I_0\alpha \qquad M_0 = Fr \qquad F < F_{\max} = \mu N$$

where $F$ = friction force
      $N$ = normal force on the body
      $\mu$ = coefficient of friction

The problem of plane motion of a rigid body may be solved by using inertia forces and moments. It is recommended that the beginning reader formulate the problem by direct application of Newton's laws, and then use the inertia force method to verify the solutions.

A certain point on a rigid body which experiences angular acceleration about a fixed point is called the center of percussion. The sum about this point of all of the moments which act on the body is always zero. The center of percussion is located at the distance $q$ from the fixed point, where

$$q = r_c + \frac{I_0}{mr_c}$$

The distance between the fixed point and the center of mass is $r_c$. The center of percussion, the center of mass, and the fixed point all lie on the same straight line.

# CHAPTER 19
# Work-Energy Methods for Particles and Rigid Bodies

## 19.1 WORK OF A FORCE, COUPLE OR TORQUE, OR MOMENT, ENERGY OF A MASS PARTICLE DUE TO POSITION OR MOTION, POTENTIAL AND KINETIC ENERGIES

**19.1** **(a)** What is the *work done* when a force acts on a particle which moves through an infinitesimal translation displacement?

**(b)** What are the fundamental units of work?

**(c)** Is work a scalar or a vector quantity?

▮ **(a)** Figure 19.1 shows a force $F$ which acts on a particle that is constrained to move along the $s$ axis. The infinitesimal work $dW$ done *by* the force *on* the particle, as this element translates through the distance $ds$, is defined to be

$$dW = (F \cos \beta)\, ds$$

The above equation is the *fundamental definition* of the work done when a force acts through a displacement.

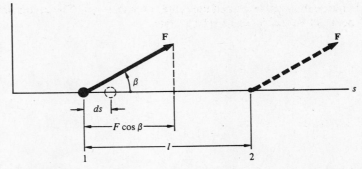

**Fig. 19.1**

**(b)** The units of work are the product of force and length. In U.S. Customary System (USCS) units, typical units for work would be the foot-pound (ft·lb) or pound-foot (lb·ft). Both ways of writing these work units are correct, but the foot-pound, or inch-pound (in·lb), is preferred in this book. In SI units, the basic unit of work is the product of newtons and meters, abbreviated N·m. It may be observed that the units of work are the same as those of couple, or torque, or moment. In the case of SI Units, the distinction between work units and moment units is clearly made. A magnitude of work of one newton-meter is defined to be one *joule*, with the symbol J. Thus, *all work and energy quantities in SI units are in joules*. No such distinction between work and moment units exists in USCS units. The conversion factor between work, or energy, in USCS units and SI units is 1 ft·lb = 1.356 J.

**(c)** Since a description of work done requires a statement of *magnitude* only, work is a scalar quantity.

**19.2** **(a)** Find the work done if a force of constant magnitude, direction, and sense acts on a particle which moves through a finite translational displacement.

**(b)** What is the difference between positive and negative work?

**(c)** What is the work done if the force is normal to the displacement of the particle?

▮ **(a)** The force in Fig. 19.1 is now assumed to be of *constant* magnitude, direction, and sense. As the point of application of this force on the particle moves along the $s$ axis in the positive coordinate sense, through distance $l$, the total work $W_{12}$ done by the force on the particle is

$$W_{12} = (F \cos \beta)l$$

where 1 and 2 represent the endpoints of the interval $l$.

(*b*)   The equation in part (*a*) contains an implicit sign convention for work. If $-90° < \beta < 90°$, the component of force is in the direction of the motion of the particle, and the work term is positive. If $90° < \beta < 270°$, then the senses of the force component and of the displacement are opposite, and the work term is negative. It follows that the work done by a force on a particle is *positive* if the sense of the component of force in the direction of motion is the same as the sense of motion of the particle.

(*c*)   It follows from the equation in part (*a*) that the work done by the force is zero if $\beta = 90°$. Thus, *if the force is normal to the path of motion of the particle, this force does no work on the particle.* A common example of this statement occurs in the case of a particle which slides along a surface. In this case, the normal contact forces *do no work* since these forces, by definition, are normal to the direction of the sliding motion. Thus, only the friction forces may do work on the particle.

**19.3**   How is work defined if the magnitude or direction of the force which acts on the particle is not constant?

▮   If either the force acting on the particle shown in Fig. 19.1, or the direction between this force and the displacement of the particle, is not constant, then the expression for the work done on the particle has the general form

$$W_{12} = \int_1^2 F \cos \beta \, ds$$

where 1 and 2 are the endpoints of the interval of motion. This equation may be solved if the force $F$ is a known function of the displacement $s$ of the particle. An example of this is seen in Prob. 19.6.

**19.4**   A particle is located at a distance $h$ above the ground, as shown in Fig. 19.4. All air resistance effects may be neglected.

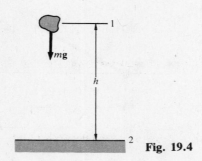

**Fig. 19.4**

(*a*)   Find the work done if the particle is allowed to fall to the ground.

(*b*)   Find the magnitude of the work done if the weight of the particle is 2.8 lb and $h = 100$ ft.

▮   (*a*)   The force acting on the particle is the static weight force $mg$, which has constant magnitude, direction, and sense. This weight force acts along the vertical direction, and there is no component of force normal to this direction. The work done by the force is

$$W_{12} = (F \cos \beta)l = mg(\cos 0°)h = mgh$$

(*b*)   For the numerical values of this problem,

$$W_{12} = 2.8(100) = 280 \text{ ft} \cdot \text{lb}$$

**19.5**   The particle in Prob. 19.4 is now allowed to slide down the frictionless inclined plane shown in Fig. 19.5*a*. Find the work done on the particle as this element travels through the distance *l*.

▮   The components of the weight force parallel and normal to the inclined plane are shown in Fig. 19.5*b*. The normal component of the weight force does no work, since this force is normal to the displacement of the particle. The work done as the particle slides down the plane is

$$W_{12} = (F \cos \beta)l = (mg \sin \xi)l$$

From Fig. 19.5*a*,

$$l \sin \xi = h$$

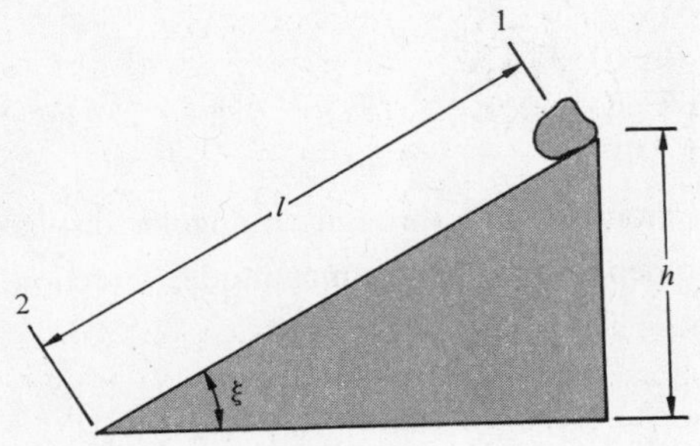

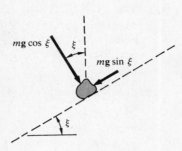

Fig. 19.5a

Fig. 19.5b

The final expression for the work done on the particle is

$$W_{12} = mg(l \sin \xi) = mgh$$

It may be seen that this result is exactly the same as that for the case, in Prob. 19.4, of free fall of the particle. These two cases are examples of a *conservative force field*, and this concept will be considered further in Prob. 19.23. It should also be noted that Fig. 19.5b is *not* a free-body diagram of the particle, but rather depicts the components of the applied force acting on the particle.

**19.6**  Figure 19.6a shows a helical spring which is acted on by a force $F$. The spring is assumed to be linear. This means that the spring constant $k$, which is a measure of the stiffness of the spring, is a constant with units of force divided by length. When $F = 0$, the end of the spring is at position 1. The spring is assumed to be massless, so that all acceleration effects may be neglected.

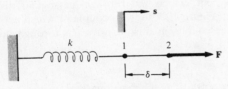

Fig. 19.6a

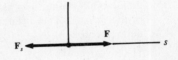

Fig. 19.6b

(*a*)  Find the work done by the force $F$ on the spring as the end of the spring moves through the distance $\delta$ from 1 to 2.

(*b*)  Find the work done by the internal force in the spring for the motion of part (*a*).

(*c*)  Find the magnitude of the work term in part (*a*) if $k = 5,250$ N/m and $\delta = 85$ mm.

▌ (*a*)  The general force relationship on the tip particle of the spring is shown in Fig. 19.6b. The applied force $F$ is opposed by the internal spring force $F_s$. The equilibrium requirement of this particle is

$$\sum F = 0 \qquad F - F_s = 0 \qquad F = F_s$$

Since the spring is linear,

$$F_s = ks$$

and therefore

$$F = ks$$

The work $W_{12}$ done by the force $F$ as the tip of the spring is stretched from position 1 to position 2 is

$$W_{12} = \int_1^2 (F \cos \beta)\, ds = \int_0^\delta (F \cos 0°)\, ds = \int_0^\delta F\, ds = \int_0^\delta ks\, ds = \frac{ks^2}{2}\Big|_0^\delta = \frac{1}{2}k\delta^2$$

(*b*)  The spring force $F_s$ acts in a sense which is opposite to the sense of motion of the tip of the spring. The work done by this force is, then,

$$W_s = \int_1^2 (F \cos \beta)\, ds = \int_0^\delta (F_s \cos 180°)\, ds = -\int_0^\delta F_s\, ds = -\int_0^\delta ks\, ds = -\frac{ks^2}{2}\Big|_0^\delta = -\frac{1}{2}k\delta^2 = -W_{12}$$

It may be seen that the work done by the spring force is the *negative* of the work done by the applied force.

(c)  Using  $k = 5,250 \, \text{N/m}$  and  $\delta = 85 \, \text{mm}$,

$$W_{12} = \tfrac{1}{2} k \delta^2 = \tfrac{1}{2}(5,250)\left(\frac{85}{1,000}\right)^2 = 19.0 \, \text{N} \cdot \text{m} = 19.0 \, \text{J}$$

**19.7**  (a)  Find the work done when a couple or torque, or a moment, acts through an angular displacement.

(b)  Find the work done if the couple or torque, or moment, has constant magnitude, direction, and sense.

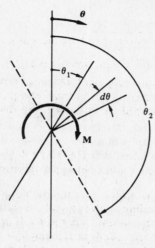

**Fig. 19.7**

❚ (a)  The work done by a couple or torque, or moment, which acts through an angular displacement is defined in a similar way as the work of a force.  Figure 19.7 shows a couple $M$ which acts in the plane of the figure.  The differential work done by the couple, as this quantity acts through the angle $d\theta$, is defined to be

$$dW = M \, d\theta$$

The equation is the *fundamental definition* of the work done when a couple or torque, or moment, acts through an angular displacement.

The work done as the couple acts through the angle between $\theta_1$ and $\theta_2$ is

$$W_{12} = \int_{\theta_1}^{\theta_2} M \, d\theta \tag{1}$$

The work quantity will be *positive* if the senses of the couple and the angular displacement are the *same*.  It may be seen that the units of the work done by a couple or torque, or moment, are again the product of force and length.

(b)  If the magnitude of the couple is *constant*, then Eq. (1) may be written as

$$W_{12} = \int_{\theta_1}^{\theta_2} M \, d\theta = M \int_{\theta_1}^{\theta_2} d\theta = M\theta \Big|_{\theta_1}^{\theta_2} = M(\theta_2 - \theta_1)$$

**19.8**  A machinist uses a T-handle wrench to finish-ream a large drilled hole.  The machinist applies equal and opposite forces of 16 N, at 150 mm from the center, as shown in Fig. 19.8.  Find the work done when the reamer has rotated through 375°.

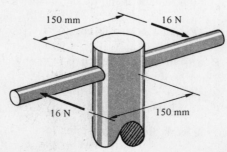

150 mm       16 N

16 N       150 mm

**Fig. 19.8**

▌ The forces applied to the handle are assumed to be constant, and the couple applied to the reamer is

$$M = 16\left(\frac{300}{1{,}000}\right) = 4.8\,\text{N} \cdot \text{m}$$

The work done during the rotation of 375° is

$$W = M(\theta_2 - \theta_1) = (4.8\,\text{N} \cdot \text{m})(375°)\,\frac{2\pi\,\text{rad}}{360°} = 31.4\,\text{J}$$

**19.9**    (*a*)   What is the definition of *energy*?

         (*b*)   In what two common forms may energy be stored in a mass particle?

▌ (*a*)   In its most fundamental conception, energy is understood to mean *the capacity for doing work*.

    (*b*)   Energy may be stored in two common forms in a mass particle. The first form is due to the *position* of the mass element, and this energy is *potential energy*. The second way that energy may be stored in a mass particle is when the body is in motion and thus has a nonzero velocity. This energy is called *kinetic energy*.

**19.10**    (*a*)   Find the potential energy of a mass particle which is raised through a vertical displacement.

         (*b*)   What is the fundamental interpretation of potential energy?

▌ (*a*)   An energy term, which is a function only of the position of a particle, is now considered. Figure 19.10*a* shows a particle resting on the ground. It is desired to raise the particle from the ground to the ledge at height *h* above the ground. A force *F* is applied to the particle, as shown in Fig. 19.10*b*, to elevate this mass element in a quasi-static manner. The term *quasi-static* means that the motion of the particle, as evidenced by the magnitude of the velocity, is so gradual that this motion may be thought of as a succession of states, in each of which the particle is in static equilibrium. Thus, all acceleration effects and the associated forces may be neglected. It follows that the force *F* must be only infinitesimally greater than the static weight *mg*, or

$$F \approx mg$$

The work done by the force *F* in raising the particle is thus

$$W_{12} = (F \cos \beta)l = (mg \cos 0°)h = mgh$$

At the conclusion of the process of raising the weight from the ground to the ledge, Fig. 19.10*c*, the work done on the weight in raising it is now "stored" in the weight, *by virtue of its position with respect to the ground from which it was raised*. This energy content is defined to be the *potential energy* of the weight.

    (*b*)   In its most basic conception, the potential energy of a body may be thought of as the ability of a body to do work when the position of the body changes. Since work is a scalar quantity, it follows that the potential energy is also a scalar quantity. In the problem in part (*a*), work was done on the weight as this element was moved from the ground to the ledge. When the weight rests on the ledge, the work done to place it in this position has been transformed to a *stored* energy in the body. If the weight is imagined to return to the ground, then the stored potential energy could be used to do work as the weight force moved downward through the distance *h*.

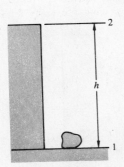

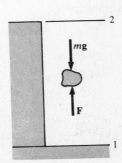

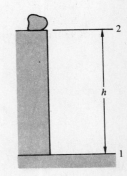

**Fig. 19.10*a***             **Fig. 19.10*b***             **Fig. 19.10*c***

**19.11**   (*a*)   Give the general form for the change in potential energy of a mass particle.

      (*b*)   What is meant by a *datum* for potential energy?

      (*c*)   Show examples of alternative choices for a datum for the potential energy of a mass particle.

      (*d*)   Find the change in potential energy when a 25-kg mass is lowered through 4.38 m.

❚   (*a*)   The symbol $V$ will be used to represent the potential energy of a mass element. In using energy methods to solve problems in dynamics, only the *change* in the potential energy, designated by $\Delta V$, has any significance. This change is defined to be

$$\Delta V = V_2 - V_1$$

where the subscripts 1 and 2 represent the initial and final endpoints, respectively, of the interval of motion of interest.

  (*b*)   A concept of fundamental importance in the definition of potential energy is that of the datum. A datum is a position of the body at which the potential energy is arbitrarily defined to be *zero*.

  (*c*)   Three possible choices of a datum for the system of Fig. 19.10*c* are shown in Figs. 19.11*a*, *b*, and *c*. In each case, the weight is raised through the distance $h$ from the ground to the ledge. Numbers 1 and 2 designate the endpoints of the interval of interest in all three cases. Three different, arbitrary datums are chosen, as shown in Figs. 19.11*a*, *b*, and *c*. The descriptions of the potential energy, and the changes in this quantity, are then as follows:

Datum at ground level (Fig. 19.11*a*):

$$V_1 = 0 \qquad V_2 = mgh \qquad \Delta V = V_2 - V_1 = mgh - 0 = mgh$$

Datum at ledge (Fig. 19.11*b*):

$$V_1 = -mgh \qquad V_2 = 0 \qquad \Delta V = V_2 - V_1 = 0 - (-mgh) = mgh$$

Datum at height $h$ above ledge (Fig. 19.11*c*):

$$V_1 = -2mgh \qquad V_2 = -mgh \qquad \Delta V = V_2 - V_1 = -mgh - (-2mgh) = mgh$$

Terms such as $V_1$ and $V_2$ are defined in terms of the arbitrarily chosen datum of the problem. The *change* in the potential energy, by comparison, is completely independent of the choice of the datum. It may be seen in the above problem that the *change* in potential energy is the same for all three cases. This is an expected result, since a single physical problem is under consideration.

  (*d*)   Using Fig. 19.11*a*, with the numbers 1 and 2 interchanged,

$$V_1 = mgh = 25(9.81)4.38 = 1{,}070 \text{ N} \cdot \text{m} = 1{,}070 \text{ J} \qquad V_2 = 0$$

The change in potential energy is

$$\Delta V = V_2 - V_1 = 0 - 1{,}070 = -1{,}070 \text{ J}$$

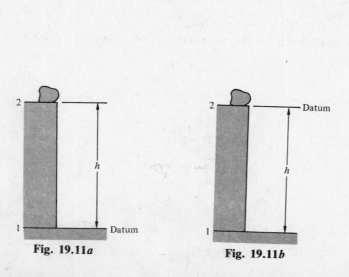

Fig. 19.11*a*                            Fig. 19.11*b*

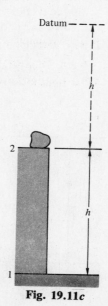

Fig. 19.11*c*

**19.12**   What are two common types of potential energy in a mechanical system?

▌ In this chapter two types of potential energy in mechanical systems are considered. The first type is referred to as *raising a weight*. If a weight is raised, work is done *on* the weight, and potential energy is stored *in* the weight. The magnitude of the increase in the potential energy is simply the product of the magnitudes of the weight and the vertical distance through which the weight is raised. If the weight is lowered, work is done *by* the weight as the potential energy decreases. The second type of potential energy is the energy stored in a spring because of the deflection of this element. The following problem illustrates this effect.

**19.13**   The linear, helical spring shown in Fig. 19.13 is initially in static equilibrium when acted on by a force $F_1 = 40$ N. The magnitude of this force is now increased until the end of the spring moves to point 2.

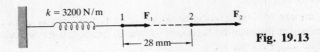

**Fig. 19.13**

(*a*)   Find the change in the potential energy stored in the spring as the end is stretched from 1 to 2.

(*b*)   Find the total potential energy stored in the spring when the end of the spring is at position 2.

▌ (*a*)   The work required to stretch a linear, helical spring through a distance $s$ was found in Prob. 19.6 to be

$$W = \tfrac{1}{2}ks^2$$

where $k$ is the spring constant, with the units of force per unit length. This work done on the spring is stored as potential energy in the spring. It may be seen that the datum for this energy quantity is implicitly given by the above equation as $s = 0$.

At position 1,

$$F_1 = ks_1 \qquad s_1 = \frac{F_1}{k} = \frac{40\ \text{N}}{3{,}200\ \text{N/m}} = 0.0125\ \text{m}$$

The displacement at point 2 is

$$s_2 = s_1 + 28\ \text{mm} = 0.0125 + \frac{28}{1{,}000} = 0.0405\ \text{m}$$

The *change* in the potential energy as the spring is stretched from 1 to 2 is

$$\Delta V = V_2 - V_1 = \tfrac{1}{2}ks_2^2 - \tfrac{1}{2}ks_1^2 = \tfrac{1}{2}k(s_2^2 - s_1^2)$$
$$= \tfrac{1}{2}(3{,}200)(0.0405^2 - 0.0125^2) = 2.37\ \text{N} \cdot \text{m} = 2.37\ \text{J}$$

(*b*)   The *total* potential energy when the spring is stretched to position 2 is

$$V_2 = \tfrac{1}{2}ks_2^2 = \tfrac{1}{2}(3{,}200)(0.0405^2) = 2.62\ \text{N} \cdot \text{m} = 2.62\ \text{J}$$

**19.14**   (*a*)   Define the *kinetic energy* of a mass particle.

(*b*)   What is the fundamental interpretation of kinetic energy?

(*c*)   What is a significant difference between the datum for potential energy and the datum for kinetic energy?

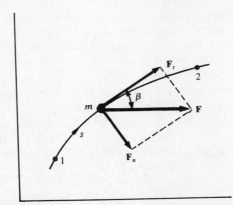

**Fig. 19.14**

**▮** (*a*)  Problem 19.10 considered a form of energy, referred to as potential energy, which is stored by virtue of the position of a particle. In this problem a second, distinct form of energy of a particle is identified. This energy quantity is kinetic energy, and it is associated with the *motion* of a particle.

Figure 19.14 shows a particle which moves along a curved path in a horizontal plane. The coordinate of length along the curved path is *s*. Since the path of the particle is horizontal, there is no change in the elevation of the particle with respect to a datum plane. Thus, there is no change in the potential energy of the particle. The particle is acted on by a resultant force $F$, with the normal and tangential components $F_n$ and $F_t$. Newton's second law in the tangential direction is written as

$$F_t = ma_t \tag{1}$$

The tangential acceleration $a_t$ may be written as

$$a_t = \frac{dv}{dt} \tag{2}$$

where $v$ is the scalar magnitude of the velocity, or speed, of the particle.

The force $F_t$ may be written as

$$F_t = F \cos \beta \tag{3}$$

Using Eqs. (2) and (3) in Eq. (1) results in

$$F \cos \beta = m \frac{dv}{dt} \tag{4}$$

The term $v$ is defined in terms of the length coordinate $s$ along the curve by

$$v = \frac{ds}{dt}$$

from which

$$dt = \frac{ds}{v} \tag{5}$$

The term $dt$ is eliminated between Eqs. (4) and (5), with the result

$$F \cos \beta = m \frac{dv}{(ds/v)} = mv \frac{dv}{ds} \qquad (F \cos \beta)\, ds = mv\, dv \tag{6}$$

Equation (6) is now integrated between the two positions 1 and 2 shown in Fig. 19.14, and the result is

$$\int_1^2 (F \cos \beta)\, ds = \int_1^2 mv\, dv = m \int_1^2 v\, dv = m \left. \frac{v^2}{2} \right|_{v_1}^{v_2} = \tfrac{1}{2}mv_2^2 - \tfrac{1}{2}mv_1^2 \tag{7}$$

The term on the left side of the above equation, by comparison with the result in Prob. 19.3, is seen to be the work done on the particle between positions 1 and 2. The terms on the right side have the form $\tfrac{1}{2}mv^2$. The latter quantity is referred to as the *kinetic energy of the particle*. The symbol $T$ is used to designate this term, and

$$T = \tfrac{1}{2}mv^2 \tag{8}$$

Since work is a scalar quantity, it follows from Eq. (7) that kinetic energy is also a scalar quantity.

The kinetic energy of a particle is an energy which the particle possesses because of its motion. Since Newton's second law was used to derive the above result, the term $v$ is an *absolute velocity* which must be defined, or measured, in an inertial coordinate system.

The path of the particle shown in Fig. 19.14 was assumed to lie in a horizontal plane. Thus, there are no changes in the potential energy of the particle. This fact does not limit the generality of the final result obtained in Eq. (7). If the path of the particle did not lie in a horizontal plane, part of the applied force would be used to change the potential energy, while the remainder would be used to change the kinetic energy of the particle.

(*b*)  In its most basic conception, the kinetic energy of a body may be thought of as the ability of a body to do work when the velocity of the body changes.

(*c*)  In the case of potential energy, a datum is chosen and the potential energy is arbitrarily defined to be zero at this point. The criterion for the selection of this datum is solely convenience. It may be seen that the kinetic energy of a particle is zero *only* when the velocity of the particle is zero, and at no other time. Thus, the datum for all kinetic energy terms is the condition of zero velocity.

**19.15**   A 3600-lb automobile moves along a straight horizontal roadway.

(a) Find the change in the kinetic energy of the automobile when its speed changes from 30 mi/h to 40 mi/h.

(b) Do the same as in part (a), if the speed changes from 40 mi/h to 50 mi/h.

▌ (a) The automobile moves as a mass particle in rectilinear translation. The speeds of the automobile are expressed in the units of feet per second as

$$30 \text{ mi/h} = 44.0 \text{ ft/s} \qquad 40 \text{ mi/h} = 58.7 \text{ ft/s} \qquad 50 \text{ mi/h} = 73.3 \text{ ft/s}$$

The change in kinetic energy has the general form

$$\Delta T = \tfrac{1}{2} m (v_2^2 - v_1^2)$$

The change in kinetic energy as the speed changes from 30 to 40 mi/h is

$$\Delta T = \frac{1}{2} \left( \frac{3{,}600}{32.2} \right) (58.7^2 - 44.0^2) = 84{,}400 \text{ ft} \cdot \text{lb}$$

(b) For a change in speed of 40 to 50 mi/h

$$\Delta T = \frac{1}{2} \left( \frac{3{,}600}{32.2} \right) (73.3^2 - 58.7^2) = 108{,}000 \text{ ft} \cdot \text{lb}$$

**19.16** The plane sliding mechanism of Prob. 16.42 is shown in Fig. 19.16. The velocity of block $B$ was found in that problem to be $v_B = 7.48 \text{ m/s}$. $m_A = 2.4 \text{ kg}$, and $m_B = 3.2 \text{ kg}$, and the mass of the connecting link may be neglected. Find the kinetic energy of the system at the instant shown in the figure.

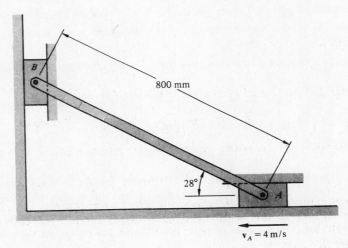

800 mm

28°

$v_A = 4 \text{ m/s}$

**Fig. 19.16**

▌ The kinetic energy of the two moving blocks is

$$T = \tfrac{1}{2} m_A v_A^2 + \tfrac{1}{2} m_B v_B^2 = \tfrac{1}{2}(2.4)4^2 + \tfrac{1}{2}(3.2)7.48^2 = 109 \text{ J}$$

**19.17** What is the relationship between the work done on a mass particle and the change in kinetic energy of this particle?

▌ When a force acts on a particle, from Eq. (7) in Prob. 19.14, the work done on the particle produces a change in kinetic energy. If the component of the force acts in the sense of the motion, the particle will experience an increase in velocity. If the particle is imagined to slow down to a lower velocity, then the stored kinetic energy of the particle could be used to do work.

If the force acting on the particle has *constant* magnitude, direction, and sense, then the work done is the product of the component of this force in the direction of motion and the displacement of the particle. If the force is variable, then the integral form on the left side of Eq. (7) in Prob. 19.14 must be used.

The relationship between the work done on the particle and the change in kinetic energy may be written as

$$W_{12} = \Delta T = T_2 - T_1 = \tfrac{1}{2} m v_2^2 - \tfrac{1}{2} m v_1^2$$

In the above equation, $W_{12}$ is the work done on the particle and $\Delta T$ is the change in the kinetic energy of this element.

**19.18**  A particle of mass 2.4 kg is at rest in position 1 on the frictionless, horizontal track shown in Fig. 19.18.  If a 50-N force at a constant inclination of 40° to the track is applied to the particle, find the speed of the particle as it passes position 2.

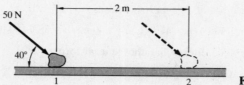

Fig. 19.18

▮  The work done on the particle as it moves from 1 to 2 is

$$W_{12} = (F \cos \beta)l = (50 \cos 40°)2 = 76.6 \text{ N} \cdot \text{m} = 76.6 \text{ J}$$

The speed $v_2$ at position 2 is found from

$$W_{12} = \tfrac{1}{2}mv_2^2 - \tfrac{1}{2}mv_1^2 \qquad 76.6 = \tfrac{1}{2}(2.4)v_2^2 \qquad v_2 = 7.99 \text{ m/s}$$

**19.19**  The particle shown in Fig. 19.19a is acted on by the system of forces shown.  It moves with rectilinear translation in a frictionless guide.

(a)  Find the work done on the particle when this element moves through a distance of 2 ft.

(b)  Find the velocity of the particle when this element has moved through a distance of 2 ft, starting from rest.

(c)  Find the change in potential energy of the particle after this element has moved through a distance of 2 ft.

(d)  Do the same as in parts (a) through (c), if the particle is acted on by the system of forces shown in Fig. 19.19b.

▮  (a)  The forces which act on the particle are shown in Fig. 19.19c.  The work done on the particle is given by

$$W_{12} = \sum (F \cos \beta)l = (2 \cos 15° + 4 \cos 60°)2 = 7.86 \text{ ft} \cdot \text{lb}$$

(b)  The velocity of the particle is found from

$$W_{12} = \tfrac{1}{2}mv_2^2 \qquad 7.86 = \frac{1}{2}\left(\frac{8}{32.2}\right)v_2^2 \qquad v_2 = 7.96 \text{ ft/s}$$

It is left as an exercise for the reader to verify the above result by using the value $a = 15.8 \text{ ft/s}^2$, found in Prob. 15.17, together with the equations for motion with constant acceleration.

(c)  The change in potential energy is zero, since the particle moves along a horizontal path.

(d)  The results are the same as in parts (a) through (c), since the 3-lb force has no component in the direction of motion.

Fig. 19.19a

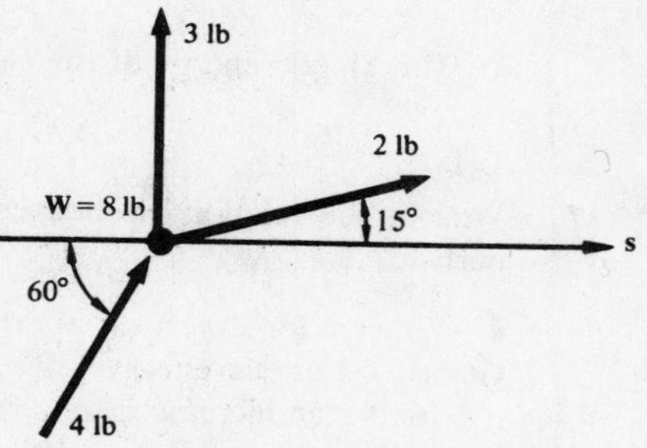

Fig. 19.19b

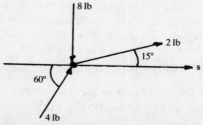

Fig. 19.19c

**19.20** Do the same as in Prob. 19.19, parts (a) through (c), for the particle in Fig. 19.20a.

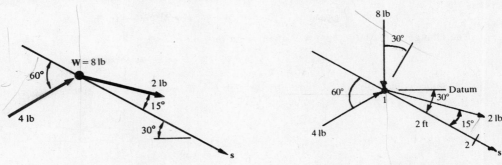

**Fig. 19.20a**                          **Fig. 19.20b**

▌ **(a)** Figure 19.20b shows the forces which act on the particle. The work done on the particle is given by

$$W_{12} = \sum (F \cos \beta)l = (2 \cos 15° + 8 \sin 30° + 4 \cos 60°)2 = 15.9 \text{ ft} \cdot \text{lb}$$

**(b)** The velocity of the particle is found from

$$W_{12} = \tfrac{1}{2}mv_2^2 \qquad 15.9 = \frac{1}{2}\left(\frac{8}{32.2}\right)v_2^2 \qquad v_2 = 11.3 \text{ ft/s}$$

The result $a = 31.9 \text{ ft/s}^2$, found in Prob. 15.18, may be used to verify the above result.

**(c)** The endpoints 1 and 2 are shown in Fig. 19.20b. The datum for potential energy is chosen to be point 1. $V_1$ and $V_2$ have the forms

$$V_1 = 0 \qquad V_2 = -8(2 \sin 30°) = -8 \text{ ft} \cdot \text{lb}$$

The change in potential energy is given by

$$\Delta V = V_2 - V_1 = -8 \text{ ft} \cdot \text{lb}$$

**19.21** Do the same as in Prob. 19.19, parts (a) through (c), for the particle in Fig. 19.21a.

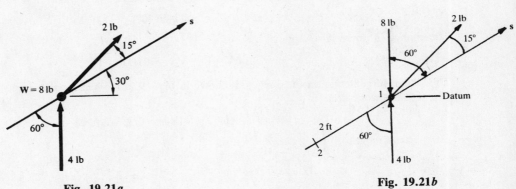

**Fig. 19.21a**                          **Fig. 19.21b**

▌ **(a)** The forces which act on the particle are shown in Fig. 19.21b. The work done on the particle has the form

$$W_{12} = \sum (F \cos \beta)l = (2 \cos 15° - 8 \cos 60° + 4 \cos 60°)2 = -0.136 \text{ ft} \cdot \text{lb}$$

It may be concluded from the above result that the particle moves in the *negative s* sense.

**(b)** The minus sign on $W_{12}$ is dropped, and the velocity is found from

$$W_{12} = \tfrac{1}{2}mv_2^2 \qquad 0.136 = \frac{1}{2}\left(\frac{8}{32.2}\right)v_2^2 \qquad v_2 = 1.05 \text{ ft/s}$$

The particle moves downward and to the left.

The solution above may be verified by using the value of $a$ found in Prob. 15.19.

(c) Figure 19.21b shows the endpoints 1 and 2 of the interval of motion. The datum for potential energy is chosen to be point 1. $V_1$ and $V_2$ have the forms

$$V_1 = 0 \qquad V_2 = -8(2\cos 60°) = -8 \text{ ft} \cdot \text{lb}$$

The change in potential energy is then given by

$$\Delta V = V_2 - V_1 = -8 \text{ ft} \cdot \text{lb}$$

**19.22** (a) Do the same as in Prob. 19.20, parts (a) through (c), if both forces are removed from the particle.

(b) Do the same as in part (a), for Prob. 19.21.

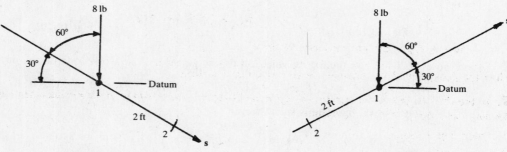

**Fig. 19.22a**                                          **Fig. 19.22b**

▌ (a) Figure 19.22 shows the weight force which acts on the particle. The work done is

$$W_{12} = (F\cos\beta)l = (8\cos 60°)2 = 8 \text{ ft} \cdot \text{lb}$$

The particle moves in the positive s sense, and

$$W_{12} = \tfrac{1}{2}mv_2^2 \qquad 8 = \frac{1}{2}\left(\frac{8}{32.2}\right)v_2^2 \qquad v_2 = 8.02 \text{ ft/s}$$

The endpoints 1 and 2 of the interval of motion are shown in Fig. 19.22a, and

$$V_1 = 0 \qquad V_2 = -8(2\sin 30°) = -8 \text{ ft} \cdot \text{lb}$$

The change in potential energy is

$$\Delta V = V_2 - V_1 = -8 \text{ ft} \cdot \text{lb}$$

(b) The weight force acting on the particle is shown in Fig. 19.22b. The particle moves down and to the left, and the work done is

$$W_{12} = (F\cos\beta)l = (8\cos 60°)2 = 8 \text{ ft} \cdot \text{lb}$$

The particle moves in the negative s sense, and

$$W_{12} = \tfrac{1}{2}mv_2^2 \qquad 8 = \frac{1}{2}\left(\frac{8}{32.2}\right)v_2^2$$
$$v_2 = 8.02 \text{ ft/s}$$

The solution above may be verified by using the value of the acceleration of the particle which was found in Prob. 15.19. The endpoints 1 and 2 of the interval of motion are shown in Fig. 19.22b, and

$$V_1 = 0 \qquad V_2 = -8(2\sin 30°) = -8 \text{ ft} \cdot \text{lb} \qquad \Delta V = V_2 - V_1 = -8 \text{ ft} \cdot \text{lb}$$

It may be observed that $v_2$ and $\Delta V$ have the same values as those in part (a).

## 19.2 CONSERVATION OF ENERGY, THE WORK-ENERGY METHOD FOR A PARTICLE

**19.23** (a) What two general types of forces are considered in problems in dynamics?

(b) Give examples of *conservative* and *nonconservative* forces.

(c) What type of force may be related to the change in potential energy of a particle?

▮ (a) In problems in dynamics it is convenient to identify two types of forces, referred to as conservative and nonconservative forces. *Conservative* forces produce changes in potential and kinetic energy which are fully recovered if the system is imagined to operate in reverse until the original condition is reached. An alternative statement of this effect is that the *conservative forces produce no permanent energy losses*. The energy changes caused by the application of *nonconservative* forces, by comparison, are not fully recovered if the system is imagined to operate in reverse.

(b) Examples of conservative forces are weight forces and spring forces. Applied forces and friction forces are examples of nonconservative forces.

(c) The work done by the conservative forces may be expressed in terms of changes in the potential energy of a particle. The work done by the nonconservative forces may *not* be expressed in this fashion. The nonconservative work, designated by the symbol $W_{NC}$, *must* be found from an equation of the form

$$W_{NC} = \int (F_{NC} \cos \beta) \, ds$$

where $F_{NC}$ is the nonconservative force.

**19.24** Give the definition of the principle of conservation of energy.

▮ In the absence of nuclear effects, *energy in the physical universe can be neither created nor destroyed.* Energy may exist in one of four basic forms:

1. Mechanical energy
2. Electrical energy
3. Chemical energy
4. Heat energy

Potential and kinetic energies of a mass particle are forms of mechanical energy. An example of electrical energy is a charged capacitor. Chemical energy may be released during a chemical reaction, such as an explosion or a combustion process. Finally, the effect of heat energy is to change the temperature of a substance.

The relationships among potential and kinetic energies of mass particles, and work, heat, and chemical energy, are studied in detail in thermodynamics. The central objective in thermodynamics is the efficient transformation of heat energy into mechanical work. In this text, only the mechanical potential and kinetic energies of mass elements will be considered.

**19.25** (a) Derive the work-energy method for the motion of a particle.

(b) Give examples of the exchange of potential and kinetic energies in a physical system.

(c) What is an important limiting characteristic of the work-energy method of problem solution?

(d) What are two useful applications of the work-energy method?

▮ (a) When a particle is acted on by a force, the particle's potential energy, kinetic energy, or both may be changed. A general formulation of this problem will be developed now. This technique rests on the basic premise that matter can be neither created nor destroyed, and thus all the energy in the system must be accounted for.

The most general statement of the conservation of energy in a mechanical system may now be written as

$$\begin{array}{c} \text{Mechanical} \\ \text{energy in} \\ \text{system in} \\ \text{position 1} \end{array} + \begin{array}{c} \text{nonconservative} \\ \text{work done on} \\ \text{the particle} \end{array} = \begin{array}{c} \text{mechanical} \\ \text{energy in} \\ \text{system in} \\ \text{position 2} \end{array} \qquad (1)$$

Figure 19.25 shows a general case of a particle which moves along a path from 1 to 2 and is acted on by a force F. Locations 1 and 2 are the endpoints of the interval of interest. At the initial position 1, the total mechanical energy content of the particle is described by the sum of the potential energy $V_1$ and the kinetic energy $T_1$. The nonconservative work, *if any*, done on the particle between positions 1 and 2 is designated $W_{NC}$. When the particle is at position 2, the total mechanical energy content is $V_2$ plus $T_2$.

By using the above notations, Eq. (1) may be written as

$$(T_1 + V_1) + W_{NC} = T_2 + V_2$$

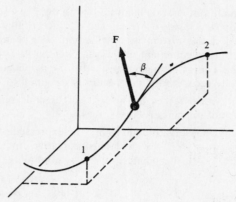

**Fig. 19.25**

This equation may be rearranged to the form

$$W_{NC} = (T_2 - T_1) + (V_2 - V_1) \qquad (2)$$

Using the notations

$$\Delta T = T_2 - T_1 \qquad \Delta V = V_2 - V_1$$

the final form of Eq. (2) is

$$W_{NC} = \Delta T + \Delta V \qquad (3)$$

This equation is a concise, formal statement of the *conservation of energy for a particle in motion.* This equation states that the nonconservative work done on a particle results in a change in kinetic energy, a change in potential energy, or both.

In many problems in dynamics no nonconservative forces act on the particle. For this case, the statement of conservation of energy reduces to

$$
\begin{aligned}
0 &= \Delta T + \Delta V \qquad && \text{conservative forces only} \\
\Delta T &= -\Delta V \qquad && \text{conservative forces only}
\end{aligned}
\qquad (4)
$$

Equation (4) has a very simple physical interpretation. A positive change, or increase, in the kinetic energy of the particle corresponds to a negative change, or decrease, of the potential energy. Thus, although the *total* energy of the particle must remain constant, potential energy may be converted to kinetic energy, and vice versa.

In Prob. 19.17 the relationship

$$W_{12} = \Delta T = T_2 - T_1$$

was introduced. $W_{12}$ is the work done on the particle, and $\Delta T$ is the change in kinetic energy. The above equation is equivalent to Eq. (3), if the conservative work is expressed in terms of the change in potential energy. It is left as an exercise for the reader to verify the results in Probs. 19.19 through 19.22 by direct use of the work-energy method.

(*b*)  Familiar examples of energy exchanges are a falling object, which exchanges potential energy for kinetic energy; a ball thrown up in the air, which exchanges kinetic energy for potential energy; and an automobile suspension spring, which stores the kinetic energy of the wheel hitting a hole in the road as potential energy in the spring.

(*c*)  The work-energy method considers states, or conditions, of motion *only at the beginning and end of the time interval* of interest in the problem. The solution yields no information whatsoever about the conditions *during* this interval. Also, when the work-energy method is used, a free-body diagram of the mass element is not necessarily required, and *the acceleration of the mass element never enters directly into the analysis. The work-energy problem is formulated solely in terms of position coordinates and velocities.*

(*d*)  The work-energy method has two major useful applications. It may be used either to obtain results which are directly useful, or to check the correctness of results obtained by the use of Newton's second law. Both of these outcomes will be illustrated in the problems in this chapter.

**19.26**  The system of Prob. 14.44 is shown in Fig. 19.26*a*. The particle is initially at rest and is dropped from a height *h* to the ground.

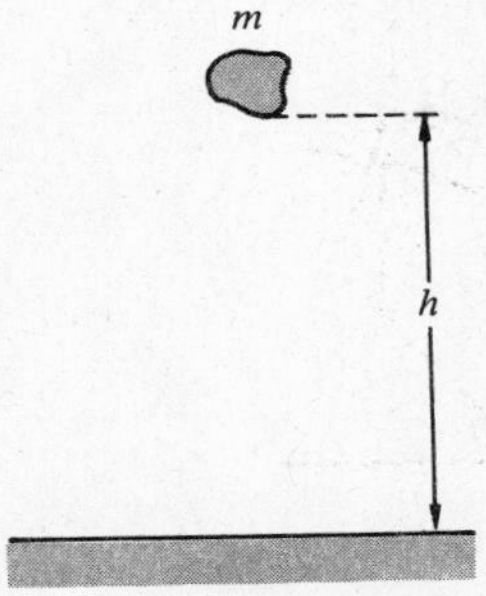

Fig. 19.26a

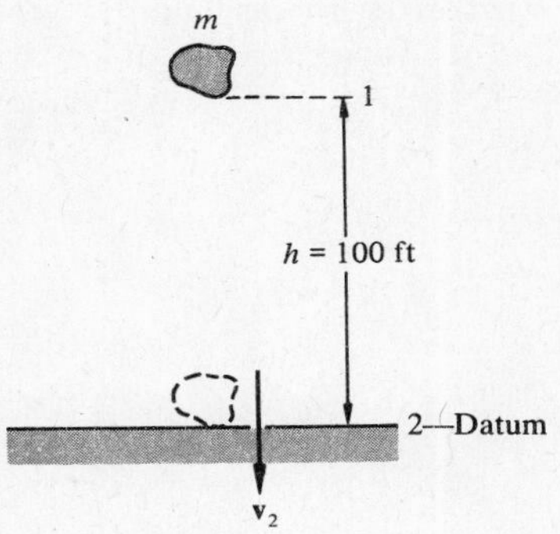

Fig. 19.26b

(a)  Use the work-energy method to find the general result for the velocity with which the particle strikes the ground.

(b)  Find the value of this velocity which corresponds to a height  $h = 100$ ft.

❙ (a)  This system is redrawn in Fig. 19.26b.  The initial position is designated 1, and the ground is designated 2.  The ground is taken to be the datum for potential energy, and there is no nonconservative work term.

The energy terms at the endpoints of the time interval are

$$T_1 = 0 \quad V_1 = mgh \quad T_2 = \tfrac{1}{2}mv_2^2 \quad V_2 = 0$$

The statement of work energy is thus

$$0 = \Delta T + \Delta V \quad 0 = (T_2 - T_1) + (V_2 - V_1) \quad 0 = (\tfrac{1}{2}mv_2^2 - 0) + (0 - mgh) \quad v_2 = \sqrt{2gh}$$

It may be observed that this result is *independent* of the mass of the particle.

(b)  For the numerical values of this example,

$$v = \sqrt{2gh} = \sqrt{2(32.2)(100)} = 80.2 \text{ ft/s}$$

This is the result obtained in Prob. 14.44 for a particle with constant acceleration.  This problem was also considered in Prob. 19.4.  It is left as an exercise for the reader to show that the result for $W_{12}$ in this latter problem could be used with Eq. (7) in Prob. 19.14 to obtain the above value of velocity.

**19.27**  A particle is projected downward from a height of 100 ft with a velocity of 35 ft/s.  Find the velocity with which it strikes the ground.

❙ Figure 19.27 shows the endpoints of the interval.  The potential and kinetic energy terms have the forms

$$T_1 = \tfrac{1}{2}mv_1^2 \quad T_2 = \tfrac{1}{2}mv_2^2 \quad V_1 = mgh \quad V_2 = 0$$

The work-energy equation is written as

$$0 = \Delta T + \Delta V = (T_2 - T_1) + (V_2 - V_1) = (\tfrac{1}{2}mv_2^2 - \tfrac{1}{2}mv_1^2) + (0 - mgh) \quad v_2 = \sqrt{v_1^2 + 2gh}$$

Using  $v_1 = 35$ ft/s  and  $h = 100$ ft,

$$v_2 = \sqrt{35^2 + 2(32.2)100} = 87.5 \text{ ft/s}$$

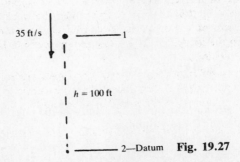

**Fig. 19.27**

**19.28**    An archer shoots an arrow vertically upward.    If the arrow ascends to a height of 90 ft, find the required value of the initial velocity of the arrow.

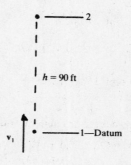

Fig. 19.28

▌ The endpoints of the interval are shown in Fig. 19.28.    The forms for $T$ and $V$ are

$$T_1 = \tfrac{1}{2}mv_1^2 \qquad T_2 = 0 \qquad V_1 = 0 \qquad V_2 = mgh$$

The work-energy equation appears as

$$0 = \Delta T + \Delta V = (T_2 - T_1) + (V_2 - V_1) = (0 - \tfrac{1}{2}mv_1^2) + (mgh = 0) \qquad v_1 = \sqrt{2gh}$$

For   $h = 90$ ft,

$$v_1 = \sqrt{2(32.2)90} = 76.1 \text{ ft/s}$$

The solution above is the same as that found in Prob. 14.48 by direct use of the equations for motion with constant acceleration.

**19.29**    The particle in Fig. 19.29a is projected upward from location $a$ with an initial velocity of 12 m/s.

(a)    Find the maximum height above the ground that the particle attains.

(b)    Find the velocity with which the particle passes point $b$.

(c)    Find the velocity with which the particle strikes the ground.

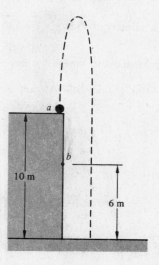

Fig. 19.29a

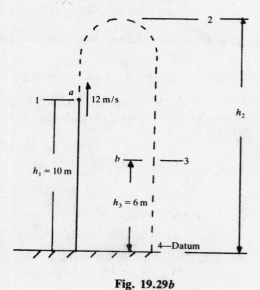

Fig. 19.29b

▌ (a)    Figure 19.29b shows points 1, 2, 3, and 4 in the interval of motion.    The kinetic and potential energies for points 1 and 2 are written as

$$T_1 = \tfrac{1}{2}mv_1^2 \qquad T_2 = 0 \qquad V_1 = mgh_1 \qquad V_2 = mgh_2$$

The work-energy equation between 1 and 2 has the form

$$0 = \Delta T + \Delta V = (T_2 - T_1) + (V_2 - V_1) = (0 - \tfrac{1}{2}mv_1^2) + (mgh_2 - mgh_1) \qquad h_2 = h_1 + \frac{v_1^2}{2g}$$

For the numerical values of the problem,

$$h_2 = 10 + \frac{12^2}{2(9.81)} = 17.3 \text{ m}$$

(b)  The forms of $T$ and $V$ at point 3, or $b$, are

$$T_3 = \tfrac{1}{2}mv_3^2 \qquad V_3 = mgh_3$$

Using the equation of work-energy between 1 and 3,

$$0 = \Delta T + \Delta V = (T_3 - T_1) + (V_3 - V_1) = (\tfrac{1}{2}mv_3^2 - \tfrac{1}{2}mv_1^2) + (mgh_3 - mgh_1)$$
$$\tfrac{1}{2}mv_3^2 = \tfrac{1}{2}mv_1^2 + mg(h_1 - h_3)$$
$$v_3 = \sqrt{v_1^2 + 2g(h_1 - h_3)} \tag{1}$$

Using  $v_1 = 12 \text{ m/s}$,  $h_1 = 10 \text{ m}$,  and  $h_3 = 6 \text{ m}$,

$$v_3 = \sqrt{12^2 + 2(9.81)(10 - 6)} = 14.9 \text{ ms/s}$$

(c)  $T_4$ and $V_4$ are written as

$$T_4 = \tfrac{1}{2}mv_4^2 \qquad V_4 = 0$$

From comparison with Eq. (1), with  $h_3 \to h_4 = 0$,

$$v_4 = \sqrt{v_1^2 + 2gh_1} = \sqrt{12^2 + 2(9.81)10} = 18.4 \text{ m/s}$$

**19.30**   Figure 19.30a shows a ball rebounding from a pavement.   Find the maximum height to which the ball will rise.

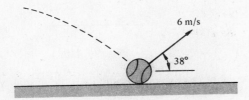

Fig. 19.30a

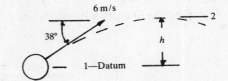

Fig. 19.30b

▌ Figure 19.30b shows the endpoints of the interval of motion.   Only the vertical component of velocity contributes to the increase in height.   For motion in the vertical direction, the kinetic and potential energy terms have the forms

$$T_1 = \tfrac{1}{2}m(6\sin 38°)^2 \qquad T_2 = 0 \qquad V_1 = 0 \qquad V_2 = mgh$$

The work-energy equation appears as

$$0 = \Delta T + \Delta V = (T_2 - T_1) + (V_2 - V_1)$$

For the numerical values of this problem,

$$0 = [0 - \tfrac{1}{2}m(6\sin 38°)^2] + (mgh - 0) \qquad \tfrac{1}{2}(6\sin 38°)^2 = 9.81h \qquad h = 0.695 \text{ m} = 695 \text{ mm}$$

The solution above may be compared with that in Prob. 14.71, where the problem was solved by direct use of the equations for plane projectile motion.

**19.31**   Use the work-energy method to verify the equation given in Prob. 14.79.

▌ Figure 19.31 shows the endpoints of the interval of motion of one typical particle of mass $m$.   The forms for $T$ and $V$ are

$$T_1 = 0 \qquad T_2 = \tfrac{1}{2}mv_2^2 \qquad V_1 = mgh \qquad V_2 = 0$$

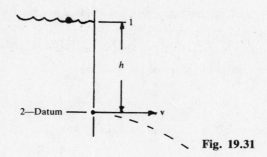

**Fig. 19.31**

Using the work-energy equation,

$$0 = = \Delta T + \Delta V = (T_2 - T_1) + (V_2 - V_1) = (\tfrac{1}{2}mv_2^2 - 0) + (0 - mgh) \qquad v_2 = \sqrt{2gh}$$

The above result is the same as that given in Prob. 14.79.

**19.32** In the position shown in Fig. 19.32a, the block just contacts the free end of the spring.

(a) If the block is released instantaneously with zero initial velocity, find the maximum deflection of the spring.

(b) Do the same as in part (a), if the block is very gradually lowered onto the spring.

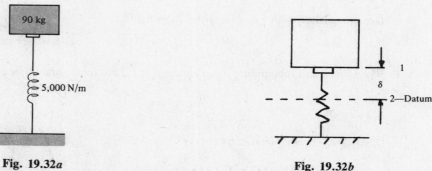

**Fig. 19.32a**                                    **Fig. 19.32b**

**❙** (a) Figure 19.32b shows the endpoints of the interval of motion, and position 2 is the position of maximum spring deflection. When the block is released instantaneously,

$$T_1 = 0 \qquad T_2 = 0 \qquad V_1 = mg\delta \qquad V_2 = \tfrac{1}{2}k\delta^2$$

$$0 = \Delta T + \Delta V = (T_2 - T_1) + (V_2 - V_1) = (0 - 0) + (\tfrac{1}{2}k\delta^2 - mg\delta) \qquad (\tfrac{1}{2}k\delta - mg)\delta = 0$$

The two solutions to the above equation are

$$\delta = 0 \qquad \tfrac{1}{2}k\delta - mg = 0$$

Since $\delta \neq 0$, the solution for the maximum displacement of the spring is

$$\delta = \frac{2mg}{k} = \frac{2(90)9.81}{5,000} = 0.353 \text{ m} = 353 \text{ mm} \qquad (1)$$

(b) When the block is gradually lowered onto the spring, the block comes to rest when the weight force is equal to the spring force. This result is written as

$$F = k\delta = mg$$

$$\delta = \frac{mg}{k} = \frac{90(9.81)}{5,000} = 0.177 \text{ m} = 177 \text{ mm} \qquad (2)$$

It may be seen that the spring deflection given by Eq. (1), for the case where the block is dropped on the spring, is *twice* that given by Eq. (2), for the case where the block is gradually lowered onto the spring. It was shown in Prob. 19.6 that the force in the spring is directly proportional to the spring deflection. Thus, the maximum spring force is *doubled* when the weight is "dropped" onto the spring. This problem forms the basis for the interpretation of *dynamic load* on a structure or machine versus *static load*, where Eq. (1) is the result for dynamic load and Eq. (2) is the result for static load.

**19.33** (*a*) A weight is dropped with zero initial velocity onto the springs shown in Fig. 19.33*a*. If all three springs have the same dimensions, find the maximum values of the deflections of the springs.

(*b*) Do the same as in part (*a*), if the weight has a downward velocity of 60 in/s in the position shown in Fig. 19.33*a*.

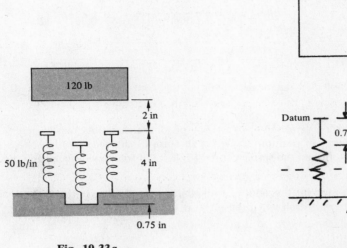

**Fig. 19.33*a***

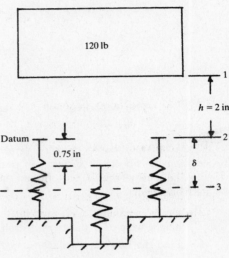

**Fig. 19.33*b***

▌ (*a*) Figure 19.33*b* shows three points in the interval of motion. Point 1 is the initial position, point 2 is the datum, and point 3 is the position of maximum spring displacement. The values of $T$ and $V$ at points 1 and 2 are

$$T_1 = 0 \qquad T_2 = \frac{1}{2}\left(\frac{120}{386}\right)v_2^2 \qquad V_1 = 120(2) \text{ in} \cdot \text{lb} \qquad V_2 = 0$$

The work-energy equation is written as

$$0 = \Delta T + \Delta V = (T_2 - T_1) + (V_2 - V_1) = \left[\frac{1}{2}\left(\frac{120}{386}\right)v_2^2 - 0\right] + [0 - 120(2)] \qquad v_2 = 39.3 \text{ in/s}$$

The above value is the velocity of the weight as it contacts the springs. The kinetic and potential energies at point 3 are

$$T_3 = 0 \qquad V_3 = 2[\tfrac{1}{2}(50)\delta^2] + \tfrac{1}{2}(50)(\delta - 0.75)^2 - 120\delta$$

The work-energy equation between points 2 and 3 has the form

$$0 = \Delta T + \Delta V = (T_3 - T_2) + (V_3 - V_2)$$

$$= \left[0 - \frac{1}{2}\left(\frac{120}{386}\right)39.3^2\right] + [(2(\tfrac{1}{2})50\delta^2 + \tfrac{1}{2}(50)(\delta - 0.75)^2 - 120\delta) - 0]$$

$$50\delta^2 + 25(\delta - 0.75)^2 - 120\delta - 240 = 0$$

$$75\delta^2 - 158\delta - 226 = 0$$

$$\delta = \frac{158 \pm \sqrt{(-158)^2 - 4(75)(-226)}}{2(75)} = \frac{158 \pm 305}{2(75)}$$

The positive root is used, and

$$\delta = 3.09 \text{ in}$$

The deflections of the outer springs are 3.09 in, and the deflection of the inner spring is $3.09 - 0.75 = 2.34$ in. This problem could have been solved directly by writing the work-energy equation between points 1 and 3 in the form

$$0 = \Delta T + \Delta V = (T_3 - T_1) + (V_3 - V_1)$$

This technique is used in the solution in part (*b*).

(*b*)  The values of $T$ and $V$ at points 1 and 3 are

$$T_1 = \frac{1}{2}\left(\frac{120}{386}\right)60^2 \qquad T_3 = 0 \qquad V_1 = 120(2) \text{ in} \cdot \text{lb} \qquad V_3 = 2[\tfrac{1}{2}(50)\delta^2] + \tfrac{1}{2}(50)(\delta - 0.75)^2 - 120\delta$$

The work-energy equation between points 1 and 3 has the form

$$0 = \Delta T + \Delta V = (T_3 - T_1) + (V_3 - V_1)$$

$$= \left[0 - \frac{1}{2}\left(\frac{120}{386}\right)60^2\right] + [(2(\tfrac{1}{2})50\delta^2 + \tfrac{1}{2}(50)(\delta - 0.75)^2 - 120\delta) - 240]$$

$$75\delta^2 - 158\delta - 786 = 0 \qquad \delta = \frac{158 \pm \sqrt{(-158)^2 - 4(75)(-786)}}{2(75)} = \frac{158 \pm 511}{2(75)} = 4.46 \text{ in}$$

The deflections of the outer and inner springs are 4.46 in and $4.46 - 0.75 = 3.71$ in, respectively.

**19.34**  Figure 19.34*a* shows a model of a spring-operated toy pistol which shoots plastic pellets. In the uncocked position the tip of the spring is at position 2. When the pistol is loaded, the tip of the spring is at position 1. The mass of the pellet is $m$, and the spring constant is $k$. All frictional effects, and the mass of the spring, may be neglected.

(*a*)  Derive a general expression for the velocity with which the pellet leaves the spring, and for the maximum height which the pellet will reach when the pistol is fired.

(*b*)  Find the numerical values for part (*a*) if the pellet weighs 15 g, $k = 70$ N/m, and $\delta = 100$ mm.

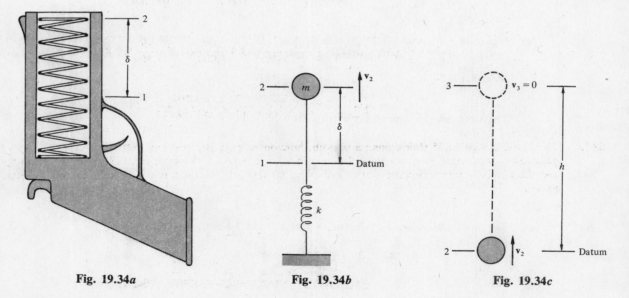

**Fig. 19.34*a***       **Fig. 19.34*b***       **Fig. 19.34*c***

▌ (*a*)  The model of the pistol spring is shown in Fig. 19.34*b*. Since both the spring and the mass may store potential energy, the potential energy terms for each of these elements will be written separately. Position 1 is the datum for potential energy of the mass. The subscripts $m$ and $s$ are used to designate the pellet mass and the spring, respectively. The required energy terms are

$$W_{\text{NC}} = 0 \qquad T_1 = 0 \qquad V_{1m} = 0 \qquad V_{1s} = \tfrac{1}{2}k\delta^2$$
$$T_2 = \tfrac{1}{2}mv_2^2 \qquad V_{2m} = mg\delta \qquad V_{2s} = 0$$

The work-energy equation has the form

$$W_{\text{NC}} = \Delta T + \Delta V = (T_2 - T_1) + (V_{2m} - V_{1m}) + (V_{2s} - V_{1s})$$
$$0 = (\tfrac{1}{2}mv_2^2 - 0) + (mg\delta - 0) + (0 - \tfrac{1}{2}k\delta^2)$$
$$\tfrac{1}{2}mv_2^2 = \tfrac{1}{2}k\delta^2 - mg\delta \qquad\qquad (1)$$

$$v_2 = \sqrt{\frac{\delta}{m}(k\delta - 2mg)} \qquad\qquad (2)$$

It is interesting to note that a limiting condition occurs if

$$2mg = k\delta \qquad mg = \tfrac{1}{2}k\delta \tag{3}$$

For this case the velocity of the mass at position 2 would be zero, and the pellet would not leave the spring. The corresponding weight of the mass, from Eq. (3), would be *exactly one half* of the force required to compress the spring by the amount $\delta$. If $2mg > k\delta$, the pellet would not reach endpoint 2.

When the pellet leaves the spring, it has the velocity given by Eq. (2). The model of the rise of the pellet to its maximum height $h$ is shown in Fig. 19.34c. Position 2 is defined to be a new datum, and the required energy terms are

$$T_2 = \tfrac{1}{2}mv_2^2 \qquad T_3 = 0 \qquad V_2 = 0 \qquad V_3 = mgh$$

From Eq. (1),

$$\tfrac{1}{2}mv_2^2 = T_2 = \tfrac{1}{2}k\delta^2 - mg\delta$$

The work-energy equation between points 2 and 3 appears as

$$W_{NC} = \Delta T + \Delta V \qquad 0 = (T_3 - T_2) + (V_3 - V_2)$$
$$0 = [0 - (\tfrac{1}{2}k\delta^2 - mg\delta)] + (mgh - 0) = -\tfrac{1}{2}k\delta^2 + mg\delta + mgh$$
$$h = \frac{\tfrac{1}{2}k\delta^2 - mg\delta}{mg} = \delta\left(\frac{k\delta}{2mg} - 1\right) \tag{4}$$

(*b*) For the present problem,

$$m = 15\,\text{g} \qquad k = 70\,\text{N/m} \qquad \delta = 100\,\text{mm}$$

Using Eq. (2), we get

$$v_2 = \sqrt{\frac{\delta}{m}(k\delta - 2mg)} = \sqrt{\frac{100/1{,}000}{15/1{,}000}\left[70\left(\frac{100}{1{,}000}\right) - 2\left(\frac{15}{1{,}000}\right)(9.81)\right]} = 6.69\,\text{m/s}$$

From Eq. (4),

$$h = \delta\left(\frac{k\delta}{2mg} - 1\right) = \frac{100}{1{,}000}\left[\frac{70(100/1{,}000)}{2(15/1{,}000)(9.81)} - 1\right] = 2.28\,\text{m}$$

**19.35** The block in Fig. 19.35 slides along a smooth, horizontal track at a constant velocity of 50 ft/s. At a certain point the smooth track joins a section of rough, horizontal track. The coefficient of friction between the block and the rough track is estimated to be 0.24. How far along the rough track will the block slide before coming to rest?

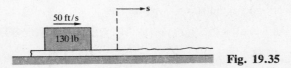

**Fig. 19.35**

▌ The distance traveled by the block along the rough track is $l$, between endpoints 1 and 2. The nonconservative force is the friction force $F$, given by

$$F = \mu mg$$

The nonconservative work is given by

$$W_{NC} = \int_0^l (F\cos\beta)\,ds$$

Since the friction force acts *opposite* to the motion of the block,

$$\cos\beta = 180°$$

and

$$W_{NC} = \int_0^l (F\cos 180°)\,ds = \int_0^l (-\mu mg)\,ds = -\mu mgl$$

Since there are no elevation changes of the block, the potential energy change $\Delta V$ of the block is zero. The equation of work-energy has the form

$$W_{NC} = \Delta T + \Delta V = T_2 - T_1 \qquad -\mu mgl = 0 - \tfrac{1}{2}mv_1^2 \qquad l = \frac{v_1^2}{2\mu g} \tag{1}$$

It may be observed that this result is independent of the mass of the block. For the present case, $v_1 = 50$ ft/s and $\mu = 0.24$. Equation (1) then appears as

$$l = \frac{v_1^2}{2\mu g} = \frac{50^2}{2(0.24)(32.2)} = 162 \text{ ft}$$

This is the result obtained in Prob. 15.29 by direct use of Newton's second law.

**19.36** A block slides along a horizontal plane with a constant value of deceleration, as shown in Fig. 19.36*a*. The speed of the block decreases from 35 in/s to zero over a length of 100 in.

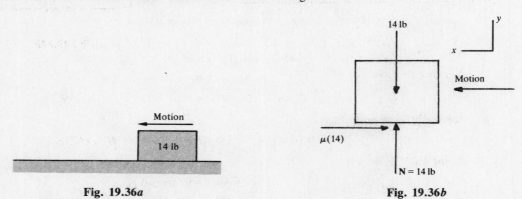

Fig. 19.36*a*                    Fig. 19.36*b*

(*a*) Find the value of the coefficient of friction.

(*b*) A second block, weighing 6 lb, is attached to the top of the 14-lb block. Find the distance through which the system of blocks will move before coming to rest, if the initial speed is 35 in/s.

▮ (*a*) Figure 19.36*b* shows the free-body diagram of the block. Point 1 is the beginning of the interval and point 2 is the end of the interval. The forms for $T$ and $V$ are

$$T_1 = \frac{1}{2}\left(\frac{14}{386}\right)35^2 \qquad T_2 = 0 \qquad V_1 = 0 \qquad V_2 = 0$$

The work done is

$$W_{NC} = -\mu(14)100 \text{ in} \cdot \text{lb}$$

The work-energy equation is written as

$$W_{NC} = \Delta T + \Delta V = (T_2 - T_1) + (V_2 - V_1) \qquad -\mu(14)100 = \left[0 - \frac{1}{2}\left(\frac{14}{386}\right)35^2\right] + (0 - 0) \qquad \mu = 0.0159$$

(*b*) The solution above is independent of the mass of the block, and is the same as that found in Prob. 15.30 by direct use of Newton's second law. Since this problem is equivalent to that in part (*a*), the blocks will move through 100 in before coming to rest.

**19.37** A constant force $P$ is applied to the crate shown in Fig. 19.37*a* when this element is moving to the right at 1.2 m/s. After the crate has moved through 4 m, the velocity is 3.7 m/s.

(*a*) Show that the 15-N · m couple has no effect on the translational motion of the crate.

(*b*) Find the magnitude of the force $P$.

▮ (*a*) Figure 19.37*b* shows the free-body diagram of the crate. For rotational equilibrium about the center of mass,

$$\sum M_0 = I_0 \alpha = 0$$

$$-15 - (0.2N_a + 0.2N_b)\frac{100}{1,000} - N_a\left(\frac{160}{1,000}\right) + N_b\left(\frac{160}{1,000}\right) = 0$$

$$-15 - 0.18N_a + 0.14N_b = 0 \tag{1}$$

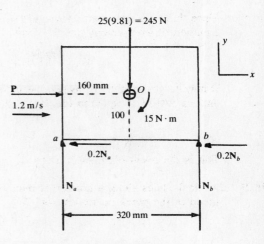

**Fig. 19.37a**            **Fig. 19.37b**

For force equilibrium,

$$\sum F_y = 0 \qquad N_a - 245 + N_b = 0 \qquad N_b = 245 - N_a \qquad (2)$$

Using Eq. (2) in Eq. (1),

$$-15 - 0.18N_a + 0.14(245 - N_a) = 0 \qquad N_a = 60.3 \text{ N}$$

Using Eq. (2),

$$N_b = 245 - 60.3 = 185 \text{ N}$$

The resultant friction force $F_T$ is given by

$$F_T = 0.2N_a + 0.2N_b = 0.2(60.3 + 185) = 49.1 \text{ N}$$

If the 15-N · m couple is removed, Eq. (1) has the form

$$-0.18N_a + 0.14N_b = 0 \qquad N_b = 1.29N_a \qquad (3)$$

Equation (2) is unchanged, with the form

$$N_b = 245 - N_a \qquad (4)$$

$N_b$ is eliminated between Eqs. (3) and (4), to obtain

$$1.29N_a = 245 - N_a \qquad N_a = 107 \text{ N} \qquad N_b = 138 \text{ N}$$

The resultant friction force is

$$F_T = 0.2N_a + 0.2N_b = 0.2(107) + 0.2(138) = 49.0 \approx 49.1 \text{ N}$$

The effect of the applied couple is to change the distribution of the total normal force between $N_a$ and $N_b$. This distribution has no effect on the total normal force $N = N_a + N_b$, or on the resultant friction force $F_T = \mu(N_a + N_b)$.

**(b)** Point 1 is the initial condition and point 2 is the final condition. The displacement, velocity, and energy terms have the values

$$s_1 = 0 \qquad v_1 = 1.2 \text{ m/s} \qquad T_1 = \tfrac{1}{2}(25)1.2^2 \qquad V_1 = 0$$
$$s_2 = 4 \text{ m} \qquad v_2 = 3.7 \text{ m/s} \qquad T_2 = \tfrac{1}{2}(25)3.7^2 \qquad V_2 = 0$$

The magnitude of the work done is

$$W_{NC} = (P - 49.0)4$$

The work-energy equation has the form

$$W_{NC} = \Delta T + \Delta V = (T_2 - T_1) + (V_2 - V_1)$$
$$(P - 49.0)4 = \tfrac{1}{2}(25)3.7^2 - \tfrac{1}{2}(25)1.2^2 \qquad P = 87.3 \text{ N}$$

**19.38**    At the instant shown in Fig. 19.38a, the block has a velocity of 8 ft/s.

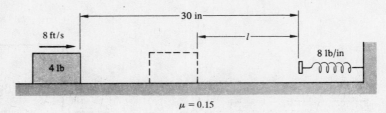

Fig. 19.38a

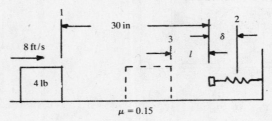

Fig. 19.38b

(a) Find the maximum value of the deflection of the spring.

(b) Find the distance $l$ through which the block travels, after rebound, before it comes to rest.

▌ (a) Figure 19.38b shows points 1, 2, and 3 in the interval of motion. Point 1 is the initial position of the block. Point 2 is the position of maximum deflection of the spring with the block at rest. Point 3 is the rest position of the block after rebound. The energy terms at positions 1 and 2 are

$$T_1 = \frac{1}{2}\left(\frac{4}{386}\right)[8(12)]^2 \qquad T_2 = 0 \qquad V_1 = 0 \qquad V_2 = \frac{1}{2}(8)\delta^2$$

The work done between 1 and 2 is given by

$$W_{NC} = -0.15(4)(30 + \delta)$$

Using the work-energy equation,

$$W_{NC} = \Delta T + \Delta V = (T_2 - T_1) + (V_2 - V_1) \qquad -0.15(4)(30 + \delta) = \left[0 - \frac{1}{2}\left(\frac{4}{386}\right)96^2\right] + \left[\frac{1}{2}(8)\delta^2 - 0\right]$$

$$4\delta^2 + 0.6\delta - 29.8 = 0 \qquad \delta = \frac{-0.6 \pm \sqrt{0.6^2 - 4(4)(-29.8)}}{2(4)} = \frac{-0.6 \pm 21.8}{2(4)}$$

The positive root is used, and

$$\delta = 2.65 \text{ in}$$

(b) The maximum travel of the block after rebound is found next. The energy terms at positions 2 and 3 are

$$T_2 = 0 \qquad T_3 = 0 \qquad V_2 = \frac{1}{2}(8)\delta^2 = \frac{1}{2}(8)2.65^2 \qquad V_3 = 0$$

The work done between 2 and 3 is

$$W_{NC} = -0.15(4)(2.65 + l)$$

The work-energy equation is written as

$$W_{NC} = \Delta T + \Delta V = (T_3 - T_2) + (V_3 - V_2) \qquad -0.15(4)(2.65 + l) = (0 - 0) + [0 - \frac{1}{2}(8)2.65^2]$$
$$l = 44.2 \text{ in}$$

It may be seen that the block comes to rest to the left of its initial position shown in Fig. 19.38a.

19.39 (a) The free length of the spring in Fig. 19.39a is 4 in. Find the distance $l$ through which the block will move when this element is released from rest in the position shown in the figure.

(b) Do the same as in part (a), for the arrangement of the system shown in Fig. 19.39b.

▌ (a) Figure 19.39c shows the endpoints 1 and 2 of the interval of motion. The energy terms at 1 and 2 are

$$T_1 = 0 \qquad T_2 = 0 \qquad V_1 = \frac{1}{2}(20)(4 - 2.6)^2 \qquad V_2 = 0$$

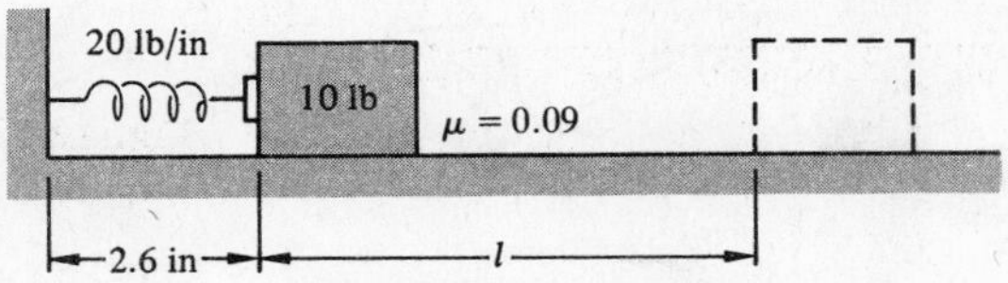

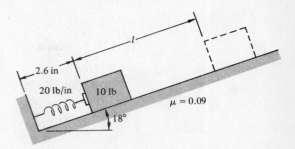

Fig. 19.39a

Fig. 19.39b

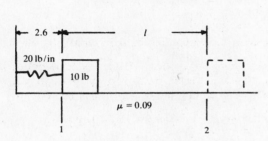

Fig. 19.39c

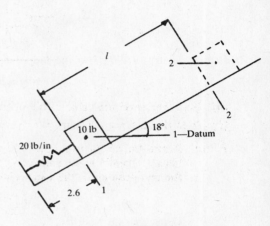

Fig. 19.39d

The work done between 1 and 2 is

$$W_{NC} = -0.09(10)l$$

Using the work-energy equation,

$$W_{NC} = \Delta T + \Delta V = (T_2 - T_1) + (V_2 - V_1) \qquad -0.09(10)l = (0 - 0) + [0 - \tfrac{1}{2}(20)1.4^2]$$
$$l = 21.8 \text{ in}$$

(b)  Figure 19.39d shows the endpoints 1 and 2 for the case where the surface is inclined.   Endpoint 1 is taken as the datum for potential energy of the mass of the block.   The required energy terms are

$$T_1 = 0 \qquad T_2 = 0 \qquad V_1 = \tfrac{1}{2}(20)(4 - 2.6)^2 \qquad V_2 = 10(l \sin 18°)$$

The work done between 1 and 2 is

$$W_{NC} = -0.09(10 \cos 18°)l$$

The work-energy equation has the form

$$W_{NC} = \Delta T + \Delta V = (T_2 - T_1) + (V_2 - V_1)$$
$$-0.09(10 \cos 18°)l = (0 - 0) + [(10 \sin 18°)l - \tfrac{1}{2}(20)1.4^2] \qquad l = 4.97 \text{ in}$$

The value for $l$ is significantly smaller for the case where the block moves up the incline.   The reason is that part of the initial potential energy of the compressed spring is used to raise the weight in the part (b) solution, while in the part (a) solution this spring energy is needed only to overcome the sliding friction force.

19.40  (a)  The velocity of the block in the position shown in Fig. 19.40a is  $v = 20$ in/s,  and  $\beta = 15°$.  How much further will the block travel before it comes to rest?

(b)  It is desired to have the block in Fig. 19.40a come to rest at position a.  If  $\beta = 15°$,  find the required value of $v$.

(c)  If  $v = 20$ in/s,  find the required value of $\beta$ for the block to come to rest at position a.

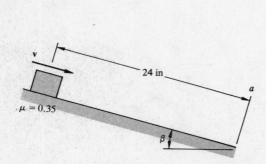

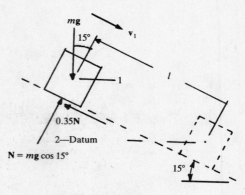

Fig. 19.40a                                    Fig. 19.40b

▌ (a)  Figure 19.40b shows the endpoints 1 and 2 of the interval of motion.   The energy terms have the forms

$$T_1 = \tfrac{1}{2}m(20)^2 \qquad T_2 = 0 \qquad V_1 = mgl \sin 15° \qquad V_2 = 0$$

The work-energy equation is written as

$$W_{NC} = \Delta T + \Delta V = (T_2 - T_1) + (V_2 - V_1) \qquad -(0.35mg \cos 15°)l = [0 - \tfrac{1}{2}m(20)^2] + [0 - mgl \sin 15°]$$
$$l(386)[0.35 \cos 15° - \sin 15°] = \tfrac{1}{2}(20)^2 \qquad\qquad (1)$$
$$l = 6.54 \text{ in}$$

(b)  Equation (1) in part (a), with   $l = 24$ in,   may be written in terms of $\beta$ and $v_1$ as

$$24(386)[0.35 \cos \beta - \sin \beta] = \tfrac{1}{2}v_1^2 \qquad\qquad (2)$$

Using   $\beta = 15°$   in Eq. (2),

$$24(386)[0.35 \cos 15° - \sin 15°] = \tfrac{1}{2}v_1^2 \qquad v_1 = 38.3 \text{ in/s}$$

(c)  Using   $v_1 = 20$ in/s   in Eq. (2) in part (b),

$$24(386)[0.35 \cos \beta - \sin] = \tfrac{1}{2}(20)^2$$
$$0.35 \cos \beta - \sin \beta = 0.0216 \qquad 0.35 \cos \beta = 0.0216 + \sin \beta \qquad\qquad (3)$$

Both sides of Eq. (3) are squared, and the term $\cos^2 \beta$ is replaced by the term $1 - \sin^2 \beta$.   The result is

$$0.35^2(1 - \sin^2 \beta) = 0.0216^2 + 2(0.0216) \sin \beta + \sin^2 \beta \qquad 1.12 \sin^2 \beta + 0.0432 \sin \beta - 0.122 = 0$$

The solution to the above quadratic equation is

$$\sin \beta = \frac{-0.0432 \pm \sqrt{0.0432^2 - 4(1.12)(-0.122)}}{2(1.12)} = \frac{-0.0432 \pm 0.741}{2(1.12)}$$

Since $\beta$ must be positive, the positive root in the above equation is used, so that

$$\sin \beta = \frac{-0.0432 + 0.741}{2(1.12)} \qquad \beta = 18.2°$$

**19.41**  The block in Fig. 19.41a is released from rest in the position shown in the figure.   The coefficient of friction over length $ab$ is 0.22; and over length $bc$ this quantity has the value 0.16.   Find the velocity with which the block passes position $c$.

▌ Figure 19.41b shows the endpoints 1 and 2 of the interval of motion, and point 2 is the datum.   The kinetic and potential energies are

$$T_1 = 0 \qquad T_2 = \tfrac{1}{2}mv_2^2 \qquad V_1 = mg(5 \sin 24°) \qquad V_2 = 0$$

The work-energy equation is written as

$$W_{NC} = \Delta T + \Delta V = (T_2 - T_1) + (V_2 - V_1)$$
$$-0.22(mg \cos 24°)3 - 0.16(mg \cos 24°)2 = (\tfrac{1}{2}mv_2^2 - 0) + [0 - mg(5 \sin 24°)]$$
$$-9.81[0.22(3) + 0.16(2)] \cos 24° = \tfrac{1}{2}v_2^2 - 9.81(5 \sin 24°) \qquad v_2 = 4.73 \text{ m/s}$$

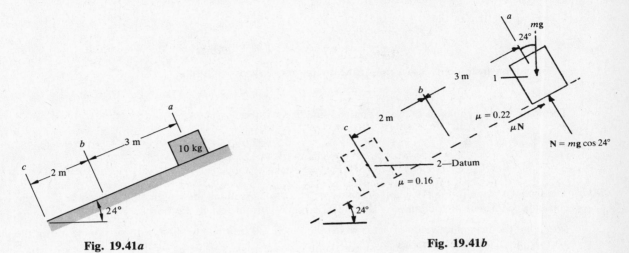

Fig. 19.41a                                Fig. 19.41b

**19.42** A force $P = 40\,\text{lb}$ is applied to the initially stationary block in Fig. 19.42a. When the block has covered a distance of 12 ft, the force is removed. Find the velocity with which the block passes its initial position.

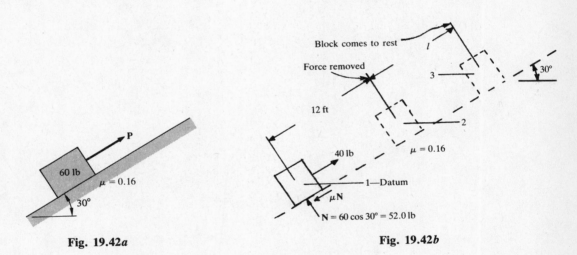

Fig. 19.42a                                Fig. 19.42b

▌ Point 1 in Fig. 19.42b is the initial position of the block, and point 2 is the position where the force is removed. The energy terms at 1 and 2 are

$$T_1 = 0 \qquad T_2 = \tfrac{1}{2}mv_2^2 \qquad V_1 = 0 \qquad V_2 = 60(12\sin 30°)$$

The nonconservative work done between points 1 and 2 is given by

$$W_{\text{NC}} = 40(12) - 0.16(52.0)12 = 380\,\text{ft}\cdot\text{lb}$$

The work-energy equation is written as

$$W_{\text{NC}} = \Delta T + \Delta V = (T_2 - T_1) + (V_2 - V_1) \qquad 380 = \left[\frac{1}{2}\left(\frac{60}{32.2}\right)v_2^2 - 0\right] + [60(12\sin 30°) - 0]$$

$$v_2^2 = 21.5\,\text{ft}^2/\text{s}^2 \qquad v_2 = 4.63\,\text{ft/s} \tag{1}$$

The sense of $v_2$ is up and to the right.
  The block comes to rest in position 3 in Fig. 19.42b. The energy terms at 3 are

$$T_3 = 0 \qquad V_3 = 60(12 + l)\sin 30°$$

The work-energy equation is

$$W_{\text{NC}} = \Delta T + \Delta V = (T_3 - T_2) + (V_3 - V_2)$$
$$-0.16(52.0)l = (0 - \tfrac{1}{2}mv_2^2) + [60(12 + l)\sin 30° - 60(12)\sin 30°] \tag{2}$$

Using Eq. (1) in Eq. (2),

$$-0.16(52.0)l = -\frac{1}{2}\left(\frac{60}{32.2}\right)21.5 + 60(\cancel{12} + l - \cancel{12})\sin 30° \qquad l = 0.523 \text{ ft}$$

For motion from point 3 to point 1, the energy terms have the forms

$$T_1 = \frac{1}{2}mv_1^2 \qquad T_3 = 0 \qquad V_1 = 0 \qquad V_3 = 60(12 + 0.523)\sin 30° = 376 \text{ ft} \cdot \text{lb}$$

The work-energy equation between points 1 and 3 is written as

$$W_{\text{NC}} = \Delta T + \Delta V = (T_1 - T_3) + (V_1 - V_3) \qquad -0.16(52.0)(12 + 0.523) = \left[\frac{1}{2}\left(\frac{60}{32.2}\right)v_1^2 - 0\right] + [0 - 376]$$

$$v_1 = 17.1 \text{ ft/s}$$

**19.43** The inclined plane in Fig. 19.43a is assumed to be frictionless along the length ab. Length bc is rough, with a constant coefficient of friction μ. The block is released from rest at position a.

(a) Find the maximum value of μ for which the block will continue to slide, without tipping, down the plane.

(b) Find the velocity of the block at position c, for the conditions of part (a).

(c) For what length bc would the block come to rest, for the conditions of part (a)?

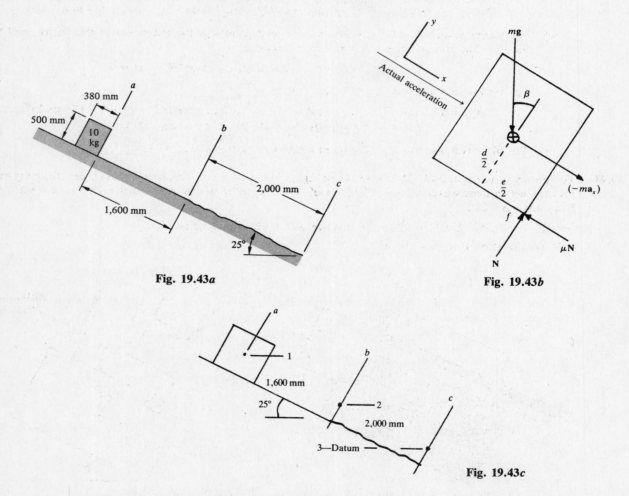

Fig. 19.43a    Fig. 19.43b    Fig. 19.43c

❙ (a) The free-body diagram of the block, with the inertia force, is shown in Fig. 19.43b. When tipping is impending, the reaction force acts at edge f. For equilibrium of the block,

$$\sum F_y = 0 \qquad N = mg\cos\beta$$

$$\sum F_x = 0 \qquad mg\sin\beta + (-ma_x) - \mu(mg\cos\beta) = 0 \qquad a_x = g(\sin\beta - \mu\cos\beta) \qquad (1)$$

$$\sum M_f = 0 \qquad -(mg \sin \beta) \frac{d}{2} - (-ma_x) \frac{d}{2} + (mg \cos \beta) \frac{e}{2} = 0 \qquad (2)$$

Using Eq. (1) in Eq. (2),

$$\frac{a}{2} (g \sin \beta) - \frac{a}{2} (-\mu g \cos \beta + g \sin \beta) - \frac{b}{2} (g \cos \beta) = 0 \qquad \mu = \frac{e}{d}$$

If $\mu \leq e/d$, the block will continue to slide without tipping. Thus,

$$\mu_{max} = \frac{e}{d} = \frac{380}{500} = 0.76$$

(b) Figure 19.43c shows three points in the interval of motion. Point 1 is the initial position of the block, point 2 is the beginning of the length of rough surface, and point 3 is the end of the rough surface, and the datum. The energy terms at 1 and 3 are

$$T_1 = 0 \qquad T_3 = \tfrac{1}{2} m v_3^2 \qquad V_1 = mg(3.6 \sin 25°) \qquad V_3 = 0$$

The work-energy equation is

$$W_{NC} = \Delta T + \Delta V = (T_3 - T_1) + (V_3 - V_1)$$

Using $\mu = 0.76$ from part (a), the above equation has the form

$$-0.76(mg \cos 25°)2 = (\tfrac{1}{2} m v_c^2 - 0) + [0 - mg(3.6 \sin 25°)] \qquad v_c = 1.68 \text{ m/s}$$

(c) The length bc in Fig. 19.43a is designated l, and point 4 is the lower end of this length, and the datum. The energy terms at 1 and 4 have the forms

$$T_1 = 0 \qquad T_4 = 0 \qquad V_1 = mg(l + 1.6) \sin 25° \qquad V_4 = 0$$

The work-energy equation is written as

$$W_{NC} = \Delta T + \Delta V = (T_4 - T_1) + (V_4 - V_1) \qquad -0.76(mg \cos 25°)l = -mg(l + 1.6) \sin 25°$$
$$l(0.76 \cos 25° - \sin 25°) = 1.6 \sin 25° \qquad l = 2.54 \text{ m} = 2,540 \text{ mm}$$

The block will come to rest 540 mm past position c shown in Fig. 19.43a.

**19.44** The conveyor system of Prob. 18.81 is shown in Fig. 19.44a. At the instant shown in the figure, the plate has a velocity, with a sense which is down and to the left, of 8 ft/s. The weight of the plate is 750 lb. $a = 6$ ft, $b = 3$ ft, and $\beta = 28°$.

(a) If $\mu = 0.15$, find how far the plate moves before its velocity is 10.5 ft/s.

(b) Do the same as in part (a), if $\mu = 0$.

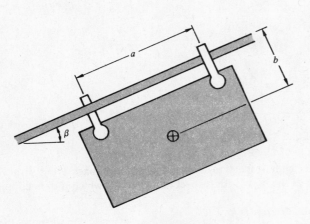

Fig. 19.44a

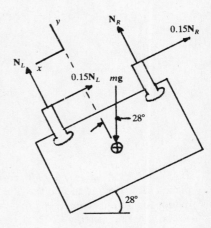

Fig. 19.44b

▮ (a) The free-body diagram of the plate is shown in Fig. 19.44b. For equilibrium in the y direction,

$$\sum F_y = 0 \qquad N_L + N_R = mg \cos 28°$$

Point 1 is the initial condition, with $x = 0$ and $v_1 = 8$ ft/s. This point is chosen to be the datum, so that

$$V_1 = 0$$

Point 2 is the final condition, with $x = l$ and $v_2 = 10.5$ ft/s. The kinetic and potential energy terms at 1 and 2 are

$$T_1 = \tfrac{1}{2}m(8)^2 \qquad T_2 = \tfrac{1}{2}m(10.5)^2 \qquad V_1 = 0 \qquad V_2 = -mgl \sin 28°$$

The work done between 1 and 2 is

$$W_{NC} = -0.15(N_L + N_R)l$$

The work-energy equation is written as

$$W_{NC} = \Delta T + \Delta V = (T_2 - T_1) + (V_2 - V_1)$$
$$-0.15(mg \cos 28°)l = [\tfrac{1}{2}m(10.5^2) - \tfrac{1}{2}m(8^2)] + [-mgl \sin 28° - 0] \qquad l = 2.13 \text{ ft}$$

A check may be made on the above result. From Prob. 18.89,

$$a_x = 10.9 \text{ ft/s}^2$$

Using $v_0 = 8$ ft/s, $v = 10.5$ ft/s, and $s = l$, the length $l$ is found from

$$v^2 = v_0^2 + 2a(s - s_0) \qquad 10.5^2 = 8^2 + 2(10.9)l \qquad l = 2.12 \text{ ft} \approx 2.13 \text{ ft}$$

(b) If $\mu = 0$, the work done during motion from 1 to 2 is zero. The work-energy equation then has the form

$$0 = \Delta T + \Delta V = (T_2 - T_1) + (V_2 - V_1) = [\tfrac{1}{2}m(10.5)^2 - \tfrac{1}{2}m(8)^2] + [-m(32.2)l \sin 28° - 0]$$
$$l = 1.53 \text{ ft}$$

It may be seen that, in the absence of friction, the velocity of 10.5 ft/s is reached in a shorter length of motion.

## 19.3 POTENTIAL ENERGY AND KINETIC ENERGY OF A RIGID BODY IN PLANE MOTION, THE WORK-ENERGY METHOD FOR A RIGID BODY IN PLANE MOTION

19.45 Find the general expression for the potential energy of a rigid body which moves in a vertical plane.

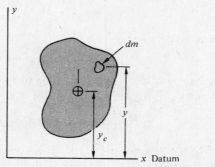

**19.45**

▌ Figure 19.45 shows a rigid body positioned with respect to a set of coordinate axes lying in a vertical plane. It is desired to find the potential energy of the body with respect to the $x$ axis, which is defined to be the datum for potential energy.

The potential energy $dV$ of a typical mass element $dm$ is

$$dV = dm\, gy$$

where $dm\, g$ is the weight of the element. The total potential energy is found as

$$V = \int_{vol} dm\, gy = g \int_{vol} y\, dm \tag{1}$$

The definition of the centroidal coordinate $y_c$ is

$$y_c = \frac{\int_{\text{vol}} y \, dm}{m} \quad \text{or} \quad \int_{\text{vol}} y \, dm = my_c \tag{2}$$

The term $\int_{\text{vol}} y \, dm$ is eliminated between Eqs. (1) and (2), with the result.

$$V = g(my_c) = (mg)y_c \tag{3}$$

The term $mg$ is the *total weight* of the rigid body. Thus, the potential energy of a body may be found by imagining all the weight to be concentrated at the center of mass (CM) and treating this point as a particle in translation.

**19.46**  (*a*)  Find the general expression for the kinetic energy of a rigid body which rotates about a fixed axis.

(*b*)  Find the general expression for the kinetic energy of a rigid body which moves in plane motion.

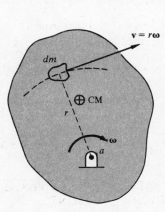

Fig. 19.46*a*

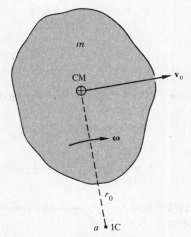

Fig. 19.46*b*

▌ (*a*)  Figure 19.46*a* shows a rigid body which is in plane rotational motion about a *fixed axis*.
The *absolute* velocity of the typical mass element $dm$ is

$$v = r\omega$$

The kinetic energy of this element is

$$dT = \tfrac{1}{2}(dm)v^2 = \tfrac{1}{2}(dm)(r\omega)^2$$

The total kinetic energy of the body is found by summing the contributions of all the individual mass elements, in the form

$$T = \int dT = \int_{\text{vol}} \tfrac{1}{2}(dm)(r\omega)^2 = \tfrac{1}{2}\omega^2 \int_{\text{vol}} r^2 \, dm \tag{1}$$

The integral in Eq. (1) may be recognized as the *mass moment of inertia* of the rigid body about the fixed point $a$, given by

$$I_a = \int_{\text{vol}} r^2 \, dm$$

The final form for the kinetic energy $T$ is

$$T = \tfrac{1}{2}I_a\omega^2 \tag{2}$$

(*b*)  Figure 19.46*b* shows a rigid body which moves with general plane motion. It was shown in Prob. 16.40 that, when a body rotates with general plane motion, there is a fixed point about which the body may be assumed to be instantaneously rotating. This point is the instant center (IC) shown in the figure. The velocity $v_0$ of the CM of the body is an *absolute* velocity, so that the IC is a point which is *fixed* with respect to a set of inertial coordinates. The kinetic energy of the body in general plane rotation about a fixed point, following Eq. (2) in Prob. 19.46, is

$$T = \tfrac{1}{2}I_a\omega^2$$

For the present problem, $I_a$ is the mass moment of inertia of the body about the fixed point which is the instant center. Using the parallel-axis theorem for moments of inertia, we get

$$I_a = I_0 + mr_0^2$$

where $I_0$ is the mass moment of inertia of the body about the CM. The kinetic energy term now has the form

$$T = \tfrac{1}{2}I_a\omega^2 = \tfrac{1}{2}(I_0 + mr_0^2)\omega^2 = \tfrac{1}{2}I_0\omega^2 + \tfrac{1}{2}mr_0^2\omega^2 \tag{3}$$

Since $v_0 = r_0\omega$, the final form of Eq. (3) is

$$T = \tfrac{1}{2}mv_0^2 + \tfrac{1}{2}I_0\omega^2 \tag{4}$$

It may be seen that the kinetic energy of a body in general plane motion has *two* distinct components. The first is the term $\tfrac{1}{2}mv_0^2$. This term represents the *translational* kinetic energy due to the absolute velocity $v_0$ of the CM. The second term, $\tfrac{1}{2}I_0\omega^2$, is the *rotational* kinetic energy due to the angular velocity $\omega$ of the body.

*Equations (2) and (4) are of fundamental importance when the work-energy method is used to solve problems of plane motion of rigid bodies.*

**19.47**  A disk of diameter 250 mm and thickness 35 mm is made of steel, with $\rho = 7{,}830 \text{ kg/m}^3$. The disk moves in general plane motion. The velocity diagram of the center of mass is shown in Fig. 19.47a, and the angular velocity diagram is shown in Fig. 19.47b. Find the values of the kinetic energy of the disk at the times 3 s, 4 s, and 10 s.

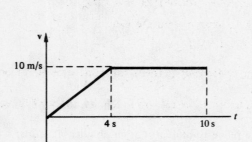

Fig. 19.47a

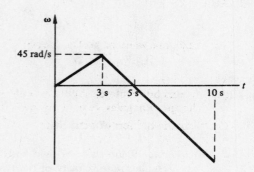

Fig. 19.47b

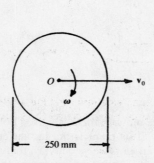

Fig. 19.47c

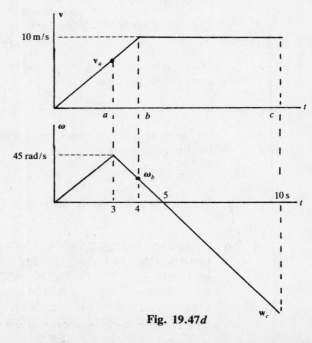

Fig. 19.47d

❚ Figure 19.47$c$ shows the translational velocity $v$, and the rotational velocity $\omega$, of the disk. The mass of the disk is given by

$$m = \rho t A = 7{,}830(35)\,\frac{\pi(250)^2}{4}\left(\frac{1}{1{,}000}\right)^3 = 13.5\,\text{kg}$$

The mass moment of inertia of the disk is found as

$$I_0 = \tfrac{1}{8}md^2 = \tfrac{1}{8}(13.5)\left(\frac{250}{1{,}000}\right)^2 = 0.105\,\text{kg}\cdot\text{m}^2$$

The kinetic energy of the disk has the form

$$T = \tfrac{1}{2}mv_0^2 + \tfrac{1}{2}I_0\omega^2$$

From the similar triangles in Fig. 19.47$d$,

$$\frac{v_a}{3} = \frac{10}{4} \qquad v_a = 7.5\,\text{m/s} \qquad \frac{\omega_b}{5-4} = \frac{45}{5-3} = \frac{\omega_c}{10-5}$$

$$\omega_b = 22.5\,\text{rad/s} \qquad \omega_c = 113\,\text{rad/s}$$

The kinetic energy at time $a$ has the value

$$T_a = \tfrac{1}{2}mv_a^2 + \tfrac{1}{2}I_0\omega_a^2 = \tfrac{1}{2}(13.5)7.5^2 + \tfrac{1}{2}(0.105)45^2 = 486\,\text{J}$$

At time $b$,

$$T_b = \tfrac{1}{2}mv_b^2 + \tfrac{1}{2}I_0\omega_b^2 = \tfrac{1}{2}(13.5)10^2 + \tfrac{1}{2}(0.105)22.5^2 = 702\,\text{J}$$

At time $c$,

$$T_c = \tfrac{1}{2}mv_c^2 + \tfrac{1}{2}I_0\omega_c^2 = \tfrac{1}{2}(13.5)10^2 + \tfrac{1}{2}(0.105)113^2 = 1{,}350\,\text{J}$$

The maximum value of kinetic energy is seen to occur at time $c$, with the value

$$T_{\max} = T_c = 1{,}350\,\text{J}$$

**19.48** (*a*) A rigid body rotates about a fixed axis. Find the relationship between the work done on the body and the change in angular velocity of the body.

(*b*) What is the form of the result in part (*a*), if the moment effect which acts on the body is constant?

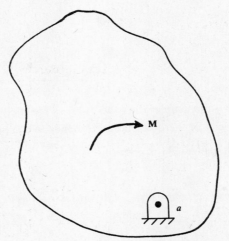

**Fig. 19.48**

❚ (*a*) Figure 19.48 shows a rigid body which moves in plane rotational motion about a fixed axis. $M$ is the *resultant* couple or torque, or moment, about point $a$, which acts on the body. Newton's second law for rotational motion of the body about $a$ is given by

$$M = I_a\alpha = I_a\,\frac{d\omega}{dt} \tag{1}$$

where $\omega$ is the angular velocity of the body. This latter term may be written as

$$\omega = \frac{d\theta}{dt} \qquad dt = \frac{d\theta}{\omega} \tag{2}$$

where $\theta$ is the angular displacement of the body.   The term $dt$ is eliminated between Eqs. (1) and (2), with the result

$$M\,d\theta = I_a\,\omega\,d\omega$$

The above equation is integrated between the two endpoints 1 and 2 of an interval of motion, to obtain

$$\int_1^2 M\,d\theta = \int_1^2 I_a\,\omega\,d\omega = I_a \left.\frac{\omega^2}{2}\right|_1^2 = \tfrac{1}{2}I_a\omega_2^2 - \tfrac{1}{2}I_a\omega_1^2 \tag{3}$$

The term on the left side of Eq. (3) is the work done on the body.   The two terms on the right side give the change in kinetic energy of the body between points 1 and 2.   Thus, *the work done on the rotating body is equal to the change in the kinetic energy of the body.*

Equation (3) may be written in the compact form

$$W_{12} = \Delta T$$

where $W_{12}$ is the work done and $\Delta T$ is the change in kinetic energy.

(*b*)   If the couple or torque, or moment, which acts on the body is constant, Eq. (3) may be written as

$$\int_1^2 M\,d\theta = M \int_1^2 d\theta = \left.M\theta\right|_1^2 = M(\theta_2 - \theta_1) = \tfrac{1}{2}I_a\omega_2^2 - \tfrac{1}{2}I_a\omega_1^2$$

The above equation may be written as

$$W_{12} = M\,\Delta\theta = \Delta T$$

If several moment effects act on the body, the above equation has the form

$$W_{12} = \sum M\,\Delta\theta = \Delta T$$

**19.49**   The disk in Fig. 19.49 is acted upon by a couple $M$ with a constant magnitude of $10\,\text{N}\cdot\text{m}$.   The mass of the disk is $5\,\text{kg}$.

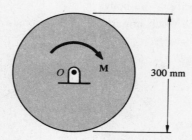

**Fig. 19.49**

(*a*)   Find the work done by the couple when the disk has completed two revolutions, starting from rest.

(*b*)   Find the angular velocity of the disk that corresponds to the position in part (*a*).

▌ (*a*)   The work done by the couple is given by

$$W_{12} = M\,\Delta\theta = 10(2)2\pi = 126\,\text{J}$$

(*b*)   The angular velocity $\omega_2$ of the disk after two revolutions is found from

$$W_{12} = \Delta T = \tfrac{1}{2}I_0\omega_2^2$$

Using   $I_0 = \tfrac{1}{8}md^2$   in the above equation results in

$$126 = \frac{1}{2}\left[\frac{1}{8}\,(5)\left(\frac{300}{1{,}000}\right)^2\right]\omega_2^2 \qquad \omega_2 = 66.9\,\text{rad/s}$$

The solution above is the same as that found in Prob. 18.6 by direct use of Newton's second law.

**19.50**   Figure 19.50 shows a heavy cast-iron flywheel that rotates at a constant speed of $150\,\text{r/min}$.   A friction braking system is used to bring the flywheel to rest.   When the flywheel is braked, it experiences a constant retarding moment of $5\,\text{N}\cdot\text{m}$.   The mass moment of inertia of the flywheel about its center axis is $6.25\,\text{kg}\cdot\text{m}^2$.

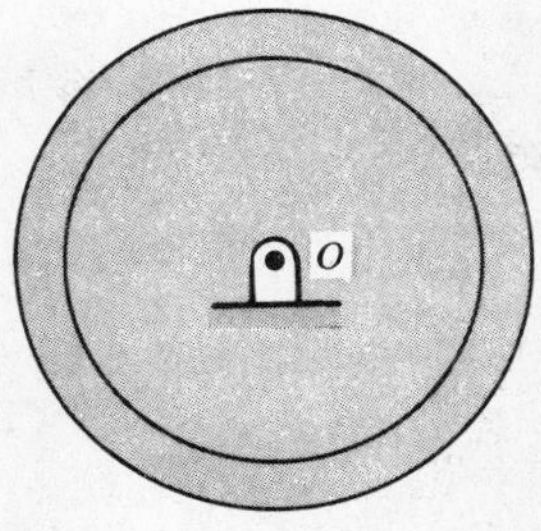

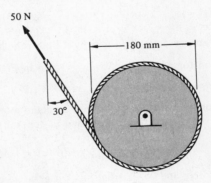

**Fig. 19.50**

Determine how many revolutions the flywheel rotates through before coming to rest.

▌ The work done is equated to the change in kinetic energy, with the result

$$W_{12} = \Delta T \qquad M(\theta_2 - \theta_1) = \tfrac{1}{2} I_0 (\omega_2^2 - \omega_1^2)$$

Using

$$\theta_2 = \theta \qquad \theta_1 = 0 \qquad \omega_2 = 0 \qquad \omega_1 = 150 \, \text{r/min} = 15.7 \, \text{rad/s} \qquad M = -5 \, \text{N} \cdot \text{m}$$

in the above equation results in

$$-5(\theta - 0) = \tfrac{1}{2}(6.25)(0 - 15.7^2) \qquad \theta = 154 \, \text{rad} = 24.5 \, \text{r}$$

The solution above may be compared with that found in Prob. 18.9, where the problem was solved by direct use of Newton's second law.

19.51 The 4-kg disk shown in Fig. 19.51 is initially at rest. A force of 50 N is applied to the thin, inextensible cable wrapped around the disk.

50 N

180 mm

30°

**Fig. 19.51**

(*a*) Find the angular velocity of the disk after this element has completed four revolutions.

(*b*) Do the same as in part (*a*) if, in addition to the cable force, the disk is acted on by a counterclockwise couple of magnitude 2 N · m. The couple acts in the plane of the disk.

▌ (*a*) The work done is equal to the change in kinetic energy, with the form

$$W_{12} = M \, \Delta\theta = \Delta T \qquad M(\theta_2 - \theta_1) = \tfrac{1}{2} I_0 \omega_2^2$$

Using $I_0 = \tfrac{1}{8} m d^2$ in the above equation results in

$$50\left(\frac{90}{1,000}\right) 4(2\pi) = \frac{1}{2}\left[\frac{1}{8}(4)\left(\frac{180}{1,000}\right)^2\right]\omega_2^2 \qquad \omega_2 = 118 \, \text{rad/s}$$

The above result was obtained in Prob. 18.19, part (*c*), by direct use of Newton's second law.

(*b*) Equating the work done to the change in kinetic energy results in

$$W_{12} = \sum M \, \Delta\theta = \Delta T \qquad M(\theta_2 - \theta_1) = \tfrac{1}{2} I_0 \omega_2^2$$

$$\left[50\left(\frac{90}{1,000}\right) - 2\right]4(2\pi) = \frac{1}{2}\left[\frac{1}{8}(4)\left(\frac{180}{1,000}\right)^2\right]\omega_2^2 \qquad \omega_2 = 88.1 \, \text{rad/s} = 14.0 \, \text{r/s}$$

The result above is the same as 88.0 rad/s ≈ 88.1 rad/s, obtained in Prob. 18.19, part (*e*), by the direct use of Newton's second law.

**19.52**   A 6-kg cylinder rests in a trough, as shown in Fig. 19.52.   A couple of constant magnitude   $M = 3.8\,\text{N} \cdot \text{m}$   is applied to the cylinder.   Through how many revolutions has the cylinder rotated at the instant that its angular velocity is 110 rad/s?

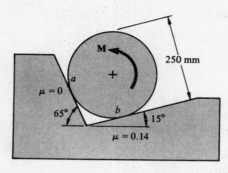

**Fig. 19.52**

▮ This problem was considered earlier in Prob. 18.22.   Using the free-body diagram in Fig. 18.22b, the friction force at $b$ was found to be   $0.14N_b = 0.14(52.9)\,\text{N}$,   acting in a sense to oppose the motion of the cylinder.   The work done on the cylinder results in a change in kinetic energy of this element, with the form

$$W_{12} = \sum M\,\Delta\theta = \Delta T = \tfrac{1}{2}I_0\omega_2^2$$

Using   $I_0 = \tfrac{1}{8}md^2$   in the above equation gives

$$\left[3.8 - 0.14(52.9)\left(\frac{125}{1,000}\right)\right]\theta = \frac{1}{2}\left[\frac{1}{8}(6)\left(\frac{250}{1,000}\right)^2\right]110^2 \qquad \theta = 98.7\,\text{rad} = 15.7\,\text{r}$$

The above result may be checked by using the value   $\alpha = 61.3\,\text{rad/s}^2$,   found in Prob. 18.22, in the equation

$$\omega^2 = \omega_0^2 + 2\alpha(\theta - \theta_0) \qquad 110^2 = 2(61.3)\theta \qquad \theta = 98.7\,\text{rad}$$

**19.53**   Give the form of the work-energy equation for a rigid body which moves in plane motion.

▮ The work-energy equation for a mass particle was given in Prob. 19.25 as

$$W_{\text{NC}} = \Delta T + \Delta V \tag{1}$$

This equation is also valid for the case of a rigid body in plane motion.   The term $W_{\text{NC}}$ is the nonconservative work done on the rigid body, and it may be expressed in general symbolic form as

$$W_{\text{NC}} = \int (F_{\text{NC}} \cos\beta)\,ds + \int M_{\text{NC}}\,d\theta$$

$F_{\text{NC}}$ is the nonconservative force, and $M_{\text{NC}}$ is the nonconservative couple or torque, or moment, acting on the rigid body.   The kinetic energy terms are found by using Eq. (2) or Eq. (4) in Prob. 19.46.   The potential energy terms due to elevation changes of the body are found by using Eq. (3) in Prob. 19.45.
   In Prob. 19.48 the relationship

$$W_{12} = \Delta T = T_2 - T_1$$

was introduced.   $W_{12}$ is the work done on the body and $\Delta T$ is the change in kinetic energy of the body.   The above equation is equivalent to Eq. (1) above if the conservative work is expressed in terms of the change in potential energy.   It is left as an exercise for the reader to verify the results in Probs. 19.49 through 19.52 by direct use of the work-energy method.

**19.54**   The system of Prob. 18.17 is shown in Fig. 19.54.   The weight of the disk is 20 lb.   A thin, inextensible string is wrapped around the outside of the disk.   The disk is initially at rest when a constant force of 10 lb is applied to the string.   Find the angular velocity of the disk, in revolutions per minute, when 42 in of string has been unwound from the disk.   The mass moment of inertia of the disk about its center was found in Prob. 18.17 to be $0.415\,\text{lb} \cdot \text{s}^2 \cdot \text{in}$.

▮ The angular displacement of the disk which corresponds to unwinding 42 in of string is found from

$$s = r\theta \qquad 42 = 4(\theta) \qquad \theta = 10.5\,\text{rad}$$

The string force is a nonconservative force, and the nonconservative moment is

$$M_{\text{NC}} = 10(4) = 40\,\text{in} \cdot \text{lb}$$

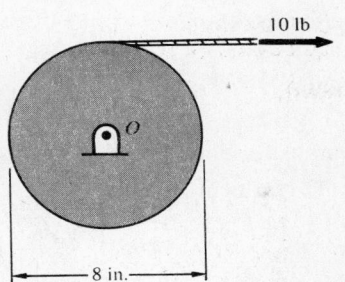

Fig. 19.54

Since this moment is constant, the equation for the nonconservative work has the form

$$W_{NC} = M_{NC}(\theta_2 - \theta_1) = 40(10.5 - 0) = 420 \text{ in} \cdot \text{lb}$$

The required energy terms are

$$T_1 = 0 \qquad T_2 = \tfrac{1}{2}I\omega_2^2 = \tfrac{1}{2}(0.415)\omega_2^2 = 0.208\omega_2^2$$
$$V_1 = 0 \qquad V_2 = 0$$

The work-equation energy appears as

$$W_{NC} = \Delta T + \Delta V = (T_2 - T_1) + (V_2 - V_1)$$

$$420 = (0.208\omega_2^2 - 0) + (0 - 0) \qquad \omega = 44.9 \text{ rad/s} = 44.9\left(\frac{60}{2\pi}\right) = 429 \text{ r/min}$$

This is the result, within computational roundoff error, which was obtained in Prob. 18.17 by direct use of Newton's second law.

19.55   Figure 19.55a shows a slender rod of mass $m$ and length $l$. The rod is released with zero initial velocity from the position $\theta = 0°$.

(a)   Find the angular velocity of the rod as a function of $\theta$.

(b)   Find the kinetic energy of the rod as a function of $\theta$.

(c)   Find the numerical results for parts (a) and (b), if $l = 1,200$ mm and $m = 2.4$ kg.

(d)   Find the maximum value of the angular velocity of the rod.

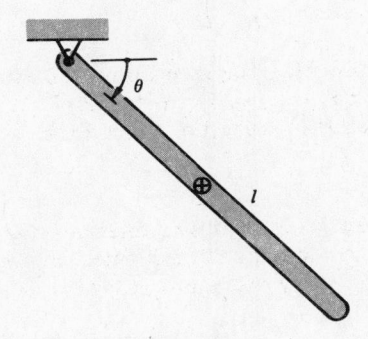

Fig. 19.55a

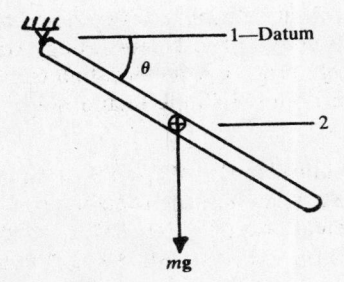

Fig. 19.55b

▌ (a)   Figure 19.55b shows the endpoints of an interval of motion of the rod. Point 1 is the datum, and this point has a fixed location. Point 2 has a variable location, which is a function of angle $\theta$. The forms for $T$ and $V$ are

$$T_1 = 0 \qquad V_1 = 0$$
$$T_2 = T = \tfrac{1}{2}I\omega^2 = \tfrac{1}{2}(\tfrac{1}{3}ml^2)\omega^2 = \tfrac{1}{6}ml^2\omega^2 \qquad (1)$$

$$V_2 = V = -mg\left(\frac{l}{2}\right)\sin\theta \qquad (2)$$

The work-energy equation has the form

$$0 = \Delta T + \Delta V = (T_2 - T_1) + (V_2 - V_1) = (\tfrac{1}{6}ml^2\omega^2 - 0) + \left(-mg\,\frac{l}{2}\sin\theta - 0\right) \qquad \omega = \sqrt{\frac{3g\sin\theta}{l}}$$

$$(3)$$

The above result is seen to be independent of the mass of the rod.

(b) Using Eq. (3) in Eq. (1) results in

$$T = \tfrac{1}{6}ml^2\left(\frac{3g\sin\theta}{l}\right) = \tfrac{1}{2}mgl\sin\theta$$

It may be observed that the above result is equal and opposite to the result in Eq. (2). This is an expected result, since the rod exchanges its initial potential energy for the kinetic energy due to rotation.

(c) Using $l = 1,200$ mm and $m = 2.4$ kg in the above equations results in

$$\omega = \sqrt{\frac{3g\sin\theta}{l}} = \sqrt{\frac{3(9.81)\sin\theta}{1.2}} = 4.95\sqrt{\sin\theta}\ \text{rad/s}$$

$$T = \tfrac{1}{2}mgl\sin\theta = \tfrac{1}{2}(2.4)9.81(1.2)\sin\theta = 14.1\sin\theta\ \text{J}$$

(d) $\omega$ has a maximum when $\theta = 90°$, so that

$$\omega_{max} = 4.95\sqrt{\sin 90°} = 4.95\ \text{rad/s}$$

19.56 The yo-yo of Prob. 18.38 is shown in Fig. 19.56a. This element is modeled as a homogeneous disk of mass $m$ connected to an inextensible string.

(a) If the yo-yo is released from rest, find the general forms for the translational and rotational velocities of this element after the center has moved downward through distance $h$.

(b) Find the numerical values for part (a) if $r = 1.3$ in and $h = 30$ in.

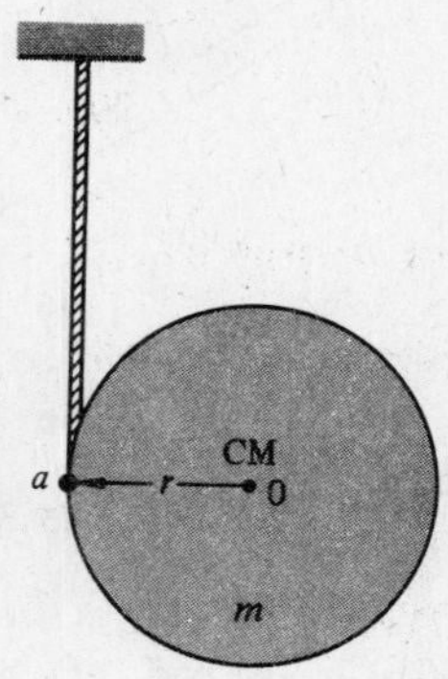

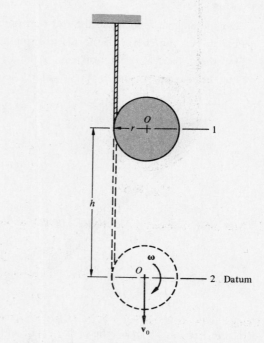

Fig. 19.56a

Fig. 19.56b

▮ (a) The initial and final positions are designated 1 and 2, as shown in Fig. 19.56b, and position 2 is taken as the datum.

The required energy terms in the initial and final positions are

$$T_1 = 0 \qquad T_2 = \tfrac{1}{2}mv_0^2 + \tfrac{1}{2}I_0\omega^2 \qquad V_1 = mgh \qquad V_2 = 0$$

By using $v_0 = r\omega$ and $I_0 = \frac{1}{2}mr^2$, the kinetic energy in the final position may be written as

$$T_2 = \frac{1}{2}mv_0^2 + \frac{1}{2}\left(\frac{1}{2}mr^2\right)\left(\frac{v_0^2}{r^2}\right) = \frac{3}{4}mv_0^2$$

There is no nonconservative work term, and the work-energy equation has the form

$$0 = \Delta T + \Delta V = (T_2 - T_1) + (V_2 - V_1) = \left(\frac{3}{4}mv_0^2 - 0\right) + (0 - mgh) \qquad v_0 = \sqrt{\frac{4}{3}gh}$$

$$\omega = \frac{v_0}{r} = \frac{\sqrt{\frac{4}{3}gh}}{r}$$

(b) For the numerical values of the problem,

$$v_0 = \sqrt{\frac{4}{3}gh} = \sqrt{\frac{4}{3}(32.2)\left(\frac{30}{12}\right)} = 10.4 \text{ ft/s} \qquad \omega = \frac{v_0}{r} = \frac{10.4}{1.3/12} = 96 \text{ rad/s}$$

The above results were obtained in Prob. 18.38 by direct use of Newton's second law. It may again be observed that these results are independent of the mass of the body.

19.57    Figure 19.57a shows the 0.15-lb yo-yo of Prob. 18.40, modeled as a cylindrical disk connected to an inextensible cable. At the instant shown in the figure, the yo-yo has a counterclockwise angular velocity of 30 rad/s. Find the increase in height that the yo-yo will attain before it comes to rest.

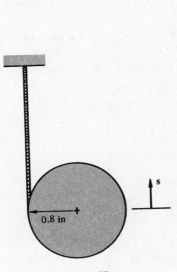

Fig. 19.57a

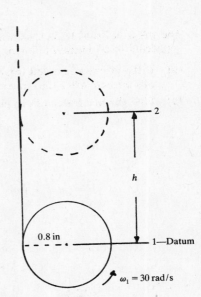

Fig. 19.57b

▌ The mass moment of inertia of the disk about its center axis is given by

$$I_0 = \frac{1}{8}md^2 = \frac{1}{8}\left(\frac{0.15}{386}\right)1.6^2 = 1.24 \times 10^{-4} \text{ lb} \cdot \text{s}^2 \cdot \text{in}$$

The endpoints of the interval of motion are shown in Fig. 19.57b. The kinetic energy at 1 is given by

$$T_1 = \frac{1}{2}mv_1^2 + \frac{1}{2}I_0\omega_1^2$$

Using $v_1 = r\omega_1 = 0.8(30) = 24 \text{ in/s}$ in the above equation results in

$$T_1 = \frac{1}{2}\left(\frac{0.15}{386}\right)24^2 + \frac{1}{2}(1.24 \times 10^{-4})30^2 = 0.168 \text{ in} \cdot \text{lb}$$

The remaining energy terms have the forms

$$V_1 = 0 \qquad T_2 = 0 \qquad V_2 = 0.15h$$

The work-energy equation is written as

$$0 = \Delta T + \Delta V = (T_2 - T_1) + (V_2 - V_1) = (0 - 0.168) + (0.15h - 0) \qquad h = 1.12 \text{ in}$$

The above result was obtained in Prob. 18.40 by direct use of the equations of motion with constant acceleration.

**19.58**  The cylinder in Fig. 19.58a rolls without slipping on the inclined surface.   It is released from rest in position a.

(a)  Find the kinetic energy of the cylinder when it reaches position b, in terms of the translational velocity at b.

(b)  Find the numerical values of the translational and rotational velocities of the cylinder, and the kinetic energy, when the cylinder reaches position b.

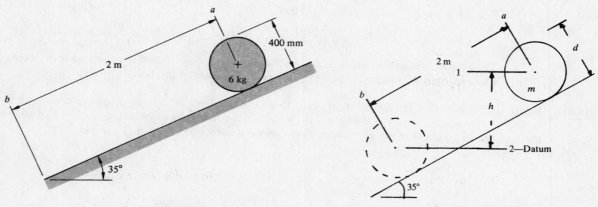

Fig. 19.58a                                                    Fig. 19.58b

▌  (a)  Figure 19.58b shows the endpoints of the interval of motion.   The forms for $T_1$, $V_1$, and $V_2$ are

$$T_1 = 0 \qquad V_1 = mgh \qquad V_2 = 0$$

$T_2$ has the form

$$T_2 = \tfrac{1}{2}mv_2^2 + \tfrac{1}{2}I_0\omega_2^2$$

Using   $I_0 = \tfrac{1}{8}md^2 = \tfrac{1}{2}mr^2$   and   $v_2 = r\omega_2$   in the above equation results in

$$T_2 = \tfrac{1}{2}mv_2^2 + \tfrac{1}{2}(\tfrac{1}{2}mr^2)\left(\frac{v_2}{r}\right)^2 = \tfrac{3}{4}mv_2^2 \qquad (1)$$

(b)  The work-energy equation has the form

$$0 = \Delta T + \Delta V = (T_2 - T_1) + (V_2 - V_1) = (\tfrac{3}{4}mv_2^2 - 0) + (0 - mgh) \qquad (2)$$

$$v_2 = \sqrt{\tfrac{4}{3}gh} = 1.15\sqrt{gh} = 1.15\sqrt{9.81(2\sin 35°)} = 3.86 \text{ m/s}$$

It may be observed that the result for $v_2$ is independent of the mass and diameter of the cylinder. The angular velocity at position b is found to be

$$\omega_2 = \frac{v_2}{r} = \frac{3.86}{200/1,000} = 19.3 \text{ rad/s}$$

The kinetic energy of the cylinder at position b, from Eq. (1), is

$$T_2 = \tfrac{3}{4}mv_2^2$$

From comparison of the above result with Eq. (2),

$$T_2 = mgh = 6(9.81)2\sin 35° = 67.5 \text{ J}$$

It may be observed that the kinetic energy of the cylinder in position b is exactly equal to the value of the potential energy of this element in position a.   It is left as an exercise for the reader to explain why this is an expected result.

**19.59**  (a)  Do the same as in Prob. 19.58 if the cylinder is hollow, as shown in Fig. 19.59.

(b)  Do the same as in Prob. 19.58, if the cylinder is replaced by a homogeneous sphere with the same outside diameter and mass.

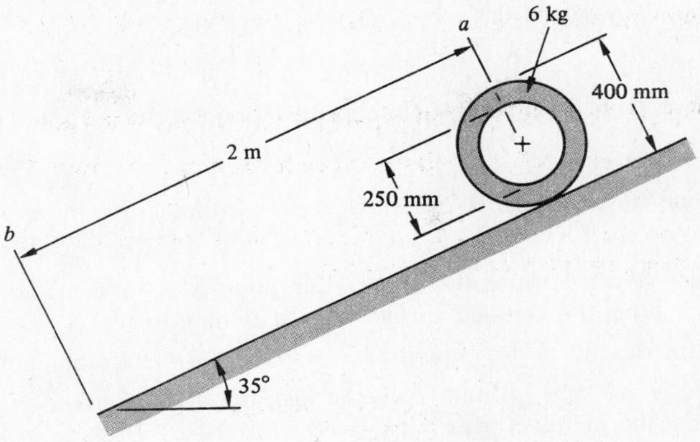

**Fig. 19.59**

**❙ (a)** The endpoints 1 and 2 shown in Fig. 19.58*b* are used. From Case 5 in Table 17.13,

$$I_0 = \tfrac{1}{8}m(d_0^2 + d_i^2) = \tfrac{1}{8}m(400^2 + 250^2)\left(\frac{1}{1,000}\right)^2 = 0.0278m \quad \text{kg} \cdot \text{m}^2$$

The energy terms $T_1$, $V_1$, and $V_2$ are written as

$$T_1 = 0 \qquad V_1 = mgh \qquad V_2 = 0$$

The translational and rotational velocities at point 2 have the relationship

$$v_2 = r\omega_2 \qquad \omega_2 = \frac{v_2}{r} = \frac{v_2}{200/1,000} = 5v_2$$

The kinetic energy at 2 has the form

$$T_2 = \tfrac{1}{2}mv_2^2 + \tfrac{1}{2}I_0\omega_2^2 = \tfrac{1}{2}mv_2^2 + \tfrac{1}{2}(0.0278m)(5v_2)^2 = 0.848mv_2^2$$

It may be observed that the above value for $T_2$ is greater than the value of $T_2$ for the solid cylinder in Prob. 19.58. Using the work-energy equation,

$$0 = \Delta T + \Delta V = (T_2 - T_1) + (V_2 - V_1) = (0.848mv_2^2 - 0) + (0 - mgh)$$
$$v_2 = 1.09\sqrt{gh} = 1.09\sqrt{9.81(2\sin 35°)} = 3.66 \text{ m/s}$$

It may be observed that the above value of $v_2$ is less than the value of $v_2$ in Prob. 19.58. The angular velocity at position *b* is given by

$$\omega_2 = \frac{v_2}{r} = \frac{3.66}{200/1,000} = 18.3 \text{ rad/s}$$

The kinetic energy of the cylinder at position *b* is

$$T_2 = 0.848mv_2^2 = mgh = 6(9.81)(2\sin 35°) = 67.5 \text{ J}$$

The above value is the same as that found in Prob. 19.58, which is an expected result.

**(b)** From Case 1 in Table 17.13,

$$I_0 = \tfrac{1}{10}md^2 = \tfrac{2}{5}mr^2$$

The required terms have the forms

$$T_1 = 0 \qquad V_1 = mgh \qquad v_2 = r\omega_2 \qquad \omega_2 = \frac{v_2}{r}$$

$$T_2 = \tfrac{1}{2}mv_2^2 + \tfrac{1}{2}I_0\omega_2^2 = \tfrac{1}{2}mv_2^2 + \tfrac{1}{2}(\tfrac{2}{5}mr^2)\left(\frac{v_2}{r}\right)^2 = \tfrac{7}{10}mv_2^2$$

The above value for $T_2$ is less than the value of $T_2$ in Prob. 19.58.
Using the work-energy equation,

$$0 = \Delta T + \Delta V = (T_2 - T_1) + (V_2 - V_1) = (\tfrac{7}{10}mv_2^2 - 0) + (0 - mgh)$$
$$v_2 = 1.20\sqrt{gh} = 1.20\sqrt{9.81(2\sin 35°)} = 4.03 \text{ m/s}$$

The above value for $v_2$ is greater than the value of $v_2$ for the solid cylinder in Prob. 19.58.

$$\omega_2 = \frac{v_2}{r} = \frac{4.03}{200/1,000} = 20.2 \text{ rad/s}$$

The kinetic energy of the sphere at position $b$ is

$$T_2 = \tfrac{7}{10}mv_2^2 = mgh = 6(9.81)(2\sin 35°) = 67.5 \text{ J}$$

The above value for $T_2$ is the same as that found in part (a), and in Prob. 19.58, which is an expected result.

**19.60** A homogeneous disk of diameter $d$ and mass $m$ rolls without slipping up an inclined surface, as shown in Fig. 19.60a.

(a) If the angular velocity of the disk at a certain point is $\omega_1$, find the additional distance $l$ through which the disk will roll along the inclined surface before coming to rest.

(b) Find the numerical value for part (a) if $\beta = 26°$, $d = 200$ mm, and $\omega_1 = 80$ r/min.

(c) Do the same as in parts (a) and (b) if the disk has a set of holes drilled on a concentric circle about the center, so that the radius of gyration $k_0$ is equal to $0.4d$. Both holes lie on the same diameter of the disk.

(d) For what theoretical symmetrical mass distribution of the disk would the distance $l$ found in part (a) be maximum?

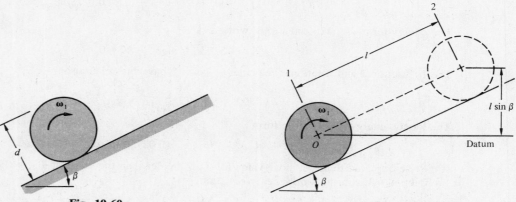

**Fig. 19.60a**

**Fig. 19.60b**

❚ (a) The initial and final locations are designated 1 and 2, as shown in Fig. 19.60b. The required energy terms are

$$T_1 = \tfrac{1}{2}mv_1^2 + \tfrac{1}{2}I_0\omega_1^2 \qquad T_2 = 0 \qquad V_1 = 0 \qquad V_2 = mgl\sin\beta$$

There is no nonconservative work done, and the work-energy equation has the form

$$0 = \Delta T + \Delta V = (T_2 - T_1) + (V_2 - V_1) = [0 - (\tfrac{1}{2}mv_1^2 + \tfrac{1}{2}I_0\omega_1^2)] + (mgl\sin\beta - 0) \qquad (1)$$

With a view toward the solution of part (c), the mass moment of inertia will be expressed in terms of the radius of gyration, with the form

$$I_0 = k_0^2 m$$

Using $v_1 = r\omega_1$, Eq. (1) appears as

$$0 = -[\tfrac{1}{2}m(r\omega_1)^2 + \tfrac{1}{2}(k_0^2 m)\omega_1^2] + mgl\sin\beta \qquad l = \frac{r^2 + k_0^2}{2g\sin\beta}\,\omega_1^2 \qquad (2)$$

It is interesting to observe that this result is independent of the value of the mass of the disk.

(b) The radius of gyration of a solid homogeneous disk is found as

$$k_0 = \sqrt{\frac{I_0}{m}} = \sqrt{\frac{\tfrac{1}{2}mr^2}{m}} = \frac{r}{\sqrt 2} = 0.707r$$

Using

$$\omega_1 = 80\left(\frac{2\pi}{60}\right) = 8.38 \text{ rad/s} \qquad \text{and} \qquad r = \tfrac{1}{2}(200) = 100 \text{ mm}$$

in Eq. (2) results in

$$l = \frac{r^2 + (0.707r)^2}{2g\sin\beta}\,\omega_1^2 = \frac{(1 + 0.707^2)(100)^2}{2(9.81)\sin 26°}(8.38)^2\left(\frac{1 \text{ m}}{1,000 \text{ mm}}\right)^2 = 0.122 \text{ m}$$

(*c*) For the case where the radius of gyration of the disk is $0.4d = 0.8r$, the distance traveled by the disk is

$$l = \frac{(r^2 + k_0^2)}{2g \sin \beta}\, \omega_1^2 = \frac{(1 + 0.8^2)(100)^2}{2(9.81) \sin 26°}\, (8.38)^2 \left(\frac{1}{1,000}\right)^2 = 0.134 \text{ m}$$

(*d*) From consideration of Eq. (2), it may be seen that the distance $l$ will be maximum when $k_0$ has its maximum value. The theoretical limiting maximum value of $k_0$ corresponds to the case where all the mass of the disk is assumed to be concentrated along the rim. For this case, $k_0 = r$, and the theoretical maximum value of $l$ is

$$l_{\max} = \frac{(1 + 1)(100)^2}{2(9.81) \sin 26°}\, (8.38)^2 \left(\frac{1}{1,000}\right)^2 = 0.163 \text{ m}$$

**19.61**  A homogeneous cylinder is released from rest at position *a* in Fig. 19.61*a*. It rolls without sliding until it reaches position *b*. Length *bc* of the inclined plane is contaminated with lubricant, and, for the purpose of this problem, the coefficient of friction of this surface may be assumed to be zero.

(*a*)  Find the angular velocity, and the velocity of the center, when the cylinder reaches position *b*.

(*b*)  Do the same as in part (*a*), when the cylinder reaches position *c*.

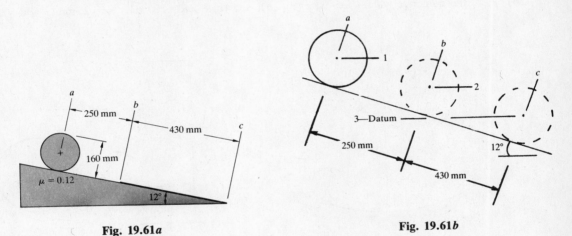

Fig. 19.61*a*                      Fig. 19.61*b*

▮ (*a*)  Figure 19.61*b* shows three points in the interval of motion. Using the result in Prob. 19.58, part (*b*),

$$v_2 = 1.15\sqrt{gh} = 1.15 \sqrt{(9.81)\left(\frac{250}{1,000}\right) \sin 12°} = 0.821 \text{ m/s}$$

$$\omega_2 = \frac{v_2}{r} = \frac{0.821}{80/1,000} = 10.3 \text{ rad/s}$$

The above results—0.821 m/s ≈ 0.824 m/s for $v_2$ and for $\omega_2$—are the same as those obtained in Prob. 18.52 by direct use of Newton's second law.

(*b*)  From Prob. 19.58,

$$T_2 = \tfrac{3}{4} m v_2^2$$

The potential energy at 2 is given by

$$V_2 = mg\left(\frac{430}{1,000}\right) \sin 12°$$

The kinetic energy at 3 has the form

$$T_3 = \tfrac{1}{2} m v_3^2 + \tfrac{1}{2} I_0 \omega_3^2$$

Using $I_0 = \tfrac{1}{2} mr^2$, $\omega_3 = \omega_2$, and $\omega_2 = v_2/r$ in the above equation results in

$$T_3 = \tfrac{1}{2} m v_3^2 + \tfrac{1}{2}\left(\tfrac{1}{2} mr^2\right)\left(\frac{v_2^2}{r}\right) = \tfrac{1}{2} m(v_3^2 + \tfrac{1}{2} v_2^2)$$

The potential energy at 3 is given by $V_3 = 0$. The work-energy equation between points 2 and 3 has the form

$$0 = \Delta T + \Delta V = (T_3 - T_2) + (V_3 - V_2) = [\tfrac{1}{2}m(v_3^2 + \tfrac{1}{2}v_2^2) - \tfrac{3}{4}mv_2^2] + \left[0 - mg\left(\frac{430}{1,000}\right)\sin 12°\right]$$

$$= \tfrac{1}{2}v_3^2 - \tfrac{1}{2}v_2^2 - g\left(\frac{430}{1,000}\right)\sin 12° = \tfrac{1}{2}v_3^2 - \tfrac{1}{2}(0.821)^2 - 9.81\left(\frac{430}{1,000}\right)\sin 12°$$

$$v_3 = 1.56 \text{ m/s} \qquad \omega_3 = \omega_2 = 10.3 \text{ rad/s}$$

The result for $v_3$ is that obtained in Prob. 18.52 by direct use of Newton's second law.

**19.62**  Figure 19.62*a* shows a cylinder of mass $m$ and a track with the shape of a circular arc.  The radius of the cylinder is $r$, and the radius of the track is $R$.  The cylinder is released from rest at the position $\beta$.

(*a*)  Find the angular velocity of the cylinder and the velocity of the center of this element, at position *b*, if the cylinder is assumed to roll without sliding.

(*b*)  Do the same as in part (*a*), if the cylinder is assumed to slide without rolling.

(*c*)  Find the normal acceleration of the center of the cylinder when it passes through position *b*, for the conditions of parts (*a*) and (*b*).

(*d*)  Find the numerical values for parts (*a*), (*b*), and (*c*) if  $R = 200$ mm,  $r = 25$ mm,  and  $\beta = 35°$.

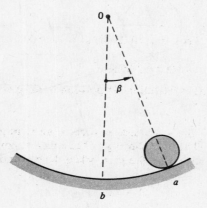

**Fig. 19.62*a***

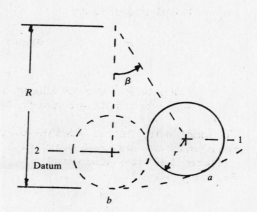

**Fig. 19.62*b***

▌ (*a*)  Figure 19.62*b* shows the endpoints of the interval of motion.  For the case of rolling without sliding,

$$T_1 = 0 \qquad T_2 = \tfrac{1}{2}mv_2^2 + \tfrac{1}{2}I_0\omega_2^2$$
$$V_1 = mg(R - r)(1 - \cos\beta) \qquad V_2 = 0$$

Using  $I_0 = \tfrac{1}{8}md^2 = \tfrac{1}{2}mr^2$  and  $\omega_2 = v_2/r$  in the above equation results in

$$T_2 = \tfrac{3}{4}mv_2^2$$

The work-energy equation has the form

$$0 = \Delta T + \Delta V = (T_2 - T_1) + (V_2 - V_1) = (\tfrac{3}{4}mv_2^2 - 0) + [0 - mg(R - r)(1 - \cos\beta)]$$

The translational and rotational velocities of the cylinder at position *b* are

$$v_2 = \sqrt{\tfrac{4}{3}g(R - r)(1 - \cos\beta)} \qquad \omega_2 = \frac{v_2}{r}$$

(*b*)  For the case of sliding without rolling, the terms $T$, $V_1$, and $V_2$ are the same as in part (*a*).  The term $T_2$ has the form

$$T_2 = \tfrac{1}{2}mv_2^2$$

The work-energy equation is written as

$$0 = \Delta T + \Delta V = (T_2 - T_1) + (V_2 - V_1) = (\tfrac{1}{2}mv_2^2 - 0) + [0 - mg(R - r)(1 - \cos\beta)]$$

The velocities at $b$ are

$$v_2 = \sqrt{2g(R-r)(1-\cos\beta)} \qquad \omega_2 = 0$$

It may be seen that $v_2$ for the case of sliding is greater than $v_2$ for the case of rolling.

(c) The normal acceleration of the center of the cylinder at position $b$ has the form

$$a_n = \frac{v^2}{\rho} = \frac{v_2^2}{R-r}$$

For rolling without sliding,

$$a_n = \frac{\frac{4}{3}g(R-r)(1-\cos\beta)}{R-r} = \frac{4}{3}g(1-\cos\beta)$$

For sliding without rolling,

$$a_n = \frac{2g(R-r)(1-\cos\beta)}{R-r} = 2g(1-\cos\beta)$$

(d) For the case of rolling,

$$v_2 = \sqrt{\frac{4}{3}(9.81)\left(\frac{200-25}{1,000}\right)(1-\cos35°)} = 0.643 \text{ m/s} \qquad \omega_2 = \frac{v_2}{r} = \frac{0.643}{25/1,000} = 25.7 \text{ rad/s}$$

$$a_n = \frac{4}{3}(9.81)(1-\cos35°) = 2.37 \text{ m/s}^2$$

For the case of sliding,

$$v_2 = \sqrt{2(9.81)\left(\frac{200-25}{1,000}\right)(1-\cos35°)} = 0.788 \text{ m/s} \qquad \omega_2 = 0$$

$$a_n = 2(9.81)(1-\cos35°) = 3.55 \text{ m/s}^2$$

It may be seen that the velocity, and the normal acceleration, at position $b$ are greater for the case where the cylinder is assumed to slide without rolling.

**19.63** The spring in Fig. 19.63 is connected to a 12-lb cylindrical roller which is in equilibrium in position 1. The plane surface is sufficiently rough so that the roller will roll without slipping. The roller is moved to position 2 and released with zero initial velocity. Find the angular velocity of the roller when the center of this element is 3 in from position 2.

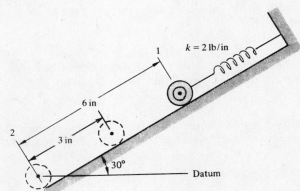

Fig. 19.63

▌ The spring is in equilibrium in position 1, with a spring force $F_s$ equal to the component of the weight force along the inclined surface, given by

$$F_s = mg\sin30° = 12\sin30° = 6\text{ lb}$$

The deflection of the spring in position 1 is found from

$$\delta_1 = \frac{F_s}{k} = \frac{6\text{ lb}}{2\text{ lb/in}} = 3\text{ in}$$

The potential and kinetic energy terms at position 2 are

$$T_2 = 0 \qquad V_2 = \tfrac{1}{2}k\delta_2^2 = \tfrac{1}{2}(2)(6+3)^2 = 81\text{ in}\cdot\text{lb}$$

The roller is released from position 2. At position 3, which is 3 in from position 2, the energy terms are

$$V_3 = mg(3)\sin 30° + \tfrac{1}{2}k(9-3)^2 = 12(3)\sin 30° + \tfrac{1}{2}(2)6^2 = 54.0 \text{ in} \cdot \text{lb}$$
$$T_3 = \tfrac{1}{2}mv_3^2 + \tfrac{1}{2}I_0\omega_3^2$$

Using

$$v_3 = r\omega_3 \qquad \omega_3 = \frac{v_3}{r}$$

$$I_0 = \tfrac{1}{8}md^2 = \tfrac{1}{2}mr^2 \qquad T_3 = \tfrac{1}{2}mv_3^2 + \tfrac{1}{2}(\tfrac{1}{2}mr^2)\left(\frac{v_3}{r}\right)^2 = \tfrac{3}{4}mv_3^2$$

The work-energy equation between points 2 and 3 has the form

$$0 = \Delta T + \Delta V = (T_3 - T_2) + (V_3 - V_2) = \left[\frac{3}{4}\left(\frac{12}{386}\right)v_3^2 - 0\right] + (54.0 - 81)$$

$$v_3 = 34.0 \text{ in/s} \qquad \omega_3 = \frac{v_3}{r} = \frac{34.0}{1} = 34.0 \text{ rad/s}$$

## 19.4  THE WORK-ENERGY METHOD FOR CONNECTED BODIES, THE WORK-ENERGY METHOD USED TO FIND THE NORMAL ACCELERATION OF A PARTICLE

19.64   (a)   How may the work-energy method be used to solve for the motion of connected bodies?

(b)   Give an example of a connected-body problem for which the work-energy method is not a convenient form of solution.

▌ (a)   The case of motion of connected bodies was first presented in Prob. 15.40. The technique of solution was shown to consist of drawing a free-body diagram of each mass element and writing Newton's second law for each of these elements. In the process of doing this, the cable or link forces between the bodies were treated as unknown quantities. The resulting system of equations was then solved simultaneously to obtain the accelerations and the cable or link forces.

The motion of certain types of systems of connected bodies also may be obtained by using the work-energy method. The solution technique consists of expressing the potential energies of all the bodies in terms of position coordinates measured from a datum. The kinetic energy of the system is then expressed in terms of the velocities of the mass elements. If both angular and translational motions occur in the system, the velocities may be related by equations of the form

$$v = r\omega$$

If the nonconservative work is zero, the problem may be solved directly by applying the work-energy equation in the form

$$0 = \Delta T + \Delta V$$

If there is nonconservative work and this quantity may be expressed in a form *which does not include the unknown cable or link forces*, then the work-energy equation has the form

$$W_{NC} = \Delta T + \Delta V$$

(b)   An example of a problem of motion of connected bodies which cannot be solved conveniently by using the work-energy method is shown in Fig. 19.64. This system was considered in Prob. 15.51. As the

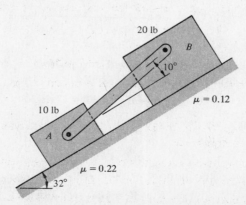

Fig. 19.64

pair of connected blocks slides down the plane, nonconservative work is done as the friction forces on the blocks act through a distance. These friction forces are functions of the normal forces acting on the blocks. These normal forces are functions of the *unknown force* in the link which connects the two blocks. Since this link force is not known at the outset of the problem, this problem cannot be solved directly by using the work-energy method.

This problem can be solved by the energy method, however, if a free-body diagram of each block is drawn, as was done in Fig. 15.51, and the work-energy equation written for *each* mass element. This operation yields two equations in two unknowns. One unknown is the link force, and the other is the particular position coordinate or velocity which is to be found. For a case such as this, it may be more convenient to solve the problem by the direct use of Newton's second law, as was done in Prob. 15.51.

**19.65**  The system of Prob. 15.41 is shown in Fig. 19.65a. If the blocks are released from rest in the position shown, find the velocity of each block when these elements have moved through a displacement of 1.5 m. The cable is inextensible, the pulley pin is assumed to be frictionless, and the mass moment of inertia of the pulley may be neglected.

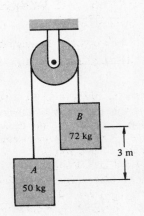

**Fig. 19.65a**

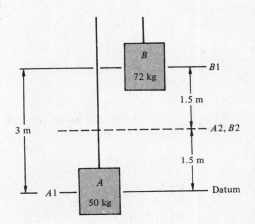

**Fig. 19.65b**

❚  The position chosen for the datum for potential energy is shown in Fig. 19.65b. The required energy terms are

$$T_{A1} = 0 \qquad T_{B1} = 0 \qquad T_{A2} = \tfrac{1}{2} m_A v_A^2 \qquad T_{B2} = \tfrac{1}{2} m_B v_B^2$$
$$V_{A1} = 0 \qquad V_{B1} = 72(9.81)3 = 2{,}120 \text{ N} \cdot \text{m} = 2{,}120 \text{ J}$$
$$V_{A2} = 50(9.81)1.5 = 736 \text{ N} \cdot \text{m} = 736 \text{ J} \qquad V_{B2} = 72(9.81)1.5 = 1{,}060 \text{ N} \cdot \text{m} = 1{,}060 \text{ J}$$

Since the cable is inextensible,

$$v_{A2} = v_{B2} = v_2$$

and

$$T_2 = T_{A2} + T_{B2} = \tfrac{1}{2}(m_A + m_B)v_2^2 = \tfrac{1}{2}(50 + 72)v_2^2 = 61v_2^2$$

There is no nonconservative work done, and the work-energy equation has the form

$$0 = \Delta T + \Delta V = (T_2 - T_1) + (V_2 - V_1) = [(T_{A2} + T_{B2}) - (T_{A1} + T_{B1})] + [(V_{A2} + V_{B2}) - (V_{A1} + V_{B1})]$$
$$= [61v_2^2 - (0 + 0)] + [(736 + 1{,}060) - (0 + 2{,}120)]$$
$$v_2 = 2.30 \text{ m/s}$$

This is the value  $2.31 \text{ m/s} \approx 2.30 \text{ m/s}$  which was obtained in Prob. 15.41 by direct use of Newton's second law.

**19.66**  At the instant shown in Fig. 19.66a, block A has an upward velocity of 1.85 m/s. The cable is inextensible, and all friction effects may be neglected.

(a)  Find the additional upward motion of block A before this element comes to rest.

(b)  Do the same as in part (a), if a mass of 1 kg is added to block B.

(c)  Do the same as in part (a), if a mass of 1 kg is added to block A.

**Fig. 19.66a**

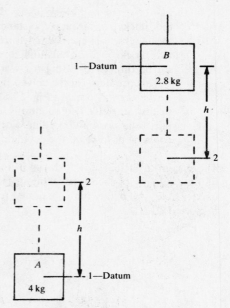

**Fig. 19.66b**

(a) Figure 19.66b shows the datum for each block and the endpoints of the interval of motion. The required energy terms are

$$T_1 = \tfrac{1}{2}(4 + 2.8)1.85^2 \qquad T_2 = 0 \qquad V_1 = 0$$
$$V_2 = 4(9.81)h - 2.8(9.81)h = 1.2(9.81)h$$

The work-energy equation has the form

$$0 = \Delta T + \Delta V = (T_2 - T_1) + (V_2 - V_1) = [0 - \tfrac{1}{2}(6.8)1.85^2] + [1.2(9.81)h - 0]$$
$$h = 0.988 \text{ m} = 988 \text{ mm}$$

(b) Figure 19.66b is used, with 1 kg of mass added to block B, so that

$$m_B = 2.8 + 1 = 3.8 \text{ kg}$$

The energy terms have the forms

$$T_1 = \tfrac{1}{2}(4 + 3.8)1.85^2 \qquad T_2 = 0 \qquad V_1 = 0$$
$$V_2 = 4(9.81)h - 3.8(9.81)h = 0.2(9.81)h$$

The work-energy equation is written as

$$0 = \Delta T + \Delta V = (T_2 - T_1) + (V_2 - V_1) = [0 - \tfrac{1}{2}(7.8)1.85^2] + [0.2(9.81)h - 0]$$
$$h = 6.80 \text{ m}$$

(c) Using Fig. 19.66b, with 1 kg of mass added to block A,

$$m_A = 4 + 1 = 5 \text{ kg} \qquad T_1 = \tfrac{1}{2}(5 + 2.8)1.85^2 \qquad V_1 = 0$$
$$T_2 = 0 \qquad V_2 = 5(9.81)h - 2.8(9.81)h = 2.2(9.81)h$$
$$0 = \Delta T + \Delta V = (T_2 - T_1) + (V_2 - V_1) = [0 - \tfrac{1}{2}(7.8)1.85^2] + [2.2(9.81)h - 0]$$
$$h = 0.618 \text{ m} = 618 \text{ mm}$$

A comparison may be made between the above solutions and those in Probs. 15.43 and 15.44. In these latter two problems, the *times* for the blocks to come to rest were found. In the present problem, the distances moved by the blocks before coming to rest are found. The times to come to rest may *not* be found by using the work-energy method.

It is left as an exercise for the reader to use the acceleration values in Probs. 15.43 and 15.44 to verify the results found above for the motion of the blocks.

**19.67** The system in Fig. 19.67a is released from rest. $W = 100 \text{ lb}$ and the coefficient of friction between the block and the plane is 0.3.

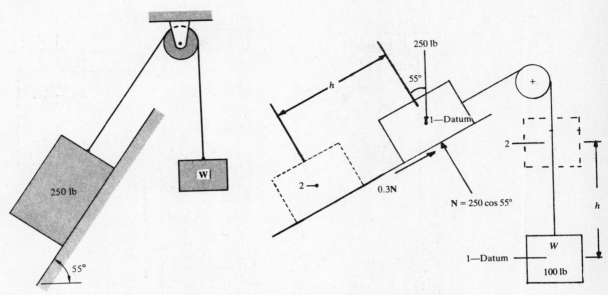

**Fig. 19.67a**                                    **Fig. 19.67b**

(*a*)  Find the distance through which the blocks must move to give these elements a velocity of 4 ft/s.

(*b*)  Do the same as in part (*a*), if  $W = 300$ lb.

▌ (*a*)  Figure 19.67*b* shows the endpoints of the interval of motion.  The 100-lb block is assumed to move upward.  The energy terms at the endpoints of the interval, using  $v_2 = 4$ ft/s,  are

$$T_1 = 0 \qquad T_2 = \frac{1}{2}\left(\frac{250 + 100}{32.2}\right)4^2 \qquad V_1 = 0 \qquad V_2 = 100h - 250(h \sin 55°)$$

The work done on the 250-lb block is given by

$$W_{\text{NC}} = -0.3(250 \cos 55°)h$$

The work-energy equation has the form

$$W_{\text{NC}} = \Delta T + \Delta V = (T_2 - T_1) + (V_2 - V_1)$$
$$-0.3(250 \cos 55°)h = \left[\frac{1}{2}\left(\frac{350}{32.2}\right)4^2 - 0\right] + [100h - 250(h \sin 55°) - 0] \qquad h = 1.41 \text{ ft}$$

Since *h* is a positive quantity, the initial assumption of upward motion of the 100-lb block is correct.

(*b*)  The 300-lb block is assumed to move downward.  Figure 19.67*b* is used, with the sense of the friction force 0.3*N* changed.  The energy terms at 1 and 2, using  $v_2 = 4$ ft/s,  are

$$T_1 = 0 \qquad T_2 = \frac{1}{2}\left(\frac{250 + 300}{32.2}\right)4^2 \qquad V_1 = 0 \qquad V_2 = -300h + 250(h \sin 55°)$$

The work done is

$$W_{\text{NC}} = -0.3(250 \cos 55°)h$$

Using the work-energy equation,

$$W_{\text{NC}} = \Delta T + \Delta V = (T_2 - T_1) + (V_2 - V_1)$$
$$-0.3(250 \cos 55°)h = \left[\frac{1}{2}\left(\frac{550}{32.2}\right)4^2 - 0\right] + [-300h + 250(h \sin 55°) - 0] \qquad h = 2.62 \text{ ft}$$

The results above may be verified by using the values of the acceleration of the blocks, found in Prob. 15.45, together with the equations for motion with constant acceleration.

19.68    Find the velocity of the blocks shown in Fig. 19.68*a* after these elements have moved through 2 m.  The blocks are initially at rest, and the mass of block *B* is 160 kg.

▌ Figure 19.68*b* shows the initial position 1 of the blocks.  Position 2 is where the 1,500-N block has moved 2 m down the plane.  The energy terms for the system of blocks are

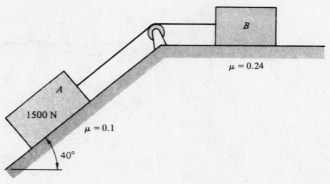

**Fig. 19.68a**

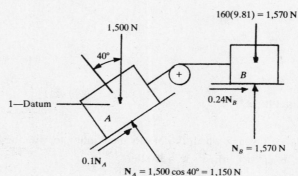

**Fig. 19.68b**

$$T_1 = 0 \qquad T_2 = \frac{1}{2}\left(\frac{1{,}500}{9.81} + 160\right)v_2^2 \qquad V_1 = 0 \qquad V_2 = -1{,}500(2\sin 40°)$$

Using the work-energy equation,

$$W_{NC} = \Delta T + \Delta V = (T_2 - T_1) + (V_2 - V_1)$$

$$-[0.1(1{,}150) + 0.24(1{,}570)]2 = \left[\frac{1}{2}\left(\frac{1{,}500}{9.81} + 160\right)v_2^2 - 0\right] + [-1{,}500(2\sin 40°) - 0] \qquad v_2 = 2.46 \text{ m/s}$$

The above result is the same as that found in Prob. 15.46 by direct use of Newton's second law.

It is left as an exercise for the reader to use the result $a = 1.51 \text{ m/s}^2$, found in Prob. 15.46 as the acceleration of the blocks, to verify the above result for $v_2$.

**19.69** For what range of values of weight of block $A$ in Fig. 19.69a will the velocity of the blocks be less than, or equal to, 2 ft/s after the blocks have moved through a distance of 4 ft, starting from rest.

❚ It is first assumed that block $B$ moves to the right. Figure 19.69b shows this condition, and points 1 and 2 are the endpoints of the interval of motion. $v_2$ is chosen to be equal to 2 ft/s. The required energy terms are

$$T_1 = 0 \qquad T_2 = \frac{1}{2}\left(\frac{W_A + 300}{32.2}\right)2^2 \qquad V_1 = 0 \qquad V_2 = W_A(4\sin 55°)$$

The work done on the two blocks is given by

$$W_{NC} = 100(4) - 36(4) - 0.2(W_A \cos 55°)4$$

The work-energy equation has the form

$$W_{NC} = \Delta T + \Delta V = (T_2 - T_1) + (V_2 - V_1)$$

$$100(4) - 36(4) - 0.2(W_A \cos 55°)4 = \left[\frac{1}{2}\left(\frac{W_A + 300}{32.2}\right)2^2 - 0\right] + [W_A(4\sin 55°) - 0] \qquad W_A = 62.5 \text{ lb}$$

It is now assumed that block $B$ moves to left. Figure 19.69b, with opposite senses of the friction forces, is used. $v_2 = 2$ ft/s, and the energy terms have the forms

$$T_1 = 0 \qquad T_2 = \frac{1}{2}\left(\frac{W_A + 300}{32.2}\right)2^2 \qquad V_1 = 0 \qquad V_2 = -W_A(4\sin 55°)$$

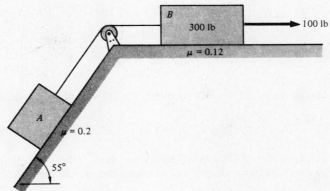

Fig. 19.69a

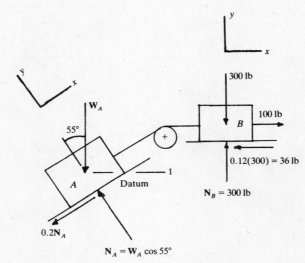

Fig. 19.69b

The work done is given by

$$W_{NC} = -100(4) - 0.12(300)4 - 0.2(W_A \cos 55°)4$$

The work-energy equation is written as

$$W_{NC} = \Delta T + \Delta V = (T_2 - T_1) + (V_2 - V_1)$$

$$-100(4) - 0.12(300)4 - 0.2(W_A \cos 55°)4 = \left[ \frac{1}{2} \left( \frac{W_A + 300}{32.2} \right) 2^2 - 0 \right] + [-W_A(4 \sin 55°) - 0] \qquad W_A = 204 \text{ lb}$$

From the above results, to satisfy the velocity requirements of the problem,

$$W_A \geq 62.5 \text{ lb} \qquad W_A \leq 204 \text{ lb}$$

It may be observed that there is a range of values of $W_A$ for which the blocks will not move at all, because of the friction forces.

For impending motion of block $B$ to the right,

Block $A$: $\qquad\qquad\qquad \sum F_x = 0 \qquad T - W_A \sin 55° - 0.2W_A \cos 55° = 0$

Block $B$: $\qquad\qquad\qquad\qquad \sum F_x = 0 \qquad 100 - 36 - T = 0$

$T$ is eliminated between the above equations, to obtain

$$W_A = 68.5 \text{ lb}$$

For impending motion of block $B$ to the left,

Block $A$: $\qquad\qquad\qquad \sum F_x = 0 \qquad -W_A \sin 55° + T + 0.2W_A \cos 55° = 0$

Block $B$: $\qquad\qquad\qquad\qquad \sum F_x = 0 \qquad -T + 100 + 36 = 0$

$T$ is eliminated, and

$$W_A = 193 \text{ lb}$$

The necessary condition for the blocks to move is

$$W_A \leq 68.5 \text{ lb} \qquad W_A \geq 193 \text{ lb}$$

The range of values of weight of block $A$ for which the velocity of the blocks will be less than, or equal to, 2 ft/s is then stated as

$$62.5 \text{ lb} \leq W_A \leq 68.5 \text{ lb} \qquad \text{or} \qquad 193 \text{ lb} \leq W_A \leq 204 \text{ lb}$$

**19.70** The system of blocks in Fig. 19.70a is released from rest. The coefficient friction on all sliding surfaces is 0.08. Find the velocity of block $B$ after this element has moved through 20 in.

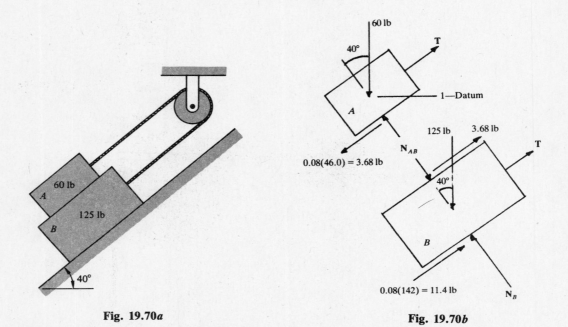

Fig. 19.70a

Fig. 19.70b

▌ Figure 19.70b shows the free-body diagrams of the two blocks. Block $B$ is assumed to move down the plane, so that the friction forces have the senses shown. The normal forces acting on the blocks have the values

$$N_{AB} = 60 \cos 40° = 46.0 \text{ lb} \qquad N_B = N_{AB} + 125 \cos 40° = 142 \text{ lb}$$

The initial position is designated 1, and this point is also the datum for potential energy. The energy terms are

$$T_1 = 0 \qquad T_2 = \frac{1}{2}\left(\frac{60+125}{386}\right)v_2^2 \qquad V_1 = 0 \qquad V_2 = (60-125)20 \sin 40°$$

The work-energy equation is written as

$$W_{NC} = \Delta T + \Delta V = (T_2 - T_1) + (V_2 - V_1)$$

$$-3.68(20) - (3.68 + 11.4)20 = \left[\frac{1}{2}\left(\frac{60+125}{386}\right)v_2^2 - 0\right] + [(60-125)20 \sin 40° - 0] \qquad v_2 = 43.8 \text{ in/s} = 3.65 \text{ ft/s}$$

Since $v_2$ is a positive number, the original assumption of the sense of the motion of the blocks was correct. Had this assumption been incorrect, $v_2$ would have been found as an imaginary number. It is left as an exercise for the reader to use the result 4.02 ft/s² for the acceleration of the blocks, found in Prob. 15.48, to verify the above value for $v_2$.

**19.71** (a) The system shown in Fig. 19.71a is initially at rest. A second block of mass 3.5 kg is attached to block $B$. Find the angular velocity of the pulley when block $A$ has moved through a distance of 1.75 m. Neglect the mass of the pulley.

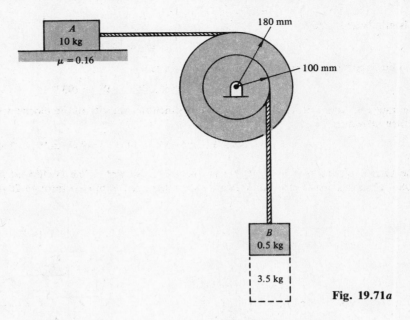

**Fig. 19.71*a***

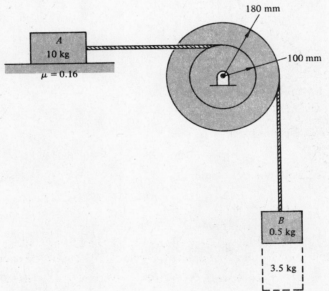

**Fig. 19.71*b***

(*b*)  Do the same as in part (*a*), if the system is rearranged as shown in Fig. 19.71*b*.

▌ (*a*)  Figure 19.71*c* shows the normal and friction forces acting on block *A*, and the endpoints of the interval of motion.  The translational and angular motions are related by

$$s = r\theta \qquad s_A = 180\theta \tag{1}$$
$$s_B = 100\theta \tag{2}$$

Dividing Eq. (1) by Eq. (2) results in

$$\frac{s_A}{s_B} = \frac{180}{100} = 1.8 \qquad s_A = 1.75 \text{ m} \qquad s_B = \frac{s_A}{1.8} = \frac{1.75}{1.8} \qquad \frac{v_A}{v_B} = 1.8 \qquad v_A = 1.8 v_B$$

The energy terms at the endpoints of the interval of motion are

$$T_1 = 0 \qquad T_2 = \tfrac{1}{2} m_A v_A^2 + \tfrac{1}{2} m_B v_B^2 = \tfrac{1}{2}(10)(1.8 v_B)^2 + \tfrac{1}{2}(4)v_B^2$$

$$V_1 = 0 \qquad V_2 = -4(9.81)\left(\frac{1.75}{1.8}\right)$$

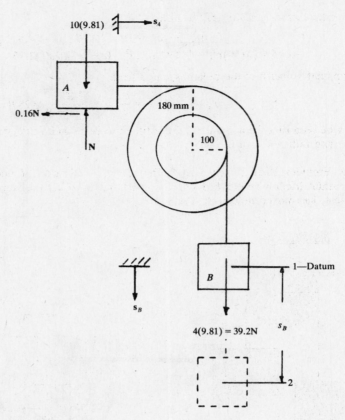

**Fig. 19.71c**

The work done on block $A$ is

$$W_{\text{NC}} = -0.16(10)9.81s_A = -0.16(10)9.81(1.75) = -27.5 \text{ J}$$

Using the work-energy equation,

$$W_{\text{NC}} = \Delta T + \Delta V = (T_2 - T_1) + (V_2 - V_1)$$

$$-27.5 = [\tfrac{1}{2}(10)(1.8v_B)^2 + \tfrac{1}{2}(4)v_B^2 - 0] + \left[-4(9.81)\frac{1.75}{1.8} - 0\right] \qquad v_B = 0.765 \text{ m/s}$$

The angular velocity of the pulley is found from

$$v_B = r\omega \qquad \omega = \frac{v_B}{r} = \frac{0.765}{100/1,000} \qquad \omega = 7.65 \text{ rad/s}$$

(*b*) Figure 19.71c is used, with the 180-mm and 100-mm dimensions interchanged.

$$s = r\theta \qquad s_A = 100\theta \tag{3}$$
$$s_B = 180\theta \tag{4}$$

Dividing Eq. (3) by Eq. (4) gives

$$\frac{s_A}{s_B} = \frac{100}{180} = \frac{1}{1.8} \qquad s_A = 1.75 \text{ m}$$

$$s_B = 1.8s_A = 1.8(1.75) \text{ m} = 3.15 \text{ m} \qquad \frac{v_B}{v_A} = \frac{180}{100} = 1.8 \qquad v_B = 1.8v_A$$

The required energy terms are

$$T_1 = 0 \qquad T_2 = \tfrac{1}{2}m_A v_A^2 + \tfrac{1}{2}m_B v_B^2 = \tfrac{1}{2}(10)v_A^2 + \tfrac{1}{2}(4)(1.8v_A)^2$$
$$V_1 = 0 \qquad V_2 = -4(9.81)3.15$$

The work done is given by

$$W_{\text{NC}} = -0.16(10)9.81s_A = -0.16(10)9.81(1.75) = -27.5 \text{ J}$$

The work-energy equation has the form

$$W_{NC} = \Delta T + \Delta V = (T_2 - T_1) + (V_2 - V_1)$$
$$-27.5 = [\tfrac{1}{2}(10)v_A^2 + \tfrac{1}{2}(4)(1.8v_A)^2 - 0] + [-4(9.81)3.15 - 0] \qquad v_A = 2.89 \text{ m/s}$$

The angular velocity of the pulley is found from

$$v_A = r\omega \qquad \omega = \frac{v_A}{r} = \frac{2.89}{100/1,000} = 28.9 \text{ rad/s}$$

It may be seen that the angular velocity of the pulley is significantly greater for the arrangement of the blocks and pulley shown in Fig. 19.71b.

**19.72** The system of Problem 18.56 is repeated in Fig. 19.72a. Because of poor lubrication, the pulley must overcome a constant friction moment of 12 N · m as it rotates. Find the velocity of the blocks after block $B$, starting from rest, has moved downward 1 m.

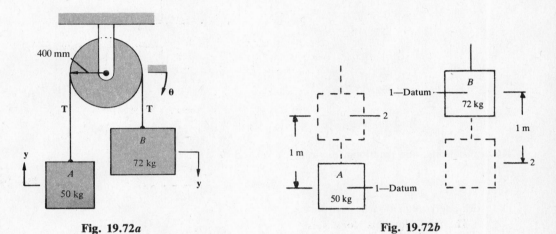

Fig. 19.72a                                            Fig. 19.72b

❚ Figure 19.72b shows the endpoints of the interval of motion. The energy terms are

$$T_1 = 0 \qquad T_2 = \tfrac{1}{2}(50 + 72)v_2^2$$
$$V_1 = 0 \qquad V_2 = 50(9.81)1 - 72(9.81)1 = -216 \text{ J} \qquad W_{NC} = -12\theta$$

The translational and rotational motions are related by

$$y = r\theta \qquad 1 = \left(\frac{400}{1,000}\right)\theta \qquad \theta = 2.5 \text{ rad}$$

The nonconservative work done is

$$W_{NC} = -12(2.5) = -30 \text{ J}$$

Using the equation of work-energy,

$$W_{NC} = \Delta T + \Delta V = (T_2 - T_1) + (V_2 - V_1) \qquad -30 = [\tfrac{1}{2}(122)v_2^2 - 0] + [-216 - 0] \qquad v_2 = 1.75 \text{ m/s}$$

The answer above may be compared with the result 1.71 m/s ≈ 1.75 m/s, obtained in Prob. 18.56 by direct use of Newton's second law. The percent difference between these two values is 2.3 percent, and this large difference is due to an accumulation of computational roundoff error.

**19.73** The system of Prob. 18.59 is shown in Fig. 19.73a. The diameter of the disk is 150 mm, the mass of the block is 600 g, and the mass moment of inertia of the disk about its center axis was found to be $2.76 \times 10^{-3}$ kg · m². If the system is released from rest, find the angular velocity of the disk after the block has moved through a displacement of 1,400 mm.

❚ Figure 19.73b shows the endpoints of the interval of motion. The energy terms are

$$T_1 = 0 \qquad T_2 = \frac{1}{2}\left(\frac{600}{1,000}\right)v_2^2 + \tfrac{1}{2}I_0\omega_2^2 = \frac{1}{2}\left(\frac{600}{1,000}\right)v_2^2 + \tfrac{1}{2}(2.76 \times 10^{-3})\omega_2^2$$

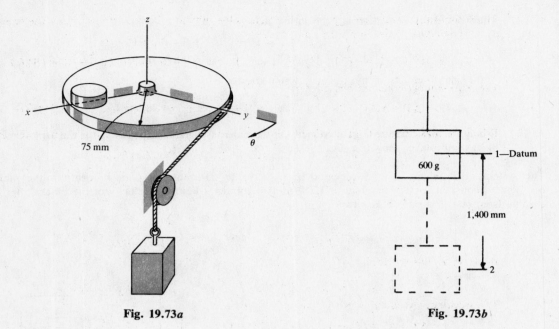

**Fig. 19.73a**

**Fig. 19.73b**

$$V_1 = 0 \qquad V_2 = -\left(\frac{600}{1,000}\right)9.81\left(\frac{1,400}{1,000}\right) = -8.24 \, \text{J}$$

The translational and rotational motions are related by

$$v_2 = \frac{75}{1,000} \, \omega_2$$

The work-energy equation has the form

$$0 = \Delta T + \Delta V = (T_2 - T_1) + (V_2 - V_1) = \left[\frac{1}{2}\left(\frac{600}{1,000}\right)\left(\frac{75}{1,000}\,\omega_2\right)^2 + \frac{1}{2}(2.76 \times 10^{-3})\omega_2^2 - 0\right] + (-8.24 - 0)$$

$$\omega_2 = 51.8 \, \text{rad/s}$$

It is left as an exercise for the reader to confirm the above result by using the value $\alpha = 72.0 \, \text{rad/s}^2$, found in Prob. 18.59, for the angular acceleration of the disk.

**19.74** Figure 19.74 shows a system of weights and pulleys. The cable is assumed to not slip on the pulleys. The masses and mass moments of inertia are $m_A = 20 \, \text{kg}$, $m_B = 14 \, \text{kg}$, $I_{OC} = 0.005 \, \text{kg} \cdot \text{m}^2$, and $I_{OD} = 0.010 \, \text{kg} \cdot \text{m}^2$.

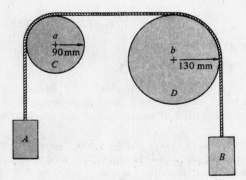

**Fig. 19.74**

(a) Find the velocities of the weights and pulleys when weight $B$, starting from rest, has moved upward 900 mm. Neglect the mass moments of inertia of the pulleys.

(b) Do the same as in part (a), but include the mass moments of inertia of the pulleys.

**(a)** The datum for potential energy for each weight is the initial position of the weight. The energy terms at the endpoints of the interval of motion are

$$T_1 = 0 \qquad T_2 = \tfrac{1}{2}(20+14)v_2^2 \qquad V_1 = 0 \qquad V_2 = (-20+14)9.81\left(\frac{900}{1{,}000}\right) = -53.0\,\text{J}$$

The work-energy equation has the form

$$0 = \Delta T + \Delta V = (T_2 - T_1) + (V_2 - V_1) = [\tfrac{1}{2}(34)v_2^2 - 0] + [-53.0] \qquad v_2 = 1.77\,\text{m/s}$$

**(b)** When the mass moments of inertia of the pulleys are included, the energy terms have the forms

$$T_1 = 0 \qquad T_2 = \tfrac{1}{2}(20+14)v_2^2 + \tfrac{1}{2}I_{0C}\omega_C^2 + \tfrac{1}{2}I_{0D}\omega_D^2$$

$$V_1 = 0 \qquad V_2 = (-20+14)9.81\left(\frac{900}{1{,}000}\right) = -53.0\,\text{J}$$

Using

$$\omega_C = \frac{v_2}{90/1{,}000} = 11.1v_2 \quad\text{and}\quad \omega_D = \frac{v_2}{130/1{,}000} = 7.69v_2$$

the term $T_2$ has the form

$$T_2 = \tfrac{1}{2}(34)v_2^2 + \tfrac{1}{2}(0.005)(11.1v_2)^2 + \tfrac{1}{2}(0.010)(7.69v_2)^2 = 17.6v_2^2$$

The work-energy equation is written as

$$0 = \Delta T + \Delta V = (T_2 - T_1) + (V_2 - V_1) = (17.6v_2^2 - 0) + (-53.0 - 0) \qquad v_2 = 1.74\,\text{m/s}$$

The angular velocities of the pulleys are then found to be

$$\omega_C = \frac{1.74}{90/1{,}000} = 19.3\,\text{rad/s} \qquad \omega_D = \frac{1.74}{130/1{,}000} = 13.4\,\text{rad/s}$$

The results in parts (a) and (b) are the same as those found in Prob. 18.60 by direct use of Newton's second law.

**19.75** Do the same as in Prob. 19.74, for the system in Fig. 19.75. Pulleys $D$ and $E$ have the same mass properties.

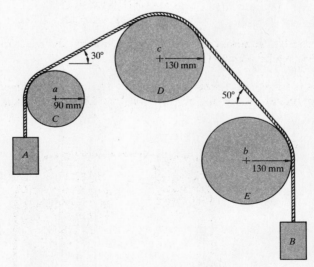

**Fig. 19.75**

**(a)** When the mass moments of inertia of the pulleys are neglected, the result for the velocity of the blocks is the same as that found in part (a) of Prob. 19.74.

**(b)** The terms $T_1$, $V_1$, and $V_2$ are the same as those in part (b) of Prob. 19.74. The term $T_2$ is written as

$$T_2 = \tfrac{1}{2}(20+14)v_2^2 + \tfrac{1}{2}I_{0C}\omega_C^2 + \tfrac{1}{2}I_{0D}\omega_D^2 + \tfrac{1}{2}I_{0E}\omega_E^2$$

Using $\omega_C = 11.1v_2$ and $\omega_D = \omega_E = 7.69v_2$, $T_2$ has the form

$$T_2 = \tfrac{1}{2}(34)v_2^2 + \tfrac{1}{2}(0.005)(11.1v_2)^2 + \tfrac{1}{2}(0.010)(7.69v_2)^2 + \tfrac{1}{2}(0.010)(7.69v_2)^2 = 17.9v_2^2$$

The work-energy equation is written as

$$0 = \Delta T + \Delta V = (T_2 - T_1) + (V_2 - V_1) = (17.9v_2^2 - 0) + (-53.0 - 0) \qquad v_2 = 1.72 \text{ m/s}$$

The angular velocities of the pulleys are then given by

$$\omega_C = \frac{1.72}{90/1,000} = 19.1 \text{ rad/s} \qquad \omega_D = \omega_E = \frac{1.72}{130/1,000} = 13.2 \text{ rad/s}$$

The above results are the same as those found in Prob. 18.61 by direct use of Newton's second law.

**19.76**   Figure 19.76a shows a rack-and-pinion gear arrangement. The rack may be approximated as a steel rod of 0.5-in by 0.5-in square cross section, and the pinion may be approximated as a steel disk of 0.5-in thickness. All frictional effects may be neglected. The specific weight of steel is 489 lb/ft³. The system is released from rest in the position shown in the figure.

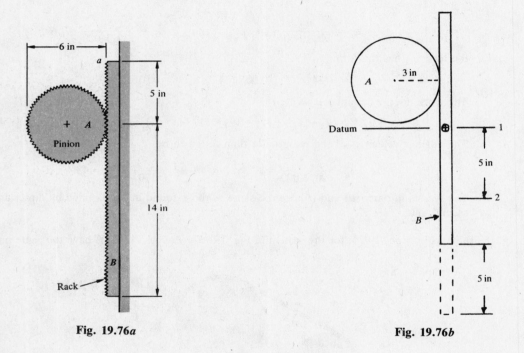

Fig. 19.76a                                        Fig. 19.76b

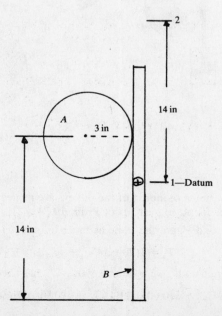

Fig. 19.76c

(a) Find the velocity of the rack at the position where the rack loses contact with the pinion.

(b) Do the same as in part (a), if the pinion is assumed to have zero mass.

(c) Do the same as in parts (a) and (b), if the pinion must overcome a constant friction moment of $1.2$ in · lb.

(d) Find the minimum required value of upward velocity of the rack, in the position shown in Fig. 19.76a, for which the rack will just loose contact with the pinion. Neglect all friction effects.

▮ (a) From Prob. 18.63,

$$W_A = 4.00 \text{ lb} \qquad W_B = 1.34 \text{ lb}$$

Figure 19.76b shows the endpoints of the interval of motion. The required energy terms are

$$T_1 = 0 \qquad T_2 = \tfrac{1}{2}m_B v_2^2 + \tfrac{1}{2}I_A \omega_2^2 \qquad V_1 = 0 \qquad V_2 = -1.34(5) = -6.7 \text{ in · lb}$$

The translational and rotational velocities are related by

$$v_2 = r\omega_2 \qquad v_2 = 3\omega_2$$

Using $I_A = I_0 = \tfrac{1}{8}md^2$, the term $T_2$ has the form

$$T_2 = \frac{1}{2}\left(\frac{1.34}{386}\right)v_2^2 + \frac{1}{2}\left[\frac{1}{8}\left(\frac{4.00}{386}\right)6^2\right]\left(\frac{v_2}{3}\right)^2 = 0.00433v_2^2$$

The work-energy equation has the form

$$0 = \Delta T + \Delta V = (T_2 - T_1) + (V_2 - V_1) = [0.00433v_2^2 - 0] + [-6.7 - 0] \qquad v_2 = 39.3 \text{ in/s}$$

The above value for $v_2$ may be compared with the result $39.4 \text{ in/s} \approx 39.3 \text{ in/s}$, found in Prob. 18.63, part (c), by direct use of Newton's second law.

(b) If the pinion has zero mass, $I_A = 0$. $T_1$, $V_1$, and $V_2$ are the same as in part (a), and $T_2$ has the form

$$T_2 = \tfrac{1}{2}m_B v_2^2 = \frac{1}{2}\left(\frac{1.34}{386}\right)v_2^2$$

The work-energy equation is written as

$$W_{NC} = \Delta T + \Delta V = (T_2 - T_1) + (V_2 - V_1) \qquad 0 = \left[\frac{1}{2}\left(\frac{1.34}{386}\right)v_2^2 - 0\right] + (-6.7 - 0) \qquad v_2 = 62.1 \text{ in/s}$$

The above result was found in Prob. 18.63, part (d), by direct use of Newton's second law.

(c) For the case of the pinion with mass, $T_1$, $T_2$, $V_1$, and $V_2$ are the same as in part (a). The translational and rotational motions are related by

$$s = r\theta \qquad 5 = 3\theta \qquad \theta = \tfrac{5}{3} \text{ rad}$$

The work-energy equation has the form

$$W_{NC} = \Delta T + \Delta V = (T_2 - T_1) + (V_2 - V_1) \qquad -1.2(\tfrac{5}{3}) = [0.00433v_2^2 - 0] + (-6.7 - 0)$$
$$v_2 = 33.0 \text{ in/s}$$

If the pinion has zero mass, $I_A = 0$, $T_1$, $T_2$, $V_1$, and $V_2$ are the same as in part (b). The velocity $v_2$ is found from

$$W_{NC} = \Delta T + \Delta V = (T_2 - T_1) + (V_2 - V_1) \qquad -1.2\left(\frac{5}{3}\right) = \left[\frac{1}{2}\left(\frac{1.34}{386}\right)v_2^2 - 0\right] + (-6.7 - 0)$$
$$v_2 = 52.1 \text{ in/s}$$

The results above may be verified by comparison with the values of $v_2$ obtained in Prob. 18.64.

(d) $v_1$ is the initial upward velocity. When $v_1$ is minimum, the velocity of the rack will be zero when this element has moved upward 14 in. Figure 19.76c shows the endpoints of the interval of motion. $T_1$ has the form of the term $T_2$ in part (a), so that

$$T_1 = 0.00433v_1^2$$

The remaining energy terms are

$$T_2 = 0 \qquad V_1 = 0 \qquad V_2 = 1.34(14) = 18.8 \text{ in · lb}$$

The work-energy equation has the form

$$0 = \Delta T + \Delta V = (T_2 - T_1) + (V_2 - V_1) = (0 - 0.00433v_1^2) + (18.8 - 0) \qquad v_1 = 65.9 \text{ in/s}$$

**19.77** The gears in Fig. 19.77a are steel, with $\gamma = 489\,\text{lb/ft}^3$, $d_1 = 4\,\text{in}$, and $d_2 = 10\,\text{in}$. The thickness of the gears is 1 in. An 8-lb weight is attached to the cable. Find the velocity of the weight after this element, starting from rest, has lowered through a distance of 20 in.

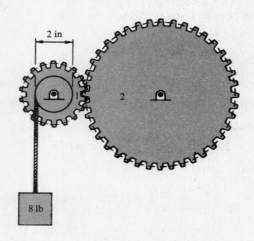

Fig. 19.77a

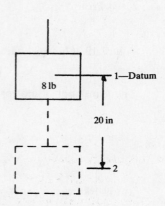

Fig. 19.77b

From Prob. 18.70, part (a),

$$I_{01} = 0.0184\,\text{lb}\cdot\text{s}^2\cdot\text{in} \qquad I_{02} = 0.720\,\text{lb}\cdot\text{s}^2\cdot\text{in}$$

Figure 19.77b shows the endpoints of the interval of motion of the weight. The required energy terms have the forms

$$T_1 = 0 \qquad T_2 = \tfrac{1}{2}mv^2 + \tfrac{1}{2}I_{01}\omega_1^2 + \tfrac{1}{2}I_{02}\omega_2^2 \qquad V_1 = 0 \qquad V_2 = -8(20) = -160\,\text{in}\cdot\text{lb}$$

where $v$, $\omega_1$, and $\omega_2$ are the translational and rotational velocities at point 2. The translational and rotational motions are related by

$$v = r\omega = (1)\omega_1 \qquad \omega_1 = v$$

The gear velocities are in the ratio

$$\frac{\omega_1}{\omega_2} = \frac{10}{4} \qquad \omega_2 = \tfrac{4}{10}\omega_1 = \tfrac{4}{10}v$$

Using the above results, $T_2$ has the form

$$T_2 = \frac{1}{2}\left(\frac{8}{386}\right)v^2 + \frac{1}{2}(0.0184)v^2 + \frac{1}{2}(0.720)(\tfrac{4}{10}v)^2 = 0.0772v^2$$

The work-energy equation is written as

$$0 = \Delta T + \Delta V = (T_2 - T_1) + (V_2 - V_1) = (0.0772v_2^2 - 0) + (-160 - 0) \qquad v_2 = 45.5\,\text{in/s}$$

The result above was found in Prob. 18.70, part (b), by direct use of Newton's second law.

**19.78** Do the same as in Prob. 19.77, if the gears and weight are arranged as shown in Fig. 19.78.

The terms $I_{01}$, $I_{02}$, $T_1$, $V_1$, and $V_2$ are the same as in Prob. 19.77, and Fig. 19.77b is used. $T_2$ has the form

$$T_2 = \tfrac{1}{2}mv^2 + \tfrac{1}{2}I_{01}\omega_1^2 + \tfrac{1}{2}I_{02}\omega_2^2$$

The relationships among the velocities are given by

$$v = r\omega = 1(\omega_2) \qquad \omega_2 = v \qquad \frac{\omega_1}{\omega_2} = \frac{10}{4} \qquad \omega_1 = \tfrac{10}{4}v$$

$T_2$ then has the form

$$T_2 = \frac{1}{2}\left(\frac{8}{386}\right)v^2 + \tfrac{1}{2}(0.0184)(\tfrac{10}{4}v)^2 + \tfrac{1}{2}(0.720)v^2 = 0.428v^2$$

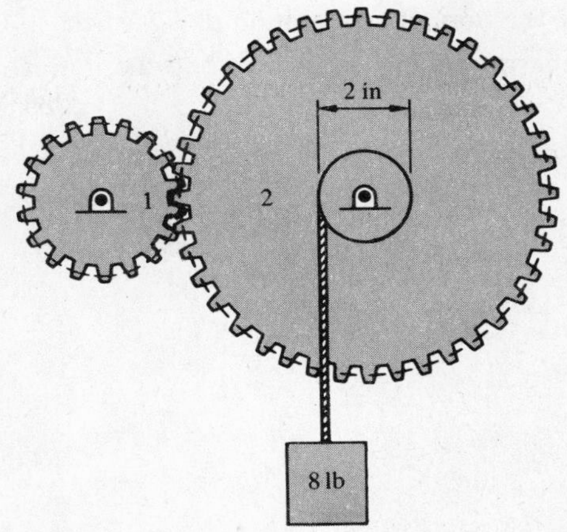

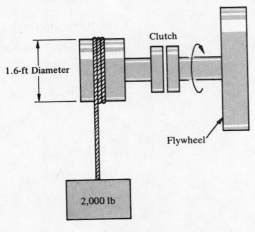

Fig. 19.78

The work-energy equation is written as

$$0 = \Delta T + \Delta V = (T_2 - T_1) + (V_2 - V_1) = (0.428 v^2 - 0) + (-160 - 0) \qquad v = 19.3 \, \text{in/s}$$

The above result was found in Prob. 18.71, part (a), by direct use of Newton's second law.

**19.79**  The flywheel in Fig. 19.79a rotates at 100 r/min.   When the clutch is engaged, the stored energy of the flywheel may be used to raise the weight.   The mass moment of inertia of the flywheel is $5.9 \times 10^5 \, \text{lb} \cdot \text{in}^2$,   and the mass moment of inertia of the drum on which the cable winds may be neglected.   It may be assumed that there is a loss of 5 percent of the initial energy as a result of slippage during the clutch engagement.   Find the distance through which the weight is raised when the clutch is engaged.

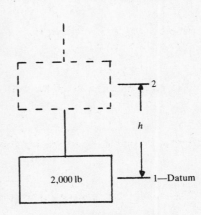

Fig. 19.79a

Fig. 19.79b

▎ Figure 19.79b shows the endpoints of the interval of motion.   The energy terms have the forms

$$T_1 = \tfrac{1}{2} I_0 \omega_1^2 \qquad T_2 = 0 \qquad V_1 = 0 \qquad V_2 = mgh$$

For the case of no energy losses, the work-energy equation has the form

$$0 = \Delta T + \Delta V = (T_2 - T_1) + (V_2 - V_1) = (0 - \tfrac{1}{2} I_0 \omega_1^2) + (mgh - 0) \qquad mgh = \tfrac{1}{2} I_0 \omega_1^2$$

For the case of a 5 percent energy loss, the above equation has the form

$$mgh = 0.95(\tfrac{1}{2} I_0 \omega_1^2)$$

Using   $\omega_1 = 100 \, \text{r/min} = 10.5 \, \text{rad/s}$,   the distance $h$ is found from

$$2,000h = 0.95 \left[ \frac{1}{2} \left( \frac{5.9 \times 10^5}{386} \right) 10.5^2 \right] \qquad h = 40.0 \, \text{in} = 3.33 \, \text{ft}$$

**19.80**  (*a*)  Give the equation for the normal acceleration of a particle.

(*b*)  How may the work-energy method be used to find the normal acceleration of a particle?

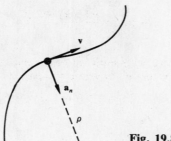

**Fig. 19.80**

▮  (*a*)  Figure 19.80 shows a particle which moves along a plane curve.  The normal component $a_n$ of the acceleration of the particle was shown in Prob. 14.56 to have the form

$$a_n = \frac{v^2}{\rho}$$

where $\rho$ is the radius of curvature and $v$ is the speed of the particle along the curve.  The physical interpretation of normal acceleration is that this quantity is required to change the direction of the velocity of the particle so that this element will follow the curved path of motion.

(*b*)  In many problems it is particularly convenient to use the work-energy method to directly obtain the value of $v$ in the equation $a_n = v^2/\rho$.  The technique will be illustrated in the following problems.

**19.81**  A frictionless particle of weight 2 lb slides down a track in a vertical plane, as shown in Fig. 19.81*a*.  At the lowest point in the track, the particle has an apparent weight that is 50 percent greater than its actual weight.  Find the height $h$ from which the particle was released.

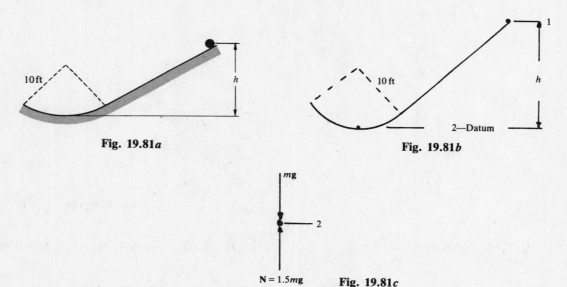

**Fig. 19.81*a***          **Fig. 19.81*b***

**Fig. 19.81*c***

▮  Figure 19.81*b* shows the endpoints of the interval of motion.  The energy terms at 1 and 2 are

$$T_1 = 0 \qquad T_2 = \tfrac{1}{2}mv_2^2 \qquad V_1 = mgh \qquad V_2 = 0$$

The work-energy equation has the form

$$0 = \Delta T + \Delta V = (T_2 - T_1) + (V_2 - V_1) = (\tfrac{1}{2}mv_2^2 - 0) + (0 - mgh) \qquad v_2^2 = 2gh \qquad (1)$$

The free-body diagram of the particle, at the lowest point in the track, is shown in Fig. 19.81*c*.  Newton's second law has the form

$$\sum F_n = ma_n \qquad 1.5mg - mg = ma_n = m\,\frac{v_2^2}{\rho} \tag{2}$$

Using $v_2^2$ from Eq. (1) in Eq. (2) results in

$$0.5mg = \frac{m(2gh)}{\rho} \qquad h = \tfrac{1}{4}\rho = \tfrac{1}{4}(10) = 2.5\ \text{ft}$$

**19.82** A mass particle moves along a smooth curved track that lies in a vertical plane, as shown in Fig. 19.82a. Determine if there is any point where the particle will lose contact with the track.

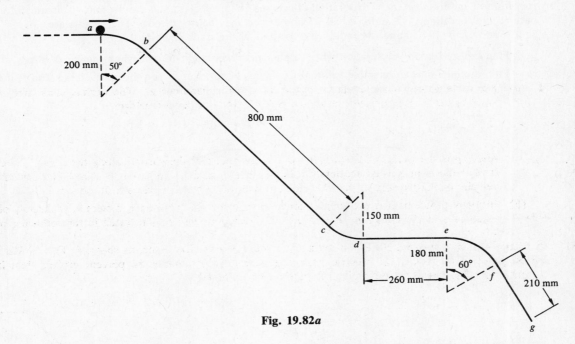

Fig. 19.82a

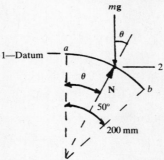

Fig. 19.82b

**❘** Track section $ab$ is considered first. The work-energy equation is written between the endpoints 1 and 2 shown in Fig. 19.82b. The $T$ and $V$ terms have the forms

$$T_1 = 0 \qquad T_2 = \tfrac{1}{2}mv_2^2 \qquad V_1 = 0 \qquad V_2 = -mg\left(\frac{200}{1{,}000}\right)(1 - \cos\theta)$$

The work-energy equation is written as

$$0 = \Delta T + \Delta V = (T_2 - T_1) + (V_2 - V_1) = (\tfrac{1}{2}mv_2^2 - 0) + [-mg(0.2)(1 - \cos\theta) - 0]$$
$$v_2^2 = 2g(0.2)(1 - \cos\theta) \tag{1}$$

The forces which act on the particle are shown in Fig. 19.82b. Newton's second law is written as

$$\sum F_n = ma_n \qquad mg\cos\theta - N = ma_n = m\,\frac{v_2^2}{\rho} \tag{2}$$

The limiting condition for which the particle will lose contact with the track is given by $N = 0$. Using this result, $v_2^2$ is eliminated between Eqs. (1) and (2), to obtain

$$\left[1 + \frac{2(0.2)}{0.2}\right]\cos\theta = 2 \qquad \cos\theta = \tfrac{2}{3} \qquad \theta = 48.2°$$

The particle will lose contact with the track just before point $b$, at $\theta = 48.2°$.

**19.83** Figure 19.83$a$ shows a preliminary engineering design for a roller coaster track to be built for an amusement park. The car is towed up a ramp to position 1. A slight leftward horizontal force is then applied to the car, and it rolls down the track. All the dimensions of the car are small compared with the length and radii of curvature of the track. Thus, all rotation effects may be neglected, and the car may be assumed to move as a particle in translation. In addition, all friction effects may be neglected. The heights are $h_1 = 60$ ft, $h_2 = 25$ ft, and $h_3 = 40$ ft, and the radius of curvature of the track at point 2 is $h_1 - h_2$.

(*a*) Find the maximum acceleration which a passenger would experience when the car is in position 2.

(*b*) Find the minimum permissible value of the radius of curvature when the car is in position 3, if the wheels of the car are not to lose contact with the track at this location.

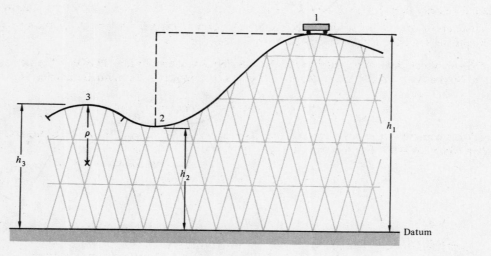

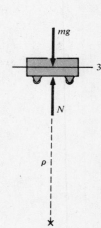

Fig. 19.83$a$        Fig. 19.83$b$

▌ (*a*) The velocity at position 2 will be found first. The ground is chosen to be the datum, and

$$T_1 = 0 \qquad T_2 = \tfrac{1}{2}mv_2^2 \qquad V_1 = mgh_1 \qquad V_2 = mgh_2$$

The work-energy equation has the form

$$0 = \Delta T + \Delta V = (T_2 - T_1) + (V_2 - V_1) = (\tfrac{1}{2}mv_2^2 - 0) + (mgh_2 - mgh_1) \qquad v_2^2 = 2g(h_1 - h_2) \quad (1)$$

The normal acceleration of the car at position 2 is

$$a_n = \frac{v_2^2}{\rho} \qquad (2)$$

$v_2^2$ from Eq. (1) is substituted into Eq. (2). Using $\rho = h_1 - h_2$ for the radius of curvature results in

$$a_n = \frac{2g(h_1 - h_2)}{h_1 - h_2} = 2g$$

At position 2 the passenger would experience an acceleration which is twice the value of the gravitational acceleration.

(*b*) The free-body diagram of the car in position 3 is shown in Fig. 19.83$b$. Newton's second law has the form

$$\sum F_n = ma_n \qquad mg - N = ma_n = \frac{mv_3^2}{\rho} \qquad (3)$$

The work-energy equation is written between positions 1 and 3, with the result

$$0 = \Delta T + \Delta V = (T_3 - T_1) + (V_3 - V_1) = (\tfrac{1}{2}mv_3^2 - 0) + (mgh_3 - mgh_1) \qquad v_3^2 = 2g(h_1 - h_3) \quad (4)$$

The car will lose contact with the track when $N = 0$. Using this result, $v_3^2$ is eliminated between Eqs. (3) and (4), to obtain

$$mg = \frac{2mg(h_1 - h_3)}{\rho} \qquad \rho = 2(h_1 - h_3) = 2(60 - 40) = 40 \text{ ft}$$

It may be seen that $\rho$ is equal to $h_3$, and this result is the *minimum* permissible value for the radius of curvature at position 3.

**19.84** Figure 19.84a shows a circular hoop which lies in a vertical plane. The internal surface of the hoop is a slightly grooved, frictionless track. Initially, a small particle of mass $m$ rests on the lowest portion of the track. A spring-loaded arm is used to impart an initial velocity to the mass element. This mass has a magnitude of 35 g, and the diameter of the hoop is 400 mm.

(a) Find the minimum required value of the initial velocity of the particle at position 1 if, when this element is in position 2, it is not to lose contact with the track.

(b) If the initial velocity $v_1$ is 2.80 m/s, find the position where the particle loses contact with the track.

(c) For the position of part (b), find the tangential acceleration of the particle and discuss the magnitude, direction, and sense of the total acceleration of the particle.

▮ (a) The free-body diagram of the mass particle in position 2 is shown in Fig. 19.84b. The normal force exerted by the track on the particle is $N$. Newton's second law in the normal direction is written as

$$\sum F_n = ma_n \qquad N + mg = ma_n = m\frac{v_2^2}{\rho} \qquad (1)$$

Since $N$ is a compressive reaction force, it may have only positive values. Thus, a limiting condition is reached when $N = 0$. For this case, Eq. (1) appears as

$$0 + mg = \frac{mv_2^2}{d/2} \qquad v_2^2 = \tfrac{1}{2}dg \qquad (2)$$

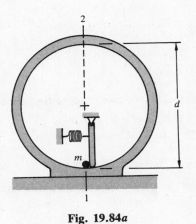

**Fig. 19.84a**

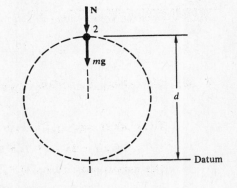

**Fig. 19.84b**

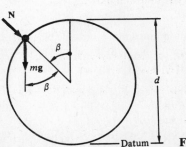

**Fig. 19.84c**

The work-energy equation is now written between positions 1 and 2. The required energy terms are

$$T_1 = \tfrac{1}{2}mv_1^2 \qquad T_2 = \tfrac{1}{2}mv_2^2 \qquad V_1 = 0 \qquad V_2 = mgd \tag{3}$$

and

$$0 = \Delta T + \Delta V = (T_2 - T_1) + (V_2 - V_1) = (\tfrac{1}{2}mv_2^2 - \tfrac{1}{2}mv_1^2) + (mgd - 0) \qquad v_2^2 = v_1^2 - 2gd \tag{4}$$

The term $v_2^2$ is eliminated between Eqs. (2) and (4), with the result

$$v_1^2 = \tfrac{5}{2}dg \qquad v_1 = \sqrt{\tfrac{5}{2}dg}$$

Using $d = 400$ mm,

$$v_1 = \sqrt{\tfrac{5}{2}\left(\tfrac{400}{1,000}\right)9.81} = 3.13 \text{ m/s}$$

If the initial velocity of the mass particle is 3.13 m/s, the element will just lose contact with the track at position 2. It may be observed that this result is independent of the mass.

(b) Figure 19.84c shows the free-body diagram when the mass particle is at an arbitrary position defined by angle $\beta$. Newton's second law in the normal direction has the form

$$\sum F_n = ma_n \qquad N + mg\cos\beta = ma_n = \frac{mv^2}{d/2} \tag{5}$$

The kinetic and potential energies, when the particle is in the position defined by angle $\beta$, are

$$T = \tfrac{1}{2}mv^2 \qquad V = mg\left(\frac{d}{2}\right)(1 + \cos\beta)$$

The initial energy terms $T_1$ and $V_1$ are given by Eq. (3) in part (a). The work-energy equation is written as

$$0 = \Delta T + \Delta V = (T - T_1) + (V - V_1) = (\tfrac{1}{2}mv^2 - \tfrac{1}{2}mv_1^2) + \left[mg\left(\frac{d}{2}\right)(1 + \cos\beta) - 0\right] \tag{6}$$

The position at which the particle loses contact with the track is defined by $N = 0$. Using this condition, $v^2$ is eliminated between Eqs. (5) and (6) to obtain

$$\cos\beta = \frac{2}{3dg}(v_1^2 - dg)$$

With the values $v_1 = 2.80$ m/s and $d = 400$ mm,

$$\cos\beta = \frac{2}{3(400/1,000)(9.81)}\left[2.80^2 - \frac{400}{1,000}(9.81)\right] = 0.667 \qquad \beta = 48.3°$$

(c) From consideration of Fig. 19.84c, Newton's second law in the normal and tangential directions has the forms

$$\sum F_n = ma_n \qquad mg\cos\beta = ma_n \qquad a_n = g\cos\beta$$
$$\sum F_t = ma_t \qquad mg\sin\beta = ma_t \qquad a_t = g\sin\beta$$

It may be seen that $a_n$ and $a_t$ are the components of the gravitational acceleration vector $g$. Thus, at the instant when the particle loses contact with the track, the total acceleration of the particle is equal to $g$.

19.85 A small particle is at rest on top of a fixed cylinder, as shown in Fig. 19.85a. The surface of the cylinder is assumed to be frictionless.

(a) If the mass is given an infinitesimal lateral displacement to set it into motion, find the position at which this particle loses contact with the cylindrical surface.

(b) How would the result in part (a) change if the surface of the cylinder were not frictionless?

▮ (a) The initial position is designated 1, and an arbitrary subsequent position is designated 2, as shown in Fig. 19.85b. The radius of the cylinder is $r$, and the datum for potential energy is at position 2. The required energy terms are

$$T_1 = 0 \qquad T_2 = \tfrac{1}{2}mv^2 \qquad V_1 = 0 \qquad V_2 = -mgr(1 - \cos\beta)$$

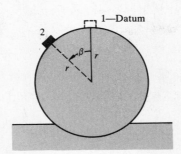

Fig. 19.85a          Fig. 19.85b

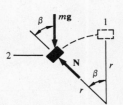

Fig. 19.85c

The work-energy equation is then

$$0 = \Delta T + \Delta V = (T_2 - T_1) + (V_2 - V_1) = (\tfrac{1}{2}mv^2 - 0) + [-mgr(1 - \cos \beta) - 0] \qquad (1)$$

The free-body diagram of the mass particle in position 2 is shown in Fig. 19.85c. Newton's second law for the particle has the form

$$\sum F_n = ma_n \qquad mg \cos \beta - N = ma_n = m\,\frac{v^2}{r} \qquad (2)$$

The limiting condition $N = 0$ corresponds to loss of contact between the mass element and the cylinder. Using this condition, and eliminating $v^2$ from Eqs. (1) and (2), results in

$$\cos \beta = \tfrac{2}{3} \qquad \beta = 48.2°$$

It is interesting to note that the above result is independent of *both* the mass of the particle and the dimensions of the cylinder. It may also be observed that the above value $48.2° \approx 48.3°$ for $\beta$ is the same as the value obtained in Prob. 19.84. It is left as an exercise for the reader to show that these are equivalent problems.

(*b*) If sliding friction exists between the particle and the cylinder, there is a *negative* nonconservative work term, since the friction forces act opposite to the actual sense of the motion. The work-energy equation has the form

$$W_{\mathrm{NC}} = \Delta T + \Delta V \qquad W_{\mathrm{NC}} < 0$$

The terms $\Delta T$ and $\Delta V$ are the same as before, so that

$$W_{\mathrm{NC}} = \tfrac{1}{2}mv^2 - mgr(1 - \cos \beta) \qquad (3)$$

The term $v^2$ is eliminated between Eqs. (2) and (3). Using the condition $N = 0$ and defining $\beta_f$ to be the position, for the case of sliding friction, at which contact is lost, we have

$$\cos \beta_f = \frac{2 + [2/(rmg)]W_{\mathrm{NC}}}{3}$$

Since $W_{\mathrm{NC}}$ is a *negative* quantity, it follows that

$$\cos \beta_f < \cos \beta \qquad \beta_f > \beta$$

It may be seen that, when sliding friction is present, the particle has a larger arc of motion before it loses contact with the cylindrical surface.

**19.86**  The track of parabolic shape in Fig. 19.86a lies in a vertical plane, and is defined by $y = 5 + 0.3x^2$, where $x$ and $y$ are in inches. All friction effects may be neglected.

(*a*)  Find the normal acceleration, in $\mathrm{in/s^2}$, of the particle as it passes through position $b$. The particle starts from rest at position $a$.

(*b*)  Do the same as in part (*a*) if the particle has a velocity of 20 in/s as it passes through position $a$.

(*c*)  Do the same as in parts (*a*) and (*b*), if the track is inverted, as shown in Fig. 19.86b. The particle is assumed to not lose contact with the track.

▋ (*a*)  The radius of curvature $\rho_b$ at point $b$ is found first. The equation of the parabolic curve is

$$y = 5 + 0.3x^2$$

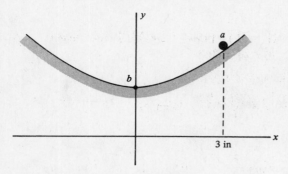

Fig. 19.86a

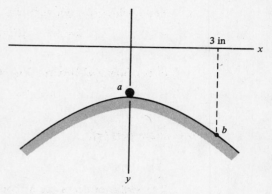

Fig. 19.86b

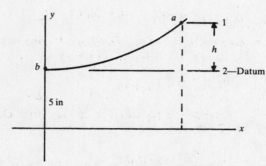

Fig. 19.86c

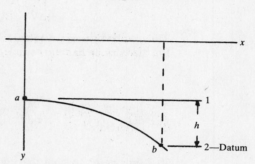

Fig. 19.86d

The endpoints of this curve, and the values of the first and second derivatives, are

$$y|_{x=0} = 5 \text{ in} \qquad y|_{x=3 \text{ in}} = 5 + 0.3(3)^2 = 7.70 \text{ in}$$

$$\frac{dy}{dx} = 2(0.3)x = 0.6x \qquad \frac{d^2y}{dx^2} = 0.6$$

The radius of curvature of a plane curve, from Prob. 14.63, has the form

$$\frac{1}{\rho} = \frac{\pm d^2y/dx^2}{[1 + (dy/dx)^2]^{3/2}}$$

For the curve in Fig. 19.86a,

$$\frac{1}{\rho_b} = \frac{1}{\rho}\bigg|_{x=0} = \frac{\pm 0.6}{(1 + 0^2)^{3/2}} = 0.6 \qquad \rho_b = 1.67 \text{ in}$$

Figure 19.86c shows the endpoints of the interval of motion. Using $h = 7.70 - 5 = 2.70$ in, the energy terms have the forms

$$T_1 = 0 \qquad T_2 = \tfrac{1}{2}mv_2^2 \qquad V_1 = mgh \qquad V_2 = 0$$

The work-energy equation is written as

$$0 = \Delta T + \Delta V = (T_2 - T_1) + (V_2 - V_1) = (\tfrac{1}{2}mv_2^2 - 0) + (0 - mgh) \qquad v_2^2 = 2gh \qquad (1)$$

The normal acceleration of the particle at point $b$ is given by

$$a_n = \frac{v_2^2}{\rho_b} \qquad (2)$$

$v_2^2$ is eliminated between Eqs. (1) and (2), with the result

$$a_n = \frac{2gh}{\rho_b} = \frac{2(386)2.70}{1.67} = 1{,}250 \text{ in/s}^2$$

(b) For the case where the particle has $v_1 = 20$ in/s at point $a$,

$$T_1 = \tfrac{1}{2}m(20)^2 \qquad T_2 = \tfrac{1}{2}mv_2^2 \qquad V_1 = mgh \qquad V_2 = 0$$

The work-energy equation has the form

$$0 = \Delta T + \Delta V = (T_2 - T_1) + (V_2 - V_1) = [\tfrac{1}{2}mv_2^2 - \tfrac{1}{2}m(20)^2] + (0 - mgh)$$
$$v_2^2 = 20^2 + 2gh \tag{3}$$

$v_2^2$ is eliminated between Eqs. (2) and (3), to obtain

$$a_n = \frac{v_2^2}{\rho_b} = \frac{20^2 + 2(386)2.70}{1.67} = 1{,}490 \text{ in/s}^2$$

(c) From part (a),

$$\frac{dy}{dx} = 0.6x \qquad \frac{d^2y}{dx^2} = 0.6$$

The radius of curvature at point $b$ is found as

$$\frac{1}{\rho_b} = \frac{1}{\rho}\bigg|_{x=3 \text{ in}} = \frac{\pm 0.6}{[1 + [0.6(3)]^2]^{3/2}} = 0.0687 \qquad \rho_b = 14.6 \text{ in}$$

The endpoints of the interval of motion are shown in Fig. 19.86d. The $T$ and $V$ terms at 1 and 2 are the same as in part (a). Using $v_2^2 = 2gh$ and $h = 2.70$ in in Eq. (2) results in

$$a_n = \frac{v_2^2}{\rho_b} = \frac{2gh}{\rho_b} = \frac{2(386)2.70}{14.6} = 143 \text{ in/s}^2$$

For the case where $v_1 = 20$ in/s, the $T$ and $V$ terms are the same as in part (b). Using $v_2^2 = 20^2 + 2gh$ and $h = 2.70$ in in Eq. (2) results in

$$a_n = \frac{v_2^2}{\rho_b} = \frac{20^2 + 2(386)2.70}{14.6} = 170 \text{ in/s}^2$$

**19.87** In the proposed system shown in Fig. 19.87a it is desired to have the small spherical mass particle, which is released from rest at position $a$, fall into an opening at position $b$.

(a) As one extreme condition, all friction effects may be neglected and it may be assumed that the particle slides, without rolling, along the track. Find the corresponding value of $h$.

(b) As the other extreme condition, it may be assumed that there is sufficient friction so that the particle rolls without sliding. Find the corresponding value of $h$.

❚ (a) The required value of velocity as the particle exits the curved track is found first. Figure 19.87b shows the track exit velocity $v_0$ as the initial velocity for a plane projectile motion problem. The $x$ motion of the particle is given by

$$x = v_{0x}t$$

Using $v_{0x} = v_0 \cos 30°$ in the above equation results in

$$3 = (v_0 \cos 30°)t \qquad t = \frac{3}{v_0 \cos 30°} \tag{1}$$

The $y$ motion of the particle is given by

$$y = v_{0y}t - \tfrac{1}{2}gt^2 = (v_{0y} - \tfrac{1}{2}gt)t = 0 \qquad v_{0y} - \tfrac{1}{2}gt = 0$$

Using Eq. (1) and $v_{0y} = v_0 \sin 30°$ in the above equation results in

$$v_0 \sin 30° - \tfrac{1}{2}(9.81)\left(\frac{3}{v_0 \cos 30°}\right) = 0 \qquad v_0 = 5.83 \text{ m/s}$$

Figure 19.87c shows the endpoints of the interval of motion. For the case where the particle slides without rolling the energy terms are

$$T_1 = 0 \qquad T_2 = \tfrac{1}{2}mv_2^2 = \tfrac{1}{2}m(5.83)^2 \qquad V_1 = mgh \qquad V_2 = 0$$

The work-energy equation has the form

$$0 = \Delta T + \Delta V = (T_2 - T_1) + (V_2 - V_1) = [\tfrac{1}{2}m(5.83)^2 - 0] + [0 - m(9.81)h] \qquad h = 1.73 \text{ m}$$

(b) The case is now considered where the particle rolls without sliding. Using Case 1 in Table 17.13,

$$I_0 = \tfrac{1}{10}md^2$$

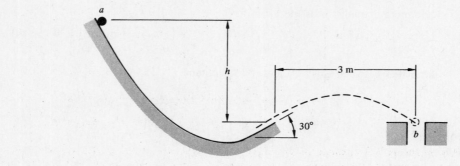

Fig. 19.87a

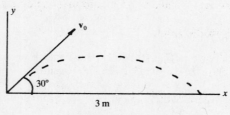

Fig. 19.87b

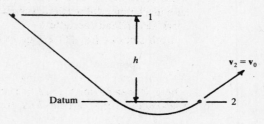

Fig. 19.87c

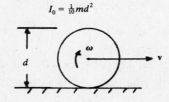

Fig. 19.87d

The energy terms have the forms

$$T_1 = 0 \qquad T_2 = \tfrac{1}{2}mv_2^2 + \tfrac{1}{2}I_0\omega_2^2 \qquad V_1 = mgh \qquad V_2 = 0$$

Using $v_2 = (d/2)\omega_2$ and $\omega_2 = 2v_2/d$ in the above equation results in

$$T_2 = \tfrac{1}{2}mv_2^2 + \tfrac{1}{2}\left(\tfrac{1}{10}md^2\right)\left(\frac{2v_2}{d}\right)^2 = 0.7mv_2^2 = 0.7m(5.83)^2$$

The work-energy equation is written as

$$0 = \Delta T + \Delta V = (T_2 - T_1) + (V_2 - V_1) = [0.7m(5.83)^2 - 0] + [0 - m(9.81)h] \qquad h = 2.43 \text{ m}$$

The required height $h$ is greater for the case of rolling of the particle, since additional energy is required to produce this rolling motion. The percent difference between the two height values is

$$\%\text{D} = \frac{2.43 - 1.73}{1.73}(100) = 40.5\%$$

## 19.5  POWER AS THE WORK DONE, OR ENERGY EXPENDED, PER UNIT TIME

**19.88**  Figure 19.88 shows a particle which is acted on by a resultant force $F$ and which is constrained to move along the $s$ axis.

(**a**)  What power is expended as the particle moves through the displacement $ds$ in time $dt$?

(**b**)  What is the definition of the term *power*?

(**c**)  What are the basic units of power?

(**d**)  What is meant by the term power *efficiency*?

▌ (**a**)  The fundamental definition of the work $dW$ done on the particle, as this element translates through the displacement $ds$, was given in Prob. 19.1 by

$$dW = (F \cos \beta)\, ds \qquad\qquad (1)$$

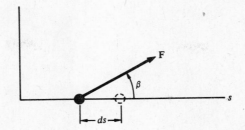

**Fig. 19.88**

This result is *independent of the time it took for the event to occur*. The term *power*, which characterizes the time required for a given amount of work to be done, will now be introduced. In its most fundamental definition, power is the work done per unit time. The distinguishing characteristic of a more powerful machine versus a less powerful machine is that the former can do a given amount of work faster than the latter.

Both sides of Eq. (1) are divided by $dt$, with the result

$$P = \frac{dW}{dt} = (F \cos \beta) \frac{ds}{dt} \tag{2}$$

where $P$ is the symbol used to represent power. The term $ds/dt$ is the magnitude of the velocity, or the speed, of the particle, and this quantity may be written as

$$\frac{ds}{dt} = v \tag{3}$$

Equations (2) and (3) are combined, to obtain

$$P = (F \cos \beta)v \tag{4}$$

(*b*) It may be seen from Eq. (4) that the *product of force and velocity is the basic definition of power*. It follows from this equation that the power will be constant only if both the component of force in the direction of motion, and the velocity, are constant.

(*c*) The basic units of power are energy divided by time. In USCS units, typical power units are foot-pounds per second or foot-pounds per minute. A very useful unit is the *horsepower*, designated hp, and defined by

$$1 \text{ hp} = 550 \text{ ft} \cdot \text{lb/s} = 33{,}000 \text{ ft} \cdot \text{lb/min}$$

In SI units, the basic unit of power is $1 \text{ N} \cdot \text{m/s} = 1 \text{ J/s}$. A power of one joule per second is defined to be one *watt*, designated by the symbol W. Thus,

$$1 \text{ W} = 1 \frac{\text{J}}{\text{s}}$$

The conversion relationship between power in USCS and SI units is

$$1 \text{ hp} = 746 \text{ W}$$

(*d*) In any physical, mechanical system there are certain inherent power losses due to friction, hysteresis losses in materials, and resistance effects in electric conductors. A convenient macroscopic representation of all these losses is contained in a term which is referred to as efficiency. This quantity is represented by the sumbol $\eta$ and is defined to be

$$\eta = \frac{\text{energy, or power, output}}{\text{energy, or power, input}}$$

For a system which has no losses of any kind, $\eta = 1$. This case represents a theoretical upper limit for any real system. Thus, for actual mechanical systems,

$$\eta < 1$$

**19.89** Figure 19.89 shows a shaft which rotates with angular velocitiy $\omega$ and transmits a torque, or moment, of magnitude $M$. What power is transmitted through the shaft?

▌ By using an analysis similar to that for the case of a force acting on a particle, it can be shown that the power transmitted through the shaft is

$$P = M\omega \tag{1}$$

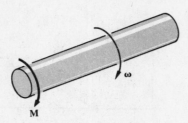

**Fig. 19.89**

The units of power in the rotational case are the same as those in the translational case. A very convenient relationship among the torque, angular speed, and power, in USCS units, is given by

$$M = 63,000\left(\frac{P}{N}\right) \tag{2}$$

In Eq. (2), $T$ is in inch-pounds, $N$ is in revolutions per minute, and $P$ is in horsepower.

In SI units, the relationship among torque, angular speed, and power is

$$M = \frac{60}{2\pi}\left(\frac{P}{N}\right) = 9.55\left(\frac{P}{N}\right)$$

where $M$ is in newton-meters, $P$ is in watts, and $N$ is in revolutions per minute.

**19.90**   Derive the equation given as Eq. (2) in Prob. 19.89, which relates torque, angular speed, and power.

❚ The power $P$ transmitted through a shaft which rotates with angular speed $\omega$ was given by Eq. (1) in Prob. 19.89 as

$$P = M\omega \tag{1}$$

The units chosen for the terms in this equation are horsepower for $P$, inch-pounds for $M$, and revolutions per minute for $\omega$. Equation (1) is written in terms of its units as

$$P(\text{hp}) = \frac{M(\text{in}\cdot\text{lb})N\left(\dfrac{\text{r}}{\text{min}}\right)\left(\dfrac{1\,\text{ft}}{12\,\text{in}}\right)\left(\dfrac{2\pi\,\text{rad}}{1\,\text{r}}\right)}{33,000\,\dfrac{\text{ft}\cdot\text{lb}}{\text{min}}\Big/1\,\text{hp}}$$

The above equation is solved for $M$, with the result

$$M = \frac{33,000}{\frac{1}{12}(2\pi)}\left(\frac{P}{N}\right) = 63,025\left(\frac{P}{N}\right)$$

The constant in the above equation is rounded to three significant figures, with the final result

$$M = 63,000\left(\frac{P}{N}\right)$$

**19.91**   Into what two convenient groups may problems in power be divided?

❚ For ease of interpretation, problems in power may be classified as one of two general types. In the first type, the power output is *constant*. A typical example would be an electric motor driving a blower. The required torque output of the motor to drive the blower is constant, and the blower operates at constant speed. In the second type either the force (or moment) required, or the velocity, or both, are *variable*. For this case, Eq. (4) in Prob. 19.88 or Eq. (1) in Prob. 19.89 are still valid, but the power must be thought of in terms of *instantaneous values*. A typical example of this type of problem is an automobile in motion. Because of changing conditions such as speed, roadway inclination, and wind conditions, the power requirements of the vehicle are changing almost continually. Both of the above cases will be illustrated in the following problems.

**19.92**   Figure 19.92 shows a forklift truck. The fork elevates a crate of mass 820 kg uniformly through a vertical distance of 2 m in 10 s. The truck is powered by a battery-operated motor. The total power efficiency of the motor and lift assembly is 0.62. Find the required electrical power input to the system in watts and in horsepower.

❚ The weight force of the crate is

$$W = mg = 820(9.81) = 8.040\,\text{N}$$

Fig. 19.92

and the constant vertical velocity of this element is

$$v = \frac{2 \text{ m}}{10 \text{ s}} = 0.2 \frac{\text{m}}{\text{s}}$$

The theoretical power $P_t$ required to raise the load is

$$P_t = Wv = 8,040(0.2) = 1,610 \text{ J/s} = 1,610 \text{ W}$$

The actual power $P_a$ required to operate the system is

$$P_a = \frac{P_t}{\eta} = \frac{1,610}{0.62} = 2,600 \text{ W} = 2,600 \text{ W} \left( \frac{1 \text{ hp}}{746 \text{ W}} \right) = 3.49 \text{ hp}$$

**19.93**   Figure 19.93 shows a man in an elevator.  The weight of the man is 180 lb and the weight of the elevator cage is 1,450 lb.  The elevator uses a 10-hp motor.  If the overall efficiency of the elevator power system is 75 percent, find the maximum value of the constant speed with which the elevator can move upward.

❚   Using the definition of power,

$$P = Fv \qquad 0.75(10)(\text{hp}) = \frac{(1450 + 180)(\text{lb})v}{550(\text{ft} \cdot \text{lb/s})/\text{hp}} \qquad v = 2.53 \text{ ft/s}$$

Fig. 19.93

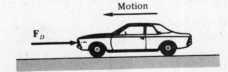

Fig. 19.94

**19.94**   Figure 19.94 shows an automobile which moves with constant speed along a horizontal road surface.  As the vehicle moves, it must displace stationary air.  The total resisting force of the air on the vehicle is represented by the resultant force $F_D$.  This force is referred to as a *drag force*.  It is shown in Prob. 21.13 that a description of this force is given by

$$F_D = C_D(\tfrac{1}{2} \rho v^2) A$$

where $\rho$ is the mass density of the air, $A$ is the projected area of the vehicle on a plane which is normal to the direction of motion, $v$ is the speed, and $C_D$ is a constant called the *drag coefficient*.  The specific weight of the air is $0.07637 \text{ lb/ft}^3$, the projected area of the automobile is $20.6 \text{ ft}^2$, and the drag coefficient is assumed to have the value. 0.68.

(a) Find the power output of the automobile engine required to overcome the air resistance if the vehicle moves with a constant speed of 50 mi/h.

(b) Do the same as in part (a) if the speed of the vehicle is 70 mi/h.

▌ (a) At 50 mi/h,

$$v = 50 \frac{\text{mi}}{\text{h}} \left( \frac{5{,}280 \text{ ft/mi}}{3{,}600 \text{ s/h}} \right) = 73.3 \frac{\text{ft}}{\text{s}}$$

The drag force is

$$F_D = C_D(\tfrac{1}{2}\rho v^2)A = 0.68 \left[ \frac{1}{2} \left( \frac{0.07637}{32/2} \right)(73.3)^2 \right] 20.6 = 89.3 \text{ lb}$$

The required power is then

$$P = F_D v = 89.3(73.3)\left( \frac{1 \text{ hp}}{550 \text{ (ft} \cdot \text{lb/s)/hp}} \right) = 11.9 \text{ hp}$$

(b) At 70 mi/h,

$$v = 70 \left( \frac{5{,}280}{3{,}600} \right) = 103 \text{ ft/s}$$

The drag force is

$$F_D = C_D(\tfrac{1}{2}\rho v^2)A = 0.68 \left[ \frac{1}{2} \left( \frac{0.07637}{32.2} \right)(103)^2 \right] 20.6 = 176 \text{ lb}$$

and the required power is

$$P = F_D v = \frac{176(103)}{550} = 33.0 \text{ hp}$$

It may be seen that a 40 percent increase in the speed of the vehicle almost triples the magnitude of the required power. These results illustrate the effects which are responsible for the drastic increase in fuel consumption when an automobile operates at high speed.

**19.95** Figure 19.95a shows a shaft with four gears mounted on it. The system is driven through gear B, and power is taken out through the remaining three gears. Find the torque transmitted through shaft sections AB, BC, and CD. The rotational speed of the shaft is 2,800 r/min.

▌ Gear A and shaft section AB are isolated from the system, as shown in Fig. 19.95b. Power from gear B enters end B of shaft length AB and leaves at end A through gear A. The speed of the shaft is constant. It may thus be concluded, from Eq. (2) in Prob. 19.89, that the torque in shaft length AB is constant. This value is

$$M_{AB} = 63{,}000\left( \frac{\text{hp}_{AB}}{N} \right) = 63{,}000\left( \frac{28}{2{,}800} \right) = 630 \text{ in} \cdot \text{lb}$$

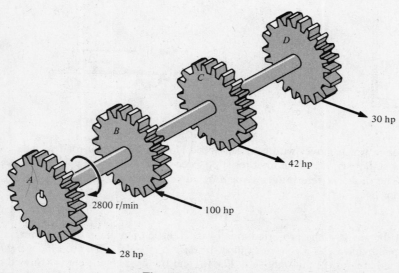

Fig. 19.95a

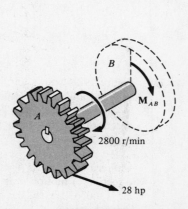

Fig. 19.95b

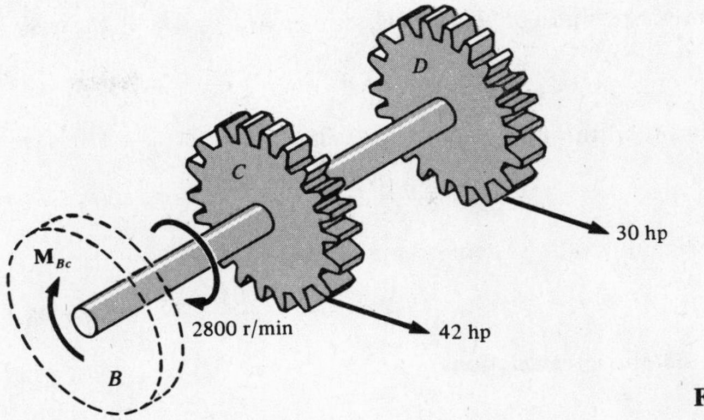

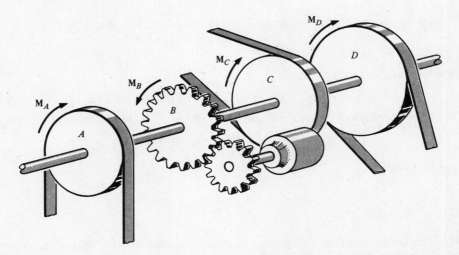

**Fig. 19.95c**

Gears $C$ and $D$ and shaft length $BCD$ are isolated from the system, as shown in Fig. 19.95c. The torque transmitted through shaft length $BC$ is then

$$M_{BC} = 63,000\left(\frac{\text{hp}_{BC}}{N}\right) \tag{1}$$

The power transmitted through shaft length $BC$ must satisfy the output requirements of gears $C$ and $D$. Thus,

$$\text{hp}_{BC} = 42 + 30 = 72 \text{ hp}$$

Equation (1) now appears as

$$M_{BC} = 63,000\left(\frac{72}{2,800}\right) = 1,620 \text{ in} \cdot \text{lb}$$

By using a similar analysis, it can be shown that the torque transmitted through the shaft length $CD$ is

$$M_{CD} = 63,000\left(\frac{30}{2,800}\right) = 675 \text{ in} \cdot \text{lb}$$

**19.96** The moment input $M_B$ through gear $B$ to the shaft in Fig. 19.96 is $310 \text{ N} \cdot \text{m}$. The shaft rotates at $1,500 \text{ r/min}$. The output moments are $M_A = 100 \text{ N} \cdot \text{m}$, $M_C = 80 \text{ N} \cdot \text{m}$, and $M_D = 130 \text{ N} \cdot \text{m}$. Find the power transmitted through shaft lengths $AB$, $BC$, and $CD$ of the shaft.

**Fig. 19.96**

▮ The power transmitted through shaft length $AB$, using $M_A = M_{AB} = 100 \text{ N} \cdot \text{m}$ and $N = 1,500 \text{ r/min}$ in Eq. (3) of Prob. 19.89, is

$$P_{AB} = \frac{M_{AB}N}{9.55} = \frac{100(1,500)}{9.55} = 15,700 \text{ W} = 15.7 \text{ kW}$$

The power transmitted through length $BC$, using $M_{BC} = M_C + M_D = 80 + 130 = 210\,\text{N} \cdot \text{m}$, is given by

$$P_{BC} = \frac{M_{BC}N}{9.55} = \frac{210(1,500)}{9.55} = 33,000\,\text{W} = 33.0\,\text{kW}$$

The power transmitted through length $CD$, using $M_D = M_{CD} = 130\,\text{N} \cdot \text{m}$, is found from

$$P_{CD} = \frac{M_{CD}N}{9.55} = \frac{130(1,500)}{9.55} = 20,400\,\text{W} = 20.4\,\text{kW}$$

The total power input to the system occurs at gear $B$, with

$$P_B = \frac{M_B N}{9.55} = \frac{310(1,500)}{9.55} = 48,700\,\text{W} = 48.7\,\text{kW}$$

As a check on the above calculations,

$$P_B \stackrel{?}{=} P_{AB} + P_{BC} \qquad 48.7 \stackrel{?}{=} 15.7 + 33.0 \qquad 48.7 \equiv 48.7$$

**19.97** Figure 19.97 shows a blade used to stir a thick liquid solution. The blade rotates at 4 rad/s and overcomes a fluid resistance of 250 ft · lb. If the electrical power input to the motor is 2 hp, find the efficiency of the motor.

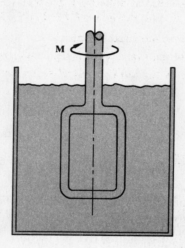

**Fig. 19.97**

▐ The angular speed of the blade is expressed in revolutions per minute as

$$\omega = 4\,\frac{\text{rad}}{\text{s}}\left(\frac{60\,\text{s}}{1\,\text{min}}\right)\left(\frac{1\,\text{r}}{2\pi\,\text{rad}}\right) = 38.2\,\text{r/min}$$

The resisting torque is

$$M = 250\,\text{ft} \cdot \text{lb} = 250(12) = 3,000\,\text{in} \cdot \text{lb}$$

The actual power required to stir the liquid is designated $P_a$. Using $M = 63,000(P_a/N)$,

$$P_a = \frac{MN}{63,000} = \frac{3,000(38.2)}{63,000} = 1.82\,\text{hp}$$

The efficiency of the motor is found from

$$\eta = \frac{P_a}{P_{\text{input}}} = \frac{1.82}{2} = 0.91 = 91\%$$

**19.98** Figure 19.98 shows a model of a disk-brake system. The shaft speed is 600 r/min. Find the power dissipated by the brake when the disk-pad forces $P$ have the magnitude 20 kN. The coefficient of friction between the pads and the disk is 0.6.

▐ $M$ is the total moment of the friction forces which act on the disk, with the magnitude

$$M = 2(\mu N)\frac{d}{2} = 2(0.6)20,000\left(\frac{150}{1,000}\right) = 3,600\,\text{N} \cdot \text{m}$$

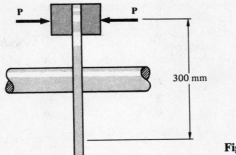

**Fig. 19.98**

The power dissipated by the brake is found from

$$P = \frac{MN}{9.55} = \frac{3,600(600)}{9.55} = 226,000 \text{ W} = 226 \text{ kW}$$

**19.99** The weight in Fig. 19.99a is lowered at a constant speed of 3.7 m/s. The coefficient of friction between the brake shoe and the drum is 0.36, and the mass of the drum is 10 kg.

(a) Find the power dissipated at the brake shoe.

(b) Find the required value of the brake lever force $P$, and the lever and drum pin forces at $a$ and $b$.

(c) Express the forces acting on the lever in formal vector notation.

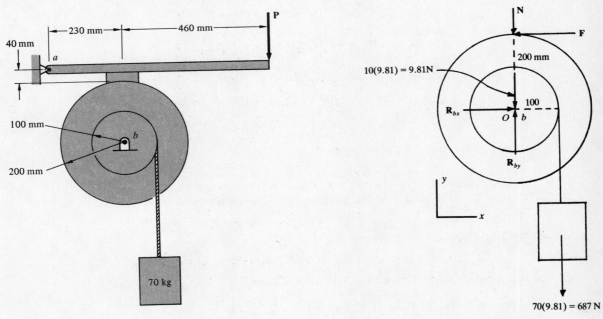

**Fig. 19.99a**

**Fig. 19.99b**

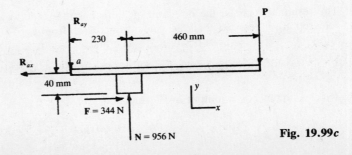

**Fig. 19.99c**

▌ (a) Figure 19.99b shows the forces exerted by the brake shoe on the drum, and the weight force of the 70-kg mass. Since the weight is lowered at constant speed, the mass and the drum are in static equilibrium. For moment equilibrium of the drum,

$$\sum M_b = 0 \qquad F(200) - 687(100) = 0 \qquad F = 344 \text{ N}$$

The angular velocity of the drum is found from

$$v = r\omega \qquad 3.7 = \left(\frac{100}{1,000}\right)\omega \qquad \omega = 37 \text{ rad/s}$$

The power dissipated at the brake shoe is then given by

$$P = M\omega = Fr\omega = 344\left(\frac{200}{1,000}\right)37 = 2,550 \text{ W}$$

As an alternative solution to this problem, the power dissipated at the brake shoe is equal to the power output of the moving block. This power is given by

$$P = Fv = 687(3.7) = 2,540 \text{ W} \approx 2,550 \text{ W}$$

(b) Figure 19.99c shows the free-body diagram of the brake lever. The normal force N is found from

$$F = 344 = \mu N = 0.36 N \qquad N = 956 \text{ N}$$

For equilibrium of the lever,

$$\sum M_a = 0 \qquad 344(40) + 956(230) - P(230 + 460) = 0 \qquad P = 339 \text{ N}$$

$$\sum F_x = 0 \qquad -R_{ax} + 344 = 0 \qquad R_{ax} = 344 \text{ N}$$

$$\sum F_y = 0 \qquad -R_{ay} + 956 - 339 = 0 \qquad R_{ay} = 617 \text{ N}$$

The pin force at a is given by

$$R_a = \sqrt{R_{ax}^2 + R_{ay}^2} = \sqrt{344^2 + 617^2} = 706 \text{ N}$$

For force equilibrium of the drum, from Fig. 19.99b,

$$\sum F_x = 0 \qquad R_{bx} - F = R_{bx} - 344 = 0 \qquad R_{bx} = 344 \text{ N}$$

$$\sum F_y = 0 \qquad R_{by} - N - 98.1 - 687 = R_{by} - 956 - 98.1 - 687 = 0 \qquad R_{by} = 1,740 \text{ N}$$

The pin force at b has the value

$$R_b = \sqrt{R_{bx}^2 + R_{by}^2} = \sqrt{344^2 + 1,740^2} = 1,770 \text{ N}$$

(c) The forces which act on the lever, in formal vector notation, are as follows.

Applied force: $\qquad\qquad\qquad \mathbf{P} = -P\mathbf{j} = -339\mathbf{j} \qquad$ N

Drum force: $\qquad\qquad \mathbf{F} = F\mathbf{i} + N\mathbf{j} = 344\mathbf{i} + 956\mathbf{j} \qquad$ N

Pin force: $\qquad\qquad \mathbf{R}_a = -R_{ax}\mathbf{i} - R_{ay}\mathbf{j} = -344\mathbf{i} - 617\mathbf{j} \qquad$ N

**19.100** The system of Probs. 19.4 and 19.26 is shown in Fig. 19.100.

(a) Find the maximum power output of the mass particle as it falls freely in the gravitational field.

(b) Express the answer to part (a) in SI units.

▌ (a) The resultant force in the direction of motion is the static weight force $W = 2.8$ lb. The particle is released from rest, so that the velocity increases from zero to the maximum value $v_2 = \sqrt{2gh} = \sqrt{2g(100)} = 80.2$ ft/s at position 2. The maximum power output occurs at position 2, with the value

$$P_{max} = Wv_2 = 2.8 \text{ lb}\left(80.2 \frac{\text{ft}}{\text{s}}\right) = 225 \frac{\text{ft} \cdot \text{lb}}{\text{s}} = 225 \frac{\text{ft} \cdot \text{lb}}{\text{s}}\left(\frac{1 \text{ hp}}{550 \text{ ft} \cdot \text{lb/s}}\right) = 0.409 \text{ hp} \qquad (1)$$

(b) The maximum power output in SI units is

$$P_{max} = 0.409 \text{ hp}\left(\frac{746 \text{ W}}{1 \text{ hp}}\right) = 305 \text{ W} \qquad (2)$$

W = 2.8 lb

100 ft

$v_2 = 80.2$ ft/s          **Fig. 19.100**

In this problem the power varies continually with time. Thus, Eqs. (1) and (2) represent the *instantaneous* power output of the mass particle at position 2. As a comparison with the results in Eqs. (1) or (2), the instantaneous power output at the initial position is zero, since the particle is at rest in this position.

**19.101**  An automobile weighs 4,000 lb. The coefficient of friction between the tires and the roadway is 0.6.

(*a*)  Find the instantaneous power dissipation when the driver locks the brakes and starts to slide, when the vehicle is traveling at a speed of 60 mi/h, and find the energy expended in bringing the automobile to rest.

(*b*)  Do the same as in part (*a*), if the initial speed of the automobile is 70 mi/h.

(*c*)  Do the same as in part (*a*), if the initial speed of the automobile is 80 mi/h.

**❙** (*a*)  The friction force of the roadway on the automobile is given by

$$F = \mu N = \mu mg = \mu W = 0.6(400) = 2,400 \text{ lb}$$

Using  60 mi/h = 88.0 ft/s,  the power dissipated when the automobile starts to slide is

$$P = Fv = \frac{2,400(88.0)}{550} = 384 \text{ hp}$$

The energy expended in bringing the automobile to rest is

$$T = \tfrac{1}{2}mv^2 = \frac{1}{2}\left(\frac{4,000}{32.2}\right)88.0^2 = 4.81 \times 10^5 \text{ ft} \cdot \text{lb}$$

(*b*)  The power dissipated, using  70 mi/h = 103 ft/s,  is given by

$$P = Fv = \frac{2,400(103)}{550} = 449 \text{ hp}$$

The energy expended is

$$T = \tfrac{1}{2}mv^2 = \frac{1}{2}\left(\frac{4,000}{32.2}\right)103^2 = 6.59 \times 10^5 \text{ ft} \cdot \text{lb}$$

(*c*)  Using  80 mi/h = 117 ft/s,  the power dissipated is

$$P = Fv = \frac{2,400(117)}{550} = 511 \text{ hp}$$

The energy expended is

$$T = \tfrac{1}{2}mv^2 = \frac{1}{2}\left(\frac{4,000}{32.2}\right)117^2 = 8.50 \times 10^5 \text{ ft} \cdot \text{lb}$$

The percent increase in speed from 60 to 80 mi/h is 33.3 percent. The corresponding increase in maximum power dissipation is also 33.3 percent. The corresponding increase in energy expended to bring the vehicle to rest is

$$\%\text{D} = \frac{8.50 \times 10^5 - 4.81 \times 10^5}{4.81 \times 10^5}\, 100 = 76.7\%$$

**19.102** (a) The block in Fig. 19.102a moves downward with a constant velocity of 7.5 m/s, and $\beta = 21.8°$. Find the power used to overcome the friction forces.

(b) The angle of the inclined surface is increased to 40°, and the block has a velocity of 7.5 m/s at the instant that it passes position a. Find the instantaneous power dissipation when the block is at position b.

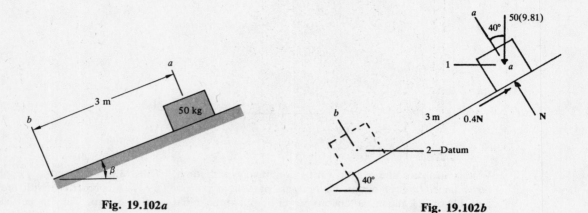

Fig. 19.102a            Fig. 19.102b

❚ (a) Since the block moves with constant velocity, this element is in static equilibrium. The friction force $\mu(mg \cos \beta)$ is equal to the component $mg \sin \beta$ of the weight force, so that

$$\mu = \tan \beta = \tan 21.8° = 0.4$$

The friction force is given by

$$F = \mu(mg \cos \beta) = 0.4(50)9.81 \cos 21.8° = 182 \text{ N}$$

The power required to overcome the friction force is found from

$$P = Fv = 182(7.5) = 1,370 \text{ W}$$

An alternative solution to this problem is as follows. The power expended to overcome the friction force comes from the component of the weight force parallel to the plane. With this interpretation,

$$P = Fv = (mg \sin \beta)v = [50(9.81) \sin 21.8°]7.5 = 1,370 \text{ W}$$

(b) Figure 19.102b shows the endpoints of the interval. The energy terms at points 1 and 2 are

$$T_1 = \tfrac{1}{2}(50)7.5^2 = 1,410 \text{ J} \qquad T_2 = \tfrac{1}{2}(50)v_2^2$$
$$V_1 = 50(9.81)(3 \sin 40°) = 946 \text{ J} \qquad V_2 = 0$$

Using $\mu = 0.4$ from part (a), the nonconservative work done is given by

$$W_{NC} = -0.4(50)9.81 \cos 40°(3) = -451 \text{ J}$$

The work-energy equation has the form

$$W_{NC} = \Delta T + \Delta V = (T_2 - T_1) + (V_2 - V_1) \qquad -451 = [\tfrac{1}{2}(50)v_2^2 - 1,410] + (0 - 946)$$
$$v_2 = 8.73 \text{ m/s}$$

The instantaneous power dissipation when the block is at position b is given by

$$P = Fv = [50(9.81) \sin 40° - 0.4(50)9.81 \cos 40°]8.72 = 1,440 \text{ W}$$

**19.103** (a) The drum and weight system in Fig. 19.103a is released from rest. The mass moment of inertia of the drum is $0.61 \text{ lb} \cdot \text{s}^2 \cdot \text{in}$, and the coefficient of friction between the brake shoe and the drum is 0.4. When the weight has lowered 3 ft, a force $P_a = 45 \text{ lb}$ is applied to the brake lever. Find how much farther the weight will lower before it comes to rest.

(b) Find the maximum value of the instantaneous power dissipation during the braking process.

❚ (a) The velocity after the weight has lowered 3 ft is found first. Figure 19.103b shows the endpoints 1 and 2 of this interval of motion. The energy terms at 1 and 2 have the forms

$$T_1 = 0 \qquad T_2 = \tfrac{1}{2}mv_2^2 + \tfrac{1}{2}I_0\omega_2^2 \qquad V_1 = 40(3)12 \qquad V_2 = 0$$

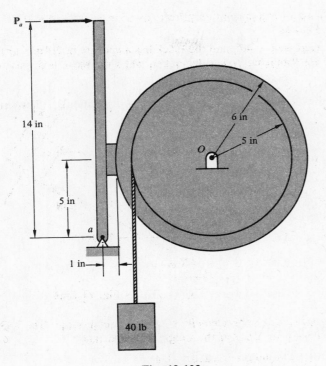

**Fig. 19.103a**

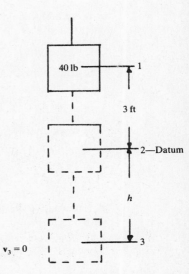

**Fig. 19.103b**

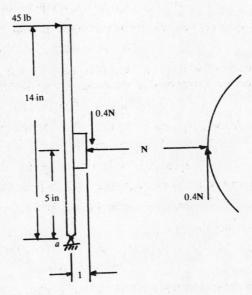

**Fig. 19.103c**

Using $v_2 = r\omega_2$ and $\omega_2 = v_2/r = v_2/5$ in the above equation results in

$$T_2 = \frac{1}{2}\left(\frac{40}{386}\right)v_2^2 + \frac{1}{2}(0.61)\left(\frac{v_2}{5}\right)^2 = 0.0640v_2^2$$

The work-energy equation has the form

$$0 = \Delta T + \Delta V = (T_2 - T_1) + (V_2 - V_1) = (0.0640v_2^2 - 0) + (0 - 1{,}440) \qquad v_2 = 150 \text{ in/s}$$

Figure 19.103c shows the components of the force transmitted between the brake shoe and the drum. For moment equilibrium of the lever,

$$\sum M_a = 0 \qquad -45(14) - 0.4N(1) + N(5) = 0 \qquad N = 137 \text{ lb}$$

The friction force has the value

$$F = 0.4(137) = 54.8 \text{ lb}$$

$M$ is the moment acting on the drum due to the friction force, with the value

$$M = 54.8(6) = 329 \text{ in} \cdot \text{lb}$$

$h$ is the additional distance that the block moves before coming to rest at position 3. The energy terms at position 3 are

$$T_3 = 0 \qquad V_3 = -mgh = -40h \qquad \text{in} \cdot \text{lb}$$

The translational and angular motions are related by

$$h = r\theta \qquad \theta = \frac{h}{r} = \frac{h}{5} \qquad \text{rad}$$

The work-energy equation has the form

$$W_{NC} = \Delta T + \Delta V = (T_3 - T_2) + (V_3 - V_2)$$

$$-329\theta = -329\left(\frac{h}{5}\right) = [0 - 0.0640(150^2)] + (-40h - 0) \qquad h = 55.8 \text{ in} = 4.65 \text{ ft}$$

An alternative method of solution is to use the work-energy equation directly between points 1 and 3, with the form

$$W_{NC} = \Delta T + \Delta V = (T_3 - T_1) + (V_3 - V_1)$$

$$-329\left(\frac{h}{5}\right) = (0 - 0) + (-40h - 1,400) \qquad h = 55.8 \text{ in} = 4.65 \text{ ft}$$

It may be observed that this method does not yield $v_2$, which is needed in the part $(b)$ solution.

$(b)$ The power dissipation is given by

$$P = Fv$$

The friction force which acts on the drum is

$$F = 54.8 \text{ lb} = \text{constant}$$

The power dissipation $P$ will be maximum when $v$ is maximum. The maximum value of $v$ is

$$v_{max} = v_2 = 150 \text{ in/s}$$

The maximum value of the instantaneous power dissipation during the braking process is

$$P_{max} = Fv_{max} = \frac{54.8(150)}{12(550)} = 1.25 \text{ hp}$$

The above power dissipation occurs at position 2, at the onset of braking.

**19.104** Do the same as in Prob. 19.103, if the system is rearranged as shown in Fig. 19.104.

$(a)$ This problem is the same as Prob. 19.103, except that the sense of the friction force $0.4\,N$ changes. Figure 19.103$b$ is used, and $v_2 = 150$ in/s, as in Prob. 19.103. Using Fig. 19.103$c$, with the sense of the friction force $0.4\,N$ changed,

$$\sum M_a = 0 \qquad -45(14) + 0.4N(1) + N(5) = 0 \qquad N = 117 \text{ lb}$$

The friction force is given by

$$F = 0.4(117) = 46.8 \text{ lb}$$

The friction moment is expressed as

$$M = 46.8(6) = 281 \text{ in} \cdot \text{lb}$$

The energy terms at points 1 and 3 are the same as in Prob. 19.103, and

$$W_{NC} = -281\theta$$

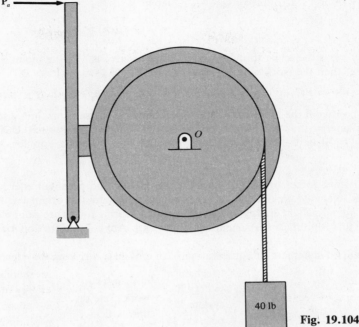

**Fig. 19.104**

The work-energy equation between 1 and 3 has the form

$$W_{NC} = \Delta T + \Delta V = (T_3 - T_1) + (V_3 - V_1) \qquad -281\left(\frac{h}{5}\right) = (0 - 0) + (-40h - 1{,}440)$$

$$h = 88.9 \text{ in} = 7.41 \text{ ft}$$

(*b*)  The maximum value of the instantaneous power dissipation is given by

$$P = Fv$$

Using  $F = 46.8 \text{ lb} = \text{constant}$  and  $v_{max} = v_2 = 150 \text{ in/s},$  the power term is

$$P_{max} = \frac{46.8(150)}{12(550)} = 1.06 \text{ hp}$$

The above power dissipation occurs at the onset of braking.

**19.105**  Give a summary of the basic concepts of work-energy methods for particles and rigid bodies.

▮  In its most fundamental conception energy is understood to mean the capacity for doing work.  Energy may be stored in a mass particle by virtue of the motion of this element.  This form of energy is called kinetic energy, with the basic form

$$T = \tfrac{1}{2}mv^2$$

Energy may also be stored in a mass particle by virtue of the position of the particle.  This form of energy is called potential energy.  When motion of a particle occurs, potential energy may be converted to kinetic energy, and kinetic energy may be converted to potential energy.

The work done by a force is defined to be

$$W_{12} = \int_1^2 (F \cos \beta) \, ds$$

where 1 and 2 represent the endpoints of the distance through which the point of application of the force moves.  If both the magnitude and the orientation of the force with respect to the displacement are constant, then the work may be expressed as

$$W_{12} = (F \cos \beta)l$$

The work done by a couple or torque, or moment, is

$$W_{12} = \int_{\theta_1}^{\theta_2} M \, d\theta$$

where $\theta_2 - \theta_1$ is the angle through which the moment acts. If the moment has constant magnitude, direction, and sense, then the work may be expressed as

$$W_{12} = M(\theta_2 - \theta_1)$$

Work is positive if the sense of the force or moment is the same as the sense of the displacement. The units of work or energy are the product of force and length. The usual USCS units are inch-pounds or foot-pounds. In SI units, the fundamental work and energy units are joules, written as J, where

$$1 \, J = 1 \, N \cdot m$$

Conservative forces produce changes in the kinetic and potential energy which are fully recoverable if the system is imagined to operate in reverse and return to the original condition. Thus, conservative forces produce no permanent energy losses. Weight and spring forces are examples of conservative forces. Nonconservative forces produce permanent energy losses. Applied and friction forces are examples of nonconservative forces.

The general statement of conservation of energy in a mechanical system is

$$\begin{matrix} \text{Mechanical} \\ \text{energy in} \\ \text{system in} \\ \text{position 1} \end{matrix} + \begin{matrix} \text{nonconservative} \\ \text{work done on} \\ \text{the particle} \end{matrix} = \begin{matrix} \text{mechanical} \\ \text{energy in} \\ \text{system in} \\ \text{position 2} \end{matrix} \qquad (1)$$

This equation may be written in the form

$$W_{NC} = \Delta T + \Delta V$$

where $W_{NC}$ is the nonconservative work, $\Delta T$ is the change in kinetic energy, and $\Delta V$ is the change in potential energy. If there is no nonconservative work done, the work-energy equation has the form

$$0 = \Delta T + \Delta V = (T_2 - T_1) + (V_2 - V_1)$$

The potential energy of a rigid body may be found by imagining all the mass of the body to be concentrated at the center of mass and treating this point as a particle in translation. The kinetic energy of a rigid body which rotates about a fixed point $a$ with angular velocity $\omega$ is

$$T = \tfrac{1}{2} I_a \omega^2$$

where $I_a$ is the mass moment of inertia about the fixed point $a$. If the rigid body moves in general plane motion, the kinetic energy has the form

$$T = \tfrac{1}{2} m v_c^2 + \tfrac{1}{2} I_0 \omega^2$$

where $v_c$ is the absolute velocity of the center of mass and $I_0$ is the mass moment of inertia of the body about this point.

The work-energy method considers states, or conditions, at only the endpoints of the time interval of interest. The solutions yield no information whatsoever about the conditions during the interval. The acceleration terms, and time, never enter directly into these solutions.

The normal acceleration of a particle which moves in plane curvilinear translation is given by

$$a_n = \frac{v^2}{\rho}$$

where $\rho$ is the radius of curvature of the path. In many problems the path of the motion lies in a vertical plane. For such cases the work-energy method may be used to find the velocity at the point of interest. This value may then be used in the above equation to find the normal acceleration of the particle.

Power is defined to be the work done unit time. In translational systems, the power is given by

$$P = (F \cos \beta) v$$

In rotational systems, the power is expressed as

$$P = M \omega$$

The fundamental units of power are the units of energy divided by time. In USCS units, power is expressed in foot-pounds per second or foot-pounds per minute. The horsepower unit is defined by

$$1 \text{ hp} = 550 \text{ ft} \cdot \text{lb}/\text{s} = 33{,}000 \text{ ft} \cdot \text{lb}/\text{min}$$

In SI units power is expressed in newton-meters per second, where

$$1 \text{ N} \cdot \text{m}/\text{s} = 1 \text{ J}/\text{s} = 1 \text{ W}$$

The symbols J and W in the above equation represent joules and watts, respectively. The conversion relationship for power in the two systems of units is

$$1 \text{ hp} = 746 \text{ W}$$

The efficiency of a mechanical system is defined by

$$\eta = \frac{\text{energy, or power, output}}{\text{energy, or power, input}}$$

Since all real systems have power losses, it follows that

$$\eta < 1$$

# CHAPTER 20
# Impulse-Momentum Methods for Particles and Rigid Bodies

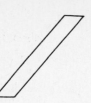

## 20.1 IMPULSE OF A FORCE AND LINEAR MOMENTUM OF A PARTICLE

**20.1**  Summarize the two general methods of solving problems in dynamics: by the use of Newton's second law and by the use of the work-energy method.

❙  In the preceding chapters two methods of solving problems in dynamics have been used.  The general characteristics of these methods are summarized here.

### Direct Application of Newton's Second Law

In solving a problem by the direct application of Newton's second law, a necessary preliminary step is to draw a free-body diagram of the body under consideration.  The resultant force, or couple, torque, or moment, which acts on the body is then used in equations of the form

$$F = ma$$

for cases of translation, or

$$M = I\alpha$$

for cases of rotation.  These equations are solved, and the basic quantity found is the *acceleration* of the body.  If the resultant force, or couple, torque, or moment, which acts on the body is *constant*, then the acceleration of the body will be *constant*.  This value may then be used in known kinematic equations with the forms

$$v = v_0 + at \qquad \omega = \omega_0 + \alpha t$$
$$v^2 = v_0^2 + 2a(s - s_0) \qquad \omega^2 = \omega_0^2 + 2\alpha(\theta - \theta_0)$$
$$s = s_0 + v_0 t + \tfrac{1}{2}at^2 \qquad \theta = \theta_0 + \omega_0 t + \tfrac{1}{2}\alpha t^2$$

The above equations relate the acceleration to the displacements, velocities, and times of the problem.  The results of these computations are a *complete time history* of the motion of the body under consideration.  The solution of problems in dynamics by the direct use of Newton's second law yields the *maximum* possible amount of information about the motion of the body.

### Solutions Using Energy Methods

When using the energy method, the problem is described in terms of position coordinates, and velocities, at the endpoints of the time interval of interest.  Thus, the solution obtained must be in terms of a displacement coordinate or a velocity.

There are two distinct steps in solving problems in dynamics by using energy methods.  One step is to express the *potential energy* of the body with respect to some arbitrarily chosen datum.  The other is to express the velocity of the body in terms of coordinates describing the motion of the body.  This velocity is then used to find the kinetic energy of the body.  The potential and kinetic energy terms are then used in an equation which is a basic statement of conservation of energy.  The acceleration of the body, and the time, never enter directly into the solution, and the work-energy method provides information about *only* the endpoints of the time interval under consideration.  Thus, the solution is inherently incomplete, since it does not give a complete time history of the motion.  In many problems, however, only the states of motion at these endpoints of the interval are of interest.  The lack of a complete time history of motion is the characteristic which distinguishes work-energy solutions from solutions obtained by the direct application of Newton's second law.

**20.2**  Figure 20.2*a* shows a sphere which is dropped from a height *h* onto a horizontal plate.  It is desired to find the maximum height to which the sphere will rebound.  Show that this problem cannot be solved by using either of the two methods described in Prob. 20.1.

❙  This problem may be divided into three regimes of motion.  In the first regime, the sphere is in free fall until it contacts the plate.  In the second, the sphere is in contact with the plate.  In the third, the sphere moves upward with constant deceleration until the final height is reached.

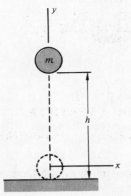

Fig. 20.2*a*

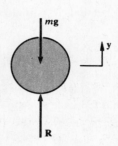

Fig. 20.2*b*

The motion in the first and third regimes may be found readily by using either Newton's second law or the work-energy method. The free-body diagram of the sphere during the second regime is shown in Fig. 20.2*b*, and *R* is the force exerted by the plate on the sphere. The equation of motion for this case is

$$R - mg = m\ddot{y} \qquad (1)$$

From physical reasoning, it may be deduced that the reaction force *R* is some variable function of time. It follows from Eq. (1) that the acceleration is *not* constant. The symbol $\ddot{y}$, rather than $a_y$, is used to emphasize this effect. Both the reaction force *R* and the acceleration $\ddot{y}$ of the sphere are unknown. Thus, Eq. (1) is one equation in two unknowns, and this equation cannot be solved directly with the information given.

The thought might occur to the reader that a second, independent equation could be written that is a statement of conservation of energy, which would relate the kinetic energies, and thus the velocities, of the sphere before and after its contact with the plate. Unfortunately this cannot be done, since experimental evidence shows that the sphere will rebound to a height which is *less than* the height from which it was dropped. Thus, there is an energy loss of unknown magnitude in the problem.

This example typifies a class of problems in which one body impacts a second body. It is in this type of problem that the impulse-momentum methods presented in this chapter find their maximum utility. It will be seen in Prob. 20.16 that the problem considered above may be treated by the introduction of a term which is referred to as the *coefficient of restitution*.

**20.3** A particle which moves in rectilinear translation is acted on by a resultant force *R*.

(*a*) Give the form of the *impulse-momentum equation* for the particle.

(*b*) What is the *linear momentum* of the particle?

(*c*) What are the units of linear momentum?

(*d*) What is the graphical interpretation of an impulse?

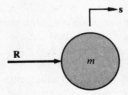

Fig. 20.3

▌ (*a*) Figure 20.3 shows the particle. The displacement coordinate of the particle is defined to be *s*, with the associated velocity and acceleration *v* and *a*. As before, the implication of particle motion is that there are no rotation effects.

Newton's second law may be written as

$$R = ma \qquad (1)$$

Since

$$a = \frac{dv}{dt} \qquad (2)$$

Eq. (1) may be written as

$$R = m \frac{dv}{dt} \qquad (3)$$

Equation (3) is multiplied by $dt$ and integrated, with the result

$$\int_1^2 R \, dt = \int_1^2 m \, dv = mv \Big|_1^2 = m(v_2 - v_1) \qquad (4)$$

where 1 and 2 designate the endpoints of the time interval of interest. The term on the left side of Eq. (4) is defined to be the *impulse* of the force $R$. The impulse of a force is a vector quantity which has the same direction as the force $R$. If $R$ is a known function of time, this integral can be evaluated directly. In the general case, typified by Prob. 20.2, $R$ is *not* known. For this case the quantity

$$I' = \int_1^2 R \, dt \qquad (5)$$

is a symbolic term which is used to represent the impulse $I'$ of the force $R$. This integral is usually not evaluated directly in the solution of problems. It may be seen from Eq. (5) that the units of the impulse of a force are the product of force and time.

The numerical value of an impulse is not in itself a particularly useful quantity. The force associated with the impulse is the desired quantity. In order to obtain the value of this force, it is necessary to know the duration of the impulse and the variation of the force of the impulse with respect to time. In the majority of actual problems this latter information is not known, nor can it be obtained easily.

The numbers 1 and 2 designate the endpoints of the time interval of interest in the problem. The quantity $v_2 - v_1$ represents the *change in the velocity* of the particle, while the quantity $m(v_2 - v_1)$ represents the *change in the linear momentum* of the particle. The general conclusion may now be drawn from Eq. (4) that the effect of an impulse acting on a particle is to produce a change in the velocity of the particle.

Equation (4) is referred to as the impulse-momentum equation for a particle.

(b) The two terms on the right side of Eq. (4) are the product of the mass of the particle and the velocity of this element. The product $mv$ is defined to be the linear momentum of the particle, and this quantity is of fundamental importance in engineering dynamics. Since the linear momentum is expressed in terms of the velocity $v$, it follows that *linear momentum is a vector quantity*.

(c) It may be seen that the units of linear momentum are the product of mass and velocity. From consideration of Eq. (4), it follows that these units are the same as the units of impulse. The units of both of the above quantities are the product of force and time. These units are typically pound-seconds in USCS units and newton-seconds in SI units.

(d) The impulse given by Eq. (5) may be interpreted as the area under the force-time curve of the resultant force acting on the particle. If this curve is known, the area under this curve may then be used directly to find the velocity *change* of the particle.

20.4 What is the form of the impulse-momentum equation for a particle in rectilinear translation if the force which acts on the particle has a constant value?

❚ If the resultant force acting on the particle is constant, then the impulse of this force may be written as

$$I' = \int_1^2 R \, dt = R \int_1^2 dt = Rt \Big|_1^2 = R(t_2 - t_1)$$

This result is combined with Eq. (4) in Prob. 20.3, to obtain

$$R(t_2 - t_1) = m(v_2 - v_1)$$

The above equation is an elementary relationship between the times and velocities at the endpoints of the interval of interest. It should be noted that this equation is true *only if the resultant force which acts on the particle has a constant value*.

20.5 An automobile moves along a straight roadway with a constant speed of 40 mi/h. A constant friction force exerted by the roadway on the tires of the automobile acts for 5 s, until the automobile reaches a speed of 50 mi/h. The weight of the automobile is 3,600 lb.

(*a*)  Find the required value of the constant friction force that will produce this change of velocity.

(*b*)  Do the same as in part (*a*), if the speed of the car changes from 50 mi/h to 60 mi/h in 5 s.

**▮** (*a*)  The velocities at the endpoints of the interval are

$$40 \text{ mi/h} = 58.7 \text{ ft/s} \qquad 50 \text{ mi/h} = 73.3 \text{ ft/s}$$

The impulse-momentum equation has the form

$$\int_1^2 R \, dt = R(t_2 - t_1) = m(v_2 - v_1)$$

The friction force is $F$, and

$$F(5) = \frac{3{,}600}{32.2}(73.3 - 58.7) \qquad F = 326 \text{ lb}$$

(*b*)  When the automobile changes speed from 50 to 60 mi/h,

$$50 \text{ mi/h} = 73.3 \text{ ft/s} \qquad 60 \text{ mi/h} = 88.0 \text{ ft/s}$$

$$\int_1^2 R \, dt = R(t_2 - t_1) = m(v_2 - v_1)$$

$$F(5) = \frac{3{,}600}{32.2}(88.0 - 73.3) \qquad F = 329 \text{ lb} \approx 326 \text{ lb}$$

It may be seen that the required friction force is a function of the *change* in velocity, rather than the value of the velocity.  The above result may be verified by using the values obtained in Prob. 15.13 for the acceleration of the automobile.

**20.6**  The particle in Fig. 20.6 is acted on by the system of forces shown.  It moves with rectilinear translation in a frictionless guide.  At $t = 0$ the particle is at rest.  Find the velocity when $t = 2$ s.

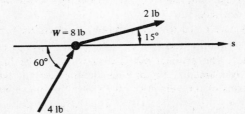

**Fig. 20.6**

**▮** The resultant force which acts on the particle is constant.  The impulse-momentum equation is written as

$$\int_1^2 R \, dt = R(t_2 - t_1) = m(v_2 - v_1)$$

$$(4 \cos 60° + 2 \cos 15°)2 = \frac{8}{32.2}(v - 0) \qquad v = 31.7 \text{ ft/s}$$

The result  $v = 31.7 \approx 31.6 \text{ ft/s}$  is the same as that obtained in Prob. 15.17 by direct use of Newton's second law.

**20.7**  The system of Prob. 15.29 is repeated in Fig. 20.7.  A block weighing 130 lb slides along a smooth, horizontal track with a velocity of 50 ft/s.  At a certain point the smooth track joins a section of rough horizontal track.  The coefficient of friction between the block and the rough track is 0.24.  How long will it take the block to come to rest after it contacts the rough track?

**▮** State 1 occurs when the block first contacts the rough horizontal track, and state 2 occurs when the block comes to rest.  The constant force which acts on the block is the friction force, given by

$$F = -\mu N = -0.24(130) = -31.2 \text{ lb}$$

The minus sign is required in the above equation since the friction force acts in a sense which is opposite to that of the motion.  The remaining terms which are required are

$$t_1 = 0 \qquad t_2 = t \qquad v_1 = 50 \text{ ft/s} \qquad v_2 = 0$$

$$m = \frac{130}{32.2} = 4.04 \text{ lb} \cdot \text{s}^2/\text{ft}$$

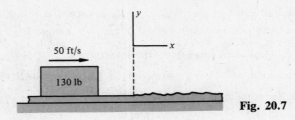

Fig. 20.7

The impulse-momentum equation, for constant force, now appears as

$$\int_1^2 R\,dt = R(t_2 - t_1) = m(v_2 - v_1) \qquad -31.2(t - 0) = 4.04(0 - 50) \qquad t = 6.47\,\text{s}$$

This is the result previously obtained in Prob. 15.29, where this problem was solved by direct application of Newton's second law.

**20.8**   The block in Fig. 20.8a is released from rest at $t = 0$.

(a)   Find the velocity of the block at the end of 3 s.

(b)   Do the same as in part (a) if, at the instant shown in the figure, the block has a velocity of 6 ft/s up the plane.

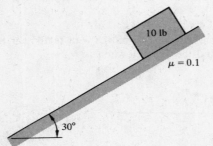

Fig. 20.8a

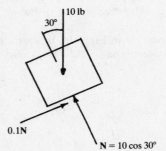

Fig. 20.8b

▌ (a)   Figure 20.8b shows the free-body diagram of the block.   The impulse-momentum equation has the form

$$\int_1^2 R\,dt = m(v_2 - v_1) \qquad [10 \sin 30° - 0.1(10 \cos 30°)]3 = \frac{10}{32.2}\,(v_2 - 0) \qquad v_2 = 39.9\,\text{ft/s}$$

(b)   When the block moves up the plane, the sense of the friction force is opposite that shown in Fig. 20.8b.   $t_1 = 0$   is defined to be the initial time of the problem, with   $v_1 = 6$ ft/s,   and $t_2$ is the time when the block comes to rest, with   $v_2 = 0$.   For motion up the plane

$$\int_1^2 R\,dt = m(v_2 - v_1) \qquad [-10 \sin 30° - 0.1(10 \cos 30°)]t_2 = \frac{10}{32.2}\,(0 - 6) \qquad t_2 = 0.318\,\text{s}$$

Time $t_3$ is defined by,

$$t_3 = 3 - t_2 = 3 - 0.318 = 2.68\,\text{s}$$

For motion down the plane, the sense of the friction force is that shown in Fig. 20.8b.   The impulse-momentum equation is written as

$$\int_2^3 R\,dt = m(v_3 - v_2) \qquad [10 \sin 30° - 0.1(10 \cos 30°)]2.68 = \frac{10}{32.2}\,(v_3 - 0) \qquad v_3 = 35.7\,\text{ft/s}$$

**20.9**   A block rests on a horizontal surface, as shown in Fig. 20.9a.   It is acted on by an applied force $P$, and the variation of this force with time is shown in Fig. 20.9b.

(a)   If the block is initially at rest, find the velocity of the block when   $t = 5$ s.

(b)   Verify the result in part (a) by direct use of Newton's second law.

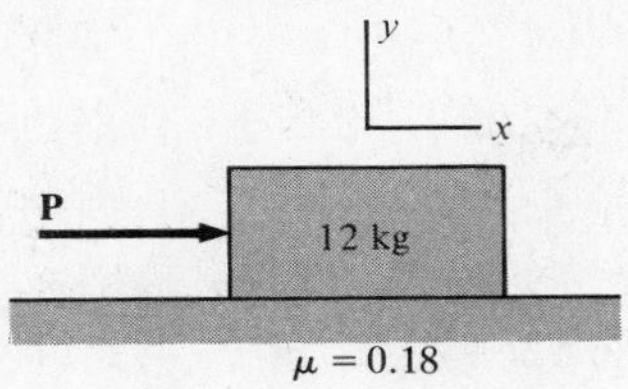

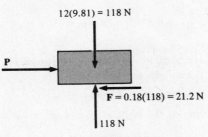

Fig. 20.9a

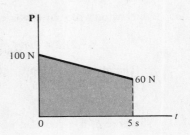

Fig. 20.9b

12(9.81) = 118 N

P

F = 0.18(118) = 21.2 N

118 N

Fig. 20.9c

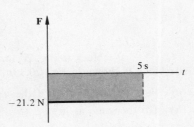

Fig. 20.9d

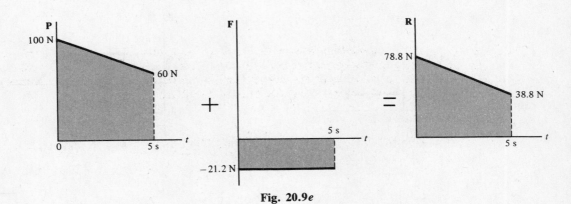

Fig. 20.9e

**(a)** The free-body diagram of the block is shown in Fig. 20.9c. The force-time diagram of the friction force is shown in Fig. 20.9d. The force-time curve of the *resultant force* in the $x$ direction, which acts on the block, is the sum of the curves in Figs. 20.9b and 20.9d, and this result is shown in Fig. 20.9e.

The equation of impulse momentum now appears as

$$I' = \int_1^2 R\, dt = m(v_2 - v_1)$$

$\int_1^2 R\, dt$ is the area under the $R - t$ curve. The above equation has the form

$$\frac{78.8 + 38.8}{2}(5) = 12(v_2 - 0) \qquad v_2 = 24.5\,\text{m/s}$$

**(b)** The equation of the $P - t$ curve is

$$P = mt + b$$

Using the endpoint conditions $t = 0$, $P = 100\,\text{N}$; $t = 5\,\text{s}$, $P = 60\,\text{N}$, the constants $m$ and $b$ are found as

$$m = -8\,\text{N/s} \qquad b = 100\,\text{N}$$

The force $P$ is then

$$P = -8t + 100\,\text{N}$$

Newton's second law is written as

$$P - 21.2 = ma = m\ddot{x} \quad (-8t + 100) - 21.2 = 12\ddot{x} \quad \ddot{x} = \frac{d^2x}{dt^2} = -0.667t + 6.57 \quad \text{m/s}^2$$

The above equation is integrated, to obtain

$$\frac{dx}{dt} = v = -0.667\frac{t^2}{2} + 6.57t + c_1 \quad \text{m/s} \tag{1}$$

where $c_1$ is a constant of integration. Using the initial condition $t = 0$, $v = 0$ in Eq. (1) gives

$$c_1 = 0$$

At $t = 5$ s,

$$v = -0.667\frac{(5)^2}{2} + 6.57(5) = 24.5 \text{ m/s}$$

20.10 At $t = 0$ the block in Fig. 20.10a is moving to the left at 10 ft/s. The time variation of the force that acts on the block is shown in Fig. 20.10b, and this force is positive in the sense shown in the figure.

(a) Find the value of $P_1$ if the block is to move leftward at 26 ft/s, when $t = 10$ s.

(b) Do the same as in part (a), if the block is moving to the right at 10 ft/s at $t = 0$.

▌ (a) Figure 20.10c shows the conditions at $t = 0$. Initially, force $P$ and the friction force oppose motion. The first step is to find the time $t_2$ when $v_2 = 0$. Using $t_1 = 0$, the impulse-momentum equation has the form

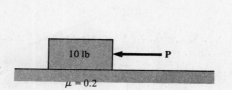

Fig. 20.10a

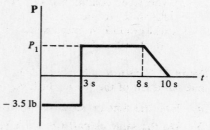

Fig. 20.10b

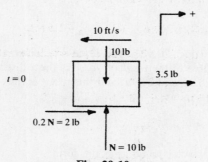

Fig. 20.10c

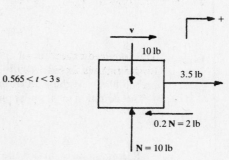

Fig. 20.10d

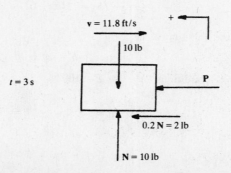

Fig. 20.10e

$$\int_1^2 R\,dt = m(v_2 - v_1) \qquad (3.5+2)t_2 = \frac{10}{32.2}[0-(-10)] \qquad t_2 = 0.565\,\text{s}$$

The conditions for $0.565 < t < 3\,\text{s}$ are shown in Fig. 20.10d. The velocity $v_3$ when $t_3 = 3\,\text{s}$ is found next. Using the impulse-momentum equation,

$$\int_2^3 R\,dt = m(v_3 - v_2) \qquad (3.5-2)(3-0.565) = \frac{10}{32.2}(v_3 - 0) \qquad v_3 = 11.8\,\text{ft/s} \quad \text{(to the right)}$$

The condition at $t=3\,\text{s}$ is shown in Fig. 20.10e. When $t=3\,\text{s}$, from Fig. 20.10b, the force $P_1$ starts to act to the left on the the block. The time $t_4$, when $v_4 = 0$, is found from

$$\int_3^4 R\,dt = m(v_4 - v_3) \qquad (P_1+2)(t_4-3) = \frac{10}{32.2}[0-(-11.8)] \qquad t_4 = \frac{10(11.8)}{32.2(P_1+2)} + 3 \qquad (1)$$

At time $t_5 = 10\,\text{s}$, $v_5 = 26\,\text{ft/s}$ to the left. For $t_4 < t \leqslant 10\,\text{s}$, the friction force opposes force $P$. The impulse-momentum equation has the form

$$\int_4^5 R\,dt = m(v_5 - v_4) \qquad P_1(8-t_4) + \tfrac{1}{2}P_1(10-8) - 2(10-t_4) = \frac{10}{32.2}(26-0)$$

$$9P_1 + (2-P_1)t_4 = 20 + \frac{10(26)}{32.2} \qquad (2)$$

Using $t_4$ from Eq. (1) in Eq. (2) gives

$$P_1^2 - 2.29P_1 - 6.14 = 0 \qquad P_1 = \frac{2.29 \pm \sqrt{(2.29)^2 - 4(1)(-6.14)}}{2(1)} = \frac{2.29 \pm 5.46}{2(1)} = 3.88\,\text{lb}$$

(b) The given condition is that the block is moving to the right at 10 ft/s at $t=0$. From part (a), $v_3 = 11.8\,\text{ft/s}$ to the right at $t=3\,\text{s}$. The force $P = P_1$, from Fig. 20.10b, is positive for $t>3\,\text{s}$ and will produce an increase in the rightward velocity of the block. Thus, there is no positive value of $P_1$ which will satisfy the problem statement.

20.11  Figure 20.11a shows a heavy cabinet. All friction effects, and the mass of the wheels, may be neglected. The time variation of the force which acts on the cabinet is shown in Fig. 20.11b.

(a) Find the time at which the velocity of the cabinet is 1.2 m/s to the right, if the cabinet is initially at rest.

(b) Do the same as part (a), for a velocity of 1.2 m/s to the left.

(c) Do the same as in parts (a) and (b), if the cabinet moves along the incline shown in Fig. 20.11c.

▮ (a) The velocity at $t=2\,\text{s}$ is found first. $P$ and $v$ have the same senses, and

$$\int_1^2 R\,dt = m(v_2 - v_1) \qquad 2(150) = 200(v_2 - 0) \qquad v_2 = 1.5\,\text{m/s}$$

The required value $v = 1.2\,\text{m/s}$ will occur at $t_3 < 2\,\text{s}$. The time $t_3$ at which the velocity is 1.2 m/s to the right is found from

$$\int_1^3 R\,dt = m(v_3 - v_1) \qquad 150t_3 = 200(1.2 - 0) \qquad t_3 = 1.6\,\text{s}$$

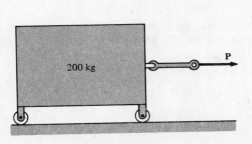

Fig. 20.11a

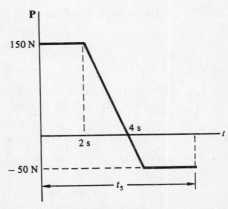

Fig. 20.11b

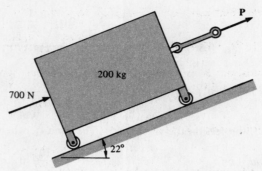

Fig. 20.11c

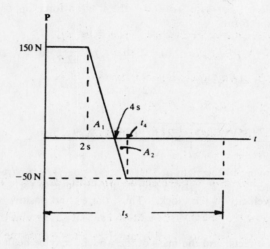

Fig. 20.11d

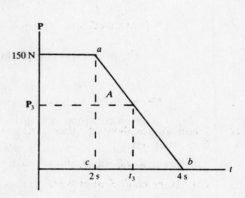

Fig. 20.11e

(b) $P$ and $v$ have opposite senses. From the similar triangles $A_1$ and $A_2$ in Fig. 20.11d,

$$\frac{150}{4-2} = \frac{50}{t_4-4} \qquad t_4 = 4.67 \text{ s}$$

When $t = t_5$, $v_5 = -1.2 \text{ m/s}$. The impulse-momentum equation has the form

$$\int_1^5 R \, dt = m(v_5 - v_1) \qquad 2(150) + \tfrac{1}{2}(2)150 - \tfrac{1}{2}(50)(4.67 - 4) - 50(t_5 - 4.67) = 200(-1.2 - 0)$$

$$t_5 = 18.1 \text{ s}$$

(c) $v = 1.2 \text{ m/s}$ up the incline. $R$ is the resultant force in the positive $P$ sense, given by

$$R = P - 200(9.81) \sin 22° + 700 = P - 35.0 \text{ N}$$

The velocity $v_2$ at $t = 2 \text{ s}$ is found from

$$\int_1^2 R \, dt = m(v_2 - v_1) \qquad \int_0^2 (P - 35) \, dt = m(v_2 - v_1)$$

$$2(150) - 35(2) = 200(v_2 - 0) \qquad v_2 = 1.15 \text{ m/s}$$

It may be seen that the required velocity $v_3 = 1.2 \text{ m/s}$ will occur at $t_1 > 2 \text{ s}$.
From Fig. 20.11e, triangle $A$ is similar to triangle $abc$, so that

$$\frac{150 - P_3}{t_3 - 2} = \frac{150}{4 - 2} \qquad P_3 = -75t_3 + 300 \tag{1}$$

The impulse-momentum equation is written as

$$\int_1^3 R \, dt = m(v_3 - v_1) \qquad \int_0^3 P \, dt - 35t_3 = 200(v_3 - 0) \qquad 150(2) + \left(\frac{150 + P_3}{2}\right)(t_3 - 2)$$

$$-35t_3 = 200(1.2 - 0) \tag{2}$$

$P_3$ is eliminated between Eqs. (1) and (2), to obtain

$$t_3^2 - 7.07t_3 + 10.4 = 0 \qquad t_3 = \frac{7.07 \pm \sqrt{7.07^2 - 4(1)(10.4)}}{2(1)} = \frac{7.07 \pm 2.90}{2(1)} = 2.09, \; 4.99 \text{ s}$$

A determination is next made of which of the above two roots is the solution to the problem. At $t_3 = 2.09$ s, using Eq. (1),

$$P_3 = 300 - 75(2.09) = 143 \text{ N}$$

At $t_3 = 4.99$ s, using Eq. (1),

$$P_3 = 300 - 75(4.99) = -74.3 \text{ N}$$

Since $P_3$ must be positive, the time $t_3$ is given by

$$t_3 = 2.09 \text{ s}$$

Time $t = t_5$ corresponds to $v = 1.2$ m/s, down the incline. From part (a),

$$t_4 = 4.67 \text{ s}$$

It is assumed that $t_5 > 4.67$ s. The impulse-momentum equation has the form

$$\int_1^5 R \, dt = m(v_5 - v_1) \qquad 150(2) + \tfrac{1}{2}(2)150 + \tfrac{1}{2}(4.67 - 4)50$$

$$-50(t_5 - 4.67) - 35t_1 = 200(-1.2) \qquad t_5 = 10.7 \text{ s}$$

Since $10.7 \text{ s} > 4.66 \text{ s}$, the above assumption was correct.

20.12  Figure 20.12a shows the time variation of a force which acts on the block shown in Fig. 20.12b. The block is initially at rest.

(a)  If the surface is frictionless, find the velocity of the block at the end of 1 s, and at the end of 3 s.

(b)  If the coefficient of friction between the block and the plane is 0.15, find the time at which the block starts to move.

(c)  Do the same as in part (a), for the condition of part (b).

❚  (a)  The area under the $P - t$ curve between $t = 0$ and $t = 1$ s is represented by the two areas $A_1$ and $A_2$ in Fig. 20.12c. $v_1$ is the velocity at the end of 1 s. The impulse-momentum equation has the form

$$\int_0^1 R \, dt = m(v_1 - v_0) \qquad A_1 + A_2 = m(v_1 - 0)$$

$$\tfrac{1}{2}(1)(6 - 2.8) + 1(2.8) = \frac{25}{32.2}(v_1 - 0) \qquad v_1 = 5.67 \text{ ft/s}$$

At the end of 3 s,

$$\int_0^3 R \, dt = m(v_3 - v_0)$$

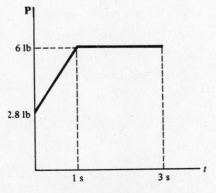

**Fig. 20.12a**

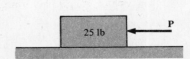

**Fig. 20.12b**

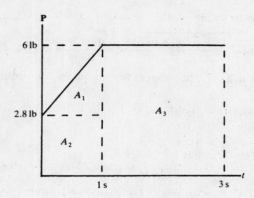

Fig. 20.12c

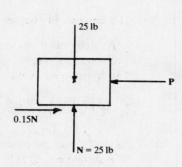

Fig. 20.12d

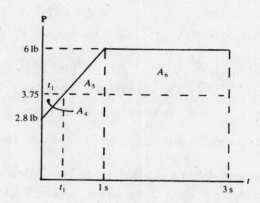

Fig. 20.12e

Using areas $A_1$, $A_2$, and $A_3$ in Fig. 20.12c,

$$A_1 + A_2 + A_3 = m(v_3 - 0) \qquad \tfrac{1}{2}(1)(6 - 2.8) + 1(2.8) + 2(6) = \frac{25}{32.2}\,(v_3 - 0) \qquad v_3 = 21.1\,\text{ft/s}$$

(b) The free-body diagram of the block is shown in Fig. 20.12d. The friction force $F$ which acts on the block is given by

$$F = 0.15\,N = 0.15(25) = 3.75\,\text{lb}$$

The block starts to move when $P = 3.75\,\text{lb}$. The corresponding time is $t_1$, shown in Fig. 20.12e. $A_4$ and $A_5$ in this figure are similar triangles, so that

$$\frac{t_1}{3.75 - 2.8} = \frac{1 - t_1}{6 - 3.75} \qquad t_1 = 0.297\,\text{s}$$

The block starts to move when $t_1 = 0.297\,\text{s}$.

(c) At the end of 1 s,

$$\int_0^1 R\,dt = m(v_1 - v_0) \qquad A_5 = \frac{25}{32.2}\,(v_1 - 0)$$

$$\tfrac{1}{2}(1 - 0.297)(6 - 3.75) = \frac{25}{32.2}\,v_1 \qquad v_1 = 1.02\,\text{ft/s}$$

At the end of 3 s,

$$\int_0^3 R\,dt = m(v_3 - v_0) \qquad A_5 + A_6 = \frac{25}{32.2}\,(v_3 - 0) \qquad \tfrac{1}{2}(1 - 0.297)(6 - 3.75) + 2(6 - 3.75) = \frac{25}{32.2}\,v_3$$

$$v_3 = 6.81\,\text{ft/s}$$

20.13 (a) Do the same as in parts (a) and (c) of Prob. 20.12 if, at $t = 0$, the block has a velocity of 15 ft/s to the right.

(**b**)  The block in Prob. 20.12 has a velocity $v$, to the right, at $t = 0$. For what value of $v$ would the velocity of the block be zero at $t = 3$ s? Consider the two cases with $\mu = 0$ and $\mu = 0.15$.

▌ (**a**)  The impulse-momentum equation has the form

$$\int_0^1 R\, dt = m(v_1 - v_0)$$

The initial velocity $v$ and the force $P$ have opposite senses, so that

$$v_0 = -15 \text{ ft/s}$$

Using Fig. 20.12$c$, the velocity at the end of 1 s is found from

$$A_1 + A_2 = m[v_1 - (-15)] \qquad \tfrac{1}{2}(1)(6 - 2.8) + 1(2.8) = \frac{25}{32.2}(v_1 + 15) \qquad v_1 = -9.33 \text{ ft/s}$$

The above velocity is to the right, and force $P$ and velocity $v$ still have opposite senses. At the end of 3 s,

$$\int_0^3 R\, dt = m(v_3 - v_0)$$

$$A_1 + A_2 + A_3 = m[v_3 - (-15)] = \tfrac{1}{2}(1)(6 - 2.8) + 1(2.8) + 2(6) = \frac{25}{32.2}(v_3 + 15)$$

$$v_3 = 6.12 \text{ ft/s}$$

The velocity is to the left, and force $P$ and velocity $v$ now have the same senses.

When there are friction forces between the sliding surfaces, both the friction force $F$ and the force $P$ oppose motion. The friction force, from part (*b*) of Prob. 20.12, is

$$F = 3.75 \text{ lb}$$

At the end of 1 s, using Fig. 20.12$c$,

$$\int_0^1 R\, dt = m(v_1 - v_0) \qquad A_1 + A_2 + (1)3.75 = m[v_1 - (-15)]$$

$$\tfrac{1}{2}(1)(6 - 2.8) + 1(2.8) + 3.75 = \frac{25}{32.2}(v_1 + 15) \qquad v_1 = -4.50 \text{ ft/s}$$

The above velocity is to the right, and the force $P$ and velocity $v$ have opposite senses. The friction force $F$ and the force $P$ oppose motion until $v = 0$. The corresponding time $t_1$, using Fig. 20.12$c$, is found from

$$\int_0^{t_1} R\, dt = m(v_{t_1} - v_0) \qquad A_1 + A_2 + (t_1 - 1)6 + 3.75 t_1 = m[0 - (-15)]$$

$$\tfrac{1}{2}(1)(6 - 2.8) + 1(2.8) + (t_1 - 1)6 + 3.75 t_1 = \frac{25}{32.2}(15) \qquad t_1 = 1.36 \text{ s}$$

For $t > t_1$, the forces $P$ and $F$ have opposite senses. At the end of 3 s,

$$\int_{t_1=1.36 \text{ s}}^3 R\, dt = m(v_3 - v_{t_1}) \qquad (3 - 1.36)6 - 3.75(3 - 1.36) = \frac{25}{32.2}(v_3 - 0)$$

$$v_3 = 4.75 \text{ ft/s}$$

The velocity is to the left, and force $P$ and velocity $v$ have the same senses.

(**b**)  The impulse-momentum equation is given by

$$\int_1^2 R\, dt = m(v_2 - v_1)$$

Using $v_1 = v$, $v_2 = 0$, and $\mu = 0$, and the areas shown in Fig. 20.12$c$,

$$A_1 + A_2 + A_3 = m[0 - (-v)] \qquad \tfrac{1}{2}(1)(6 - 2.8) + 1(2.8) + 2(6) = 16.4 = \frac{25}{32.2}\, v \qquad v = 21.1 \text{ ft/s}$$

It may be observed that the above value of velocity is the same as that found in part (*a*) of Prob. 20.12, for the velocity at the end of 3 s when the block moves on a frictionless surface. It is left as an exercise for the reader to show why this is an expected result. For $\mu = 0.15$, both the friction force $F$ and the force $P$ oppose motion. The friction force, from part (*b*) of Prob. 20.12, is $F = 3.75$ lb. The impulse-momentum equation, using Fig. 20.12$c$, has the form

$$\int_1^2 R\, dt = m(v_2 - v_1) \qquad A_1 + A_2 + A_3 + 3.75(3) = m[0 - (-v)] \qquad 16.4 + 3.75(3) = \frac{25}{32.2}\, v$$

$$v = 35.6 \text{ ft/s}$$

## 20.2 IMPACT, CONSERVATION OF LINEAR MOMENTUM, COEFFICIENT OF RESTITUTION

**20.14** (*a*) What is meant by the term *impact*?

(*b*) Describe the sequence of events when one body impacts a second body.

(*c*) What three effects characterize the impact problem in dynamics?

(*d*) What is the principal use of solutions to impact problems?

▐ (*a*) Impact is a term that describes the phenomenon of colliding physical bodies. Examples of impact problems abound in the physical world. Typical cases include a hammer driving a nail into a piece of wood, a ball dropped onto the ground, a bowling ball striking pins, and automobile bumpers contacting as a car maneuvers into a tight parking space.

The interactions which occur during impact are extremely complex and not fully understood. If the impacting velocity is sufficiently small, there will be no permanent deformation of the striking bodies; and these bodies, after impact, will be restored to their original shapes. If the velocity is greater than some critical, limiting value, there will be permanent deformation of one, or both, of the impacting bodies. A commonly seen example is the end of a cold chisel. The end of a new chisel is shown in Fig. 20.14*a*, while Fig. 20.14*b* shows the typical appearance of this end after considerable use.

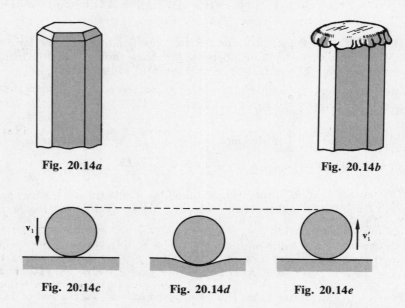

Fig. 20.14*a*                Fig. 20.14*b*

Fig. 20.14*c*        Fig. 20.14*d*        Fig. 20.14*e*

(*b*) A typical sequence of impact of one body with another body is shown in Figs. 20.14*c*, *d*, and *e*. For the purpose of this discussion, one body is assumed to be a sphere and the other body is assumed to be a stationary horizontal plate. In addition, the fundamental assumption in this text—that the bodies considered in the problems are rigid—is relaxed, so that the bodies may be imagined to deform slightly. Figure 20.14*c* shows the sphere in its undeformed condition at the instant that it first makes contact with the plate. In the following time interval, shown in Fig. 20.14*d*, a period of deformation occurs in both bodies. The amount of deformation is a function of the velocity of impact and the materials and shapes of the impacting bodies. At the end of the period of deformation, the undeformed portion of the sphere has zero velocity, and the original potential energy of the sphere, due to elevation, has been expended in compressing portions of the two bodies.

Following the regime of deformation is a regime of restoration, or restitution. During this regime the energy (due to deformation) which is stored in the bodies is converted to kinetic energies, with associated velocities. If it is assumed that no permanent deformation occurs, the appearance of the sphere at the end of restitution is as shown in Fig. 20.14*e*. The original shape of the sphere is restored, and the sphere leaves the plate with an upward velocity $v_1'$. For an idealized case in which there are no permanent energy losses during the impact process, the magnitude of the rebound velocity $v_1'$ would equal to the magnitude of the striking velocity $v_1$. In actual cases of impact of physical bodies, *there is always a permanent energy loss associated with the deformation and restitution phases of the impact*. Thus, the rebound velocity must obey the relationship

$$|v_1'| < |v_1|$$

The absolute value signs are required in the above equation because the senses of $v_1$ and $v_1'$ are opposite. A measure of the energy lost during impact, and the corresponding value of the rebound velocity, is given by a term called the *coefficient of restitution*. This term is defined in Prob. 20.16.

As a consistent usage throughout this chapter, a *prime* will be used to designate a velocity *after* impact. Thus, if $v_A$ and $v_B$ are the velocities of two bodies before they impact each other, their velocities after impact are written as $v_A'$ and $v_B'$.

(c) The following three effects characterize the impact problem in engineering dynamics.

1. The duration of the impact is much less than any other observed times in the problem. The case of dropping a sphere onto the horizontal plate may be simulated by dropping a ball onto the ground. It may be observed that the actual time that the ball is in contact with the ground is much less than either the time that the ball is falling toward the ground or the time that the ball is rebounding away from the ground.

2. Because of the short time duration of the impact, *the impacting bodies are assumed to experience an instantaneous change in velocity, with no change in displacement.* In the problem of the sphere falling on the horizontal plate, shown in Fig. 20.2a, the total impact phenomenon is assumed to occur at a *constant* value of displacement of the sphere given by $y = 0$. The velocity, by comparison, changes from a *downward* value of $v_1$ to an *upward* value of $v_1'$ during the same infinitesimal time duration of the impact.

3. During the impact process, contact reaction forces *of extremely large magnitude* exist for very short time intervals. These forces are assumed to be much greater than any other forces, such as weight, spring, or friction forces, which act on the body.

(d) The principal use of solutions to problems of impact is to obtain a set of initial velocities for the regime of motion following the impact. This effect is illustrated in several problems in this chapter.

**20.15** (a) Give the equation for the conservation of linear momentum of two particles which move in rectilinear translation.

(b) What is a significant limitation of the equation found in part (a)?

▌ (a) Figure 20.15a shows two spheres which move in translation along the same axis. The spheres have the constant velocities $v_A$ and $v_B$. It will be assumed that $v_A > v_B$, so that sphere $A$ will eventually collide with sphere $B$. Figure 20.15b shows the free-body diagrams during the collision, or impact, of the two spheres. During this period of contact there exists a compressive force $R$ between the two spheres and with it an associated impulse $I' = \int_1^2 R\,dt$. From Newton's third law, the impulses acting on the two spheres must have opposite senses. At the cessation of the impact process, the two spheres have the new velocities $v_A'$ and $v_B'$ shown in Fig. 20.15c. All velocities in this problem are considered to be positive quantities if they act in the arbitrarily chosen positive sense shown in the figure.

The equations of impulse-momentum for each sphere are

Sphere $A$:
$$-\int_1^2 R\,dt = m_A(v_A' - v_A) \tag{1}$$

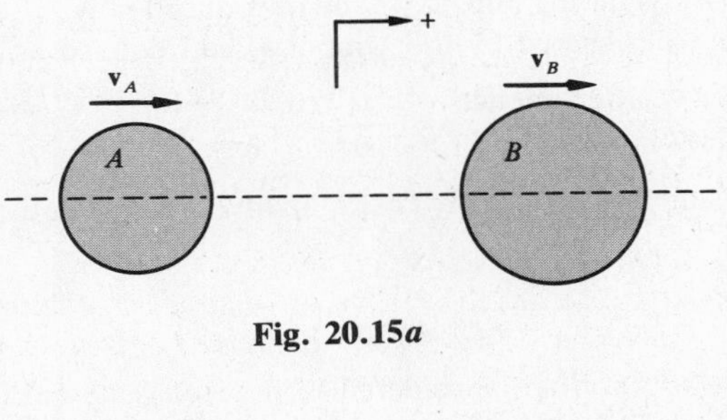

Fig. 20.15a

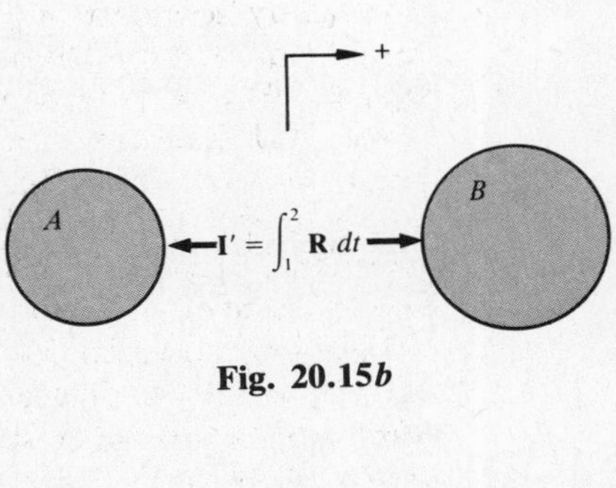

Fig. 20.15b

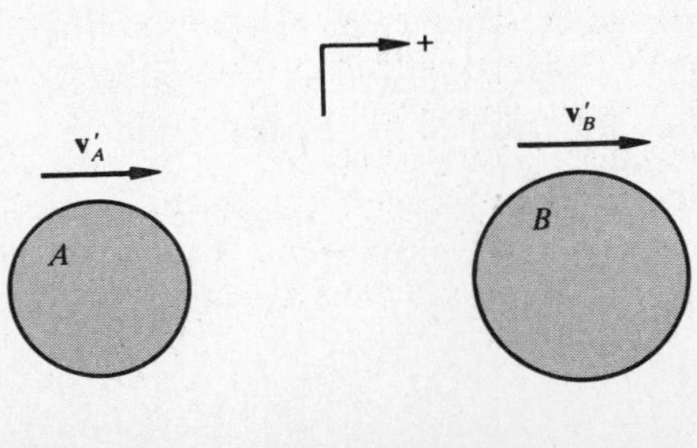

Fig. 20.15c

Sphere $B$:

$$\int_1^2 R\, dt = m_B(v_B' - v_B) \tag{2}$$

The minus sign is required in Eq. (1) to conform with the positive sense of motion shown in the figure. Equations (1) and (2) are added, with the result

$$0 = m_A(v_A' - v_A) + m_B(v_B' - v_B)$$

This equation may be written in the form

$$m_A v_A + m_B v_B = m_A v_A' + m_B v_B' \tag{3}$$

The left side of Eq. (3) is the combined initial momentum which the two spheres possessed before the impact, and the right side of the equation expresses the combined momentum of the two spheres after impact. The impulse which one sphere exerts on the other sphere during the time of impact is an internal force effect within the system of the two spheres. Equation (3) now leads to a conclusion of fundamental importance in dynamics: For colliding bodies, in the absence of external forces, the total linear momentum of the system of bodies is constant. An alternative statement of this effect is that the *linear momentum of the system is conserved.*

In the problem illustrated in Fig. 20.15a, a desired solution would be the values of the two velocities $v_A'$ and $v_B'$ of the spheres after impact. It may be seen that Eq. (3) is a single equation in these two unknowns. Thus, this equation by itself is not sufficient to solve the problem. In the following problem a second independent equation, which relates the velocities before and after impact, will be developed in terms of the coefficient of restitution. With this latter equation, the complete solution may be obtained for the case of the two colliding spheres.

(*b*)  A very important observation about Eq. (3) is that it may be used only to find the new velocities of the bodies after impact. It does *not* yield information about the duration of the impact or the magnitude of the impact forces.

20.16   (*a*)  Give the definition of the *coefficient of restitution*.

(*b*)  What is *plastic impact*?

(*c*)  What is *elastic impact*?

▮ (*a*)  When two physical bodies collide, there is a permanent energy loss as a result of effects occurring in the material during the deformation and restitution phases of the impact process. This energy loss manifests itself in lower values of velocity after impact than would have been the case had there been no energy loss.

In the initial condition *before* impact, as shown in Fig. 20.15a, the relative velocity of the two spheres may be expressed as

$$v_r = v_A - v_B$$

The relative velocity immediately after impact may be written as

$$v_r' = v_A' - v_B'$$

A term which is a measure of the energy loss during impact will be introduced now. This quantity is referred to as the coefficient of restitution, designated by the symbol $e$, and defined as

$$e = -\frac{\text{relative velocity after impact}}{\text{relative velocity before impact}} = -\frac{v_r'}{v_r} = -\frac{v_A' - v_B'}{v_A - v_B} \tag{1}$$

If the numerator and denominator of the above equation are multiplied by $-1$, this equation may also be written in the form

$$e = -\frac{v_B' - v_A'}{v_B - v_A}$$

Two limiting values of the coefficient of restitution are considered in parts (*b*) and (*c*) below.

(*b*)  One extreme value of the coefficient of restitution occurs during *plastic impact*. In a plastic impact, the colliding bodies are assumed to physically adhere to each other and move with a common velocity after impact. For this case,

$$v_A' = v_B'$$

and, from Eq. (1),

$$e = 0$$

The plastic impact case corresponds to the *maximum* energy loss which may occur between the two impacting bodies.

(c) In an *elastic impact* the assumption is made that no permanent energy loss results from the impact process. Thus, all the energy is recovered, and elastic impact may be considered to be a theoretical limiting case. It can be shown that, for the case of elastic impact, $e = 1$. It then follows that the range of possible values of $e$ is given by

$$0 \le e \le 1$$

If an impact is not elastic, it is referred to as *inelastic*. The actual value of the coefficient of restitution must be determined experimentally. In a given problem, a value may be estimated for this quantity.

20.17 A ball weighing 0.5 lb is dropped onto the ground from a height of 4 ft, as shown in Fig. 20.17. Find the height to which the ball will rebound, and the energy loss associated with the impact, if the coefficient of restitution is assumed to be 0.85.

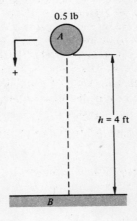

0.5 lb

$h = 4$ ft

Fig. 20.17

▋ The velocity $v_A$ of the ball just before impact with the ground is

$$v_A = \sqrt{2gh} = \sqrt{2(32.2)(4)} = 16.0 \, \text{ft/s}$$

The ground is body $B$, and

$$v_A = 16.0 \, \text{ft/s} \qquad v_B = 0 \qquad v'_B = 0$$

The rebound velocity $v'_A$ of the ball is found from

$$e = -\frac{v'_A - v'_B}{v_A - v_B} \qquad 0.85 = -\frac{v'_A - 0}{16.0 - 0} \qquad v'_A = -13.6 \, \text{ft/s}$$

The minus sign in the result for $v'_A$ indicates that the sense of the velocity is upward. The equation of impulse momentum is not required in this problem, since the velocity of the ground is zero throughout the impact process. Also, the above result is independent of the mass of the ball.

The final height $h'$ attained by the ball after impact is found from

$$h' = \frac{(v'_A)^2}{2g} = \frac{(-13.6)^2}{2(32.2)} = 2.87 \, \text{ft}$$

The initial potential energy is

$$V_1 = mgh = 0.5(4) = 2 \, \text{ft} \cdot \text{lb}$$

The final potential energy is

$$V_2 = mgh' = 0.5(2.87) = 1.44 \, \text{ft} \cdot \text{lb}$$

The permanent loss in energy may be stated in the form

$$\%D = \frac{V_1 - V_2}{V_1}(100) = \frac{2 - 1.44}{2}(100) = 28\%$$

It may be seen that, due to the impact, there is a permanent loss of 28 percent of the initial energy in the system.

**20.18**   A solid metal sphere of mass 0.6 kg is dropped onto a rigid surface from a height   $h = 1.5\,\text{m}$,   as shown in Fig. 20.18a.   The coefficient of restitution is 0.81.

(**a**)   Find the maximum height $h'$ attained by the sphere after the impact.

(**b**)   Find the energy loss that occurs during the impact.

(**c**)   Find the value of the impulse that occurs during the impact.

(**d**)   To what height $h''$ would the sphere rebound if it impacted the ground with a velocity of 8 m/s?

(**e**)   Find the energy loss for the motion of part (d).

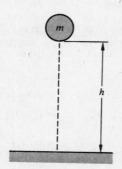

Fig. 20.18a

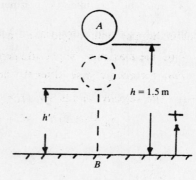

Fig. 20.18b

Fig. 20.18c

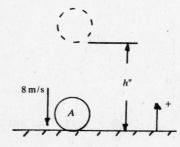

Fig. 20.18d

(**a**)   Figure 20.18b shows the positive sense of motion, and the initial and final heights.   The impact velocity of the sphere with the surface is given by

$$v_A = -\sqrt{2gh} = -\sqrt{2(9.81)1.5} = -5.42\,\text{m/s}$$

Using the definition of the coefficient of restitution,

$$e = -\frac{v'_B - v'_A}{v_B - v_A} = -\frac{v'_A}{v_A} \qquad v'_A = -ev_A = -0.81(-5.42) = 4.39\,\text{m/s}$$

The rebound height $h'$ is found from

$$h' = \frac{v'^2_A}{2g} = \frac{4.39^2}{2(9.81)} = 0.982\,\text{m} = 982\,\text{mm}$$

(**b**)   The energy loss during the impact is given by

$$\Delta T = \tfrac{1}{2}m(v'^2_A - v^2_A) = \tfrac{1}{2}(0.6)(4.39^2 - 5.42^2) = -3.03\,\text{J}$$

(**c**)   Figure 20.18c shows the impulse $I'$ acting on the sphere during the impact.   The value of $I'$ is found from

$$I' = \int_1^2 R\,dt = m(v'_A - v_A) = 0.6[4.39 - (-5.42)] = 5.89\,\text{N}\cdot\text{s}$$

(**d**)   Figure 20.18d shows the case where the sphere impacts the ground with a velocity of 8 m/s.   The impact velocity is stated as

$$v_A = -8\,\text{m/s}$$

From the definition of the coefficient of restitution,

$$e = -\frac{(v_B' - v_A')}{v_B - v_A} = -\frac{v_A'}{v_A} \qquad v_A' = -ev_A = -0.81(-8) = 6.48 \text{ m/s}$$

The height $h''$ of rebound is found from

$$h'' = \frac{v_A'^2}{2g} = \frac{6.48^2}{2(9.81)} = 2.14 \text{ m}$$

(e)  The energy loss during the impact is given by

$$\Delta T = \tfrac{1}{2}m(v_A'^2 - v_A^2) = \tfrac{1}{2}(0.6)(6.48^2 - 8^2) = -6.60 \text{ J}$$

**20.19**  A ball is dropped onto a rigid floor, as shown in Fig. 20.19a.  The coefficient of restitution is 0.87.

(a)  Find the maximum height attained by the ball after the first, second, and third impacts.

(b)  Find the energy losses after the first, second, and third impacts.

❙  (a)  The velocity, and height, after the first impact are found first.  The results are

$$v_A = -\sqrt{2gh} = -\sqrt{2(32.2)10} = -25.4 \text{ ft/s} \qquad v_A' = -ev_A = -0.87(-25.4) = 22.1 \text{ ft/s}$$

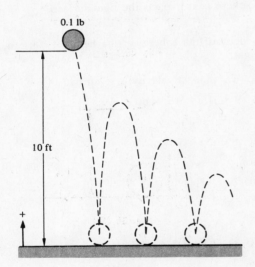

Fig. 20.19a

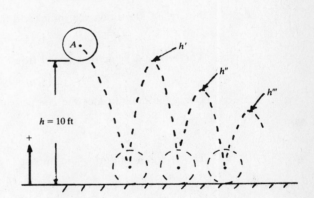

Fig. 20.19b

The height $h'$, shown in Fig. 20.19b, is found as

$$h' = \frac{v_A'^2}{2g} = \frac{22.1^2}{2(32.2)} = 7.58 \text{ ft}$$

The above result may be generalized to the form

$$h' = \frac{v'^2}{2g} = \frac{(ev)^2}{2g} = \frac{e^2 v^2}{2g} = \frac{e^2(2gh)}{2g} = e^2 h$$

In the above equation $h$ is the height from which the ball is dropped before impact, and $h'$ is the rebound height after impact.  The results for the problem in Fig. 20.19a are then as follows.
After the first impact,

$$h' = e^2 h = 0.87^2(10) = 7.57 \approx 7.58 \text{ ft}$$

After the second impact,

$$h'' = e^2 h' = 0.87^2(7.57) = 5.73 \text{ ft}$$

After the third impact,

$$h''' = e^2 h'' = 0.87^2(5.73) = 4.34 \text{ ft}$$

(*b*)   The energy loss during an impact has the form

$$\Delta V = mg(h - h')$$

After the first impact,

$$\Delta V = mg(h - h') = 0.1(10 - 7.57) = 0.243 \text{ ft} \cdot \text{lb}$$

After the second impact,

$$\Delta V = mg(h' - h'') = 0.1(7.57 - 5.73) = 0.184 \text{ ft} \cdot \text{lb}$$

After the third impact,

$$\Delta V = mg(h'' - h''') = 0.1(5.73 - 4.34) = 0.139 \text{ ft} \cdot \text{lb}$$

It may be seen from the above results that the energy loss decreases with each successive impact.

**20.20**   A solid sphere is dropped from a height of 1,800 mm onto a rigid surface.   The maximum height of rebound of the sphere is 1,510 mm, and   $m = 400$ g.

(*a*)   Find the value of the coefficient of restitution.

(*b*)   Find the energy loss associated with the impact.

(*c*)   Find the maximum height of the rebound after the sphere next impacts the rigid surface.

(*d*)   Find the impulses of the two impacts.

(*e*)   Find the required value of the initial downward velocity of the sphere, at the height 1,800 mm, if the rebound height is to be the same as the original height.

▌ (*a*)   Figure 20.20*a* shows the initial and final heights.   The impact velocity is given by

$$v_A = -\sqrt{2gh} = -\sqrt{2(9.81)1.8} = -5.94 \text{ m/s}$$

The rebound velocity is found from

$$v_A' = \sqrt{2gh_1} = \sqrt{2(9.81)1.51} = 5.44 \text{ m/s}$$

Using the definition of the coefficient of restitution,

$$e = -\frac{v_B' - v_A'}{v_B - v_A} = -\frac{v_A'}{v_A} = -\frac{5.44}{(-5.94)} = 0.916$$

(*b*)   The energy loss during the impact is given by

$$\Delta V = mg(h - h') = 0.4(9.81)(1.8 - 1.51) = 1.14 \text{ J}$$

(*c*)   The velocity $v_A''$ after the second impact is given by

$$v_A'' = -ev_A' = -0.916(-5.44) = 4.98 \text{ m/s}$$

The height attained after the second impact is

$$h'' = \frac{v_A''^2}{2g} = \frac{4.98^2}{2(9.81)} = 1.26 \text{ m} = 1,260 \text{ mm}$$

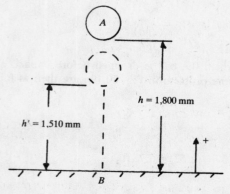

**Fig. 20.20*a***

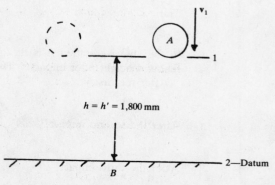

**Fig. 20.20*b***

(*d*)  The value of the impulse during the first impact is

$$I' = m(v'_A - v_A) = 0.4[5.44 - (-5.94)] = 4.55\,\text{N}\cdot\text{s}$$

During the second impact, the impulse is

$$I'' = m(v''_A - v'_A) = 0.4[4.98 - (-5.44)] = 4.17\,\text{N}\cdot\text{s}$$

(*e*)  Figure 20.20*b* shows the endpoints of the interval of motion from the initial position to the ground.  The energy terms at these points are

$$T_1 = \tfrac{1}{2}mv_1^2 \qquad T_2 = \tfrac{1}{2}mv_2^2 \qquad V_1 = mgh \qquad V_2 = 0$$

The work-energy equation has the form

$$0 = \Delta T + \Delta V = (\tfrac{1}{2}mv_2^2 - \tfrac{1}{2}mv_1^2) + (0 - mgh)$$
$$v_2 = \sqrt{v_1^2 + 2gh} \tag{1}$$

The value 0.916 for the coefficient of restitution, found in part (*a*), is used, so that

$$v'_2 = 0.916 v_2$$

The height and velocity after rebound are related by

$$h' = \frac{v_2'^2}{2g} = \frac{(0.916 v_2)^2}{2g} \tag{2}$$

Using $v_2$ from Eq. (1) in Eq. (2) results in

$$h' = (0.916)^2\,\frac{v_1^2 + 2gh}{2g} \qquad 1.8 = 0.916^2\!\left[\frac{v_1^2}{2(9.81)} + 1.8\right] \qquad v_1 = 2.60\,\text{m/s}$$

**20.21**  Figure 20.21*a* shows two prismatic rods which move along a common straight line on a plane surface.  The masses of the two rods are equal, and the coefficient of friction between the rods and the surface is 0.28.  Rod *B* is initially at rest.

(*a*)  Find the velocities of the two rods after impact, in terms of $v_A$ and $e$.

(*b*)  Describe the velocities after impact, if elastic impact, with $e = 1$, is assumed.

(*c*)  Do the same as in part (*b*), if plastic impact, with $e = 0$, is assumed.

Fig. 20.21*a*                                   Fig. 20.21*b*

▌ (*a*)  The configuration after impact is shown in Fig. 20.21*b*.  The impulse-momentum equation is

$$m_A v_A + m_B v_B = m_A v'_A + m_B v'_B \qquad mv_A + m(0) = mv'_A + mv'_B \qquad v'_A + v'_B = v_A \tag{1}$$

The equation which relates the relative velocities before and after impact is

$$e = -\frac{v'_A - v'_B}{v_A - v_B} = -\frac{v'_A - v'_B}{v_A - 0} \qquad v'_A - v'_B = -ev_A \tag{2}$$

Equations (1) and (2) are solved simultaneously, and the result is

$$v'_A = \frac{1-e}{2}\,v_A \qquad v'_B = \frac{1+e}{2}\,v_A$$

(*b*)  For the case of elastic impact, $e = 1$ and

$$v'_A = \frac{1-1}{2}\,v_A = 0 \qquad v'_B = \frac{1+1}{2}\,v_A = v_A$$

For this case, it may be seen that the two rods *exchange* their velocities.  After the impact, rod *A* is at rest and rod *B* has the original velocity of rod *A*.

An approximation of the above phenomenon may be observed with impacting billiard balls. This latter probem is not exactly the same as the problem of the impacting rods, because of the rotation effects of the balls.

(c)  For the case of plastic impact, $e = 0$ and

$$v_A' = \frac{1-0}{2}\, v_A = \tfrac{1}{2} v_A \qquad v_B' = \frac{1+0}{2}\, v_A = \tfrac{1}{2} v_A$$

The two rods adhere to each other and move with a common velocity which is *one-half* of the original impact velocity.

It may be seen that the friction forces do not enter into this problem. This is because of the earlier assumption that the magnitudes of forces such as weight, or friction, forces are negligible compared to the magnitudes of impact forces. These friction forces, however, affect the motion of the rods after impact.

**20.22**  Figure 20.22*a* shows two rods which move along a common line on a plane surface. $m_A = 2$ kg, $m_B = 1$ kg, $v_B = 3$ m/s, and $e = 0.65$.

(a)  Find the general expression for the velocities of the two rods after impact.

(b)  Find the value of the initial velocity $v_A$ which will cause rod $A$ to be at rest after the impact, and the corresponding velocity of rod $B$ after the impact.

(c)  Find the magnitude of the impulse for the conditions of part (b).

(d)  Find the percent decrease in energy which corresponds to the impact in part (b).

Fig. 20.22*a*                   Fig. 20.22*b*

(a)  The velocities of the two rods after impact are shown in Fig. 20.22*b*. Using Eq. (3) in Prob. 20.15, the equation of impulse momentum is

$$m_A v_A + m_B(-v_B) = m_A v_A' + m_B v_B' \tag{1}$$

The equation which relates the relative velocities is

$$e = -\frac{v_A' - v_B'}{v_A - (-v_B)} \tag{2}$$

Equations (1) and (2) are solved simultaneously, and the results are

$$v_A' = \frac{(m_A - e m_B)v_A - (1 + e)m_B v_B}{m_A + m_B} \tag{3}$$

$$v_B' = \frac{(1 + e)m_A v_A - (m_B - e m_A)v_B}{m_A + m_B} \tag{4}$$

Equations (3) and (4) are an extremely useful set of results for the velocities of two bodies after impact, and these equations will be used repeatedly in subsequent problems in this chapter. The reader is urged to study these equations carefully, together with Figs. 20.22*a* and *b*, which show the positive senses of the four velocities $v_A$, $v_B$, $v_A'$, and $v_B'$. It should be noted that the positive senses of $v_A'$ and $v_B'$ are always the same as the positive sense of $v_A$. If $v_B$ is negative, a necessary condition for impact to occur is $v_A > |v_B|$.

(b)  Using the values $m_A = 2$ kg, $m_B = 1$ kg, $v_B = 3$ m/s, and $e = 0.65$, with $v_A' = 0$, Eq. (3) appears as

$$v_A' = 0 = [2 - 0.65(1)]v_A - (1 + 0.65)(1)(3) = 3.67 \text{ m/s}$$

The corresponding value of $v_B'$ is found from Eq. (4) as

$$v_B' = \frac{(1 + 0.65)(2)(3.67) - [1 - 0.65(2)]3}{2 + 1} = 4.34 \text{ m/s}$$

(c)  The impulse which occurs during the impact is

$$I' = m_A(v_A' - v_A) = 2(0 - 3.67) = -7.34 \text{ N} \cdot \text{s}$$

The minus sign indicates that the sense of this impulse is to the left on rod $A$. From Newton's third law, it may be concluded that an impulse of the same magnitude acts to the right on rod $B$. It may be observed that the numerical value of the impulse has no particular useful application in this problem.

(d)  The initial kinetic energy of the system is

$$T_1 = \tfrac{1}{2}m_A v_A^2 + \tfrac{1}{2}m_B v_B^2 = \tfrac{1}{2}(2)(3.67^2) + \tfrac{1}{2}(1)(3^2) = 18.0 \text{ N} \cdot \text{m} = 18.0 \text{ J}$$

The kinetic energy after impact is

$$T_2 = \tfrac{1}{2}m_B(v_B')^2 = \tfrac{1}{2}(1)(4.34^2) = 9.42 \text{ J}$$

The percent decrease in energy is

$$\%\text{D} = \frac{T_1 - T_2}{T_1}(100) = \frac{18.0 - 9.42}{18.0}(100) = 47.7\%$$

20.23  The two blocks in Fig. 20.23a approach each other at constant velocity. After impact, block $A$ has a velocity of 3.1 m/s to the left.

(a)  Find the value of the coefficient of restitution, and the velocity of block $B$ after the impact.

(b)  Find the value of the impulse.

(c)  Find the percent loss in energy during the impact.

(d)  What value of initial velocity of block $B$ will cause the velocity after impact of this block to be zero?

(e)  Find the corresponding value of the velocity of block $A$ after impact, for the condition of part (d).

▌ (a)  The known values of velocity are

$$v_A' = -3.1 \text{ m/s} \qquad v_A = 2 \text{ m/s} \qquad v_B = 3 \text{ m/s}$$

The masses of the two blocks are

$$m_A = 0.62 \text{ kg} \qquad m_B = 0.84 \text{ kg}$$

Using Eqs. (3) and (4) in Prob. 20.22,

$$v_A' = \frac{(m_A - em_B)v_A - (1 + e)m_B v_B}{m_A + m_B} \qquad (1)$$

$$-3.1 = \frac{[0.62 - e(0.84)]2 - (1 + e)0.84(3)}{0.62 + 0.84} \qquad e = 0.773$$

$$v_B' = \frac{(1 + e)m_A v_A - (m_B - em_A)v_B}{m_A + m_B} = \frac{(1 + 0.773)0.62(2) - [0.84 - 0.773(0.62)]3}{0.62 + 0.84} = 0.765 \text{ m/s} \quad (2)$$

Fig. 20.23a

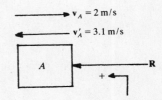

Fig. 20.23b

(b)  The value of the impulse, using Fig. 20.23b, is given by

$$I' = \int_1^2 R \, dt = m(v_A' - v_A) = 0.62[3.1 - (-2)] = 3.16 \text{ N} \cdot \text{s}$$

(c)  The values of the kinetic energy of the system of two blocks, before and after impact, are

$$T_1 = \tfrac{1}{2}m_A v_A^2 + \tfrac{1}{2}m_B v_B^2 = \tfrac{1}{2}(0.62)2^2 + \tfrac{1}{2}(0.84)3^2 = 5.02 \text{ J}$$

$$T_2 = \tfrac{1}{2}m_A v_A'^2 + \tfrac{1}{2}m_B v_B'^2 = \tfrac{1}{2}(0.62)3.1^2 + \tfrac{1}{2}(0.84)0.765^2 = 3.22 \text{ J}$$

The percent loss in energy during the impact is

$$\%D = \frac{3.22 - 5.02}{5.02}(100) = -35.9\%$$

(d) The conditions of the problem are

$$v_A = 2\,\text{m/s} \qquad v_B' = 0$$

Using Eq. (2), and the value of $e$ found in part (a),

$$v_B' = 0 = \frac{(1 + 0.773)0.62(2) - [0.84 - 0.773(0.62)]v_B}{0.62 + 0.84} \qquad v_B = 6.09\,\text{m/s}$$

(e) Using Eq. (1) in part (a),

$$v_A' = \frac{[0.62 - 0.773(0.84)]2 - (1 + 0.773)0.84(6.09)}{0.62 + 0.84} = -6.25\,\text{m/s}$$

After the impact, block $A$ moves to the left and block $B$ has zero velocity.

20.24   Block $B$ in Fig. 20.24 is initially at rest, and the plane surface is assumed to be frictionless.

(a)   Find the velocities after impact.  Also find the displacement of each block and the separation distance between the two blocks 2 s after impact, if elastic impact is assumed.

(b)   Do the same as in part (a), if a coefficient of restitution of 0.8 is assumed.

(c)   Do the same as in part (a), if a coefficient of restitution of 0.2 is assumed.

(d)   Do the same as in part (a), if plastic impact is assumed.

(e)   Compare the results found in parts (a) through (d).

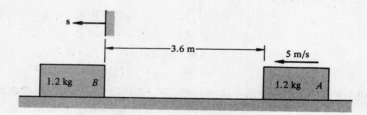

Fig. 20.24

▌ (a)   The conditions just before impact are

$$v_A = 5\,\text{m/s} \qquad v_B = 0$$

Using Eqs. (3) and (4) in Prob. 20.22,

$$v_A' = \frac{(m_A - em_B)v_A - (1 + e)m_B v_B}{m_A + m_B} = \frac{[1.2 - (1)1.2]5 - (1 + 1)1.2(0)}{1.2 + 1.2} = 0 \qquad (1)$$

$$v_B' = \frac{(1 + e)m_A v_A - (m_B - em_A)v_B}{m_A + m_B} = \frac{(1 + 1)1.2(5) - [1.2 - (1)1.2]0}{1.2 + 1.2} = 5\,\text{m/s} \qquad (2)$$

The motion of the blocks after impact is described by

$$t = 0 \qquad s_A = 0 \qquad v_A = 0$$
$$t = 2\,\text{s} \qquad s_A = 0$$
$$t = 0 \qquad s_B = 0 \qquad v_B = 5\,\text{m/s}$$
$$t = 2\,\text{s} \qquad s_B = v_B t = 5(2) = 10\,\text{m}$$

The separation distance between the blocks 2 s after impact is

$$\Delta s = s_B - s_A = 10 - 0 = 10\,\text{m}$$

(b)   Using equations of the form of Eq. (1) and Eq. (2), with $e = 0.8$, results in

$$v_A' = \frac{[1.2 - 0.8(1.2)]5 - 0}{1.2 + 1.2} = 0.5\,\text{m/s} \qquad v_B' = \frac{(1 + 0.8)1.2(5) - 0}{1.2 + 1.2} = 4.5\,\text{m/s}$$

The motion after impact is given by

$$t = 0 \qquad s_A = 0 \qquad v_A = 0.5 \, \text{m/s}$$
$$t = 2 \, \text{s} \qquad s_A = v_A t = 0.5(2) = 1 \, \text{m}$$
$$t = 0 \qquad s_B = 0 \qquad v_B = 4.5 \, \text{m/s}$$
$$t = 2 \, \text{s} \qquad s_B = v_B t = 4.5(2) = 9 \, \text{m}$$

The separation distance 2 s after impact is

$$\Delta s = s_B - s_A = 9 - 1 = 8 \, \text{m}$$

(c)  For  $e = 0.2$,  and using equations of the form of Eq. (1) and Eq. (2),

$$v_A' = \frac{[1.2 - 0.2(1.2)]5 - 0}{1.2 + 1.2} = 2 \, \text{m/s} \qquad v_B' = \frac{(1 + 0.2)1.2(5) - 0}{1.2 + 1.2} = 3 \, \text{m/s}$$

The motion after impact is described by

$$t = 0 \qquad s_A = 0 \qquad v_A = 2 \, \text{m/s}$$
$$t = 2 \, \text{s} \qquad s_A = v_A t = 2(2) = 4 \, \text{m}$$
$$t = 0 \qquad s_B = 0 \qquad v_B = 3 \, \text{m/s}$$
$$t = 2 \, \text{s} \qquad s_B = v_B t = 3(2) = 6 \, \text{m}$$

The separation distance after 2 s is

$$\Delta s = s_B - s_A = 6 - 4 = 2 \, \text{m}$$

(d)  For plastic impact, with  $e = 0$,  Eqs. (1) and (2) have the forms

$$v_A' = \frac{[1.2 - 0]5 - 0}{1.2 + 1.2} = 2.5 \, \text{m/s} \qquad v_B' = \frac{(1 + 0)1.2(5) - 0}{1.2 + 1.2} = 2.5 \, \text{m/s}$$

The motion after impact is given by

$$t = 0 \qquad s_A = 0 \qquad v_A = 2.5 \, \text{m/s}$$
$$t = 2 \, \text{s} \qquad s_A = v_A t = 2.5(2) = 5 \, \text{m}$$
$$t = 0 \qquad s_B = 0 \qquad v_B = 2.5 \, \text{m/s}$$
$$t = 2 \, \text{s} \qquad s_B = v_B t = 2.5(2) = 5 \, \text{m}$$

As an expected result in plastic impact, the separation distance is expressed by

$$\Delta s = s_B - s_A = 5 - 5 = 0$$

(e)  The results in parts (a) through (d) are shown in Table 20.1.  As $e$ decreases from 1 to zero, the following effects may be observed.

1.  $v_A'$ increases from zero to 2.5 m/s.
2.  $v_B'$ decreases from 5 m/s to 2.5 m/s.
3.  The separation distance 2 s after impact decreases from 10 m to zero.

**TABLE 20.1**

| | | | $t = 2$ s | | |
|---|---|---|---|---|---|
| $e$ | $v_A'$ | $v_B'$ | $s_A$ | $s_B$ | $\Delta s$ |
| 1 | 0 | 5 | 0 | 10 | 10 |
| 0.8 | 0.5 | 4.5 | 1 | 9 | 8 |
| 0.2 | 2 | 3 | 4 | 6 | 2 |
| 0 | 2.5 | 2.5 | 5 | 5 | 0 |

**20.25**  Two blocks approach each other on a plane horizontal surface, as shown in Fig. 20.25a.  At  $t = 0$  the separation distance between the blocks is 3 m.  The coefficient of friction between the blocks and the surface is 0.15, and elastic impact is assumed.  $m_A = 2 \, \text{kg}$,  $m_B = 3 \, \text{kg}$,  $v_A = 4 \, \text{m/s}$,  and  $v_B = 6.5 \, \text{m/s}$.

(**a**)  Find the time of impact.

(**b**)  Find the velocities after impact.

(**c**)  Find the magnitude of the impulse.

(**d**)  Find the separation distance between the blocks when these elements come to rest.

❚ (**a**)  For motion before impact, the displacement coordinates $s_A$ and $s_B$ in Fig. 20.25a are used.  The initial velocities at $t = 0$  are

$$v_{Ai} = 4\ \text{m/s} \qquad v_{Bi} = 6.5\ \text{m/s}$$

The acceleration of the blocks is given by

$$a = -\mu g = a_A = a_B = -0.15(9.81) = -1.47\ \text{m/s}^2$$

The displacements $s_A$ and $s_B$ are related by

$$s_A + s_B = 3$$

Using  $s = s_0 + v_0 t + \frac{1}{2}at^2$  in the above equation results in

$$[4t + \tfrac{1}{2}(-1.47)t^2] + [(6.5t + \tfrac{1}{2}(-1.47)t^2] = 3 \qquad -1.47t^2 + 10.5t - 3 = 0$$

$$t = \frac{-10.5 \pm \sqrt{10.5^2 - 4(-1.47)(-3)}}{2(-1.47)} = \frac{-10.5 \pm 9.62}{-2(1.47)} = 0.299,\ 6.84\ \text{s}$$

The use of  $t = 6.84\ \text{s}$  yields negative values for $v_A$ and $v_B$.  Thus,

$$t = 0.299\ \text{s}$$

Using

$$v = v_0 + at$$

the velocities of the blocks just prior to impact are

$$v_A = 4 - 1.47(0.299) = 3.56\ \text{m/s} \qquad v_B = 6.5 - 1.47(0.299) = 6.06\ \text{m/s}$$

(**b**)  Using Eqs. (3) and (4) in Prob. 20.22, with  $e = 1$,  results in

$$v_A' = \frac{[2 - (1)3]3.56 - (1 + 1)3(6.06)}{2 + 3} = -7.98\ \text{m/s}$$

$$v_B' = \frac{(1 + 1)2(3.56) - [3 - (1)2]6.06}{2 + 3} = 1.64\ \text{m/s}$$

Block $A$ moves to the right, and block $B$ moves to the left.

(**c**)  Figure 20.25b shows the impulsive force $R$ which acts on block $A$.  The impulse of this force is given by

$$I' = \int_1^2 R\,dt = m_A(v_A' - v_A) = 2[7.98 - (-3.56)] = 23.1\ \text{N} \cdot \text{s}$$

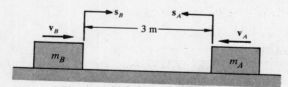

**Fig. 20.25a**

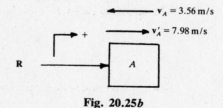

**Fig. 20.25b**

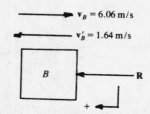

**Fig. 20.25c**

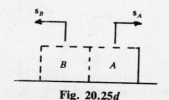

**Fig. 20.25d**

Figure 20.25c shows the impulsive force $R$ acting on block $B$. The reader may use this figure to verify the above result for $I'$.

(d)  Figure 20.25d shows the displacement coordinates of the blocks after impact. In terms of these coordinates,

$$v_A = 7.98 \text{ m/s} \qquad v_B = 1.64 \text{ m/s}$$

The acceleration of the blocks after the impact is the same as in part (a). The displacements of the blocks, when these elements come to rest, are found from

$$v^2 = v_0^2 + 2a(s - s_0)$$
$$0 = 7.98^2 + 2(-1.47)s_A \qquad s_A = 21.7 \text{ m}$$
$$0 = 1.64^2 + 2(-1.47)s_B \qquad s_B = 0.915 \text{ m}$$

The separation distance when the blocks come to rest is given by

$$\Delta s = s_A + s_B = 22.6 \text{ m}$$

**20.26**  (a)  Do the same as in Prob. 20.25, if the coefficient of restitution is 0.8, and find the energy loss during the impact.

(b)  Do the same as in part (a), if the coefficient of restitution is 0.2.

▌ (a)  The impact occurs at $t = 0.229$ s, and the velocities just before impact are

$$v_A = 3.56 \text{ m/s} \qquad v_B = 6.06 \text{ m/s}$$

as found in part (a) of Prob. 20.25. Using Eqs. (3) and (4) in Prob. 20.22, with $e = 0.8$, gives

$$v_A' = \frac{[2 - 0.8(3)]3.56 - (1 + 0.8)3(6.06)}{2 + 3} = -6.83 \text{ m/s} \qquad (1)$$

$$v_B' = \frac{(1 + 0.8)2(3.56) - [3 - 0.8(2)]6.06}{2 + 3} = 0.866 \text{ m/s} \qquad (2)$$

Block $A$ moves to the right and block $B$ moves to the left.

Figure 20.26a shows the impulsive force which acts on block $A$. The impulse of this force is given by

$$I' = \int_1^2 R \, dt = m_A(v_A' - v_A) = 2[6.83 - (-3.56)] = 20.8 \text{ N} \cdot \text{s}$$

The displacement coordinates $s_A$ and $s_B$ of the blocks after impact are seen in Fig. 20.25d. In terms of these coordinates,

$$v_A = 6.83 \text{ m/s} \qquad v_B = 0.866 \text{ m/s}$$

The acceleration of the blocks after the impact is the same as that before the impact, given by

$$a = a_A = a_B = -1.47 \text{ m/s}^2$$

The displacements when the blocks come to rest are found from

$$v^2 = v_0^2 + 2a(s - s_0)$$
$$0 = 6.83^2 + 2(-1.47)s_A \qquad s_A = 15.9 \text{ m}$$
$$0 = 0.866^2 + 2(-1.47)s_B \qquad s_B = 0.255 \text{ m}$$

The separation distance between the blocks when these elements come to rest is given by

$$\Delta s = s_A + s_B = 16.2 \text{ m}$$

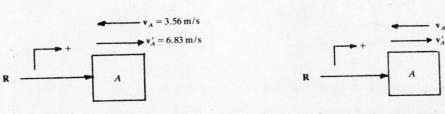

Fig. 20.26a                                    Fig. 20.26b

The initial and final kinetic energies of the system of two blocks are expressed as

$$T_1 = \tfrac{1}{2}m_A v_A^2 + \tfrac{1}{2}m_B v_B^2 = \tfrac{1}{2}(2)3.56^2 + \tfrac{1}{2}(3)6.06^2 = 67.8 \text{ J}$$
$$T_2 = \tfrac{1}{2}m_A v_A'^2 + \tfrac{1}{2}m_B v_B'^2 = \tfrac{1}{2}(2)6.83^2 + \tfrac{1}{2}(3)0.866^2 = 47.8 \text{ J}$$
$$\Delta T = T_2 - T_1 = -20.0 \text{ J}$$

The energy loss due to the impact is 20.0 J.

(b) The solution follows that in part (a). The velocities prior to impact are

$$v_A = 3.56 \text{ m/s} \qquad v_B = 6.06 \text{ m/s}$$

Using the forms of Eqs. (1) and (2) in part (a),

$$v_A' = \frac{[2-(0.2)3]3.56 - (1+0.2)3(6.06)}{2+3} = -3.37 \text{ m/s}$$

$$v_B' = \frac{(1+0.2)2(3.56) - [3-0.2(2)]6.06}{2+3} = -1.44 \text{ m/s}$$

The impulsive force $R$ which acts on block $A$ is seen in Fig. 20.26b. The magnitude of the impulse is given by

$$I' = \int_1^2 R \, dt = m_A(v_A' - v_A) = 2[3.37 - (-3.56)] = 13.9 \text{ N} \cdot \text{s}$$

In terms of the $s_A$ and $s_B$ coordinates in Fig. 20.25d,

$$v_A = 3.37 \text{ m/s} \qquad v_B = -1.44 \text{ m/s}$$

The displacements when the blocks come to rest are found from

$$v^2 = v_0^2 + 2a(s - s_0)$$
$$0 = 3.37^2 + 2(-1.47)s_A \qquad s_A = 3.86 \text{ m}$$
$$0 = (-1.44)^2 + 2(-1.47)s_B \qquad s_B = 0.705 \text{ m}$$

The final separation distance is given by

$$\Delta s = s_A - s_B = 3.86 - 0.705 = 3.16 \text{ m}$$

From part (a),

$$T_1 = 67.8 \text{ J}$$

The kinetic energy after impact is

$$T_2 = \tfrac{1}{2}m_A v_A'^2 + \tfrac{1}{2}m_B v_B'^2 = \tfrac{1}{2}(2)3.37^2 + \tfrac{1}{2}(3)1.44^2 = 14.5 \text{ J}$$
$$\Delta T = T_2 - T_1 = -53.3 \text{ J}$$

The loss of energy due to the impact is 53.3 J.

**20.27** Block $B$ in Fig. 20.27a is initially at rest on a spring. Block $A$ is dropped, with zero initial velocity, and plastic impact between the two blocks is assumed. $m_A = 0.3$ kg and $m_B = 0.2$ kg.

(a) Find the maximum value of the deflection of the spring from the position shown in the figure.

(b) Do the same as part (a) if, in the position shown in the figure, block $A$ has a downward velocity of 2 m/s.

$\blacksquare$ (a) The velocity of block $A$ when it impacts block $B$ is given by

$$v_A = \sqrt{2gh} = \sqrt{2(9.81)1.4} = 5.24 \text{ m/s}$$

Using Eq. (4) in Prob. 20.22, with $e = 0$ and $v_B = 0$, gives

$$v_B' = \frac{(1+e)m_A v_A - (m_B - em_A)v_B}{m_A + m_B} = \frac{(1+0)0.3(5.24) - 0}{0.3 + 0.2} = 3.14 \text{ m/s} \qquad (1)$$

Since the impact is plastic,

$$v_A' = v_B' = 3.14 \text{ m/s}$$

$\delta_0$ is the initial spring compression due to the static weight of $m_B$. The value of this term is

$$\delta_0 = \frac{m_B g}{k} = \frac{0.2(9.81)}{800} = 0.00245 \text{ m}$$

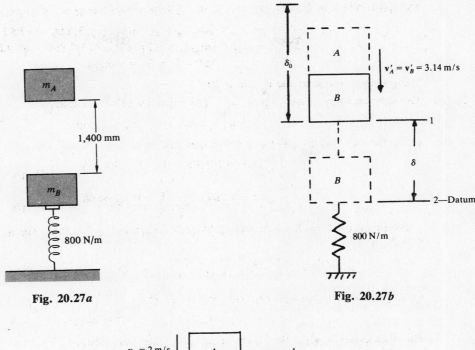

Fig. 20.27a

Fig. 20.27b

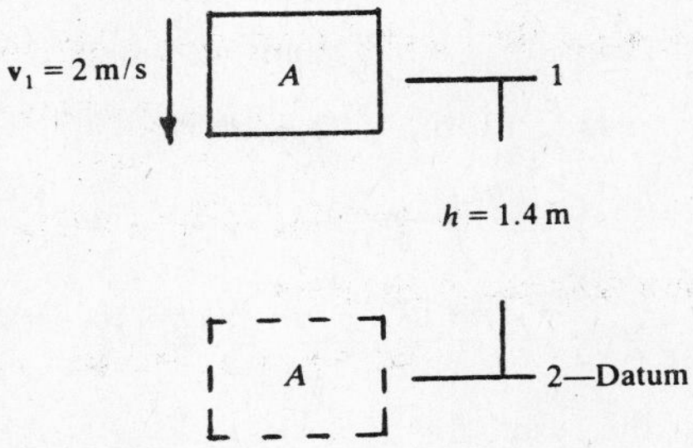

Fig. 20.27c

Figure 20.27b shows the endpoints 1 and 2 of the interval of motion of the combined system of blocks $A$ and $B$. The energy terms at 1 and 2 are

$$T_1 = \tfrac{1}{2}(0.3 + 0.2)3.14^2 \qquad V_1 = \tfrac{1}{2}k\delta_0^2 + m_B g\delta = \tfrac{1}{2}(800)0.00245^2 + 0.2(9.81)\delta$$

$$T_2 = 0 \qquad V_2 = \tfrac{1}{2}k(\delta + \delta_0)^2 = \tfrac{1}{2}k(\delta^2 + 2\delta_0\delta + \delta_0^2) = \tfrac{1}{2}(800)\delta^2 + 800(0.00245)\delta + \tfrac{1}{2}(800)0.00245^2$$

The work-energy equation has the form

$$0 = \Delta T + \Delta V = (T_2 - T_1) + (V_2 - V_1)$$

$$0 = [0 - \tfrac{1}{2}(0.5)3.14^2] + [\tfrac{1}{2}(800)\delta^2 + 800(0.00245)\delta + \tfrac{1}{2}\cancel{(800)0.00245^2}$$
$$- (\tfrac{1}{2}\cancel{(800)0.00245^2} + 0.2(9.81)\delta)]$$

$$\tfrac{1}{2}(800)\delta^2 + 800(0.00245\delta) - 0.2(9.81)\delta - \tfrac{1}{2}(0.5)3.14^2 = 0$$

$$400\delta^2 + \cancel{1.96\delta} - \cancel{1.96\delta} - 2.46 = 0 \qquad \delta = 0.0784 \text{ m} = 78.4 \text{ mm}$$

(b) Block $A$ has an initial velocity of $2 \text{ m/s}$. The impact velocity of this block with block $B$ is next found. Figure 20.27c shows the endpoints of the interval of motion before impact. The energy terms are

$$T_1 = \tfrac{1}{2}m_A v_1^2 \qquad T_2 = \tfrac{1}{2}m_A v_2^2 \qquad V_1 = m_A gh \qquad V_2 = 0$$

The work-energy equation is written as

$$0 = \Delta T + \Delta V = (T_2 - T_1) + (V_2 - V_1) = (\tfrac{1}{2}m_A v_2^2 - \tfrac{1}{2}m_A v_1^2) + (0 - m_A gh)$$

$$v_2 = \sqrt{v_1^2 + 2gh} = v_A = \sqrt{2^2 + 2(9.81)1.4} = 5.61 \text{ m/s}$$

Using the form of Eq. (1) in part (a),

$$v_B' = \frac{(1 + 0)0.3(5.61) - 0}{0.3 + 0.2} = 3.37 \text{ m/s}$$

Since the impact is plastic,

$$v'_A = v'_B = 3.37 \text{ m/s}$$

The kinetic energy at point 1 in Fig. 20.27$b$ is given by

$$T_1 = \tfrac{1}{2}(0.3 + 0.2)3.37^2$$

$V_1$, $T_2$, and $V_2$ have the same functional forms as in part ($a$). The work-energy equation has the form

$$0 = \Delta T + \Delta V = (T_2 - T_1) + (V_2 - V_1) \qquad 400\delta^2 - 2.84 = 0 \qquad \delta = 0.0843 \text{ m} = 84.3 \text{ mm}$$

**20.28** In the position shown in Fig. 20.28$a$ block $B$ just contacts the spring. Block $A$ impacts block $B$ with a velocity of 8 ft/s. $W_A = 2$ lb and $W_B = 3$ lb.

**Fig. 20.28$a$**

($a$) Find the velocity of block $A$ after the impact and the maximum distance through which the spring is compressed, if the impact is elastic and the plane surface is frictionless.

($b$) Do the same as in part ($a$), if the impact is assumed to be plastic.

($c$) Do the same as in part ($a$), if $e = 0.9$ and $\mu = 0.1$.

($d$) Do the same as in part ($a$), if $e = 0.5$ and $\mu = 0.25$.

▮ ($a$) Using Eqs. (3) and (4) in Prob. 20.22, with $e = 1$, $v_A = 8$ ft/s, and $v_B = 0$, gives

$$v'_A = \frac{(m_A - em_B)v_A - (1 + e)m_Bv_B}{m_A + m_B} = \frac{[(2/g) - (1)(3/g)]8 - 0}{2/g + 3/g} = -1.6 \text{ ft/s}$$

$$v'_B = \frac{(1 + e)m_Av_A - (m_B - em_A)v_B}{m_A + m_B} = \frac{(1 + 1)(2/g)(8) - 0}{2/g + 3/q} = 6.4 \text{ ft/s}$$

The minus sign on the result for $v'_A$ indicates that block $A$ rebounds away from the spring.

Figure 20.28$b$ shows the endpoints of the interval of motion following the impact. The energy terms at 1 and 2 are

$$T_1 = \tfrac{1}{2}m_Bv'^2_B = \frac{1}{2}\left(\frac{3}{386}\right)[6.4(12)]^2 \text{ in} \cdot \text{lb} \qquad T_2 = 0$$

$$V_1 = 0 \qquad V_2 = \tfrac{1}{2}(1)\delta^2$$

The work-energy equation has the form

$$0 = \Delta T + \Delta V = (T_2 - T_1) + (V_2 - V_1) = \left[0 - \frac{1}{2}\left(\frac{3}{386}\right)[6.4(12)]^2\right] + [\tfrac{1}{2}(1)\delta^2 - 0] \qquad \delta = 6.77 \text{ in}$$

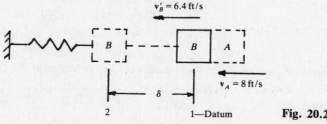

**Fig. 20.28$b$**

($b$) Using the equations for $v'_A$ and $v'_B$ in part ($a$), with $e = 0$, gives

$$v'_A = \frac{[(2/g) - (0)(3/g)]8 - 0}{2/g + 3/g} = 3.2 \text{ ft/s} \qquad v'_B = \frac{(1 + 0)(2/g)(8) - 0}{2/g + 3/g} = 3.2 \text{ ft/s}$$

Both blocks move as a single unit of weight 5 lb, with velocity 3.2 ft/s to the left in Fig. 20.28$a$. The deflection of the spring is found using

$$T_1 = \tfrac{1}{2}(m_A + m_B)v_B'^2 = \tfrac{1}{2}\left(\frac{2+3}{386}\right)[3.2(12)]^2 \qquad T_2 = 0$$

$$V_1 = 0 \qquad V_2 = \tfrac{1}{2}(1)\delta^2$$

$$0 = \Delta T + \Delta V = (T_2 - T_1) + (V_2 - V_1) = \left[0 - \frac{1}{2}\left(\frac{5}{386}\right)[3.2(12)]^2\right] + [\tfrac{1}{2}(1)\delta^2 - 0] \qquad \delta = 4.37\,\text{in}$$

(c)  Using $e = 0.9$ in the equations for $v_A'$ and $v_B'$ in part (a) gives

$$v_A' = \frac{[(2/g) - 0.9(3/g)]8 - 0}{2/g + 3/g} = -1.12\,\text{ft/s}$$

$$v_B' = \frac{(1+0.9)(2/g)(8) - 0}{2/g + 3/g} = 6.08\,\text{ft/s}$$

It may be seen that block $A$ moves to the right after the impact.
The energy terms at the endpoints of the interval are

$$T_1 = \tfrac{1}{2}m_B v_B'^2 = \frac{1}{2}\left(\frac{3}{386}\right)[6.08(12)]^2 \qquad T_2 = 0 \qquad V_1 = 0 \qquad V_2 = \tfrac{1}{2}(1)\delta^2$$

The work-energy equation has the form

$$W_{\text{NC}} = \Delta T + \Delta V = (T_2 - T_1) + (V_2 - V_1)$$

$$-\mu N\delta = -0.1(3)\delta = \left[0 - \frac{1}{2}\left(\frac{3}{386}\right)[6.08(12)]^2\right] + [\tfrac{1}{2}(1)\delta^2 - 0] \qquad 0.5\delta^2 + 0.3\delta - 20.7 = 0$$

$$\delta = \frac{-0.3 \pm \sqrt{0.3^2 - 4(0.5)(-20.7)}}{2(0.5)} = \frac{-0.3 \pm 6.44}{1}$$

The positive root is used, so that

$$\delta = 6.14\,\text{in}$$

(d)  Using $e = 0.5$ in the equations for $v_A'$ and $v_B'$ in part (a) gives

$$v_A' = \frac{[(2/g) - 0.5(3/g)]8 - 0}{2/g + 3/g} = 0.8\,\text{ft/s} \qquad v_B' = \frac{(1+0.5)(2/g)(8) - 0}{2/g + 3/g} = 4.8\,\text{ft/s}$$

Since $v_A'$ is positive, block $A$ moves to the left.   Thus, after impact, block $A$ continues to move toward block $B$.   The two blocks experience a second impact at a later time, but that problem is not considered in this solution.
The deflection of the spring is found using

$$T_1 = \tfrac{1}{2}m_B v_B'^2 = \frac{1}{2}\left(\frac{3}{386}\right)[4.8(12)]^2 \qquad T_2 = 0 \qquad V_1 = 0 \qquad V_2 = \tfrac{1}{2}(1)\delta^2$$

$$W_{\text{NC}} = \Delta T + \Delta V = (T_2 - T_1) + (V_2 - V_1)$$

$$-0.25(3)\delta = \left[0 - \frac{1}{2}\left(\frac{3}{386}\right)[4.8(12)]^2\right] + [\tfrac{1}{2}(1)\delta^2 - 0] \qquad 0.5\delta^2 + 0.75\delta - 12.9 = 0$$

$$\delta = \frac{-0.75 \pm \sqrt{0.75^2 - 4(0.5)(-12.9)}}{2(0.5)} = \frac{-0.75 \pm 5.13}{1}$$

Using the positive root,

$$\delta = 4.38\,\text{in}$$

**20.29**  At $t = 0$, both blocks are released from the positions shown in Fig. 20.29.   When the blocks collide, elastic impact is assumed.

(a)  Find the time at which impact occurs, and the velocities after impact.

(b)  Do the same as in part (a), if the coefficient of restitution is 0.8.

(c)  Do the same as in part (a), if the coefficient of restitution is 0.2.

(d)  Compare the results in parts (a) through (c).

▎ (a)  Sliding motion of the blocks is impending when $\mu = \tan 15° = 0.268$.   Since $0.268 < \mu_B = 0.3$, block $B$ is stationary until impact occurs.   The acceleration of block $A$ is given by

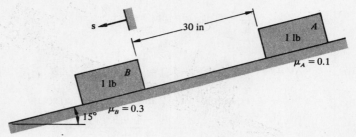

**Fig. 20.29**

$$a_A = (\sin \beta - \mu_A \cos \beta)g = (\sin 15° - 0.1 \cos 15°)32.2 = 5.22 \text{ ft/s}^2$$

The time for block $A$ to reach block $B$, using the conditions at $t = 0$ of $v_0 = 0$, $s = 0$, and $s_0 = -30$ in, is found from

$$s = s_0 + v_0 t + \tfrac{1}{2}at^2 \qquad 0 = -\tfrac{30}{12} + \tfrac{1}{2}(5.22)t^2 \qquad t = 0.979 \text{ s}$$

The velocity with which block $A$ impacts block $B$ is found from

$$v^2 = v_0^2 + 2a(s - s_0) \qquad v_A^2 = 2(5.22)\tfrac{30}{12} \qquad v_A = 5.11 \text{ ft/s}^2$$

Using Eqs. (3) and (4) in Prob. 20.22,

$$v_A' = \frac{(m_A - em_B)v_A - (1 + e)m_B v_B}{m_A + m_B} = \frac{[(1/g) - (1)(1/g)]5.11 - 0}{1/g + 1/g} = 0 \qquad (1)$$

$$v_B' = \frac{(1 + e)m_A v_A - (m_B - em_A)v_B}{m_A + m_B} = \frac{(1 + 1)(1/g)(5.11) - 0}{1/g + 1/g} = 5.11 \text{ m/s} \qquad (2)$$

(b) The time for block $A$ to reach block $B$, and the velocity of block $A$ before impact, are the same as in part (a). Using Eqs. (1) and (2) in part (a), with $e = 0.8$ and $v_A = 5.11$ ft/s,

$$v_A' = \frac{[(1/g) - 0.8(1/g)]5.11 - 0}{1/g + 1/g} = 0.511 \text{ ft/s}$$

$$v_B' = \frac{(1 + 0.8)(1/g)(5.11) - 0}{1/g + 1/g} = 4.60 \text{ ft/s}$$

(c) The time, and velocity of block $A$ before impact, are the same as in part (a). Using Eqs. (1) and (2) in part (a), with $e = 0.2$ and $v_A = 5.11$ ft/s,

$$v_A' = \frac{[(1/g) - 0.2(1/g)]5.11 - 0}{1/g + 1/g} = 2.04 \text{ ft/s}$$

$$v_B' = \frac{(1 + 0.2)(1/g)(5.11) - 0}{1/g + 1/g} = 3.07 \text{ ft/s}$$

(d) The results for parts (a) through (c) are shown in Table 20.2. It may be seen that as $e$ decreases, $v_A'$ increases and $v_B'$ decreases. In all cases, the velocities of the blocks after impact are in the sense of positive $s$ shown in Fig. 20.29.

**TABLE 20.2**

| $e$ | $v_A'$, m/s | $v_B'$, m/s |
|-----|-------------|-------------|
| 1 | 0 | 5.11 |
| 0.8 | 0.511 | 4.60 |
| 0.2 | 2.04 | 3.07 |

**20.30** At the instant shown in Fig. 20.30a, block $B$ is released with zero initial velocity. $v_A$ is constant, and the blocks collide at a position 20 in from the original position of block $B$. The impact is assumed to be elastic. $W_A = 10$ lb and $W_B = 12$ lb.

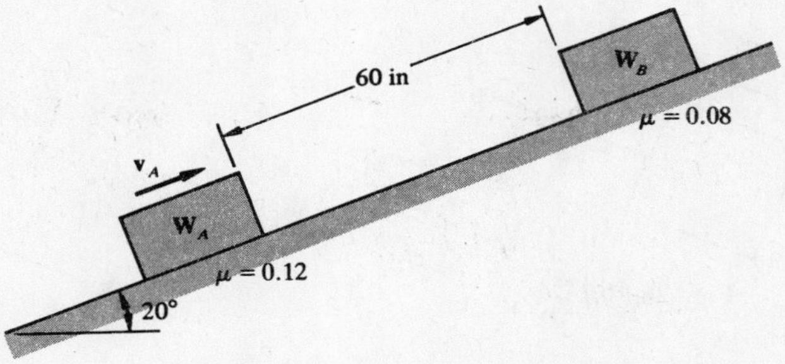

**Fig. 20.30a**

(a) Find the velocities of the blocks after impact, and the value of the impulse.

(b) Do the same as in part (a), if the coefficient of restitution is 0.8, and find the percent loss in energy due to the impact.

(c) Do the same as in part (a), if the coefficient of restitution is 0.2, and find the percent loss in energy due to the impact.

(d) Compare the results in parts (a) through (c).

▌ (a) The time and velocity when block B has moved 20 in is found first. The acceleration of block B is

$$a_B = (\sin \beta - \mu \cos \beta)g = (\sin 20° - 0.08 \cos 20°)386 = 103 \text{ in/s}^2$$

The velocity of block B is found from

$$v^2 = v_0^2 + 2a(s - s_0) \qquad v_B^2 = 2(103)20 \qquad v_B = 64.2 \text{ in/s}$$

The time which corresponds to the above velocity is found from

$$v = v_0 + at \qquad 64.2 = 103t \qquad t = 0.623 \text{ s}$$

The velocity of block A at the above time is found from

$$s = v_0 t \qquad 40 = v_A(0.623) \qquad v_A = 64.2 \text{ in/s}$$

The velocities of the two blocks just before impact are

$$v_A = 64.2 \text{ in/s} \qquad v_B = 64.2 \text{ in/s}$$

Using Eqs. (3) and (4) in Prob. 20.22,

$$v_A' = \frac{(m_A - em_B)v_A - (1 + e)m_B v_B}{m_A + m_B} = \frac{[(10/g) - (1)(12/g)]64.2 - (1 + 1)(12/g)(64.2)}{10/g + 12/g} = -75.9 \text{ in/s}$$
$$(1)$$

$$v_B' = \frac{(1 + e)m_A v_A - (m_B - em_A)v_B}{m_A + m_B} = \frac{(1 + 1)(10/g)(64.2) - [(12/g) - (1)(10/g)]64.2}{10/g + 12/g} = 52.5 \text{ in/s}$$
$$(2)$$

The velocity of block A after impact is down the plane, and the velocity of block B is up the plane. The velocities before and after the impact are shown in Figs. 20.30b and c. The value of the impulse is given by

$$I' = \int_1^2 R \, dt = m_B(v_B' - v_B) = \frac{12}{386}[52.5 - (-64.2)] = 3.63 \text{ lb} \cdot \text{s}$$

As a check on the above calculation,

$$I' \overset{?}{=} m(v_A' - v_A) \qquad 3.63 \overset{?}{=} \frac{10}{386}[75.9 - (-64.2)] \qquad 3.63 \equiv 3.63$$

(b) The velocities before impact are the same as in part (a). Using Eqs. (1) and (2) in part (a), with e = 0.8, gives

$$v_A' = \frac{[(10/g) - (0.8)(12/g)]64.2 - (1 + 0.8)(12/g)(64.2)}{10/g + 12/g} = -61.9 \text{ in/s}$$

$$v_B' = \frac{(1 + 0.8)(10/g)(64.2) - [(12/g) - (0.8)(10/g)]64.2}{10/g + 12/g} = 40.9 \text{ in/s}$$

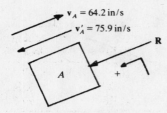

**Fig. 20.30b**

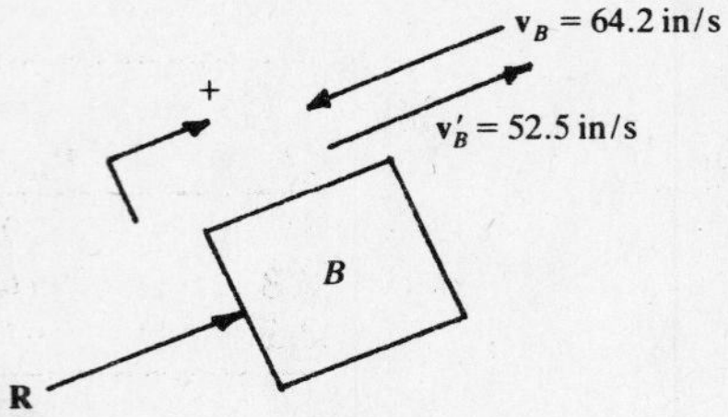

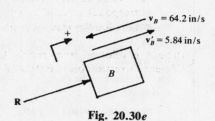

**Fig. 20.30c**

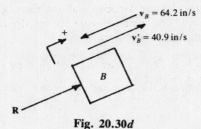

**Fig. 20.30d**

**Fig. 20.30e**

The velocities of the blocks after impact, as in part (a), have opposite senses.
    The impulse shown in Fig. 20.30d as acting on block B is given by

$$I' = \int_1^2 R\,dt = m_B(v_B' - v_B) = \frac{12}{386}[40.9 - (-64.2)] = 3.27\,\text{lb}\cdot\text{s}$$

The energy of the system of two blocks before impact is

$$T_1 = \tfrac{1}{2}m_A v_A^2 + \tfrac{1}{2}m_B v_B^2 = \frac{1}{2}\left(\frac{10}{386}\right)64.2^2 + \frac{1}{2}\left(\frac{12}{386}\right)64.2^2 = 117\,\text{in}\cdot\text{lb}$$

The energy after impact is

$$T_2 = \tfrac{1}{2}m_A v_A'^2 + \tfrac{1}{2}m_B v_B'^2 = \frac{1}{2}\left(\frac{10}{386}\right)(-61.9)^2 + \frac{1}{2}\left(\frac{12}{386}\right)40.9^2 = 75.6\,\text{in}\cdot\text{lb}$$

The percent loss in energy due to the impact is expressed as

$$\%\text{D} = \frac{T_1 - T_2}{T_1}(100) = \frac{117 - 75.6}{117}(100) = 35.4\%$$

An alternative statement of the above result is that the percent change in the energy of the system of two blocks, due to the impact, is −35.4%.

(c)    The velocities before impact are the same as in part (a).   Using   e = 0.2 in,   Eqs. (1) and (2) in part (a) give

$$v_A' = \frac{[(10/g) - (0.2)(12/g)]64.2 - (1 + 0.2)(12/g)(64.2)}{10/g + 12/g} = -19.8\,\text{in/s}$$

$$v_B' = \frac{(1 + 0.2)(10/g)(64.2) - [(12/g) - (0.2)(10/g)]64.2}{10/g + 12/g} = 5.84\,\text{in/s}$$

Using Fig. 20.30e, the impulse is

$$I' = m_B(v_B' - v_B) = \frac{12}{386}[5.84 - (-64.2)] = 2.18\,\text{lb}\cdot\text{s}$$

The energy of the system before impact, found in part (b), is

$$T_1 = 117\,\text{in}\cdot\text{lb}$$

The energy of the system after impact is

$$T_2 = \frac{1}{2}\left(\frac{10}{386}\right)(-19.8)^2 + \frac{1}{2}\left(\frac{12}{386}\right)5.84^2 = 5.61\,\text{in}\cdot\text{lb}$$

The loss in energy due to the impact is

$$\%\text{D} = \frac{T_1 - T_2}{T_1}(100) = \frac{117 - 5.61}{117}(100) = 95.2\%$$

**TABLE 20.3**

| $e$ | $v'_A$, in/s | $v'_B$, in/s | $I'$, lb·s | energy loss, % |
|-----|-----|-----|-----|-----|
| 1 | −75.9 | 52.5 | 3.63 | 0 |
| 0.8 | −61.9 | 40.9 | 3.27 | 35.4 |
| 0.2 | −19.8 | 5.84 | 2.18 | 95.2 |

(d)  A comparison of the results in parts (a) through (c) is given in Table 20.3.  It may be seen that, as the coefficient of restitution decreases, the magnitudes of the velocity and the impulse decrease.  The corresponding trend in energy loss is an increase.  It is left as an exercise for the reader to find the results for the case of plastic impact of the blocks.

**20.31**  Particle A is released from rest in the position shown in Fig. 20.31.  The mass of each particle is 100 g, the coefficient of restitution is 0.92, and all friction effects may be neglected.

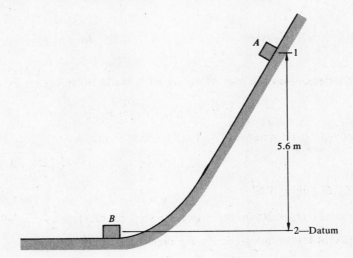

**Fig. 20.31**

(a)  Find the velocity of each particle after the impact, and the percent change in energy due to the impact.

(b)  Do the same as in part (a), if the coefficient of restitution is 0.5.

(c)  Do the same as in part (a) if, because of sliding friction effects, there is a 10 percent loss in energy as particle A slides down the track.

(d)  Do the same as in part (a), for the conditions of parts (b) and (c).

**I**  (a)  The velocity of block A before impact it given by

$$v_A = \sqrt{2gh} = \sqrt{2(9.81)5.6} = 10.5 \text{ m/s}$$

Using Eqs. (3) and (4) in Prob. 20.22,

$$v'_A = \frac{(m_A - em_B)v_A - (1+e)m_Bv_B}{m_A + m_B} = \frac{[0.1 - 0.92(0.1)]10.5 - 0}{0.1 + 0.1} = 0.42 \text{ m/s} \qquad (1)$$

$$v'_B = \frac{(1+e)m_Av_A - (m_B - em_A)v_B}{m_A + m_B} = \frac{(1+0.92)0.1(10.5) - 0}{0.1 + 0.1} = 10.1 \text{ m/s} \qquad (2)$$

Both blocks move to the left after the impact.  The energy of the system of two blocks in the initial position and the final position is given by

$$T_1 = mgh = 0.1(9.81)5.6 = 5.49 \text{ J}$$
$$T_2 = \tfrac{1}{2}m_Av'^2_A + \tfrac{1}{2}m_Bv'^2_B = \tfrac{1}{2}(0.1)0.42^2 + \tfrac{1}{2}(0.1)10.1^2 = 5.11 \text{ J}$$

The percent change in energy due to the impact is

$$\%\text{D} = \frac{T_2 - T_1}{T_1} (100) = \frac{5.11 - 5.49}{5.49} (100) = -6.92\%$$

(**b**)  Using   $e = 0.5$   and   $v_A = 10.5 \, \text{m/s}$   in Eqs. (1) and (2) in part (*a*) gives

$$v'_A = \frac{[0.1 - 0.5(0.1)]10.5 - 0}{0.1 + 0.1} = 2.63 \, \text{m/s} \qquad v'_B = \frac{(1 + 0.5)0.1(10.5) - 0}{0.1 + 0.1} = 7.88 \, \text{m/s}$$

From part (*a*),

$$T_1 = 5.49 \, \text{J}$$

The energy after impact is given by

$$T_2 = \tfrac{1}{2}m_A v'^2_A + \tfrac{1}{2}m_B v'^2_B = \tfrac{1}{2}(0.1)2.63^2 + \tfrac{1}{2}(0.1)7.88^2 = 3.45 \, \text{J}$$

The percent change in energy is

$$\%D = \frac{T_2 - T_1}{T_1}(100) = \frac{3.45 - 5.49}{5.49}(100) = -37.2\%$$

(**c**)  From part (*a*),

$$T_1 = 5.49 \, \text{J}$$

Because of the 10 percent energy loss due to sliding friction, the energy at position 2 is

$$T_2 = 0.9T_1 = 0.9(5.49)$$

The velocity of block *A* just before impact is found from

$$T_2 = \tfrac{1}{2}mv^2_A \qquad 0.9(5.49) = \tfrac{1}{2}(0.1)v^2_A \qquad v_A = 9.94 \, \text{m/s}$$

Using Eqs. (1) and (2) in part (*a*),

$$v'_A = \frac{[0.1 - 0.92(0.1)]9.94 - 0}{0.1 + 0.1} = 0.398 \, \text{m/s} \qquad v'_B = \frac{(1 + 0.92)0.1(9.94) - 0}{0.1 + 0.1} = 9.54 \, \text{m/s}$$

Using

$$T_1 = 0.9(5.49) = 4.94 \, \text{J} \qquad \text{and} \qquad T_2 = \tfrac{1}{2}m_A v'^2_A + \tfrac{1}{2}m_B v'^2_B = \tfrac{1}{2}(0.1)0.398^2 + \tfrac{1}{2}(0.1)9.54^2 = 4.56 \, \text{J}$$

the percent energy change is found as

$$\%D = \frac{T_2 - T_1}{T_1}(100) = \frac{4.56 - 4.94}{4.94}(100) = -7.69\%$$

(**d**)  From part (*c*),

$$v_A = 9.94 \, \text{m/s}$$

Using Eqs. (1) and (2) in part (*a*), with   $e = 0.5$,   gives

$$v'_A = \frac{[0.1 - 0.5(0.1)]9.94 - 0}{0.1 + 0.1} = 2.49 \, \text{m/s} \qquad v'_B = \frac{(1 + 0.5)0.1(9.94) - 0}{0.1 + 0.1} = 7.46 \, \text{m/s}$$

The energy terms before and after impact are

$$T_1 = 4.94 \, \text{J} \qquad T_2 = \tfrac{1}{2}m_A v'^2_A + \tfrac{1}{2}m_B v'^2_B = \tfrac{1}{2}(0.1)2.49^2 + \tfrac{1}{2}(0.1)7.46^2 = 3.09 \, \text{J}$$

The percent change in energy is found as

$$\%D = \frac{T_2 - T_1}{T_1}(100) = \frac{3.09 - 4.94}{4.94}(100) = -37.4\%$$

**20.32**  Block *B* in Fig. 20.32 is initially at rest.   The coefficient of friction between the blocks and the plane surface is 0.2, and the impact is assumed to be elastic.  $W_A = 20 \, \text{lb}$   and   $W_B = 10 \, \text{lb}$.

(**a**)  Find the minimum required value of the velocity $v_A$, for the position of block *B* shown in the figure, that will push this block off the plane surface.

(**b**)  Do the same as in part (*a*), if   $W_A = 10 \, \text{lb}$   and   $W_B = 20 \, \text{lb}$.

(**c**)  Do the same as in part (*a*), if   $e = 0.9$.

(**d**)  Do the same as in part (*a*), if   $e = 0.5$.

(**e**)  Compare the results in parts (*a*), (*c*), and (*d*).

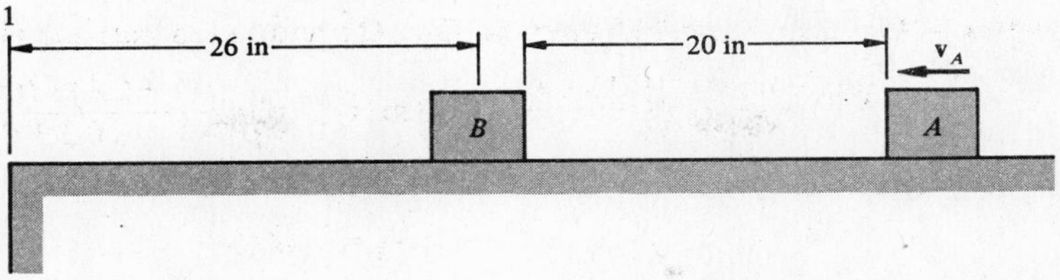

**Fig. 20.32**

**▌(a)** Using Newton's second law, the acceleration of block $B$ is found as

$$F = ma \qquad -2 = \frac{10}{32.2}\, a_B \qquad a_B = -6.44 \text{ ft/s}^2$$

The required initial velocity $v_B'$ to reach position 1 in Fig. 20.32 is found from

$$v^2 = v_0^2 + 2a(s - s_0) \qquad 0 = v_B'^2 + 2(-6.44)\frac{26}{12} \qquad v_B' = 5.28 \text{ ft/s}$$

$v_B'$ is the required velocity of block $B$ after the impact.

The velocity of block $A$ just before impact is designated $v_{A,\text{impact}}$. Using Eq. (4) in Prob. 20.22,

$$v_B' = \frac{(1+e)m_A v_A - (m_B - em_A)v_B}{m_A + m_B} \tag{1}$$

$$5.28 = \frac{(1+1)(20/g)v_{A,\text{impact}} - 0}{20/g + 10/g} \qquad v_{A,\text{impact}} = 3.96 \text{ ft/s}$$

The velocity $v_A$ of block $A$ in the initial position shown in Fig. 20.32 in found next. $a_A = a_B = -6.44 \text{ ft/s}^2$, and $v_A$ is found from

$$v^2 = v_0^2 + 2a(s - s_0) \qquad 3.96^2 = v_A^2 + 2(-6.44)\frac{20}{12} \qquad v_A = 6.09 \text{ ft/s}$$

**(b)** The weights of the two blocks are

$$W_A = 10 \text{ lb} \qquad W_B = 20 \text{ lb}$$

and

$$e = 1$$

From part (a),

$$v_B' = 5.28 \text{ ft/s}$$

Using Eq. (1) in part (a),

$$5.28 = \frac{(1+1)(10/g)v_{A,\text{impact}} - 0}{10/g + 20/g} \qquad v_{A,\text{impact}} = 7.92 \text{ ft/s}$$

The initial velocity of $A$ is found from

$$v^2 = v_0^2 + 2a(s - s_0) \qquad 7.92^2 = v_A^2 + 2(-6.44)\frac{20}{12} \qquad v_A = 9.18 \text{ ft/s}$$

The above value may be compared with the result $v_A = 6.09 \text{ ft/s}$ found in part (a). It may be seen that a larger initial velocity $v_A$ is required when the 10-lb block impacts the stationary 20-lb block.

**(c)** The weights of the two blocks are

$$W_A = 20 \text{ lb} \qquad W_B = 10 \text{ lb}$$

and

$$e = 0.9$$

From part (a),

$$v_B' = 5.28 \text{ ft/s}$$

Using Eq. (1) in part (a),

$$v_B' = 5.28 = \frac{(1+0.9)(20/g)v_{A,\text{impact}} - 0}{20/g + 10/g} \qquad v_{A,\text{impact}} = 4.17 \text{ ft/s}$$

The initial velocity of $A$ is found from

$$v^2 = v_0^2 + 2a(s - s_0) \qquad 4.17^2 = v_A^2 + 2(-6.44)\frac{20}{12} \qquad v_A = 6.23 \text{ ft/s}$$

(*d*) The weights of the two blocks are

$$W_A = 20\,\text{lb} \qquad W_B = 10\,\text{lb}$$

and
$$e = 0.5$$

From part (*a*),

$$v_B' = 5.28\,\text{ft/s}$$

Using Eq. (1) in part (*a*),

$$v_B' = 5.28 = \frac{(1+0.5)(20/g)v_{A,\text{impact}} - 0}{20/g + 10/g} \qquad v_{A,\text{impact}} = 5.28\,\text{ft/s}$$

$v_A$ is found from

$$v^2 = v_0^2 + 2a(s - s_0) \qquad 5.28^2 = v_A^2 + 2(-6.44)\frac{20}{12} \qquad v_A = 7.02\,\text{ft/s}$$

(*e*) Table 20.4 shows the results for parts (*a*), (*c*), and (*d*).
It may be seen that as the coefficient of restitution decreases, the required value of the initial velocity of block *A* increases.

**TABLE 20.4**

| $e$ | $v_A$, ft/s |
| --- | --- |
| 1 | 6.09 |
| 0.9 | 6.23 |
| 0.5 | 7.02 |

**20.33** A 600-g block rests on the edge of a table, as shown in Fig. 20.33. A second block, weighing 400 g and moving with velocity $v_A$, strikes the first block and causes the trajectory shown in the figure. The impact is assumed to be "nearly elastic," with an assumed value of the coefficient of restitution of 0.95. Find the initial velocity $v_A$ and the final velocity $v_A'$ of the striking block.

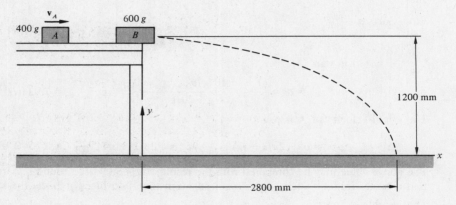

**Fig. 20.33**

▎ The equation of motion of block *B* in the *y* direction is

$$y = y_0 + v_{0y}t - \tfrac{1}{2}gt^2$$

The time *t* at which block *B* strikes the ground at $y = 0$ is found from
$$0 = 1.2 + 0 - \tfrac{1}{2}(9.81)t^2 \qquad t = 0.495\,\text{s}$$

The equation of motion of block *B* in the *x* direction is

$$x = x_0 + v_{0x}t$$

The *x* component of the initial velocity of block *B* is found from

$$2.8 = 0 + v_{0x}(0.495) \qquad v_{0x} = 5.66\,\text{m/s}$$

The above quantity is equal to $v_B'$, the velocity of block *B* after the impact.

Using Eq. (4) in Prob. 20.22,

$$v_B' = \frac{(1+e)m_A v_A - (m_B - em_A)v_B}{m_A + m_B} \qquad 5.66 = \frac{(1+0.95)(400/1,000)v_A - 0}{400/1,000 + 600/1,000} \qquad v_A = 7.26 \text{ m/s}$$

The velocity $v_A'$ is found from Eq. (3) in Prob. 20.22 as

$$v_A' = \frac{(m_A - em_B)v_A - (1+e)m_B v_B}{m_A + m_B} = \frac{[400/1,000 - 0.95(600/1,000)]7.26 - 0}{400/1,000 + 600/1,000} = -1.23 \text{ m/s}$$

The minus sign in the above result indicates that block $A$ moves leftward after the impact.

**20.34** Block $B$ in Fig. 20.34$a$ is initially at rest. When it is struck by block $A$, it describes the trajectory shown in the figure. Both blocks weigh 2 lb.

(a) Find the required value of $v_A$, the velocity of block $A$ in the position shown in the figure, if the plane is frictionless and elastic impact is assumed.

(b) Do the same as in part (a), if the coefficient of friction between the block and the plane is 0.1 and the coefficient of restitution is 0.9.

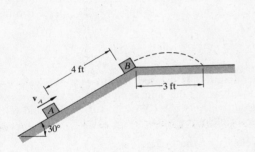

**Fig. 20.34$a$**

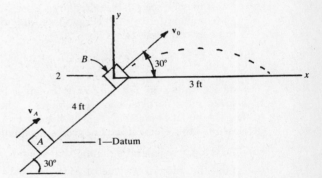

**Fig. 20.34$b$**

▌ (a) Figure 20.34$b$ shows the initial velocity $v_0$ of block $B$ after the impact. The $x$ component of this motion is described by

$$x = v_{0x}t \qquad 3 = (v_0 \cos 30°)t \qquad t = \frac{3}{v_0 \cos 30°} \tag{1}$$

The $y$ component of the motion is given by

$$y = y_0 + v_{0y}t - \tfrac{1}{2}gt^2 \qquad 0 = (v_0 \sin 30°)t - \tfrac{1}{2}(32.2)t^2 = [v_0 \sin 30° - \tfrac{1}{2}(32.2)t]t \tag{2}$$

Using $t$ from Eq. (1) in Eq. (2) gives

$$v_0 \sin 30° = \frac{\tfrac{1}{2}(32.2)3}{v_0 \cos 30°} \qquad v_0^2 \sin 30° \cos 30° = \tfrac{1}{2}(32.2)3 \qquad v_0 = 10.6 \text{ ft/s}$$

The above value $v_0 = 10.6$ ft/s is $v_B'$, the required velocity of block $B$ after the impact. The initial velocities of the blocks before impact are

$$v_B = 0 \qquad v_A = v_{A,\text{impact}}$$

Using Eq. (4) in Prob. 20.22,

$$v_B' = \frac{(1+e)m_A v_A - (m_B - em_A)v_B}{m_A + m_B} \tag{3}$$

$$10.6 = \frac{(1+1)(2/g)v_{A,\text{impact}} - 0}{2/g + 2/g} \qquad v_{A,\text{impact}} = 10.6 \text{ ft/s}$$

The endpoints 1 and 2 of the interval of motion before the impact are shown in Fig. 20.34$b$. The energy terms at these points are

$$T_1 = \tfrac{1}{2}m_A v_A^2 \qquad T_2 = \tfrac{1}{2}m_A(10.6)^2 \qquad V_1 = 0 \qquad V_2 = m_A g(4 \sin 30°)$$

The work-energy equation has the form

$$0 = \Delta T + \Delta V = (T_2 - T_1) + (V_2 - V_1) = \left[\frac{1}{2}\left(\frac{2}{32.2}\right)10.6^2 - \frac{1}{2}\left(\frac{2}{32.2}\right)v_A^2\right] + [2(4\sin 30°) - 0]$$

$$v_A = 15.5 \text{ ft/s}$$

(b)  $v_B'$ has the same value as in part (a), given by  $v_B' = 10.6$ ft/s.  Using  $e = 0.9$,  and Eq. (3) in part (a), gives

$$v_B' = 10.6 = \frac{(1 + 0.9)(2/g)v_{A,\text{impact}} - 0}{2/g + 2/g} \qquad v_{A,\text{impact}} = 11.2 \text{ ft/s}$$

The energy terms at 1 and 2 are

$$T_1 = \tfrac{1}{2}m_A v_A^2 \qquad T_2 = \tfrac{1}{2}m_A(11.2)^2 \qquad V_1 = 0 \qquad V_2 = m_A g(4 \sin 30°)$$

The work-energy equation has the form

$$W_{NC} = \Delta T + \Delta V = (T_2 - T_1) + (V_2 - V_1)$$

$$-0.1(2\cos 30°)4 = \left[\frac{1}{2}\left(\frac{2}{32.2}\right)11.2^2 - \frac{1}{2}\left(\frac{2}{32.2}\right)v_A^2\right] + [2(4\sin 30°) - 0] \qquad v_A = 16.6 \text{ ft/s}$$

It may be seen that the required value of $v_A$ in the problem in part (b) is larger than that in part (a), to compensate for the energy losses due to sliding friction and inelastic impact.  The percent difference between these two values of $v_A$ is

$$\%D = \frac{16.6 - 15.5}{15.5}(100) = 7.1\%$$

**20.35**    The device shown in Fig. 20.35a is called a ballistic pendulum.  It is used to measure the velocity of a high-speed mass particle, such as a bullet.  The box is filled with sand and supported by thin, inextensible cables.  A bullet of mass $m_A$ approaches the box with constant velocity $v_A$.  When it strikes the sand-filled box of mass $m_B$, this element experiences a maximum increase in elevation given by $h$.

(a)    Develop the general equation that relates $h$ to the velocity of the bullet.

(b)    Find the numerical value of $h$, if  $m_A = 14$ g,  $m_B = 5$ kg,  and  $v_A = 500$ m/s.

▌ (a)    For conservation of linear momentum,

$$m_A v_A + m_B v_B = m_A v_A' + m_B v_B'$$

$v_B = 0$,   and   $v_B'$ is found from

$$m_A v_A + 0 = (m_A + m_B)v_B' \qquad v_B' = \left(\frac{m_A}{m_A + m_B}\right)v_A \qquad\qquad (1)$$

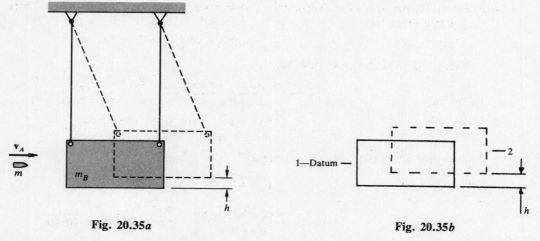

Fig. 20.35a                                                              Fig. 20.35b

Figure 20.35b shows the endpoints of the interval of motion of the sand-filled box.  The energy terms at 1 and 2 are

$$T_1 = \tfrac{1}{2}mv_B'^2 \qquad T_2 = 0 \qquad V_1 = 0 \qquad V_2 = mgh$$

The work-energy equation is written as

$$0 = \Delta T + \Delta V = (T_2 - T_1) + (V_2 - V_1) = (0 - \tfrac{1}{2} m v_B'^2) + (mgh - 0)$$

$$h = \frac{v_B'^2}{2g} \tag{2}$$

$v_B'$ from Eq. (1) is used in Eq. (2), with the result

$$h = \frac{1}{2g} \left( \frac{m_A}{m_A + m_B} \right)^2 v_A^2$$

(b) The numerical value of $h$ is given by

$$h = \frac{1}{2g} \left( \frac{m_A}{m_A + m_B} \right)^2 v_A^2 = \frac{1}{2(9.81)} \left( \frac{0.014}{0.014 + 5} \right)^2 500^2 = 0.0993 \text{ m} = 99.3 \text{ mm}$$

**20.36** The simple pendulum in Fig. 20.36a is released from rest at position $\theta$. Mass $A$ moves down, and there is an inelastic impact with mass $B$.

(a) Find the general expressions for the velocities of masses $A$ and $B$ after impact.

(b) Find the numerical value of the result in part (a), if $m = 150$ g, $l = 900$ mm, $e = 0.92$, and $\theta = 20°$, and find the percent loss of energy during the impact.

(c) Do the same as in part (b), if $e = 0.4$.

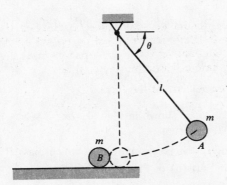

Fig. 20.36a

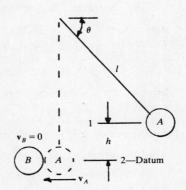

Fig. 20.36b

$\blacksquare$ (a) Figure 20.36b shows the endpoints of the interval of motion. The velocity of mass $A$ just before impact is given by

$$v_A = \sqrt{2gh} = \sqrt{2gl(1 - \sin \theta)} \tag{1}$$

Using Eqs. (3) and (4) in Prob. 20.22,

$$v_A' = \frac{(m_A - em_B)v_A - (1 + e)m_B v_B}{m_A + m_B} = \frac{(m_A - em_B)v_A - 0}{m_A + m_B} \tag{2}$$

$$v_B' = \frac{(1 + e)m_A v_A - (m_B - em_A)v_B}{m_A + m_B} = \frac{(1 + e)m_A v_A - 0}{m_A + m_B} \tag{3}$$

Using Eq. (1) in Eqs. (2) and (3) results in

$$v_A' = \left( \frac{1 - e}{2} \right) \sqrt{2gl(1 - \sin \theta)} \tag{4}$$

$$v_B' = \frac{(1 + e)m\sqrt{2gl(1 - \sin \theta)}}{2m} = \left( \frac{1 + e}{2} \right) \sqrt{2gl(1 - \sin \theta)} \tag{5}$$

(b) The numerical values of $v_A'$ and $v_B'$ are found to be

$$v_A' = \left( \frac{1 - e}{2} \right) \sqrt{2gl(1 - \sin \theta)} = \left( \frac{1 - 0.92}{2} \right) \sqrt{2(9.81)0.9(1 - \sin 20°)} = 0.136 \text{ m/s}$$

$$v_B' = \left(\frac{1+e}{2}\right)\sqrt{2gl(1-\sin\theta)} = \left(\frac{1+0.92}{2}\right)\sqrt{2(9.81)0.9(1-\sin 20°)} = 3.27\ \text{m/s}$$

Both of the above velocities are to the left in Fig. 20.36a.

The initial energy of the system of two masses is

$$T_1 = mgh = 0.150(9.81)0.9(1-\sin 20°) = 0.871\ \text{J}$$

The final energy of the system is

$$T_2 = \tfrac{1}{2}m_A v_A'^2 + \tfrac{1}{2}m_B v_B'^2 = \tfrac{1}{2}(0.15)0.136^2 + \tfrac{1}{2}(0.15)3.27^2 = 0.803\ \text{J}$$

The percent energy loss due to the impact is

$$\%D = \frac{T_1 - T_2}{T_1}(100) = \frac{0.871 - 0.803}{0.871}(100) = 7.8\%$$

(c)   $v_A$ is the same as in part (a).   Using Eqs. (4) and (5) in part (a),

$$v_A' = \left(\frac{1-0.4}{2}\right)\sqrt{2(9.81)0.9(1-\sin 20°)} = 1.02\ \text{m/s}$$

$$v_B' = \left(\frac{1+0.4}{2}\right)\sqrt{2(9.81)0.9(1-\sin 20°)} = 2.39\ \text{m/s}$$

From part (b),

$$T_1 = 0.871\ \text{J}$$

The energy at 2 is given by

$$T_2 = \tfrac{1}{2}m_A v_A'^2 + \tfrac{1}{2}m_B v_B'^2 = \tfrac{1}{2}(0.15)1.02^2 + \tfrac{1}{2}(0.15)2.39^2 = 0.506\ \text{J}$$

The percent loss of energy is

$$\%D = \frac{T_1 - T_2}{T_1}(100) = \frac{0.871 - 0.506}{0.871}(100) = 41.9\%$$

**20.37**   A second pendulum is added to the system of Prob. 20.36, as shown in Fig. 20.37a.   Mass A is released from rest at the position θ.   The impact is inelastic, and   $l_1 = 680$ mm.

(a)   Find the general expression for $\theta_1$, corresponding to the maximum height which mass B attains.

(b)   Find the numerical value of $\theta_1$, using the data in part (b) of Prob. 20.36.

(c)   Find the position of maximum elevation of mass A after the impact.

(d)   Find the percent loss of energy during the impact.

▌ (a)   Figure 20.37b shows the endpoints of the interval of motion after the impact.   The kinetic energy of mass B after the impact is

$$T_1 = \tfrac{1}{2}m v_B'^2$$

Using Eq. (5) from Prob. 20.36 in the above equation,

$$T_1 = \tfrac{1}{2}m\left(\frac{1+e}{2}\right)^2 2gl(1-\sin\theta)$$

The remaining energy terms are

$$V_1 = 0 \qquad V_2 = mgh_1 = mgl_1(1-\sin\theta_1) \qquad T_2 = 0$$

The work-energy equation has the form

$$0 = \Delta T + \Delta V = (T_2 - T_1) + (V_2 - V_1) = \left[0 - \tfrac{1}{2}m\left(\frac{1+e}{2}\right)^2 2gl(1-\sin\theta)\right] + [mgl_1(1-\sin\theta_1) - 0]$$

$$\sin\theta_1 = 1 - \frac{(1+e)^2}{4}\left(\frac{l}{l_1}\right)(1-\sin\theta)$$

(b)   Using   $l = 900$ mm,   $e = 0.92$,   $\theta = 20°$,   and   $l_1 = 680$ mm,

$$\sin\theta_1 = 1 - \frac{(1+e)^2}{4}\left(\frac{l}{l_1}\right)(1-\sin\theta) = 1 - \frac{(1+0.92)^2}{4}\left(\frac{900}{680}\right)(1-\sin 20°) \qquad \theta_1 = 11.4°$$

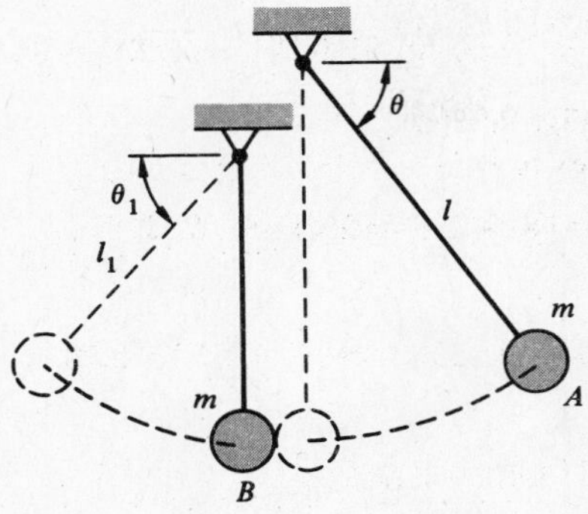

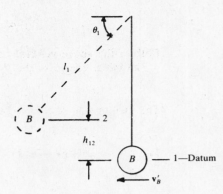

Fig. 20.37a

Fig. 20.37b

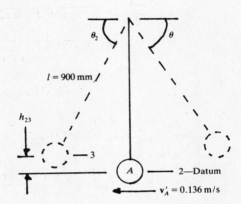

Fig. 20.37c

(c) Figure 20.37c shows angle $\theta_2$ corresponding to the maximum elevation of mass $A$ after the impact. The velocity of block $A$ after impact, from part (b) of Prob. 20.36, is

$$v_A' = 0.136 \text{ m/s}$$

The energy terms at the endpoints of the interval of motion are

$$T_2 = \tfrac{1}{2}mv_A'^2 = \tfrac{1}{2}m(0.136)^2 \qquad T_3 = 0 \qquad V_2 = 0 \qquad V_3 = mgh_{23} = mg(0.9)(1 - \sin\theta_2)$$

The work-energy equation has the form

$$0 = \Delta T + \Delta V = (T_3 - T_2) + (V_3 - V_2) = [0 - \tfrac{1}{2}m(0.136)^2] + [m(9.81)0.9(1 - \sin\theta_2) - 0]$$
$$\theta_2 = 87.4°$$

(d) The percent loss in energy during the impact is 7.8 percent, as found in part (b) of Prob. 20.36.

**20.38** The pendulum of Prob. 20.36 is arranged as shown in Fig. 20.38a. Mass $A$ is released from rest at position $\theta$, and it strikes mass $B$ at a later time. The dimensions of the masses are small compared to $l$. Using the numerical data in part (b) of Prob. 20.36, find the values of $x$ and $y$ that define the trajectory of mass $B$ after impact.

▌ The velocity $v_A$ of mass $A$ just before impact is found first. Figure 20.38b shows the endpoints of the interval of motion from the release of mass $A$ to impact of mass $A$ with mass $B$. The energy terms at 1 and 2 are

$$T_1 = 0 \qquad T_2 = \tfrac{1}{2}0.150v_A^2 \qquad V_1 = 0.150(9.81)(0.9)(1 - \sin 20°) \qquad V_2 = 0.150(9.81)(0.9)(1 - \cos 18°)$$

The work-energy equation has the form

$$0 = \Delta T + \Delta V = (T_2 - T_1) + (V_2 - V_1) = [\tfrac{1}{2}(0.150)v_A^2 - 0] + [0.150(9.81)(0.9)(1 - \cos 18°)$$
$$- 0.150(9.81)(0.9)(1 - \sin 20°)]$$
$$v_A = 3.28 \text{ m/s}$$

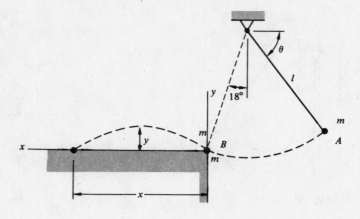

Fig. 20.38a

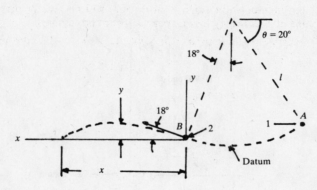

Fig. 20.38b

Using Eq. (4) in Prob. 20.22,

$$v_B' = \frac{(1+e)m_A v_A - (m_B - e m_A)v_B}{m_A + m_B} = \frac{(1+0.92)0.150(3.28) - 0}{0.150 + 0.150} = 3.15 \text{ m/s}$$

The initial velocity of the plane projectile motion of mass $B$ after the impact is

$$v_0 = v_B' = 3.15 \text{ m/s}$$

The $y$-component motion is given by

$$v_y^2 = v_{0y}^2 - 2g(y - y_0) \qquad 0 = (3.15 \sin 18°)^2 - 2(9.81)y \qquad y = 0.0483 \text{ m} = 48.3 \text{ mm}$$

The time of the projectile motion is found from

$$y = y_0 + v_{0y}t - \tfrac{1}{2}gt^2 \qquad 0 = (3.15 \sin 18°)t - \tfrac{1}{2}(9.81)t^2 \qquad t = 0, 0.198 \text{ s}$$

$t = 0$   is the initial time, at endpoint 2 in Fig. 20.38b, so that the time of the projectile motion is 0.198 s.   The $x$-component motion is given by

$$x = v_{0x}t = (3.15 \cos 18°)0.198 = 0.593 \text{ m} = 593 \text{ mm}$$

20.39    All four pendulums in Fig. 20.39a are the same.   Mass $A$ is released from rest at position $\theta$.

(a)    Describe what occurs following the impact of mass $A$ with mass $B$.

(b)    Find the position $\theta_1$ corresponding to the maximum elevation of mass $D$.

(c)    Find the values of $\theta_1$, for   $\theta = 20°$   and values of $e$ of 0.95, 0.85, 0.65, and 0.45.

(d)    Do the same as in parts (a) and (b), if pendulum system $C$ is removed from the problem, so that mass $B$ contacts mass $D$ in the rest position.

❚ (a)    Mass $A$ impacts mass $B$.   There is no change in displacement of either mass, and the velocity of mass $B$ after the impact is $v_B'$.   Mass $B$ impacts mass $C$.   There is no change in displacement of either mass, and the velocity of mass $C$ after impact is $v_C'$.   Mass $C$ impacts mass $D$.   There is no change in displacement of either mass, and the velocity of mass $D$ after impact is $v_D'$.   The above effects are shown in Fig. 20.39b.

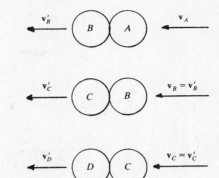

**Fig. 20.39a**                    **Fig. 20.39b**

If $e < 1$, the final-position angle $\theta_1$ of mass $D$ will be greater than the initial-position angle $\theta$ of mass $A$, because of the energy loss associated with each impact.

(b)   Equation (3) in Prob. 20.36 is used, with $m_A = m_B = m$. For the impact of masses $A$ and $B$,

$$v_B' = \left(\frac{1+e}{2}\right)v_A$$

For the impact of masses $B$ and $C$,

$$v_C' = \left(\frac{1+e}{2}\right)v_B \qquad (1)$$

Using

$$v_B = v_B' = \left(\frac{1+e}{2}\right)v_A$$

in Eq. (1) gives

$$v_C' = \left(\frac{1+e}{2}\right)^2 v_A$$

For the impact of masses $C$ and $D$,

$$v_D' = \left(\frac{1+e}{2}\right)v_C \qquad (2)$$

Using

$$v_C = v_C' = \left(\frac{1+e}{2}\right)^2 v_A$$

in Eq. (2) gives

$$v_D' = \left(\frac{1+e}{2}\right)^3 v_A \qquad (3)$$

The velocity of mass $A$ just before impact is given by Eq. (1) in part (a) of Prob. 20.36 as

$$v_A = \sqrt{2gl(1 - \sin \theta)} \qquad (4)$$

Figure 20.39c shows the endpoints of the interval of motion of mass $D$. The required energy terms are

$$T_1 = \tfrac{1}{2}mv_D'^2 \qquad T_2 = 0 \qquad V_1 = 0 \qquad V_2 = mgl(1 - \sin \theta_1)$$

The work-energy equation has the form

$$0 = \Delta T + \Delta V = (T_2 - T_1) + (V_2 - V_1) = (0 - \tfrac{1}{2}mv_D'^2) + [mgl(1 - \sin \theta_1) - 0]$$

$$\sin \theta_1 = 1 - \frac{v_D'^2}{2gl} \qquad (5)$$

Using Eq. (3) in Eq. (5) gives

$$\sin \theta_1 = 1 - \left(\frac{1+e}{2}\right)^6 v_A^2 \qquad (6)$$

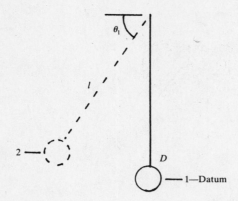

**Fig. 20.39c**

Using Eq. (4) in Eq. (6) results in

$$\sin \theta_1 = 1 - \left(\frac{1+e}{2}\right)^6 \frac{2gl}{2gl} (1 - \sin \theta) = 1 - \left(\frac{1+e}{2}\right)^6 (1 - \sin \theta) \tag{7}$$

$\theta_1$ may be found from the above equation, and this angle defines the maximum elevation of mass $D$ after the impact.

(c)   Using   $\theta = 20°$   and   $e = 0.95$   in Eq. (7) results in

$$\sin \theta_1 = 1 - \left(\frac{1+0.95}{2}\right)^6 (1 - \sin 20°) \qquad \theta_1 = 25.8°$$

For   $e = 0.85$,   using Eq. (7),

$$\sin \theta_1 = 1 - \left(\frac{1+0.85}{2}\right)^6 (1 - \sin 20°) \qquad \theta_1 = 36.0°$$

For   $e = 0.65$,   with Eq. (7),

$$\sin \theta_1 = 1 - \left(\frac{1+0.85}{2}\right)^6 (1 - \sin 20°) \qquad \theta_1 = 52.4°$$

For   $e = 0.45$,   using Eq. (7),

$$\sin \theta_1 = 1 - \left(\frac{1+0.45}{2}\right)^6 (1 - \sin 20°) \qquad \theta_1 = 64.7°$$

(d)   If mass $C$ is removed from the system, mass $D$ plays the role of mass $C$.   Using the solutions in part (a),

$$v_D' = v_C' = \left(\frac{1+e}{2}\right)^2 v_A \qquad \sin \theta_1 = 1 - \left(\frac{1+e}{2}\right)^4 (1 - \sin \theta)$$

## 20.3   DIRECT AND OBLIQUE CENTRAL IMPACT, IMPULSIVE FORCES

**20.40**   (a)   What is the definition of *central impact*?

(b)   What are the definitions of *direct* central impact and *oblique* central impact?

(c)   For the case of oblique central impact what is the significant characteristic of the components of velocity normal to the line of impact?

▌   (a)   Figure 20.40a shows two bodies at the instant before impact.   Points $a$ and $b$ are coincident points on the surfaces of the bodies at the location at which impact occurs.   If the common normal line to the surfaces at points $a$ and $b$ also intersects the centers of mass of both bodies, the impact is referred to as central impact.   In central impact, there is no tendency of the bodies to rotate.   If this condition is not fulfilled, one or both of the bodies will possess angular velocity after the impact.

(b)   If the velocities of the two bodies are *collinear* with the common normal line, the impact is described as *direct central impact*.   All the impact problems considered so far in this chapter have been cases of direct central impact.   If one or both of the velocities of the bodies are *not* collinear with the common normal line, the impact is referred to as *oblique central impact*.

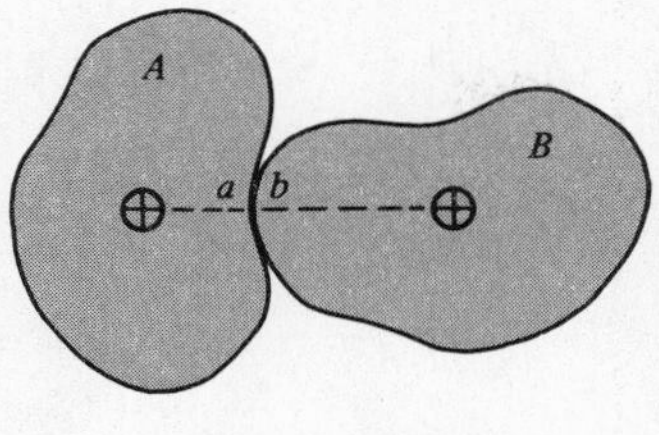

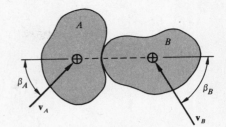

Fig. 20.40a

Fig. 20.40b

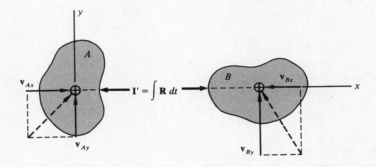

Fig. 20.40c

(c)  Figure 20.40b shows two bodies in a general configuration of oblique central impact. The initial velocities $v_A$ and $v_B$, with the directions $\beta_A$ and $\beta_B$, are known, and it is desired to find the magnitudes and directions of the velocities of the two bodies after impact.

The free-body diagrams during impact of the two bodies are shown in Fig. 20.40c. Because of the definition of *central impact*, the direction of the impulse $I'$ which acts between the two bodies must be along the $x$ axis. Thus, *there is no component of this impulse along the $y$ axis*. The equation of impulse momentum in the $y$ direction is written for each body, with the result

$$0 = m_A(v'_{Ay} - v_{Ay}) \qquad v'_{Ay} = v_{Ay}$$
$$0 = m_B(v'_{By} - v_{By}) \qquad v'_{By} = v_{By}$$

The very significant conclusion is now arrived at that, in oblique central impact, *the components of the velocities of the two impacting bodies normal to the line of impact do not change*. For the case considered above, the $x$ components of the velocities are then used as if the problem were one of direct central impact along the $x$ axis.

**20.41**  Two spheres approach each other with constant velocity along rectilinear paths, as shown in Fig. 20.41a. The coefficient of restitution is assumed to be 0.85.

(a)  Find the magnitudes and directions of the velocities of the two spheres after impact, and the percent loss of energy, for the impact configuration shown in Fig. 20.41b.

(b)  Do the same as in part (a) for the impact configurations shown in Fig. 20.41c.

▌ (a)  The impact configuration for Fig. 20.41b is shown in Fig. 20.41d. The components of initial velocity are

$$v_{Ax} = 50 \sin 40° = 32.1 \text{ in/s} \qquad v_{Bx} = -80 \text{ in/s}$$
$$v_{Ay} = -50 \cos 40° = -38.3 \text{ in/s} \qquad v_{By} = 0$$

The components of velocity in the $y$ direction are unchanged, so that

$$v'_{Ay} = v_{Ay} = -38.3 \text{ in/s} \qquad v'_{By} = v_{By} = 0$$

The equation of impulse momentum in the $x$ direction is

$$m_A v_{Ax} + m_B v_{Bx} = m_A v'_{Ax} + m_B v'_B x \qquad \frac{1.5}{g}(32.1) + \frac{2}{g}(-80) = \frac{1.5}{g}v'_{Ax} + \frac{2}{g}v'_{Bx} \qquad (1)$$

The equation relating the relative velocities is

$$e = -\frac{v'_{Bx} - v'_{Ax}}{v_{Bx} - v_{Ax}} \qquad 0.85 = \frac{-(v'_{Bx} - v'_{Ax})}{-80 - 32.1} \qquad (2)$$

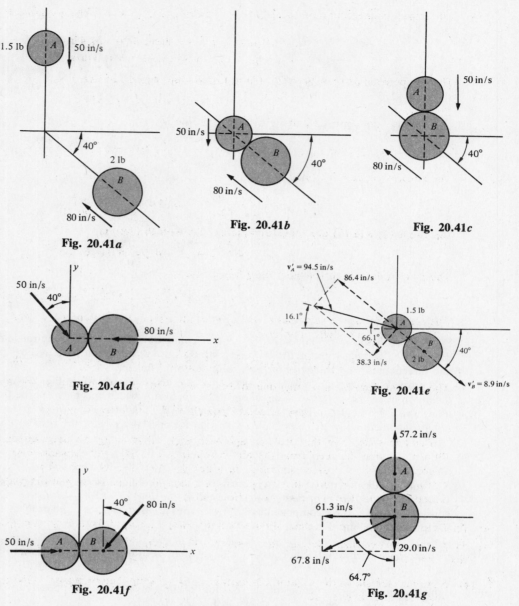

Fig. 20.41a

Fig. 20.41b

Fig. 20.41c

Fig. 20.41d

Fig. 20.41e

Fig. 20.41f

Fig. 20.41g

Equations (1) and (2) are solved simultaneously, with the results

$$v'_{Ax} = -86.4 \text{ in/s} \qquad v'_{Bx} = 8.9 \text{ in/s}$$

The magnitudes of the two velocities after impact are

$$v'_A = \sqrt{(v'_{Ax})^2 + (v'_{Ay})^2} = \sqrt{(-86.4)^2 + (-38.3)^2} = 94.5 \text{ in/s} \qquad v'_B = v'_{Bx} = 8.9 \text{ in/s}$$

The orientations of the post-impact velocities are shown in Fig. 20.41e.
The initial kinetic energy is

$$T_1 = \tfrac{1}{2}m_A v_A^2 + \tfrac{1}{2}m_B v_B^2 = \frac{1}{2}\left(\frac{1.5}{386}\right)(50^2) + \frac{1}{2}\left(\frac{2}{386}\right)(80^2) = 21.4 \text{ in} \cdot \text{lb} \qquad (3)$$

The value of the kinetic energy after impact is

$$T_2 = \tfrac{1}{2}m_A(v'_A)^2 + \tfrac{1}{2}m_B(v'_B)^2 = \frac{1}{2}\left(\frac{1.5}{386}\right)(94.5^2) + \frac{1}{2}\left(\frac{2}{386}\right)(8.9^2) = 17.6 \text{ in} \cdot \text{lb}$$

The percent decrease in energy due to the impact is

$$\%\text{D} = \frac{T_1 - T_2}{T_1}(100) = \frac{21.4 - 17.6}{21.4}(100) = 17.8\%$$

(b)   The impact configuration of Fig. 20.41c is redrawn in Fig. 20.41f.   The components of initial velocity are

$$v_{Ax} = 50 \text{ in/s} \qquad v_{Bx} = -80 \sin 40° = -51.4 \text{ in/s}$$
$$v_{Ay} = 0 \qquad v_{By} = -80 \cos 40° = -61.3 \text{ in/s}$$

The components of velocity in the $y$ direction do not change, so that

$$v'_{Ay} = v_{Ay} = 0 \qquad v'_{By} = v_{By} = -61.3 \text{ in/s}$$

The impulse-momentum equation in the $x$ direction is

$$m_A v_{Ax} + m_B v_{Bx} = m_A v'_{Ax} + m_B v'_{Bx} \qquad \frac{1.5}{g}(50) + \frac{2}{g}(-51.4) = \frac{1.5}{g} v'_{Ax} + \frac{2}{g} v'_{Bx} \qquad (4)$$

The equation of relative velocities is

$$e = -\frac{v'_{Bx} - v'_{Ax}}{v_{Bx} - v_{Ax}} \qquad 0.85 = \frac{-(v'_{Bx} - v'_{Ax})}{-51.4 - 50} \qquad (5)$$

Equations (4) and (5) are solved simultaneously, with the results

$$v'_{Ax} = -57.2 \text{ in/s} \qquad v'_{Bx} = 29.0 \text{ in/s}$$

The magnitudes of the velocities after impact are

$$v'_A = v'_{Ax} = -57.2 \text{ in/s} \qquad v'_B = \sqrt{(v'_{Bx})^2 + (v'_{By})^2} = \sqrt{29.0^2 + (-61.3)^2} = 67.8 \text{ in/s}$$

The orientation of the velocities of the spheres after impact is shown in Fig. 20.41g.
    The kinetic energy after impact is

$$T_2 = \tfrac{1}{2} m_A (v'_A)^2 + \tfrac{1}{2} m_B (v'_B)^2 = \frac{1}{2}\left(\frac{1.5}{386}\right)(57.2^2) + \frac{1}{2}\left(\frac{2}{386}\right)(67.8^2) = 18.3 \text{ in} \cdot \text{lb}$$

The percent decrease in energy due to the impact, using the initial value in Eq. (3), is

$$\%\text{D} = \frac{T_1 - T_2}{T_1}(100) = \frac{21.4 - 18.3}{21.4}(100) = 14.5\%$$

The two spheres in this problem approach each other along the fixed directions shown in Fig. 20.41a.  It may be seen from the above solutions that *the final velocities after impact are heavily dependent on the precise configuration in which the spheres strike each other.*
    It may also be observed that the post-impact velocities in parts (a) and (b) could have been found directly by using Eqs. (3) and (4) in Prob. 20.22.

**20.42**   (a)   Find the magnitude and direction of the velocities, after impact, of the two spheres shown in Fig. 20.42a.
    (b)   Find the energy loss during the impact.  $W_A = 6$ lb, $W_B = 10.5$ lb,  $v_A = 60$ in/s,  $v_B = 72$ in/s, and  $e = 0.84$.

$\blacksquare$   (a)   Figure 20.42b shows the $x$ and $y$ components of the velocities before impact.   The $y$ components of the velocities after impact are

$$v'_{Ay} = v_{Ay} = 0 \qquad v'_{By} = v_{By} = -36 \text{ in/s}$$

Using Eqs. (3) and (4) in Prob. 20.22,

$$v'_{Ax} = \frac{(m_A - e m_B)v_{Ax} - (1 + e)m_B v_{Bx}}{m_A + m_B} = \frac{[(6/g) - 0.84(10.5/g)]60 - (1 + 0.84)(10.5/g)(62.4)}{6/g + 10.5/g}$$

$$= -83.3 \text{ in/s}$$

$$v'_{Bx} = \frac{(1 + e)m_A v_{Ax} - (m_B - e m_A)v_{Bx}}{m_A + m_B} = \frac{(1 + 0.84)(6/g)60 - [(10.5/g) - 0.84(6/g)]62.4}{6/g + 10.5/g}$$

$$= 19.5 \text{ in/s}$$

Figure 20.42c shows the velocities after impact.   The magnitudes of these velocities are

$$v'_A = 83.3 \text{ in/s} \qquad \text{to the left}$$

$$v'_B = \sqrt{v'^2_{Bx} + v'^2_{By}} = \sqrt{19.5^2 + 36^2} = 40.9 \text{ in/s}$$

The direction $\theta$ of $v'_B$ is found from

$$\tan \theta = \frac{36}{19.5} \qquad \theta = 61.6°$$

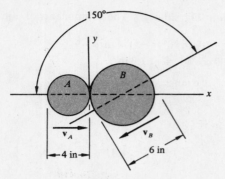

Fig. 20.42a

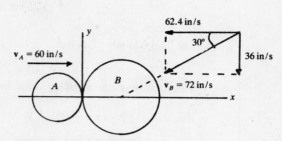

Fig. 20.42b

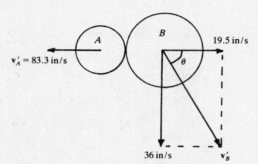

Fig. 20.42c

(b)  The kinetic energy of the two spheres before impact is

$$T_1 = \tfrac{1}{2}m_A v_A^2 + \tfrac{1}{2}m_B v_B^2 = \frac{1}{2}\left(\frac{6}{386}\right)60^2 + \frac{1}{2}\left(\frac{10.5}{386}\right)72^2 = 98.5 \text{ in} \cdot \text{lb}$$

The kinetic energy after impact is

$$T_2 = \tfrac{1}{2}m_A v_A'^2 + \tfrac{1}{2}m_B v_B'^2 = \frac{1}{2}\left(\frac{6}{386}\right)83.3^2 + \frac{1}{2}\left(\frac{10.5}{386}\right)40.9^2 = 76.7 \text{ in} \cdot \text{lb}$$

The percent loss in energy due to the impact is

$$\%\text{D} = \frac{T_1 - T_2}{T_1}\,(100) = \frac{98.5 - 76.7}{98.5}\,(100) = 22.1\%$$

**20.43**  Two identical 2.5-kg spherical masses move along rectilinear paths with constant velocity, as shown in Fig. 20.43a.

(a)  Find the magnitude and direction of the velocities of the spheres after impact, for the impact configuration shown in Fig. 20.43b.  The impact is assumed to be elastic.

(b)  Do the same as in part (a), if the velocities are interchanged, so that $v_A = 2.44$ m/s  and  $v_B = 1.64$ m/s.

(c)  Do the same as in part (a), if the coefficient of restitution is 0.9.

(d)  Find the percent energy loss associated with the impact in part (c).

▌ (a)  Figure 20.43c shows the impact configuration referenced to the $x_1 y_1$ coordinates.  The angle $\theta$ between the $x$ and $x_1$ axes is found from

$$\sin \theta = \frac{r}{2r} \qquad \theta = 30°$$

The $y_1$ components of the velocity after impact are

$$v_{Ay_1}' = v_{Ay_1} = -0.82 \text{ m/s} \qquad v_{By_1}' = v_{By_1} = -2.11 \text{ m/s}$$

Using Eqs. (3) and (4) in Prob. 20.22,

$$v_{Ax_1}' = \frac{(m_A - em_B)v_{Ax_1} - (1+e)m_B v_{Bx_1}}{m_A + m_B} = \frac{[m - (1)m]1.42 - (1+1)m(1.22)}{m + m} = -1.22 \text{ m/s} \qquad (1)$$

Fig. 20.43a

Fig. 20.43b

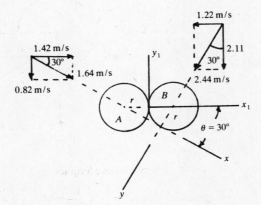

Fig. 20.43c

$$v'_{Bx_1} = \frac{(1+e)m_A v_{Ax_1} - (m_B - em_A)v_{Bx_1}}{m_A + m_B} = \frac{(1+1)m(1.42) - [m - (1)m]1.22}{m + m} = 1.42 \text{ m/s} \quad (2)$$

The velocities after impact are seen to be independent of the masses of the two elements $A$ and $B$. Figures 20.43$d$ and $e$ show the velocity components of mass $A$ and mass $B$ after the impact.

Using Fig. 20.43$d$, the velocity components of mass $A$ after impact, in the $x$ and $y$ directions, are found as

$$v'_{Ax} = 0.82 \sin 30° - 1.22 \sin 60° = -0.647 \text{ m/s}$$
$$v'_{Ay} = 0.82 \cos 30° + 1.22 \cos 60° = 1.32 \text{ m/s}$$

The magnitude and direction of $v'_A$ are given by

$$v'_A = \sqrt{v'^2_{Ax} + v'^2_{Ay}} = \sqrt{0.647^2 + 1.32^2} = 1.47 \text{ m/s} \qquad \tan \theta = \frac{1.32}{0.647} \qquad \theta = 63.9°$$

$\theta$ is the angle between $v'_A$ and the $x$ axis. The velocity components of mass $B$ after the impact, using Fig. 20.43$e$, are

$$v'_{Bx} = 1.42 \cos 30° + 2.11 \cos 60° = 2.28 \text{ m/s} \qquad v'_{By} = 2.11 \cos 30° - 1.42 \sin 30° = 1.12 \text{ m/s}$$

The magnitude and direction of $v'_B$ are found from

$$v'_B = \sqrt{v'^2_{Bx} + v'^2_{By}} = \sqrt{2.28^2 + 1.12^2} = 2.54 \text{ m/s} \qquad \tan \theta = \frac{1.12}{2.28} \qquad \theta = 26.2°$$

Figure 20.43$f$ shows the velocities, before and after impact, of both masses.

(b) Figure 20.43$g$ shows the case where the initial velocities are interchanged, so that

$$v_A = 2.44 \text{ m/s} \qquad v_B = 1.64 \text{ m/s}$$

The $y_1$ components of the velocity after impact are

$$v'_{Ay_1} = v_{Ay_1} = -1.22 \text{ m/s} \qquad v'_{By_1} = v_{By_1} = -1.42 \text{ m/s}$$

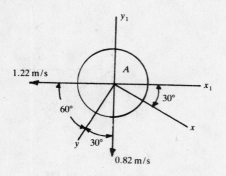

Fig. 20.43d

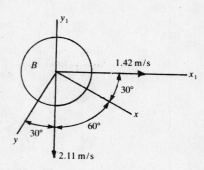

Fig. 20.43e

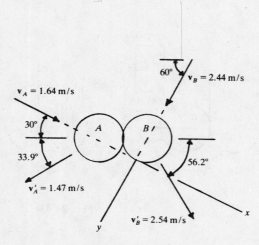

Fig. 20.43f

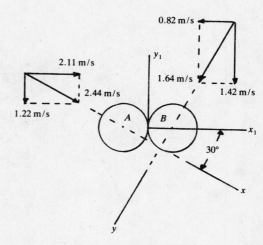

Fig. 20.43g

Using Eqs. (1) and (2) in part (a),

$$v'_{Ax_1} = \frac{[m-(1)m]2.11 - (1+1)m(0.82)}{m+m} = -0.82 \text{ m/s}$$

$$v'_{Bx_1} = \frac{(1+1)m(2.11) - [m-(1)m]0.82}{m+m} = 2.11 \text{ m/s}$$

Figures 20.43h and i show the velocities of masses A and B after the impact. The velocity components in the x and y directions after impact are

$$v'_{Ax} = 1.22 \sin 30° - 0.82 \sin 60° = -0.100 \text{ m/s}$$
$$v'_{Ay} = 1.22 \cos 30° + 0.82 \cos 60° = 1.47 \text{ m/s}$$

The magnitude and direction of $v'_A$ are found as

$$v'_A = \sqrt{v'^2_{Ax} + v'^2_{Ay}} = \sqrt{(-0.100)^2 + 1.47^2} = 1.47 \text{ m/s} \qquad \tan \theta = \frac{1.47}{0.100} \qquad \theta = 86.1°$$

Using Fig. 20.43i, the velocity components of mass B after the impact are

$$v'_{Bx} = 2.11 \cos 30° + 1.42 \sin 30° = 2.54 \text{ m/s}$$
$$v'_{By} = 1.42 \cos 30° - 2.11 \sin 30° = 0.175 \text{ m/s}$$

The magnitude and direction of $v'_B$ are given by

$$v'_B = \sqrt{v'^2_{Bx} + v'^2_{By}} = \sqrt{2.54^2 + 0.175^2} = 2.55 \text{ m/s} \qquad \tan \theta = \frac{0.175}{2.54} \qquad \theta = 3.94°$$

The velocities of the two masses before and after impact are shown in Fig. 20.43j.

A comparison may be made of the results shown in Figs. 20.43f and j. It may be seen that the velocities of masses A and B after impact are the same in both cases. The directions of these velocities, however, are interchanged.

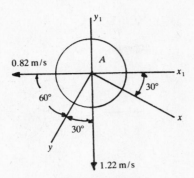

**Fig. 20.43h**

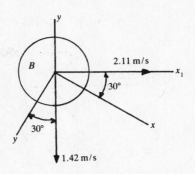

**Fig. 20.43i**

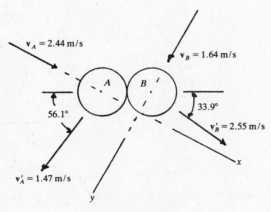

**Fig. 20.43j**

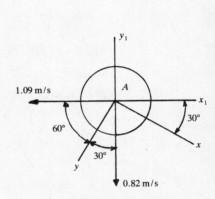

**Fig. 20.43k**

(c)    The solution for the case where the coefficient of restitution is 0.9 is similar to that in part (a).    The steps in this solution are shown below.

Using Fig. 20.43c,

$$v'_{Ay_1} = v_{Ay_1} = -0.82 \text{ m/s} \qquad v'_{By_1} = v_{By_1} = -2.11 \text{ m/s}$$

Using Eqs. (1) and (2) in part (a),

$$v'_{Ax_1} = \frac{[m - (0.9)m]1.42 - (1 + 0.9)m(1.22)}{m + m} = -1.09 \text{ m/s}$$

$$v'_{Bx_1} = \frac{(1 + 0.9)m(1.42) - [m - (0.9)m]1.22}{m + m} = 1.29 \text{ m/s}$$

The results above for the velocity components are shown in Figs. 20.43k and l.    The components, magnitudes, and directions of $v'_A$ and $v'_B$ are found as

$$v'_{Ax} = 0.82 \sin 30° - 1.09 \sin 60° = -0.534 \text{ m/s} \qquad v'_{Ay} = 0.82 \cos 30° + 1.09 \cos 60° = 1.26 \text{ m/s}$$

$$v'_A = \sqrt{v'^2_{Ax} + v'^2_{Ay}} = \sqrt{0.534^2 + 1.26^2} = 1.37 \text{ m/s} \qquad \tan \theta = \frac{1.26}{0.534} \qquad \theta = 67.0°$$

$$v'_{Bx} = 1.29 \cos 30° + 2.11 \cos 60° = 2.17 \text{ m/s} \qquad v'_{By} = 2.11 \sin 60° - 1.29 \sin 30° = 1.18 \text{ m/s}$$

$$v'_B = \sqrt{v'^2_{Bx} + v'^2_{By}} = \sqrt{2.17^2 + 1.18^2} = 2.47 \text{ m/s} \qquad \tan \theta = \frac{1.18}{2.17} \qquad \theta = 28.5°$$

The velocities of the two masses before and after impact are shown in Fig. 20.43m.    The results in this figure may be compared with those for the case of elastic impact, shown in Fig. 20.43f.    It may be seen that the effect of the inelastic impact is to change both the magnitudes and directions of the velocities of A and B after impact.

(d)    The kinetic energy before impact is

$$T_1 = \tfrac{1}{2}mv_A^2 + \tfrac{1}{2}mv_B^2 = \tfrac{1}{2}(2.5)(1.64)^2 + \tfrac{1}{2}(2.5)(2.44)^2 = 10.8 \text{ J}$$

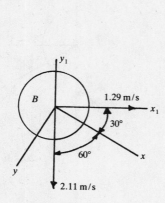

**Fig. 20.43*l***

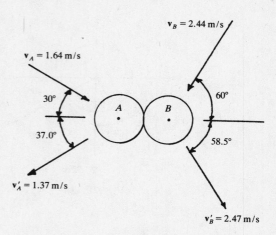

**Fig. 20.43*m***

The kinetic energy after impact is given by

$$T_2 = \tfrac{1}{2}mv_A'^2 + \tfrac{1}{2}mv_B'^2 = \tfrac{1}{2}(2.5)(1.37)^2 + \tfrac{1}{2}(2.5)(2.47)^2 = 9.97 \text{ J}$$

The percent loss of energy due to the impact is

$$\%\text{D} = \frac{T_1 - T_2}{T_1}\,(100) = \frac{10.8 - 9.97}{10.8}\,(100) = 7.69\%$$

**20.44** Do the same as in Prob. 20.43, part (*a*), for the impact configuration shown in Fig. 20.44*a*.

❚ The $x_1$ and $y_1$ components of the velocity before impact are shown in Fig. 20.44*b*. The $y_1$ components of velocity after impact are

$$v_{Ay_1}' = v_{Ay_1} = -1.16 \text{ m/s} \qquad v_{By_1}' = v_{By_1} = -1.73 \text{ m/s}$$

Using Eqs. (1) and (2) in Prob. 20.43,

$$v_{Ax_1}' = \frac{[m - (1)m]1.16 - (1+1)m(1.73)}{m + m} = -1.73 \text{ m/s}$$

$$v_{Bx_1}' = \frac{(1+1)m(1.16) - [m - (1)m]1.73}{m + m} = 1.16 \text{ m/s}$$

The velocities of masses $A$ and $B$ after impact are shown in Figs. 20.44*c* and *d*. The components, magnitudes, and directions of $v_A'$ and $v_B'$ are found as

$$v_{Ax}' = 1.16 \sin 45° - 1.73 \sin 45° = -0.403 \text{ m/s} \qquad v_{Ay}' = 1.16 \cos 45° + 1.73 \cos 45° = 2.04 \text{ m/s}$$

$$v_A' = \sqrt{v_{Ax}'^2 + v_{Ay}'^2} = \sqrt{(-0.403)^2 + 2.04^2} = 2.08 \text{ m/s} \qquad \tan\theta = \frac{2.04}{0.403} \qquad \theta = 78.8°$$

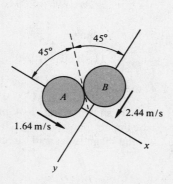

**Fig. 20.44*a***

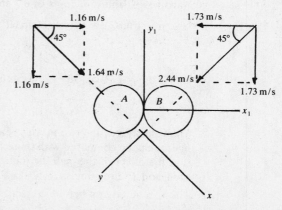

**Fig. 20.44*b***

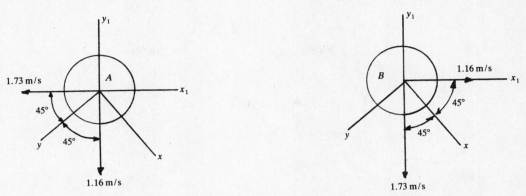

Fig. 20.44c                                    Fig. 20.44d

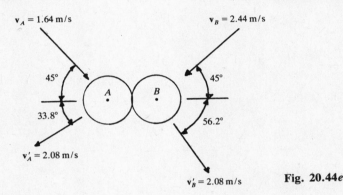

Fig. 20.44e

$$v'_{Bx} = 1.16 \cos 45° + 1.73 \cos 45° = 2.04 \text{ m/s} \qquad v'_{By} = 1.73 \sin 45° - 1.16 \sin 45° = 0.403 \text{ m/s}$$

$$v'_B = \sqrt{v'^2_{Bx} + v'^2_{By}} = \sqrt{2.04^2 + 0.43^2} = 2.08 \text{ m/s} \qquad \tan \theta = \frac{0.403}{2.04} \qquad \theta = 11.2°$$

The velocities of masses $A$ and $B$ before and after impact are seen in Fig. 20.44e. It is interesting to observe that the magnitudes of $v'_A$ and $v'_B$ are the same, but their directions are significantly different.

**20.45**   A sphere of weight $W_A = 1$ lb is released from rest in the position shown in Fig. 20.45a, and impacts an identical sphere that is at rest. The track is frictionless, and the coefficient of restitution is 0.92.

(a)   Find the magnitude, direction, and sense of the velocities of the two spheres after impact.

(b)   Find the value of the impulse that occurs during the impact.

(c)   Find the percent loss in energy due to the impact.

(d)   Do the same as in parts (a), (b), and (c) if, at the instant of impact shown in Fig. 20.45a, sphere $B$ has a rightward velocity in the $y$ direction of 5 ft/s.

▌ (a)   Since the track is frictionless, sphere $A$ slides down the track without rolling. The velocity of sphere $A$ before impact is given by

$$v_A = \sqrt{2gh} = \sqrt{2(32.2)2} = 11.3 \text{ ft/s}$$

Figure 20.45b shows the impact configuration of the two spheres. Angle $\theta$ is found from

$$\sin \theta = \frac{r}{2r} \qquad \theta = 30°$$

The velocity components in the $y_1$ direction are

$$v'_{Ay_1} = v_{Ay_1} = 5.65 \text{ ft/s} \qquad v'_{By_1} = v_{By_1} = 0$$

Using Eqs. (3) and (4) in Prob. 20.22,

$$v'_{Ax_1} = \frac{(m_A - em_B)v_{Ax_1} - (1 + e)m_Bv_{Bx_1}}{m_A + m_B} = \frac{(m - 0.92m)9.79 - 0}{m + m} = 0.392 \text{ ft/s} \qquad (1)$$

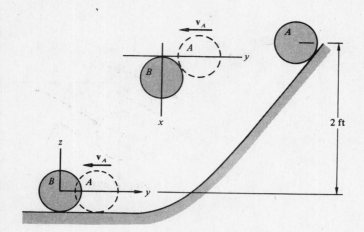

**Fig. 20.45a**

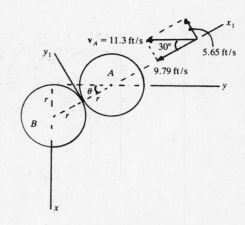

**Fig. 20.45b**

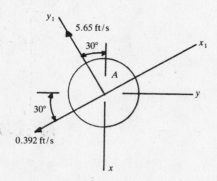

**Fig. 20.45c**

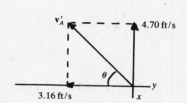

**Fig. 20.45d**

$$v'_{Bx_1} = \frac{(1+e)m_A v_{Ax_1} - (m_B - em_A)v_{Bx_1}}{m_A + m_B} = \frac{(1+0.92)m(9.79) - 0}{m + m} = 9.40 \text{ ft/s} \qquad (2)$$

Figure 20.45c shows the velocity components of sphere $A$ after the impact. The components of velocity in the $x$ and $y$ directions are

$$v'_{Ax} = -5.65 \cos 30° + 0.392 \sin 30° = -4.70 \text{ ft/s}$$
$$v'_{Ay} = -0.392 \cos 30° - 5.65 \sin 30° = -3.16 \text{ ft/s}$$

The above two components of velocity of sphere $A$ are shown in Fig. 20.45d. The magnitude and direction of $v'_A$ are found as

$$v'_A = \sqrt{v'^2_{Ax} + v'^2_{Ay}} = \sqrt{(-4.70)^2 + (-3.16)^2} = 5.66 \text{ ft/s}$$

$$\tan \theta = \frac{4.70}{3.16} \qquad \theta = 56.1°$$

The magnitude and direction of the velocity of sphere $B$ after the impact are

$$v'_B = 9.40 \text{ ft/s} \qquad \theta = 30°$$

These results are shown in Fig. 20.45e.

(b)  Figure 20.45f shows the impulsive force $R$ which acts on sphere $B$. The value of the impulse that occurs during the impact is given by

$$I' = \int_1^2 R \, dt = m_B(v'_B - v_B) = \frac{1}{32.2} (9.40 - 0) = 0.292 \text{ lb} \cdot \text{s}$$

(c)  The initial energy $V_1$ of the system of two spheres is

$$V_1 = mgh = 1(2) = 2 \text{ ft} \cdot \text{lb}$$

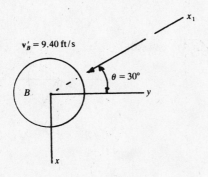

Fig. 20.45e

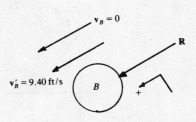

Fig. 20.45f

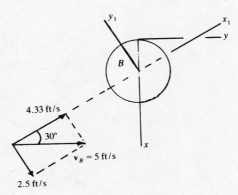

Fig. 20.45g

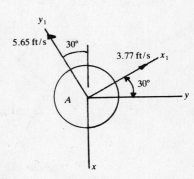

Fig. 20.45h

The energy $T_2$ after impact is

$$T_2 = \tfrac{1}{2}m_A v_A'^2 + \tfrac{1}{2}m_B v_B'^2 = \frac{1}{2}\left(\frac{1}{32.2}\right)5.66^2 + \frac{1}{2}\left(\frac{1}{32.2}\right)9.40^2 = 1.87 \text{ ft} \cdot \text{lb}$$

The percent loss of energy due to the impact is

$$\%\text{D} = \frac{V_1 - T_2}{V_1}\,(100) = \frac{2 - 1.87}{2}\,(100) = 6.5\%$$

(d) From part (a),

$$v_A = 11.3 \text{ ft/s}$$

Figure 20.45g shows the velocity components of sphere $B$, in the $x_1$ and $y_1$ directions, before impact. The velocity components, after impact, in the $y_1$ direction are

$$v_{Ay_1}' = v_{Ay_1} = 5.65 \text{ ft/s} \qquad v_{By_1}' = v_{By_1} = -2.5 \text{ ft/s}$$

Using Eqs. (3) and (4) in Prob. 20.22,

$$v_{Ax_1}' = \frac{(m_A - em_B)v_{Ax_1} - (1+e)m_B v_{Bx_1}}{m_A + m_B} = \frac{(m - 0.92m)9.79 - (1+0.92)m(4.33)}{m + m} = -3.77 \text{ ft/s}$$

$$v_{Bx_1}' = \frac{(1+e)m_A v_{Ax_1} - (m_B - em_A)v_{Bx_1}}{m_A + m_B} = \frac{(1+0.92)m(9.79) - (m - 0.92m)4.33}{m + m} = 9.23 \text{ ft/s}$$

Figure 20.45h shows the velocity components of sphere $A$, after the impact, in the $x_1$ and $y_1$ directions. The components of velocity in the $x$ and $y$ directions are

$$v_{Ax}' = -5.65\cos 30^\circ - 3.77\sin 30^\circ = -6.78 \text{ ft/s}$$
$$v_{Ay}' = -5.65\sin 30^\circ + 3.77\cos 30^\circ = 0.440 \text{ ft/s}$$

The above results are shown in Fig. 20.45i, and the magnitude and direction of $v_A'$ are found as

$$v_A' = \sqrt{v_{Ax}'^2 + v_{Ay}'^2} = \sqrt{(-6.78)^2 + (0.440)^2} = 6.79 \text{ ft/s} \qquad \tan\theta = \frac{6.78}{0.440} \qquad \theta = 86.3^\circ$$

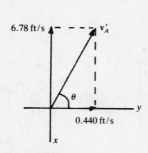

Fig. 20.45i

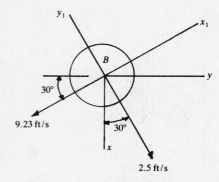

Fig. 20.45j

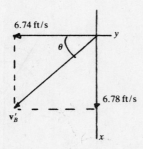

Fig. 20.45k

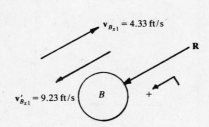

Fig. 20.45l

The velocity components of sphere $B$ after the impact, in the $x_1$ and $y_1$ directions, are shown in Fig. 20.45j. The components of $v_B'$ in the $x$ and $y$ directions are

$$v_{Bx}' = 9.23 \sin 30° + 2.5 \cos 30° = 6.78 \text{ ft/s}$$
$$v_{By}' = 2.5 \sin 30° - 9.23 \cos 30° = -6.74 \text{ ft/s}$$

The above results are shown in Fig. 20.45k. The magnitude and direction of $v_B'$ are found to be

$$v_B' = \sqrt{v_{Bx}'^2 + v_{By}'^2} = \sqrt{6.78^2 + (-6.74)^2} = 9.56 \text{ ft/s} \qquad \tan \theta = \frac{6.78}{6.74} \qquad \theta = 45.2°$$

The impulsive force $R$ that acts on sphere $B$ is shown in Fig. 20.45l. The value of the impulse is given by

$$I' = \int_1^2 R \, dt = m_B(v_{Bx_1}' - v_{Bx_1}) = \frac{1}{32.2}[9.23 - (-4.33)] = 0.421 \text{ lb} \cdot \text{s}$$

The energy terms before and after impact are given by

$$T_1 + V_1 = \tfrac{1}{2}m_B v_B^2 + m_A gh = \frac{1}{2}\left(\frac{1}{32.2}\right)5^2 + 1(2) = 2.39 \text{ ft} \cdot \text{lb}$$

$$T_2 = \tfrac{1}{2}m_A v_A'^2 + \tfrac{1}{2}m_B v_B'^2 = \frac{1}{2}\left(\frac{1}{32.2}\right)6.79^2 + \frac{1}{2}\left(\frac{1}{32.2}\right)9.56^2 = 2.14 \text{ ft} \cdot \text{lb}$$

The percent loss of energy during the impact is found to be

$$\%\text{D} = \frac{T_1 - T_2}{T_1}(100) = \frac{2.39 - 2.14}{2.39}(100) = 10.5\%$$

**20.46** (a) Sphere $A$ in Fig. 20.46a moves with a constant velocity $v_A = 20 \text{ in/s}$. Find the magnitude and direction of the velocities of spheres $B$ and $C$ after impact. Each sphere has a weight of 1.4 lb and a diameter of 2.4 in, and the impact is assumed to be elastic.

(b) Do the same as in part (a), if the coefficient of restitution is 0.85, and find the percent loss of energy due to the impact.

(c) Do the same as in part (a), for the configuration of the spheres shown in Fig. 20.46b.

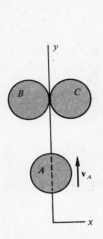

**Fig. 20.46a**

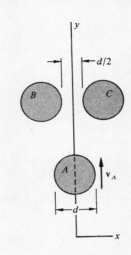

**Fig. 20.46b**

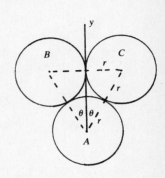

**Fig. 20.46c**

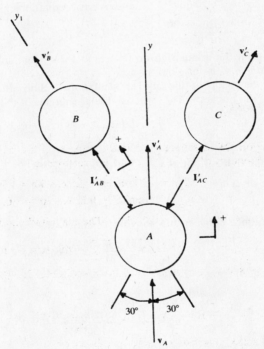

**Fig. 20.46d**

▌ (*a*) From the impact configuration in Fig. 20.46*c*,

$$\sin \theta = \frac{r}{2r} \qquad \theta = 30°$$

Figure 20.46*d* shows the velocities of spheres *B* and *C* after impact, and the impulses which act on the spheres. From symmetry,

$$I'_{AB} = I'_{AC} \qquad v'_B = v'_C$$

For motion of sphere *A* along the *y* axis, the impulse-momentum equation has the form

$$I' = \int_1^2 R \, dt = m(v' - v) \qquad -2I'_{AB} \cos 30° = m(v'_A - v_A) \qquad (1)$$

For motion of sphere *B* along the $y_1$ axis, the impulse-momentum equation is written as

$$I' = \int_1^2 R \, dt = m(v' - v) \qquad I'_{AB} = m(v'_B - 0) \qquad (2)$$

Using $I'_{AB}$ from Eq. (2) in Eq. (1) gives

$$-2(mv'_B)\cos 30° = m(v'_A - v_A) \tag{3}$$

The coefficient of restitution $e$ is defined by

$$e = -\frac{v'_B - v'_A}{v_B - v_A}$$

All velocities in the above equation are interpreted as velocity components along the line of impact. For the present problem,

$$e = \frac{-(v'_B - v'_A \cos 30°)}{0 - v_A \cos 30°} \qquad -ev_A \cos 30° = -v'_B + v'_A \cos 30° \tag{4}$$

$v'_A$ is eliminated between Eqs. (3) and (4) to obtain

$$v'_B = \frac{(1+e)v_A \cos 30°}{2\cos^2 30° + 1} \tag{5}$$

From Eq. (3),

$$v'_A = v_A - 2v'_B \cos 30° \tag{6}$$

Using $e = 1$ and $v_A = 20$ in/s in the above two equations gives

$$v'_B = \frac{(1+1)20(\cos 30°)}{2\cos^2 30° + 1} = 13.9 \text{ in/s} \qquad v'_A = 20 - 2(13.9)\cos 30° = -4.08 \text{ in/s}$$

It may be seen from the minus sign on the result for $v'_A$ that sphere $A$ rebounds in a sense which is opposite to the sense of its original motion.

A check may be made of the above calculations. Since the impact is elastic, energy must be conserved during the impact. The initial and final energies of the system of three spheres are given by

$$T_1 = \frac{1}{2}\left(\frac{1.4}{386}\right)(20)^2 = 0.725 \text{ in} \cdot \text{lb}$$

$$T_2 = \tfrac{1}{2}m_A v'^2_A + \tfrac{1}{2}m_B v'^2_B + \tfrac{1}{2}m_C v'^2_C = \frac{1}{2}\left(\frac{1.4}{386}\right)[4.08^2 + 2(13.9)^2] = 0.731 \text{ in} \cdot \text{lb}$$

For conservation of energy,

$$T_1 \overset{?}{=} T_2$$

The percent difference between the above two values, due to roundoff error, is

$$\frac{0.731 - 0.725}{0.731}(100) = 0.8\%$$

so that $0.725 \approx 0.731$ and $T_1 \approx T_2$.

(b) Using Eq. (5) in part (a), with $e = 0.85$,

$$v'_B = \frac{(1 + 0.85)20(\cos 30°)}{2\cos^2 30° + 1} = 12.8 \text{ in/s}$$

Using the above result in Eq. (6) in part (a),

$$v'_A = 20 - 2(12.8)\cos 30° \qquad v_A = -2.17 \text{ in/s}$$

The energy of the system before impact, from part (a), is

$$T_1 = 0.725 \text{ in} \cdot \text{lb}$$

The energy after impact is

$$T_2 = \frac{1}{2}\left(\frac{1.4}{386}\right)[2.17^2 + 2(12.8)^2] = 0.603 \text{ in} \cdot \text{lb}$$

The percent loss of energy that occurs during the impact is

$$\%\text{D} = \frac{T_1 - T_2}{T_1}(100) = \frac{0.725 - 0.603}{0.725}(100) = 16.8\%$$

(c) Figure 20.46e shows the impact configuration. Using Fig. 20.46f, angle $\theta$ is found from

$$\sin \theta = \frac{\frac{3}{4}d}{d} = \frac{3}{4} \qquad \theta = 48.6°$$

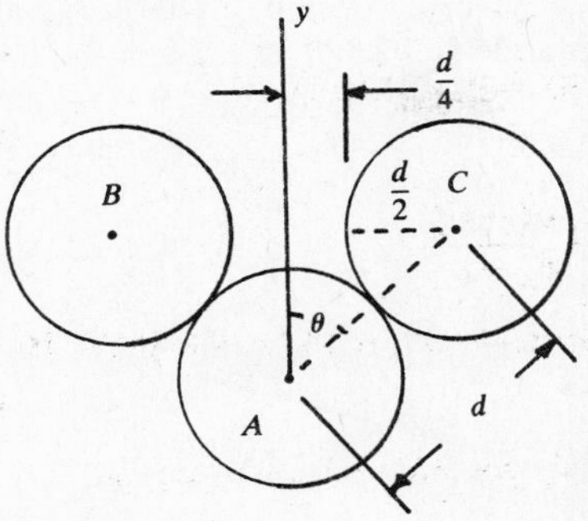

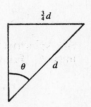

**Fig. 20.46e**

**Fig. 20.46f**

Using Eq. (5) in part (a), with $e = 1$ and $30° \rightarrow 48.6°$, gives

$$v_B' = \frac{(1+1)20(\cos 48.6°)}{2 \cos^2 48.6° + 1} = 14.1 \text{ in/s}$$

The above result is used in Eq. (6) in part (a), with the result

$$v_A' = 20 - 2(14.1) \cos 48.6° = 1.35 \text{ in/s}$$

It may be observed in this case that, since $v_A'$ is positive, sphere $A$ does not rebound after the impact. A check may now be made on the above results. From part (a),

$$T_1 = 0.725 \text{ in} \cdot \text{lb}$$

The energy after impact is given by

$$T_2 = \tfrac{1}{2}m[v_A'^2 + 2v_B'^2] = \tfrac{1}{2}m[1.35^2 + 2(14.1)^2] = 0.724 \text{ in} \cdot \text{lb}$$

Since the impact is elastic,

$$T_1 \overset{?}{=} T_2 \qquad 0.725 \approx 0.724 \text{ in} \cdot \text{lb}$$

**20.47**    What is the difference between *impulsive* forces and impact forces?

▌ A situation which is very similar to the case of impact is the problem of *impulsive* forces. Impulsive forces are forces of very large magnitude which occur for very short durations. The distinguishing characteristic between impulsive forces and impact forces is that *impulsive forces do not occur because of the impact of physical bodies*. These forces are typically caused by explosive effects, such as the firing of a bullet from a gun or the detonation process in an internal-combustion engine, where the fuel mixture in the cylinder ignites and exerts momentary forces of very large magnitudes on the face of the piston. Since impact is not present, there is no need for a quantity such as the coefficient of restitution. The equation which expresses the conservation of linear momentum is still valid, and this equation may be used directly to solve the problem.

**20.48**    The rifle shown in Fig. 20.48a weighs 8 lb. It fires a bullet, weighing 0.8 oz, which leaves the rifle with a muzzle velocity of 2,000 ft/s. Find the recoil velocity of the rifle.

▌ The initial velocities of the rifle and the bullet are zero. The model of the system when the rifle is fired is shown in Fig. 20.48b. $I'$ is the impulse which acts between the rifle and the bullet. This is an example of impulsive motion, and the equation of impulse momentum appears as

$$m_A v_A + m_B v_B = m_A v_A' + m_B v_B' \qquad 0 + 0 = \frac{8}{g} v_A' + \frac{0.8(\frac{1}{16})}{g}(2,000) \qquad v_A' = -12.5 \text{ ft/s}$$

The minus sign on $v_A'$ indicates that the recoil velocity of the rifle is to the left.

**Fig. 20.48a**

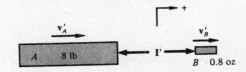

**Fig. 20.48b**

**20.49** Figure 20.49 shows a model of an 8-in howitzer used on a naval battleship. The 8-in-diameter shell weighs 200 lb. The absolute velocity of the shell as it leaves the muzzle is 2,200 ft/s. When the howitzer is fired, it is observed that the barrel has a recoil velocity of 60 ft/s.

(a) Find the weight of the howitzer barrel.

(b) Find the kinetic energy of the shell as it leaves the howitzer.

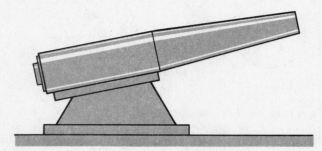

Fig. 20.49

▮ (a) $A$ and $B$ are the designations of the barrel and shell, respectively. The impulse-momentum equation has the form

$$m_A v_A + m_B v_B = m_A v_A' + m_B v_B' \qquad 0 + 0 = \frac{W_A}{g}(-60) + \frac{200}{g}(2,200) \qquad W_A = 7,330 \text{ lb}$$

(b) The kinetic energy of the shell as it leaves the barrel is given by

$$T = \tfrac{1}{2} m v^2 = \frac{1}{2}\left(\frac{200}{32.2}\right)2,200^2 = 1.50 \times 10^7 = 15,000,000 \text{ ft} \cdot \text{lb}$$

**20.50** Figure 20.50a shows an artillery cannon. It fires a 90-mm shell that weighs 4.41 lb. The muzzle velocity is 1,100 ft/s, and the weight of the cannon barrel is 160 lb. A nest of eight equal springs, each of stiffness $k$, is used to limit the recoil displacement. Find the value of $k$, if the displacement of the cannon barrel after firing must not exceed 10 in. Neglect the effect of the weight of the cannon barrel on the springs.

▮ The cannon is designated $A$, and the shell is $B$. The impulse-momentum equation is written as

$$m_A v_A + m_B v_B = m_A v_A' + m_B v_B' \qquad 0 + 0 = \frac{160}{g} v_A' + \frac{4.41}{g}(1,100) \qquad v_A' = -30.3 \text{ ft/s}$$

Figure 20.50b shows the endpoints of the interval of recoil motion. The energy terms have the forms

$$T_1 = \tfrac{1}{2} m v_A'^2 = \frac{1}{2}\left(\frac{160}{32.2}\right)30.3^2 \qquad T_2 = 0 \qquad V_1 = 0 \qquad V_2 = 8[\tfrac{1}{2} k (10)^2 \text{ in} \cdot \text{lb}]\left(\frac{1 \text{ ft}}{12 \text{ in}}\right)$$

$k$ is the stiffness of one spring, and the factor of 8 in the above equation represents the combined stiffness of the 8 springs.

The work-energy equation has the form

$$0 = \Delta T + \Delta V = (T_2 - T_1) + (V_2 - V_1) = \left[0 - \frac{1}{2}\left(\frac{160}{32.2}\right)30.3^2\right] + \{8[\tfrac{1}{2}k(10)^2](\tfrac{1}{12}) - 0\}$$

$$k = 68.4 \text{ lb/in}$$

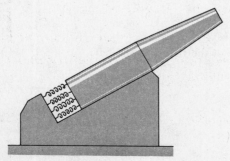

Fig. 20.50a

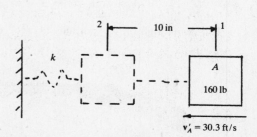

Fig. 20.50b

## 20.4  ANGULAR MOMENTUM, AND IMPULSE-MOMENTUM EQUATION, OF A RIGID BODY IN PLANE MOTION, IMPACT OF RIGID BODIES IN PLANE MOTION, IMPACT AT THE CENTER OF PERCUSSION

**20.51**   (*a*)   Give the form of the *angular momentum* of a rigid body which rotates about a fixed axis.

   (*b*)   Is angular momentum a vector or a scalar quantity?

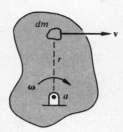

**Fig. 20.51**

▌ (*a*)   Figure 20.51 shows a rigid body which rotates about a fixed point *a* with angular velocity $\omega$. A typical mass element *dm* in the body has the velocity *v* shown in the figure. The linear momentum of this mass element is $dm\,v$. If this linear momentum of the mass element is multiplied by *r*, the distance from the linear momentum vector to the axis of rotation, the product is referred to as the angular momentum, or *moment of momentum*, of the particle. This latter quantity is designated by the symbol *H*, and

$$dH = (v\,dm)r \tag{1}$$

Equation (1) is now integrated over all the mass elements of the body, with the result

$$H = \int dH = \int_{\text{vol}} (v\,dm)r \tag{2}$$

Since

$$v = r\omega$$

Eq. (2) may be written as

$$H = \int_{\text{vol}} (r\omega\,dm)r = \int_{\text{vol}} r^2\omega\,dm$$

The angular velocity $\omega$ is the same for all mass elements of the body. Thus, this quantity may be moved outside the integral sign, and the result is

$$H = \omega \int_{\text{vol}} r^2\,dm$$

The integral may now be identified as the *mass moment of inertia of the body about the fixed point a*. This quantity is designated $I_a$, and the final form of the angular momentum of the body is

$$H = I_a\omega$$

   (*b*)   The angular momentum is a vector quantity which has the direction of the angular velocity vector. For the case of a rigid body which moves with plane motion, this direction will always be normal to the plane of the motion. The effect on the motion of the body when the *direction* of the angular momentum vector is allowed to change is considered in Probs. 22.41 to 22.47.

**20.52**   Give the form of the impulse-momentum equation for a rigid body which rotates about a fixed axis.

▌ The rigid body in Fig. 20.52 is acted on by the applied moment $M_a$ about the fixed point *a*. Newton's second law for rotation of the body about the fixed point is

$$M_a = I_a\alpha \tag{1}$$

$I_a$ is the mass moment of inertia of the body about the fixed point, and $\alpha$, the angular acceleration of the body, may be expressed as

$$\alpha = \frac{d\omega}{dt} \tag{2}$$

**Fig. 20.52**

Equations (1) and (2) are combined and integrated over a time interval with the endpoints 1 and 2. The result is

$$\int_1^2 M_a \, dt = \int_1^2 I_a \, d\omega = I_a \int_1^2 d\omega = I_a(\omega_2 - \omega_1) = H_2 - H_1 \qquad (3)$$

The quantity $\int_1^2 M \, dt$ is the *impulse of the moment* $M_a$. It may be seen from Eq. (3) that the effect of this impulse is to give the rigid body a change in angular momentum, with a corresponding change in angular velocity. The magnitude of the impulse of the moment also may be identified as the area under the moment-time curve between times $t_1$ and $t_2$.

If the moment which acts on the rigid body is *constant*, then Eq. (3) reduces to the elementary form

$$M_a(t_2 - t_1) = I_a(\omega_2 - \omega_1)$$

**20.53** The electric motor shown in Fig. 20.53*a* has the two-stage starting torque-time curve shown in Fig. 20.53*b*. The armature, and other rotating parts of the motor, may be approximated as a disk of mass 25 kg and diameter 400 mm.

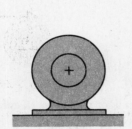

**Fig. 20.53*a***

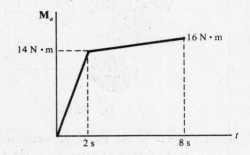

**Fig. 20.53*b***

(*a*) Find the steady-state angular velocity of the motor, in revolutions per minute, at the end of the startup phase of motion if no load is connected to the motor.

(*b*) A magnetic brake, which exerts a constant braking torque, is later used to stop the motor. If the motor is observed to come to rest in 2 s, find the value of the braking torque.

❚ (*a*) The mass moment of inertia of the rotating parts is

$$I_a = \tfrac{1}{8}md^2 = \tfrac{1}{8}(25)\left(\frac{400}{1,000}\right)^2 = 0.5 \text{ kg} \cdot \text{m}^2$$

The impulse of the starting torque is equal to the area $A$ under the moment-time curve in Fig. 20.53*b*, given by

$$A = \tfrac{1}{2}(2)14 + \frac{14 + 16}{2}(8 - 2) = 104 \text{ N} \cdot \text{s}$$

The impulse-momentum equation has the form

$$\int_1^2 M_a \, dt = I_a(\omega_2 - \omega_1)$$

$$104 \text{ N} \cdot \text{s} = (0.5 \text{ kg} \cdot \text{m}^2)(\omega_2 - 0) \qquad \omega_2 = 208 \text{ rad/s} = 208\left(\frac{60}{2\pi}\right) = 1,990 \text{ r/min}$$

(b) The braking of the motor occurs with a constant value of braking torque $M_B$. Thus,

$$M_B(t_2 - t_1) = I_a(\omega_2 - \omega_1)$$

By using $t_2 - t_1 = 2\,\text{s}$, $\omega_1 = 208\,\text{rad/s}$, and $\omega_2 = 0$, the above equation appears as

$$M_B(2) = 0.5(0 - 208) \qquad M_B = -52\,\text{N}\cdot\text{m}$$

The minus sign in the result for $M_B$ indicates that the sense of the braking torque is opposite to the sense of rotation.

**20.54** Figure 20.54a shows two elements in a high-speed mechanism. Cam $A$ rotates clockwise at 2,500 r/min, and cam $B$ is stationary. The shaft of cam $A$ is momentarily lowered into the position shown, so that the projection on this cam can strike the projection on cam $B$. After the impact, cam $B$ has a counterclockwise angular velocity, and the shaft of cam $A$ is raised to its original position. $I_A = 6 \times 10^{-4}\,\text{lb}\cdot\text{s}^2\cdot\text{in}$ and $I_B = 9.8 \times 10^{-6}\,\text{lb}\cdot\text{s}^2\cdot\text{in}$.

(a) If the impact of the cams is assumed to be plastic, find the velocities of both cams after impact.

(b) Find the percent loss of energy caused by the impact.

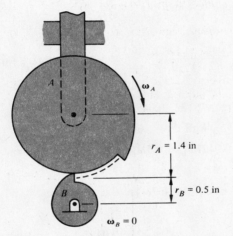

**Fig. 20.54a**

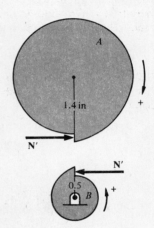

**Fig. 20.54b**

▌ (a) Figure 20.54b shows the tangential force impulse $N'$ which acts on both cams and the two assumed positive senses of rotation. The equations of impulse momentum are

Cam $A$: $\qquad\qquad\qquad -N'(1.4) = I_A(\omega_A' - \omega_A)$       (1)

Cam $B$: $\qquad\qquad\qquad N'(0.5) = I_B(\omega_B' - \omega_B)$       (2)

$N'$ is eliminated between Eqs. (1) and (2), with the result

$$\frac{I_A(\omega_A' - \omega_A)}{-1.4} = \frac{I_B(\omega_B' - \omega_B)}{0.5}$$

Using $\omega_A = 2{,}500\,\text{r/min} = 262\,\text{rad/s}$ and $\omega_B = 0$, we get

$$\frac{6 \times 10^{-4}(\omega_A' - 262)}{-1.4} = \frac{9.8 \times 10^{-6}\omega_B'}{0.5} \qquad \omega_A' - 262 = -0.0457\omega_B' \tag{3}$$

Since plastic impact is assumed, the tangential velocity of the two cams at the point of contact must be the same after the impact. This kinematic relationship is

$$r_A\omega_A' = r_B\omega_B' \qquad 1.4\omega_A' = 0.5\omega_B' \tag{4}$$

Equations (3) and (4) are solved simultaneously, and the final results are

$$\omega_B' = 650\,\text{rad/s} = 6{,}210\,\text{r/min} \qquad \omega_A' = \frac{0.5}{1.4}\,\omega_B' = 232\,\text{rad/s} = 2{,}220\,\text{r/min}$$

(b) The initial kinetic energy is

$$T_1 = \tfrac{1}{2}I_A\omega_A^2 = \tfrac{1}{2}(6 \times 10^{-4})(262^2) = 20.6\,\text{in}\cdot\text{lb}$$

The kinetic energy of the system after the impact is

$$T_2 = \tfrac{1}{2}I_A(\omega_A')^2 + \tfrac{1}{2}I_B(\omega_B')^2 = \tfrac{1}{2}(6 \times 10^{-4})(232^2) + \tfrac{1}{2}(9.8 \times 10^{-6})(650^2) = 18.2 \text{ in} \cdot \text{lb}$$

The percent loss of energy is

$$\%\text{D} = \frac{T_1 - T_2}{T_1}\,(100) = \frac{20.6 - 18.2}{20.6}\,(100) = 11.7\%$$

**20.55** The impact of the two elements in the high-speed mechanism in Prob. 20.54 is assumed to be inelastic.   A term $e$, which represents a coefficient of restitution effect, of the ratio of the relative tangential velocities before and after impact, is assumed to have the form

$$e = -\left(\frac{r_A\omega_A' - r_B\omega_B'}{r_A\omega_A - r_B\omega_B}\right)$$

(a)   Find the velocities of both cams after impact, if $e = 0.9$.

(b)   Do the same as in part (a), if $e = 0.75$.

(c)   Do the same as in part (a), if $e = 0.5$.

(d)   Compare all of the results in parts (a) through (c), and in Prob. 20.54.

❚ (a)   The coefficient of restitution term has the form

$$e = -\frac{r_A\omega_A' - r_B\omega_B'}{r_A\omega_A - r_B\omega_B} = \frac{-(1.4\omega_A' - 0.5\omega_B')}{1.4(262) - 0} \qquad 1.4(262)e = -1.4\omega_A' + 0.5\omega_B' \qquad (1)$$

Equation (3) in Prob. 20.54 is

$$\omega_A' - 262 = -0.0457\omega_B' \qquad (2)$$

$\omega_A'$ is eliminated between Eqs. (1) and (2), with the result

$$1.4(262)e = -1.4(262 - 0.0457\omega_B') + 0.5\omega_B' \qquad \omega_B' = (1 + e)650 \text{ rad/s} \qquad (3)$$

From Eq. (2),

$$\omega_A' = 262 - 0.0457\omega_B' \text{ rad/s} \qquad (4)$$

Using $e = 0.9$ in Eqs. (3) and (4),

$$\omega_B' = (1 + 0.9)650 = 1,240 \text{ rad/s} \qquad \omega_A' = 262 - 0.0457(1,240) = 205 \text{ rad/s}$$

(b)   Using $e = 0.75$,

$$\omega_B' = (1 + 0.75)650 = 1,140 \text{ rad/s} \qquad \omega_A' = 262 - 0.0457(1,140) = 210 \text{ rad/s}$$

(c)   Using $e = 0.5$,

$$\omega_B' = (1 + 0.5)650 = 975 \text{ rad/s} \qquad \omega_A' = 262 - 0.0457(975) = 217 \text{ rad/s}$$

(d)   The results for $\omega_A'$ and $\omega_B'$, for $e = 0$ to $e = 0.9$, are shown in Table 20.5.   It may be seen that, as the coefficient of restitution decreases, $\omega_A'$ increases and $\omega_B'$ decreases.

### TABLE 20.5

| $e$ | $\omega_A'$, rad/s | $\omega_B'$, rad/s |
|-----|--------------------|--------------------|
| 0.9 | 205 | 1,240 |
| 0.75 | 210 | 1,140 |
| 0.50 | 217 | 975 |
| 0 | 232 | 650 |

**20.56** Give the equations which govern the case of impact of rigid bodies in plane motion.

❚ For the case of impact of rigid bodies in plane motion, two independent effects occur.

**Translation of the Center of Mass**

For this case, the equation

$$I' = \int_1^2 R\,dt = m(v' - v) \tag{1}$$

may be written.  $I'$ is the impulse of the resultant force which acts on the body, and $v$ and $v'$ are the initial and final values, respectively, of the velocity of the *center of mass*.

### Rotation of the Body about Its Center of Mass

For this case, the equation

$$\int_1^2 M_0\,dt = I_0(\omega' - \omega) \tag{2}$$

may be written.  The mass moment of inertia $I_0$ is with respect to the center of mass, and the resultant moment $M_0$ is with respect to the same point.  This moment will be expressed in terms of the components of the impulse $I'$, from Eq. (1), of the resultant force which acts on the body.

When Eqs. (1) and (2) are used, additional assumptions must be introduced which determine whether the impact is plastic or elastic, and whether slipping motion is impending.  These effects will be illustrated in the following problems.  In all cases, it is assumed that the impacting bodies possess a common plane of symmetry in which act all the impulsive forces and moments.  Thus, there is no tendency for the bodies to move out of this plane of motion after the impact.

20.57   The disk shown in Fig. 20.57a rolls without slipping along a horizontal surface until it impacts the edge of a rigid step.

(a)   Find the motion of the disk after impact and the energy loss, if plastic impact is assumed and the friction force between the disk and the edge is sufficiently large so that slipping motion is not impending.  Find the numerical values of the velocity components after impact, if $v_c = 10$ m/s,  $m = 5$ kg,  $\beta = 30°$,  and  $r = 160$ mm.

(b)   Find the necessary condition if slipping motion during impact is not impending.

(c)   Do the same as in part (a) if slipping motion during impact is impending.  Find the numerical values of the velocities after impact if  $\mu = 0.05$.

(d)   Do the same as in part (a) if the edge of the step is assumed to be frictionless.

(e)   Summarize the results for the velocities after impact, for parts (a), (c), and (d).

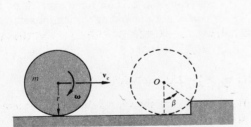

**Fig. 20.57a**

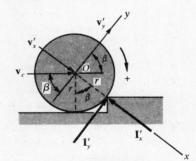

**Fig. 20.57b**

▌ (a)   The configuration of the disk and the edge during impact is shown in Fig. 20.57b.  Since the disk initially rolls without slipping

$$v_c = r\omega \tag{1}$$

The components $I'_x$ and $I'_y$ of the total impulse exerted by the edge on the disk are shown in the figure in their assumed *actual* senses.  These quantities will be considered to be *positive* if they act in the positive coordinate senses.  $I'_x$ must be a compressive effect, and the sense of $I'_y$ is chosen as shown, since this quantity is a friction force effect which must resist the initial angular velocity $\omega$.  The positive sense of rotation is as shown in Fig. 20.57b.

The equations of translational and rotational impulse momentum now appear as

$$-I'_x = m(v'_x - v_c \sin \beta) \tag{2}$$

$$I'_y = m(v'_y - v_c \cos \beta) \tag{3}$$

$$-I'_y r = I_0(\omega' - \omega) \tag{4}$$

Plastic impact is assumed, so that

$$v'_x = 0 \tag{5}$$

Since slipping motion is not impending during impact, the contact point between the disk and the edge is an instant center of rotation of the disk. Thus,

$$v'_y = r\omega' \tag{6}$$

$I'_y$ is eliminated from Eqs. (3) and (4), and the result is used with Eqs. (1) and (5), together with $I_0 = \frac{1}{2}mr^2$, to obtain

$$v'_y = \frac{v_c}{3}(1 + 2\cos\beta) \tag{7}$$

The remaining two components of velocity after impact, using Eqs. (5) and (6), are

$$v'_x = 0 \qquad \omega' = \frac{v'_y}{r} = \frac{v_c}{3r}(1 + 2\cos\beta) \tag{8}$$

$v'_x$ and $v'_y$ are positive when acting in the positive $x$ and $y$ coordinate senses in Fig. 20.57b, and the positive sense of $\omega'$ is clockwise. The numerical values for part (a) are

$$v_c = 10\,\text{m/s} \qquad v'_x = 0 \qquad v'_y = \tfrac{10}{3}(1 + 2\cos 30°) = 9.11\,\text{m/s}$$

$$\omega = \frac{v_c}{r} = \frac{10}{0.16} = 62.5\,\text{rad/s} \qquad \omega' = \frac{v'_y}{r} = \frac{9.11}{0.16} = 56.9\,\text{rad/s}$$

The mass moment of inertia of the disk about its center is

$$I_0 = \tfrac{1}{2}mr^2 = \tfrac{1}{2}(5)(0.16)^2 = 0.064\,\text{kg}\cdot\text{m}^2$$

The initial kinetic energy is

$$T_1 = \tfrac{1}{2}mv_c^2 + \tfrac{1}{2}I_0\omega^2 = \tfrac{1}{2}(5)(10^2) + \tfrac{1}{2}(0.064)(62.5^2) = 375\,\text{J}$$

The kinetic energy after the impact is

$$T_2 = \tfrac{1}{2}m(v'_y)^2 + \tfrac{1}{2}I_0(\omega')^2 = \tfrac{1}{2}(5)(9.11^2) + \tfrac{1}{2}(0.064)(56.9^2) = 311\,\text{J}$$

The percent loss in energy due to the impact is

$$\%\text{D} = \frac{T_1 - T_2}{T_1}(100) = \frac{375 - 311}{375}(100) = 17.1\%$$

(b)  The impulse component $I'_y$ in Fig. 20.57b is caused by the friction force exerted by the edge on the disk. The necessary condition, if slipping motion is not impending during impact, is

$$\mu I'_x > I'_y \qquad \mu > \frac{I'_y}{I'_x} \tag{9}$$

where $\mu$ is the coefficient of friction.

Using $v'_x = 0$ and $v'_y$ from Eq. (7) in Eqs. (2) and (3), we get

$$I'_x = mv_c\sin\beta \qquad I'_y = \frac{mv_c}{3}(1 - \cos\beta)$$

The limiting condition, Eq. (9), then has the form

$$\mu > \frac{I'_y}{I'_x} = \frac{(mv_c/3)(1 - \cos\beta)}{mv_c\sin\beta} = \frac{1 - \cos\beta}{3\sin\beta} \tag{10}$$

It is interesting to note that the result above is independent of the mass and velocities, and is a function only of angle $\beta$. The limiting value of $\mu$, for $\beta = 30°$, is

$$\mu_{\text{min}} = \frac{1 - \cos 30°}{3\sin 30°} = 0.089$$

(c)  For the case where slipping motion is impending during the plastic impact, Eqs. (2) through (5) are still valid. The condition of impending slipping motion is reflected in the equation

$$I'_y = \mu I'_x \tag{11}$$

The solution to this system of equations, given by Eqs. (2) through (5), and Eq. (11), is

$$v'_x = 0 \qquad v'_y = v_c(\cos\beta + \mu\sin\beta) \qquad \omega' = \omega - \frac{rmv_c\mu\sin\beta}{I_0} \tag{12}$$

The numerical values, with $\mu = 0.05$, are

$$v_x' = 0 \qquad v_y' = 10(\cos 30° + 0.05 \sin 30°) = 8.91 \text{ m/s}$$

$$\omega' = 62.5 - \frac{0.016(5)(10)(0.05) \sin 30°}{0.064} = 59.4 \text{ rad/s}$$

The kinetic energy after the impact is

$$T_2 = \tfrac{1}{2}mv_y'^2 + \tfrac{1}{2}I_0\omega'^2 = \tfrac{1}{2}(5)(8.91^2) + \tfrac{1}{2}(0.064)(59.4^2) = 311 \text{ J}$$

The percent loss in energy due to the impact is

$$\%\text{D} = \frac{T_1 - T_2}{T_1}(100) = \frac{375 - 311}{375}(100) = 17.1\%$$

(d)  If the edge is frictionless,

$$I_y' = 0$$

From Eq. (3),

$$v_y' - v_c \cos \beta = 0 \qquad v_y' = v_c \cos \beta = 10 \cos 30° = 8.66 \text{ m/s}$$

From Eq. (4),

$$0 = I_0(\omega' - \omega) \qquad \omega' = \omega = 62.5 \text{ rad/s}$$

The kinetic energy after impact is

$$T_2 = \tfrac{1}{2}mv_y'^2 + \tfrac{1}{2}I_0\omega'^2 = \tfrac{1}{2}(5)(8.66^2) + \tfrac{1}{2}(0.064)(62.5^2) = 312 \text{ J}$$

It may be seen that the values in parts (a), (c) and (d), for the kinetic energy after impact, are the same, within three-significant-figure accuracy.

The very important conclusion may now be reached that *the energy loss is a function only of the impact phenomenon*, and not of the value of the friction force between the disk and the edge. The reason is that *the friction forces do not act through a distance and thus do not produce any work.*

(e)  The velocities after impact for the three cases considered are summarized in Fig. 20.57c. It may be seen that the case of slipping motion not impending produces the minimum change in translational velocity and the maximum change in rotational velocity. The frictionless case exhibits the opposite trend.

In the following two problems, the subsequent motion of the disk as it moves along a trajectory and then impacts the ground will be studied.

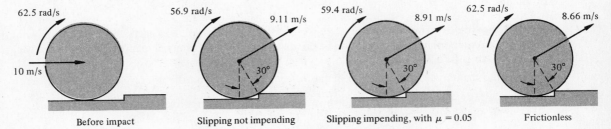

Fig. 20.57c

**20.58**  Find the complete time history of the motion of the disk of Prob. 20.57 from the time it leaves the edge of the step until it makes contact again with the ground. Assume that slipping motion is not impending during the impact with the edge.

❙ Figure 20.58a shows the configuration of the disk at the cessation of impact with the edge. A new set of $xy$ coordinates, which are attached to the ground, are shown in the figure. These coordinates will be used as reference axes for the subsequent plane projectile motion of the disk.

The initial velocity components at $t = 0$ are

$$v_{0x} = v_y' \cos \beta = 9.11 \cos 30° = 7.89 \text{ m/s} \qquad v_{0y} = v_y' \sin \beta = 9.11 \sin 30° = 4.56 \text{ m/s}$$

The total time of the flight is found from

$$y = y_0 + v_{0y}t - \tfrac{1}{2}gt^2 \qquad 0 = 0 + 4.56t - \tfrac{1}{2}(9.81)t^2 \qquad t = 0, \quad 0.0930 \text{ s}$$

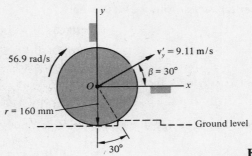

Fig. 20.58a

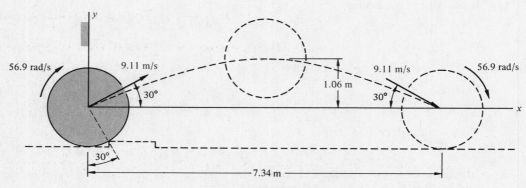

Fig. 20.58b

The first solution corresponds to the initial time, and the time of the flight is 0.930 s. The horizontal distance traveled by the disk is

$$x = v_{0x}t = 7.89(0.930) = 7.34 \text{ m}$$

The maximum height attained by the disk is given by

$$v_y^2 = v_{0y}^2 - 2g(y - y_0) \qquad 0 = 4.56^2 - 2(9.81)(y_{max} - 0) \qquad y_{max} = 1.06 \text{ m}$$

Since all air resistance effects are neglected, there is no loss in kinetic energy. Thus, the magnitudes of the initial and final velocities are the same. The details of the trajectory motion are shown in Fig. 20.58b. It also may be observed that there is no change in the angular velocity of the disk.

20.59 (a) Find the velocity components of the disk of Prob. 20.58 after it impacts the ground following its free flight. Assume that slipping motion of the disk during impact is not impending and that the coefficient of restitution is 0.75.

(b) Find the magnitude and direction of the rebound velocity, and the energy loss due to impact.

▌ (a) The impact configuration is shown in Fig. 20.59a, together with a new position of the xy coordinates. The equations of impulse momentum are

$$I_x' = m(v_x' - v_c \cos \beta) \qquad (1)$$
$$I_y' = m[v_y' - (-v_c \sin \beta)] \qquad (2)$$
$$-I_x'r = I_0(\omega' - \omega) \qquad (3)$$

The minus sign in the parentheses in Eq. (2) indicates that the y component of the initial velocity acts in the negative coordinate sense. The condition of slipping motion not impending requires that

$$v_x' = r\omega' \qquad (4)$$

The coefficient of restitution is defined by

$$e = -\frac{v_A' - v_B'}{v_A - v_B} \qquad (5)$$

The ground is assumed to be body B, and

$$v_B = 0 \qquad v_B' = 0$$

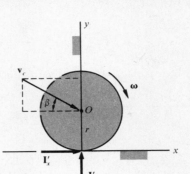

Fig. 20.59*a*

The translational velocity of the disk in the $y$ direction, before impact, is

$$v_A = -v_c \sin \beta$$

and the velocity of this element in the $y$ direction, after impact, is

$$v_A' = v_y'$$

Equation (5) then has the form

$$e = -\frac{v_y' - 0}{-v_c \sin \beta - 0} \qquad v_y' = ev_c \sin \beta \tag{6}$$

Equations (1) through (4) and (6) are five equations in the five unknowns $v_x'$, $v_y'$, $\omega'$, $I_x'$, and $I_y'$.
    The solutions for the velocities after impact are

$$v_x' = \frac{rmv_c \cos \beta + I_0}{rm + I_0/r} \qquad v_y' = ev_c \sin \beta \qquad \omega' = \frac{v_x'}{r} \tag{7}$$

The numerical values of this problem are

$$v_c = 9.11 \text{ m/s} \qquad \omega = 56.9 \text{ rad/s} \qquad \beta = 30° \qquad r = 160 \text{ mm}$$
$$m = 5 \text{ kg} \qquad I_0 = 0.064 \text{ kg} \cdot \text{m}^2 \qquad e = 0.75$$

The values of the velocity components after impact are

$$v_x' = \frac{0.16(5)(9.11) \cos 30° + 0.064(56.9)}{0.16(5) + 0.064/0.16} = 8.29 \text{ m/s}$$

$$v_y' = 0.75(9.11) \sin 30° = 3.42 \text{ m/s} \qquad \omega' = \frac{v_x'}{r} = \frac{8.29}{0.16} = 51.8 \text{ rad/s}$$

These velocities are shown in Fig. 20.59*b*.

**(b)**  The magnitude $v_c'$ and direction $\theta$ of the resultant velocity, shown in Fig. 20.59*b*, are given by

$$v_c' = \sqrt{v_x'^2 + v_y'^2} = \sqrt{8.29^2 + 3.42^2} = 8.97 \text{ m/s} \qquad \theta = \tan^{-1} \frac{3.42}{8.29} = 22.4°$$

The kinetic energy before impact is

$$T_1 = \tfrac{1}{2}mv_c^2 + \tfrac{1}{2}I_0\omega^2 = \tfrac{1}{2}(5)(9.11^2) + \tfrac{1}{2}(0.064)(56.9^2) = 311 \text{ J}$$

The kinetic energy after impact is

$$T_2 = \tfrac{1}{2}m(v_c')^2 + \tfrac{1}{2}I_0(\omega')^2 = \tfrac{1}{2}(5)(8.97^2) + \tfrac{1}{2}(0.064)(51.8^2) = 287 \text{ J}$$

The percent loss in energy is

$$\%D = \frac{T_1 - T_2}{T_1} (100) = \frac{311 - 287}{311} (100) = 7.7\%$$

It may be seen that after impact both the translational and rotational velocities decrease. It is left as an exercise for the reader to verify that if $e = 1$, the energy loss would have been zero.
    In addition to being the particular impact conditions in the present problem, the configuration shown in Fig. 20.59*a* may be viewed as a general case of a spinning disk impacting a plane. An interesting result is obtained if $\omega$ is assumed to have a sense which is opposite that shown in the figure. If the magnitude of $\omega$ is sufficiently large, Eq. (7) shows that (recalling that $\omega$ is now negative) $v_x'$ may have a

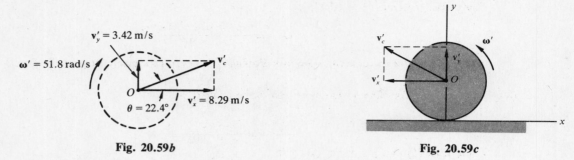

**Fig. 20.59b**                    **Fig. 20.59c**

negative value.   From Eq. (6), $v'_y$ is always positive, and the direction of the rebound velocity $v'_c$ for this case would be as shown in Fig. 20.59c.   The disk, after impact, would rebound in the general direction from which it had come.   In addition, the angular velocity $\omega'$, after impact, would have the sense shown in the figure.

This effect may be demonstrated readily with a rubber ball. The ball is held between a person's thumb and middle finger and given an initial spin as it is tossed to the ground.   The sense of the spin is such that the top surface of the ball moves toward the person.   With a minimum amount of practice, it should be possible to catch the ball on the rebound!

**20.60**  (*a*)  Find the velocities, after impact, of the disk in Prob. 20.59, if slipping motion during impact is impending.

(*b*)  Find the limiting value of the coefficient of friction that determines whether slipping motion is impending.

(*c*)  Find the numerical results for part (*b*), using the data of Prob. 20.59, for $e = 0.9$ and for $e = 0.6$.

(*d*)  Find the numerical values of the velocities of the disk after impact, for $e = 0.9$ and $e = 0.6$, and with $\mu = 0.6$.

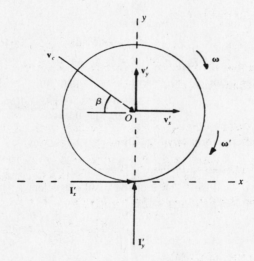

**Fig. 20.60**

▌  (*a*)  Figure 20.60 shows the velocity and impulse components. Equations (1) through (3), and (6), in Prob. 20.59 are still valid, and those results are repeated below.

$$I'_x = m(v'_x - v_c \cos \beta) \tag{1}$$

$$I'_y = m(v'_y + v_c \sin \beta) \tag{2}$$

$$-I'_x r = I_0(\omega' - \omega) \tag{3}$$

$$v'_y = ev_c \sin \beta \tag{4}$$

When slipping motion is impending,

$$I'_x = \mu I'_y \tag{5}$$

Using Eqs.(1), (2), and (4) in Eq. (5),

$$v'_x = v_c[\cos \beta + \mu(1 + e) \sin \beta] \tag{6}$$

Using Eqs. (1) and (6) in Eq. (3),

$$\omega' = \omega - \frac{rmv_c\mu(1+e)\sin\beta}{I_0} \tag{7}$$

$v_x'$, $v_y'$, and $\omega'$, for the case of impending slipping motion during impact, are given by Eqs. (4), (6), and (7).

(b)  The condition for slipping motion to not be impending is

$$\mu > \frac{I_x'}{I_y'} \tag{8}$$

Using Eqs. (1) and (2) in Eq. (8) gives

$$\mu_{\min} = \frac{m(v_x' - v_c\cos\beta)}{m(v_y' + v_c\sin\beta)} \tag{9}$$

Using Eqs. (7) in Prob. 20.59 in Eq. (9) gives

$$\mu_{\min} = \frac{I_0\omega - (I_0/r)v_c\cos\beta}{(rm + I_0/r)v_c(1+e)\sin\beta} \tag{10}$$

(c)  The numerical values of this problem are

$$v_c = 9.11\,\text{m/s} \qquad \omega = 56.9\,\text{rad/s} \qquad \beta = 30° \qquad r = 160\,\text{mm}$$
$$m = 5\,\text{kg} \qquad I_0 = 0.064\,\text{kg}\cdot\text{m}^2$$

The result for $\mu_{\min}$, using $e = 0.9$, is found as

$$\mu_{\min} = \frac{0.064(56.9) - (0.064/0.160)(9.11)\cos 30°}{[0.160(5) + 0.064/0.160]9.11(1 + 0.9)\sin 30°} \tag{11}$$

$$\mu_{\min} = 0.0468 \tag{12}$$

For $e = 0.6$, from Eq. (12),

$$\mu_{\min} = 0.0468\left(\frac{1 + 0.9}{1 + 0.6}\right) = 0.0556 \tag{13}$$

It may be seen that the value of $\mu_{\min}$ increases with decreasing values of $e$.

(d)  For $e = 0.9$, from Eq. (12),

$$\mu_{\min} = 0.0468$$

Since $0.6 > 0.0468$, slipping motion of the disk is not impending. Using Eqs. (7) in Prob. 20.59, the velocities after impact have the forms

$$v_x' = \frac{rmv_c\cos\beta + I_0\omega}{rm + I_0/r} = \frac{0.160(5)9.11\cos 30° + 0.064(56.9)}{0.160(5) + 0.064/0.160} = 8.29\,\text{m/s} \tag{14}$$

$$v_y' = ev_c\sin\beta = 0.9(9.11)\sin 30° = 4.10\,\text{m/s} \qquad v' = \sqrt{v_x'^2 + v_y'^2} = \sqrt{8.29^2 + 4.10^2} = 9.25\,\text{m/s}$$

$$\omega' = \frac{v_x'}{r} = \frac{8.29}{0.160} = 51.8\,\text{rad/s} \tag{15}$$

For $e = 0.6$, from Eq. (13),

$$\mu_{\min} = 0.0556$$

Since $0.6 > 0.0556$, slipping motion of the disk is not impending. $v_x'$ and $\omega'$ are given by Eqs. (14) and (15) as

$$v_x' = 8.29\,\text{m/s} \qquad \omega' = 51.8\,\text{rad/s}$$

$v_y'$ is found from

$$v_y' = ev_c\sin\beta = 0.6(9.11)\sin 30° = 2.73\,\text{m/s} \qquad v' = \sqrt{v_x'^2 + v_y'^2} = \sqrt{8.29^2 + 2.73^2} = 8.73\,\text{m/s}$$

**20.61**  (a)  Do the same as in Prob. 20.59, for the case of a homogeneous sphere of mass $m$ impacting the ground. Use the numerical constants of that problem.

(b)  Do the same as in part (a), if the sphere of mass $m$ is hollow, with an inside diameter 0.82 times the magnitude of the outside diameter.

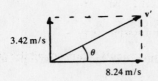

Fig. 20.61a

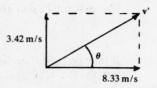

Fig. 20.61b

(a) From Case 1 in Table 17.13,

$$I_0 = \tfrac{1}{10}md^2 = \tfrac{1}{10}(5)\left(\frac{320}{1,000}\right)^2 = 0.0512\ \text{kg}\cdot\text{m}^2$$

Equations (7) in Prob. 20.59 have the forms

$$v_x' = \frac{rmv_c\cos\beta + I_0\omega}{rm + I_0/r} = \frac{0.160(5)9.11\cos 30° + 0.0512(56.9)}{0.160(5) + 0.0512/0.160} = 8.24\ \text{m/s}$$

$$v_y' = ev_c\sin\beta = 0.75(9.11)\sin 30° = 3.42\ \text{m/s}$$

$$\omega' = \frac{v_x'}{r} = \frac{8.24}{0.160} = 51.5\ \text{rad/s} \qquad v' = \sqrt{v_x'^2 + v_y'^2} = \sqrt{8.24^2 + 3.42^2} = 8.92\ \text{m/s}$$

The direction $\theta$ of the velocity after impact, shown in Fig. 20.61a, is found from

$$\tan\theta = \frac{3.42}{8.24} \qquad \theta = 22.5°$$

The kinetic energy before impact is given by

$$T_1 = \tfrac{1}{2}mv_c^2 + \tfrac{1}{2}I_0\omega^2 = \tfrac{1}{2}(5)9.11^2 + \tfrac{1}{2}(0.0512)56.9^2 = 290\ \text{J}$$

The kinetic energy after impact is

$$T_2 = \tfrac{1}{2}mv'^2 + \tfrac{1}{2}I_0\omega'^2 = \tfrac{1}{2}(5)8.92^2 + \tfrac{1}{2}(0.0512)51.5^2 = 267\ \text{J}$$

The percent loss in energy due to the impact is given by

$$\%\mathrm{D} = \frac{T_1 - T_2}{T_1}(100) = \frac{290 - 267}{290}(100) = 7.9\%$$

(b) The moment of inertia of the hollow sphere, using Case 2 in Table 17.13, is

$$I_0 = \tfrac{1}{10}m\left(\frac{d_0^5 - d_i^5}{d_0^3 - d_i^3}\right)$$

Using $d_i = 0.82d_0$ in the above equation results in

$$I_0 = \tfrac{1}{10}m\left(\frac{d_0^5 - (0.82d_0)^5}{d_0^3 - (0.82d_0)^3}\right) = 0.140md_0^2 \qquad (1)$$

Using $d_0 = 2(0.160) = 0.320\ \text{m}$ in Eq. (1) gives

$$I_0 = 0.140(5)(0.320)^2 = 0.0717\ \text{kg}\cdot\text{m}^2$$

Equations (7) in Prob. 20.59 are used, and the results are

$$v_x' = \frac{rmv_c\cos\beta + I_0\omega}{rm + I_0/r} = \frac{0.160(5)9.11\cos 30° + 0.0717(56.9)}{0.160(5) + 0.0717/0.160} = 8.33\ \text{m/s}$$

$$v_y' = ev_c\sin\beta = 0.75(9.11)\sin 30° = 3.42\ \text{m/s}$$

$$\omega' = \frac{v_x'}{r} = \frac{8.33}{0.160} = 52.1\ \text{rad/s} \qquad v' = \sqrt{v_x'^2 + v'_y^2} = \sqrt{8.33^2 + 3.42^2} = 9.00\ \text{m/s}$$

Figure 20.61b shows the direction $\theta$ of $v'$, and the value of this quantity is found from

$$\tan\theta = \frac{3.42}{8.33} \qquad \theta = 22.3°$$

The kinetic energies before and after impact are

$$T_1 = \tfrac{1}{2}mv_c^2 + \tfrac{1}{2}I_0\omega^2 = \tfrac{1}{2}(5)9.11^2 + \tfrac{1}{2}(0.0717)56.9^2 = 324\ \text{J} \qquad T_2 = \tfrac{1}{2}(5)9.00^2 + \tfrac{1}{2}(0.0717)52.1^2 = 300\ \text{J}$$

The percent loss in energy during the impact is found to be

$$\%D = \frac{324 - 300}{324}\,(100) = 7.4\%$$

It may be seen that the percent energy loss for the hollow sphere impacting the ground is less than the loss for the solid sphere impacting the ground.

**20.62** Find the relationship between $v_c$ and $\omega$, of the disk in Prob. 20.59, if the direction of the rebound velocity is to be collinear with the direction of $v_c$.

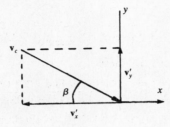

**Fig. 20.62**

▮ The components of the rebound velocity must be equal and opposite to the components of $v_c$, as shown in Fig. 20.62, so that

$$v'_x = -v_c \cos \beta \qquad v'_y = -v_c \sin \beta \qquad \frac{v'_y}{v'_x} = -\tan \beta$$

Using Eqs. (7) in Prob. 20.59 in the above equation with $\tan \beta$ results in

$$\frac{ev_c \sin \beta}{(rmv_c \cos \beta + I_0\omega)/(rm + I_0/r)} = -\tan \beta \qquad (1)$$

For a solid disk, using Case 4 in Table 17.13,

$$I_0 = \tfrac{1}{8}md^2 = \tfrac{1}{2}mr^2 \qquad (2)$$

Using Eq. (2) in Eq. (1) gives

$$\omega = -3(e + \tfrac{2}{3})(\cos \beta)\,\frac{v_c}{r}$$

It may be seen from the above equation that $\omega$ is negative. Thus, the sense of $\omega$ in Fig. 20.59a must be counterclockwise.

Using the values from Prob. 20.59 of $e = 0.75$, $\beta = 30°$, $v_c = 9.11$ m/s, and $r = 0.160$ m,

$$\omega = -3(0.75 + \tfrac{2}{3})(\cos 30°)\,\frac{9.11}{0.160} = -210\,\text{rad/s}$$

**20.63** A disk rolls along a plane surface, as shown in Fig. 20.63a. Plastic impact is assumed, and the coefficient of friction between the disk and the edge of the step is 0.10. $W = 16$ lb, $r = 3$ in, and $\omega = 75$ rad/s.

(a) Find the translational and angular velocities of the disk after impact, the energy loss and the percent loss of energy that occurs during impact.

(b) Find the horizontal distance through which the center of the disk moves, after impact, before this element contacts the upper plane surface.

(c) Do the same as in parts (a) and (b), if the coefficient of friction is 0.20.

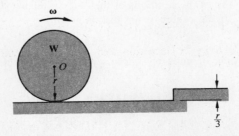

**Fig. 20.63a**

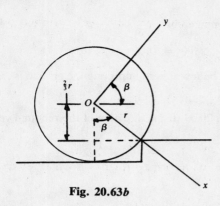

**Fig. 20.63b**

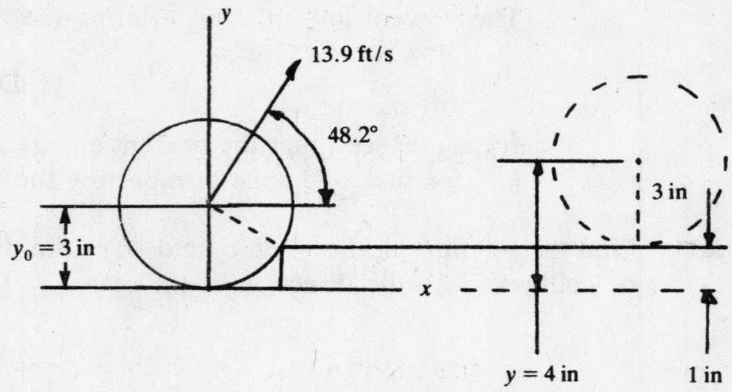

**Fig. 20.63c**

(a) Figure 20.63b shows the impact configuration. Angle $\beta$ is found from

$$\cos \beta = \frac{\frac{2}{3}r}{r} = \frac{2}{3} \qquad \beta = 48.2° \qquad \sin \beta = 0.745$$

Using Eq. (9) in part (b) of Prob. 20.57,

$$\mu_{min} = \frac{1 - \cos \beta}{3 \sin \beta} = \frac{1 - \frac{2}{3}}{3(0.745)} = 0.149$$

The necessary condition is that $\mu > 0.149$ if slipping motion is not impending during the impact. In the present problem, $\mu = 0.10 < 0.149$, so that slipping motion during impact is impending.

Using Eqs. (10) in part (c) of Prob. 20.57,

$$v'_x = 0 \qquad v'_y = v_c(\cos \beta + \mu \sin \beta) \qquad \omega' = \omega - \frac{rmv_c\mu \sin \beta}{I_0}$$

The translational velocity of the disk is given by

$$v_c = r\omega = \tfrac{3}{12}(75) = 18.8 \text{ ft/s}$$

Using the above result, and $\beta = 48.2°$, $\mu = 0.10$, and $d = 6$ in, the velocities after impact are found to be

$$v'_x = 0 \qquad v'_y = 18.8(\cos 48.2° + 0.10 \sin 48.2°) = 13.9 \text{ ft/s}$$

$$\omega' = 75 - \frac{\frac{3}{12}(16/32.2)18.8(0.10) \sin 48.2°}{\frac{1}{8}(16/32.2)(0.5)^2} = 63.8 \text{ rad/s}$$

The kinetic energy before and after impact is

$$T_1 = \tfrac{1}{2}mv_c^2 + \tfrac{1}{2}I_0\omega^2 = \frac{1}{2}\left(\frac{16}{32.2}\right)18.8^2 + \frac{1}{2}\left[\frac{1}{8}\left(\frac{16}{32.2}\right)(0.5)^2\right]75^2 = 131 \text{ ft} \cdot \text{lb}$$

$$T_2 = \tfrac{1}{2}mv'^2_y + \tfrac{1}{2}I_0\omega'^2 = \frac{1}{2}\left(\frac{16}{32.2}\right)13.9^2 + \frac{1}{2}\left[\frac{1}{8}\left(\frac{16}{32.2}\right)(0.5)^2\right]63.8^2 = 79.6 \text{ ft} \cdot \text{lb}$$

The loss in energy during impact is

$$\Delta T = T_1 - T_2 = 51.4 \text{ ft} \cdot \text{lb}$$

The percent loss in energy is given by

$$\%D = \frac{T_1 - T_2}{T_1}(100) = \frac{51.4}{131} = 39.2\%$$

(b) Figure 20.63c shows the positions of the disk at impact, and when it contacts the upper plane surface. The $y$ motion of the center of the disk is given by

$$y = y_0 + v_{0y}t - \tfrac{1}{2}gt^2$$

$t = 0$ is the time of impact. Using $y = 4$ in, $y_0 = 3$ in, and $v_0 = 13.9$ ft/s in the above equation results in

$$\tfrac{4}{12} = \tfrac{3}{12} + (13.9 \sin 48.2°)t - \tfrac{1}{2}(32.2)t^2 \qquad 16.1t^2 - 10.4t + 0.0833 = 0$$

$$t = \frac{10.4 \pm \sqrt{10.4^2 - 4(16.1)0.0833}}{2(16.1)} = \frac{10.40 \pm 10.14}{2(16.1)} = 0.00807, \ 0.638 \text{ s}$$

$t = 0.00807$ s   corresponds to upward motion, at the position   $y = 4$ in,   and, when the disk contacts the upper plane surface,

$$t = 0.638 \text{ s}$$

The $x$ motion is given by

$$x = v_{0x}t = (13.9 \cos 48.2°)0.638 = 5.91 \text{ ft}$$

(c)  From part (a),

$$\mu_{\min} = 0.149$$

For this problem   $\mu = 0.20$,   so that slipping motion during impact is not impending.   Using Eqs. (7) and (8) in Prob. 20.57,

$$v'_x = 0 \qquad v'_y = \frac{v_c}{3}(1 + 2\cos\beta) \qquad \omega' = \frac{v_c}{3r}(1 + 2\cos\beta)$$

Using   $v_c = 18.8$ ft/s   and   $\beta = 48.2°$   in the above equations results in

$$v'_x = 0 \qquad v'_y = \frac{18.8}{3}(1 + 2\cos 48.2°) = 14.6 \text{ ft/s}$$

$$\omega' = \frac{18.8}{3(\frac{3}{12})}(1 + 2\cos 48.2°) = 58.5 \text{ rad/s}$$

From part (a),

$$T_1 = 131 \text{ ft} \cdot \text{lb}$$

The kinetic energy after the impact is

$$T_2 = \tfrac{1}{2}mv'^2_y + \tfrac{1}{2}I_0\omega'^2 = \frac{1}{2}\left(\frac{16}{32.2}\right)14.6^2 + \frac{1}{2}\left[\frac{1}{8}\left(\frac{16}{32.2}\right)(0.5)^2\right]58.5^2 = 79.5 \text{ ft} \cdot \text{lb}$$

The loss, and percent loss, in energy due to the impact are found to be

$$\Delta T = T_1 - T_2 = 51.5 \text{ ft} \cdot \text{lb} \qquad \%\text{D} = \frac{T_1 - T_2}{T_1}(100) = \frac{51.5}{131}(100) = 39.3\%$$

It may be observed that the result above is the same as that found in part (a), within three-significant-figure accuracy.  This is an expected result, since the energy loss is not caused by the friction force but is due only to the plastic impact.

The solution for the motion after impact follows that in part (b).  Using   $y = 4$ in,   $y_0 = 3$ in,   and   $v_0 = 14.6$ ft/s,

$$y = y_0 + v_{0y}t - \tfrac{1}{2}gt^2 \qquad \tfrac{4}{12} = \tfrac{3}{12} + (14.6 \sin 48.2°)t - \tfrac{1}{2}(32.2)t^2 \qquad 16.1t^2 - 10.9t + 0.0833 = 0$$

$$t = \frac{10.9 \pm \sqrt{10.9^2 - 4(16.1)0.0833}}{2(16.1)} = \frac{10.90 \pm 10.65}{2(16.1)} = 0.00776, \ 0.669 \text{ s}$$

The $x$ motion of the disk is given by

$$x = v_{0x}t = (14.6 \cos 48.2°)0.669 = 6.51 \text{ ft}$$

It may be observed that the disk travels further in the horizontal direction if slipping motion during impact is not impending.

**20.64**   The low plane surface in Fig. 20.64a is lubricated, so that the coefficient of friction between it and the disk is effectively zero.  The disk impacts the edge of the step with translational velocity $v_c$ and zero angular velocity.  Plastic impact is assumed, and the coefficient of friction between the disk and the edge of the step is 0.28.   $r = 200$ mm,   $m = 3$ kg,   and   $v_c = 6$ m/s.

(a)  Find the translational and angular velocities of the disk after impact, if slipping motion is assumed to not be impending.

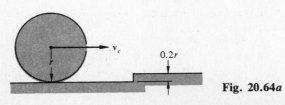

Fig. 20.64a

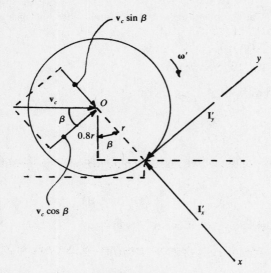

**Fig. 20.64b**

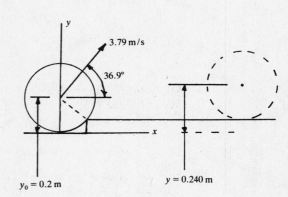

**Fig. 20.64c**

(*b*)  Do the same as in part (*a*), if slipping motion is assumed to be impending.

(*c*)  Find the energy loss that occurs during impact.

(*d*)  Find the horizontal distances through which the center of the disk moves, after impact, before this element contacts the upper plane surface.

▌  (*a*)  Figure 20.64*b* shows the impulse components $I'_x$ and $I'_y$ in their assumed actual senses. It is assumed that slipping motion is not impending during the impact. The impulse-momentum equations have the forms

$$-I'_x = m(v'_x - v_c \sin \sin \beta) \tag{1}$$
$$-I'_y = m(v'_y - v_c \cos \beta) \tag{2}$$
$$I'_y r = I_0(\omega' - 0) \tag{3}$$

Since the impact is assumed to be plastic,

$$v'_x = 0 \tag{4}$$

For the assumption that slipping motion is not impending,

$$v'_y = r\omega' \tag{5}$$

$I'_y$ is eliminated between Eqs. (2) and (3), and $\omega'$ is eliminated from the resulting equation by using Eq. (5) and the result $I_0 = \frac{1}{2}mr^2$, to obtain

$$v'_y = \tfrac{2}{3}v_c \cos \beta \tag{6}$$

Using Eq. (5),

$$\omega' = \frac{2v_c \cos \beta}{3r} \tag{7}$$

The velocity components after impact, for the case of slipping motion not impending, are given by Eqs. (4), (6), and (7).

(*b*)  The condition for slipping motion to not be impending is

$$\mu \geq \frac{I'_y}{I'_x} = \frac{m(v_c \cos \beta - \tfrac{2}{3}v_c \cos \beta)}{mv_c \sin \beta}$$

$$\mu \geq \frac{\tfrac{1}{3}\cos \beta}{\sin \beta} = \frac{1}{3}\left(\frac{\cos 36.9°}{\sin 36.90°}\right) = 0.444 \tag{8}$$

In this problem  $\mu = 0.28$,  so that slipping motion of the disk is not impending.

When slipping motion is impending, Eqs. (1) through (3) are still valid and, for plastic impact,

$$v'_x = 0$$

The condition of impending slipping motion requires that

$$I'_y = \mu I'_x \tag{9}$$

$I'_x$ and $I'_y$ from Eqs. (1) and (2) are used in Eq. (9), to obtain

$$m(v_c \cos \beta - v'_y) = \mu m(0 + v_c \sin \beta) \qquad v'_y = v_c(\cos \beta - \mu \sin \beta) \tag{10}$$

From Eq. (3), using Eqs. (2) and (10), and the result $I_0 = \tfrac{1}{2}mr^2$,

$$\omega' = \frac{2\mu v_c \sin \beta}{r} \tag{11}$$

The velocity components after impact, for the case of impending slipping motion, are given by Eqs. (4), (10), and (11).

A check may now be made on the above results.   If

$$\mu = \frac{\tfrac{1}{3}\cos \beta}{\sin \beta} \tag{12}$$

then $\omega'$ given by Eq. (11) should be equal to $v'_y/r$, as the condition of slipping motion not impending.   Using Eq. (12) in Eq. (10) gives

$$v'_y = v_c\left[\cos \beta - \left(\frac{\tfrac{1}{3}\cos \beta}{\sin \beta}\right)\sin \beta\right] = \tfrac{2}{3}v_c \cos \beta$$

Using Eq. (12) in Eq. (11) gives

$$\omega' = \frac{2v_c \sin \beta}{r}\left(\frac{\tfrac{1}{3}\cos \beta}{\sin \beta}\right) = \frac{2}{3}\frac{v_c}{r}\cos \beta$$

Using the condition of no slipping,

$$\omega' \overset{?}{=} \frac{v'_y}{r}$$

$$\frac{2}{3}\frac{v_c}{r}\cos \beta \overset{?}{=} \frac{\tfrac{2}{3}v_c \cos \beta}{r} \qquad 1 \equiv 1$$

From Fig. 20.64$b$,

$$\cos \beta = \frac{0.8r}{r} = 0.8 \qquad \beta = 36.9°$$

Equation (8) is written as

$$\mu_{\min} = \tfrac{1}{3}\cot \beta = \tfrac{1}{3}\cot 36.9° = 0.444$$

For the present problem,

$$\mu = 0.28 < 0.444$$

so that slipping motion is not impending.   Using $\mu = 0.28$, $r = 200$ mm, $\beta = 36.9°$, $m = 3$ kg, and $v_c = 6$ m/s, the numerical values of the velocities after impact may be found.   Using Eq. (4),

$$v'_x = 0$$

Using Eq. (10),

$$v'_y = v_c(\cos \beta - \mu \sin \beta) = 6(\cos 36.9° - 0.28 \sin 36.9°) = 3.79 \text{ m/s}$$

Using Eq. (11),

$$\omega' = \frac{2\mu v_c \sin \beta}{r} = \frac{2(0.28)6 \sin 36.9°}{0.2} = 10.1 \text{ rad/s}$$

(c)   The initial and final values of the kinetic energy are

$$T_1 = \tfrac{1}{2}mv_c^2 = \tfrac{1}{2}(3)6^2 = 54.0 \text{ J}$$
$$T_2 = \tfrac{1}{2}mv'^2_y + \tfrac{1}{2}I_0\omega'^2 = \tfrac{1}{2}(3)3.79^2 + \tfrac{1}{2}[\tfrac{1}{2}(3)0.2^2]10.1^2 = 24.6 \text{ J}$$

The loss in energy during the impact is

$$\Delta T = T_1 - T_2 = 54.0 - 24.6 = 29.4 \text{ J}$$

The percent loss in energy is given by

$$\%D = \frac{\Delta T}{T_1}(100) = \frac{29.4}{54.0}(100) = 54.4\%$$

(d) Figure 20.64c shows the positions of the disk at impact and when it contacts the upper plane surface. The y motion of the center of the disk is given by

$$y = y_0 + v_{0y}t - \tfrac{1}{2}gt^2$$

$t = 0$ is the time of impact. Using $y = 0.240\,\text{m}$, $y_0 = 0.2\,\text{m}$, and $v_0 = 3.79\,\text{m/s}$ in the above equation results in

$$0.240 = 0.2 + (3.79\sin 36.9°)t - \tfrac{1}{2}(9.81)t^2 \qquad 4.91t^2 - 2.28t + 0.04 = 0$$

$$t = \frac{2.28 \pm \sqrt{2.28^2 - 4(4.91)0.04}}{2(4.91)} = \frac{2.28 \pm 2.10}{2(4.91)} = 0.0183,\ 0.446\,\text{s}$$

$t = 0.446\,\text{s}$ is used, and the x motion is given by

$$x = v_{0x}t = (3.79\cos 36.9°)0.446 = 1.35\,\text{m}$$

20.65   (a) Find the general forms for the velocities of the disk in Prob. 20.57 after impact. The impact is assumed to be inelastic, with a coefficient of restitution $e$, and impending motion of the disk is assumed not to occur during the impact.

  (b) Find the necessary condition for impending motion not to occur during the impact.

  (c) Do the same as in part (a), if motion is impending during impact.

  (d) Find the numerical values of the velocities of the disk, and the energy loss, after impact. Use the numerical data of Prob. 20.57, together with $e = 0.6$ and $\mu = 0.04$.

  (e) Do the same as in part (d), if $e = 0.6$ and $\mu = 0.06$.

  (f) Compare the results found in parts (d) and (e).

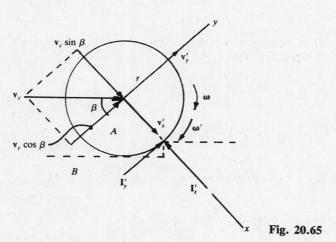

Fig. 20.65

▌ (a) Figure 20.65 shows the impact configuration. The impact motion is defined by Eqs. (2) through (4) in Prob. 20.57, with the forms

$$-I'_x = m(v'_x - v_c \sin \beta) \tag{1}$$

$$I'_y = m(v'_y - v_c \cos \beta) \tag{2}$$

$$-I'_y r = I_0(\omega' - \omega) \tag{3}$$

For the case of inelastic impact,

$$e = -\frac{v'_A - v'_B}{v_A - v_B} \tag{4}$$

The disk is body $A$ and the ground is body $B$. The velocities before and after impact have the forms

$$v_B = v'_B = 0 \qquad v'_A = v'_x \qquad v_A = v_c \sin \beta$$

The above terms are used in Eq. (4), with the result

$$e = -\frac{(v_x' - 0)}{v_c \sin \beta - 0} \qquad v_x' = -ev_c \sin \beta \tag{5}$$

For the case where slipping motion is not impending,

$$v_y' = r\omega' \tag{6}$$

$I_y'$ is eliminated between Eqs. (2) and (3), to obtain

$$m(v_y' - v_c \cos \beta) = -\frac{I_0}{r}(\omega' - \omega) \tag{7}$$

Using Eq. (6), and $v_c = r\omega$, in Eq. (7) gives

$$m(v_y' - v_c \cos \beta) = -\frac{I_0}{r}\left(\frac{v_y'}{r} - \frac{v_c}{r}\right) \tag{8}$$

Using $I_0 = \frac{1}{2}mr^2$ in Eq. (8) gives

$$v_y' = \frac{v_c}{3}(1 + 2\cos \beta) \tag{9}$$

The above equation is the same as Eq. (7) in Prob. 20.57, and this is an expected result. $v_x'$ is given by Eq. (5). $\omega'$ is found by using Eq. (9) in Eq. (6). The result is

$$\omega' = \frac{v_y'}{r} = \frac{v_c}{3r}(1 + 2\cos \beta) \tag{10}$$

(b) The condition for slipping motion to not be impending is

$$\mu > \frac{I_y'}{I_x'}$$

Using Eqs. (1) and (2) in the above equation results in

$$\mu > \frac{I_y'}{I_x'} = \frac{m(v_y' - v_c \cos \beta)}{-m(v_x' - v_c \sin \beta)} = \frac{(v_c/3)(1 + 2\cos \beta) - v_c \cos \beta}{-(-ev_c \sin \beta - v_c \sin \beta)} = \frac{1 - \cos \beta}{3(1 + e)\sin \beta} \tag{11}$$

(c) When slipping motion is impending, Eqs. (1) through (3), and Eq. (5), are still valid. The condition of impending slipping motion is

$$I_y' = \mu I_x' \tag{12}$$

Using Eqs. (1) and (2) in Eq. (12),

$$m(v_y' - v_c \cos \beta) = \mu m(v_c \sin \beta - v_x') \tag{13}$$

Using Eq. (5) in Eq. (13) gives

$$v_y' - v_c \cos \beta = \mu(v_c \sin \beta + ev_c \sin \beta) \qquad v_y' = v_c[\cos \beta + \mu(1 + e)\sin \beta] \tag{14}$$

The $x$ component of velocity is given by Eq. (5) as

$$v_x' = -ev_c \sin \beta$$

Using $I_y'$ from Eq. (2) in Eq. (3) gives

$$-[m(v_y' - v_c \cos \beta)]r = I_0(\omega' - \omega) \tag{15}$$

Using Eq. (9) in Eq. (15) results in

$$-[\mu(1 + e)v_c \sin \beta + v_c \cos \beta - v_c \cos \beta] = \frac{I_0}{mr}(\omega' - \omega) \qquad \omega' = \omega - \frac{rm(1 + e)\mu v_c \sin \beta}{I_0} \tag{16}$$

(d) The numerical constants of the problem are

$$e = 0.6 \qquad \mu = 0.04 \qquad v_c = 10 \text{ m/s} \qquad m = 5 \text{ kg} \qquad \beta = 30° \qquad r = 0.16 \text{ m}$$
$$\omega = 62.5 \text{ rad/s} \qquad I_0 = 0.064 \text{ kg} \cdot \text{m}^2$$

The limiting condition for the coefficient of friction is given by Eq. (11) as

$$\mu_{min} = \frac{1 - \cos \beta}{3(1 + e)\sin \beta} = \frac{1 - \cos 30°}{3(1 + 0.6)\sin 30°} = 0.0558$$

For the present problem, $\mu = 0.04 < 0.0558$, so that slipping motion is impending. Using Eqs. (5), (14), and (16),

$$v_x' = -ev_c \sin \beta = -0.6(10) \sin 30° = -3 \text{ m/s}$$

$$v_y' = v_c[\cos \beta + \mu(1 + e) \sin \beta] = 10[\cos 30° + 0.04(1 + 0.6) \sin 30°] = 8.98 \text{ m/s}$$

$$\omega' = \omega - \frac{rm(1 + e)\mu v_c \sin \beta}{I_0} = 62.5 - \frac{0.16(5)(1 + 0.6)0.04(10) \sin 30°}{0.064} = 58.5 \text{ rad/s}$$

The kinetic energy before impact, from part (a) of Prob. 21.57, is

$$T_1 = 375 \text{ J}$$

The kinetic energy after impact is

$$T_2 = \tfrac{1}{2}mv'^2 + \tfrac{1}{2}I_0\omega'^2 = \tfrac{1}{2}(5)[(-3)^2 + 8.98^2] + \tfrac{1}{2}(0.064)58.5^2 = 334 \text{ J}$$

The energy loss due to the impact is given by

$$\Delta T = T_1 - T_2 = 41 \text{ J}$$

The percent loss in energy is found to be

$$\%D = \frac{\Delta T}{T_1} (100) = \frac{41}{375} (100) = 10.9\%$$

(e) The numerical values of part (d), together with $\mu = 0.06$, are used. Since $\mu = 0.06 > 0.0558$, slipping motion is not impending. Equations (5), (9), and (10) have the forms

$$v_x' = -ev_c \sin \beta = -0.6(10) \sin 30° = -3 \text{ m/s}$$

$$v_y' = \frac{v_c}{3} (1 + 2 \cos \beta) = \tfrac{10}{3}(1 + 2 \cos 30°) = 9.11 \text{ m/s} \qquad \omega' = \frac{v_y'}{r} = \frac{9.11}{0.16} = 56.9 \text{ rad/s}$$

The kinetic energy terms before and after impact are

$$T_1 = 375 \text{ J} \qquad T_2 = \tfrac{1}{2}mv'^2 + \tfrac{1}{2}I_0\omega'^2 = \tfrac{1}{2}(5)[(-3)^2 + 9.11^2] + \tfrac{1}{2}(0.064)56.9^2 = 334 \text{ J}$$

The loss, and percent loss, of energy are found to be

$$\Delta T = T_1 - T_2 = 41 \text{ J} \qquad \%D = \frac{\Delta T}{T_1} (100) = \frac{41}{375} (100) = 10.9\%$$

The above results are the same as in part (d). This is an expected result, since the energy loss is a function only of the inelastic impact.

(f) Table 20.6 shows the velocity components for the cases where slipping motion is, and is not, impending. It may be seen that the effect of the larger value of $\mu$ in part (e), so that slipping motion is not impending, is to produce, after impact, a larger value of $v_y'$ and a smaller value of $\omega'$.

### TABLE 20.6

| velocity component | part (a), slipping motion impending | part (e), slipping motion not impending |
|---|---|---|
| $v_x'$, m/s | −3 | −3 |
| $v_y'$, m/s | 8.98 | 9.11 |
| $\omega'$, rad/s | 58.5 | 56.9 |

20.66 (a) A solid, homogeneous sphere of mass $m$ rolls along a plane surface and impacts a step, as shown in Fig. 20.66a. Find the limiting value of the coefficient of friction, between the edge of the step and the sphere, that determines whether slipping motion of the sphere during impact is impending. The impact is assumed to be plastic.

(b) Do the same as in part (a), if the sphere of mass $m$ is hollow, with the inside diameter 0.82 times the magnitude of the outside diameter.

▮ (a) Equations (2) through (6) of Prob. 20.57, for the case of slipping motion not impending, are used. $I_y'$ is eliminated between Eqs. (3) and (4) of that problem, to obtain

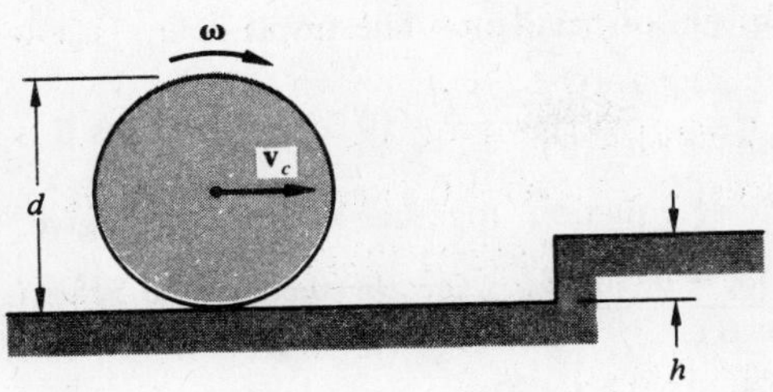

Fig. 20.66a

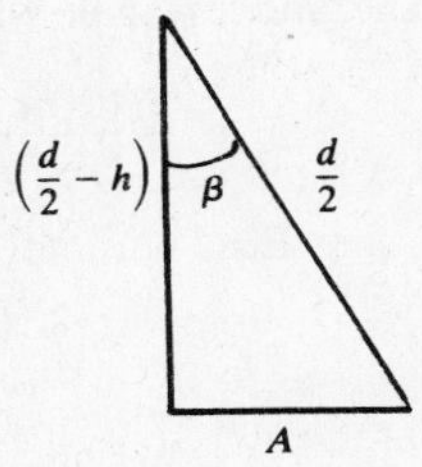

Fig. 20.66b

$$m(v_y' - v_c \cos \beta) = -\frac{I_0}{r}(\omega' - \omega)$$

Using Eqs. (1) and (6) in Prob. 20.57 in the above equation gives

$$m(v_y' - v_c \cos \beta) = -\frac{I_0}{r}\left(\frac{v_y'}{r} - \frac{v_c}{r}\right) \qquad v_y' = \frac{v_c(I_0 + mr^2 \cos \beta)}{I_0 + mr^2} \qquad (1)$$

For a solid sphere, from Case 1 in Table 17.13,

$$I_0 = \tfrac{1}{10}md^2 = \tfrac{2}{5}mr^2$$

The form of $v_y'$ is then

$$v_y' = \frac{v_c(\tfrac{2}{5}mr^2 + mr^2 \cos \beta)}{\tfrac{2}{5}mr^2 + mr^2} = \tfrac{1}{7}v_c(2 + 5 \cos \beta) \qquad (2)$$

The condition for slipping motion not to be impending is

$$\mu > \frac{I_y'}{I_x'}$$

Using Eqs. (2) and (3) in Prob. 20.57 in the above equation, together with Eq. (2), results in

$$\mu_{\min} = \frac{m[\tfrac{1}{7}v_c(2 + 5 \cos \beta) - v_c \cos \beta]}{-m[0 - v_c \sin \beta]} = \frac{2(1 - \cos \beta)}{7 \sin \beta} \qquad (3)$$

If

$$\mu > \frac{2(1 - \cos \beta)}{7 \sin \beta}$$

slipping motion of the sphere is not impending.

Angle $\beta$ is now found in terms of the dimensions $d$ and $h$ in Fig. 20.66a. Using Fig. 20.66b,

$$A = \sqrt{\left(\frac{d}{2}\right)^2 - \left(\frac{d}{2} - h\right)^2} = \sqrt{dh - h^2}$$

$$\cos \beta = \frac{(d/2 - h)}{d/2} = 1 - \frac{2h}{d} \qquad (4)$$

$$\sin \beta = \frac{\sqrt{dh - h^2}}{d/2} = 2\sqrt{\frac{h}{d} - \left(\frac{h}{d}\right)^2} \qquad (5)$$

Eqs. (4) and (5) are used in Eq. (3), with the result

$$\mu_{\min} = \frac{2h}{7d\sqrt{(h/d)(1 - h/d)}} = \frac{0.286h}{d\sqrt{(h/d)(1 - h/d)}}$$

(b) From Case 2 in Table 17.13,

$$I_0 = \tfrac{1}{10}m\left(\frac{d_0^5 - d_i^5}{d_0^3 - d_i^3}\right)$$

Using $d_0 = d$ and $d_i = 0.82d$ in the above equation results in

$$I_0 = \tfrac{1}{10}m\left[\frac{d^5 - (0.82d)^5}{d^3 - (0.82d)^3}\right] = 0.140md^2$$

The above value, with $r = d/2$, is used in Eq. (1) of part (a). The result is

$$v_y' = \frac{v_c(I_0 + mr^2 \cos \beta)}{I_0 + mr^2} = \frac{v_c[0.140md^2 + m(d/2)^2 \cos \beta]}{0.140md^2 + m(d/2)^2} = (0.359 + 0.641 \cos \beta)v_c \qquad (6)$$

Using Eqs. (2) and (3) in Prob. 20.57 in the above equation, together with Eq. (6), gives

$$\mu_{min} = \frac{m[(0.359 + 0.641 \cos \beta)v_c - v_c \cos \beta]}{-m(0 - v_c \sin \beta)} = \frac{0.359 - 0.359 \cos \beta}{\sin \beta}$$

$$\mu_{min} = \frac{0.359(1 - \cos \beta)}{\sin \beta} = \frac{0.359\{1 - [1 - 2(h/d)]\}}{2\sqrt{(h/d) - (h/d)^2}} = \frac{0.359h}{d\sqrt{(h/d)(1 - h/d)}}$$

The above value of $\mu_{min}$ is seen to be larger than the corresponding value of $\mu_{min}$ for the solid sphere given in part (a).

20.67 A yo-yo modeled as a homogeneous disk of mass 1.4 kg is released from rest in the position shown in Fig. 20.67a. When the center of the yo-yo is 1 m from the ground, the string is cut.

(a) Find the velocities of the disk after impact with the ground, if $e = 1$ and $\mu = 0.3$, and the radius is 50 mm.

(b) Do the same as in part (a), if $e = 1$ and $\mu = 0.15$.

(c) Do the same as in part (a), if $e = 0.5$ and $\mu = 0.3$.

(d) Do the same as in part (a), if $e = 0.5$ and $\mu = 0.15$.

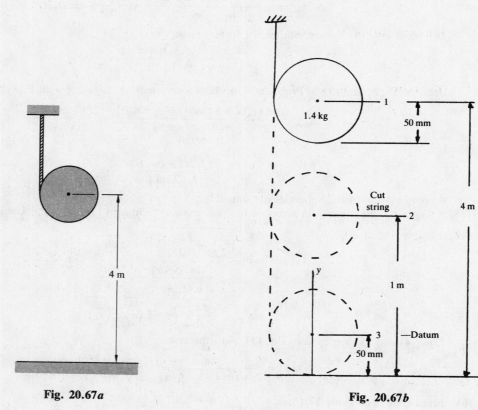

Fig. 20.67a          Fig. 20.67b

▌ (a) Figure 20.67b shows points 1, 2, and 3 in the interval of motion after the yo-yo is released from the initial position. The energy terms at 1 and 2 are

$$T_1 = 0$$
$$V_1 = 1.4(9.81)(4 - 0.050)$$

Using

$$\omega_2 = \frac{v_2}{r} = \frac{v_2}{0.050} = 20v_2 \quad \text{and} \quad I_0 = \tfrac{1}{8}md^2$$

the energy terms at 2 are

$$T_2 = \tfrac{1}{2}mv_2^2 + \tfrac{1}{2}I_0\omega_2^2 = \tfrac{1}{2}(1.4)v_2^2 + \tfrac{1}{2}\left[\tfrac{1}{8}(1.4)\left(\frac{100}{1,000}\right)^2\right](20v_2)^2 = 1.05v_2^2 \qquad V_2 = 1.4(9.81)(1-0.050)$$

The work-energy equation has the form

$$0 = \Delta T + \Delta V = (T_2 - T_1) + (V_2 - V_1) = (1.05v_2^2 - 0) + [1.4(9.81)(1-0.050) - 1.4(9.81)(4-0.050)]$$
$$v_2^2 = 39.2 \text{ m}^2/\text{s}^2 \qquad v_2 = 6.26 \text{ m/s}$$
$$\omega_2^2 = (20v_2)^2 = 400(39.2) \qquad \omega_2 = 125 \text{ rad/s}$$

The kinetic energy at 2 has the value

$$T_2 = 1.05v_2^2 = 1.05(39.2) = 41.2 \text{ J}$$

At the instant that the string is cut at position 2, the disk has angular velocity $\omega_2$. This velocity is constant until the disk impacts the ground at position 3, so that $\omega_3 = \omega_2$. The energy terms at 3 are

$$T_3 = \tfrac{1}{2}mv_3^2 + \tfrac{1}{2}I_0\omega_2^2 = \tfrac{1}{2}(1.4)v_3^2 + \tfrac{1}{2}\left[\tfrac{1}{8}(1.4)\left(\frac{100}{1,000}\right)^2\right]400(39.2) = 0.7v_3^2 + 13.7 \qquad V_3 = 0$$

The work-energy equation for motion between 2 and 3 has the form

$$0 = \Delta T + \Delta V = (T_3 - T_2) + (V_3 - V_2) = [(0.7v_3^2 + 13.7) - 41.2] + [0 - 1.4(9.81)(1-0.050)]$$

$$v_3 = 7.61 \text{ m/s} \qquad \omega_3 = \omega_2 = \frac{v_2}{r} = \frac{6.26}{0.050} = 125 \text{ rad/s}$$

Equation (10) in Prob. 20.60 has the form

$$\mu_{\min} = \frac{I_0\omega - (I_0/r)v_c \cos\beta}{(rm + I_0/r)v_c(1+e)\sin\beta} \qquad (1)$$

The numerical values of the problem are

$$\beta = 90° \qquad v_c = v_3 = 7.61 \text{ m/s} \qquad \omega = \omega_3 = 125 \text{ rad/s} \qquad r = 0.050 \text{ m} \qquad m = 1.4 \text{ kg}$$

$$I_0 = \tfrac{1}{8}md^2 = \tfrac{1}{8}(1.4)\left(\frac{100}{1,000}\right)^2 = 0.00175 \text{ kg} \cdot \text{m}^2$$

Using $e = 1$ and $\mu = 0.3$ in Eq. (1) results in

$$\mu_{\min} = \frac{0.00175(125) - 0}{[0.050(1.4) + 0.00175/0.050]7.61(1+1)\sin 90°} = 0.137$$

Since $\mu = 0.3 > 0.137$, slipping motion of the disk is not impending during the impact. Using Eqs. (7) in Prob. 20.59,

$$v_x' = \frac{rmv_c\cos\beta + I_0\omega}{rm + I_0/r} = \frac{0.050(1.4)7.61\cos 90° + 0.00175(125)}{0.050(1.4) + 0.00175/0.050} = 2.08 \text{ m/s} \qquad (2)$$

$$v_y' = ev_c\sin\beta = 1(7.61)\sin 90° = 7.61 \text{ m/s} \qquad (3)$$

$$\omega' = \frac{v_x'}{r} = \frac{2.08}{0.050} = 41.6 \text{ rad/s} \qquad (4)$$

(b) For the case of $e = 1$ and $\mu = 0.15$, use of Eq. (1) yields the same result, $\mu_{\min} = 0.137$, found in part (a). Thus, the results for the velocity are the same as those found in part (a).

(c) Using $e = 0.5$ and $\mu = 0.3$ in Eq. (1) results in

$$\mu_{\min} = \frac{0.00175(125) - 0}{[0.050(1.4) + 0.00175/0.050]7.61(1+0.5)\sin 90°} = 0.183$$

Since $\mu = 0.3 > 0.183$, slipping motion of the disk is not impending during the impact, and the solution is given by Eqs. (2) through (4) in part (a). From Eq. (2),

$$v_x' = 2.08 \text{ m/s}$$

From Eq. (3),

$$v_y' = ev_c\sin\beta = 0.5(7.61)\sin 90° = 3.81 \text{ m/s}$$

From Eq. (4),

$$\omega' = 41.6 \text{ rad/s}$$

(d) For the case of $e = 0.5$ and $\mu = 0.15$, the use of Eq. (1) yields the same result, $\mu_{\text{min}} = 0.183$, found in part (c). Since $\mu = 0.15 < 0.183$, slipping motion is impending. Using Eqs. (4), (6), and (7) in Prob. 20.60,

$$v'_x = v_c[\cos \beta + \mu(1 + e) \sin \beta] = 7.61[\cos 90° + 0.15(1 + 0.5) \sin 90°] = 1.71 \text{ m/s}$$
$$v'_y = ev_c \sin \beta = 0.5(7.61) \sin 90° = 3.81 \text{ m/s}$$
$$\omega' = \omega - \frac{rmv_c\mu(1 + e) \sin \beta}{I_0} = 125 - \frac{0.050(1.4)7.61(0.15)(1 + 0.5) \sin 90°}{0.00175} = 56.5 \text{ rad/s}$$

**20.68** A thin, rigid prismatic rod of mass $m$ and length $2l$, with zero angular velocity, impacts a plane surface at angle $\theta$, as shown in Fig. 20.68a. The velocity of the rod just before impact is $v_c$. The coefficient of restitution is $e$, and the coefficient of friction between the rod and the surface is $\mu$.

(a) Find the $x$ and $y$ components of the velocity of the center of mass, and the angular velocity of the rod, after impact.

(b) Write the equations of motion for the problem in part (a) if, in addition to the vertical velocity $v_c = v_{cy}$ just before impact, the rod has a velocity component $v_{cx}$ in the positive $x$-coordinate sense and a clockwise angular velocity $\omega_0$.

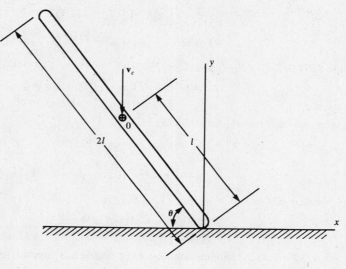

Fig. 20.68a

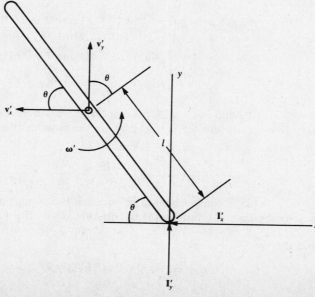

Fig. 20.68b

▎ (a) Figure 20.68b shows the components $I_x'$ and $I_y'$ of the impulse, and the velocity components $v_x'$, $v_y'$, and $\omega'$ after impact. All these effects are shown in their assumed actual senses. The impulse-momentum equations have the forms

$$I_x' = m(v_x' - 0) \tag{1}$$

$$I_y' = m[v_y' - (-v_c)] \tag{2}$$

$$I_y' l \cos\theta - I_x' l \sin\theta = I_0(\omega' - 0) \tag{3}$$

where $I_0$ is the mass moment of inertia of the rod about the center of mass. The coefficient of restitution, from Prob. 20.16, has the form

$$e = -\frac{v_A' - v_B'}{v_A - v_B}$$

The rod is considered to be body A and the ground is body B, and $v_B = v_B' = 0$. Using the velocity components along the line between the center of mass and the contact point of the rod with the plane surface,

$$e = -\frac{[(v_x' \cos\theta + v_y' \sin\theta) - 0]}{-v_c \sin\theta - 0} = \frac{v_x' \cos\theta + v_y' \sin\theta}{v_c \sin\theta} \tag{4}$$

Equations (1) through (4) are four equations in the five unknowns $v_x'$, $v_y'$, $\omega'$, $I_x'$, and $I_y'$. A fifth equation, which is a function of whether or not slipping motion is impending, is developed next.

## Case A: Slipping Motion during Impact Not Impending

For this case the impact point is an instant center of rotation of the rod with respect to the plane surface. The velocity relationship, from Fig. 20.68b, is

$$l\omega' = v_x' \sin\theta - v_y' \cos\theta \tag{5}$$

$I_x'$ and $I_y'$ from Eqs. (1) and (2) are used in Eq. (3), with the result

$$m(v_y' + v_c)l \cos\theta - mlv_x' \sin\theta = I_0\omega' \tag{6}$$

From Eq. (4),

$$v_x' = \frac{ev_c \sin\theta - v_y' \sin\theta}{\cos\theta} \tag{7}$$

From Eq. (5),

$$\omega' = \frac{v_x' \sin\theta - v_y' \cos\theta}{l} \tag{8}$$

$v_x'$ is eliminated between Eqs. (7) and (8), to obtain

$$\omega' = \frac{ev_c \sin^2\theta - v_y'}{l \cos\theta} \tag{9}$$

$v_x'$ from Eq. (7) and $\omega'$ from Eq. (9) are next used in Eq. (6), together with

$$I_0 = k_0^2 m \qquad k_0^2 = \frac{I_0}{m}$$

where $k_0$ is the radius of gyration of the rod about the center of mass. The result is

$$v_y' = v_c\left(e \sin^2\theta - \frac{l^2}{l^2 + k_0^2} \cos^2\theta\right) \tag{10}$$

Using Eq. (10) in Eq. (7) results in

$$v_x' = v_c\left(e + \frac{l^2}{l^2 + k_0^2}\right)\sin\theta \cos\theta \tag{11}$$

$v_y'$ from Eq. (10) is used in Eq. (9), to obtain

$$\omega' = \frac{v_c}{l}\left(\frac{l^2}{l^2 + k_0^2}\right)\cos\theta \tag{12}$$

Equations (10) through (12) give the velocity components after impact, for case A, where slipping motion during impact is not impending. The subscript A is used on these terms to indicate case A motion, and the final forms are

$$v_{xA}' = v_c\left(e + \frac{l^2}{l^2 + k_0^2}\right)\sin\theta \cos\theta \tag{13}$$

$$v'_{yA} = v_c\left(e\sin^2\theta - \frac{l^2}{l^2 + k_0^2}\cos^2\theta\right) \tag{14}$$

$$\omega'_A = \frac{v_c}{l}\left(\frac{l^2}{l^2 + k_0^2}\right)\cos\theta \tag{15}$$

It may be observed from Eq. (15) that the value of $\omega'_A$ is independent of the value of $e$, the coefficient of restitution. The necessary condition for case A operation is

$$\mu > \frac{I'_x}{I'_y}$$

Equations (1) and (2) are used in the above equation, with the result

$$\mu_{min} = \frac{I'_x}{I'_y} = \frac{mv'_x}{m(v'_y + v_c)} \tag{16}$$

Using $v'_x$ and $v'_y$ from Eqs. (10) and (11) in Eq. (16) gives

$$\mu_{min} = \frac{[e(l^2 + k_0^2) + l^2]\sin\theta\cos\theta}{(1 + e)(l^2 + k_0^2)\sin^2\theta + k_0^2\cos^2\theta} \tag{17}$$

If $\mu > \mu_{min}$, slipping motion during impact is not impending, and the velocity components after impact are given by Eqs. (13) through (15).

## Case B: Slipping Motion during Impact Impending

Equations (1) through (4) are still valid. The condition of impending slipping motion requires

$$I'_x = \mu I'_y \tag{18}$$

Using Eqs. (1) and (2) in Eq. (18) gives

$$v'_x = \mu(v'_y + v_c) \tag{19}$$

Equations (4) and (19) are solved simultaneously, and the result is

$$v'_y = \frac{v_c(e\sin\theta - \mu\cos\theta)}{\sin\theta + \mu\cos\theta} \tag{20}$$

Using Eq. (20) in Eq. (19) gives

$$v'_x = \frac{\mu v_c(1 + e)\sin\theta}{\sin\theta + \mu\cos\theta} \tag{21}$$

Using Eqs. (1) and (2) in Eq. (3) gives

$$m(v'_y + v_c)l\cos\theta - mv'_x l\sin\theta = I_0\omega' \tag{22}$$

$v'_x$ and $v'_y$ are eliminated from Eq. (22), using Eqs. (20) and (21). The result is

$$\omega' = \frac{v_c l(1 + e)\sin\theta(\cos\theta - \mu\sin\theta)}{k_0^2(\sin\theta + \mu\cos\theta)} \tag{23}$$

Equations (20), (21), and (23) give the velocity components after impact for case B, where slipping motion is impending. The subscript $B$ is used on these terms to indicate case B motion, and the final forms are

$$v'_{xB} = \frac{\mu v_c(1 + e)\sin\theta}{\sin\theta + \mu\cos\theta} \tag{24}$$

$$v'_{yB} = \frac{v_c(e\sin\theta - \mu\cos\theta)}{\sin\theta + \mu\cos\theta} \tag{25}$$

$$\omega'_B = \frac{v_c l(1 + e)\sin\theta(\cos\theta - \mu\sin\theta)}{k_0^2(\sin\theta + \mu\cos\theta)} \tag{26}$$

The necessary condition for case B motion is

$$\mu \leq \mu_{min}$$

where $\mu_{min}$ is given by Eq. (17).

(*b*)  The rod has general plane motion just before impact, with the velocities

$$v_y = v_{cy} \qquad \text{downward}$$
$$v_x = v_{cx} \qquad \text{to the right}$$
$$\omega = \omega_0 \qquad \text{clockwise}$$

The impulse-momentum equations have the forms

$$I'_x = m[v'_x - (-v_{cx})] \qquad (27)$$
$$I'_y = m[v'_y - (-v_{cy})] \qquad (28)$$
$$I'_y l \cos \theta - I'_x l \sin \theta = I_0[\omega' - (-\omega_0)] \qquad (29)$$

The equation defining the coefficient of restitution is

$$e = \frac{-[(v'_x \cos \theta + v'_y \sin \theta) - 0]}{[-(v_{cy} \sin \theta + v_{cx} \cos \theta) - 0]} \qquad (30)$$

Equations (27) through (30) are the counterparts of Eqs. (1) through (4) in part (*a*).  The remainder of the solution to the problem follows the same steps as in part (*a*).

**20.69**  The rod in Prob. 20.68 has a length of 900 mm and it impacts the plane surface with a velocity $v_c = 6$ m/s.  Find the values of $\mu_{\min}$, and the velocity components after impact, for the following values of $\theta$, $e$, and $\mu$,

$$\theta = 25°, 50°, 75° \qquad e = 1, 0.5, 0.2 \qquad \mu = 0, 0.3$$

▌  $l$ and $k_0$ are found from

$$2l = 900 \qquad l = 450 \text{ mm} = 0.45 \text{ m} \qquad I_0 = k_0^2 m$$

$$\tfrac{1}{12} m(2l)^2 = k_0^2 m \qquad k_0 = \frac{l}{\sqrt{3}} = \frac{0.45}{\sqrt{3}} = 0.260 \text{ m}$$

Table 20.7 shows the results for $\mu_{\min}$, and $v'_x$, $v'_y$, and $\omega'$, for the required values of $\theta$, $e$, and $\mu$.  The following effects may be observed from the data in this table.

1.  Case B, with impending slipping motion, occurs at smaller values of $\theta$.  Case A, where slipping motion is not impending, occurs at larger values of $\theta$.  It should be noted that the practical range of $\theta$ is $0 \le \theta \le 90°$.

**TABLE 20.7**

| $\theta$ | $e$ | $\mu$ | $\mu_{\min}$ | case A | | | case B | | |
|---|---|---|---|---|---|---|---|---|---|
| | | | | $v'_{xA}$, m/s | $v'_{yA}$, m/s | $\omega'_A$, rad/s | $v'_{xB}$, m/s | $v'_{yB}$, m/s | $\omega'_B$, rad/s |
| | 1 | 0 | 1.75 | . . . | . . . | . . . | 0 | 6.00 | 145 |
| | 1 | 0.3 | 1.75 | . . . | . . . | . . . | 2.19 | 1.30 | 75.9 |
| | 0.5 | 0 | 1.65 | . . . | . . . | . . . | 0 | 3.00 | 109 |
| 25° | 0.5 | 0.3 | 1.65 | . . . | . . . | . . . | 1.64 | −0.524 | 56.9 |
| | 0.2 | 0 | 1.55 | . . . | . . . | . . . | 0 | 1.20 | 87.0 |
| | 0.2 | 0.3 | 1.55 | . . . | . . . | . . . | 1.31 | −1.62 | 45.5 |
| | 1 | 0 | 0.786 | . . . | . . . | . . . | 0 | 6.0 | 103 |
| | 1 | 0.3 | 0.786 | . . . | . . . | . . . | 2.88 | 3.59 | 52.8 |
| | 0.5 | 0 | 0.768 | . . . | . . . | . . . | 0 | 3.00 | 77.1 |
| 50° | 0.5 | 0.3 | 0.768 | . . . | . . . | . . . | 2.16 | 1.19 | 39.6 |
| | 0.2 | 0 | 0.751 | . . . | . . . | . . . | 0 | 1.20 | 61.7 |
| | 0.2 | 0.3 | 0.751 | . . . | . . . | . . . | 1.73 | −0.248 | 31.7 |
| | 1 | 0 | 0.257 | . . . | . . . | . . . | 0 | 6.00 | 41.4 |
| | 1 | 0.3 | 0.257 | 2.88 | 5.23 | 1.59 | . . . | . . . | . . . |
| | 0.5 | 0 | 0.253 | . . . | . . . | . . . | 0 | 3.00 | 31.1 |
| 75° | 0.5 | 0.3 | 0.253 | 2.13 | 2.43 | 1.59 | . . . | . . . | . . . |
| | 0.2 | 0 | 0.250 | . . . | . . . | . . . | 0 | 1.20 | 24.8 |
| | 0.2 | 0.3 | 0.250 | 1.68 | 0.749 | 1.59 | . . . | . . . | . . . |

**2.** The values of $\mu_{min}$ decrease with increasing $\theta$.

**3.** $v'_{xB}$ is zero for all cases with $\mu = 0$, which is an expected result.

**4.** For $\mu = 0$, $v'_{yB}$ is a function of $e$ only, and is not a function of $\theta$.

**5.** $v'_x$ and $v'_y$, for both cases A and B, decrease with decreasing $e$. For certain cases $v'_{yB}$ changes from a positive value (upward motion) to a negative value (downward motion) as $e$ decreases. An example of this effect is seen in the comparison of the two cases, with $\theta = 50°$ and $\mu = 0.3$. For $e = 0.5$, $v'_{yB} = 1.19\,\text{m/s}$ and, for $e = 0.2$, $v'_{yB} = -0.248\,\text{m/s}$.

**6.** For given values of $e$ and $\mu$, $\omega'_B$ decreases with increasing $\theta$.

**7.** For given values of $\theta$ and $e$, $\omega'_B$ decreases with increasing $\mu$.

**8.** For given values of $\theta$ and $\mu$, $\omega'_B$ decreases with decreasing $e$.

**9.** For case A, $\omega'_A$ is a constant which is independent of $e$. From the result in Eq. (15), this is an expected result.

**20.70**   Figure 20.70 shows the hands of a ball player on a baseball bat. The bat is assumed to rotate counterclockwise in the plane, and the location of the player's hands is considered to be an instantaneous fixed point about which the bat rotates. The bat has angular velocity $\omega$ as it strikes the ball in the position shown in the figure, and the ball exerts the impulse $I'_x$ on the bat. Show that, for certain locations of the impact of the ball on the bat, the player will feel no impulsive force reaction on his or her hands as the bat strikes the ball.

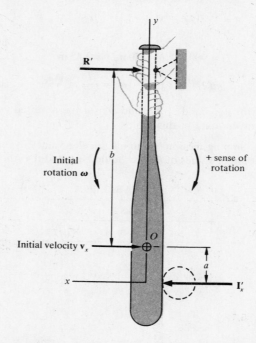

**Fig. 20.70**

▮   The impulse of the ball on the bat is $I'_x$, as shown in the figure, and $R'$ is the impulse exerted by the player's hands on the bat. The impulse $I'_x$ occurs at the arbitrary distance $a$ from the CM of the bat.

The equations of impulse momentum, for translation of the CM and rotation about this point, are

$$I'_x - R' = m[v'_x - (-v_x)] \qquad I'_x a + R'b = I_0[\omega' - (-\omega)]$$

where $I_0$ is the mass moment of inertia of the bat about its center of mass. The translational and rotational velocities are related by

$$v_x = b\omega \qquad v'_x = b\omega'$$

The above set of equations is solved for $R'$, with the result

$$R' = \frac{(I_0 - abm)(\omega' + \omega)}{a + b}$$

If $R' = 0$, the player will feel no force reaction on his or her hands. The necessary condition for this, from the above equation, is

$$I_0 - abm = 0$$

$I_0$ may be expressed in terms of the centroidal radius of gyration $k_0$ as

$$I_0 = k_0^2 m$$

Using this result, the equation above appears as

$$k_0^2 m - abm = 0 \qquad a = \frac{k_0^2}{b}$$

The length $a$ given by the above equation locates the *center of percussion* (CP) of the bat. This term was defined in Prob. 18.98. It may be concluded that the maximum comfort of the player will be obtained if the ball is hit at the center of percussion of the bat.

**20.71**  (*a*)  The slender rod in Fig. 20.71*a* is released from rest. It rotates until it strikes a bumper. Find where the bumper should be placed to minimize the impulsive force experienced by the hinge pin.

(*b*)  Do the same as in part (*a*), for the arrangement of the rod shown in Fig. 20.71*b*.

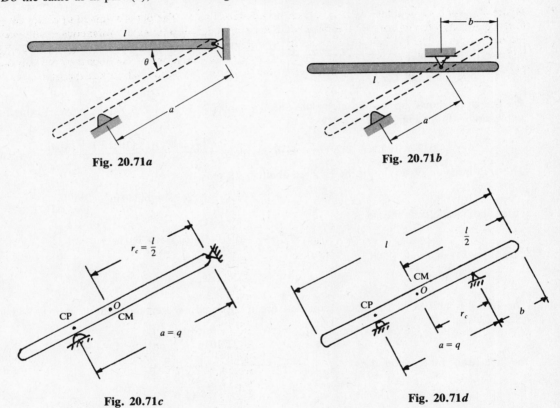

**Fig. 20.71*a***　　　　**Fig. 20.71*b***

**Fig. 20.71*c***　　　　**Fig. 20.71*d***

▌ (*a*)  The bumper should be located at the center of percussion of the rod. Figure 20.71*c* shows the locations of the CP and the CM of the rod. The nomenclature of Prob. 18.98 is used. With $I_0 = \frac{1}{12}ml^2$ the dimension $a$ is found as

$$a = q = r_c + \frac{I_0}{mr_c} = \frac{l}{2} + \frac{\frac{1}{12}ml^2}{m(l/2)} = \frac{2}{3}l$$

(*b*)  The locations of the CM and CP are shown in Fig. 20.71*d*. Using $I_0 = \frac{1}{12}ml^2$ and $r_c = l/2 - b$, the dimension $a$ is found from

$$a = q = r_c + \frac{I_0}{mr_c} = \left(\frac{l}{2} - b\right) + \frac{\frac{1}{12}ml^2}{m(l/2 - b)} = \frac{2(l^2 - 3bl + 3b^2)}{3(l - 2b)}$$

**20.72**  Figure 20.72*a* shows a storage canister with a hinged lid. In use, after material is removed from the canister, the lid is allowed to drop to the closed position. Two rubber stops, located distance $a$ from the hinge, are to be used to cushion the closing action. Find the optimum spacing $a$ for these stops that will result in the "smoothest" closing action.

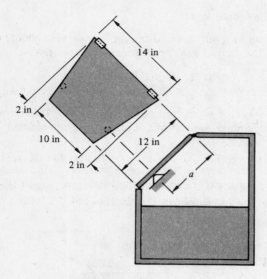

**Fig. 20.72a**

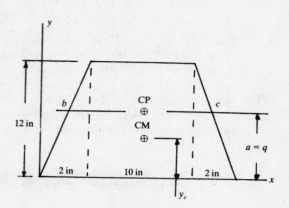

**Fig. 20.72b**

▌ The stops should be located along the line $bc$, shown in Fig. 20.72$b$, which passes through the center of percussion. From Fig. 20.72$b$,

$$A = \tfrac{1}{2}(2)12 + 10(12) + \tfrac{1}{2}(2)12 = 144 \text{ in}^2 \qquad y_c = \frac{\frac{12}{3}[\tfrac{1}{2}(2)12] + \frac{12}{2}(10)12 + \frac{12}{3}[\tfrac{1}{2}(2)12]}{144} = 5.67 \text{ in}$$

The area moment of inertia of the lid area about the $x$ axis is

$$I_x = \tfrac{1}{12}(2)12^3 + \tfrac{1}{3}(10)12^3 + \tfrac{1}{12}(2)12^3 = 6{,}340 \text{ in}^4$$

Using the parallel-axis theorem,

$$I_{0x} = I_x - A y_c^2 = 6{,}340 - 144(5.67)^2 = 1{,}710 \text{ in}^4$$

Using Eq. (4) in Prob. 18.98,

$$q = r_c + \frac{I_0}{mr_c}$$

The terms $I_0$ and $m$ in the above equation are interpreted as area moment of inertia, and area, respectively. Using

$$r_c = y_c = 5.67 \text{ in} \qquad I_0 = I_{0x} = 1{,}710 \text{ in}^4 \qquad \text{and} \qquad A = 144 \text{ in}^2$$

the final result for $a$ is found as

$$a = q = r_c + \frac{I_0}{A r_c} = 5.67 + \frac{1{,}710}{144(5.67)} = 7.76 \text{ in}$$

**20.73**  Give a summary of the basic concepts of impulse-momentum methods for particles and rigid bodies.

▌ The impulse $I'$ of a resultant force $R$ which acts on a particle is

$$I' = \int_1^2 R \, dt$$

The linear momentum $mv$ of a particle is the product of the mass and the velocity of the particle. The linear momentum is a vector quantity. The impulse-momentum equation for a single particle is

$$I' = \int_1^2 R \, dt = m(v_2 - v_1)$$

where 1 and 2 are the endpoints of the time interval of interest. The effect of an impulse acting on a particle is to change the velocity of the particle. The units of impulse and linear momentum are the product of force and time. The USCS units are pound-seconds, and the SI units are newton-seconds. The area under the force-time curve over a given time interval is equal to the change in linear momentum during this time interval. If the resultant force $R$ which acts on the particle is constant, the impulse-momentum equation has the form

$$R(t_2 - t_1) = m(v_2 - v_1)$$

Impact is a term which describes the phenomenon of colliding physical bodies. An irreversible energy loss is always associated with the impact of physical bodies. The impact problem has three characteristics:

1. The duration of the impact is much less than the other observed times in the problem.

2. Because of the short duration of the impact, the impacting bodies are assumed to experience an instantaneous change in velocity with no change in displacement.

3. During the impact process, reaction forces between the bodies, of extremely large magnitude, exist for very short time intervals.

The principal use of solutions to problems of impact is to obtain a set of initial velocities for the regime of motion following the impact.

For colliding bodies in the absence of external forces, the total linear momentum of the system of bodies is constant. For two colliding bodies, the equation of conservation of linear momentum has the form

$$m_A v_A + m_B v_B = m_A v_A' + m_B v_B'$$

where primes are used to designate velocities after impact.

The coefficient of restitution is defined by

$$e = -\frac{v_A' - v_B'}{v_A - v_B}$$

In plastic impact, $e = 0$ and the colliding bodies adhere to each other after impact. The case of plastic impact results in the maximum possible energy loss. The case of elastic impact, with $e = 1$, is a theoretical limiting case in which the energy loss is zero.

In central impact, the common normal line to the surfaces at the contact point passes through the centers of mass of both bodies. In central impact there is no tendency of the bodies to rotate. Direct central impact occurs if the velocities of the two bodies are collinear with the common normal line. If one or both of the velocities of the bodies are not collinear with the common normal line, the impact is referred to as oblique central impact. In oblique central impact, the components of the velocities of the two impacting bodies normal to the line of impact do not change during the impact.

Impulsive forces are forces of very large magnitude which exist for very short periods. The difference between impulsive forces and impact forces is that impulsive forces do not occur because of the impact of physical bodies. These forces are typically caused by explosive effects.

The angular momentum $H$ of a body is a vector quantity defined by

$$H = I\omega$$

where $I$ is the mass moment of inertia about the reference point of interest. The rotational form of the equation of impulsive momentum is

$$\int_1^2 M\, dt = H_2 - H_1 = I(\omega_2 - \omega_1)$$

If the moment which acts on the rigid body is constant, the impulse-momentum equation has the form

$$M(t_2 - t_1) = I(\omega_2 - \omega_1)$$

Two distinct effects occur in the case of impact of bodies in general plane motion, as described below.

1. Translation of the center of mass, with the equation of motion

$$I' = \int_1^2 R\, dt = m(v' - v)$$

where $v$ and $v'$ are the initial and final values, respectively, of the velocity of the center of mass.

2. Rotation of the body about its center of mass, with the equation of motion

$$\int_1^2 M_0\, dt = I_0(\omega' - \omega)$$

where $\omega$ and $\omega'$ are the initial and final values, respectively, of the angular velocities, and $I_0$ is the mass moment of inertia about the center of mass.

# CHAPTER 21
# Rectilinear Motion of a Body with Resisting, or Drag, Forces

## 21.1 CONSTANT DRAG FORCE, DRAG FORCE DIRECTLY PROPORTIONAL TO VELOCITY, LINEAR RESISTANCE LAW

**21.1**    (*a*)   Summarize the results for the motion of a particle or rigid body which is acted on by a constant force and moves in rectilinear translation.

(*b*)   What two general types of resultant forces may act on a particle or rigid body which moves in rectilinear translation?

(*c*)   What three common types of drag forces may act on a rigid body which moves in rectilinear translation?

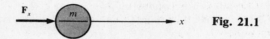

**Fig. 21.1**

**▌** (*a*)   In Chaps. 14 and 15 the kinematics and dynamics of a particle in rectilinear motion were studied. Figure 21.1 shows the typical appearance of such a particle. The particle is constrained to move along the $x$ axis in rectilinear motion. $F_x$ is the component of the *resultant* force, in the $x$ direction, of all the forces acting on the particle. Newton's second law for this problem has the elementary form

$$F_x = m\ddot{x} \tag{1}$$

In the previous treatment in Prob. 15.9, the case was considered where the magnitude of the resultant force had a *constant* value. For this case, from Eq. (1), the acceleration has a constant value. The equations which relate the displacement, velocity, acceleration, and time, for the case of constant acceleration, were found to be

$$v = v_0 + at \qquad x = x_0 + v_0 t + \tfrac{1}{2}at^2 \qquad v^2 = v_0^2 + 2a(x - x_0)$$

where $v_0$ and $x_0$ are the initial velocity and displacement, respectively, at $t = 0$. It should be emphasized that these equations are true *only* for the case where the particle is acted on by a resultant force of *constant* magnitude which produces a *constant* acceleration. In many actual problems the resultant force which acts on a body is constant, so that the above set of equations finds widespread application in the solution of practical engineering problems.

It was subsequently shown in Prob. 18.27 that the center of mass of a rigid body moves as a particle in translation which is acted on by the resultant of all the forces acting on the body. Thus, in the following discussion, *all the results apply equally to particles, and to the centers of mass of rigid bodies, in translation*. The analysis of the rectilinear motion of a body will be extended in this chapter to include the cases where the body is acted on by resisting, or drag, forces.

(*b*)   As a preliminary consideration, the resultant force which acts on a particle or rigid body may be envisioned to be one of two general types. The first type has the same sense as the velocity and may be thought of as an effect which "assists" motion, with the net effect of increasing the velocity of the particle or body. Typical examples of this type of force are the weight force of a falling body and a cable tensile force which acts on a body. The second type of force that may act on a body has a sense which is opposite to that of the velocity and may be thought of as a force which "opposes" motion, with the effect of decreasing the magnitude of the velocity. One type of force in this latter category is referred to as a *dissipative force*. Dissipative forces always transform a part of the kinetic energy of the particle, through frictional dissipative effects, to heat energy, with an unrecoverable decrease in the velocity. Typical examples include friction forces which exist when one body slides along the surface of a second body, and the drag forces which act on a body, such as an automobile, airplane, or boat, as it moves through a fluid medium.

It is interesting to note that a particular force in a given problem may act so as to either assist or oppose motion. In the case of a body which is projected upward from the ground, for example, the same weight force *opposes* motion during the ascent of the body and *assists* the motion during descent

588

of the body. In the following problems a very important consideration is whether a force tends to assist the motion or to oppose it.

(c) For convenience the resisting, or drag, forces which act on a body may be separated into three groups:

1. The drag force is a constant.

2. The drag force is directly proportional to velocity.

3. The drag force is directly proportional to velocity squared.

The motion caused by each type of drag force will be considered in the following problems in this chapter.

**21.2** Give an example of rectilinear motion of a particle with constant drag force.

▌ A common example of rectilinear motion of a particle with constant drag force is sliding coulomb friction. Here, the general form of the resisting force $F_D$ is

$$F_D = -\mu N = \text{constant}$$

where $N$ is the normal contact force between the sliding surfaces and $\mu$ is the coefficient of friction. The particle is assumed to move in the positive sense of the $x$ axis, and the drag force $F_D$ is defined to act in this positive sense. Thus, the minus sign in the above equation indicates that this sliding friction drag force *always acts to oppose the motion.*

Several problems in Chap. 15 illustrated the techniques of solution for the motion of particles which are acted on by *constant* drag forces. It may be observed for this case that, since the drag force is constant, the resulting acceleration (in this case, actually a deceleration) of the particle or body must also be constant.

**21.3** (a) What type of motion results in a drag force which is directly proportional to velocity?

(b) Give the general form of a drag force which is directly proportional to velocity.

(c) Write the equations of motion for a particle which is acted on by a constant force with a sense the same as that of the motion, and a viscous drag force, and find the velocity and displacement of the particle.

(d) What are the forms of the velocity and displacement if the particle in part (c) is acted on by only a viscous drag force?

(e) Summarize the equations for the velocity and displacement of a particle acted on by a constant force, with a sense the same as that of the motion, and a viscous drag force.

▌ (a) The general term used to describe the case where the drag force is directly proportional to velocity is *viscous resistance.* This type of motion is observed both in cases where surfaces which move with low velocity relative to one another are separated by a thin layer of lubricant, and in cases involving the motion of submerged bodies at very low velocities in a viscous fluid, such as when very fine particles in a container of liquid settle slowly to the bottom. The concept of a viscous resisting force is also used extensively in vibration analysis to characterize the dissipative effect of an element which is referred to as a *viscous damper.* The concept of a viscous damper may be used to represent the dissipative effects in a vibration system.

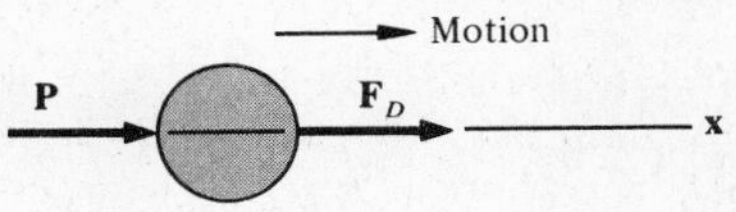

Fig. 21.3

(b) Figure 21.3 shows the forces acting on a particle in rectilinear translation with viscous drag forces. $P$ is a constant external force which acts on the particle, in the sense of the motion. In the following analysis, it will be assumed that $P > 0$. The viscous drag force $F_D$ is written as

$$F_D = -cv \qquad (1)$$

where $v = dx/dt$ is the velocity of the particle and $c$ is a constant. It may be observed from Eq. (1) that the mathematical structure of force $F_D$ is such that it *automatically* adjusts the sense of the drag force to oppose the velocity. If $v$ is positive, so that the particle moves to the right in Fig. 21.3, then $F_D$ is negative and acts to the left. If $v$ is negative, which characterizes leftward motion of the particle, then force $F_D$ is positive and acts to the right on the particle, to oppose its motion. This characteristic of the viscous drag force of automatically adjusting its sense finds widespread application in the solution of vibration problems of oscillatory motion about an equilibrium position.

(c)    The equation of motion of the particle in Fig. 21.3 is

$$\sum F_x = ma_x \qquad P + F_D = m\,\frac{dv}{dt}$$

$$P - cv = m\,\frac{dv}{dt} \qquad \frac{dv}{dt} = \frac{P}{m} - \frac{cv}{m}$$

With the notations

$$\xi = \frac{P}{m} > 0 \qquad \eta = \frac{c}{m} \tag{2}$$

the equation of motion may be written as

$$\frac{dv}{dt} = \xi - \eta v \tag{3}$$

An immediate conclusion may be drawn from Eq. (3).   When the right side of this equation is equal to zero,

$$\frac{dv}{dt} = 0 \qquad v = \text{constant} = v_T$$

This limiting value of constant velocity $v_T$ of the particle is called the *terminal velocity*.   This quantity is found by setting the right side of Eq. (3) equal to zero, with the result

$$\xi - \eta v_T = 0 \tag{4}$$

$$v_T = \frac{\xi}{\eta} = \frac{P/m}{c/m} = \frac{P}{c} \tag{5}$$

When the particle reaches its terminal velocity, the resultant force which acts on it is identically zero, since the magnitude of the applied force $P$ is exactly equal to the magnitude $cv_T$ of the viscous drag force.   It may also be concluded from Eq. (5) that the range of velocities in the problem must satisfy $v < \xi/\eta$.   It is left as an exercise for the reader to show that, if this latter inequality is not true, the viscous drag forces would increase the velocity of the particle.

The initial conditions for this problem are $t = 0$, $v = v_0$, and $x = x_0$.   At the end of the time interval of consideration, the motion is described by $x$, $v$, and $t$.

Equation (3) may be written in the form

$$dt = \frac{dv}{\xi - \eta v} \tag{6}$$

Equation (6) is integrated, with the result

$$\int_0^t dt = \int_{v_0}^v \frac{dv}{\xi - \eta v}$$

This equation has the form

$$\int_1^2 \frac{du}{u} = \ln u \Big|_1^2$$

and it may be written as

$$\int_0^t dt = \frac{1}{-\eta}\int_{v_0}^v \frac{-\eta\,dv}{\xi - \eta v} \qquad t = \Big|_0^t = \frac{-1}{\eta}\ln(\xi - \eta v)\Big|_{v_0}^v = -\frac{1}{\eta}\big[\ln(\xi - \eta v) - \ln(\xi - \eta v_0)\big]$$

$$t = -\frac{1}{\eta}\ln\left[\frac{\xi - \eta v}{\xi - \eta v_0}\right] \qquad e^{-\eta t} = \frac{\xi - \eta v}{\xi - \eta v_0}$$

This equation is solved for $v$, with the result

$$v = \frac{\xi}{\eta} - \left(\frac{\xi}{\eta} - v_0\right)e^{-\eta t} \tag{7}$$

By using $v_T = \xi/\eta$ from Eq. (5), Eq. (7) has the final form

$$v = v_T - (v_T - v_0)\,e^{-\eta t} \tag{8}$$

The velocity and displacement of a particle in rectilinear translation obey the fundamental relationship

$$v = \frac{dx}{dt} \qquad dx = v\,dt \qquad \int_0^x dx = \int_0^t v\,dt \tag{9}$$

Equation (8) is now used in the right side of Eq. (9) to obtain

$$\int_0^x dx = \int_0^t [v_T - (v_T - v_0) e^{-\eta t}]\, dt \qquad x\Big|_0^x = \left[v_T t - \left(\frac{1}{-\eta}\right)(v_T - v_0) e^{-\eta t}\right]_0^t$$

$$x = v_T t + \frac{1}{\eta}(v_0 - v_T)(1 - e^{-\eta t}) \tag{10}$$

Equations (8) and (10) are the results for the velocity and displacement of the particle at time $t$.

(d) If the particle is acted on by only the viscous drag force, then the equations in part ($c$) have simpler forms. For this case,

$$P = 0 \qquad v_T = \frac{P}{c} = 0$$

and the velocity and displacement are given by

$$v = v_0 e^{-\eta t} \qquad x = \frac{v_0}{\eta}(1 - e^{-\eta t}) \tag{11}$$

An interesting consequence of the assumption of viscous drag force may be observed in the above equations. From Eqs. (11), it may be concluded that theoretically it takes infinite time for the particle to attain zero velocity. The corresponding displacement $x_m$ of the particle, by comparison, is finite and is found to be

$$x_m = \lim_{t \to \infty} x = \lim_{t \to \infty} \frac{v_0}{\eta}(1 - e^{-\eta t}) = \frac{v_0}{\eta} \tag{12}$$

The apparent inconsistency of the above mathematical results with the physically observed motion of bodies may be explained as follows. At very small values of velocity of the particle or body, the constant sliding coulomb friction drag forces, which were neglected in this analysis of the motion, predominate over the viscous drag forces. It is these sliding friction forces which eventually bring the body to rest.

(e) The results for the case where the applied constant force has the *same* sense as the velocity are summarized below.

$$v_T = \frac{P}{c} \qquad \eta = \frac{c}{m} \tag{13}$$

$$v = v_T - (v_T - v_0) e^{-\eta t} \tag{14}$$

$$x = v_T t + \frac{1}{\eta}(v_0 - v_T)(1 - e^{-\eta t}) \tag{15}$$

$$v = v_0 e^{-\eta t} \qquad \qquad P = 0 \tag{16}$$

$$x = \frac{v_0}{\eta}(1 - e^{-\eta t}) \qquad \therefore v_T = 0 \tag{17}$$

$$x_m = \frac{v_0}{\eta} \tag{18}$$

**21.4** (a) The rigid arm in Fig. 21.4$a$ is connected to two viscous dampers. $c_A = 0.70\,\text{N}\cdot\text{s/m}$ and $c_B = 0.58\,\text{N}\cdot\text{s/m}$. A force $P$ is applied to the arm and end $d$ moves downward with an initial velocity of 4 m/s. Find the initial force in each damper.

(b) Find the value of force $P$, and the hinge-pin force, for the conditions of part ($a$).

(c) Find the initial velocity of end $d$ of the arm in Fig. 21.4$a$, if a vertical force of 2.5 N is applied to the arm at point $c$.

(d) A vertical force $P$ is to be applied to the arm in Fig. 21.4$a$. How far from the hinge pin should this force be applied, if the initial value of the hinge-pin force is to be zero?

(e) Find the initial forces in dampers $A$ and $B$, if a 4-N force is applied at the location found in part ($d$).

▌ (a) The translational velocity of a point on the arm is related to the angular velocity of the arm by

$$v = r\omega \qquad \frac{v}{r} = \text{constant}$$

The velocities of points $b$ and $c$ are found from

$$\frac{4}{2{,}150} = \frac{v_c}{1{,}400} = \frac{v_b}{800} \qquad v_b = 1.49\,\text{m/s} \qquad v_c = 2.60\,\text{m/s}$$

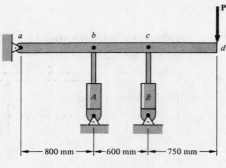

**Fig. 21.4a**

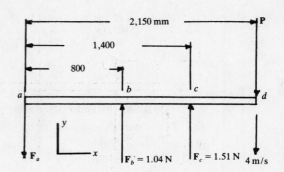

**Fig. 21.4b**

The initial forces in dampers $A$ and $B$ are found as

$$F = cv \qquad F_A = F_b = c_A v_b = 0.70(1.49) = 1.04 \text{ N} \qquad F_B = F_c = c_B v_c = 0.58(2.60) = 1.51 \text{ N}$$

(*b*)   Figure 21.4*b* shows the free-body diagram of the arm.   For equilibrium

$$\sum M_a = 0 \qquad 1.04(800) + 1.51(1,400) - P(2,150) = 0 \qquad P = 1.37 \text{ N}$$

$$\sum F_y = 0 \qquad -F_a + 1.04 + 1.51 - 1.37 = 0 \qquad F_a = 1.18 \text{ N}$$

(*c*)   The velocities of points $b$, $c$, and $d$ are related by

$$\frac{v_b}{800} = \frac{v_c}{1,400} = \frac{v_d}{2,150} \qquad v_c = \frac{1,400}{800} v_b = 1.75 v_b \tag{1}$$

The forces in dampers $A$ and $B$ are expressed in terms of $v_b$ as

$$F_b = c_A v_b = 0.70 v_b \qquad F_c = c_B v_c = 0.58 v_c = 0.58(1.75 v_b) = 1.02 v_b$$

These damper forces are shown in the free-body diagram in Fig. 21.4*c*.
  For equilibrium of the arm,

$$\sum M_a = 0 \qquad (0.70 v_b)800 + (1.02 v_b)1,400 - 2.5(1,400) = 0 \qquad v_b = 1.76 \text{ m/s}$$

Using Eq. (1),

$$v_d = \frac{2,150}{800} \qquad v_b = \frac{2,150}{800}(1.76) = 4.73 \text{ m/s}$$

(*d*)   From part (*c*),

$$F_b = 0.70 v_b \qquad F_c = 1.02 v_b$$

$v_b$ is eliminated between the above two equations, with the result

$$\frac{F_b}{0.70} = \frac{F_c}{1.02} \qquad F_b = 0.686 F_c \tag{2}$$

  Figure 21.4*d* shows the free-body diagram of the arm, with the vertical force $P$ acting at distance $x$ from point $a$.   For equilibrium,

$$\sum M_{b'} = 0 \qquad F_c(1,400 - x) - F_b(x - 800) = 0 \tag{3}$$

Using Eq. (2) in Eq. (3) gives

$$F_c(1,400 - x) - 0.686 F_c(x - 800) = 0 \qquad x = 1,160 \text{ mm}$$

(*e*)   From Fig. 21.4*d*, with   $P = 4 \text{ N}$,

$$\sum F_y = 0 \qquad F_b - 4 + F_c = 0 \tag{4}$$

Use of Eq. (2) in Eq. (4) results in

$$0.686 F_c - 4 + F_c = 0 \qquad F_c = 2.37 \text{ N}$$

Using Eq. (2), $F_b$ is found as

$$F_b = 0.686 F_c = 0.686(2.37) = 1.63 \text{ N}$$

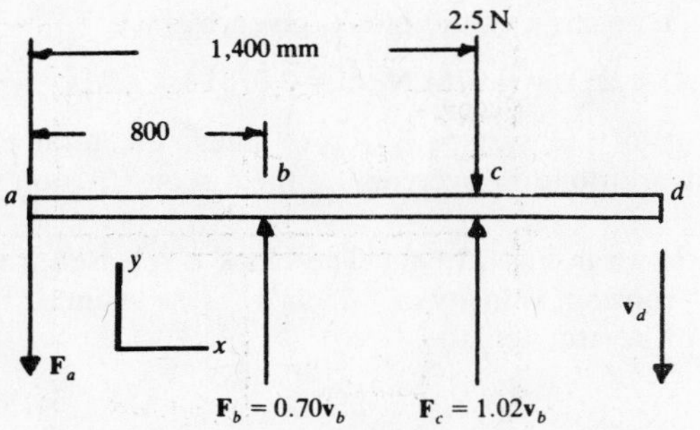

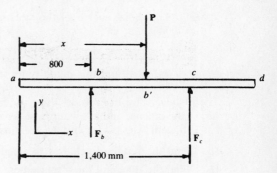

$$F_b = 0.70v_b \qquad F_c = 1.02v_b$$

**Fig. 21.4c**

**Fig. 21.4d**

**21.5** A 100-g steel block rests on a plane horizontal surface, shown in Fig. 21.5, which is coated with a thick oil. A second block, shown as the dashed outline in the figure, impacts the steel block and causes it to move rightward with an initial velocity of 1.25 m/s. The drag force on the block is assumed to be viscous, and from an earlier measurement the constant $c$ was found to have the value $0.350\ \text{N} \cdot \text{s/m}$.

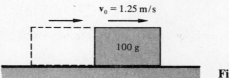

**Fig. 21.5**

(**a**) Find the distance through which the block moves before coming to rest.

(**b**) Find the time at which the velocity of the block is reduced to 50 percent of its initial value.

(**c**) Do the same as in part (b), for a reduction of the velocity to 1 percent of its initial value.

(**d**) Find the maximum value of the drag force which acts on the block.

(**e**) Find the total energy loss during the sliding motion of the block.

▮ (**a**) The block starts with an initial velocity $v_0 = 1.25\ \text{m/s}$ and later comes to rest. The constant $\eta$, from Eq. (13) in Prob. 21.3, is

$$\eta = \frac{c}{m} = \frac{0.350}{0.1} = 3.5\ \text{s}^{-1} \tag{1}$$

The total distance $x_m$ through which the block moves before coming to rest, using Eq. (18) in Prob. 21.3, is

$$x_m = \frac{v_0}{\eta} = \frac{1.25}{3.5} = 0.357\ \text{m}$$

(**b**) Equation (16) in Prob. 21.3 may be written in the form

$$\frac{v}{v_0} = e^{-\eta t} = e^{-3.5t}$$

When the velocity of the block is 50 percent of its initial value,

$$\frac{v}{v_0} = 0.50 = e^{-3.5t} \qquad \ln 0.50 = -3.5t \qquad t = \frac{-\ln 0.50}{3.5} = 0.198\ \text{s}$$

(**c**) When the velocity of the block is 1 percent of its original value

$$\frac{v}{v_0} = 0.01 = e^{-3.5t} \qquad t = 1.32\ \text{s}$$

(**d**) The velocity of the block decreases from the initial maximum velocity $v_0$ to zero velocity. The maximum value of the drag force, from Eq. (1) in Prob. 21.3, occurs at maximum velocity, so that

$$F_{D,\text{max}} = -cv_{\text{max}} = -cv_0 = -0.350(1.25) = -0.438\ \text{N}$$

(e) The initial energy of the system is the kinetic energy of the block, given by

$$T_1 = \tfrac{1}{2}mv_0^2 = \tfrac{1}{2}(0.1)(1.25)^2 = 0.0781 \text{ N} \cdot \text{m} = 0.0781 \text{ J}$$

When the block comes to rest, the energy of the system is zero. Thus, the total energy loss is 0.0781 J, and this quantity represents the work done in overcoming the viscous friction forces.

**21.6** The plane surface in Prob. 21.5 is now inclined, as shown in Fig. 21.6a. The block is released from rest at the top of the incline, and at a later time it moves with a constant velocity of 1.25 m/s. It is assumed that the plane has sufficient length to allow the terminal velocity to be attained.

(a) Find the required value of the angle $\beta$.

(b) Find the time and the displacement of the block when the velocity of this element has reached 50 percent, 90 percent, and 99 percent of the value of the terminal velocity.

(c) If the block had been set into motion with an initial velocity, would the result for part (a) have been different?

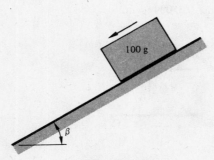

**Fig. 21.6a**

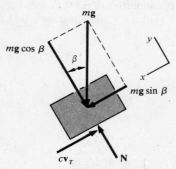

**Fig. 21.6b**

▌ (a) The free-body diagram of the block, when this element moves with its constant terminal velocity, is shown in Fig. 21.6b. For force equilibrium of the block,

$$\sum F_x = 0 \qquad mg \sin \beta - cv_T = 0 \qquad (1)$$

$$0.1(9.81) \sin \beta - 0.350(1.25) = 0 \qquad \beta = 26.5°$$

(b) The velocity of the block, using Eq. (14) in Prob. 21.3, with $v_0 = 0$, is

$$v = v_T(1 - e^{-\eta t}) \qquad \frac{v}{v_T} = 1 - e^{-\eta t}$$

The displacement of the block, from Eq. (15) in Prob. 21.3, with $v_0 = 0$, is

$$x = v_T\left[t - \frac{1}{\eta}(1 - e^{-\eta t})\right]$$

$\eta$ has the value $3.5 \text{ s}^{-1}$ given by Eq. (1) in Prob. 21.5. When the velocity of the block has reached 50 percent of its terminal value,

$$\frac{v}{v_T} = 0.50 = 1 - e^{-3.5t} \qquad t = 0.198 \text{ s} \qquad x = 1.25[0.198 - (1/3.5)(1 - e^{-3.5(0.198)})] = 0.0690 \text{ m} = 69 \text{ mm}$$

By using the above equations, it can be shown that when the block reaches 90 percent of its terminal velocity,

$$t = 0.658 \text{ s} \qquad x = 0.501 \text{ m} = 501 \text{ mm}$$

At 99 percent of the terminal velocity, the values are

$$t = 1.32 \text{ s} \qquad x = 1.30 \text{ m} = 1,300 \text{ mm}$$

(c) It may be seen from Eq. (1) that $\beta$ is a function of only $mg$, $c$, and $v_T$. Since none of these three terms is a function of the initial velocity $v_0$, it follows that the value of $\beta$ is *independent* of the initial velocity. Setting the block into motion with an initial velocity would, however, change the value of $t$ and $x$ found in part (b).

**21.7** The inclined plane surface in Fig. 21.7a is coated with a heavy lubricant, and viscous resistance is assumed. At $t = 0$ the 5-lb block is given an initial velocity $v_0 = 8$ ft/s down the plane.

(a) Find the value of the terminal velocity.

(b) Find the velocity and displacement of the block when $t = 2.6$ s.

(c) Find the energy loss at the time in part (b).

(d) Find the maximum value of the drag force in the interval $0 \le t \le 2.6$ s.

(e) Find the time when the velocity of the block is 98 percent of the terminal velocity.

▌ (a) The force $P$ in the sense of the motion is given by

$$P = 5 \sin 15° = 1.29 \text{ lb}$$

Using Eq. (13) in Prob. 21.3,

$$\eta = \frac{c}{m} = \frac{0.03(32.2)}{5} = 0.193 \text{ s}^{-1} \qquad v_T = \frac{P}{c} = \frac{1.29}{0.03} = 43.0 \text{ ft/s}$$

(b) Using Eqs. (14) and (15) in Prob. 21.3, with $t = 2.6$ s,

$$v = v_T - (v_T - v_0) e^{-\eta t} = 43.0 - (43.0 - 8) e^{-0.193(2.6)} = 21.8 \text{ ft/s} \qquad (1)$$

$$x = v_T t + \frac{1}{\eta} (v_0 - v_T)(1 - e^{-\eta t}) = 43(2.6) + \frac{1}{0.193} (8 - 43.0)(1 - e^{-0.193(2.6)}) = 40.2 \text{ ft}$$

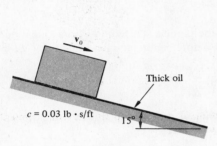

**Fig. 21.7a**

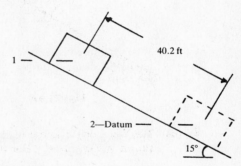

**Fig. 21.7b**

(c) Figure 21.7b shows the endpoints of the interval of motion. The work-energy equation is written as

$$W_{NC} = \Delta T + \Delta V = (T_2 - T_1) + (V_2 - V_1) = \left[ \frac{1}{2} \left( \frac{5}{32.2} \right) 21.8^2 - \frac{1}{2} \left( \frac{5}{32.2} \right) 8^2 \right] + [0 - 5(40.2 \sin 15°)]$$
$$= -20.1 \text{ ft} \cdot \text{lb}$$

The energy loss may be stated as

$$\text{Energy loss} = |W_{NC}| = 20.1 \text{ ft} \cdot \text{lb}$$

(d) The magnitude of the drag force is given by

$$F = cv$$

This force will be maximum when the velocity is maximum. This condition occurs at $t = 2.6$ s, with $v = 21.8$ ft/s, and the maximum value of the drag force is

$$F_{max} = cv_{max} = 0.03(21.8) = 0.654 \text{ lb}$$

(e) Equation (1) in part (b) may be written as

$$e^{-\eta t} = \frac{v_T - v}{v_T - v_0} = \frac{v_T - 0.98 v_T}{v_T - v_0} \qquad e^{-0.193t} = \frac{43.0 - 0.98(43.0)}{43.0 - 8} = 0.02457$$

$$-0.193t = \ln 0.02457 \qquad t = 19.2 \text{ s}$$

**21.8** The 0.25-lb block shown in Fig. 21.8 is at rest on a horizontal plane surface coated with a thick oil. The drag force on the block is assumed to be viscous. At $t = 0$, the horizontal force shown in the figure is applied to the block. This force is removed 1.8 s later.

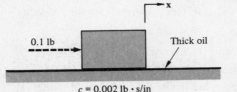

$$c = 0.002 \text{ lb} \cdot \text{s/in}$$ **Fig. 21.8**

(a) Find the velocity and displacement of the block at the instant that the force is removed.

(b) Find the additional distance through which the block moves before it comes to rest.

(c) Find the maximum value of the drag force that acts on the block.

(d) Find the energy loss at $t = 1.8 \text{ s}$, and the total energy loss during the motion of the block.

▌(a) Using Eq. (13) in Prob. 21.3,

$$\eta = \frac{c}{m} = \frac{0.002(386)}{0.25} = 3.09 \text{ s}^{-1} \qquad v_T = \frac{P}{c} = \frac{0.1}{0.002} = 50 \text{ in/s}$$

Using Eqs. (14) and (15) in Prob. 21.3, with $v_0 = 0$ and $t = 1.8 \text{ s}$, gives

$$v = v_T - (v_T - v_0) e^{-\eta t} = 50 - 50 e^{-3.09(1.8)} = 49.8 \text{ in/s}$$

$$x = v_T t + \frac{1}{\eta}(v_0 - v_T)(1 - e^{-\eta t}) = 50(1.8) + \frac{1}{3.09}(-50)(1 - e^{-3.09(1.8)}) = 73.9 \text{ in}$$

(b) When the force is removed, $v_0 = 49.8 \text{ in/s}$ and the particle is acted on by only the drag force. The additional distance through which the block moves before coming to rest, using Eq. (18) in Prob. 2.13, is given by

$$x_m = \frac{v_0}{\eta} = \frac{49.8}{3.09} = 16.1 \text{ in}$$

(c) The maximum value of the drag force occurs with the maximum velocity 49.8 in/s at $t = 1.8 \text{ s}$. The value of this force is

$$F_{\max} = cv_{\max} = 0.002(49.8) = 0.0996 \text{ lb}$$

(d) The work done by the applied force is

$$W_{12} = Px = 0.1(73.9) = 7.39 \text{ in} \cdot \text{lb}$$

The kinetic energy of the block at the end of 1.8 s is

$$T_2 = \tfrac{1}{2}mv^2 = \frac{1}{2}\left(\frac{0.25}{386}\right)49.8^2 = 0.803 \text{ in} \cdot \text{lb}$$

The energy loss at the end of 1.8 s is

$$\Delta W = W_{12} - T_2 = 7.39 - 0.803 = 6.59 \text{ in} \cdot \text{lb}$$

The total energy loss during the motion of the block is

$$\Delta W = W_{12} = 7.39 \text{ in} \cdot \text{lb}$$

since $v = 0$ at the beginning and end of the motion of the block.

**21.9** Figure 21.9a shows a model of a proposed shock-absorbing device. A weight falls vertically and imparts an initial velocity to a platform. The platform is connected to the ground by a telescoping cylinder arrangement which is filled with oil. Plastic impact between the weight and the platform is assumed, and the resisting force exerted by the cylinder rod is assumed to be viscous. The weight of the platform and rod may be neglected.

(a) Find the initial velocity of the platform.

(b) Find the velocity and displacement of the platform at the end of 0.1, 0.5, and 1 s.

(c) Find the energy dissipated by the cylinder at the times in part (b).

(d) Find the maximum height from which the weight may be dropped onto the platform, if the maximum force which may be exerted on the platform is 90 lb.

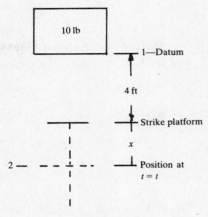

**Fig. 21.9a**                    **Fig. 21.9b**

(a)   The velocity with which the weight impacts the platform is

$$v_0 = \sqrt{2gh} = \sqrt{2(32.2)4} = 16.0 \text{ ft/s}$$

(b)   Since the impact is plastic, $v_0$ is the initial velocity of the platform.   The velocity and displacement of the platform are given by Eqs. (14) and (15) in Prob. 21.3 as

$$v = v_T - (v_T - v_0)\, e^{-\eta t} \qquad x = v_T t + \frac{1}{\eta}\,(v_0 - v_T)(1 - e^{-\eta t})$$

Using Eq. (13) in Prob. 21.3,

$$v_T = \frac{P}{c} = \frac{10}{2.5} = 4 \text{ ft/s} \qquad \eta = \frac{c}{m} = \frac{2.5(32.2)}{10} = 8.05 \text{ s}^{-1}$$

At   $t = 0.1$ s,

$$v = 4 - (4 - 16)\, e^{-8.05(0.1)} = 9.37 \text{ ft/s} \qquad x = 4(0.1) + \frac{1}{8.05}\,(16 - 4)(1 - e^{-8.05(0.1)}) = 1.22 \text{ ft}$$

At   $t = 0.5$ s,

$$v = 4 - (4 - 16)\, e^{-8.05(0.5)} = 4.21 \text{ ft/s} \qquad x = 4(0.5) + \frac{1}{8.05}\,(16 - 4)(1 - e^{-8.05(0.5)}) = 3.46 \text{ ft}$$

At   $t = 1$ s,

$$v = 4 - (4 - 16)\, e^{-8.05(1)} = 4.00 \text{ ft/s} \qquad x = 4(1) + \frac{1}{8.05}\,(16 - 4)(1 - e^{-8.05(1)}) = 5.49 \text{ ft}$$

It may be seen that the terminal velocity   $v_T = 4$ ft/s   has been reached at   $t = 1$ s.

(c)   Figure 21.9b shows the endpoints 1 and 2 of an interval of motion.   Point 1 is the initial position of the weight, and point 2 is the position at time $t$.   The energy terms at 1 and 2, for   $t = 0.1$ s,   are

$$T_1 = 0 \qquad T_2 = \tfrac{1}{2} m v_2^2 = \frac{1}{2}\left(\frac{10}{32.2}\right)9.37^2 \qquad V_1 = 0 \qquad V_2 = -10(4 + 1.22)$$

The energy dissipated by the cylinder is designated $\Delta W$, and

$$\Delta W = (T_1 + V_1) - (T_2 + V_2) = 0 - \left[\frac{1}{2}\left(\frac{10}{32.2}\right)9.37^2 - 10(5.22)\right] = 38.6 \text{ ft} \cdot \text{lb}$$

At   $t = 0.5$ s,

$$T_1 = 0 \qquad T_2 = \frac{1}{2}\left(\frac{10}{32.2}\right)4.21^2 \qquad V_1 = 0 \qquad V_2 = -10(4 + 3.46)$$

$$\Delta W = 0 - \left[\frac{1}{2}\left(\frac{10}{32.2}\right)4.21^2 - 10(7.46)\right] = 71.8 \text{ ft} \cdot \text{lb}$$

At   $t = 1$ s,

$$T_1 = 0 \qquad T_2 = \frac{1}{2}\left(\frac{10}{32.2}\right)4.00^2 \qquad V_1 = 0 \qquad V_2 = -10(4 + 5.49)$$

$$\Delta W = 0 - \left[\frac{1}{2}\left(\frac{10}{32.2}\right)4.00^2 - 10(9.49)\right] = 92.4 \text{ ft} \cdot \text{lb}$$

(*d*)  The maximum force which may be exerted on the platform is

$$F_{max} = 90 \text{ lb} = c v_{max}$$

where $v_{max}$ is the maximum allowable value of the velocity of the platform.   From the above equation,

$$v_{max} = \frac{90}{2.5} = 36 \text{ fts/s}$$

$h_{max}$ is the maximum allowable height of the weight above the platform, and

$$v_{max} = \sqrt{2gh_{max}} \qquad h_{max} = \frac{v_{max}^2}{2g} = \frac{36^2}{2(32.2)} = 20.1 \text{ ft}$$

**21.10**   Figure 21.10 shows a spherical body submerged in a liquid.   It is shown in texts in fluid mechanics that if the velocity of the body is sufficiently small, the drag force is given by Stoke's law as   $F_D = -3\pi\mu v d$,   where $v$ is the velocity and $\mu$, a property of the liquid, is the absolute viscosity.   The units of this latter quantity are the product of force and time divided by area.   It can be shown that Stoke's law is valid if   $(vd\rho/\mu) < 1$,   where $\rho$ is the mass density of the liquid.   Find the viscous drag force constant for a sphere that moves with Stoke's law motion.

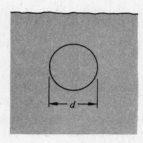

**Fig. 21.10**

❚  The viscous drag force is defined in Prob. 21.3 as

$$F_D = -cv \qquad (1)$$

Stoke's law is given by

$$F_D = -3\pi\mu v d \qquad (2)$$

Equations (1) and (2) are set equal to each other, with the result

$$c = 3\pi\mu d$$

**21.11**   Figure 21.11*a* shows a cylindrical tank filled with a heavy oil with a specific gravity of 0.92 and an absolute viscosity of $0.01 \text{ lb} \cdot \text{s/ft}^2$.   The specific weight of water is $62.4 \text{ lb/ft}^3$.   A steel sphere, of diameter   $d = 0.125 \text{ in}$   and specific weight $0.283 \text{ lb/in}^3$, is released from rest at the surface at   $t = 0$.

(*a*)   Find the terminal velocity of the sphere.

(*b*)   Find the velocity and displacement of the sphere at   $t = 1 \text{ s}$.

(*c*)   Estimate the time at which the sphere reaches the bottom of the tank.

(*d*)   Do the same as in parts (*a*), (*b*), and (*c*), if the sphere is aluminum, with a specific weight of $0.1 \text{ lb/in}^3$.

(*e*)   Do the same as in parts (*a*), (*b*), and (*c*), if the sphere is lead, with a specific weight of $0.411 \text{ lb/in}^3$.

(*f*)   Compare the results obtained in parts (*a*) through (*e*).

❚  (*a*)   It is assumed that Stoke's law motion (see Prob. 21.10) exists, so that   $vd\rho/\mu < 1$, where $\rho$ is the mass density of the liquid, $\mu$ is the absolute viscosity of the liquid, and $v$ is the velocity.   The above inequality will be tested after the value for $v$ is obtained.   The viscous drag force constant is given by

$$c = 3\pi\mu d = 3\pi(0.01)\left(\frac{0.125}{12}\right) = 9.82 \times 10^{-4} \text{ lb} \cdot \text{s/ft}$$

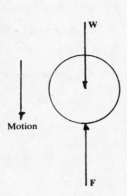

Fig. 21.11*a*

Fig. 21.11*b*

$W$ is the weight of the sphere, with the value

$$W = 0.283\left(\frac{4}{3}\right)\pi\left(\frac{0.125}{2}\right)^3 = 2.89 \times 10^{-4}\ \text{lb}$$

$F$ is the buoyant force which acts on the sphere.  The magnitude of this force is equal to the weight of liquid displaced by the sphere.  Using $62.4\ \text{lb/ft}^3$ as the specific weight of water, and $SG = 0.92$, the value of $F$ is found as

$$F = \gamma V = \frac{62.4(0.92)}{12^3}\ \frac{4}{3}\ \pi\left(\frac{0.125}{2}\right)^3 = 3.40 \times 10^{-5}\ \text{lb}$$

The resultant force which acts on the sphere is

$$P = W - F = 2.89 \times 10^{-4} - 3.40 \times 10^{-5} = 2.55 \times 10^{-4}\ \text{lb}$$

Using Eq. (5) in Prob. 21.3, the terminal velocity of the sphere is found to be

$$v_T = \frac{P}{c} = \frac{2.55 \times 10^{-4}}{9.82 \times 10^{-4}} = 0.260\ \text{ft/s}$$

As a check on the initial assumptions,

$$\frac{v d\rho}{\mu} = \frac{0.260(0.125/12)0.92(62.4/32.2)}{0.01} = 0.483 < 1$$

The sphere is seen to move with Stoke's law motion.

(*b*)  Using Eq. (2) in Prob. 21.3,

$$\eta = \frac{c}{m} = \frac{9.82 \times 10^{-4}(32.2)}{2.89 \times 10^{-4}} = 109\ \text{s}^{-1}$$

Equations (8) and (10) in Prob. 21.3 are used, with the initial condition

$$v_0 = 0$$

The results, when $t = 1\ \text{s}$,  are

$$v = v_T(1 - e^{-\eta t}) = 0.260[1 - e^{-109(1)}] = 0.260\ \text{ft/s}$$

$$x = v_T\left[t - \left(\frac{1 - e^{-\eta t}}{\eta}\right)\right] = 0.260\left[1 - \left(\frac{1 - e^{-109(1)}}{109}\right)\right] \qquad x = 0.258\ \text{ft}$$

(*c*)  An estimate is now made of the time for the sphere to reach the bottom of the tank.  At the end of 1 s, from the result in part (*a*),

$$x = 0.258\ \text{ft}$$

If the sphere had moved with the constant velocity $v = v_T = 0.260\ \text{ft/s}$, the displacement at the end of 1 s would have been

$$x = v_T t = 0.260(1) = 0.260\ \text{ft} \qquad 0.260 \approx 0.258$$

It is thus assumed that the sphere moves through the entire distance from the liquid surface to the bottom of the tank, with $v = v_T$. The approximate time for this descent is found from

$$x = v_T t \qquad 5 = 0.260t \qquad t = 19.2 \text{ s}$$

(d) Stoke's law motion is again assumed, and the solution follows that in parts (a) through (c).

$$c = 9.82 \times 10^{-4} \text{ lb} \cdot \text{s/ft} \qquad W = 0.1(\tfrac{4}{3})\pi \left(\frac{0.125}{2}\right)^3 = 1.02 \times 10^{-4} \text{ lb} \qquad F = 3.40 \times 10^{-5} \text{ lb}$$

$$P = W - F = 1.02 \times 10^{-4} - 3.40 \times 10^{-5} = 6.80 \times 10^{-5} \text{ lb}$$

$$v_T = \frac{P}{c} = \frac{6.80 \times 10^{-5}}{9.82 \times 10^{-4}} = 0.0692 \text{ ft/s}$$

As a check on the assumption of Stoke's law motion,

$$\frac{vd\rho}{\mu} = \frac{0.0692(0.125/12)0.92(62.4/32.2)}{0.01} = 0.129 < 1$$

$$\eta = \frac{c}{m} = \frac{9.82 \times 10^{-4}(32.2)}{1.02 \times 10^{-4}} = 310 \text{ s}^{-1}$$

Using $v_0 = 0$, the velocity and displacement at $t = 1$ s are

$$v = v_T(1 - e^{-\eta t}) = 0.0692[1 - e^{-310(1)}] = 0.0692 \text{ ft/s}$$

$$x = v_T \left[ t - \left( \frac{1 - e^{-\eta t}}{\eta} \right) \right] = 0.0692 \left[ 1 - \left( \frac{1 - e^{-310(1)}}{310} \right) \right] = 0.0690 \text{ ft}$$

The approximate time for the sphere to reach the bottom of the tank is found from

$$x = v_T t \qquad 5 = 0.0692t \qquad t = 72.3 \text{ s}$$

(e) The solution follows that in parts (a) through (c), and Stoke's law motion is again assumed.

$$c = 9.82 \times 10^{-4} \text{ lb} \cdot \text{s/ft} \qquad W = 0.411(\tfrac{4}{3})\pi \left(\frac{0.125}{2}\right)^3 = 4.20 \times 10^{-4} \text{ lb} \qquad F = 3.40 \times 10^{-5} \text{ lb}$$

$$P = W - F = 4.20 \times 10^{-4} - 3.40 \times 10^{-5} = 3.86 \times 10^{-4} \text{ lb} \qquad v_T = \frac{P}{c} = \frac{3.86 \times 10^{-4}}{9.82 \times 10^{-4}} = 0.393 \text{ ft/s}$$

The assumption of Stoke's law motion is verified by the result

$$\frac{vd\rho}{\mu} = \frac{0.393(0.125/12)0.92(62.4/32.2)}{0.01} = 0.730 < 1 \qquad \eta = \frac{c}{m} = \frac{9.82 \times 10^{-4}(32.2)}{4.20 \times 10^{-4}} = 75.3 \text{ s}^{-1}$$

The velocity and displacement at $t = 1$ s, using $v_0 = 0$, are

$$v = v_T(1 - e^{-\eta t}) = 0.393[1 - e^{-75.3(1)}] = 0.393 \text{ ft/s}$$

$$x = v_T \left[ t - \left( \frac{1 - e^{-\eta t}}{\eta} \right) \right] = 0.393 \left[ 1 - \left( \frac{1 - e^{-75.3(1)}}{75.3} \right) \right] = 0.388 \text{ ft}$$

The approximate time for the sphere to reach the bottom of the tank is given by

$$x = v_T t \qquad 5 = 0.393t \qquad t = 12.7 \text{ s}$$

(f) A summary of the results in parts (a) through (e) is given in Table 21.1. It may be seen that as the mass density of the material of the sphere decreases, the terminal velocity decreases and the time to reach the bottom of the tank increases. It may also be seen that, for all three cases, the terminal velocity has been reached at the end of 1 s.

### TABLE 21.1

| material of sphere | t = 1 s | | approximate time to reach bottom of tank, s | terminal velocity, ft/s |
|---|---|---|---|---|
| | v, ft/s | x, ft | | |
| Lead | 0.393 | 0.388 | 12.7 | 0.393 |
| Steel | 0.260 | 0.258 | 19.2 | 0.260 |
| Aluminum | 0.0692 | 0.0690 | 72.3 | 0.0692 |

**21.12** Figure 21.12 shows two spheres that are released from rest at the same instant at the free surface of a column of with $d_A = 0.075$ in and $\gamma_A = 0.318 \, \text{lb/in}^3$. Sphere $B$ is magnesium, with $\gamma_B = 0.0659 \, \text{lb/in}^3$. Assume that both spheres rapidly reach their terminal velocities, so that the motion of the two spheres is with constant velocity.

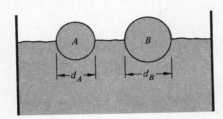

**Fig. 21.12**

(*a*) Find the required value of the diameter of sphere $B$, if the two spheres are to reach the bottom at the same time.

(*b*) Find the time required for the motion in part (*a*), if the depth of the liquid is 5 ft.

▌ (*a*) For both spheres to reach the bottom at the same time,

$$v_{T,A} = v_{T,B}$$

From Eq. (13) in Prob. 21.3,

$$\frac{P_A}{c_A} = \frac{P_B}{c_B}$$

Stoke's law motion is assumed, subject to later verification. Using the result in Prob. 21.10,

$$\frac{P_A}{3\pi\mu d_A} = \frac{P_B}{3\pi\mu d_B} \qquad \frac{P_A}{d_A} = \frac{P_B}{d_B} \qquad (1)$$

The weight of sphere $A$ is

$$W_A = \gamma_A V_A = 0.318\left(\frac{4}{3}\right)\pi\left(\frac{0.075}{2}\right)^3 = 7.02 \times 10^{-5} \, \text{lb}$$

The buoyant force acting on sphere $A$, using $\gamma = 62.4 \, \text{lb/ft}^3$ as the specific weight of water, is given by

$$F_A = \gamma V_A = \frac{62.4(0.92)}{12^3}\frac{4}{3}\pi\left(\frac{0.075}{2}\right)^3 = 7.34 \times 10^{-6} \, \text{lb}$$

The resultant force acting on sphere $A$ is

$$P_A = W_A - F_A = 7.02 \times 10^{-5} - 7.34 \times 10^{-6} = 6.29 \times 10^{-5} \, \text{lb}$$

For sphere $B$,

$$W_B = \gamma_B V_B = 0.0659\left(\frac{4}{3}\right)\pi\left(\frac{d_B}{2}\right)^3 = 0.0345 d_B^3 \qquad \text{lb}$$

$$F_B = \gamma V_B = \frac{62.4(0.92)}{12^3}\frac{4}{3}\pi\left(\frac{d_B}{2}\right)^3 = 0.0174 d_B^3 \qquad \text{lb}$$

$$P_B = W_B - F_B = 0.0345 d_B^3 - 0.0174 d_B^3 = 0.0171 d_B^3 \qquad \text{lb}$$

where $d_B$ is in inches. The above values are used in Eq. (1), with the result

$$\frac{6.29 \times 10^{-5}}{0.075} = \frac{0.0171 d_B^3}{d_B} \qquad d_B = 0.221 \, \text{in}$$

The assumption of Stoke's law motion is now verified. For sphere $B$, using the result in Prob. 21.10,

$$c_B = 3\pi(0.01)\frac{0.221}{12} = 0.00174 \, \text{lb} \cdot \text{s/ft}$$

Using Eq. (13) in Prob. 21.3,

$$v_{T,B} = \frac{P_B}{c_B} = \frac{0.0171(0.221)^3}{0.00174} = 0.106 \text{ ft/s} \qquad \frac{vd\rho}{\mu} = \frac{0.106(0.221/12)0.92(62.4/32.2)}{0.01} = 0.348 < 1$$

The above result shows that Stoke's law motion occurs.  For sphere $A$,

$$\frac{vd\rho}{\mu} < 1$$

since  $v_{T,A} = v_{T,B} = v$  and  $d_A < d_B$.

(b)  The time for the spheres to reach the bottom of the tank is found from

$$x = v_T t \qquad 5 = 0.106t \qquad t = 47.2 \text{ s}$$

## 21.2   DRAG FORCE PROPORTIONAL TO VELOCITY SQUARED, QUADRATIC RESISTANCE LAW WITH APPLIED CONSTANT FORCE WITH SAME, AND WITH OPPOSITE, SENSE AS VELOCITY

**21.13**   Figure 21.13a shows a body which moves through a fluid medium such as air or water.

(a)  Describe the force effects which resist the motion of the body.

(b)  Give the form of the drag force which is caused primarily by pressure effects.

(c)  Give typical values of the drag coefficient for bodies of simple geometry.

▌ (a)   The movement of the body through the fluid creates an unbalanced pressure distribution on the surface of the body, which tends to oppose the motion of this element.   In addition, there are internal friction effects within the fluid because it was deformed from its original shape by the passage of the body.   There are also frictional drag forces on the solid boundary, or "skin," of the body, which is in contact with the fuid.   The force $F_D$, shown in Fig. 21.13a, is the force required to overcome *all* these resisting forces as the body moves through the fluid.

In the following discussion, it is assumed that *the major component of the total drag force is due to the pressure effects*.   The physical problems which closely correspond to this assumption would be either bodies with plane or curved surfaces which are normal to the direction of the velocity of the body, or bodies which have sharp edges where the fluid loses contact with the body.   Examples of bodies where the pressure effects are *not* the major contribution to drag force are plane, or slightly curved, surfaces in which the fluid flow is along the surface, and pointed slender bodies whose axes are parallel to the velocity of the body.   Typical examples of such shapes are airplane wings, boat hulls, and other "streamlined" shapes.   The determination of the drag forces on such bodies is beyond the scope of this book, and this subject is treated in texts on fluid mechanics.

All the following analysis is for the case where the body moves with respect to a stationary fluid.   These results also may be applied to the case where the body is stationary and the fluid moves relative to it, as in a wind tunnel.   It must be emphasized that all the following results are valid *only for velocities which are much less than the speed of sound*, so that the flow field is subsonic.   For sea-level air at 60°F, this limiting sonic velocity is approximately 1,120 ft/s.

(b)   Figure 21.13b shows a body which moves in a fluid medium and experiences a force resistance primarily due to a pressure difference over the surface area of the body.   The particle is constrained to move *only along the positive x axis, with a positive velocity v*.   Thus, the body shown in this figure may move only *to the right* in the figure.   The drag force which acts on the particle is designated $F_D$, and this quantity is defined to be positive if it acts in the positive x-coordinate sense.

It is shown in texts in fluid mechanics and aerodynamics that the drag force may be expressed in the functional form

$$F_D = -C_D(\tfrac{1}{2}\rho v^2)A \qquad\qquad (1)$$

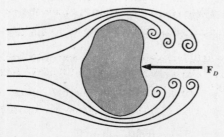

**Fig. 21.13a**

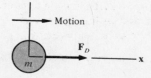

**Fig. 21.13b**

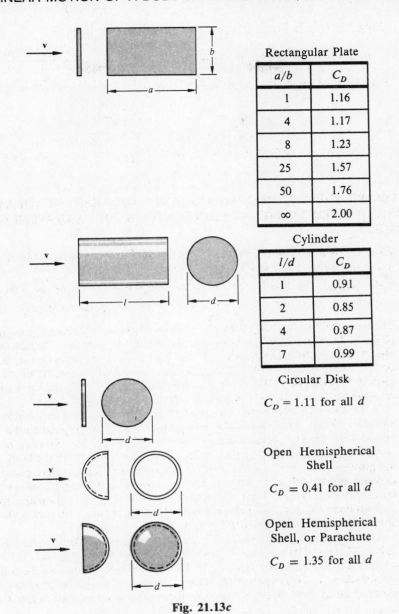

**Rectangular Plate**

| $a/b$ | $C_D$ |
|-------|-------|
| 1 | 1.16 |
| 4 | 1.17 |
| 8 | 1.23 |
| 25 | 1.57 |
| 50 | 1.76 |
| $\infty$ | 2.00 |

**Cylinder**

| $l/d$ | $C_D$ |
|-------|-------|
| 1 | 0.91 |
| 2 | 0.85 |
| 4 | 0.87 |
| 7 | 0.99 |

**Circular Disk**

$C_D = 1.11$ for all $d$

**Open Hemispherical Shell**

$C_D = 0.41$ for all $d$

**Open Hemispherical Shell, or Parachute**

$C_D = 1.35$ for all $d$

**Fig. 21.13c**

In this expression $A$ is the projected area of the body on a plane which is normal to the direction of motion. The term $\frac{1}{2}\rho v^2$, where $\rho$ is the mass density of the fluid and $v$ is the velocity of the body, is referred to as the *dynamic pressure*, and this quantity is used extensively in fluid mechanics. For the purposes of the present discussion, this term will be considered to be merely a defined quantity, having no other particular significance. The term $C_D$ is called the *drag coefficient*, and for other than very small values of the velocity, this term may be assumed to be constant. On the basis of this assumption, and from consideration of Eq. (1), it may be seen that the drag force is *proportional to the velocity squared*. This force-velocity relationship is referred to as the *quadratic resistance law*.

(c)  Figure 21.13c shows typical values of the drag coefficient for bodies of simple geometry.

21.14  Figure 21.14 shows a temporary wall at a construction site. A wind with a velocity of 50 km/h and the direction shown in the figure acts on the wall. Standard air, with a specific weight of 0.07637 lb/ft$^3$, or a density of 1.22 kg/m$^3$, is assumed. Find the resultant force that acts on the wall.

▎ The wall is a rectangular plate normal to the direction of flow. Using the nomenclature of Fig. 21.13c,

$$\frac{a}{b} = \frac{6}{4} = 1.5$$

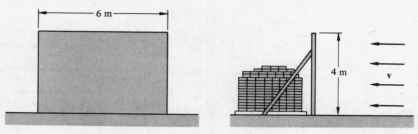

**Fig. 21.14**

It may be seen in this figure that the change in $C_D$, for a change in $a/b$ from 1 to 4, is only 0.01. It is thus assumed that $C_D \approx 1.16$. Using $50 \text{ km/h} = 13.9 \text{ m/s}$ the drag force of the wind on the wall is found as

$$F_D = C_D[\tfrac{1}{2}\rho v^2]A = 1.16[\tfrac{1}{2}(1.22)13.9^2]6(4) = 3,280 \text{ N}$$

**21.15** The packing crate with a uniform mass distribution shown in Fig. 21.15a rests on a rough ground surface. The width of the crate is 1 m. The crate is exposed to a wind of constant velocity $v$ with the direction shown in the figure. The mass of the crate is 25 kg, and standard air, with the properties given in Prob. 21.14, may be assumed. Estimate the value of $v$ that will cause tipping motion of the crate to be impending.

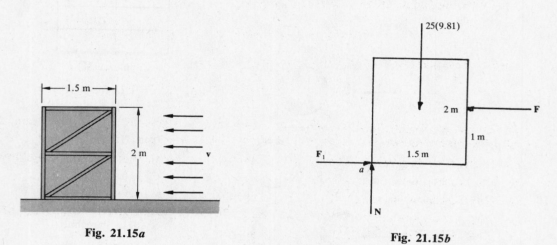

**Fig. 21.15a**          **Fig. 21.15b**

❚ Since the ground surface is rough, the crate is assumed to tip before it slides. Figure 21.15b shows the free-body diagram for impending tipping motion about edge $a$. For equilibrium,

$$\sum M_a = 0 \qquad -25(9.81)\frac{1.5}{2} + F(1) = 0 \qquad F = 184 \text{ N}$$

For a rectangular plate with $a/b = 2/1 = 2$, $C_D$ is found from Fig. 21.13c to be $C_D \approx 1.16$. Using the equation for the drag force,

$$F = C_D[\tfrac{1}{2}\rho v^2]A \qquad 184 = 1.16[\tfrac{1}{2}(1.22)v^2]2(1) \qquad v = 11.4 \text{ m/s} = 41.0 \text{ km/h}$$

**21.16** A large cylindrical storage tank is transported on a tractor-drawn flatbed trailer, as shown in Fig. 21.16. Standard air, with the properties given in Prob. 21.14, is assumed.

(a) Estimate the value of the drag force of the atmosphere on the tractor and trailer when this unit travels at 25 mi/h.

(b) Find the power loss caused by the drag force in part (a).

❚ (a) It is assumed that the plane area of the end of the tank is the major component of the resistance to motion. Thus, the problem is that of a right circular cylinder moving in the direction of its axis. The length-over-diameter ratio of the cylindrical tank is

$$\frac{l}{d} = \frac{16}{12} = 1.33$$

**Fig. 21.16**

**TABLE 21.2**

| $l/d$ | $C_D$ |
|---|---|
| 1 | 0.91 |
| 1.33 | $C_D$ |
| 2 | 0.85 |

Table 21.2 is constructed, using the values of $C_D$ for $l/d = 1$ and $l/d = 2$. Using linear interpolation,

$$\frac{2 - 1.33}{2 - 1} = \frac{C_D - 0.85}{0.91 - 0.85} \qquad C_D = 0.890$$

The velocity of the trailer is

$$25 \text{ mi/h} = 36.7 \text{ ft/s}$$

The magnitude of the drag force $F_D$ is given by

$$F_D = C_D(\tfrac{1}{2}\rho v^2)A = 0.890\left[\frac{1}{2}\left(\frac{0.07637}{32.2}\right)36.7^2\right]\frac{\pi(12)^2}{4} = 161 \text{ lb}$$

(**b**)  The power loss $P$ caused by the drag force in part (*a*) is

$$P = F_D v = \frac{161(36.7)}{550} = 10.7 \text{ hp}$$

**21.17**  The van shown in Fig. 21.17*a* travels along a horizontal roadway with constant speed.  The frontal area is 36 ft$^2$, and the weight of the van is 2,600 lb.  Standard air, with the properties in Prob. 21.14, is assumed.

(**a**)  Estimate the drag force, and the power required to overcome this force, when the van travels at 30 mi/h.

(**b**)  Do the same as in part (*a*), for a speed of 50 mi/h.

(**c**)  Do the same as in part (*a*), for a speed of 70 mi/h.

(**d**)  The van rolls down the incline shown in Fig. 21.17*b*.  Find the value of $\beta$ for which the van would move at a constant speed of 20 mi/h.

(**e**)  Find the power dissipation that corresponds to the condition in part (*d*).

**Fig. 21.17*a***

**Fig. 21.17*b***

▮ (**a**)  The van is assumed to have the approximate shape of a cylinder, with $l/d = 3$.  From linear interpolation of the table in Fig. 21.13*c*,

$$C_D = \frac{0.85 + 0.87}{2} = 0.86$$

The velocity of the van is

$$30 \text{ mi/h} = 44 \text{ ft/s}$$

The magnitude of the drag force acting on the van is given by

$$F_D = C_D(\tfrac{1}{2}\rho v^2)A = 0.86\left[\frac{1}{2}\left(\frac{0.07637}{32.2}\right)44^2\right]36 = 71.1\,\text{lb}$$

The power required to overcome the drag force is found to be

$$P = F_D v = \frac{71.1(44)}{550} = 5.69\,\text{hp}$$

(b)  Using  $50\,\text{mi/h} = 73.3\,\text{ft/s}$,  the magnitude of the drag force is found as

$$F_D = C_D(\tfrac{1}{2}\rho v^2)A = 0.86\left[\frac{1}{2}\left(\frac{0.07637}{32.2}\right)73.3^2\right]36 = 197\,\text{lb}$$

The power required to overcome the drag force is

$$P = F_D v = \frac{197(73.3)}{550} = 26.3\,\text{hp}$$

(c)  At  $70\,\text{mi/h} = 103\,\text{ft/s}$,  the drag force and power are given by

$$F_D = C_D(\tfrac{1}{2}\rho v^2)A = 0.86\left[\frac{1}{2}\left(\frac{0.07637}{32.2}\right)103^2\right]36 = 390\,\text{lb} \qquad P = F_D v = \frac{390(103)}{550} = 73.0\,\text{hp}$$

It may be seen that the drag force, and power required, increase rapidly with increasing speed.

(d)  At  $20\,\text{mi/h} = 29.3\,\text{ft/s}$,  the magnitude of the drag force is given by

$$F_D = C_D(\tfrac{1}{2}\rho v^2)A = 0.86\left[\frac{1}{2}\left(\frac{0.07637}{32.2}\right)29.3^2\right]36 = 31.5\,\text{lb}$$

Figure 21.17c shows the free-body diagram of a model of the van.  Since the van moves with constant speed, this element is in static equilibrium.  For force equilibrium in the x direction,

$$\sum F_x = 0 \qquad F_D - 2{,}600\sin\beta = 0 \qquad 31.5 = 2{,}600\sin\beta \qquad \beta = 0.694° \approx 1°$$

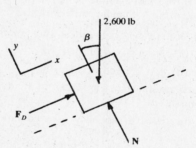

Fig. 21.17c

(e)  The power dissipated by the drag force in part (a) is given by

$$P = F_D v = \frac{31.5(29.3)}{550} = 1.68\,\text{hp}$$

21.18   Figure 21.18 shows a lightweight high-speed elevator for a high-rise building.  The top of the elevator has the form of a square with 2.5-m sides.  The total mass of the elevator and passengers is 1,000 kg.  The height-over-width ratio of the elevator is $1\tfrac{1}{3}$.  The elevator shaftway has unobstructed sides for free circulation of air.  Standard air, with the properties in Prob. 21.14, is assumed.

(a)  Estimate the drag force when the elevator moves upward at 15 m/s.

(b)  Find the power required to overcome the drag force.

(c)  Find the corresponding value of the cable force.

(d)  Find the total power required to raise the elevator at 15 m/s.

▮ (a)  The elevator is approximated as a cylinder, with  $l/d = 1\tfrac{1}{3} = \tfrac{4}{3} = 1.33$.  Table 21.3 is constructed, using values from Fig. 21.13c.  Using linear interpolation,

$$\frac{2 - 1.33}{2 - 1} = \frac{C_D - 0.85}{0.91 - 0.85} \qquad C_D = 0.89$$

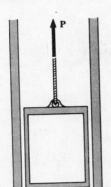

**Fig. 21.18**

**TABLE 21.3**

| $l/d$ | $C_D$ |
|---|---|
| 1 | 0.91 |
| 1.33 | $C_D$ |
| 2 | 0.85 |

The drag force which acts on the elevator is given by

$$F_D = C_D(\tfrac{1}{2}\rho v^2)A = 0.89[\tfrac{1}{2}(1.22)15^2]2.5^2 = 763 \text{ N}$$

(*b*) The power required to overcome the drag force is given by

$$P_D = F_D v = 763(15) = 11,400 \text{ W} = 11.4 \text{ kW}$$

(*c*) The force $F_C$ in the cable is the sum of the weight force and the drag force, given by

$$F_C = 1,000(9.81) + 763 = 10,600 \text{ N}$$

(*d*) The total power required to raise the elevator at 15 m/s is

$$P_{total} = F_C v = 10,600(15) = 159,000 \text{ W} = 159 \text{ kW}$$

**21.19** Figure 21.19 shows a diving bell on the floor of the ocean. The bell has a spherical shape, is 8 ft in diameter, and weighs 40 tons. The specific weight of seawater is 64 lb/ft³.

(*a*) Find the value of the drag force, when the diving bell is raised at the rate of 100 ft/min.

(*b*) Find the cable force required to raise the bell at the rate of 100 ft/min.

(*c*) Find the power required for the condition of part (*b*).

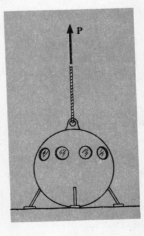

**Fig. 21.19**

▌ (*a*) The drag coefficient is approximated as the drag coefficient of an open hemispherical shell with the value $C_D = 0.41$, as given in Fig. 21.13*c*. The speed at which the diving bell is raised is $v = 100 \text{ ft/min} = 1.67 \text{ ft/s}$. The drag force is given by

$$F_D = C_D(\tfrac{1}{2}\rho v^2)A = 0.41\left[\frac{1}{2}\left(\frac{64}{32.2}\right)1.67^2\right]\frac{\pi(8)^2}{4} = 57.1 \text{ lb}$$

The static weight of the diving bell is

$$W = 40(2,000) = 80,000 \text{ lb}$$

$F$ is the buoyant force on the diving bell, and is equal to the weight of seawater displaced. The value of $F$ is given by

$$F = \gamma V = 64(\tfrac{4}{3})\pi(4)^3 = 17,200 \text{ lb}$$

The resultant cable force $F_{\text{total}}$ required to raise the diving bell at the rate of 100 ft/min is given by

$$F_{\text{total}} = W - F + F_D = 80,000 - 17,200 + 57.1 \qquad P = 62,900 \text{ lb}$$

It may be seen that the drag force is very small compared with the weight and buoyant forces acting on the diving bell.

(c) The power required to raise the diving bell is given by

$$P = F_{\text{total}}v = \frac{62,900(100)}{33,000} = 191 \text{ hp}$$

21.20 Figure 21.20 shows a proposed device to stir liquid in a shallow tank. The wheels rotate at 50 r/min and the length of each blade is 5 ft. The liquid is carbon tetrachloride, with a specific weight of 99.5 lb/ft³. Estimate the power required to stir the liquid.

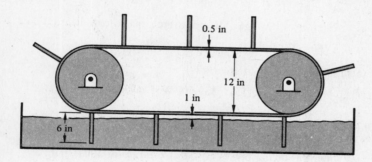

Fig. 21.20

**TABLE 21.4**

| $a/b$ | $C_D$ |
|---|---|
| 8 | 1.23 |
| 12 | $C_D$ |
| 25 | 1.57 |

❚ The power requirement is computed for the case of four blades submerged in the liquid, in the position shown in Fig. 21.20. The area of each blade is

$$A = \tfrac{5}{12}(5) = 2.08 \text{ ft}^2 \text{ per blade}$$

The length-over-width ratio of each blade is

$$\frac{a}{b} = \frac{5(12)}{5} = 12$$

The data from Fig. 21.13c, for a rectangular plate, are used. Table 21.4 shows the values of $C_D$ for $a/b$ ratios of 8 and 25. Using linear interpolation,

$$\frac{25 - 12}{25 - 8} = \frac{1.57 - C_D}{1.57 - 1.23} \qquad C_D = 1.31$$

The translational velocity of the blades is given by

$$v = r\omega = \frac{6}{12}\left[\frac{50(2\pi)}{60}\right] = 2.62 \text{ ft/s}$$

The drag force which acts on each blade is

$$F_D = C_D(\tfrac{1}{2}\rho v^2)A = 1.31\left[\frac{1}{2}\left(\frac{99.5}{32.2}\right)2.62^2\right]2.08 = 28.9 \text{ lb per blade}$$

The power required to stir the liquid is

$$P = F_D v = \frac{4(28.9)2.62}{550} = 0.551 \text{ hp}$$

21.21    Figure 21.21$a$ shows a device, referred to as a cup anenometer, which is used to measure wind velocity.  If the anenometer shaft is held stationary, with the orientation of the wind direction shown in Fig. 21.21$b$, estimate the torque, in inch-ounces, exerted by the wind forces on the shaft.  The wind velocity is 35 mi/h.  Standard air, with the properties given in Prob. 21.14, is assumed, and   16 oz = 1 lb.

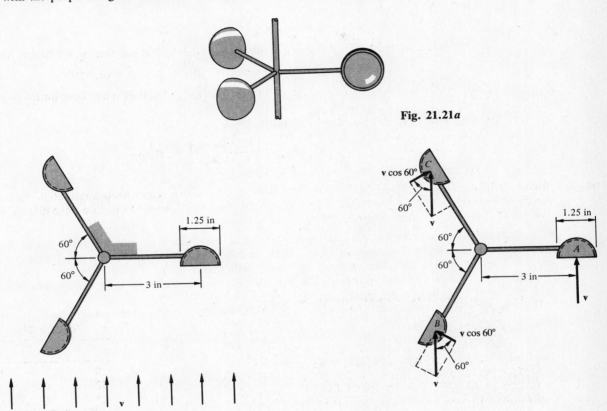

Fig. 21.21$a$

Fig. 21.21$b$

Fig. 21.21$c$

❚  The three cups are designated $A$, $B$, and $C$, as shown in Fig. 21.21$c$.  Cup $A$ behaves as an open hemispherical shell, and the drag coefficient for this shape, from Fig. 21.13$c$, is  $C_D = 1.35$.  The velocity of the wind stream past the anenometer is

$$v = \left(35\ \frac{mi}{h}\right)\left(\frac{5,280\ ft/mi}{3,600\ s/h}\right) = 51.3\ ft/s$$

The force $F_A$ which acts on cup $A$ is

$$F_A = C_D(\tfrac{1}{2}\rho v^2)A = 1.35\left[\frac{1}{2}\left(\frac{0.07637}{32.2}\right)(51.3)^2\right]\frac{\pi(1.25)^2}{4(144)} = 0.0359\ lb = 0.574\ oz$$

The velocity of the wind stream passing cups $B$ and $C$ is resolved into components along the arms and normal to these directions.  The drag forces caused by the components of the wind velocity along the arm are assumed to offer a negligible contribution to the torque which acts on the shaft.

Cups $B$ and $C$ are assumed to act as open hemispherical shells with their convex sides exposed to the flow.  From Fig. 21.13$c$,  $C_D = 0.41$  and

$$v = 51.3\cos 60° = 25.7\ ft/s$$

The drag forces on cups $B$ and $C$, acting normal to the directions of these arms, are

$$F_B = F_C = C_D(\tfrac{1}{2}\rho v^2)A = 0.41\left[\frac{1}{2}\left(\frac{0.07637}{32.2}\right)(25.7)^2\right]\frac{\pi(1.25)^2}{4(144)} = 0.00274\ lb = 0.0438\ oz$$

The torque exerted by the three cup forces on the shaft is

$$M = 0.574(3) - 2[0.0438(3)] = 1.46\ in \cdot oz$$

**21.22** Figure 21.22 shows a stirring apparatus which is used in a chemical manufacturing operation. The paddle assembly consists of four vertical steel blades which rotate about a fixed vertical axis. The entire assembly is submerged in a large tank which contains a chemical solution with a specific gravity of 1.08. Estimate the power required to drive the paddlewheel at a constant speed of 30 r/min.

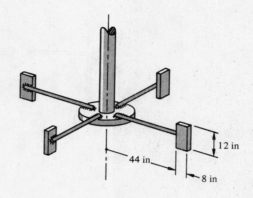

12 in

44 in

8 in

**Fig. 21.22**

❙ Each blade is analyzed as a rectangular plate, normal to the direction of motion, which moves through a fluid. It is assumed that the disturbance produced in the fluid by any one blade has no effect on the flow field about the remaining blades. The value of the drag coefficient is found from Fig. 21.13c. The $a/b$ ratio is $12/8 = 1.5$. Since this value is close to 1, and $C_D$ changes by only 0.01 for a change in $a/b$ from 1 to 4, the drag coefficient is taken as $C_D \approx 1.16$.

The velocity of each blade will be taken as the average velocity at the midpoint of the blade. Thus,

$$v = r\omega = \frac{44 + 4}{12}(30)\left(\frac{2\pi}{60}\right) = 12.6 \text{ ft/s}$$

The total drag force which acts on each blade is then

$$F_D = C_D(\tfrac{1}{2}\rho v^2)A$$

Using $62.4 \text{ lb/ft}^3$ as the specific weight of water, we get

$$F_D = 1.16\left\{\frac{1}{2}\left[\frac{1.08(62.4)}{32.2}(12.6)^2\right]\right\}\frac{12(8)}{144} = 128 \text{ lb}$$

The torque transmitted to the shaft by the force which acts on each blade is

$$M = 128(4) = 512 \text{ ft}\cdot\text{lb per blade}$$

The total power required to drive the four-bladed paddle assembly is

$$P = 4M\omega = \frac{4(512)[30(2\pi)/60]}{550} = 11.7 \text{ hp}$$

**21.23** Figure 21.23a shows a stirring blade arrangement used to agitate a water-soluble chemical solution. The constant power input into the shaft is 3,400 W.

(*a*)  Estimate the angular velocity of the shaft.

(*b*)  Estimate the force exerted on each of the disk-shaped blades.

❙ (*a*)  The chemical solution is assumed to have the density of water. Each disk is assumed to be in translational motion with an average velocity $v$ of its center. It is assumed that the disturbance produced in the fluid by any one blade has no effect on the flow field about the remaining blades. Figure 21.23b shows the force acting on a disk, and the velocity of this element.
For a disk, from Fig. 21.13c,

$$C_D = 1.11$$

The drag force on each disk, using $\rho = 1,000 \text{ kg/m}^3$ for water, is given by

$$F_D = C_D(\tfrac{1}{2}\rho v^2)A = 1.11[\tfrac{1}{2}(1,000)v^2]\pi(250/1,000)^2/4 = 27.2v^2 \qquad \text{N} \qquad\qquad (1)$$

$M$, the driving torque acting on the shaft, is given by

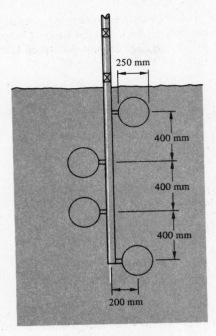

Fig. 21.23a

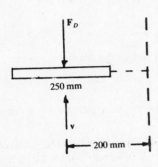

Fig. 21.23b

$$M = 4F_D \frac{200}{1,000} = 0.8F_D \qquad \text{N} \cdot \text{m} \tag{2}$$

$F_D$ is eliminated between Eqs. (1) and (2), with the result

$$M = 0.8(27.2v^2) \qquad \text{N} \cdot \text{m} \tag{3}$$

The power to drive the shaft is given by

$$P = M\omega \tag{4}$$

$M$ is eliminated between Eqs. (3) and (4), to obtain

$$P = 3,400 = 0.8(27.2v^2)\omega \tag{5}$$

Using $v = r\omega = (200/1,000)\omega = 0.2\omega$ in Eq. (5) results in

$$3,400 = 0.8(27.2)(0.2\omega)^2\omega \qquad \omega = 15.7 \, \text{rad/s} = 150 \, \text{r/min}$$

(b) Using Eq. (1),

$$F_D = 27.2v^2 = 27.2\left[\frac{200}{1,000}(15.7)\right]^2 = 268 \, \text{N}$$

**21.24** Find the velocity and displacement of a body which is acted on by a constant force with a sense the same as that of the motion, and a quadratic resisting force.

▌ In deriving the forms for the velocity and displacement for the case of a body which is acted on by quadratic resisting forces, a distinction must be made between applied forces which *assist* the motion and those which *oppose* the motion. The case is first considered where the external applied force assists the motion. The free-body diagram for this case is shown in Fig. 21.24a. The applied force $P$ has a *constant* magnitude, and the drag force $F_D$ was defined in Prob. 21.13. The equation of motion is

$$P + F_D = m\frac{dv}{dt} \qquad P - C_D(\tfrac{1}{2}\rho v^2)A = m\frac{dv}{dt}$$

$$\frac{dv}{dt} = \frac{P}{m} - \frac{C_D\rho A}{2m}v^2 \tag{1}$$

Using the notations

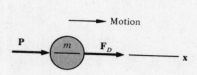

**Fig. 21.24a**

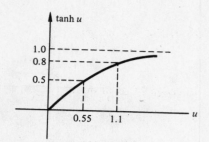

**Fig. 21.24b**

$$\xi = \frac{P}{m} \qquad \zeta = \frac{C_D \rho A}{2m} \tag{2}$$

Eq. (1) has the form

$$\frac{dv}{dt} = \xi - \zeta v^2 \tag{3}$$

The terminal velocity $v_T$ is attained when the right side of Eq. (3) is equal to zero, or

$$\xi - \zeta v_T^2 = 0 \qquad v_T = \sqrt{\frac{\xi}{\zeta}} \tag{4}$$

When the body reaches its terminal velocity $v_T$, the resultant force acting on it is zero, since the applied force $P$ is exactly balanced by the quadratic resisting force $C_D(\frac{1}{2}\rho v_T^2)A$.

Equation (3) may be written in the form

$$dt = \frac{dv}{\xi - \zeta v^2} = \frac{dv}{\xi[1 - (\zeta/\xi)v^2]} = \frac{dv}{\xi[1 - (v/v_T)^2 - 2]} \tag{5}$$

At the beginning of the time interval of interest, $t = 0$ and $v = v_0$, and $v$ and $t$ are the values at the end of this interval. Equation (5) is now integrated, with the result

$$\int_0^t dt = \frac{1}{\xi} \int_{v_0}^v \frac{dv}{1 - (v/v_T)^2} \tag{6}$$

Equation (6) is of the form

$$\int \frac{du}{1 - u^2} = \tanh^{-1} u + \text{constant} \qquad u < 1$$

where $\tanh^{-1} u$ is the inverse hyperbolic tangent and $\tanh u$ is defined by

$$\tanh u = \frac{e^u - e^{-u}}{e^u + e^{-u}}$$

A plot of $\tanh u$, for positive $u$, is shown in Fig. 21.24b. It may be seen that, for large values of $u$, $\tanh u \rightarrow 1$.

The substitution

$$u = \frac{v}{v_T} \qquad du = \frac{1}{v_T} dv$$

is next made. Since, from physical considerations, $v < v_T$, it follows that $u < 1$.

Equation (6) may be written as

$$\int_0^t dt = \frac{v_T}{\xi} \int_{v_0}^v \frac{(1/v_T)\, dv}{1 - (v/v_T)^2}$$

$$t \Big|_0^t = \frac{v_T}{\xi} \tanh^{-1} \frac{v}{v_T} \Big|_{v_0}^v \qquad t = \frac{v_T}{\xi} \left( \tanh^{-1} \frac{v}{v_T} - \tanh^{-1} \frac{v_0}{v_T} \right) \qquad v < v_T \tag{7}$$

Equation (7) may be solved for $v$, with the result

$$v = v_T \tanh \left( \frac{\xi t}{v_T} + \tanh^{-1} \frac{v_0}{v_T} \right) \qquad v < v_T$$

The displacement of the body is found from

$$dx = v\, dt$$

Using Eq. (5), we get

$$dx = \frac{v\,dv}{\xi[1-(v/v_T)^2]} \qquad \int_0^x dx = \frac{1}{\xi}\int_{v_0}^v \frac{v\,dv}{1-(v/v_T)^2} \tag{8}$$

Equation (8) is in the form

$$\int \frac{du}{u} = \ln u + \text{constant} \tag{9}$$

and it may be written as

$$\int_0^x dx = \frac{v_T^2}{(-2)\xi}\int_{v_0}^v \frac{(-2)v\,dv}{v_T^2 - v^2}$$

$$x\Big|_0^x = \frac{-v_T^2}{2\xi}\ln(v_T^2 - v^2)\Big|_{v_0}^v = -\frac{v_T^2}{2\xi}[\ln(v_T^2-v^2) - \ln(v_T^2 - v_0^2)] = \frac{v_T^2}{2\xi}\ln\frac{v_T^2 - v_0^2}{v_T^2 - v^2} \qquad v < v_T$$

If the body starts from rest, with $v_0 = 0$, the velocity and displacement have the simplified forms

$$v = v_T \tanh\frac{\xi t}{v_T} \tag{10}$$

$$x = \frac{v_T^2}{2\xi}\ln\frac{v_T^2}{v_T^2 - v^2} \tag{11}$$

Equation (10) relates the time $t$ of motion to the velocity $v$, while Eq. (11) relates the displacement $x$ to the velocity $v$. If $v$ is eliminated between these two equations, it can be shown that the result is

$$x = \frac{v_T^2}{\xi}\ln\left(\cosh\frac{\xi t}{v_T}\right) \tag{12}$$

The results for the case where the applied constant force has the same sense as the velocity are summarized below.

$$v_T = \sqrt{\frac{\xi}{\zeta}} \qquad \xi = \frac{P}{m} \qquad \zeta = \frac{C_D \rho A}{2m} \tag{13}$$

For $v_0 \neq 0$:

$$v = v_T \tanh\left(\frac{\zeta t}{v_T} + \tanh^{-1}\frac{v_0}{v_T}\right) \tag{14}$$

$$x = \frac{v_T^2}{2\xi}\ln\frac{v_T^2 - v_0^2}{v_T^2 - v^2} \tag{15}$$

For $v_0 = 0$:

$$v = v_T \tanh\frac{\xi t}{v_T} \tag{16}$$

$$x = \frac{v_T^2}{2\xi}\ln\frac{v_T^2}{v_T^2 - v^2} \tag{17}$$

$$x = \frac{v_T^2}{\xi}\ln\left(\cosh\frac{\xi t}{v_T}\right) \tag{18}$$

**21.25**  Find the velocity and displacement of a body which is acted on by a constant force with a sense opposite to that of the motion, and a quadratic resisting force.

❚ Figure 21.25 shows the free-body diagram for the case where the constant applied force $P$ opposes the motion of the body. The drag force, as before, is positive when it acts in the positive coordinate sense, with the functional form given in Prob. 21.13 by

$$F_D = -C_D(\tfrac{1}{2}\rho v^2)A$$

The equation of motion is

$$-P + F_D = m\frac{dv}{dt} \qquad -P - C_D(\tfrac{1}{2}\rho v^2)A = m\frac{dv}{dt} \qquad \frac{dv}{dt} = -\frac{P}{m} - \frac{C_D \rho A}{2m}v^2 \tag{1}$$

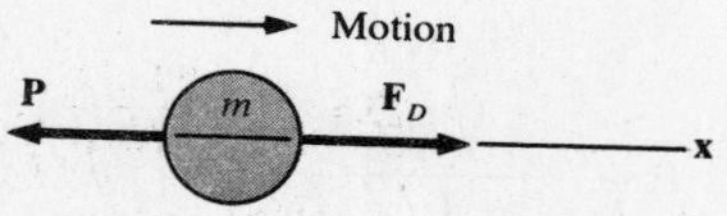

Fig. 21.25

By using the quantities $\xi$ and $\zeta$, defined in Prob. 21.24 as

$$\xi = \frac{P}{m} \qquad \zeta = \frac{C_D \rho A}{2m} \tag{2}$$

Eq. (1) appears as

$$\frac{dv}{dt} = -\xi - \zeta v^2 \tag{3}$$

Inspection of Eq. (3) leads to a very significant conclusion for the case where the applied force *opposes* the motion. Since the right side of the equation is *always* negative, never zero, it follows that

1. The rate of change of the velocity is always negative, so that the velocity may only decrease from some initial value.

2. A terminal velocity *does not exist for* this type of problem. The parameters $\xi$ and $\zeta$ were related to each other in Eq. (4) in Prob. 21.24, with the form

$$v_T = \sqrt{\frac{\xi}{\zeta}} \tag{4}$$

For the case in Prob. 21.24, where the applied force assisted motion, the quantity $v_T$ had the interpretation of a terminal velocity. Although a terminal velocity does *not* exist in the present problem, Eq. (4) will be used in the form

$$v_T^* = \sqrt{\frac{\xi}{\zeta}} \tag{5}$$

The asterisk is used to emphasize that the quantity $v_T^*$ is *not* a terminal velocity, but rather a quantity which is defined by the above equation. By using this result, Eq. (3) may be written as

$$dt = \frac{-dv}{\xi[1 + (v/v_T^*)^2]} \tag{6}$$

$$\int_0^t dt = -\frac{1}{\xi} \int_{v_0}^v \frac{dv}{1 + (v/v_T^*)^2} \tag{7}$$

Equation (7) is of the form

$$\int \frac{du}{a + bu^2} = \frac{1}{\sqrt{ab}} \tan^{-1} \sqrt{\frac{b}{a}}\, u + \text{constant} \qquad a > 0,\ b > 0$$

With the substitutions

$$a = 1 > 0 \qquad b = \frac{1}{(v_T^*)^2} > 0 \qquad u = v \qquad du = dv$$

Eq. (7) may be integrated directly, with the result

$$t \Big|_0^t = -\frac{1}{\xi} \left(\frac{1}{1/v_T^*}\right) \tan^{-1} \sqrt{\frac{1}{(v_T^*)^2}}\, v \Big|_{v_0}^v \qquad t = \frac{v_T^*}{\xi} \left(\tan^{-1} \frac{v_0}{v_T^*} - \tan^{-1} \frac{v}{v_T^*}\right)$$

This equation may be solved for $v$, with the result

$$v = v_T^* \tan\left(\tan^{-1} \frac{v_0}{v_T^*} - \frac{\xi t}{v_T^*}\right)$$

The values of the arctangent functions in the above two equations must be expressed in radians.

When the body comes to rest, $v = 0$ and the corresponding time $t_m$ may be found from the above equation as

$$t_m = \frac{v_T^*}{\xi} \tan^{-1} \frac{v_0}{v_T^*} \tag{8}$$

The displacement is found by using Eq. (6), together with

$$dx = v\, dt$$

in the form

$$dx = \frac{-v\, dv}{\xi[1 + (v/v_T^*)^2]} \tag{9}$$

Equation (9) is in the form of Eq. (8) in Prob. 21.24, and

$$\int_0^x dx = -\frac{1}{\xi}\int_{v_0}^v \frac{v\, dv}{1 + (v/v_T^*)^2} = -\frac{1}{\xi}\frac{(v_T^*)^2}{2}\int_{v_0}^v \frac{[2/(v_T^*)^2]v\, dv}{1 + (v/v_T^*)^2}$$

$$x\Big|_0^x = x = -\frac{(v_T^*)^2}{2\xi}\ln\left[1 + \left(\frac{v}{v_T^*}\right)^2\right]_{v_0}^v = -\frac{(v_T^*)^2}{2\xi}\left\{\ln\left[1 + \left(\frac{v}{v_T^*}\right)^2\right] - \ln\left[1 + \left(\frac{v_0}{v_T^*}\right)^2\right]\right\}$$

$$= \frac{(v_T^*)^2}{2\xi}\ln\frac{(v_T^*)^2 + v_0^2}{(v_T^*)^2 + v^2} \tag{10}$$

When the body comes to rest, $v = 0$ and the corresponding displacement $x_m$ is found from Eq. (10) as

$$x_m = \frac{(v_T^*)^2}{2\xi}\ln\frac{(v_T^*)^2 + v_0^2}{(v_T^*)^2} \tag{11}$$

The results for the case where the applied constant force has a sense opposite that of the velocity are summarized below.

$$v_T^* = \sqrt{\frac{\xi}{\zeta}} \qquad \xi = \frac{P}{m} \qquad \zeta = \frac{C_D\rho A}{2m} \tag{12}$$

$$v = v_T^*\tan\left(\tan^{-1}\frac{v_0}{v_T^*} - \frac{\xi t}{v_T^*}\right) \tag{13}$$

$$x = \frac{(v_T^*)^2}{2\xi}\ln\frac{(v_T^*)^2 + v_0^2}{(v_T^*)^2 + v^2} \tag{14}$$

The time and displacement when the body comes to rest are

$$t_m = \frac{v_T^*}{\xi}\tan^{-1}\frac{v_0}{v_T^*} \tag{15}$$

$$x_m = \frac{(v_T^*)^2}{2\xi}\ln\frac{(v_T^*)^2 + v_0^2}{(v_T^*)^2} \tag{16}$$

**21.26** Figure 21.26 shows a parachutist in free flight in the absence of any cross winds. The weight of the person is 225 lb, and the weight of the chute and rigging is 25 lb.

(a) Find the terminal velocity of the parachutist. Assume standard air with the properties given in Prob. 21.14.

(b) Do the same as in part (a) if a person of slight build, who weighs 100 lb, uses the parachute.

▌ (a) The weight $W$ of the person and the parachute is a force which assists the motion, so that

$$P = W$$

← 20 ft →

**Fig. 21.26**

The projected area of the parachute is

$$A = \frac{\pi (20)^2}{4} = 314 \text{ ft}^2$$

The value of the drag coefficient, from Fig. 21.13c, is 1.35. Using Eqs. (13) in Prob. 21.24,

$$\xi = \frac{P}{m} = \frac{W}{W/g} = g = 32.2 \text{ ft/s}^2 \qquad \zeta = \frac{C_D \rho A}{2m} = \frac{1.35(0.07637/32.2)(314)}{2(250/32.2)} = 0.0647 \text{ ft}^{-1}$$

$$v_T = \sqrt{\frac{\xi}{\zeta}} = \sqrt{\frac{32.2}{0.0647}} = 22.3 \text{ ft/s} = 15.2 \text{ mi/h}$$

(b) With a chutist of weight 100 lb, the total weight is 125 lb. Thus

$$\zeta = \frac{C_D \rho A}{2m} = \frac{1.35(0.07637/32.2)(314)}{2(125/32.2)} = 0.129 \text{ ft}^{-1}$$

$$v_T = \sqrt{\frac{\xi}{\zeta}} = \sqrt{\frac{32.2}{0.129}} = 15.8 \text{ ft/s} = 10.8 \text{ mi/h}$$

The terminal velocity of the heavier person is greater than that of the person of slight build by

$$\%D = \frac{22.3 - 15.8}{15.8} (100) = 41\%$$

**21.27** Figure 21.27a shows a steel boat anchor, with a density of 7,830 kg/m³ and a mass of 25 kg. The anchor is released with zero initial velocity at the surface of a lake. The depth of the lake is 10 m. It is assumed that the line attached to the anchor offers no resistance as this latter element sinks into the water. The density of water is 1,000 kg/m³.

(a) Find the velocity and displacement of the anchor at the end of 1, 2, and 3 s.
(b) Find the time at which the anchor strikes the bottom of the lake.
(c) Find the maximum value of the drag force that acts on the anchor during its descent.
(d) Do the same as in parts (a), (b), and (c) if the anchor, with the same dimensions, is made of lead with a density of 11,400 kg/m³.

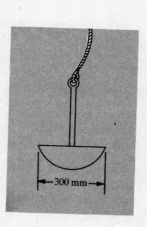

Fig. 21.27a

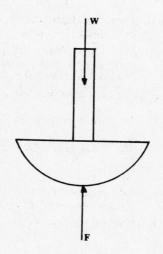

Fig. 21.27b

▮ (a) The volume of the anchor is found from

$$m = \rho V \qquad 25 = 7,830 V \qquad V = 0.00319 \text{ m}^3$$

F is the buoyant force, shown in Fig. 21.27b, which acts on the anchor. This force is equal to the weight of the water displaced, with the value

$$F = \rho g V = 1,000(9.81)(0.00319) = 31.3 \text{ N}$$

The resultant force which acts on the anchor when it is submerged is

$$P = W - F = 25(9.81) - 31.3 = 214 \text{ N} \tag{1}$$

Using Eqs. (13) from Prob. 21.24,

$$\xi = \frac{P}{m} = \frac{214}{25} = 8.56 \text{ m/s}^2 \qquad \zeta = \frac{C_D \rho A}{2m}$$

The shape of the anchor is approximated as an open hemispherical shell, shown in Fig. 21.13c, with $C_D = 0.41$. $\zeta$ then has the value

$$\zeta = \frac{0.41(1,000)[\pi(300/1,000)^2/4]}{2(25)} = 0.580 \text{ m}^{-1}$$

The terminal velocity of the anchor, using Eq. (13) in Prob. 21.24, is given by

$$v_T = \sqrt{\frac{\xi}{\zeta}} = \sqrt{\frac{8.56}{0.580}} = 3.84 \text{ m/s}$$

The velocity and displacement of the anchor, from Eqs. (16) and (17) in Prob. 21.24, have the forms

$$v = v_T \tanh \frac{\xi t}{v_T} \qquad x = \frac{v_T^2}{2\xi} \ln \left( \frac{v_T^2}{v_T^2 - v^2} \right)$$

When $t = 1$ s,

$$v = 3.84 \tanh \frac{8.56(1)}{3.84} = 3.75 \text{ m/s}$$

The above value of $v$ is 97.7 percent of the terminal velocity $v_T$.

$$x = \frac{3.84^2}{2(8.56)} \ln \left( \frac{3.84^2}{3.84^2 - 3.75^2} \right) = 2.65 \text{ m}$$

It is now assumed that $v \approx v_T$ for $t > 1$ s. At $t = 2$ s,

$$v = v_T = 3.84 \text{ m/s} \qquad x = x|_{t=1} + v_T t = 2.65 + 3.84(1) = 6.49 \text{ m}$$

At $t = 3$ s,

$$v = v_T = 3.84 \text{ m/s} \qquad x = x|_{t=2} + v_T t = 6.49 + 3.84(1) = 10.3 \text{ m}$$

(b) Since the depth of the lake is only 10 m, the anchor reaches the bottom at a time less than 3 s. Using $x = 10$ m,

$$x = x|_{t=1} + v_T(t - 1) \qquad 10 = 2.65 + 3.84(t - 1) \qquad t = 2.91 \text{ s}$$

The anchor strikes the bottom of the lake at $t = 2.91$ s.

(c) The drag force is maximum at the terminal velocity, with the value

$$F_D = C_D(\tfrac{1}{2}\rho v^2)A = 0.41[\tfrac{1}{2}(1,000)3.84^2]\left[ \frac{\pi(300/1,000)^2}{4} \right] = 214 \text{ N}$$

The above result, from comparison with Eq. (1), satisfies

$$F_D = P = 214 \text{ N}$$

This is an expected result, based on the requirement of static equilibrium of the anchor when this element moves with the constant velocity $v = v_T$.

(d) When the anchor is made of lead, the mass is

$$m = \rho V = 11,400(0.00319) = 36.4 \text{ kg}$$

The resultant force acting on the anchor is

$$P = W - F = 36.4(9.81) - 31.3 = 326 \text{ N}$$

$\xi$, $\zeta$, and $v_T$ have the values

$$\xi = \frac{P}{m} = \frac{326}{36.4} = 8.96 \text{ m/s}^2 \qquad \zeta = \frac{C_D \rho A}{2m} = \frac{0.41(1,000)[\pi(300/1,000)^2/4]}{2(36.4)} = 0.398 \text{ m}^{-1}$$

$$v_T = \sqrt{\frac{\xi}{\zeta}} = \sqrt{\frac{8.96}{0.398}} = 4.74 \text{ m/s}$$

The velocity and displacement of the anchor are given by

$$v = v_T \tanh \frac{\xi t}{v_T} \qquad x = \frac{v_T^2}{2\xi} \ln \left( \frac{v_T^2}{v_T^2 - v^2} \right)$$

When $t = 1$ s,

$$v = 4.74 \tanh \left[ \frac{8.96(1)}{4.74} \right] = 4.53 \text{ m/s} \qquad x = \frac{4.74^2}{2(8.96)} \ln \left( \frac{4.74^2}{4.74^2 - 4.53^2} \right) = 3.07 \text{ m}$$

When $t = 2$ s,

$$v = 4.74 \tanh \left[ \frac{8.96(2)}{4.74} \right] = 4.74 \text{ m/s} = v_T$$

Since $v = v_T$ Eq. (18), rather than Eq. (17), in Prob. 21.24 is used to find $x$ at $t = 2$ s. The result is

$$x = \frac{v_T^2}{\xi} \ln \left( \cosh \frac{\xi t}{v_T} \right) = \frac{4.74^2}{8.93} \ln \left( \cosh \left[ \frac{8.96(2)}{4.74} \right] \right) = 7.74 \text{ m}$$

For $t > 2$ s, $v = v_T$.
At $t = 3$ s,

$$v = v_T = 4.74 \text{ m/s} \qquad x = x|_{t=2} + v_T t = 7.74 + 4.74(1) = 12.5 \text{ m}$$

Since the depth of the lake is only 10 m, the anchor reaches the bottom at a time less than 3 s. Using $x = 10$ m,

$$x = x|_{t=2} + v_T(t-2) \qquad 10 = 7.74 + 4.74(t-2) \qquad t = 2.48 \text{ s}$$

The above time may be compared with that found in part (b). The lead anchor, made of the denser material, reaches the bottom of the lake in less time than that required for the steel anchor.

The drag force is maximum at the terminal velocity, with the value

$$F_D = C_D(\tfrac{1}{2}\rho v^2)A = 0.41[\tfrac{1}{2}(1,000)4.74^2] \left( \frac{\pi(300/1,000)^2}{4} \right) = 326 \text{ N}$$

It may be seen that the above force is equal to the resultant force $P$ which acts on the anchor.

**21.28** A ball is projected vertically upward from the ground with an initial velocity of 70 ft/s. The 2.5-in-diameter ball weighs 0.1 lb. Standard air, with the properties in Prob. 21.14, is assumed.

(a) Estimate the maximum height which the ball attains, and the time to reach this height.

(b) Do the same as in part (a) for the case where the drag force is assumed to be zero.

(c) Estimate the time for the ball to fall from the maximum height found in part (a) to the ground, and the velocity with which it strikes the ground.

(d) Do the same as in part (c) for the case where the drag force is assumed to be zero.

(e) Find the values of the drag force and the resultant force on the ball at the instant of launch, and when the ball strikes the ground.

(f) Find the total energy expended in overcoming the drag forces.

▌(a) The shape of the ball is approximated as an open hemispherical shell, shown in Fig. 21.13c, with $C_D = 0.41$. Equations (13) in Prob. 21.24 have the forms

$$\xi = \frac{P}{m} = \frac{W}{W/g} = g = 32.2 \text{ ft/s}^2 \qquad \zeta = \frac{C_D \rho A}{2m} = \frac{0.41(0.07637/g)\{\pi(2.5)^2/[4(144)]\}}{2(0.1/g)} = 5.34 \times 10^{-3} \text{ ft}^{-1}$$

Equation (13) in Prob. 21.24, and Eq. (12) in Prob. 21.25, have the form

$$v_T = v_T^* = \sqrt{\frac{\xi}{\zeta}} = \sqrt{\frac{32.2}{5.34 \times 10^{-3}}} = 77.7 \text{ ft/s}$$

Figure 21.28a shows the end of the ascent phase of the flight. From the ground to this position, the weight force opposes the motion. Using Eq. (16) in Prob. 21.25, we find the maximum height attained as

$$x_m = \frac{(v_T^*)^2}{2\xi} \ln \frac{(v_T^*)^2 + v_0^2}{(v_T^*)^2} = \frac{77.7^2}{2(32.2)} \ln \frac{77.7^2 + 70^2}{77.7^2} = 55.7 \text{ ft}$$

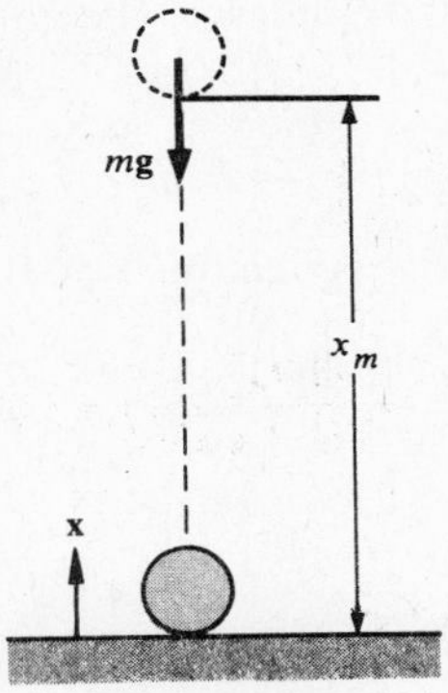

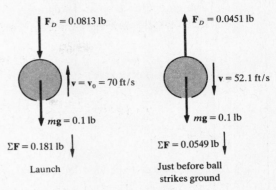

Fig. 21.28a

Fig. 21.28b

The time to reach this height is found from Eq. (15) in Prob. 21.25 as

$$t_m = \frac{v_T^*}{\xi} \tan^{-1} \frac{v_0}{v_T^*} = \frac{77.7}{32.2} \tan^{-1} \frac{70}{77.7} = 1.77 \text{ s} \tag{1}$$

(b) For the ideal case, where the drag force is absent,

$$v^2 = v_0^2 + 2ax \qquad 0 = v_0^2 - 2gx_m \qquad 0 = 70^2 - 2(32.2)x_m \qquad x_m = 76.1 \text{ ft}$$

The time required to reach this height is found from

$$v = v_0 + at \qquad 0 = v_0 - gt_m \qquad 0 = 70 - 32.2t_m \qquad t_m = 2.17 \text{ s} \tag{2}$$

Comparison of Eqs. (1) and (2) reveals that the actual time of flight during ascent is less than the theoretical time. Because of the drag forces, the ball attains a height which is only $(55.7/76.1)100 = 73.2$ percent of the height theoretically possible.

(c) As the ball returns from the maximum height to the ground, the weight force assists the motion. The ball starts this regime of motion with zero initial velocity. Using Eq. (18) in Prob. 21.24, we have

$$x = \frac{v_T^2}{\xi} \ln \left( \cosh \frac{\xi t}{v_T} \right) \qquad 55.7 = \frac{77.7^2}{32.2} \ln \left( \cosh \frac{32.2t}{77.7} \right)$$

$$\ln \left( \cosh \frac{32.2t}{77.7} \right) = 0.297 \qquad \cosh \frac{32.2t}{77.7} = e^{0.297} = 1.35 \qquad \frac{32.2t}{77.7} = 0.814 \qquad t = 1.96 \text{ s}$$

The velocity with which the ball strikes the ground is found from Eq. (16) in Prob. 21.24 as

$$v = v_T \tanh \frac{\xi t}{v_T} = 77.7 \tanh \frac{32.2(1.96)}{77.7} = 52.1 \text{ ft/s}$$

The percent difference between the initial launch velocity and the velocity when the ball returns to the ground is

$$\%\text{D} = \frac{70 - 52.1}{70} (100) = 25.6\%$$

(d) For the case where the drag forces are absent, the descent time is the *same* as the ascent time, given by Eq. (2) as 2.17 s. The velocity with which the ball strikes the ground is the same as the initial velocity of 70 ft/s with which the ball leaves the ground.

(e) The drag force on the ball, using Eq. (1) in Prob. 21.13, is

$$F_D = C_D(\tfrac{1}{2}\rho v^2)A = 0.41 \left[ \frac{1}{2} \left( \frac{0.07637}{32.2} v^2 \right) \right] \frac{\pi(2.5)^2}{4(144)} = 1.66 \times 10^{-5} v^2$$

where $F_D$ is in pounds and $v$ is in feet per second. At the instant of launch

$$v = v_0 = 70 \text{ ft/s} \qquad F_D = 1.66 \times 10^{-5}(70)^2 = 0.0813 \text{ lb}$$

Just before the ball strikes the ground,

$$v = 52.1 \text{ ft/s} \qquad F_D = 1.66 \times 10^{-5}(52.1) = 0.0451 \text{ lb}$$

The free-body diagrams for the above two positions are shown in Fig. 21.28b.

(f) The ground is taken as the datum for potential energy. When the ball leaves the ground, it has zero potential energy and an initial kinetic energy given by

$$T_1 = \tfrac{1}{2}mv_0^2 = \frac{1}{2}\left(\frac{0.1}{32.2}\right)70^2 = 7.61 \text{ ft} \cdot \text{lb}$$

When the ball contacts the ground, it again has zero potential energy, and its kinetic energy is

$$T_2 = \tfrac{1}{2}mv_f^2 = \frac{1}{2}\left(\frac{0.1}{32.2}\right)(52.1)^2 = 4.21 \text{ ft} \cdot \text{lb}$$

The difference

$$\Delta T = 7.61 - 4.21 = 3.4 \text{ ft} \cdot \text{lb}$$

is the energy expended in overcoming the drag forces on the ball. The percent loss of the initial energy is

$$\%D = \frac{3.4}{7.61}(100) = 44.7\%$$

**21.29** A man tosses a beach ball vertically upward. The ball leaves the man's hand at a position 2 m above the ground. The ball reaches a maximum height of 6 m above the ground. The man subsequently catches the ball when this element is 1.5 m above the ground. The mass of the ball is 0.14 kg and the diameter is 100 mm. Standard air, with the properties in Prob. 21.14, is assumed.

(a) Find the initial velocity of the ball, and the time at which the maximum height is reached.

(b) Find the velocity of the ball when the man catches it.

(c) Find the total time of the flight.

(d) A child drops the beach ball from a motel balcony which is 8 m above the ground. Simultaneously, another child drops a small pebble from the same balcony. Estimate the time interval between the time when the ball strikes the ground and the time when the pebble strikes the ground.

▌ (a) The shape of the ball is approximated as an open hemispherical shell, shown in Fig. 21.13c, with $C_D = 0.41$.

$$\xi = \frac{P}{m} = \frac{mg}{m} = g = 9.81 \text{ m/s}^2$$

$$\zeta = \frac{C_D \rho A}{2m} = \frac{0.41(1.22)[\pi(100/1,000)^2/4]}{2(0.14)} = 0.0140 \text{ m}^{-1}$$

$$v_T^* = \sqrt{\frac{\xi}{\zeta}} = \sqrt{\frac{9.81}{0.0140}} = 26.5 \text{ m/s}$$

The upward $x$ displacement is 4 m, and Eq. (16) in Prob. 21.25 is used to find the initial velocity $v_0$. The result is

$$x_m = \frac{(v_T^*)^2}{2\xi}\ln\left[\frac{(v_T^*)^2 + v_0^2}{(v_T^*)^2}\right] \qquad 4 = \frac{26.5^2}{2(9.81)}\ln\left[\frac{26.5^2 + v_0^2}{26.5^2}\right]$$

$$\frac{26.5^2 + v_0^2}{26.5^2} = e^{0.1118} = 1.118 \qquad v_0 = 9.10 \text{ m/s}$$

Using Eq. (15) in Prob. 21.25, the time to reach the maximum height is found to be

$$t_m = \frac{v_T^*}{\xi}\tan^{-1}\frac{v_0}{v_T^*} = \frac{26.5}{9.81}\tan^{-1}\frac{9.10}{26.5} = 0.894 \text{ s}$$

(b) The downward motion from the maximum height starts with zero initial velocity, and the downward $x$ displacement is 4.5 m. Using Eqs. (13) and (17) in Prob. 21.24,

$$v_T = v_T^* = 26.5 \text{ ms/s} \qquad x = \frac{v_T^2}{2\xi}\ln\left(\frac{v_T^2}{v_T^2 - v^2}\right) \qquad (1)$$

$$4.5 = \frac{26.5^2}{2(9.81)}\ln\left(\frac{26.5^2}{26.5^2 - v^2}\right) \qquad \left(\frac{26.5^2}{26.5^2 - v^2}\right) = e^{0.1257} = 1.134 \qquad v = 9.11 \text{ m/s}$$

(c)   The time for the ball to fall from the maximum height until the man catches it is found from Eq. (16) in Prob. 21.24 to be

$$v = v_T \tanh \frac{\xi t}{v_T} \qquad (2)$$

$$9.11 = 26.5 \tanh \frac{9.81t}{26.5} \qquad t = 0.968 \text{ s}$$

The total time of the flight is

$$t_{\text{total}} = 0.894 + 0.968 = 1.86 \text{ s}$$

(d)   The motion of the ball is considered first.   From part (a),

$$\xi = g = 9.81 \text{ m/s}^2 \qquad \zeta = 0.0140 \text{ m}^{-1}$$

From part (b),

$$v_T = 26.5 \text{ m/s}$$

Using Eq. (1) in part (b), with   $x = 8$ m,   gives

$$x = \frac{v_T^2}{2\xi} \ln\left(\frac{v_T^2}{v_T^2 - v^2}\right)$$

$$8 = \frac{26.5^2}{2(9.81)} \ln\left(\frac{26.5^2}{26.5^2 - v^2}\right) \qquad \left(\frac{26.5^2}{26.5^2 - v^2}\right) = e^{0.2235} = 1.250 \qquad v = 11.9 \text{ m/s}$$

Using Eq. (2) in part (b),

$$v = v_T \tanh \frac{\xi t}{v_T} \qquad 11.9 = 26.5 \tanh \frac{9.81t}{26.5} \qquad t = 1.306 \text{ s}$$

The ball strikes the ground 1.306 s after it is dropped from the balcony.
     The motion of the pebble is found next.   It is assumed that the drag force $F_D$ on the pebble satisfies the relationship

$$F_D \ll W$$

where $W$ is the weight of the pebble.   Thus,

$$F_D \approx 0$$

The time for the pebble to fall through a distance of 8 m is found from

$$x = \tfrac{1}{2}gt^2 \qquad 8 = \tfrac{1}{2}(9.81)t^2 \qquad t = 1.277 \text{ s}$$

The interval between the time when the ball strikes the ground and the time when the pebble strikes the ground is given by

$$\Delta t = 1.306 - 1.277 = 0.029 \text{ s}$$

21.30   Give a summary of the basic concepts of rectilinear motion of a body with resisting, or drag, forces.

❚   The sole effect of a drag force which acts on a body is to transform a part of the kinetic energy of the body to heat energy, with an unrecoverable decrease in velocity.
     If the motion of a body is resisted by sliding Coulomb friction forces as the first body slides relative to a second body, then the drag force may be assumed to be a constant, with the form

$$F_D = -\mu N$$

where $N$ is the normal contact force between the sliding surfaces and $\mu$ is the coefficient of friction.   $F_D$ is positive in the same sense as the velocity, and the minus sign indicates that the drag force opposes the motion.
     If viscous resistance to motion is assumed, the form of the drag force is

$$F_D = -cv$$

where $c$ is a constant and $v$ is the velocity of the body.   This type of motion is observed (1) where surfaces which move with low relative velocity are separated by a thin layer of lubricant and (2) in the motion of submerged bodies at low velocities.   The minus sign in the equation for $F_D$ automatically adjusts the sense of the drag force to oppose the motion.

If the predominant drag force effect is due to differences in pressure acting on the body, a quadratic resistance law may be assumed, and the drag force has the form

$$F_D = -C_D(\tfrac{1}{2}\rho v^2)A$$

$A$ is the projected area of the body on a plane which is normal to the direction of the motion, $\rho$ is the mass density of the fluid through which the body moves, and $C_D$ is the drag coefficient. A quadratic resistance law may be assumed if the body has plane or curved surfaces which are normal to the direction of motion, or sharp edges where the fluid loses contact with the body.

When the body reaches a terminal velocity, the resultant force which acts on it is zero.

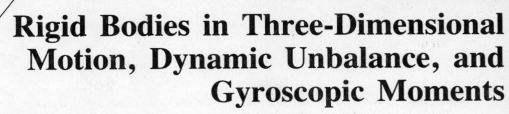

# CHAPTER 22

# Rigid Bodies in Three-Dimensional Motion, Dynamic Unbalance, and Gyroscopic Moments

## 22.1 REVIEW OF THE D'ALEMBERT, OR INERTIA, FORCE; DYNAMIC FORCES CAUSED BY ROTATING OFF-CENTER MASSES—SOLUTIONS BY DIRECT USE OF INERTIA FORCES AND BY INTEGRATION OF THE INERTIA FORCES ACTING ON THE MASS ELEMENTS

**22.1**    (*a*)    What is the definition of *plane motion* of a rigid body?

(*b*)    State the three general types of *three-dimensional motion* of a rigid body.

▌    (*a*)    A *rigid body* is defined to move with plane motion if all points in the body move in planes which are parallel to some fixed reference plane.   For this case, the forces and moments which act on the body are also assumed to lie in a plane which is parallel to the fixed reference plane *and which passes through the center of mass of the body*.   The most general description of the motion of a rigid body in plane motion characterizes the translational motion of the center of mass and rotation of the body with respect to this point.

(*b*)    In the first type of motion, the *axis* about which the rigid body rotates either is fixed in the inertial reference frame or moves so that is is always *parallel* to its original location.   An alternative description of this latter effect is that the axis about which the rigid body rotates can move with rectilinear or curvilinear translation *but cannot change its direction*.   Readers should convince themselves that the description presented thus far also applies to the case of plane motion of a rigid body presented in part (*a*).   The additional statement which distinguishes the three-dimensional problem from the case of plane motion is that, in the three-dimensional case, not all the forces, or moments, which act on the rigid body lie in planes which are normal to the axis of rotation and which pass through the center of mass of the body.

In the second type of rigid-body motion, the body possesses complete symmetry with respect to an axis about which the body rotates.   For this case, the *axis of rotation may have an arbitrary motion*.   Familiar examples of this type of motion are a spinning top and a gyroscope.   The analysis of the above two types of problems in rigid-body dynamics may be extended to include the cases where impulsive forces act on the bodies or where impact of the bodies occurs.

There is a third type of problem in rigid-body dynamics in which a rigid body of arbitrary shape can rotate about an axis which itself possesses any arbitrary motion.   This type of motion is characterized by an airplane in flight, the motion of a spinning artillery shell along its trajectory, or the orbital motion of a space vehicle.   The analysis of these types of problems is beyond the scope of this book, and they are studied extensively in texts in advanced dynamics.

**22.2**    (*a*)    Review the techniques used to show the D'Alembert, or inertia, force acting on a mass particle.

(*b*)    Define the term *centrifugal force*.

▌    (*a*)    The concept of the D'Alembert, or inertia, force was first introduced in Probs. 15.79 and 15.80.   Since the following problems in this chapter make extensive use of inertia forces, a brief review of this theory will be presented.

Figure 22.2*a* shows a particle which is assumed to move with rectilinear translation along the *s* axis.   The acceleration of the particle is *a*, and the positive sense of this term is the same as the positive sense of *s*.

The following three steps must be used when showing the inertia force which acts on a mass particle.

1.    An arrow, which represents a vector and which has the direction and the sense of the acceleration of the particle, is sketched on the particle.   This sense of the acceleration may be that which is defined by the positive sense of a displacement coordinate, such as *s*, or it may be the sense of the assumed, or actual, acceleration.

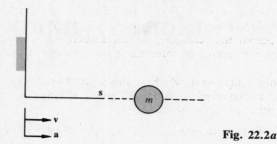

Fig. 22.2a

2. The magnitude of the vector represented by the arrow is defined to be the product of the mass and the acceleration of the particle.

3. A *minus sign* is placed in front of the magnitude which is written in step 2.

These three steps, for the particle in Fig. 22.2a, are shown in Fig. 22.2b.

In the problem above, the quantity $-ma$ is the *inertia force acting on the particle*. When the inertia force is drawn on a mass particle, the original dynamics problem is transformed to a problem in *dynamic equilibrium*, which may be solved by using the equations of static equilibrium. The inertia force is a vector quantity. When a vector quantity is multiplied by $-1$, the *net* effect is to change the sense of the vector. Thus, in Fig. 22.2b, step 3, the *actual* sense of the quantity $ma$ is to the *left*. Since an applied force, acting to the right in the positive *s*-coordinate sense, would be required to produce a positive value of $a$, the leftward-acting force mentioned above may be thought of as a fictitious "applied force" which balances the actual applied force and places the particle in a condition of *dynamic* equilibrium.

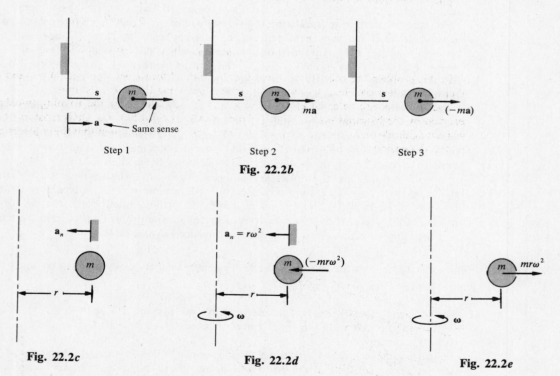

Step 1                Step 2                Step 3

Fig. 22.2b

Fig. 22.2c            Fig. 22.2d            Fig. 22.2e

(b) In the following problems of this chapter the inertia force will be required for a particle which travels in concentric circular paths about a fixed axis. Figure 22.2c shows the typical appearance of such a particle. The actual *sense* of the centripetal acceleration $a_n$ is toward the axis of rotation, as shown in this figure, and

$$a_n = r\omega^2$$

The inertia force, which is written in accordance with the rules presented above, is shown in Fig. 22.2d. If the minus sign is omitted from this force, the *sense* of the vector is changed, and this construction is shown in Fig. 22.2e. The quantity $mr\omega^2$ is called the *centrifugal force*, and this force acts *radially outward* on the particle.

Throughout the remainder of this chapter the construction shown in Fig. 22.2e will be used to represent the centrifugal force acting on a mass particle which rotates about a *fixed* axis.

**22.3**     Figure 22.3a shows a rigid shaft which rotates in fixed bearings $a$ and $b$ with constant angular velocity. Two point masses $m$ are attached to thin rigid arms, of length $r$, which form part of the shaft. The $x_0y_0z_0$ coordinates are attached to the ground.

**(a)**  Find the dynamic force effects at $a$ and $b$ due to the rotating unbalance.

**(b)**  What is an undesirable effect produced by a rotating unbalanced shaft?

▌ **(a)**  The $xyz$ coordinates are attached to the shaft and thus possess the motion of the shaft. The $x$ and $x_0$ axes are collinear, and both arms lie in the $xy$ plane. The centrifugal forces of the two masses, caused by the angular velocity of the shaft, are shown in the figure in their actual senses. It may be readily verified that the shaft assembly is in static equilibrium about the $x$ axis for any position $\theta$. The mass of the arms is neglected in this analysis.

As the shaft rotates with the two off-center masses, the centrifugal forces produce a couple $M_z$, given by

$$M_z = Fd = (mr\omega^2)d \qquad (1)$$

This couple acts in the $xy$ plane, about the $z$ axis, and is positive in the sense of the positive $z$ axis. Since the $xyz$ axes rotate with the shaft, it follows that the *couple produced by the unbalance forces also rotates with the shaft*. In order to have equilibrium of the shaft when it is rotating, the bearings must exert force reactions $R_{ay}$ and $R_{by}$ on the shaft, in the $y$ direction, as shown in the figure.

The equilibrium requirements of the shaft are

$$\sum F_y = 0 \qquad R_{ay} - R_{by} = 0 \qquad R_{ay} = R_{by} = R$$

$$\sum M_z = 0 \qquad (mr\omega^2)d - Rl = 0 \qquad R = \frac{mrd\omega^2}{l} \qquad (2)$$

In this problem the centrifugal forces are the applied forces acting on the shaft, and the resultant of these forces is a couple. The pair of reaction forces must also be a couple.

The forces $R$ are referred to as the *dynamic forces*, caused by the rotating unbalance, which are exerted by the bearings on the shaft. From Newton's third law, the shaft must exert forces $R$ of the same magnitude and opposite sense on the bearings. It may be seen that these forces are independent of the dimension $c$ and will vanish only if $r$, $d$, or $\omega^2$ is zero.

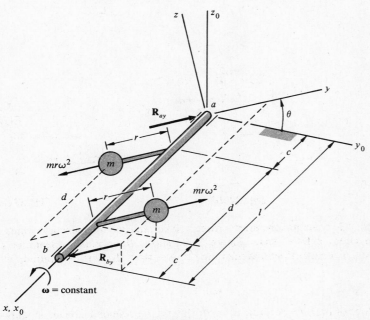

**Fig. 22.3a**

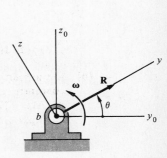

**Fig. 22.3b**

The limiting case of no rotation of the shaft is $\omega = 0$. If $r = 0$, the center of mass of the shaft assembly lies on the axis of rotation. If $d = 0$, there is no separation distance in the $x$ direction between the masses, and the problem reduces to that of plane motion of a rigid body.

(b) An undesirable effect which is produced by a shaft rotating with unbalance will now be considered. Figure 22.3b shows the force $R$ exerted by the shaft on bearing $b$. The $z_0$ axis is assumed to be normal to the ground. The components of $R$ along the $y_0$ and $z_0$ axes are

$$F_{y_0} = R \cos \theta \qquad F_{z_0} = R \sin \theta \qquad (3)$$

The equation for motion with constant angular acceleration is

$$\theta = \theta_0 + \omega_0 t + \tfrac{1}{2} \alpha t^2$$

In the present problem,

$$\alpha = 0 \qquad \omega_0 = \omega = \text{constant} \qquad \theta_0 = 0$$

and therefore

$$\theta = \omega t$$

Equations (3) are now written as

$$F_{y_0} = R \cos \omega t \qquad F_{z_0} = R \sin \omega t$$

Both the sine and cosine functions are periodic. Thus, $\sin \omega t$ and $\cos \omega t$, together with the forces $F_{y_0}$ and $F_{z_0}$, vary periodically with time. This type of motion is referred to as *vibration*. The study of vibrations is a very important part of the synthesis and analysis of mechanical systems.

It may be concluded from the above discussion that *any shaft which rotates with unbalance force will exert pulsating forces on its bearing supports.*

**22.4** The off-center masses on the shaft of Prob. 22.3 are repositioned, as shown in Fig. 22.4a. Compare the dynamic forces exerted on the bearings with those found in Prob. 22.3.

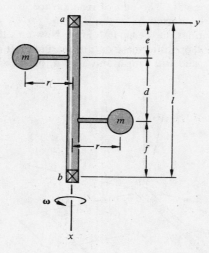

**Fig. 22.4a**

❚ The couple of the centrifugal forces, from Eq. (1) in Prob. 22.3, is a function of the spacing of the unbalance masses with respect to each other. The couple of the dynamic reaction forces of the bearings on the shaft is a function of the constant length $l$ of the shaft. Therefore, it may be concluded that the dynamic forces on the bearings are the *same* as for the case of symmetrical placement of the unbalance masses on the shaft, given by Eq. (2) in Prob. 22.3 as

$$R = \frac{mrd\omega^2}{l}$$

Several other configurations which also have the *same* dynamic bearing forces are shown in Figs. 22.4b, c, and d.

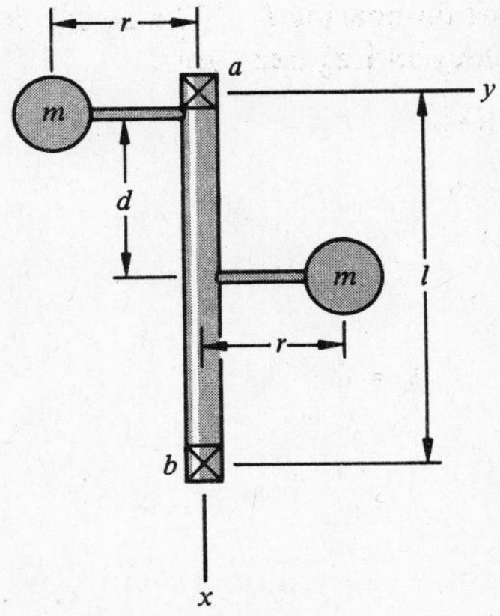

Fig. 22.4b 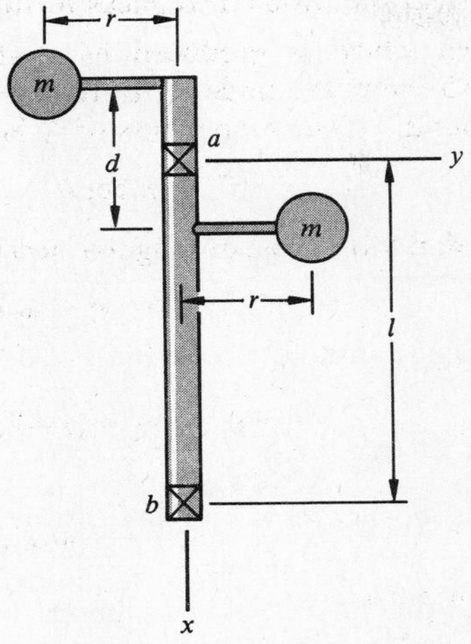 $\qquad$ Fig. 22.4c 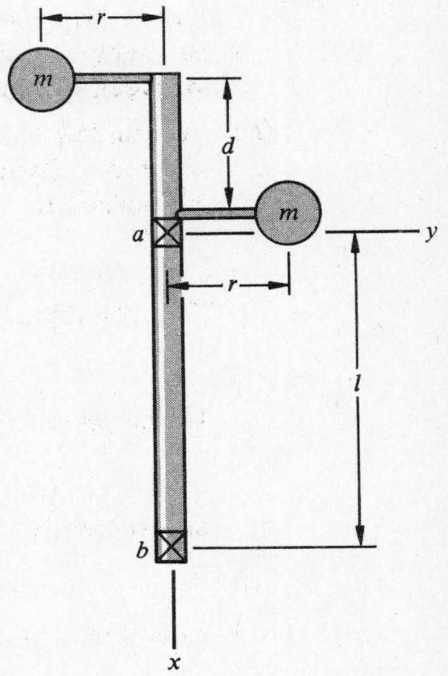 $\qquad$ Fig. 22.4d

**22.5** During the operation of the unbalanced shaft assembly in Prob. 22.3, the mass closest to the $y$ axis suddenly becomes detached and flies away from the assembly.

**(a)** What is the subsequent effect on the dynamic forces acting on bearings $a$ and $b$?

**(b)** Find the numerical values of these dynamic forces if $m = 100\,\text{g}$, $r = 20\,\text{mm}$, $c = 60\,\text{mm}$, $d = 150\,\text{mm}$, $l = 270\,\text{mm}$, and $\omega = 1,500\,\text{r/min}$.

∎ **(a)** The shaft, with the remaining mass, is shown in Fig. 22.5. The equilibrium requirements are

$$\sum M_z = 0 \qquad (mr\omega^2)(c + d) - R_{by}l = 0 \qquad R_{by} = \frac{mr(c + d)\omega^2}{l}$$

$$\sum F_y = 0 \qquad -R_{ay} + mr\omega^2 - R_{by} = 0 \qquad R_{ay} = \frac{mrc\omega^2}{l}$$

In the original condition, in Prob. 22.3, both dynamic bearing forces have the same magnitude, given by

$$R = \frac{mrd\omega^2}{l}$$

From comparison of the above results and using the relationship

$$\frac{c + d}{l} > \frac{d}{l}$$

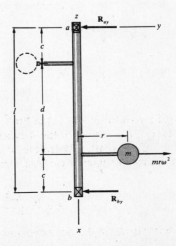

Fig. 22.5

it may be concluded that the dynamic bearing force $R_{by}$ *increases* after the mass is detached. The dynamic force $R_{ay}$ will increase if $c > d$ and decrease if $c < d$. Thus, the loss of one of the masses on the shaft produces an *increase* in at least one of the dynamic bearing forces.

(*b*)   The angular velocity of the rotating assembly is

$$\omega = 1,500\left(\frac{2\pi}{60}\right) = 157 \text{ rad/s}$$

The original values of the dynamic bearing forces are

$$R = \frac{mrd\omega^2}{l} = \frac{(100/1,000)(20/1,000)(150/1,000)(157)^2}{270/1,000} = 27.4 \text{ N}$$

The values of the bearing forces after one mass is detached are

$$R_{ay} = \frac{mrc\omega^2}{l} = \frac{100(20)60(157)^2}{270}\left(\frac{1}{1,000}\right)^2 = 11.0 \text{ N}$$

$$R_{by} = \frac{mr(c+d)\omega^2}{l} = \frac{100(20)(60+150)(157)^2}{270}\left(\frac{1}{1,000}\right)^2 = 38.3 \text{ N}$$

The percent increase in the value of the maximum dynamic force exerted on a bearing is

$$\%\text{D} = \frac{R_{by} - R}{R}(100) = \frac{38.3 - 27.4}{27.4}(100) = 40\%$$

**22.6**   The assembly shown in Fig. 22.6*a* rotates at 1,000 r/min. The two spherical masses are steel, with $\gamma = 0.283 \text{ lb/in}^3$, and these elements may be considered to be point masses attached to massless arms. At $t = 0$, the two masses lie in the $x_0$, $y_0$ plane.

(*a*)   Find the magnitude of the dynamic bearing forces.

(*b*)   Find the $z_0$ and $y_0$ components of the dynamic forces exerted by the shaft on the bearings at $t = 0.95$ s. Express these results in formal vector notation.

(*c*)   Do the same as in parts (*a*) and (*b*) if, at $t = 0$, mass $A$ suddenly becomes detached and flies away from the assembly.

(*d*)   The assembly in Fig. 22.6*a* is initially at rest. A couple about the $x$ axis, of magnitude 0.5 in · lb, is applied to the shaft at $t = 0$. Find the magnitude of the dynamic bearing forces when $t = 5$ s.

❚ (*a*)   The dynamic forces exerted by the bearings on the shaft are shown in Fig. 22.6*b*. The angular velocity of the shaft is

$$\omega = 1,000\left(\frac{2\pi}{60}\right) = 105 \text{ rad/s}$$

The magnitude of a spherical mass is given by

$$m = \frac{4}{3}\pi\left(\frac{1.25}{2}\right)^3\left(\frac{0.283}{386}\right) = 7.50 \times 10^{-4} \text{ lb} \cdot \text{s}^2/\text{in}$$

The dynamic bearing forces exerted by the bearings on the shaft, using Eq. (2) in Prob. 22.3, are

$$R = \frac{mrd\omega^2}{l} = \frac{7.50 \times 10^{-4}(4)10(105)^2}{14} = 23.6 \text{ lb} \tag{1}$$

(*b*)   The angular displacement of the shaft at $t = 0.95$ s is given by

$$\theta = \omega t = \frac{1,000}{60}0.95 = 15.83r$$

The above value corresponds to several revolutions of the assembly, and the net rotation past the $y_0$ axis at $t = 0.95$ s is found from

$$\theta = 1,000\left(\frac{2\pi}{60}\right)0.95 - 15(2\pi) = 5.236 \text{ rad} = 300°$$

The position of the assembly at $t = 0.95$ s is seen in Fig. 22.6*c*. $R^*$ is the dynamic force exerted by the shaft *on* the bearings. From Newton's third law,

$$R^* = -R = -23.6 \text{ lb}$$

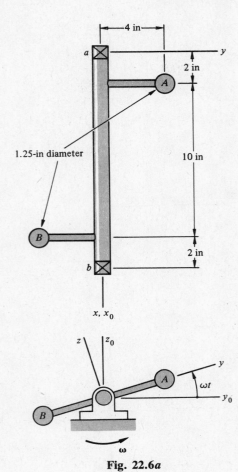

Fig. 22.6a

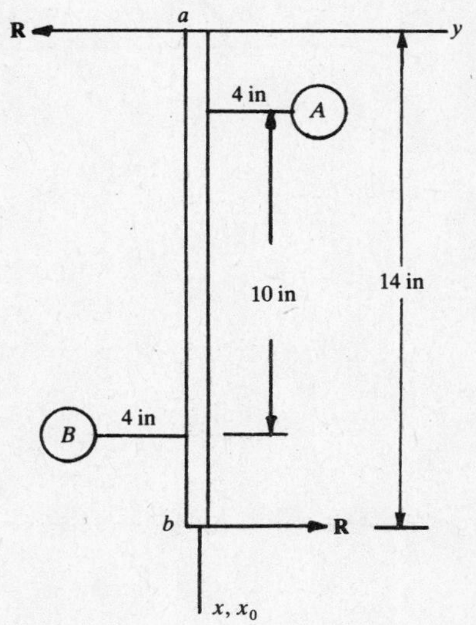

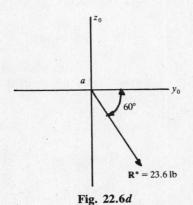

Fig. 22.6b

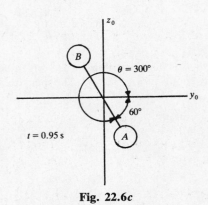

Fig. 22.6c

Fig. 22.6d

Figures 22.6d and e show the directions of $R^*$ at bearings $a$ and $b$, respectively. The components of this force at bearing $a$ are

$$R^*_{a,y_0} = 23.6 \cos 60° = 11.8 \text{ lb} \qquad R^*_{a,z_0} = -23.6 \sin 60° = -20.4 \text{ lb}$$

For this problem, $\mathbf{i}$, $\mathbf{j}$, and $\mathbf{k}$ are the unit vectors of the $x_0 y_0 z_0$ coordinate system. The dynamic force acting on the bearing at $a$ is

$$\mathbf{R}^*_a = 11.8\mathbf{j} - 20.4\mathbf{k} \qquad \text{lb}$$

The force components at bearing $b$ have the same magnitudes as those at $a$, and opposite senses. Thus, the dynamic force acting on bearing $b$ is

$$\mathbf{R}^*_b = -11.8\mathbf{j} + 20.4\mathbf{k} \qquad \text{lb}$$

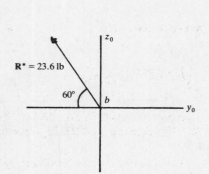

**Fig. 22.6e**

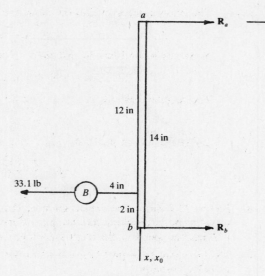

**Fig. 22.6f**

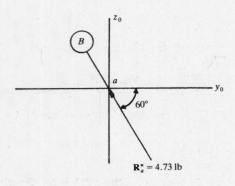

**Fig. 22.6g**

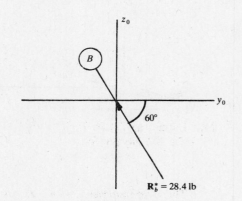

**Fig. 22.6h**

(c)  At  $t = 0$,   mass $A$ is detached from the assembly.  The centrifugal force which acts on mass $B$ is shown in the free-body diagram of the assembly shown in Fig. 22.6f.   The magnitude of this force is given by

$$F = mr\omega^2 \qquad\qquad (2)$$
$$F = 7.50 \times 10^{-4}(4)105^2 = 33.1 \text{ lb}$$

For dynamic equilibrium of the shaft,

$$\sum M_a = 0 \qquad -33.1(12) + R_b(14) = 0 \qquad R_b = 28.4 \text{ lb}$$
$$\sum M_b = 0 \qquad 33.1(2) - R_a(14) = 0 \qquad R_a = 4.73 \text{ lb}$$

It may be seen from the above results that the maximum value of the dynamic force at a bearing increases from 23.6 to 28.4 lb when mass $A$ is detached from the assembly.   From part (a),

$$\theta = 300° \qquad \text{(at } t = 0.95 \text{ s)}$$

$R_a^*$ and $R_b^*$ are the dynamic forces exerted *on* the bearings.   These forces are shown in Figs. 22.6g and h, and the components of these forces in the $y_0$ and $z_0$ directions are given by

$$R_{a,y_0}^* = -4.73\cos 60° = -2.37 \text{ lb} \qquad R_{a,z_0}^* = 4.73\sin 60° = 4.10 \text{ lb}$$
$$R_{b,y_0}^* = -28.4\cos 60° = -14.2 \text{ lb} \qquad R_{b,z_0}^* = 28.4\sin 60° = 24.6 \text{ lb}$$

The above forces may be written in formal vector notation as

$$\mathbf{R}_a^* = -2.37\mathbf{j} + 4.10\mathbf{k} \qquad \text{lb} \qquad \mathbf{R}_b^* = -14.2\mathbf{j} + 24.6\mathbf{k} \qquad \text{lb}$$

(d) The mass moment of inertia about the $x$ or $x_0$ axis is given by

$$I_x = 2mr^2 = 2(7.50 \times 10^{-4})4^2 = 0.024 \text{ lb} \cdot \text{s}^2 \cdot \text{in}$$

Newton's second law for rotational motion of the assembly about the $x$ or $x_0$ axis is

$$M_x = I_x \alpha \qquad 0.5 = 0.024\alpha \qquad \alpha = 20.8 \text{ rad/s}^2$$

The angular velocity at $t = 5$ s is found from

$$\omega = \omega_0 + \alpha t = \alpha t = 20.8(5) = 104 \text{ rad/s}$$

Using Eq. (1) in part (a),

$$R = \frac{mrd\omega^2}{l} = \frac{7.50 \times 10^{-4}(4)10(104)^2}{14} = 23.2 \text{ lb}$$

**22.7** (a) Figure 22.7 shows the construction of the bearings that support the shaft in Fig. 22.6a. The maximum force in the $y_0$ direction that the pair of mounting bolts may be subjected to is 800 lb. Find the maximum permissible speed of the shaft in revolutions per minute.

(b) Do the same as in part (a), but allow for the possibility of one of the masses becoming detached as the assembly rotates.

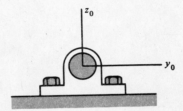

**Fig. 22.7**

▌ (a) From part (a) in Prob. 22.6, $m = 7.50 \times 10^{-4}$ lb·s²/in, $r = 4$ in, $d = 10$ in, and $l = 14$ in, and Eq. (1) has the form

$$R = \frac{mrd\omega^2}{l}$$

The maximum permissible value of $R$ is 800 lb in the $y_0$ direction. When $R$ has the direction of the $y_0$ axis,

$$R = 800 = \frac{7.50 \times 10^{-4}(4)10\omega^2}{14} \qquad \omega = 611 \text{ rad/s} = 611\left(\frac{60}{2\pi}\right) = 5,830 \text{ r/min}$$

(b) From the results in part (c) of Prob. 22.6, and using Eq. (2) and Fig. 22.6f,

$$R_b > R_a$$

For dynamic equilibrium of the assembly in Fig. 22.6f, with $F = mr\omega^2$ and $R_b = R_{b,\text{max}} = 800$ lb,

$$\sum M_a = 0 \qquad -mr\omega^2(12) + R_b(14) = -7.50 \times 10^{-4}(4)\omega^2(12) + 800(14) = 0$$

$$\omega = 558 \text{ rad/s} = 558\left(\frac{60}{2\pi}\right) = 5,330 \text{ r/min}$$

It may be seen that the maximum permissible speed of the shaft is smaller for the case where one mass becomes detached.

**22.8** (a) The assembly shown in Fig. 22.8a rotates at 50 rad/s. Find the dynamic forces at bearings $a$ and $b$.

(b) Do the same as in part (a), if mass $m_A$ becomes detached and flies away from the assembly.

(c) Do the same as in part (a), if mass $m_B$ becomes detached.

(d) Do the same as in part (a), if mass $m_C$ becomes detached.

(e) Compare the results in parts (a) through (d).

▌ (a) The centrifugal forces acting on the three masses are found from

$$F = mr\omega^2$$

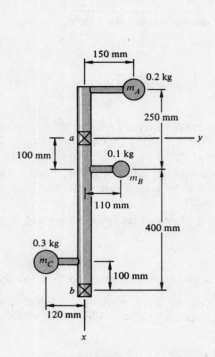

**Fig. 22.8a**

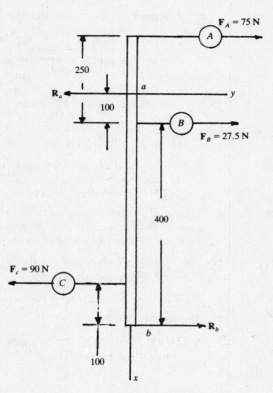

**Fig. 22.8b**

$$F_A = 0.2\left(\frac{150}{1,000}\right)50^2 = 75\text{ N} \qquad F_B = 0.1\left(\frac{110}{1,000}\right)50^2 = 27.5\text{ N} \qquad F_C = 0.3\left(\frac{120}{1,000}\right)50^2 = 90\text{ N}$$

The free-body diagram of the assembly is shown in Fig. 22.8b. For dynamic equilibrium,

$$\sum M_b = 0 \qquad 90(100) - 27.5(400) + R_a(500) - 75(650) = 0 \qquad (1)$$

$$R_a = 102\text{ N}$$

$$\sum F_y = 0 \qquad 75 - R_a + 27.5 - 90 + R_b = 0 \qquad (2)$$

$$R_b = 89.5\text{ N}$$

**(b)** $m_A$ becomes detached from the assembly. Equations (1) and (2) are used, with the term $F_A = 75\text{ N}$ omitted. The results are

$$\sum M_b = 0 \qquad 90(100) - 27.5(400) + R_a(500) = 0 \qquad R_a = 4\text{ N}$$

$$\sum F_y = 0 \qquad -R_a + 27.5 - 90 + R_b = 0 \qquad R_b = 66.5\text{ N}$$

**(c)** $m_B$ becomes detached from the assembly. Using Eqs. (1) and (2), with the term $F_B = 27.5\text{ N}$ omitted, gives

$$\sum M_b = 0 \qquad 90(100) + R_a(500) - 75(650) = 0 \qquad R_a = 79.5\text{ N}$$

$$\sum F_y = 0 \qquad 75 - R_a - 90 + R_b = 0 \qquad R_b = 94.5\text{ N}$$

**(d)** $m_C$ becomes detached from the assembly. The term $F_C = 90\text{ N}$ in Eqs. (1) and (2) is omitted, to obtain

$$\sum M_b = 0 \qquad -27.5(400) + R_a(500) - 75(650) = 0 \qquad R_a = 120\text{ N}$$

$$\sum F_y = 0 \qquad 75 - R_a + 27.5 + R_b = 0 \qquad R_b = 17.5\text{ N}$$

**(e)** Table 22.1 shows the dynamic bearing forces for all cases of mass attachment to the assembly. The maximum value of the dynamic bearing force, with all three masses attached, is $R_a = 102\text{ N}$. If mass

**TABLE 22.1**

| $m_A$, kg | $m_B$, kg | $m_C$, kg | $R_a$, N | $R_b$, N |
|-----------|-----------|-----------|----------|----------|
| 0.2 | 0.1 | 0.3 | 102 | 89.5 |
| — | 0.1 | 0.3 | 4 | 66.5 |
| 0.2 | — | 0.3 | 79.5 | 94.5 |
| 0.2 | 0.1 | — | 120 | 17.5 |

$A$ or mass $B$ becomes detached, the maximum value of the dynamic bearing force decreases. If mass $C$ becomes detached, the value of the maximum dynamic bearing force increases to $R_a = 120\,\text{N} > 102\,\text{N}$. The percent increase in the maximum dynamic bearing force for this case is

$$\%\text{D} = \frac{120 - 102}{102}\,(100) = 17.6\%$$

**22.9** Two thin rectangular brass plates are soldered to a shaft, as shown in Fig. 22.9a. The thickness of the plates is 4 mm and the density of brass is $8{,}550\,\text{kg/m}^3$.

(**a**) Find the dynamic bearing forces when the assembly rotates at 200 r/min.

(**b**) How would the results in part (*a*) change if one of the plates became detached from the shaft?

▌ (**a**) Figure 22.9b shows the free-body diagram of the plate and shaft assembly. $F$ is the centrifugal force which acts on each plate and $R$ is the dynamic force at each bearing. The angular velocity of the shaft is

$$\omega = 200\left(\frac{2\pi}{60}\right) = 20.9\,\text{rad/s}$$

The centrifugal force on each plate is found from

$$F = mr\omega^2 = 94(50)4(8{,}550)(25 + 9)(20.9)^2\left(\frac{1}{1{,}000}\right)^4 = 2.39\,\text{N}$$

The two centrifugal forces form a couple, and thus the two dynamic bearing forces must also form a couple. The quantity $R$ is found from

$$R = \frac{Fd}{l} = \frac{2.39(94)}{200} = 1.12\,\text{N}$$

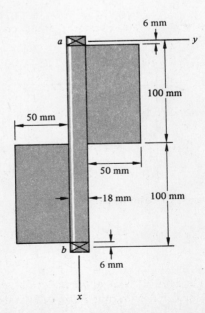

**Fig. 22.9a**

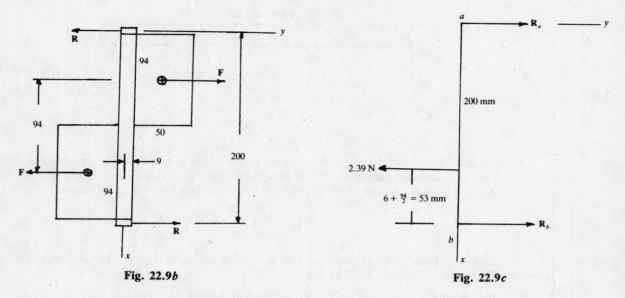

**Fig. 22.9b**                                    **Fig. 22.9c**

(*b*) Figure 22.9*c* shows the free-body diagram for the case where the upper plate becomes detached from the shaft. For dynamic equilibrium,

$$\sum M_b = 0 \qquad -R_a(200) + 2.39(53) = 0 \qquad R_a = 0.633 \text{ N}$$

$$\sum F_y = 0 \qquad R_a - 2.39 + R_b = 0 \qquad R_b = 1.76 \text{ N}$$

It may be seen that $R_b = 1.76$ N is greater than the result $R = 1.12$ N obtained in part (*a*). The result $R_a = 1.76$ N would be obtained if the lower plate became detached from the shaft.

**22.10**  The disk assembly in Fig. 22.10*a* is mounted on a shaft, as shown in Fig. 22.10*b*.

(*a*)  Find the dynamic bearing forces when the unit rotates with a speed of 3,400 r/min.

(*b*)  Do the same as in part (*a*), if, because of a faulty mounting procedure, the center of the disk is displaced 2 mm, along the line *a'a'*, from the bearing centerline.

(*c*)  Compare the results in parts (*a*) and (*b*).

(*d*)  Do the same as in part (*a*), if the disk assembly is mounted as shown in Fig. 22.10*c*.

∎ (*a*)  From Prob. 17.55,

$$m_1 = 0.725 \text{ kg} \qquad m_2 = 1.07 \text{ kg} \qquad m_3 = 6.96 \text{ kg}$$

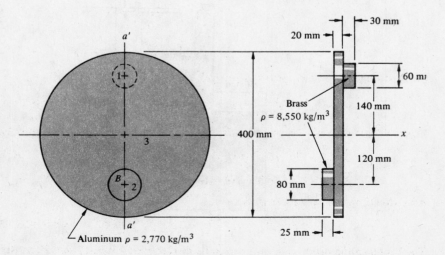

**Fig. 22.10a**

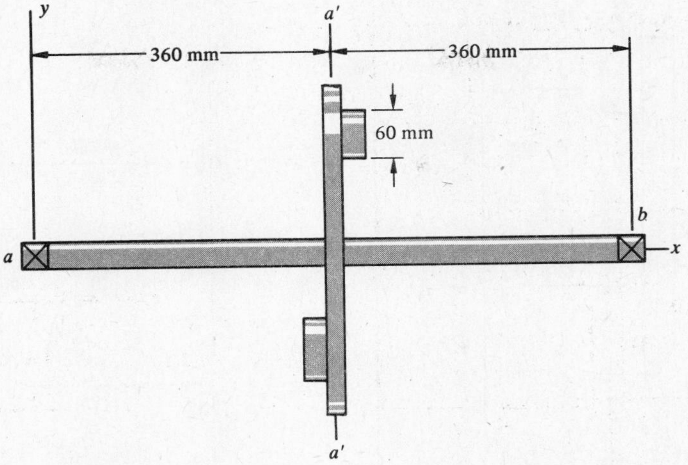

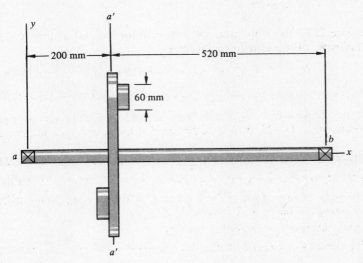

**Fig. 22.10b**

**Fig. 22.10c**

where $m_3$ is the mass of the disk. The centrifugal forces acting on each element are shown in Fig. 22.10d. These forces are given by

$$F = mr\omega^2$$

Using

$$\omega = 3,400 \text{ r/min} = 356 \text{ rad/s}$$

the centrifugal forces on elements 1 and 2 are found as

$$F_1 = 0.725\left(\frac{140}{1,000}\right)356^2 = 12,900 \text{ N} \qquad F_2 = 1.07\left(\frac{120}{1,000}\right)356^2 = 16,300 \text{ N}$$

Figure 22.10e shows the free-body diagram of the shaft. For dynamic equilibrium,

$$\sum M_a = 0 \qquad -16,300(338) + 12,900(385) + R_b(720) = 0 \qquad R_b = 754 \text{ N}$$

$$\sum M_b = 0 \qquad -R_a(720) + 16,300(382) - 12,900(335) = 0 \qquad R_a = 2,650 \text{ N}$$

As a check on the above calculations,

$$\sum F_y \stackrel{?}{=} 0 \qquad R_a + R_b + 12,900 \stackrel{?}{=} 16,300 \qquad 16,300 \equiv 16,300$$

(**b**) The center of the disk is first assumed to be displaced downward 2 mm along the line $aa$ shown in Figs. 22.10a and b. The new positions of masses 1 and 2, and the CM of the disk, mass 3, are shown in Fig.

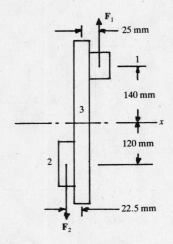

**Fig. 22.10d**

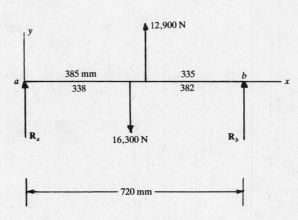

**Fig. 22.10e**

22.10$f$. The centrifugal forces acting on the three mass elements are

$$F_1 = 0.725\left(\frac{138}{1,000}\right)356^2 = 12,700 \text{ N} \qquad F_2 = 1.07\left(\frac{122}{1,000}\right)356^2 = 16,500 \text{ N}$$

$$F_3 = 6.96\left(\frac{2}{1,000}\right)356^2 = 1,760 \text{ N}$$

The free-body diagram of the shaft assembly is shown in Fig. 22.10$g$. For dynamic equilibrium,

$$\sum M_a = 0 \qquad -16,500(338) - 1,760(360) + 12,700(385) + R_b(720) = 0 \qquad R_b = 1,830 \text{ N}$$

$$\sum F_y = 0 \qquad R_a - 16,500 - 1,760 + 12,700 + R_b = 0 \qquad R_a = 3,730 \text{ N}$$

Figure 22.10$h$ shows the case where the center of the disk is displaced upward 2 mm along the line $a'a'$ shown in Figs. 22.10$a$ and $b$. The centrifugal forces are found to be

$$F_1 = 0.725\left(\frac{142}{1,000}\right)356^2 = 13,000 \text{ N} \qquad F_2 = 1.07\left(\frac{118}{1,000}\right)356^2 = 16,000 \text{ N}$$

$$F_3 = 6.96\left(\frac{2}{1,000}\right)356^2 = 1,760 \text{ N}$$

The free-body diagram of the shaft is seen in Fig. 22.10$i$. For dynamic equilibrium,

$$\sum M_a = 0 \qquad -16,000(338) + 1,760(360) + 13,000(385) - R_b(720) = 0 \qquad R_b = 320 \text{ N}$$

$$\sum F_y = 0 \qquad R_a - 16,000 + 1,760 + 13,000 - R_b = 0 \qquad R_a = 1,560 \text{ N}$$

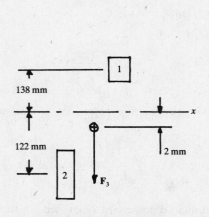

**Fig. 22.10f**

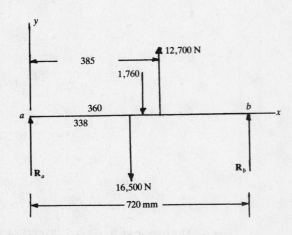

**Fig. 22.10g**

Fig. 22.10h

Fig. 22.10i

TABLE 22.2

| displacement of center of disk, mm | $R_a$, N | $R_b$, N |
|---|---|---|
| 0 | 2,650 | 754 |
| 2, downward | 3,730 | 1,830 |
| 2, upward | 1,560 | 320 |

(c)  Table 22.2 shows the bearing forces at $a$ and $b$ for the three cases considered.   It may be seen that there are significant changes in the values of the bearing forces for the given range of displacements of the center of the disk.   It may also be observed that the sense of $R_b$ in the last case in the table is opposite that of this term in the first two cases.   This effect may be seen from comparison of Figs. 22.10e, g, and i.

(d)  $F_1$ and $F_2$ are given in part (a).   Figure 22.10j shows the free-body diagram of the shaft.   For dynamic equilibrium,

$$\sum M_a = 0 \qquad -16,300(178) + 12,900(225) - R_b(720) = 0 \qquad R_b = 1.53 \text{ N} \approx 0$$

$$\sum F_y = 0 \qquad R_a - 16,300 + 12,900 - R_b = 0 \qquad R_a = 3,400 \text{ N}$$

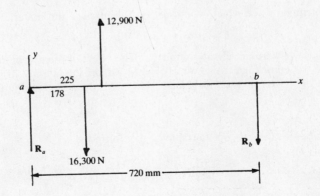

Fig. 22.10j

The problem of a rotating disk and shaft is studied further in the subject vibration analysis, when finding the *critical speed* of a shaft.

22.11    Figure 22.11a shows a cross-sectional view of a cylindrical rotor made of steel, with a specific weight of 489 lb/ft³.   The centerlines of both holes and the center axis of the rotor lie in a common plane.

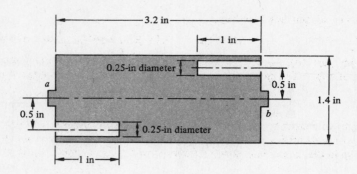

**Fig. 22.11a**

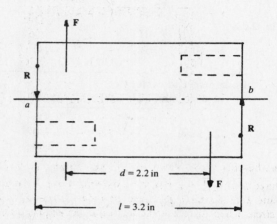

**Fig. 22.11b**

(a) Find the maximum permissible value of the angular velocity, in r/min, if the dynamic bearing forces must not exceed 6 lb.

(b) The rotor is initially at rest. A couple of magnitude 0.1 in-lb, about the *ab* axis, is applied to the rotor. For what length of time may this couple act on the rotor, if the dynamic bearing forces are not to exceed 6 lb?

▮ (a) The mass of the material which occupies one hole is given by

$$m = \frac{\pi(0.25)^2}{4} (1) \frac{489}{12^3(386)} = 3.60 \times 10^{-5} \text{ lb} \cdot \text{s}^2/\text{in}$$

The centrifugal force which would act on the mass of the material in one hole is

$$F = mr\omega^2 = 3.60 \times 10^{-5}(0.5)\omega^2 = 1.80 \times 10^{-5}\omega^2 \qquad \text{lb}$$

where $\omega$ is the angular velocity in rad/s. The effect of the *absent* mass in this problem is to change the *sense* of the centrifugal force which would act on this mass. This effect is shown in the free-body diagram of the rotor, seen in Fig. 22.11b. The dynamic bearing force is given by

$$R = \frac{Fd}{l}$$

Using $R = R_{max} = 6$ lb in the above equation gives

$$6 = \frac{1,80 \times 10^{-5}\omega^2(2.2)}{3.2} \qquad \omega = 696 \text{ rad/s} = 6{,}650 \text{ r/min}$$

(b) The mass moment of inertia of a cylinder about its center axis, from Case 4 in Table 17.13, is

$$I_0 = \tfrac{1}{8}md^2 = \tfrac{1}{32}\rho\pi d^4 h$$

where $\rho$ is the mass density. Using the above result, and the parallel-axis theorem, the mass moment of inertia of the rotor about its center axis is

$$I_0 = \frac{1}{32}\left[\frac{489}{386(12)^3}\right]\pi(1.4)^4 3.2 - 2[\tfrac{1}{8}(3.60 \times 10^{-5})0.25^2 + 3.60 \times 10^{-5}(0.5)^2] = 8.66 \times 10^{-4} \text{ lb} \cdot \text{s}^2 \cdot \text{in}$$

Newton's second law is written as

$$M_0 = I_0 \alpha \qquad 0.1 = 8.66 \times 10^{-4} \alpha \qquad \alpha = 115 \, \text{rad/s}^2$$

From part (a), $\omega = \omega_{\text{max}} = 696 \, \text{rad/s}$ corresponds to a maximum permissible dynamic bearing force of 6 lb. The maximum value of time $t$ for which the couple may act is found from

$$\omega = \omega_0 + \alpha t \qquad 696 = 115t \qquad t = 6.05 \, \text{s}$$

**22.12**   Figure 22.12a shows a cross-sectional view of a hollow cylinder supported by bearings at $a$ and $b$. A hole with a 0.200-in diameter is located at $c$, and a second hole, with a 0.284-in diameter, is located at $d$. The centerlines of both holes lie in the same plane that contains axis $ab$. The cylinder material is steel, with $\gamma = 0.283 \, \text{lb/in}^3$.

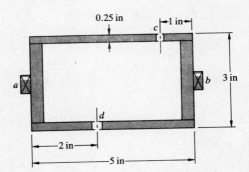

**Fig. 22.12a**

(a)   Find the dynamic bearing forces when the cylinder rotates at 1,725 r/min.

(b)   It is desired to have bearing $a$ experience no dynamic force. To accomplish this a third hole, with its axis in the same plane that contains the axes of holes $c$ and $d$, is to be drilled in the cylinder along a circumferential circle 1.6 in from bearing $a$. Find the required diameter of the hole.

▌ (a)   The centrifugal force which would act on the mass of the material which is removed from the holes is given by

$$F = mr\omega^2$$

The distance $r$ from the cylinder axis to the CM of the hole material is

$$r = 1.5 - \frac{0.25}{2} = 1.38 \, \text{in}$$

The angular velocity is

$$\omega = 1,725 \, \text{r/min} = 181 \, \text{rad/s}$$

The above results are combined, to obtain

$$F_c = \frac{\pi(0.2)^2}{4} (0.25)\left(\frac{0.283}{386}\right)(1.38)181^2 = 0.260 \, \text{lb}$$

$$F_d = \frac{\pi(0.284)^2}{4} (0.25)\left(\frac{0.283}{386}\right)(1.38)181^2 = 0.525 \, \text{lb}$$

The free-body diagram, with the reversed senses of the centrifugal forces acting on the masses of the hole material, is shown in Fig. 22.12b. For dynamic equilibrium,

$$\sum M_a = 0 \qquad F_d(2) - F_c(4) + R_b(5) = 0.525(2) - 0.260(4) + R_b(5) = 0 \qquad R_b = -0.002 \, \text{lb}$$

$R_b$ is 0.002 lb, acting downward in Fig. 22.12b.

$$\sum F_y = 0 \qquad -R_a + F_d - F_c + R_b = -R_a + 0.525 - 0.260 + (-0.002) = 0 \qquad R_a = 0.263 \, \text{lb}$$

$R_a$ is 0.263 lb, acting upward in Fig. 22.12b. It may be seen from comparison of the values above for $R_a$ and $R_b$ that $R_b \approx 0$.

(b)   Figure 22.12c shows the third hole in the cylinder. The centrifugal force that acts on the mass which would occupy the hole is

$$F_e = \frac{\pi d^2}{4} (0.25)\left(\frac{0.283}{386}\right)1.38(181)^2 = 6.51d^2$$

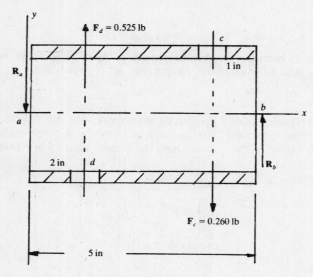

**Fig. 22.12b**

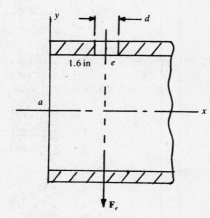

**Fig. 22.12c**

where $d$ is the diameter of the hole. For dynamic equilibrium of the cylinder, using Figs. 22.12b and c,

$$\sum M_b = 0 \qquad R_a(5) - F_d(3) + F_c(1) + F_e(3.4) = 0$$

Using the condition $R_a = 0$,

$$0(5) - 0.525(3) + 0.260(1) + 6.51d^2(3.4) = 0 \qquad d = 0.244 \text{ in}$$

**22.13** The part shown in Fig. 22.13a is fabricated from steel, with a density of 7,830 kg/m³. The part is to be chucked in a lathe for a facing operation on the flat face of the element that supports the two cylindrical stubs.

(a) Find the dynamic force exerted on the lathe chuck, and the dynamic moment about the point ($x_1 = 0$, $y_1 = 0$), if the 28-mm diameter is chucked and the part rotates about the $x_1$ axis at 740 r/min.

(b) Do the same as in part (a), if the 20-mm diameter is chucked, the part rotates about the $x$ axis at 740 r/min., and the dynamic moment is about the point ($x = 0$, $y = 0$).

(c) The part is modified to the shape shown in Fig. 22.13b. Find the dynamic force exerted on the lathe chuck if the 28-mm diameter is chucked and the part rotates about the $x_1$ axis at 740 r/min.

**❙ (a)** The angular velocity is

$$\omega = 740 \text{ r/min} = 77.5 \text{ rad/s}$$

Figure 22.13c shows the identification of the three mass elements of the part. From the solution to Prob. 17.70,

$$m_1 = 0.0492 \text{ kg} \qquad m_2 = 1.23 \text{ kg} \qquad m_3 = 0.145 \text{ kg}$$

For rotation about the $x_1$ axis, the centrifugal forces $F_1$ and $F_2$ acting on mass elements 1 and 2 have the values

$$F_1 = m_1 r\omega^2 = 0.0492\left(\frac{84}{1,000}\right)77.5^2 = 24.8 \text{ N} \qquad F_2 = m_2 r\omega^2 = 1.23\left(\frac{40}{1,000}\right)77.5^2 = 296 \text{ N}$$

The dynamic force $F$ exerted on the chuck is

$$F = F_1 + F_2 = 321 \text{ N}$$

Point $a$ in Fig. 22.13c is defined by $x_1 = y_1 = 0$. The dynamic moment $M_a$ of the forces $F_1$ and $F_2$ about this point is

$$M_a = F_1\left(\frac{10}{1,000}\right) - F_2\left(\frac{14}{1,000}\right) = 24.8\left(\frac{10}{1,000}\right) - 296\left(\frac{14}{1,000}\right) = -3.90 \text{ N} \cdot \text{m}$$

The dynamic moment $M_a$ has a clockwise sense in Fig. 22.13c. It acts in the $x_1$, $y_1$ plane, and rotates with this plane.

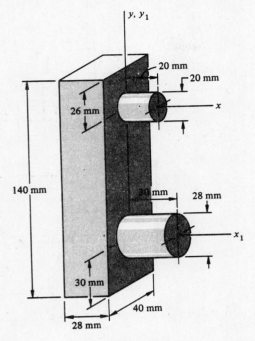

**Fig. 22.13a**

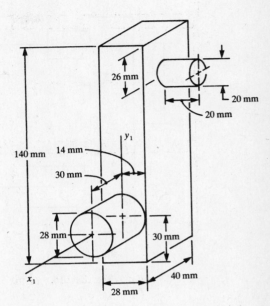

**Fig. 22.13b**

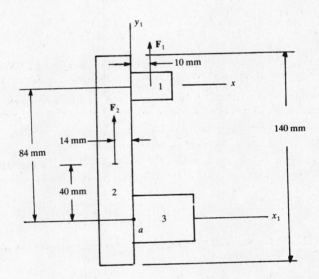

**Fig. 22.13c**

(b) Figure 22.13d shows the centrifugal forces $F_2$ and $F_3$ for the case of rotation of the part about the $x$ axis. These forces are given by

$$F_2 = m_2 r\omega^2 = 1.23\left(\frac{44}{1,000}\right)77.5^2 = 325 \text{ N} \qquad F_3 = m_3 r\omega^2 = 0.145\left(\frac{84}{1,000}\right)77.5^2 = 73.2 \text{ N}$$

Point $b$ in Fig. 22.13d is defined by $x = y = 0$. The dynamic moment $M_b$ of the forces $F_2$ and $F_3$ about this point is

$$M_b = F_2\left(\frac{14}{1,000}\right) - F_3\left(\frac{15}{1,000}\right) = 325\left(\frac{14}{1,000}\right) - 73.2\left(\frac{15}{1,000}\right) = 3.45 \text{ N} \cdot \text{m}$$

The dynamic moment $M_b$ has a counterclockwise sense in Fig. 22.13d. It lies in the $xy$ plane, and rotates with this plane.

(c) Figure 22.13e shows the centrifugal forces $F_1$ and $F_2$ when the part rotates about the $x_1$ axis. These forces have the magnitudes

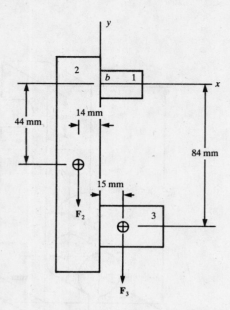

**Fig. 22.13d**

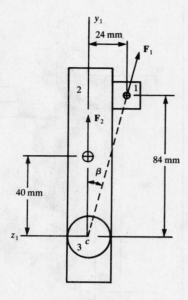

**Fig. 22.13e**

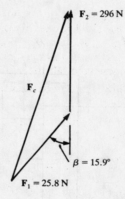

**Fig. 22.13f**

$$F_1 = m_1 r\omega^2 = 0.0492(\sqrt{24^2 + 84^2})\left(\frac{1}{1,000}\right)77.5^2 = 25.8 \text{ N}$$

$$F_2 = m_2 r\omega^2 = 1.23\left(\frac{40}{1,000}\right)77.5^2 = 296 \text{ N}$$

From Fig. 22.13e,

$$\tan \beta = \frac{24}{84} \qquad \beta = 15.9°$$

The dynamic force $F_c$ exerted on the lathe chuck is the vector sum of $F_1$ and $F_2$, and this vector addition is shown in Fig. 22.13f. Using the law of cosines

$$F_c^2 = F_1^2 + F_2^2 - 2F_1F_2 \cos(180° - 15.9°) = 25.8^2 + 296^2 - 2(25.8)296\cos(180° - 15.9°) = 321 \text{ N}$$

**22.14** The body shown in Fig. 22.14a is fabricated from 0.5-in-thick steel plate, with $\gamma = 0.283 \text{ lb/in}^3$. The ends are fitted to bearings at $a$ and $b$, and the assembly rotates about the center axis of the 20-in length.

(*a*) Describe how the dynamic bearing forces may be found.

(*b*) Find the dynamic bearing forces for the body in Fig. 22.14a, if the angular velocity of the body is 985 r/min.

(*c*) Do the same as in part (*b*), if the center axis of the 20-in length is displaced 0.050 in from the centerline of the two bearings.

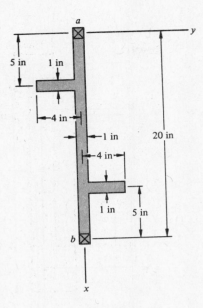

Fig. 22.14a

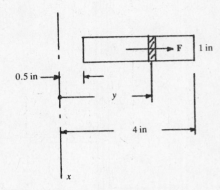

Fig. 22.14b

(a) In the preceding problems the unbalance was in the form of a *discrete mass*, and this mass was treated as a particle. The technique used to solve these problems may be extended to the case of bodies with continuous mass distribution. The centrifugal force acting on a typical differential mass element is found first. This force effect is then integrated over all the mass elements of the body to obtain the resultant force, and the moment, of all the centrifugal forces acting on the mass particles.

(b) Figure 22.14b shows the differential mass element of one of the arms. The centrifugal force $F$ which acts on this arm is given by

$$F = \int dm\, r\omega^2$$

Using $\omega = 985(2\pi/60) = 103\,\text{rad/s}$ in the above equation results in

$$F = \int_{0.5}^{4} \left[ 1(0.5)\,dy\left(\frac{0.283}{386}\right) \right] y(103)^2 = 103^2(1)0.5\left(\frac{0.283}{386}\right)\int_{0.5}^{4} y\,dy$$

$$= 103^2(1)0.5\left(\frac{0.283}{386}\right)\left(\frac{y^2}{2}\Big|_{0.5}^{4}\right) = 103^2(1)0.5\left(\frac{0.283}{386}\right)\frac{1}{2}\,(4^2 - 0.5^2) = 30.6\,\text{lb} \qquad (1)$$

Figure 22.14c shows the forces $F$, and the dynamic bearing forces $R$, which act on the body. These forces are found from

$$R = \frac{Fd}{l} = \frac{30.6(10)}{20} = 15.3\,\text{lb}$$

(c) Figure 22.14d shows the free-body diagram of the body when the center axis is displaced 0.050 in from the centerline of the bearings. Using the forms of Eq. (1) in part (a),

$$F_A = 103^2(1)0.5\left(\frac{0.283}{386}\right)\int_{0.5-0.050}^{4-0.050} y\,dy = 103^2(1)0.5\left(\frac{0.283}{386}\right)\left(\frac{y^2}{2}\Big|_{0.45}^{3.95}\right)$$

$$= 103^2(1)0.5\left(\frac{0.283}{386}\right)\frac{1}{2}\,(3.95^2 - 0.45^2) = 29.9\,\text{lb}$$

$$F_B = 103^2(1)0.5\left(\frac{0.283}{386}\right)\int_{0.5+0.050}^{4+0.050} y\,dy = 103^2(1)0.5\left(\frac{0.283}{386}\right)\left(\frac{y^2}{2}\Big|_{0.550}^{4.05}\right)$$

$$= 103^2(1)0.5\left(\frac{0.283}{386}\right)\frac{1}{2}\,(4.05^2 - 0.550^2) = 31.3\,\text{lb}$$

The centrifugal force acting on the center element of the body is

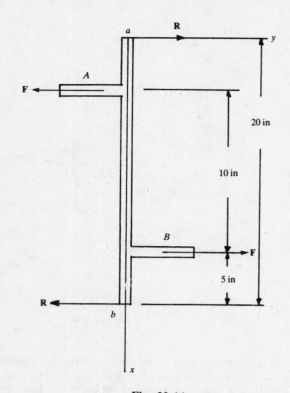

Fig. 22.14c

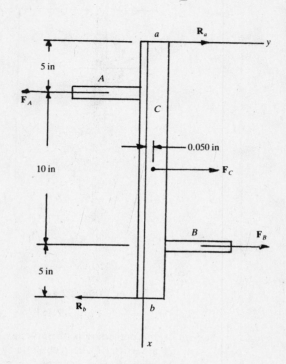

Fig. 22.14d

$$F_C = mr\omega^2 = 1(0.5)20\left(\frac{0.283}{386}\right)0.050(103)^2 = 3.89 \text{ lb}$$

For dynamic equilibrium of the body in Fig. 22.14d

$$\sum M_b = 0$$

$$-R_a(20) + F_A(15) - F_C(10) - F_B(5) = -R_a(20) + 29.9(15) - 3.89(10) - 31.3(5) = 0$$
$$R_a = 12.7 \text{ lb}$$

$$\sum F_y = 0 \qquad R_a - F_A + F_C + F_B - R_b = 12.7 - 29.9 + 3.89 + 31.3 - R_b = 0 \qquad R_b = 18.0 \text{ lb}$$

The effect of displacing the centerline of the body is to increase the value of the maximum dynamic bearing force from $R = 15.3$ lb to $R_b = 18.0$ lb.

22.15    In Fig. 22.15a a thin, rigid rod of mass $m$ is hinged to a vertical shaft which rotates with constant angular velocity $\omega$. The lower end of the rod is attached to the center axis by an inextensible cable of negligible weight, so that angle $\beta$ is constant.

(a)    Find the magnitude, direction, and location of the resultant of the *centrifugal* forces acting on the rod.

(b)    Find the tensile force in the cable, and the force exerted by the rod on the hinge pin.

(c)    Find the numerical results for parts (a) and (b), if $m = 4.2$ kg, $l = 1.35$ m, $\beta = 26.5°$, and $\omega = 825$ r/min.

❚    (a)    A set of $x_1y_1$ coordinates is positioned on the rod, as shown in Fig. 22.15b. Since the rod is thin, all points on the mass element $dm$ are at approximately the same radial distance from the axis of rotation. The differential centrifugal force $dP$ is shown in the figure in its actual sense. Since all the centrifugal forces which act on the mass elements of the rod have directions normal to the center axis of rotation, it may be concluded that these forces comprise a *parallel force system*. The mass per unit length of the rod is designated $\rho_0$, and

$$dP = dm(x_1 \sin \beta)\omega^2 = \rho_0(x_1 \sin \beta)\omega^2 \, dx_1 \qquad (1)$$

The resultant centrifugal force is then

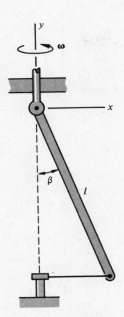

**Fig. 22.15a**

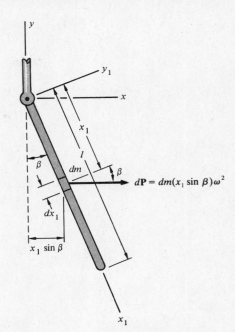

$dP = dm(x_1 \sin \beta)\omega^2$

**Fig. 22.15b**

$$P = \int dP = \int_0^l \rho_0(x_1 \sin \beta)\omega^2 \, dx_1 = \rho_0(\sin \beta)\omega^2 \int_0^l x_1 \, dx_1 = \rho_0(\sin \beta)\omega^2 \left.\frac{x_1^2}{2}\right|_0^l$$

$$= \rho_0 \frac{l^2}{2} \omega^2 \sin \beta$$

Since the mass $m$ of the rod is

$$m = \rho_0 l$$

the final value of $P$ is

$$P = m\left(\frac{l}{2} \sin \beta\right)\omega^2 \tag{2}$$

The term $(l/2) \sin \beta$ may be recognized as the radial position of the center of mass of the rod. The *magnitude* of the total centrifugal force is thus the same value which would be obtained if all the mass were imagined to be "concentrated" at the center of mass and had the acceleration of this point. The resultant centrifugal force is shown in Fig. 22.15c, and $x_0$ defines the location of the line of action of this resultant force. As will be seen in the following computation, this resultant force does *not* act through the center of mass of the rod.

The resultant moment $M_p$ of the centrifugal forces about the hinge pin can be written as

$$M_p = (P \cos \beta)x_0 = \int_0^l (dP \cos \beta)x_1 \tag{3}$$

It may be observed that the forces $P \sin \beta$ and $dP \sin \beta$, in the $x_1$ direction, pass through the hinge pin and thus contribute no moment about this point.

Equations (1) and (2) are substituted into Eq. (3) to obtain

$$m\left(\frac{l}{2} \sin \beta\right)\omega^2 x_0 = \int_0^l [\rho_0(x_1 \sin \beta)\omega^2]x_1 \, dx_1 \qquad m\left(\frac{l}{2}\right)x_0 = \rho_0 \int_0^l x_1^2 \, dx_1 = \rho_0 \left.\frac{x_1^3}{3}\right|_0^l = \frac{\rho_0 l^3}{3} \tag{4}$$

By using $m = \rho_0 l$, the final form of Eq. (4) is

$$x_0 = \tfrac{2}{3}l \tag{5}$$

This result has an interesting interpretation, since the coordinate $x_0 = \frac{2}{3}l$ locates the *center of percussion* of a slender rod which is hinged at its end. It may be recalled from Prob. 18.98 that the center of percussion is a point on a physical body about which the *resultant dynamic moment* is zero. This point, except for the limiting case shown in Prob. 18.100, is never coincident with the CM of the body. It may also be observed that the location of the line of action of the resultant of the inertia forces acting on the rod is *independent of the rotational velocity* $\omega$.

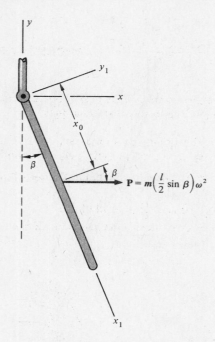

Fig. 22.15c

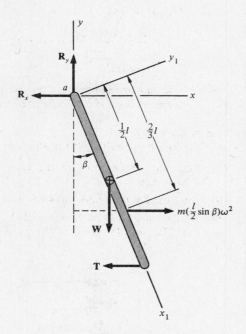

Fig. 22.15d

(*b*) The free-body diagram of the slender rod is shown in Fig. 22.15*d*. For moment equilibrium about the hinge,

$$\sum M_a = 0 \qquad m\left(\frac{l}{2} \sin \beta\right)\omega^2\left(\frac{2}{3} l \cos \beta\right) - W\left(\frac{l}{2} \sin \beta\right) - T(l \cos \beta) = 0$$

Using $m = W/g$ in the above equation gives

$$T = \frac{W}{6}\left(\frac{2\omega^2 l \sin \beta}{g} - 3 \tan \beta\right) \tag{6}$$

$T$ represents a cable tensile force, which is an inherently positive term. Thus, from Eq. (6) the magnitude of the angular velocity $\omega$ must satisfy

$$\frac{2\omega^2 l \sin \beta}{g} > 3 \tan \beta = 3 \frac{\sin \beta}{\cos \beta} \qquad \omega > \sqrt{\frac{3g}{2l \cos \beta}}$$

The remaining equilibrium requirements are

$$\sum F_x = 0 \qquad -R_x + \frac{W}{g}\left(\frac{l}{2} \sin \beta\right)\omega^2 - T = 0$$

Using $T$ from Eq. (6) in the above equation results in

$$R_x = \frac{W}{6}\left(\frac{\omega^2 l \sin \beta}{g} + 3 \tan \beta\right)$$

$$\sum F_y = 0 \qquad R_y - W = 0 \qquad R_y = W$$

The resultant force which acts on the hinge pin may then be found from

$$R = \sqrt{R_x^2 + R_y^2}$$

(*c*) $m = 4.2$ kg, $l = 1.35$ m, $\beta = 26.5°$, and $\omega = 825$ r/min $= 86.4$ rad/s. Equation (2) is used to obtain

$$P = m\left(\frac{l}{2} \sin \beta\right)\omega^2 = 4.2\left(\frac{1.35}{2} \sin 26.5°\right)86.4^2 = 9{,}440 \text{ N}$$

Using Eq. (5),

$$x_0 = \tfrac{2}{3}l = \tfrac{2}{3}(1.35) = 0.9 \text{ m}$$

Using Eq. (6)

$$T = \frac{W}{g}\left(\frac{2\omega^2 l \sin\beta}{g} - 3\tan\beta\right) = \frac{4.2(9.81)}{9.81}\left[\frac{2(86.4)^2 1.35 \sin 26.5°}{9.81} - 3\tan 26.5°\right] = 4.2(916.7 - 1.5)$$
$$= 3,840 \text{ N}$$

It may be seen that the second term in the parentheses in the above equation, due to the static weight force, is insignificant compared with the first term, due to the centrifugal force.

**22.16** The thin rod of mass 1.5 kg in Fig. 22.16a is hinged to a vertical shaft that rotates with constant angular velocity $\omega$. The lower end of the rod is attached to the center axis by an inextensible cable of negligible weight.

**(a)** Find the tensile force in the cable, and the resultant force that acts on the hinge pin, when $\omega = 1,000$ r/min.

**(b)** Find the limiting value of $\omega$ for which the cable will experience a tensile force.

**(c)** Do the same as in parts (a) and (b), if the rod is replaced by the simple pendulum of mass 1.5 kg shown in Fig. 22.16b. The mass of the pendulum rod may be neglected.

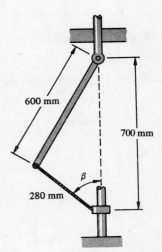

Fig. 22.16a

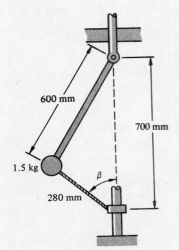

Fig. 22.16b

**(a)** Figure 22.16c shows the free-body diagram of the rod. Angle $\beta_1$ is found by using the law of cosines. The result is

$$280^2 = 700^2 + 600^2 - 2(700)600\cos\beta_1 \qquad \beta_1 = 23.3°$$

Using the law of sines,

$$\frac{600}{\sin\beta} = \frac{280}{\sin\beta_1} \qquad \beta = 57.9°$$

The location of the resultant centrifugal force on the rod is found by using Eq. (5) in Prob. 22.15. This result is shown in Fig. 22.16c.

The angular velocity of the shaft is

$$\omega = 1,000\left(\frac{2\pi}{60}\right) = 105 \text{ rad/s}$$

Using Eq. (2) from Prob. 22.15,

$$P = m\left(\frac{l}{2}\sin\beta_1\right)\omega^2 = 1.5\left[\frac{600}{2(1,000)}\sin 23.3°\right]105^2 = 1,960 \text{ N} \qquad (1)$$

For dynamic equilibrium of the rod,

$$\sum M_a = 0$$

$$-P(400\cos 23.3°) + 1.5(9.81)300\sin 23.3° + (T\cos 32.1°)(600\cos 23.3°)$$
$$+ (T\sin 32.1°)(600\sin 23.3°) = 0 \qquad (2)$$

Using $P = 1,960\,\text{N}$ in the above equation gives

$$-1,960(400\cos 23.3°) + 1.5(9.81)(300\sin 23.3°) + (T\cos 32.1°)(600\cos 23.3°)$$
$$+ (T\sin 32.1°)(600\sin 23.3°) = 0 \qquad T = 1,210\,\text{N}$$

$$\sum F_x = 0 \qquad R_{ax} - P + T\cos 32.1° = 0$$

$$R_{ax} - 1,960 + 1,210\cos 32.1° = 0 \qquad R_{ax} = 935\,\text{N}$$

$$\sum F_y = 0 \qquad R_{ay} - 1.5(9.81) - T\sin 32.1° = 0$$

$$R_{ay} - 1.5(9.81) - 1,210\sin 32.1° = 0 \qquad R_{ay} = 658\,\text{N}$$

The resultant force acting on the hinge pin is

$$R_a = \sqrt{R_{ax}^2 + R_{ay}^2} = \sqrt{935^2 + 658^2} = 1,140\,\text{N}$$

(*b*) The limiting value of the cable tensile force is $T = 0$. Equation (1) may be written as

$$P = 1.5\left[\frac{600}{2(1,000)}\sin 23.3°\right]\omega^2 = 0.178\omega^2 \qquad (3)$$

Equation (3) is used in Eq. (2), together with the condition $T = 0$. The result is

$$\sum M_a = 0 \qquad -(0.178\omega^2)(400\cos 23.3°) + 1.5(9.81)(300\sin 23.3°) = 0$$

$$\omega = 5.17\,\text{rad/s} = 49.4\,\text{r/min}$$

(*c*) Figure 22.16*d* shows the free-body diagram of the pendulum rod. The centrifugal force that acts on the pendulum mass is

$$P = mr\omega^2 = 1.5\left(\frac{600}{1,000}\sin 23.3°\right)105^2 = 3,920\,\text{N}$$

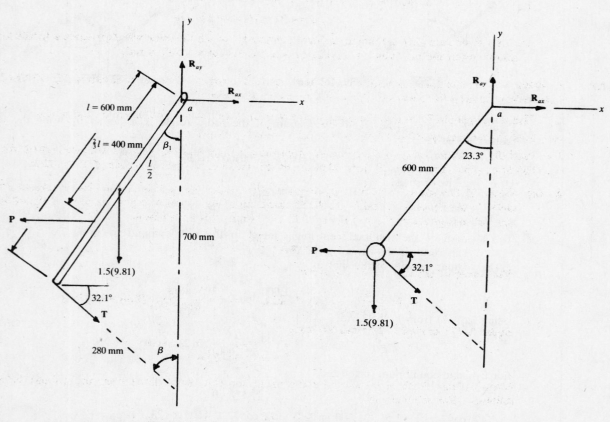

Fig. 22.16*c*                                                    Fig. 22.16*d*

For dynamic moment equilibrium of the rod,

$$\sum M_a = 0$$

$$-P(600\cos 23.3°) + 1.5(9.81)600\sin 23.3° + (T\cos 32.1°)(600\cos 23.3°)$$
$$+ (T\sin 32.1°)(600\sin 23.3°) = 0 \qquad (4)$$

Using $P = 3,920\,\text{N}$ in the above equation gives

$$-3,920(600\cos 23.3°) + 1.5(9.81)(600\sin 23.3°) + (T\cos 32.1°)(600\cos 23.3°)$$
$$+ (T\sin 32.1°)(600\sin 23.3°) = 0 \qquad T = 3,640\,\text{N}$$

It may be seen that the cable force above is greater than that in part ($a$) by a factor of $3,640/1,210 = 3.01 \approx 3$. For force equilibrium of the rod,

$$\sum F_x = 0 \qquad R_{ax} - P + T\cos 32.1° = 0$$

$$R_{ax} - 3,920 + 3,640\cos 32.1° = 0 \qquad R_{ax} = 836\,\text{N}$$

$$\sum F_y = 0 \qquad R_{ay} - 1.5(9.81) - T\sin 32.1° = 0$$

$$R_{ay} - 1.5(9.81) - 3,640\sin 32.1° = 0 \qquad R_{ay} = 1,950\,\text{N}$$

The resultant force acting on the hinge pin is

$$R_a = \sqrt{R_{ax}^2 + R_{ay}^2} = \sqrt{836^2 + 1,950^2} = 2,120\,\text{N}$$

The limiting value of the cable force occurs when $T = 0$. The centrifugal force acting on the mass may be written as

$$P = 1.5\left(\frac{600}{1,000}\sin 23.3°\right)\omega^2 = 0.356\omega^2 \qquad (5)$$

Equation (5) is used in Eq. (4), together with $T = 0$, to obtain

$$\sum M_a = 0 \qquad -(0.356\omega^2)(600\cos 23.3°) + 1.5(9.81)(600\sin 23.3°) = 0$$

$$\omega = 4.22\,\text{rad/s} = 40.3\,\text{r/min}$$

It may be seen that the limiting angular velocity for which the cable will experience a tensile force is smaller when the pendulum is a point mass at the end of a massless rod.

22.17  A slender rod of mass 1.4 kg is hinged to the center shaft of a cylindrical drum, as shown in Fig. 22.17a. The drum rotates about a fixed vertical axis with a speed of 290 r/min.

($a$)  Find the hinge-pin force and the compressive normal force exerted by the rod on the drum.

($b$)  Find the limiting speed at which the rod loses contact with the drum.

($c$)  Find the maximum value of the speed at which the assembly may be operated without causing failure of the hinge pin. The hinge-pin force that causes this element to shear is estimated to be 32 N.

❚ ($a$)  Figure 22.17b shows the free-body diagram of the rod. The rod is divided into elements 1 and 2 shown in the figure. $P_1$ and $P_2$ are the resultant centrifugal forces acting on the two rod elements, with the locations given by Eq. (5) in Prob. 22.15. Angle $\beta$ is found from

$$\sin \beta = \frac{200}{400} \qquad \beta = 30°$$

The masses of the two rod elements are

$$m_1 = \frac{400}{400 + 150}(1.4) = 1.02\,\text{kg} \qquad m_2 = \frac{150}{400 + 150}(1.4) = 0.382\,\text{kg}$$

Equation (2) in Prob. 22.15 has the form

$$P = m\left(\frac{l}{2}\sin \beta\right)\omega^2$$

where $l$ is the length of the hinged rod element and $\beta$ is the angle between the rod and the axis of rotation. For rod element 1,

$$P_1 = 1.02\left[\frac{400}{2(1,000)}\sin 30°\right]\omega^2 = 0.102\omega^2 \qquad (1)$$

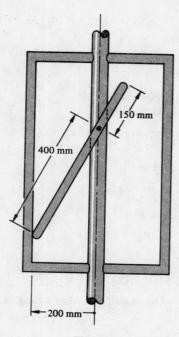

**Fig. 22.17a**

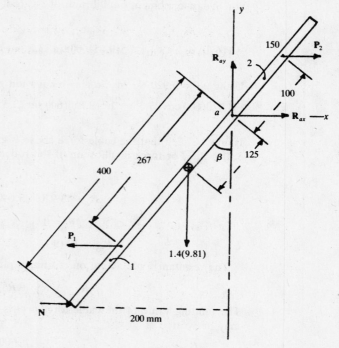

**Fig. 22.17b**

The angular velocity of the drum is

$$\omega = 290\left(\frac{2\pi}{60}\right) = 30.4 \text{ rad/s}$$

Using Eq. (1), $P_1$ has the form

$$P_1 = 0.102(30.4)^2 = 94.3 \text{ N}$$

Force $P_2$ may be written as

$$P_2 = 0.382\left[\frac{150}{2(1,000)} \sin 30°\right]\omega^2 = 0.0143\omega^2 \qquad (2)$$

$$P_2 = 0.0143(30.4)^2 = 13.2 \text{ N}$$

For dynamic equilibrium of the rod in Fig. 22.17b,

$$\sum M_a = 0$$

$$-P_2(100 \cos 30°) + 1.4(9.81)125 \sin 30° - P_1(267 \cos 30°) + N(400 \cos 30°) = -13.2(100 \cos 30°)$$
$$+ 1.4(9.81)125 \sin 30° - 94.3(267 \cos 30°) + N(400 \cos 30°) = 0 \qquad N = 63.8 \text{ N}$$

$$\sum F_x = 0 \qquad N - P_1 + R_{ax} + P_2 = 63.8 - 94.3 + R_{ax} + 13.2 = 0 \qquad R_{ax} = 17.3 \text{ N}$$

$$\sum F_y = 0 \qquad -1.4(9.81) + R_{ay} = 0 \qquad R_{ay} = 13.7 \text{ N}$$

The hinge-pin force is

$$R_a = \sqrt{R_{ax}^2 + R_{ay}^2} = \sqrt{17.3^2 + 13.7^2} = 22.1 \text{ N}$$

(*b*) When the rod loses contact with the drum, $N = 0$. For moment equilibrium of the rod, using that value of $N$,

$$\sum M_a = 0 \qquad -P_2(100 \cos 30°) + 1.4(9.81)125 \sin 30° - P_1(267 \cos 30°) = 0$$

Using Eqs. (1) and (2) in the above equation gives

$$-(0.0143\omega^2)100 \cos 30° + 1.4(9.81)125 \sin 30° - (0.102\omega^2)267 \cos 30° = 0$$
$$\omega = 5.88 \text{ rad/s} = 56.1 \text{ r/min}$$

(c)    For moment equilibrium of the rod, using Fig. 22.17b,

$$\sum M_a = 0 \qquad -P_2(100\cos 30°) + 1.4(9.81)125\sin 30° - P_1(267\cos 30°) + N(400\cos 30°) = 0$$

Using Eqs. (1) and (2) in the above equation results in

$$\sum M_a = 0$$

$$-(0.0143\omega^2)100\cos 30° + 1.4(9.81)125\sin 30° - (0.102\omega^2)267\cos 30° + N(400\cos 30°) = 0$$

$$N = 0.0717\omega^2 - 2.48$$

For force equilibrium of the rod,

$$\sum F_x = 0 \qquad N - P_1 + R_{ax} + P_2 = (0.0717\omega^2 - 2.48) - 0.102\omega^2 + R_{ax} + 0.0143\omega^2 = 0$$

$$R_{ax} = 0.016\omega^2 + 2.48$$

$$\sum F_y = 0 \qquad -1.4(9.81) + R_{ay} = 0 \qquad R_{ay} = 13.7\,\text{N}$$

The resultant force on the pin is

$$R_a = \sqrt{R_{ax}^2 + R_{ay}^2}$$

Using $R_{a,\text{max}} = 32\,\text{N}$ in the above equation gives

$$32 = \sqrt{R_{ax}^2 + R_{ay}^2} \qquad 32^2 = (0.016\omega^2 + 2.48)^2 + 13.7^2 \qquad 2.56 \times 10^{-4}\omega^4 + 0.0794\omega^2 - 830 = 0$$

$$\omega^2 = \frac{-0.0794 \pm \sqrt{0.0794^2 - 4(2.56 \times 10^{-4})(-830)}}{2(2.56 \times 10^{-4})} = \frac{-0.0794 \pm 0.925}{2(2.56 \times 10^{-4})} = 1,650$$

$$\omega = 40.6\,\text{rad/s} = 388\,\text{r/min}$$

**22.18**    The hinged rod in Prob. 22.17 is replaced by the simple pendulum arrangement shown in Fig. 22.18a.    Find the required value of $h$, if the mass is to lose contact with the drum at the same rotational speed at which the rod in Prob. 22.17 loses contact with the drum.

▐    Figure 22.18b shows the free-body diagram of the pendulum mass.    For dynamic equilibrium of the mass,

$$\sum F_y = 0 \qquad T\sin\beta - mg = 0 \tag{1}$$

$$\sum F_x = 0 \qquad -mr\omega^2 + T\cos\beta = 0 \tag{2}$$

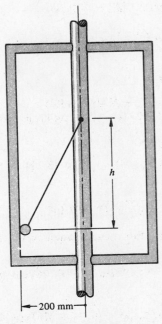

**Fig. 22.18a**

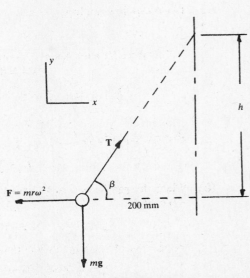

**Fig. 22.18b**

$T$ is eliminated between Eqs. (1) and (2), with the result

$$\tan \beta = \frac{g}{r\omega^2}$$

Using $\omega = 5.88 \, \text{rad/s}$, from part (b) of Prob. 22.17, in the above equation gives

$$\tan \beta = \frac{9.81}{(200/1,000)5.88^2} \qquad \beta = 54.8°$$

From Fig. 22.18b,

$$\tan \beta = \frac{h}{200} \qquad h = 284 \, \text{mm}$$

22.19    (a)   A hole is drilled in a bronze cylinder, as shown in Fig. 22.19a. Find the dynamic forces exerted on bearings $a$ and $b$ when the cylinder rotates at 3,250 r/min. The density of bronze is 8,800 kg/m³.

     (b)   Do the same as in part (a), if the orientation of the drilled hole is as shown in Fig. 22.19b. The hole may be approximated as a right circular cylinder of height $d_1$.

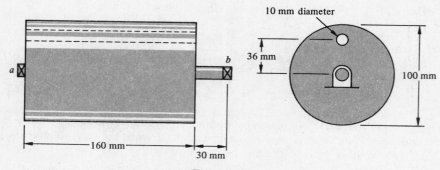

Fig. 22.19a

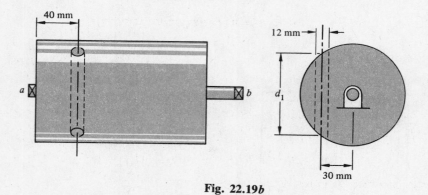

Fig. 22.19b

❚   (a)   The angular velocity of the cylinder is

$$\omega = 3,250 \, \text{r/min} = 340 \, \text{rad/s}$$

The centrifugal force which would act on the mass of the material in the hole, shown in Fig. 22.19c, is

$$F = mr\omega^2 = \frac{\pi(10)^2}{4}(160)8,800(36)(340)^2\left(\frac{1}{1,000}\right)^4 = 460 \, \text{N}$$

The free-body diagram, with the reversed sense of the centrifugal force acting on the mass of the hole material, is shown in Fig. 22.19d. For dynamic equilibrium of the cylinder and shaft unit,

$$\sum M_a = 0 \qquad -460(80) + R_b(190) = 0 \qquad R_b = 194 \, \text{N}$$

$$\sum F_y = 0 \qquad R_a - 460 + R_b = 0 \qquad R_a = 266 \, \text{N}$$

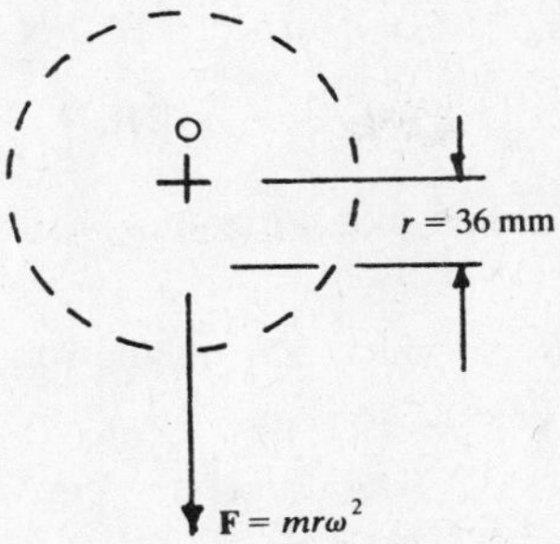

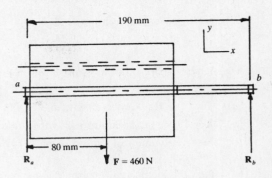

Fig. 22.19c

Fig. 22.19d

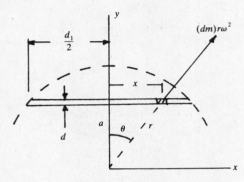

Fig. 22.19e

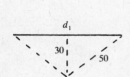

Fig. 22.19f

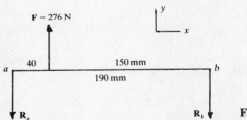

Fig. 22.19g

(*b*) Figure 22.19*e* shows the differential centrifugal force which acts on the mass of material which would fill the hole. The resultant centrifugal force in the *y* direction is

$$F_y = \int dm \, r\omega^2 \cos\theta = \int_{-d_1/2}^{d_1/2} \rho\left(\frac{\pi d^2}{4}\right) dx \, r\omega^2 \cos\theta$$

Using $r \cos\theta = a$ in the above equation gives

$$F_y = \left(\frac{\rho\pi d^2 a\omega^2}{4}\right)2\int_0^{d_1/2} dx = \frac{\rho\pi}{4}\, d^2 a\omega^2 2(x)\big|_0^{d_1/2} = \rho\left(\frac{\pi d^2}{4}\right)d_1 a\omega^2 \tag{1}$$

The above result is the same as if it were assumed that the entire mass of the hole material was located at the centroid of the hole volume. From Fig. 22.19*f*,

$$\frac{d_1}{2} = \sqrt{50^2 - 30^2} \qquad d_1 = 80 \text{ mm}$$

For the numerical values of the problem, Eq. (1) has the form

$$F_y = 8,800 \, \frac{\pi(12)^2}{4} \, (80)(30)(340)^2 \left(\frac{1}{1,000}\right)^4 = 276 \text{ N}$$

Figure 22.19*g* shows the effect of the centrifugal force of the hole material on the shaft. For dynamic equilibrium,

$$\sum M_a = 0 \qquad 276(40) - R_b(190) = 0 \qquad R_b = 58.1 \text{ N}$$

$$\sum F_y = 0 \qquad -R_a + 276 - R_b = 0 \qquad R_a = 218 \text{ N}$$

**22.20** The slender, rigid rod of mass $m$ shown in Fig. 22.20$a$ is attached to a horizontal shaft at a fixed value of angle $\beta$.

(*a*) Find the resultant moment, due to the centrifugal forces, which acts on the rod when the shaft rotates.

(*b*) Use the result found in part (*a*) to find the dynamic bearing forces.

(*c*) Find the numerical values for parts (*a*) and (*b*) if $m = 0.378$ kg and $l = 1$ m. The rotational speed is 340 r/min, and $\beta = 20°$.

(*d*) Find the factor by which the dynamic force at each bearing exceeds the static reaction force at the bearing when the rod assembly is stationary.

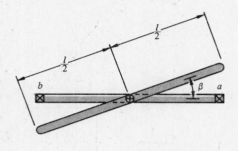

**Fig. 22.20$a$**                                    **Fig. 22.20$b$**

(*a*) The free-body diagram of the rod is shown in Fig. 22.20$b$. Each half of the rod has exactly the same configuration as a slender rod rotating with given inclination about a fixed axis. (See Prob. 22.15.) The magnitude, and the point of application, of the resultant centrifugal force which acts on half of the rod may be found from this problem. Using Eqs. (2) and (5) in Prob. 22.15 with $l \to l/2$ and $m \to m/2$, we get

$$P = m\left(\frac{l}{2}\sin\beta\right)\omega^2 \Big|_{\substack{l\to l/2\\ m\to m/2}} = \frac{m}{2}\left(\frac{l}{4}\sin\beta\right)\omega^2 = \frac{ml\omega^2 \sin\beta}{8} \qquad x_0 = \frac{2}{3}l\Big|_{l\to l/2} = \frac{l}{3}$$

The above result is shown in Fig. 22.20$b$. The dynamic moment acting on the rod, due to the resultant centrifugal forces, is

$$M = P(2x_0\cos\beta) = \frac{ml\omega^2 \sin\beta}{8}\left[2\left(\frac{l}{3}\right)\cos\beta\right] = \frac{1}{24}ml^2\omega^2 \sin 2\beta \qquad (1)$$

(*b*) Since the center of mass lies on the axis of rotation

$$R_a = R_b = R$$

For dynamic equilibrium of the rod and shaft assembly, using Fig. 22.20$b$,

$$\sum M = 0 \qquad -R(l\cos\beta) + \tfrac{1}{24}ml^2\omega^2 \sin 2\beta = 0 \qquad R = \tfrac{1}{12}ml\omega^2 \sin\beta$$

(*c*) $m = 0.378$ kg, $l = 1$ m, and $\beta = 20°$, and the angular velocity is

$$\omega = 340\left(\frac{2\pi}{60}\right) = 35.6 \text{ rad/s}$$

The dynamic moment due to the resultant centrifugal forces, from Eq. (1), is

$$M = \tfrac{1}{24}ml^2\omega^2 \sin 2\beta = \tfrac{1}{24}(0.378)(1)^2 35.6^2 \sin 2(20°) = 12.8 \text{ N} \cdot \text{m}$$

The dynamic force at each bearing is

$$R = \tfrac{1}{12}ml\omega^2 \sin\beta = \tfrac{1}{12}(0.378)(1)(35.6)^2 \sin 20° = 13.7 \text{ N}$$

(d) The weight of the rod is

$$W = mg = 0.378(9.81) = 3.71 \text{ N}$$

When the rod assembly is stationary, the static reaction force at each bearing is $3.71/2 = 1.86$ N. The factor by which the magnitude of the dynamic bearing force exceeds the static bearing force is $13.7/1.86 = 7.37$.

**22.21** (a) The cylinder in Fig. 22.21a is made of aluminum, with a specific weight of $0.1 \text{ lb/in}^3$. Two 3/16-in-diameter holes are drilled through the cylinder, and the axes of both holes lie in the same diametral plane. Find the dynamic forces at bearings $a$ and $b$ when the cylinder rotates at 6,000 r/min.

(b) Do the same as in part (a), if the holes are filled with lead with a specific weight of $710 \text{ lb/ft}^3$.

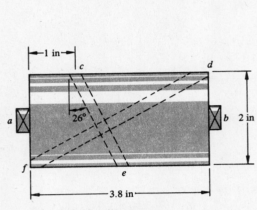

Fig. 22.21a

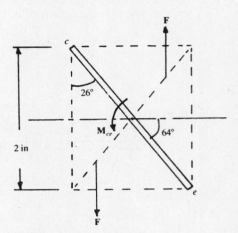

Fig. 22.21b

$\blacksquare$ (a) This problem may be solved by using the resultant moment, due to the centrifugal forces, on the mass of material which would occupy the holes. Figure 22.21b shows hole $ce$. The length $l_{ce}$ of this hole is given by

$$l_{ce} = \frac{2}{\cos 26°}$$

The mass per unit length of the hole material is

$$\rho_0 = \rho A = \frac{0.1}{386} \left[ \frac{\pi(\frac{3}{16})^2}{4} \right] = 7.15 \times 10^{-6} \text{ lb} \cdot \text{s}^2/\text{in}^2$$

The angular velocity of the cylinder is

$$\omega = 6,000 \text{ r/min} = 628 \text{ rad/s}$$

The moment due to the mass of material which would occupy hole $ce$ is given by Eq. (1) in Prob. 22.20 as

$$M = \tfrac{1}{24} m l^2 \omega^2 \sin 2\beta$$

For the numerical values of this problem,

$$M_{ce} = \tfrac{1}{24}(7.15 \times 10^{-6})\left(\frac{2}{\cos 26°}\right)\left(\frac{2}{\cos 26°}\right)^2 628^2 \sin[2(64°)] = 1.02 \text{ in} \cdot \text{lb}$$

The effect of the absent mass is to change the senses of the centrifugal forces which would act if the mass were present. The actual senses of the resultant centrifugal forces, due to the material which is removed from each half of the hole length, are shown as the forces $F$ in Fig. 22.21b. The actual sense of the dynamic moment $M_{ce}$, due to the hole mass $ce$, which acts on the cylinder is shown in Fig. 22.21b.

Hole $df$ is shown in Fig. 22.21c. From this figure

$$l_{df} = \sqrt{2^2 + 3.8^3} = 4.29 \text{ in} \qquad \tan \beta = \frac{2}{3.8} \qquad \beta = 27.8°$$

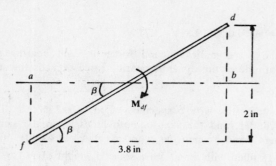

Fig. 22.21c

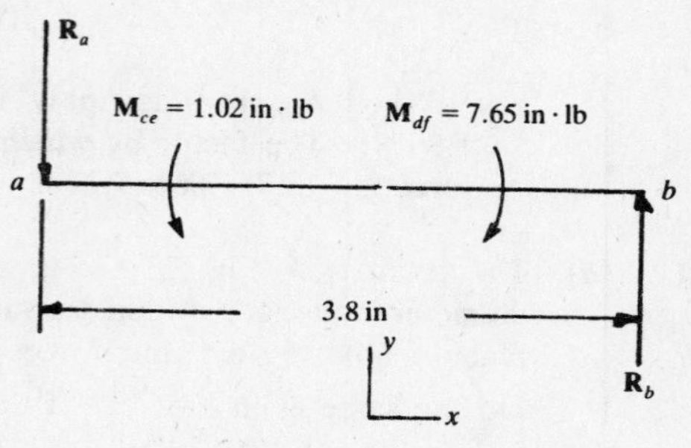

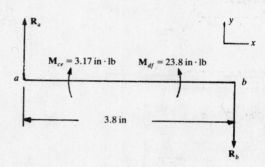

Fig. 22.21d

Fig. 22.21e

The moment due to the mass of the material in hole $ce$ is

$$M_{df} = \tfrac{1}{24}ml^2\omega^2 \sin 2\beta = \tfrac{1}{24}(7.15 \times 10^{-6})(4.29)(4.29)^2 628^2 \sin[2(27.8°)] = 7.65 \text{ in} \cdot \text{lb}$$

The actual sense of $M_{df}$ is seen in Fig. 22.21c.

The free-body diagram of shaft $ab$ is shown in Fig. 22.21d. For dynamic equilibrium of the shaft,

$$\sum M_a = 0 \qquad 1.02 - 7.65 + R_b(3.8) = 0 \qquad R_b = 1.74 \text{ lb}$$

$$\sum F_y = 0 \qquad -R_a + R_b = 0 \qquad R_a = R_b = 1.74 \text{ lb}$$

(b) The solution is similiar to that in part (a). There are, however, different values of $\rho_0$, and the senses of $M_{ce}$ and $M_{df}$ change.

The density of the hole material is greater than that of the cylinder material. Thus, the hole material acts as a slender rigid rod, as in Prob. 22.20, with a mass per unit length of

$$\rho_0 = \rho A = \left(\frac{710}{12^3} - 0.1\right)\left(\frac{1}{386}\right)\left[\frac{\pi(3/16)^2}{4}\right] = 2.22 \times 10^{-5} \text{ lb} \cdot \text{s}^2/\text{in}^2$$

The moments due to the centrifugal forces acting on the material in the holes are

$$M_{ce} = \tfrac{1}{24}(2.22 \times 10^{-5})\left(\frac{2}{\cos 26°}\right)^3 (628)^2 \sin[2(64°)] = 3.17 \text{ in} \cdot \text{lb}$$

$$M_{df} = \tfrac{1}{24}(2.22 \times 10^{-5})(4.29)^3(628)^2 \sin[2(27.8°)] = 23.8 \text{ in} \cdot \text{lb}$$

Figure 22.21e shows the actual senses of $M_{ce}$ and $M_{df}$, as dynamic moments which act on the shaft. For equilibrium of the shaft,

$$\sum M_a = 0 \qquad -3.17 + 23.8 - R_b(3.8) = 0 \qquad R_b = 5.43 \text{ lb}$$

$$\sum F_y = 0 \qquad R_a - R_b = 0 \qquad R_a = R_b = 5.43 \text{ lb}$$

22.22 The two arms in Fig. 22.22a are hinged to the rotating member and connected to each other by an inextensible, weightless cable. The arms each have a weight of 0.2 lb/in.

(a) Find the tensile force in the cable, and the forces exerted by the arms on the pin, when the assembly rotates at 850 r/min.

(b) Find the minimum value of rotational speed for which the cable force will be tensile.

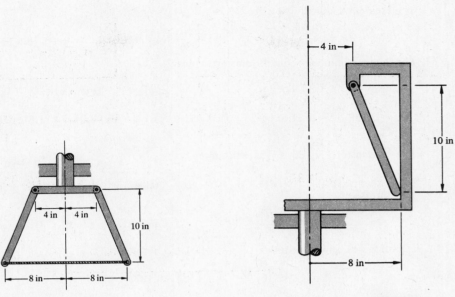

Fig. 22.22a

Fig. 22.22b

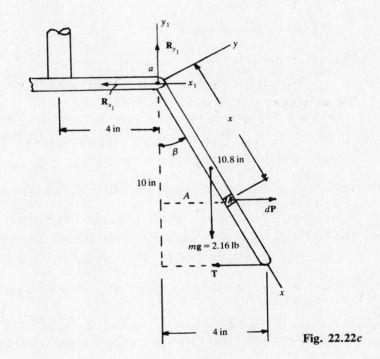

Fig. 22.22c

(c) Figure 22.22b shows a view of a cylindrical drum with a pivoted arm that rotates about a fixed axis. How could the results obtained in part (b) be used to find the speed at which the arm loses contact with the inside surface of the drum?

▌ (a) Figure 22.22c shows the centrifugal force $dP$ that acts on a differential mass element of one arm, and the cable tensile force $T$. From this figure,

$$\tan \beta = \tfrac{4}{10} \qquad \beta = 21.8° \qquad \frac{x}{A} = \frac{10.8}{4} \qquad A = 0.370x$$

The weight of the arm is

$$W = \gamma_0 l = 0.2(10.8) = 2.16 \text{ lb}$$

The elemental centrifugal force $dP$ is given by

$$dP = \left(\frac{0.2\,dx}{386}\right)(4 + A)\omega^2 = \left(\frac{0.2\,dx}{386}\right)(4 + 0.370x)\omega^2 = 0.000518\omega^2(4 + 0.370x)\,dx$$

For dynamic moment equilibrium of the arm,

$$\sum M_a = 0$$

$$-2.16\left(\frac{10.8}{2}\sin 21.8°\right) + \int_0^{10.8} [0.000518\omega^2(4 + 0.370x)\,dx]x\cos 21.8° - T(10.8\cos 21.8°) = 0$$

$$(1)$$

The second term in Eq. (1) is written as

$$0.000481\omega^2\int_0^{10.8} (4x + 0.370x^2)\,dx = 0.000481\omega^2\left[4\left(\frac{x^2}{2}\right) + 0.370\left(\frac{x^3}{3}\right)\right]\Big|_0^{10.8}$$

$$= 0.000481\omega^2\left[\tfrac{4}{2}(10.8^2 - 0) + \frac{0.370}{3}(10.8^3 - 0)\right] = 0.187\omega^2$$

Equation (1) now appears as

$$-2.16\left(\frac{10.8}{2}\sin 21.8°\right) + 0.187\omega^2 - T(10.8\cos 21.8°) = 0$$

Using

$$\omega = 850\left(\frac{2\pi}{60}\right) = 89.0\ \text{rad/s}$$

in the above equation results in

$$T = 147\ \text{lb}$$

For dynamic force equilibrium,

$$\sum F_{x_1} = 0 \qquad -R_{x_1} + \int dP - T = 0$$

$$R_{x_1} = \int_0^{10.8} 0.000518\omega^2(4 + 0.370x)\,dx - 147 = 0.000518\omega^2\left[4x + 0.370\frac{x^2}{2}\right]_0^{10.8} - 147$$

Using $\omega = 89.0\ \text{rad/s}$ in the above equation gives

$$R_{x_1} = 0.000518(89.0)^2\left[4(10.8 - 0) + \frac{0.370}{2}(10.8^2 - 0)\right] - 147 = 119\ \text{lb}$$

$$\sum F_y = 0 \qquad R_{y_1} - mg = 0 \qquad R_{y_1} = mg = 2.16\ \text{lb} \qquad R \approx R_{x_1} = 119\ \text{lb}$$

From the symmetry of construction of the system, both pins experience the same force.

(b) The minimum value of rotational speed corresponds to $T = 0$. Equation (1) is used to find $\omega$, with the result

$$-2.16\left(\frac{10.8}{2}\sin 21.8°\right) + 0.187\omega^2 = 0 \qquad \omega = 4.81\ \text{rad/s} = 45.9\ \text{r/min}$$

(c) The result is the same as in part (b). $T = 0$ is the equivalent of the normal force between the arm and the drum becoming zero when the arm loses contact with the drum.

22.23 · Figure 22.23a shows a proposed design for a device which will indicate when the angular velocity of a rotating shaft reaches a limiting, minimum value. The drum spins about a fixed vertical axis. The arm has the shape of a quarter of a thin circular ring, and it is hinged to the drum assembly at the axis of rotation. At sufficiently high speeds the centrifugal force which acts on the arm forces it against the drum.

(a) Find the magnitude and location of the resultant centrifugal force which acts on the arm.

(b) Find the normal force exerted by the drum on the arm and the force exerted by the pin on the arm.

(c) Derive a general expression for the angular velocity $\omega_0$ at which the arm loses contact with the drum.

(d) Find the numerical value of $\omega_0$ if $r = 1$ in.

▮ (a) The centrifugal force which acts on a typical mass element of the arm is shown in Fig. 22.23b. The magnitude of the resultant centrifugal force is

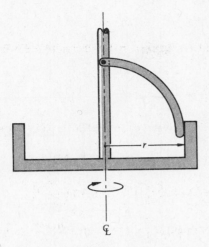

**Fig. 22.23a**

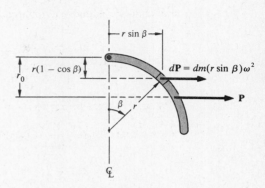

**Fig. 22.23b**

$$P = \int dP = \int dm(r \sin \beta)\omega^2 \tag{1}$$

The mass per unit length of the arm is designated $\rho_0$, so that

$$dm = \rho_0(r \, d\beta)$$

With the above result, Eq. (1) now appears as

$$P = \int_0^{\pi/2} \rho_0(r \, d\beta) r \sin \beta \omega^2 = \rho_0 r^2 \omega^2 \int_0^{\pi/2} \sin \beta \, d\beta = \rho_0 r^2 \omega^2 [-\cos \beta]_0^{\pi/2} = -\rho_0 r^2 \omega^2 [0 - 1] = \rho_0 r^2 \omega^2 \tag{2}$$

The mass $m$ of the quarter circular ring is

$$m = \frac{\rho_0 \pi r}{2} \tag{3}$$

$\rho_0$ is eliminated between Eqs. (2) and (3), with the result

$$P = \frac{2mr\omega^2}{\pi} \tag{4}$$

Figure 22.23c shows a plane curve in the form of a quarter circle. The centroidal coordinates shown in the figure may be verified by using Case 3 in Table 17.15. Equation (4) may be written as

$$P = m\left(\frac{2r}{\pi}\right)\omega^2$$

It may be concluded that the *magnitude* of the resultant centrifugal force may be found by imagining all the mass to be concentrated at the center of mass, and to have the acceleration of this point.

The location of the resultant centrifugal force, using Fig. 22.23b, is found from the resultant moment condition

$$\sum M_a = Pr_0 = \int dP \, r(1 - \cos \beta) = \int dm(r \sin \beta)\omega^2 r(1 - \cos \beta) \tag{5}$$

Using $dm = \rho_0 r d\beta$, Eq. (5) can be written as

$$Pr_0 = \rho_0 r^3 \omega^2 \int_0^{\pi/2} \sin \beta(1 - \cos \beta) \, d\beta = \rho_0 r^3 \omega^2 \left( \int_0^{\pi/2} \sin \beta \, d\beta - \int_0^{\pi/2} \sin \beta \cos \beta \, d\beta \right)$$

$$= \rho_0 r^3 \omega^2 \left( -\cos \beta \Big|_0^{\pi/2} - \frac{\sin^2 \beta}{2} \Big|_0^{\pi/2} \right) = \rho_0 r^3 \omega^2 \left[ -[0 - 1] - \frac{1}{2}(1 - 0) \right] = \frac{\rho_0 r^3 \omega^2}{2} \tag{6}$$

Using $P$ from Eq. (2) in Eq. (6) results in

$$r_0 = \frac{\rho_0 r^3 \omega^2 / 2}{\rho_0 r^2 \omega^2} = \frac{r}{2}$$

It is interesting to note that the resultant centrifugal force acts at exactly the midheight of the quarter circular arm.

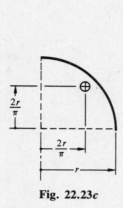

Fig. 22.23c

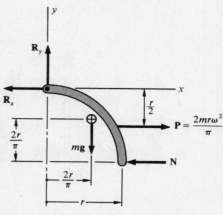

Fig. 22.23d

(b) The free-body diagram of the arm is shown in Fig. 22.23d, and $N$ is the compressive normal force between the arm and the drum. For moment equilibrium of the arm about the hinge pin

$$\sum M_a = 0 \qquad -mg\frac{2r}{\pi} + \frac{2mr\omega^2}{\pi}\frac{r}{2} - Nr = 0 \qquad N = \frac{m}{\pi}(r\omega^2 - 2g) \qquad (7)$$

For force equilibrium,

$$\sum F_x = 0 \qquad -R_x + P - N = 0 \qquad R_x = P - N = \frac{2mr\omega^2}{\pi} - \frac{m}{\pi}(r\omega^2 - 2g) = \frac{m}{\pi}(r\omega^2 + 2g)$$

$$\sum F_y = 0 \qquad R_y - mg = 0 \qquad R_y = mg$$

The force exerted by the pin on the arm is

$$R = \sqrt{R_x^2 + R_y^2}$$

(c) The limiting condition when the arm loses contact with the drum occurs when $N = 0$. The corresponding value of the angular velocity $\omega_0$ is found from Eq. (7) as

$$N = 0 \qquad r\omega_0^2 - 2g = 0 \qquad \omega_0 = \sqrt{\frac{2g}{r}}$$

It is interesting to note that this result is *completely independent* of the mass properties of the arm.

(d) The value of $\omega_0$ which corresponds to $r = 1$ in is

$$\omega_0 = \sqrt{\frac{2g}{r}} = \sqrt{\frac{2(386)}{1}} = 27.8 \text{ rad/s} = 27.8\left(\frac{60}{2\pi}\right) = 265 \text{ r/min}$$

22.24 The device in Prob. 22.23 is modified to the form shown in Fig. 22.24a. The weight of the arm is 0.025 lb.

(a) Find the magnitude and location of the resultant centrifugal force which acts on the arm.

(b) Find the normal force exerted by the curved arm on the drum, and the hinge-pin force, when $\omega = 600$ r/min.

(c) Find the general expression for the limiting value $\omega_0$ of the angular velocity at which the arm loses contact with the drum, and the numerical value of $\omega_0$, if $r = 1$ in.

▌ (a) Figure 22.24b shows the centrifugal force $dP$ which acts on a typical mass element of the arm. The resultant centrifugal force is

$$P = \int dP = \int dm\, r\omega^2 = \int_0^{\pi/2} \rho_0 r d\beta r(1 - \sin\beta)\omega^2 = \rho_0 \omega^2 r^2 \int_0^{\pi/2}(1 - \sin\beta)\,d\beta$$

$$= \rho_0\omega^2 r^2(\beta + \cos\beta)\Big|_0^{\pi/2} = \rho_0\omega^2 r^2\left[\left(\frac{\pi}{2} - 0\right) + (0 - 1)\right] = \rho_0\omega^2 r^2\left(\frac{\pi}{2} - 1\right)$$

Using $m = \rho_0\pi r/2$ and $\rho_0 = 2m/\pi r$ in the above equation results in

$$P = \frac{2m}{\pi r}\omega^2 r^2\left(\frac{\pi}{2} - 1\right) = mr\left(1 - \frac{2}{\pi}\right)\omega^2 = 0.363mr\omega^2$$

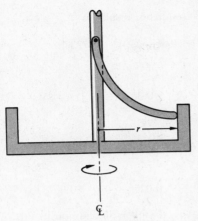

**Fig. 22.24a**

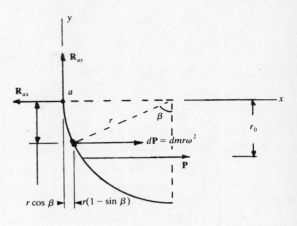

**Fig. 22.24b**

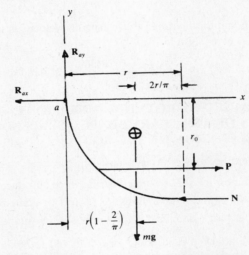

**Fig. 22.24c**

The centroidal coordinate of the plane curve shape which represents the arm, from Case 3 in Table 17.15, is shown in Fig. 22.24c. It may be seen that the magnitude of the resultant centrifugal force may be found by imagining all the mass to be concentrated at the center of mass, and to have the acceleration of this point.

The resultant centrifugal force $P$ is shown in Fig. 22.24c. The location $r_0$ of $P$ is found from the resultant moment condition

$$\sum M_a = Pr_0 = \int_0^{\pi/2} (\rho_0 r d\beta) r(1 - \sin \beta)\omega^2 r \cos \beta$$

$$\rho_0 \omega^2 r^2 \left(\frac{\pi}{2} - 1\right) r_0 = \rho_0 r^3 \omega^2 \int_0^{\pi/2} \cos \beta(1 - \sin \beta)\, d\beta \qquad (1)$$

The integral in the above equation is written as

$$\int_0^{\pi/2} \cos \beta\, d\beta - \int_0^{\pi/2} \sin \beta \cos \beta\, d\beta = \sin \beta \Big|_0^{\pi/2} - \frac{\sin^2 \beta}{2}\Big|_0^{\pi/2} = \left[(1-0) - \frac{1}{2}(1-0)\right] = \frac{1}{2}$$

Equation (1) now appears as

$$\left[\rho_0 \omega^2 r^2 \left(\frac{\pi}{2} - 1\right)\right] r_0 = \rho_0 r^3 \omega^2 \left(\frac{1}{2}\right) \qquad r_0 = \frac{r}{\pi - 2} = 0.876r$$

(b) The force $N$, shown in Fig. 22.24c, is the normal force exerted on the arm by the drum. For moment equilibrium,

$$\sum M_a = 0 \qquad -mgr\left(1 - \frac{2}{\pi}\right) + Pr_0 - Nr = 0$$

$$-mgr\left(1 - \frac{2}{\pi}\right) + mr\left(1 - \frac{2}{\pi}\right)\omega^2\left(\frac{r}{\pi-2}\right) - Nr = 0 \qquad (2)$$

$$N = \frac{m}{\pi}\left[r\omega^2 - g(\pi - 2)\right]$$

Using $\omega = 600\,\text{r/min} = 62.8\,\text{rad/s}$ in the above equation results in

$$N = \frac{0.025}{386\pi}[1(62.8)^2 - 386(\pi - 2)] = 0.0722\,\text{lb}$$

For force equilibrium

$$\sum F_x = 0 \qquad -R_{ax} + P - N = 0 \qquad -R_{ax} + 0.363\left(\frac{0.025}{386}\right)1(62.8)^2 - 0.0722 = 0$$

$$R_{ax} = 0.0205\,\text{lb}$$

$$\sum F_y = 0 \qquad R_{ay} - mg = 0 \qquad R_{ay} = 0.025\,\text{lb}$$

The resultant force which acts on the pin is

$$R_a = \sqrt{R_{ax}^2 + R_{ay}^2} = \sqrt{0.0205^2 + 0.025^2} \qquad R_a = 0.0323\,\text{lb}$$

(c) The arm loses contact with the drum when $N = 0$. Using this condition in Eq. (2) gives

$$\sum M_a = 0 \qquad -mgr\left(1 - \frac{2}{\pi}\right) + mr\left(1 - \frac{2}{\pi}\right)\omega_0^2\left(\frac{r}{\pi - 2}\right) = 0 \qquad \omega_0 = \sqrt{\frac{(\pi - 2)g}{r}}$$

It may be seen that this limiting value of angular velocity is smaller than that obtained in Prob. 22.23, by the factor $\sqrt{(\pi - 2)/2} = 0.756$. For $r = 1\,\text{in}$,

$$\omega_0 = \sqrt{\frac{(\pi - 2)386}{1}} = 21.0\,\text{rad/s} = 201\,\text{r/min}$$

## 22.2 DYNAMIC FORCES CAUSED BY ROTATING UNBALANCE—GENERAL SOLUTION FOR UNBALANCED BODIES OF ARBITRARY SHAPE, INDEPENDENCE OF DYNAMIC FORCES AND ANGULAR ACCELERATION OF THE BODY

22.25 Figure 22.25a shows a rigid body of arbitrary shape which rotates about a fixed axis.

(a) Find the general solution for the *dynamic* bearing forces produced by the rotational motion.

(b) Find the *necessary* and *sufficient* conditions for perfect dynamic balance of the body.

(c) How do the results in part (a) change if the body is a thin plane body which lies in the xy plane?

▮ (a) The shaft and the attached rigid body are considered to be an integral assembly of total mass m, and the effect of the static weight force of the assembly on the bearing forces is neglected.

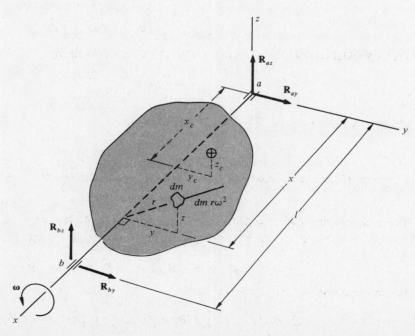

Fig. 22.25a

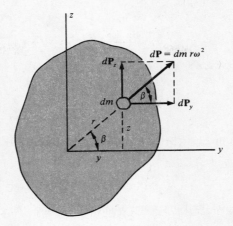

**Fig. 22.25b**

The $xyz$ axes are attached to the body, as shown in Fig. 22.25$a$, and rotate with this element. The origin of the coordinates is placed at one of the shaft bearings, designated $a$, and the $x$ axis is the axis of rotation. The spacing of the two bearings is given by $l$, and the position coordinates of the center of mass of the assembly are given by $x_c$, $y_c$, and $z_c$. The $y$ and $z$ components of the bearing forces are assumed to act in the positive coordinate senses.

The typical mass element $dm$ is acted on by the elementary centrifugal force $dP = dm\, r\omega^2$. This force and its components are shown in true view in Fig. 22.25$b$. From this figure,

$$dP_y = dP \cos \beta = (dm\, r\omega^2) \cos \beta \qquad (1)$$
$$dP_z = dP \sin \beta = (dm\, r\omega^2) \sin \beta \qquad (2)$$

Since

$$\sin \beta = \frac{z}{r} \qquad \cos \beta = \frac{y}{r}$$

Eqs. (1) and (2) may be written as

$$dP_y = (dm\, r\omega^2)\frac{y}{r} = y\omega^2\, dm \qquad dP_z = (dm\, r\omega^2)\frac{z}{r} = z\omega^2\, dm$$

The $y$ and $z$ components of the resultant centrifugal force acting on the rigid body are then

$$P_y = \int dP_y = \int_V y\omega^2\, dm = \omega^2 \int_V y\, dm \qquad (3)$$

$$P_z = \int dP_z = \int_V z\omega^2\, dm = \omega^2 \int_V z\, dm \qquad (4)$$

The centroidal coordinates $y_c$ and $z_c$ in Fig. 22.25$a$ are defined by

$$y_c = \frac{\int_V y\, dm}{m} \qquad z_c = \frac{\int_V dm}{m}$$

where $m$ is the total mass of the body. Thus,

$$\int_V y\, dm = my_c \qquad \int_V z\, dm = mz_c \qquad (5)$$

Equations (5) are substituted into Eqs. (3) and (4), with the results

$$P_y = my_c\omega^2 \qquad P_z = mz_c\omega^2$$

The requirements of force equilibrium of the body in Fig. 22.25$a$ are

$$\sum F_y = 0 \qquad R_{ay} + P_y + R_{by} = 0 \qquad R_{ay} + R_{by} = -P_y = -my_c\omega^2 \qquad (6)$$

$$\sum F_z = 0 \qquad R_{az} + P_z + R_{bz} = 0 \qquad R_{az} + R_{bz} = -P_z = -mz_c\omega^2 \qquad (7)$$

For moment equilibrium of the rigid body,

$$\sum M_y = 0 \qquad -\int_V (dP_z)x - R_{bz}l = -\int_V (z\omega^2\, dm)x - R_{bz}l = 0$$

$$R_{bz} = \frac{-\omega^2 \int_V zx \, dm}{l} \qquad (8)$$

$$\sum M_z = 0 \qquad \int_V (dP_y)x + R_{by}l = \int_V (y\omega^2 \, dm)x + R_{by}l = 0$$

$$R_{by} = \frac{-\omega^2 \int_V xy \, dm}{l} \qquad (9)$$

The numerators in Eqs. (8) and (9) may be recognized as mass products of inertia, defined in Prob. 17.93 as

$$I_{xy} = \int_V xy \, dm \qquad I_{zx} = \int_V zx \, dm$$

The final forms for the force components at bearing $b$ are

$$R_{by} = \frac{-\omega^2 I_{xy}}{l} \qquad R_{bz} = \frac{-\omega^2 I_{zx}}{l} \qquad (10)$$

Using Eqs. (6) and (7),

$$R_{ay} = -my_c\omega^2 - R_{by} = \frac{\omega^2 I_{xy}}{l} - my_c\omega^2 \qquad (11)$$

$$R_{az} = -mz_c\omega^2 - R_{bz} = \frac{\omega^2 I_{zx}}{l} - mz_c\omega^2 \qquad (12)$$

$R_{ay}$, $R_{az}$, $R_{by}$, and $R_{bz}$ are the components of the *dynamic* forces exerted on the shaft by the bearings. These terms, with opposite senses, are the dynamic forces exerted *on the bearings by the* shaft.

**(b)** The rotating assembly consisting of the rigid body and the shaft is formally defined to be in *dynamic balance if the four dynamic bearing force components in part (a) are identically zero for all values of* $\omega$. This condition requires that, from Eq. (10),

$$R_{by} = -\frac{\omega^2 I_{xy}}{l} = 0 \qquad \therefore I_{xy} = 0 \qquad (13)$$

$$R_{bx} = -\frac{\omega^2 I_{zx}}{l} = 0 \qquad \therefore I_{zx} = 0 \qquad (14)$$

and, from Eqs. (6) and (7),

$$R_{ay} + R_{by} = -my_c\omega^2 \qquad 0 + 0 = -my_c\omega^2 \qquad \therefore y_c = 0$$
$$R_{az} + R_{bz} = -mz_c\omega^2 \qquad 0 + 0 = -mz_c\omega^2 \qquad \therefore z_c = 0$$

The *necessary* condition for *static balance* of the assembly is that *the center of mass be located on the axis of rotation*. This requirement is given by

$$y_c = 0 \qquad z_c = 0$$

It may be seen that static balance is a *necessary*, but not *sufficient*, condition for dynamic balance. The additional requirement, from the results above, is

$$I_{xy} = 0 \qquad I_{zx} = 0$$

The interpretation of these equations is that the *xyz axes must be principal axes of the body.*
  In summary, for complete dynamic balance of a rotating mass assembly:

1. The center of mass of the assembly must lie on the axis of rotation.
2. The *xyz* axes which are attached to the body must be principal axes of the mass assembly.

The equations presented above are perfectly general results for the dynamic bearing forces which act on any rotating shaft-mass assembly that is supported by two bearings. As with any other general solution, the terms in the solution must be carefully interpreted in light of their original definition. The definitions which have been used in the present analysis are repeated below:

1. The *xyz* axes are *attached* to the body with the origin at one bearing, designated *a*, and the *x* axis is coincident with the axis of rotation.

**2.** The remaining bearing is designated $b$.

**3.** The axial spacing between the bearings is designated $l$, measured in the positive $x$-coordinate sense.

**4.** The centroidal coordinates and the products of inertia *are referenced to the axes which are attached to the body*.

**5.** The four components of the bearing forces exerted by the bearings on the shaft are considered to be *positive quantities* if they act in the positive coordinate senses.

(*c*) If the bodies considered are thin plane bodies which lie in the $xy$ plane, the equations presented above have simplified forms. For the assumption of thin bodies,

$$z_c \approx 0 \qquad I_{yz} \approx 0 \qquad I_{zx} \approx 0$$

Using Eqs. (10), the dynamic forces $R_{by}$ and $R_{bz}$ have the forms

$$R_{by} = -\frac{\omega^2 I_{xy}}{l} \qquad R_{bz} \approx 0 \tag{15}$$

Using Eqs. (6) and (7), the dynamic forces $R_{ay}$ and $R_{az}$ have the forms

$$R_{ay} = \frac{\omega^2 I_{xy}}{l} - m y_c \omega^2 \qquad R_{az} \approx 0 \tag{16}$$

It follows from these equations that the necessary and sufficient conditions for dynamic balance of a thin plane body which lies in the $xy$ plane are

$$y_c = 0 \qquad I_{xy} = 0$$

**22.26** Show the independence of the dynamic bearing forces and the angular acceleration of rotating, unbalanced bodies.

❚ The configuration of a rigid body of arbitrary shape which rotates about a fixed axis is shown in Fig. 22.25*a*. The assembly, consisting of the rigid body and the shaft, together with the centrifugal forces, is a three-dimensional problem in dynamic equilibrium. The necessary condition for equilibrium of this system is the satisfaction of six equations of equilibrium. In the analysis in Prob. 22.25 for the dynamic bearing forces, the four equilibrium equations

$$\sum F_y = 0 \qquad \sum F_z = 0 \qquad \sum M_y = 0 \qquad \sum M_z = 0$$

were used to find the values of the four dynamic bearing forces $R_{ay}$, $R_{az}$, $R_{by}$, and $R_{bz}$.

The remaining two equilibrium equations are

$$\sum F_x = 0 \tag{1}$$

$$\sum M_x = I\alpha \tag{2}$$

If no axial, or thrust, forces in the $x$ direction act on the assembly, Eq. (1) above is identically satisfied. If there are axial forces, a typical example of which would be the axial component of the total force on a helical gear tooth, this equation would have the form

$$F_x = R_{ax} \text{ or } R_{bx}$$

where $F_x$ is the external axial force which acts on the gear tooth and $R_{ax}$ or $R_{bx}$ is the axial component of force at a bearing. The above forces are all *static* forces and thus do not contribute to the *dynamic* bearing force.

The final equation, Eq. (2), is the rotational equation of motion about the fixed $x$ axis, and this equation may be written as

$$M_x = I_x \alpha \tag{3}$$

where $M_x$ is the resultant moment, about the $x$ axis, which acts on the body, $I_x$ is the mass moment of inertia of the body about the $x$ axis, and $\alpha$ is the angular acceleration. Since Eq. (3) is the only equilibrium equation which includes the angular acceleration, and since this equation does *not* contain any of the bearing force components, it may be concluded that the dynamic bearing forces due to rotating unbalance are independent of the angular acceleration of the rotating body.

**22.27** The system of Prob. 22.3 is redrawn in Fig. 22.27. Use the techniques presented in Prob. 22.25 to find the dynamic forces which act on the shaft.

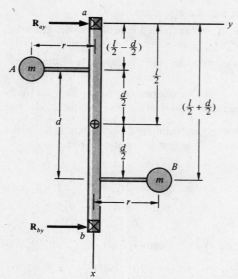

**Fig. 22.27**

▌ Since the masses are assumed to be point masses,

$$I_{xy} = \int_V xy\,dm \approx x_A y_A m + x_B y_B m$$

Using

$$x_A = \frac{l}{2} - \frac{d}{2} \qquad x_B = \frac{l}{2} + \frac{d}{2} \qquad y_A = -r \qquad y_B = r$$

the product of inertia $I_{xy}$ is found as

$$I_{xy} = \left(\frac{l}{2} - \frac{d}{2}\right)(-r)m + \left(\frac{l}{2} + \frac{d}{2}\right)rm = mrd$$

By using Eq. (10) in Prob. 22.25, the dynamic force $R_{by}$ is found as

$$R_{by} = \frac{-\omega^2 I_{xy}}{l} = \frac{-mrd\omega^2}{l}$$

With $y_c = 0$, Eq. (6) in Prob. 22.25 appears as

$$R_{ay} + R_{by} = -m\omega^2 y_c = 0 \qquad R_{ay} = -R_{by} = \frac{mrd\omega^2}{l}$$

The minus sign on the result for $R_{by}$ indicates that the actual sense of this force is to the left in Fig. 22.27. It may be observed that Eqs. (15) and (16) in Prob. 22.25 could also have been used directly to solve this problem.

**22.28**  Solve Prob. 22.8 by using the general solutions given in Prob. 22.25.

▌ (a)  Figure 22.8a is used. The product of inertia $I_{xy}$ has the form

$$I_{xy} = x_A y_A m_A + x_B y_B m_B + x_C y_C m_C = \frac{(-150)150(0.2)}{1{,}000^2} + \frac{100(110)0.1}{1{,}000^2} + \frac{400(-120)0.3}{1{,}000^2}$$

$$= -0.0178\,\text{kg} \cdot \text{m}^2$$

Using Eq. (15) in Prob. 22.25,

$$R_{by} = \frac{-\omega^2 I_{xy}}{l} = \frac{-50^2(-0.0178)}{500/1{,}000} = 89\,\text{N} \approx 89.5\,\text{N}$$

The centroidal coordinate $y_c$ is given by

$$y_c = \frac{y_A m_A + y_B m_B + y_C m_C}{m_A + m_B + m_C} = \frac{150(0.2) + 110(0.1) - 120(0.3)}{0.2 + 0.1 + 0.3} = 8.33\,\text{mm}$$

Using Eq. (16) in Prob. 22.25,

$$R_{ay} = \frac{\omega^2 I_{xy}}{l} - my_c\omega^2 = \frac{50^2(-0.0178)}{500/1,000} - 0.6\left(\frac{8.33}{1,000}\right)50^2 = -101\ N$$

When using the above general result for $R_{ay}$, positive $R_{ay}$ acts in the positive $y$-coordinate sense. Thus

$$|-101\ N| \approx 102\ N$$

(b) When mass $m_A$ becomes detached,

$$I_{xy} = x_B y_B m_B + x_C y_C m_C = \frac{100(110)0.1}{1,000^2} + \frac{400(-120)0.3}{1,000^2} = -0.0133\ kg\cdot m^2$$

$$R_{by} = \frac{-\omega^2 I_{xy}}{l} = \frac{-50^2(-0.0133)}{500/1,000} = 66.5\ N$$

$$y_c = \frac{y_B m_B + y_C m_C}{m_B + m_C} = \frac{110(0.1) - 120(0.3)}{0.1 + 0.3} = -62.5\ mm$$

$$R_{ay} = \frac{\omega^2 I_{xy}}{l} - my_c\omega^2 = \frac{50^2(-0.0133)}{500/1,000} - \frac{0.4(-62.5)50^2}{1,000} = -4\ N \qquad |-4\ N| \equiv 4\ N$$

The above results for $R_{ay}$ and $R_{by}$ were obtained in Prob. 22.8, part (b).

(c) When mass $m_B$ becomes detached,

$$I_{xy} = x_A y_A m_A + x_C y_C m_C = \frac{(-150)150(0.2)}{1,000^2} + \frac{400(-120)0.3}{1,000^2} = -0.0189\ kg\cdot m^2$$

$$R_{by} = \frac{-\omega^2 I_{xy}}{l} = \frac{-50^2(-0.0189)}{500/1,000} = 94.5\ N$$

$$y_c = \frac{y_A m_A + y_C m_C}{m_A + m_C} = \frac{150(0.2) - 120(0.3)}{0.2 + 0.3} = -12\ mm$$

$$R_{ay} = \frac{\omega^2 I_{xy}}{l} - my_c\omega^2 = \frac{50^2(-0.0189)}{500/1,000} - \frac{0.5(-12)50^2}{1,000} = -79.5\ N \qquad |-79.5\ N| \equiv 79.5\ N$$

The above results for $R_{ay}$ and $R_{by}$ were obtained in Prob. 22.8, part (c).

(d) When mass $m_C$ becomes detached,

$$I_{xy} = x_A y_A m_A + x_B y_B m_B = \frac{(-150)150(0.2)}{1,000^2} + \frac{100(110)0.1}{1,000^2} = -0.00340\ kg\cdot m^2$$

$$R_{by} = -\frac{\omega^2 I_{xy}}{l} = \frac{-50^2(-0.00340)}{500/1,000} = 17\ N \approx 17.5\ N$$

$$y_c = \frac{y_A m_A + y_B m_B}{m_A + m_B} = \frac{150(0.2) + 110(0.1)}{0.2 + 0.1} = 137\ mm$$

$$R_{ay} = \frac{\omega^2 I_{xy}}{l} - my_c\omega^2 = \frac{50^2(-0.00340)}{500/1,000} - 0.3\left(\frac{137}{1,000}\right)50^2 = -120\ N \qquad |-120\ N| \equiv 120\ N$$

The above results were obtained in Prob. 22.8, part (d).

**22.29** (a) Solve Prob. 22.14, part (b), by using the general solutions given in Prob. 22.25. Consider the two projections on the shaft to be slender rod elements.

(b) Solve Prob. 22.14, part (c), by using the general solutions given in Prob. 22.25.

▌ (a) Figures 22.14b and c are used. The product of inertia $I_{xy}$ of the body has the form

$$I_{xy} = I_{xy,A} + I_{xy,B}$$

For element $A$,

$$I_{xy,A} = \int xy\ dm = x_A \int dm$$

Using $x_A = 5$ in,

$$I_{xy,A} = 5\int_{-0.5}^{-4} (-y)\left[1(0.5)\, dy\left(\frac{0.283}{386}\right)\right] = -5(1)0.5\left(\frac{0.283}{386}\right)\int_{-0.5}^{-4} y\, dy \qquad (1)$$

The integral term in the above equation has the form

$$\int_{-0.5}^{-4} y\, dy = \frac{y^2}{2}\bigg|_{-0.5}^{-4} = \frac{1}{2}(4^2 - 0.5^2)$$

Equation (1) appears as

$$I_{xy,A} = -5(1)0.5\left(\frac{0.283}{386}\right)\left(\frac{1}{2}\right)(4^2 - 0.5^2) = -0.0144\,\text{lb}\cdot\text{s}^2\cdot\text{in} \qquad (2)$$

For element $B$,

$$I_{xy,B} = \int xy\, dm = x_B\int y\, dm$$

$x_B = 15$ in, and Eq. (2), is used, with the term $(-5)$ replaced by the term $(15)$.   The result is

$$I_{xy,B} = 15(1)0.5\left(\frac{0.283}{386}\right)\left(\frac{1}{2}\right)(4^2 - 0.5)^2 = 0.0433\,\text{lb}\cdot\text{s}^2\cdot\text{in}$$

The product of inertia of the body is

$$I_{xy} = I_{xy,A} + I_{xy,B} = -0.0144 + 0.0433 = 0.0289\,\text{lb}\cdot\text{s}^2\cdot\text{in}$$

Using Eq. (15) in Prob. 22.25,

$$R_{by} = \frac{-\omega^2 I_{xy}}{l} = \frac{-103^2(0.0289)}{20} = -15.3\,\text{lb}$$

Using Eq. (16) in Prob. 22.25, with $y_c = 0$, gives

$$R_{ay} = -R_{by} = 15.3\,\text{lb}$$

The above results were obtained in Prob. 22.14, part $(b)$.

$(b)$  Figure 22.14$d$ is used.  The product of inertia $I_{xy}$ of the body has the form

$$I_{xy} = I_{xy,A} + I_{xy,B} + I_{xy,C} \qquad (3)$$

For element $A$,

$$I_{xy,A} = \int xy\, dm = x_A\int y\, dm$$

Using the form of Eq. (1) in part $(a)$,

$$I_{xy,A} = -5(1)0.5\left(\frac{0.283}{386}\right)\int_{-(0.5-0.050)}^{-(4-0.050)} y\, dy \qquad (4)$$

The integral term is written as

$$\int_{-0.45}^{-3.95} y\, dy = \frac{y^2}{2}\bigg|_{-0.45}^{-3.95}$$

and

$$I_{xy,A} = -5(1)0.5\left(\frac{0.283}{386}\right)\left(\frac{1}{2}\right)(3.95^2 - 0.45^2) = -0.0141\,\text{lb}\cdot\text{s}^2\cdot\text{in}$$

$I_{xy,B}$ has the form

$$I_{xy,B} = \int xy\, dm = x_B\int y\, dm$$

Using the form of Eq. (4), with $x_B = 15$ in,

$$I_{xy,B} = 15(1)0.5\left(\frac{0.283}{386}\right)\int_{0.5+0.050}^{4+0.050} y\, dy$$

The integral term is written as

$$\int_{0.55}^{4.05} y\, dy = \frac{y^2}{2}\bigg|_{0.55}^{4.05}$$

and

$$I_{xy,B} = 15(1)0.5\left(\frac{0.283}{386}\right)\left(\frac{1}{2}\right)(4.05^2 - 0.55^2) = 0.0443\,\text{lb} \cdot \text{s}^2 \cdot \text{in}$$

$I_{xy,C}$ is written as

$$I_{xy,C} = x_c y_c m_C = 10(0.050)\left[(1)0.5(20)\left(\frac{0.283}{386}\right)\right] = 0.00367\,\text{lb} \cdot \text{s}^2 \cdot \text{in}$$

The product of inertia of the body, using Eq. (3), is then

$$I_{xy} = -0.0141 + 0.0443 + 0.00367 = 0.0339\,\text{lb} \cdot \text{s}^2 \cdot \text{in}$$

Using Eq. (15) in Prob. 22.25,

$$R_{by} = \frac{-\omega^2 I_{xy}}{l} = \frac{-(103)^2 0.0339}{20} = -18.0\,\text{lb}$$

Using Eq. (16) in Prob. 22.25,

$$R_{ay} = \frac{\omega^2 I_{xy}}{l} - my_c\omega^2 \qquad (5)$$

$y_c = 0.050\,\text{in}$,   and the mass of the body is

$$m = [4(1) + 20(1) + 4(1)]0.5\left(\frac{0.283}{386}\right) = 0.0103\,\text{lb} \cdot \text{s}^2/\text{in}$$

Equation (5) appears as

$$R_{ay} = \frac{103^2(0.0339)}{20} - 0.0103(0.050)103^2 = 12.5\,\text{lb} \approx 12.7\,\text{lb}$$

The results above are the same as those obtained in Prob. 22.14, part (c).

22.30  Figure 22.30a shows a stirring blade which is used in a paint manufacturing operation. The blade is formed of $\frac{3}{4}$-in-diameter steel rod. $\gamma = 0.283\,\text{lb/in}^3$, and the rotational velocity is 900 r/min. The two bearings at a and b have a maximum recommended radial load rating of 200 lb. Compare the dynamic forces which act on the bearings with the recommended load value.

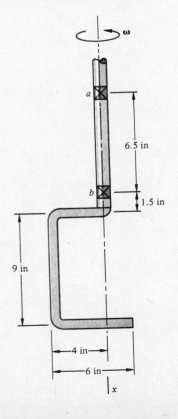

Fig. 22.30a

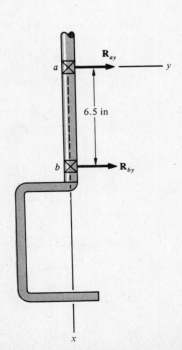

Fig. 22.30b

❚ Figure 22.30$b$ shows the dynamic forces exerted by the bearings on the blade assembly. From Prob. 17.97,

$$I_{xy} = -0.175 \, \text{lb} \cdot \text{s}^2 \cdot \text{in}$$

The centroidal coordinate $y_c$ of the blade form, using the information in Table 17.12 in Prob. 17.97, is given by

$$y_c = \frac{y_1 m_1 + y_2 m_2 + y_3 m_3}{m_1 + m_2 + m_3} = \frac{-2(0.00130) - 3.63(0.00267) - 1(0.00194)}{0.00591} = -2.41 \, \text{in}$$

Using $\omega = 900(2\pi/60) = 94.2 \, \text{rad/s}$, and Eq. (15) in Prob. 22.25, the bearing force at $b$ is

$$R_{by} = \frac{-\omega^2 I_{xy}}{l} = \frac{-94.2^2(-0.175)}{6.5} = 239 \, \text{lb}$$

The bearing force at $a$ is found from Eq. (16) in Prob. 22.25 as

$$R_{ay} = \frac{\omega^2 I_{xy}}{l} - m y_c \omega^2 = \frac{94.2^2(-0.175)}{6.5} - 0.00591(94.2)^2(-2.41) = -113 \, \text{lb}$$

It may be noted that the 239-lb magnitude of the force at bearing $b$ exceeds the maximum recommended bearing force of 200 lb by 20 percent.

**22.31** In order to reduce the the stirring blade bearing forces in Prob. 22.30, the blade is to be modified by addition of the shaded length of rod shown in Fig. 22.31$a$. Find the bearing forces for this new blade configuration, and compare them with the bearing forces in the original design.

❚ Since all the straight length elements of the blade are normal or parallel to the $xy$ axes, all centroidal products of inertia of these lengths are zero. Thus

$$I_{xy} = m x_c y_c$$

$m$ is the mass of the entire blade, and $x_c$ and $y_c$ are the centroidal coordinates, shown in Fig. 22.31$a$, of the blade form.

The blade is divided into the four straight elements shown in Fig. 22.31$b$. The centroidal coordinates of the plane curve shape of the blade are found by inspection to be

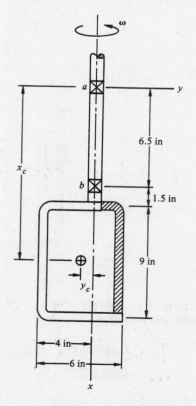

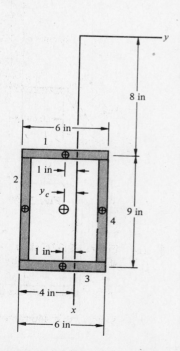

**Fig. 22.31$a$**                    **Fig. 22.31$b$**

$$x_c = 8 + 4.5 = 12.5 \text{ in} \qquad y_c = -1 \text{ in}$$

The mass of the blade, from Prob. 17.97, is

$$m = \rho_0 l = 3.24 \times 10^{-4}(6 + 8.25 + 6 + 8.25) = 9.23 \times 10^{-3} \text{ lb} \cdot \text{s}^2/\text{in}$$

The product of inertia $I_{xy}$ is then

$$I_{xy} = m x_c y_c = 9.23 \times 10^{-3}(12.5)(-1) = -0.115 \text{ lb} \cdot \text{s}^2 \cdot \text{in}$$

Using Eqs. (15) and (16) in Prob. 22.25, the bearing forces are

$$R_{by} = \frac{-\omega^2 I_{xy}}{l} = \frac{-94.2^2(-0.115)}{6.5} = 157 \text{ lb}$$

$$R_{ay} = \frac{\omega^2 I_{xy}}{l} - m\omega^2 y_c = -R_{by} - m\omega^2 y_c = -157 - 9.23 \times 10^{-3}(94.2^2)(-1) = -75.1 \text{ lb}$$

The original bearing forces, found in Prob. 22.30, were

$$R_{by} = 239 \text{ lb} \qquad R_{ay} = -113 \text{ lb}$$

It may be seen that the proposed blade modification significantly reduces the magnitude of the maximum dynamic bearing force.

**22.32** The stirring blade in Fig. 22.32a is formed of thin 8-mm-diameter bronze rod. The density of bronze is $8{,}800 \text{ kg/m}^3$.

(a) Find the dynamic forces that act on the bearings at $a$ and $b$, when the shaft rotates at 440 r/min, by integration of the centrifugal forces acting on the mass elements.

(b) Verify the results in part (a) by using the general solution, given in Prob. 22.25, for dynamic bearing forces.

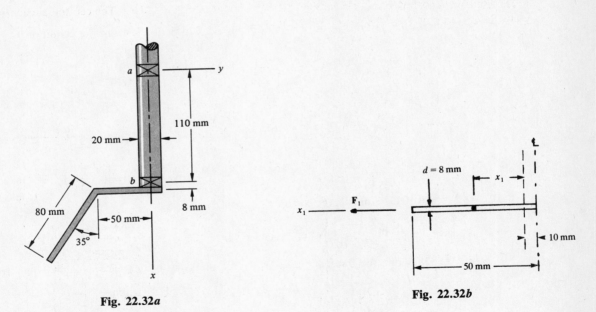

Fig. 22.32a

Fig. 22.32b

▌ (a) Figure 22.32b shows the differential mass element of the straight length of rod that extends past the shaft diameter. The centrifugal force $F_1$ which acts on this rod is

$$F_1 = \int dm\, x_1 \omega^2 = \omega^2 \int x_1 (A\, dx_1)\rho = \omega^2 \rho \left(\frac{\pi d^2}{4}\right) \int_{10}^{50} x_1\, dx_1 = \omega^2 \rho \left(\frac{\pi d^2}{4}\right) \frac{x_1^2}{2}\Big|_{10}^{50} = \frac{\pi d^2 \rho \omega^2}{8}(50^2 - 10^2)$$

Using $\omega = 440 \text{ r/min} = 46.1 \text{ rad/s}$ in the above equation gives

$$F_1 = \frac{\pi (8)^2 8{,}800(46.1)^2[(50)^2 - (10)^2]}{8(1{,}000)^4} = 1.13 \text{ N}$$

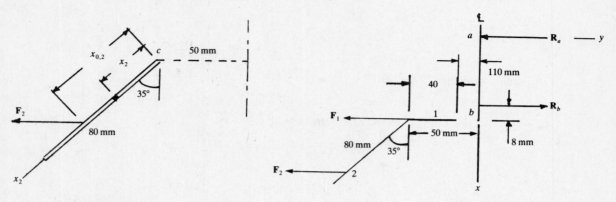

**Fig. 22.32c**

**Fig. 22.32d**

Figure 22.32c shows the differential mass element of the inclined length of rod. The resultant centrifugal force which acts on the rod is

$$F_2 = \int dm(50 + x_2 \sin 35°)\omega^2 = \omega^2 \int (50 + x_2 \sin 35°)(A\, dx_2)\rho = \omega^2 \rho\left(\frac{\pi d^2}{4}\right)\int_0^{80}(50 + x_2 \sin 35°)\, dx_2$$

$$= \omega^2 \rho\left(\frac{\pi d^2}{4}\right)\left(50x_2 + \frac{x_2^2}{2}\sin 35°\right)\Bigg|_0^{80} = \frac{\pi d^2 \rho \omega^2}{4}\left[50(80 - 0) + \frac{\sin 35°}{2}(80^2 - 0)\right]$$

$$= \frac{\pi(8)^2(8,800)46.1^2}{4(1,000)^4}\left[50(80) + \frac{\sin 35°}{2}(80)^2\right] = 5.49\ \text{N}$$

The location of $F_2$, shown as position $x_{0,2}$ in Fig. 22.32c, is found from the condition

$$\sum M_c = F_2(x_{0,2}\cos 35°) = \int dm(50 + x_2 \sin 35°)\omega^2(x_2 \cos 35°)$$

$$F_2 x_0 \cos 35° = \omega^2 \rho\left(\frac{\pi d^2}{4}\right)\int_0^{80}(50 + x_2 \sin 35°)(x_2 \cos 35°)\, dx_2 = \frac{\pi d^2 \rho \omega^2}{4}\int_0^{80}(50x_2 \cos 35°$$

$$+ x_2^2 \sin 35° \cos 35°)\, dx_2 = \frac{\pi d^2 \rho \omega^2}{4}\left[50\cos 35°\left(\frac{x_2^2}{2}\right) + \sin 35° \cos 35°\left(\frac{x_2^3}{3}\right)\right]\Bigg|_0^{80}$$

$$= \frac{\pi(8)^2 8,800(46.1)^2}{4}\left[\frac{50\cos 35°}{2}(80)^2 + \frac{\sin 35° \cos 35°}{3}(80)^3\right]\left(\frac{1}{1,000}\right)^5$$

$$= 0.1986\ \text{N}\cdot\text{m} = F_2 x_{0,2}\cos 35° = 5.49 x_{0,2}\cos 35°$$

$$x_{0,2} = 0.0442\ \text{m} = 44.2\ \text{mm}$$

Figure 22.32d shows the free-body diagram of the shaft and blade. For dynamic equilibrium,

$$\sum M_a = 0 \qquad R_b(110) - F_1(118) - F_2(118 + x_0 \cos 35°) = 0$$

$$R_b(110) - 1.13(118) - 5.49(118 + 44.2\cos 35°) = 0 \qquad R_b = 8.91\ \text{N}$$

$$\sum F_y = 0 \qquad -R_a + R_b - F_1 - F_2 = -R_a + 8.91 - 1.13 - 5.49 = 0 \qquad R_a = 2.29\ \text{N}$$

(*b*) The blade is divided into the two straight elements shown in Fig. 22.32e. The masses of the two elements are

$$m_1 = \frac{\pi d^2}{4}\, l_1 \rho = \frac{\pi(8)^2}{4}(40)8,800\left(\frac{1}{1,000}\right)^3 = 0.0177\ \text{kg}$$

$$m_2 = \frac{\pi d^2}{4}\, l_2 \rho = \frac{\pi(8)^2}{4}(80)8,800\left(\frac{1}{1,000}\right)^3 = 0.0354\ \text{kg}$$

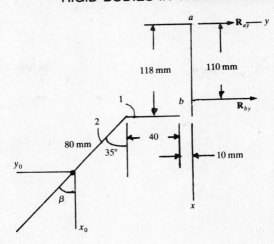

Fig. 22.32e

The centroidal product of inertia of element 2, from Prob. 17.96, has the form

$$I_{0,xy,2} = \tfrac{1}{24} m_2 l_2^2 \sin 2\beta$$

From comparison of the orientation of element 2 with respect to the $xy$ axes in Fig. 22.32e, and with respect to the $xy$ axes in Fig. 17.96a, it follows that $\beta = -35°$.

The parallel-axis theorem for product of inertia, from Prob. 17.94, is

$$I_{xy} = I_{0,xy} + m x_c y_c$$

The mass product of inertia $I_{xy}$ of the blade form is then

$$I_{xy} = (I_{0,xy,1} + m_1 x_{c1} y_{c1}) + (I_{0,xy,2} + m_2 x_{c2} y_{c2}) \tag{1}$$

The centroidal coordinates required in the above equation are

$$x_{c1} = 118 \text{ mm} \qquad y_{c1} = -(10 + \tfrac{40}{2}) = -30$$
$$x_{c2} = 118 + \tfrac{80}{2} \cos 35° = 151 \text{ mm} \qquad y_{c2} = -(50 + \tfrac{80}{2} \sin 35°) = -72.9 \text{ mm}$$

Equation (1) then has the form

$$I_{xy} = \left[ 0 + 0.0177 \frac{(118)(-30)}{1,000^2} \right] + \left[ \tfrac{1}{24}(0.0354) \frac{80^2}{1,000^2} \sin [2(-35°)] + 0.0354 \frac{(151)(-72.9)}{1,000^2} \right]$$

$$= -4.61 \times 10^{-4} \text{ kg} \cdot \text{m}^2$$

Using Eq. (15) in Prob. 22.25,

$$R_{by} = \frac{-\omega^2 I_{xy}}{l} = \frac{-46.1^2(-4.61 \times 10^{-4})}{110/1,000} = 8.91 \text{ N}$$

The centroidal coordinate of the blade form is given by

$$y_c = \frac{y_1 l_1 + y_2 l_2}{l_1 + l_2}$$

Using Fig. 22.32e,

$$y_c = \frac{-(10 + \tfrac{40}{2})(40) - (50 + \tfrac{80}{2} \sin 35°)80}{40 + 80} = -58.6 \text{ mm}$$

Using Eq. (16) in Prob. 22.25,

$$R_{ay} = \frac{\omega^2 I_{xy}}{l} - m y_c \omega^2 = \frac{46.1^2(-4.61 \times 10^{-4})}{110/1,000} - (0.0177 + 0.0354)\frac{(-58.6)}{1,000}(46.1)^2 = -2.29 \text{ N}$$

The minus sign on the above result indicates that $R_{ay}$ acts to the left in Fig. 22.32e.

**22.33** The plane blade shape in Fig. 22.33a is formed of 0.5-in-diameter steel rod. The specific weight of steel is 489 lb/ft³.

(a) Find the maximum permissible value of the rotational speed by integration of the centrifugal forces acting on the mass elements, if the dynamic bearing force is not to exceed 12 lb.

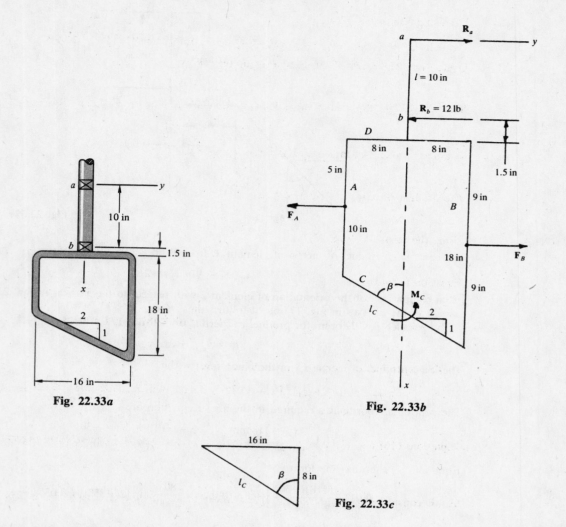

**Fig. 22.33a**

**Fig. 22.33b**

**Fig. 22.33c**

(b) Verify the results in part (a) by using the general solution, given in Prob. 22.25, for the dynamic bearing forces.

▌ (a) Figure 22.33b shows the free-body diagram of the blade and shaft. $R_b$ is assumed to have the maximum permissible value $R_{b,\text{max}} = 12$ lb. The density per unit length is given by

$$\rho_0 = \rho A = \frac{489}{12^3(386)}\left[\frac{\pi(0.5)^2}{4}\right] = 1.44 \times 10^{-4}\ \text{lb}\cdot\text{s}^2/\text{in}^2$$

The centrifugal forces acting on elements $A$ and $B$ are

$$F_A = m_A r\omega^2 = 1.44 \times 10^{-4}(10)8\omega^2 \qquad F_B = m_B r\omega^2 = 1.44 \times 10^{-4}(18)8\omega^2$$

The length and direction of element $C$, from Fig. 22.33c, are

$$l_C = \sqrt{16^2 + 8^2} = 17.9\ \text{in} \qquad \tan\beta = \tfrac{16}{8} \qquad \beta = 63.4°$$

Using Eq. (1) in Prob. 22.20,

$$M_C = \tfrac{1}{24}m_C l^2\omega^2 \sin 2\beta = \tfrac{1}{24}(1.44 \times 10^{-4})(17.9)(17.9)^2\omega^2 \sin[2(63.4°)]$$

For dynamic equilibrium of the blade and shaft assembly,

$$\sum M_a = 0$$

$$-R_b(10) - F_A(10 + 1.5 + 5) + F_B(10 + 1.5 + 9) + M_C = -12(10) - 1.44 \times 10^{-4}(10)8\omega^2(16.5)$$
$$+ 1.44 \times 10^{-4}(18)8\omega^2(20.5) + \tfrac{1}{24}(1.44 \times 10^{-4})(17.9)^3\omega^2 \sin[2(63.4°)] = -12(10) + 0.263\omega^2 = 0$$
$$\omega = 21.4\ \text{rad/s} = 204\ \text{r/min}$$

For dynamic force equilibrium,

$$\sum F_y = 0 \qquad R_a - R_b + F_B - F_A = R_a - 12 + 1.44 \times 10^{-4}(18)8\omega^2 - 1.44 \times 10^{-4}(10)8\omega^2 = 0$$

Using $\omega = 21.4 \, \text{rad/s}$ in the above equation gives

$$R_a = 7.78 \, \text{lb}$$

Since $R_a < 12 \, \text{lb}$, the maximum value of the dynamic bearing force occurs at bearing $b$, with the value

$$R_b = R_{b,\text{max}} = 12 \, \text{lb}$$

(b) The mass product of inertia of the blade shape is given by

$$I_{xy} = I_{xy,A} + I_{xy,B} + I_{xy,C} \qquad\qquad (1)$$

The product of inertia of element $C$ is

$$I_{xy,C} = I_{0,xy,C}$$

Using the result in Prob. 17.96,

$$I_{0,xy,C} = \tfrac{1}{24} m_C l_C^2 \sin 2\beta$$

Equation (1) now appears as

$$\begin{aligned}
I_{xy} &= m_A x_A y_A + m_B x_B y_B + \tfrac{1}{24} m_C l_C^2 \sin 2\beta \\
&= 1.44 \times 10^{-4}(10)(10 + 1.5 + 5)(-8) + 1.44 \times 10^{-4}(18)(10 + 1.5 + 9)8 + \tfrac{1}{24}(1.44 \times 10^{-4})(17.9) \\
&\quad \times (17.9)^2 \sin[2(63.4°)] = 0.263 \, \text{lb} \cdot \text{s}^2 \cdot \text{in}
\end{aligned}$$

The centroidal coordinate $y_c$ of the blade shape is

$$y_c = \frac{y_A l_A + y_B l_B + y_C l_C + y_D l_D}{l_A + l_B + l_C + l_D} = \frac{-8(10) + 8(18) + 0(17.9) + 0(16)}{10 + 18 + 17.9 + 16} = 1.03 \, \text{in}$$

$$l = 10 + 18 + 17.9 + 16 = 61.9 \, \text{in}$$

Equations (15) and (16) in Prob. 22.25 are next used to find the dynamic bearing forces $R_{ay}$ and $R_{by}$.

$$R_{by} = \frac{-\omega^2 I_{xy}}{l}$$

$R_{by}$ is considered to be positive if it acts in the positive $y$ sense. From Fig. 22.33b,

$$R_{by} = -12 = \frac{-\omega^2(0.263)}{10} \qquad \omega = 21.4 \, \text{rad/s} = 204 \, \text{r/min}$$

The mass of the blade is

$$m = \rho_0 l = 1.44 \times 10^{-4}(61.9) = 0.00891 \, \text{lb} \cdot \text{s}^2/\text{in}$$

$$R_{ay} = \frac{\omega^2 I_{xy}}{l} - m y_c \omega^2 = \frac{21.4^2(0.263)}{10} - 0.00891(1.03)21.4^2 = 7.84 \approx 7.78 \, \text{lb}$$

**22.34** Use the general solution for the dynamic bearing forces, given in Prob. 22.25, to verify the results found in Prob. 22.20.

▮ The inclined slender rigid rod is shown in Fig. 22.34 in its orientation with respect to the $xy$ axes of Fig. 22.25a. The product of inertia for this configuration was found in Prob. 17.96 as

$$I_{xy} = \tfrac{1}{24} m l^2 \sin 2\beta$$

The two bearing forces are shown in their functional positive senses. The general solutions for these two forces are given by Eqs. (15) and (16) in Prob. 22.25 as

$$R_{by} = \frac{-\omega^2 I_{xy}}{l} \qquad R_{ay} = \frac{\omega^2 I_{xy}}{l} - m y_c \omega^2$$

In the derivation of these equations, the term $l$ is the axial spacing between the two bearings, *measured along the positive x axis*. Thus, in this problem,

$$l \to -l \cos \beta$$

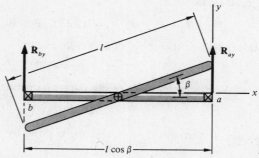

**Fig. 22.34**

The bearing force $R_{by}$ is then found as

$$R_{by} = \frac{-\omega^2 I_{xy}}{l} = \frac{-\omega^2(\frac{1}{24})ml^2 2 \sin\beta\cos\beta}{-l\cos\beta} = \tfrac{1}{12}ml\omega^2\sin\beta$$

The positive result for $R_{by}$ indicates that this force acts upward on the left end of the shaft, in the view in Fig. 22.34.

Since $y_c = 0$,

$$R_{ay} = \frac{\omega^2 I_{xy}}{l} = -R_{by} = -\tfrac{1}{12}ml\omega^2\sin\beta$$

The actual sense of the force $R_{ay}$ is downward on the right end of the shaft, as seen in Fig. 22.20b.

**22.35** In a certain experimental configuration it is desired to apply a pulsating force to a bearing. A proposed technique is shown in Fig. 22.35a. A diagonal hole is drilled through a steel cylinder. The hole lies in a plane which contains the x axis. By careful speed regulation near the nominal operating speed, the desired value of force can be obtained. If the diameter of the drilled hole is 3 mm, find the nominal speed which will produce a pulsating force of magnitude 1.2 N. The density of steel is 7,830 kg/m³.

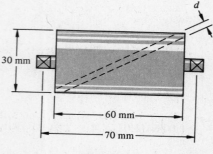

**Fig. 22.35a**

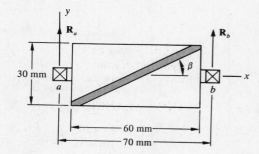

**Fig. 22.35b**

❚ The cylinder is referenced to a set of xy axes as shown in Fig. 22.35b. The product of inertia of the cylinder, considered to be a homogeneous *solid* body, is designated $I_{xy,C}$. The product of inertia of the cylinder material *which would occupy the volume of the drilled hole* is called $I_{xy,H}$. The net product of inertia $I_{xy}$ of the drilled cylinder is then

$$I_{xy} = I_{xy,C} - I_{xy,H}$$

Because of the symmetry of the cylinder with respect to the coordinate axes,

$$I_{xy,C} = 0$$

and thus

$$I_{xy} = -I_{xy,H} \tag{1}$$

From consideration of Eq. (1) above, it may be seen that the effect of the drilled hole in the cylinder is the *same* as that of a slender steel rod at a fixed angle with respect to the axis of rotation. This case was solved in Prob.

17.96, and the term $I_{xy,H}$ is given by

$$I_{xy,H} = \tfrac{1}{24}ml^2 \sin 2\beta$$

Using Eq. (1), the product of inertia of the cylinder with a hole is

$$I_{xy} = -\tfrac{1}{24}ml^2 \sin 2\beta \qquad (2)$$

The term $m$ in the above equation is the mass of the material which occupies the hole.
  From Fig. 22.35b,

$$\tan \beta = \tfrac{30}{60} \qquad \beta = 26.6°$$

The length $l_1$ of the hole is

$$l_1 = \sqrt{30^2 + 60^2} = 67.1 \text{ mm}$$

The mass of material which occupies the hole is

$$m = \frac{\pi}{4}(3^2)(67.1) \text{ mm}^3 \left(7{,}830 \, \frac{\text{kg}}{\text{m}^3}\right)\left(\frac{1 \text{ m}}{1{,}000 \text{ mm}}\right)^3 = 0.00371 \text{ kg}$$

Using Eq. (2),

$$I_{xy} = -\tfrac{1}{24}(0.00371)\left(\frac{67.1}{1{,}000}\right)^2 \sin[2(26.6°)] = -5.57 \times 10^{-7} \text{ kg} \cdot \text{m}^2 \qquad (3)$$

Equations (15) and (16) in Prob. 22.25 have the forms

$$R_{by} = -\frac{\omega^2 I_{xy}}{l} \qquad R_{ay} = \frac{\omega^2 I_{xy}}{l} - my_c\omega^2$$

where $l$ is the spacing between the bearings. The centroidal coordinate $y_c$ of the cylinder is zero, so that

$$R_{by} = -\frac{\omega^2 I_{xy}}{l} \qquad R_{ay} = -R_{by} \qquad (4)$$

The limiting value of $R_{by}$ is

$$R_{by} = 1.2 \text{ N}$$

Using this result, and Eq. (3), in Eq. (4) gives

$$1.2 = \frac{-\omega^2 I_{xy}}{l} = \frac{-\omega^2(-5.57 \times 10^{-7})}{70/1{,}000} \qquad \omega = 388 \text{ rad/s} = 3{,}710 \text{ r/min}$$

The dynamic forces on the *bearings* act down at bearing $b$ and up at bearing $a$.
  The method of solution in this problem may be compared with that in Prob. 22.21, where a similar problem was solved by direct use of the dynamic moment due to the centrifugal forces.

## 22.3 DYNAMIC BALANCING OF ROTATING BODIES

22.36  Figure 22.36a shows a rotor-shaft assembly which is supported by two bearings. The *xyz* axes are attached to the rotor, and the rotor is assumed to be in a general state of dynamic unbalance. What are the conditions for *dynamic balance* of the assembly?

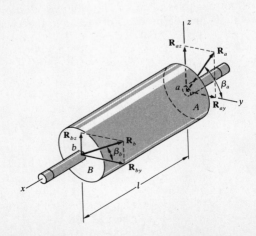

**Fig. 22.36a**

❚ If the assembly is not in a state of dynamic balance, then at least *one* of the following conditions exists.

1. The center of mass does *not* lie on the axis of rotation, and thus the assembly is *not* in static balance about this axis.

2. The $xyz$ axes are *not* principal axes of the body, and thus one or both of the products of inertia $I_{xy}$ and $I_{zx}$ are *not* zero.

The force components $R_{ay}$, $R_{az}$, $R_{by}$, and $R_{bz}$ are forces caused by the dynamic unbalance of the rotor. Those forces are assumed to be transmitted to the shaft at the locations of the end faces $A$ and $B$ of the rotor. These forces are transmitted through the shaft to the bearings, where they manifest themselves as undesirable pulsating forces acting on the bearing supports.

The force components in the figure are assumed to be positive quantities if their actual senses are the same as the senses of the positive coordinate axes. The dynamic forces, given by Eqs. (6), (7), and (10) in Prob. 22.25, are

$$R_{by} = \frac{-\omega^2 I_{xy}}{l} \qquad R_{bz} = \frac{-\omega^2 I_{zx}}{l} \qquad R_{ay} = -R_{by} - my_c\omega^2 \qquad R_{az} = -R_{bz} - mz_c\omega^2$$

In the above equations, $l$ is used as the axial spacing between the *end faces* of the rotor, as contrasted with the use of $l$ as the spacing of the bearings in Prob. 22.25.

The resultant forces at $a$ and $b$ are

$$R_a = \sqrt{R_{ay}^2 + R_{az}^2} \qquad R_b = \sqrt{R_{by}^2 + R_{bz}^2}$$

with the directions

$$\beta_a = \tan^{-1}\frac{R_{az}}{R_{ay}} \qquad \beta_b = \tan^{-1}\frac{R_{bz}}{R_{by}}$$

From physical considerations, it may be concluded that one or more of the force components $R_{ay}$, $R_{az}$, $R_{by}$, $R_{bz}$ must be negative.

It is now shown that the rotor assembly in Fig. 22.36a may be put into *dynamic balance* by the addition of correction weights to the rotor, *in two distinct parallel planes which are normal to the axis of rotation*. These planes will be chosen to be the two end planes $A$ and $B$ shown in the figure.

The true view of face $A$ is shown in Fig. 22.36b. A small mass $m_a$ is attached to the end face of the rotor, along the line of action of $R_a$, at a distance $r_a$ from the center axis. If the magnitudes of $m_a$ and $r_a$ are chosen to satisfy

$$m_a r_a \omega^2 = R_a$$

the resultant force on plane $A$ will be exactly zero. The equivalent effect may be obtained by *drilling out* a portion of the rotor material, as shown in Fig. 22.36c.

After the above correction, the rotor is acted on by only the force $R_b$ shown in Fig. 22.36a. A correction weight is now added to plane $B$, as shown in Fig. 22.36d. If $m_b$ and $r_b$ satisfy

$$m_b r_b \omega^2 = R_b$$

the resultant dynamic force on plane $B$ will be zero. This correction also could be made by drilling a hole, shown as the dashed circle in Fig. 22.36d, in face $B$ of the rotor.

When the above two correction weights are added, the rotor is in dynamic balance. The effect of the addition of the two correction weights is to *adjust the total mass distribution of the rotor*, so that the center of mass lies on the axis of rotation and the products of inertia in the $xy$ and $zx$ planes are identically zero.

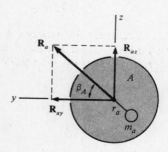

**Fig. 22.36b**

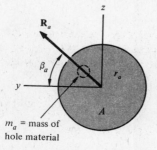

$m_a$ = mass of hole material

**Fig. 22.36c**

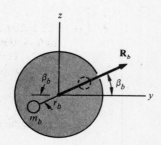

**Fig. 22.36d**

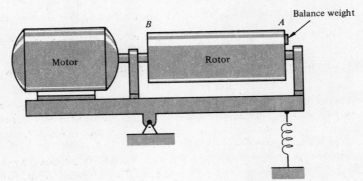

**Fig. 22.36e**

Figure 22.36e shows a schematic setup to illustrate the operation of dynamically balancing a rotor. The rotor is mounted in a fixture which is connected to the ground by a hinge pin and a spring. Plane B is directly above the hinge pin, so that any resultant centrifugal forces which act in this plane are directly reacted out to the ground. The speed of the motor is varied until a condition is reached which is described as *resonance* of the spring-mounted fixture. This condition is characterized by large angular amplitudes of the spring-mounted frame, and this motion is caused by the rotating, unbalanced force in plane A. The rotating forces in plane B have no effect on this motion. A person is imagined to place balance weights on the rotor in plane A, by trial and error, until a condition is reached where the angular vibrational motion of the fixture ceases. At this condition, *the resultant dynamic force of the rotor, at the axial location of plane A, is identically zero.*

The rotor is then turned around in the fixture, so that planes A and B exchange their positions, and the process is repeated. At the conclusion of the above operation, the rotor is dynamically balanced.

22.37   The cylindrical steel rotor shown in Fig. 22.37a has an equivalent radial unbalance force of 0.95 lb in the plane of each end face when it rotates at 1,400 r/min. It is proposed to dynamically balance the rotor by drilling $\frac{3}{8}$-in diameter holes in the end faces, on a 4-in-diameter circle, and inserting lead plugs flush with these faces.

The densities of lead and steel are 710 and 489 lb/ft³, respectively, and the drilled holes may be assumed to have the shape of a right circular cylinder.

(a)  Find the required depth of these balancing holes.

(b)  Find the required depth if drilled holes alone, at positions which are 180° from that required in part (a), are used.

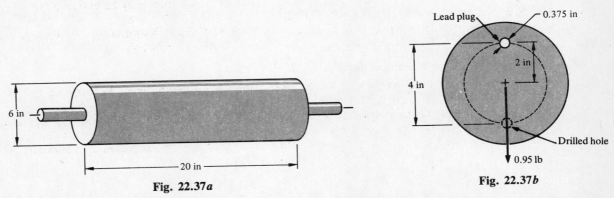

**Fig. 22.37a**                 **Fig. 22.37b**

❙ (a)  The balancing holes for the lead plugs are shown in Fig. 22.37b. The holes are assumed to be of shallow depth, so that the centroids of the hole volumes lie very nearly in the planes of the end faces. The mass density of the plug-hole combination is the difference between the densities of the lead and the steel. The rotational velocity is

$$\omega = 1,400\left(\frac{2\pi}{60}\right) = 147 \text{ rad/s}$$

The required depth $l$ of the holes is found from

$$F = mr\omega^2 \qquad 0.95 = \frac{\pi}{4}(0.375)^2 l\left(\frac{710-489}{12^3(386)}\right)(2)(147)^2 \qquad l = 0.601 \text{ in}$$

(b) For the hole alone, the density of steel is used directly, and the required hole depth is found from

$$F = mr\omega^2 \qquad 0.95 = \frac{\pi}{4}(0.375)^2 l \left(\frac{489}{[12^3(386)]}\right)(2)(147)^2 \qquad l = 0.271 \text{ in}$$

The position of this hole is shown as the dashed circle in Fig. 22.37b.

The centroid of the maximum hole volume in part (a) is located at $0.601/2 = 0.301$ in from the end faces. This dimension is $(0.301/20)100 = 1.5\%$ of the rotor length. Thus, this centroid is approximately in the plane of the end face.

**22.38** Figure 22.38a shows a shaft, with two heavy metal rotors, that rotates at 2,200 r/min. During the dynamic balancing of the assembly, correction weights were welded on faces c and d, in the same plane that contains the axis of the rotors. A steel disk 1 in thick and 1.2 in in diameter was welded to face c, on a mean diameter of 26.4 in. On face d, a steel disk 0.75 in thick and 1 in in diameter, on a mean diameter of 36.8 in, was welded on. Find the values of the dynamic bearing forces that existed before the correction weights were added. The specific weight of steel is 489 lb/ft$^3$, and the mass of the welds may be neglected.

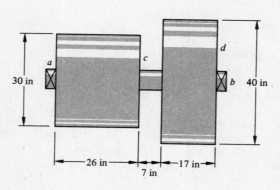

**Fig. 22.38a**

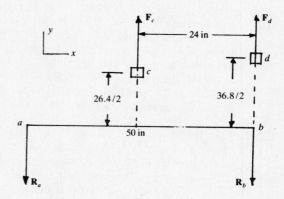

**Fig. 22.38b**

❚ Figure 22.38b shows the centrifugal forces which act on the two correction weights, and the dynamic bearing forces $R_a$ and $R_b$. $\omega = 2,200$ r/min $= 230$ rad/s, and the masses of the correction weights are given by

$$m_c = \frac{\pi(1.2^2)}{4}(1)\frac{489}{12^3(386)} = 8.29 \times 10^{-4} \text{ lb} \cdot \text{s}^2/\text{in} \qquad m_d = \frac{\pi(1)^2}{4}(0.75)\frac{489}{12^3(386)} = 4.32 \times 10^{-4} \text{ lb} \cdot \text{s}^2/\text{in}$$

The magnitudes of the centrifugal forces are

$$F = mr\omega^2 \qquad F_c = 8.29 \times 10^{-4}\left(\frac{26.4}{2}\right)230^2 = 579 \text{ lb} \qquad F_d = 4.32 \times 10^{-4}\left(\frac{36.8}{2}\right)230^2 = 420 \text{ lb}$$

Both weights lie in a single plane with the axis of rotation. For dynamic equilibrium of the assembly, using Fig. 22.38b,

$$\sum M_b = 0 \qquad R_a(50) - F_c(24) = R_a(50) - 579(24) = 0 \qquad R_a = 278 \text{ lb}$$

$$\sum F_y = 0 \qquad -R_a + F_c + F_d - R_b = -278 + 579 + 420 - R_b = 0 \qquad R_b = 721 \text{ lb}$$

$R_a$ and $R_b$ are the dynamic bearing forces which existed before the two correction weights were added to the rotors.

**22.39** The four disks in Fig. 22.39a are 200 mm in diameter and 25 mm thick. Disks B and C have 30-mm-diameter holes on a 120-mm-diameter circle. The assembly is to be dynamically balanced by drilling holes, on 110-mm-diameter circles, in disks A and D.

(a) Find the required diameters of the holes, and the positions $\beta_A$ and $\beta_D$, if all four disks are steel, with $\rho = 7,830$ kg/m$^3$.

(b) Do the same as in part (a) if disk B is aluminum, with $\rho_B = 2,770$ kg/m$^3$, disk C is bronze, with $\rho_C = 8,800$ kg/m$^3$, and disks A and D are steel.

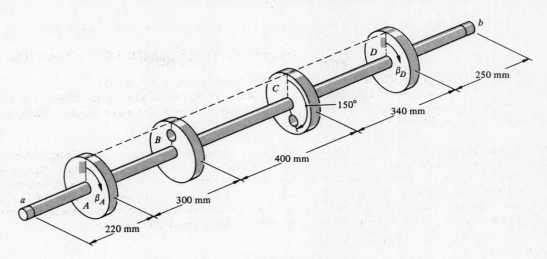

**Fig. 22.39a**

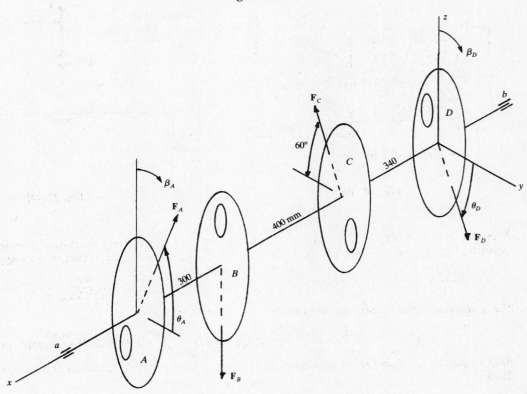

**Fig. 22.39b**

▮ (a) Figure 22.39b shows the centrifugal forces $F_B$ and $F_C$ due to the holes in disks $B$ and $C$. The centrifugal forces which result when holes are drilled in disks $A$ and $D$ are shown as $F_A$ and $F_D$, with the directions $\theta_A$ and $\theta_D$, respectively. The forces $F_A$, $F_B$, $F_C$, and $F_D$ have the forms

$$F_A = \left(\frac{\pi d_A^2}{4}\right)25(7,830)(55)(\omega^2)\left(\frac{1}{1,000}\right)^4 = 8.46 \times 10^{-6} d_A^2 \omega^2 \quad \text{N} \tag{1}$$

$$F_B = F_C = \left(\frac{\pi 30^2}{4}\right)25(7,830)60\omega^2\left(\frac{1}{1,000}\right)^4 = 0.00830\omega^2 \quad \text{N} \tag{2}$$

$$F_D = \left(\frac{\pi d_D^2}{4}\right)25(7,830)(55)(\omega^2)\left(\frac{1}{1,000}\right)^4 = 8.46 \times 10^{-6} d_D^2 \omega^2 \quad \text{N} \tag{3}$$

For moment equilibrium of the assembly,

$$\sum M_y = 0 \qquad -F_A \sin\theta_A(1{,}040) + F_B(740) - F_C \sin 60°(340) = 0$$

Equations (1), (2), and (3) are used in the above equation, with the result

$$-F_A \sin\theta_A(1{,}040) + 0.00830\omega^2(740) - 0.00830\omega^2 \sin 60°(340) = 0$$
$$F_A \sin\theta_A = 0.00356\omega^2 \tag{4}$$

$$\sum M_z = 0 \qquad F_A \cos\theta_A(1{,}040) - F_C \cos 60°(340) = 0$$

Using Eqs. (1) and (2) in the above equation gives

$$F_A \cos\theta_A(1{,}040) - 0.00830\omega^2 \cos 60°(340) = 0 \qquad F_A \cos\theta_A = 0.00136\omega^2 \tag{5}$$

Equation (4) is divided by Eq. (5), with the result

$$\tan\theta_A = 2.62 \qquad \theta_A = 69.1° \tag{6}$$

Using Eqs. (1) and (6) in Eq. (4) gives

$$8.46 \times 10^{-6}d_A^2\omega^2 \sin 69.1° = 0.00356\omega^2 \qquad d_A = 21.2 \text{ mm}$$

It may be observed that $d_A$ and $\theta_A$ are both independent of $\omega$, the angular velocity of the assembly. The angle $\beta_A$, shown in Fig. 22.39b, gives the location of the *hole* in disk A. This position is 180° from the direction of $F_A$. Thus,

$$\beta_A = 270° - \theta_A = 201°$$

For force equilibrium of the assembly,

$$\sum F_y = 0 \qquad F_A \cos\theta_A - F_C \cos 60° + F_D \cos\theta_D = 0$$

Using Eqs. (1) and (2) in the equation above results in

$$8.46 \times 10^{-6}(21.2)^2\omega^2 \cos 69.1° - 0.00830\omega^2 \cos 60° + F_D \cos\theta_D = 0$$
$$F_D \cos\theta_D = 0.00279\omega^2 \tag{7}$$

$$\sum F_z = 0 \qquad F_A \sin\theta_A - F_B + F_C \sin 60° - F_D \sin\theta_D = 0$$

Equations (1), (2), and (3) are used in the equation above, to obtain

$$8.46 \times 10^{-6}(21.2)^2\omega^2 \sin 69.1° - 0.00830\omega^2 + 0.00830\omega^2 \sin 60° - F_D \sin\theta_D = 0$$
$$F_D \sin\theta_D = 0.00244\omega^2 \tag{8}$$

Equation (8) divided by Eq. (7), with the result

$$\tan\theta_D = 0.875 \qquad \theta_D = 41.2° \tag{9}$$

Equations (3) and (9) are used in Eq. (7). The result is

$$8.46 \times 10^{-6}d_D^2\omega^2 \cos 41.2° = 0.00279\omega^2 \qquad d_D = 20.9 \text{ mm}$$

Angle $\beta_D$, seen in Fig. 22.39b, gives the position of the hole in disk D. This hole is 180° from the position of $F_D$ in Fig. 22.39b, so that

$$\beta_D = 270° + \theta_D = 311°$$

(b) Figure 22.39b is used. The centrifugal forces $F_A$ and $F_D$, found in part (a), are

$$F_A = 8.46 \times 10^{-6}d_A^2\omega^2 \qquad \text{N} \tag{10}$$
$$F_D = 8.46 \times 10^{-6}d_D^2\omega^2 \qquad \text{N} \tag{11}$$

The centrifugal forces due to the holes in disks B and C have the forms

$$F_B = \left(\frac{\pi 30^2}{4}\right)25(2{,}770)(60)(\omega^2)\left(\frac{1}{1{,}000}\right)^4 = 0.00294\omega^2 \qquad \text{N} \tag{12}$$

$$F_C = \left(\frac{\pi 30^2}{4}\right)25(8{,}800)(60)(\omega^2)\left(\frac{1}{1{,}000}\right)^4 = 0.00933\omega^2 \qquad \text{N} \tag{13}$$

For moment equilibrium,

$$\sum M_y = 0 \qquad -F_A \sin\theta_A(1{,}040) + F_B(740) - F_C \sin 60°(340) = 0$$

Equations (12) and (13) are used in the above equations, with the result

$$-F_A \sin \theta_A (1{,}040) + 0.00294\omega^2(740) - 0.00933\omega^2 \sin 60°(340) = 0$$

$$F_A \sin \theta_A = -0.000550\omega^2 \tag{14}$$

$$\sum M_z = 0 \qquad F_A \cos \theta_A (1{,}040) - F_C \cos 60°(340) = 0$$

Using Eq. (13) in the above equation gives

$$F_A \cos \theta_A (1{,}040) - 0.00933\omega^2 \cos 60°(340) = 0 \qquad F_A \cos \theta_A = 0.00153\omega^2 \tag{15}$$

Equation (14) is divided by Eq. (15), to obtain

$$\tan \theta_A = -0.359 \qquad \theta_A = -19.7° \tag{16}$$

Using Eqs. (10) and (16) in Eq. (14) gives

$$8.46 \times 10^{-6} d_A^2 \omega^2 \sin (-19.7°) = -0.000550\omega^2 \qquad d_A = 13.9 \text{ mm}$$

Angle $\beta_A$ in Fig. 22.39a is written as

$$\beta_A = 270° - \theta_A = 270° - (-19.7°) \qquad \beta_A = 290°$$

For force equilibrium of the assembly,

$$\sum F_y = 0 \qquad F_A \cos \theta_A - F_C \cos 60° + F_D \cos \theta_D = 0$$

Equations (10) and (13) are used in the above equation to obtain

$$8.46 \times 10^{-6}(13.9)^2\omega^2 \cos (-19.7°) - 0.00933\omega^2 \cos 60° + F_D \cos \theta_D = 0$$

$$F_D \cos \theta_D = 0.00313\omega^2 \tag{17}$$

$$\sum F_z = 0 \qquad F_A \sin \theta_A - F_B + F_C \sin 60° - F_D \sin \theta_D = 0$$

Using Eqs. (10), (12), and (13) in the equation above results in

$$8.46 \times 10^{-6}(13.9)^2\omega^2 \sin (-19.7°) - 0.00294\omega^2 + 0.00933\omega^2 \sin 60° - F_D \sin \theta_D = 0$$

$$F_D \sin \theta_D = 0.00459\omega^2 \tag{18}$$

Equation (18) is divided by Eq. (17), with the result

$$\tan \theta_D = 1.47 \qquad \theta_D = 55.8° \tag{19}$$

Equations (11) and (19) are used in Eq. (17). The result is

$$8.46 \times 10^{-6} d_D^2 \omega^2 \cos (55.8°) = 0.00313\omega^2 \qquad d_D = 25.7 \text{ mm}$$

Angle $\beta_D$, shown in Fig. 22.39a, has the value

$$\beta_D = 270° + \theta_D = 270° + 55.8° = 326°$$

**22.40** The rotating assembly in Fig. 22.14a is to be put into dynamic balance by welding onto the assembly the additional two arms shown in Fig. 22.40a. Find the required lengths of these two arms. The mass of the weld material may be neglected.

▮ The dynamic force $R$ which acts on each bearing, from Prob. 22.14, part (b), is

$$R = 15.3 \text{ lb}$$

The couple $M$ of the bearing forces is

$$M = R(20) = 15.3(20) = 306 \text{ in} \cdot \text{lb}$$

$d_1$ is the axial distance, shown in Fig. 22.40a, between the pieces to be welded onto the assembly. From Fig. 22.40a,

$$d_1 = 20 - 8 - 3 = 9 \text{ in}$$

Since the CM of the assembly lies on the $x$ axis,

$$l_1 = l_2$$

Let $l = l_1 + 0.5 = l_2 + 0.5$. $F$ is the centrifugal force of each piece welded on, given by

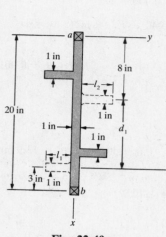

Fig. 22.40a

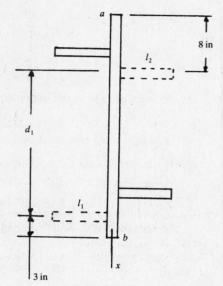

Fig. 22.40b

$$F = \int_{0.5}^{l} \left[ (1)0.5 \, dy \left( \frac{0.283}{386} \right) \right] y (103)^2 = (1)0.5 \left( \frac{0.283}{386} \right) 103^2 \int_{0.5}^{l} y \, dy = 103^2 (1)0.5 \left( \frac{0.283}{386} \right) \left( \frac{y^2}{2} \Big|_{0.5}^{l} \right)$$

$$= (103)^2 (1)0.5 \left( \frac{0.283}{386} \right) \left( \frac{1}{2} \right) (l^2 - 0.5^2)$$

The couple of the centrifugal forces of the two arms which are added is equated to the couple of the bearing forces. The result is

$$Fd_1 = M = 306 \text{ in} \cdot \text{lb} \qquad (103)^2 (1)0.5 \left( \frac{0.283}{386} \right) \left( \frac{1}{2} \right) (l^2 - 0.5^2)9 = 306 \qquad l = 4.21 \text{ in}$$

The required lengths of the arms are

$$l_1 = l_2 = l - 0.5 = 3.71 \text{ in}$$

## 22.4 DERIVATIVE OF A VECTOR WITH CONSTANT MAGNITUDE AND CHANGING DIRECTION, MOMENT EFFECTS DUE TO CHANGE IN DIRECTION OF AN AXIS OF ROTATION, THE GYROSCOPIC MOMENT

22.41     Find the time derivative of a vector with *constant* magnitude and *changing* direction.

▮ Figure 22.41a shows a vector **A**. This vector has a constant magnitude $A$, and it is constrained to move in the $xy$ plane as it rotates about the $z$ axis. It is desired to find the quantity $d\mathbf{A}/dt$.

The position of vector **A** at time $t$, designated $\mathbf{A}(t)$, is shown in Fig. 22.41b. The *direction* of this vector rotates with the angular velocity $\omega_1$, so that at a later time $t + \Delta t$ the vector $\mathbf{A}(t + \Delta t)$ has the new position shown in the figure. The change $\Delta \mathbf{A}$, by definition, is

$$\Delta \mathbf{A} = \mathbf{A}(t + \Delta t) - \mathbf{A}(t)$$

$\Delta \mathbf{A}$ is shown in the figure.

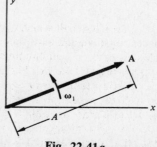

Fig. 22.41a

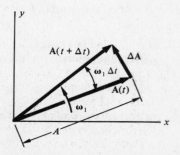

Fig. 22.41b

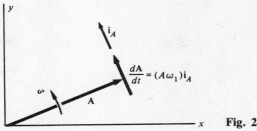

**Fig. 22.41c**

The first time derivative of **A** is defined to be

$$\frac{d\mathbf{A}}{dt} = \lim_{\Delta t \to 0} \frac{\Delta \mathbf{A}}{\Delta t}$$

The *direction* of this vector is that of the vector $\Delta \mathbf{A}$. This direction, in the limit, is *perpendicular to the direction of* **A**. The *sense* of $d\mathbf{A}/dt$ is the same as that obtained by imagining the vector **A** to be rotated 90° in the same sense as its angular velocity.

Finally, the *magnitude* of $\Delta \mathbf{A}$ is found, in the limit, as

$$\lim_{\Delta t \to 0} |\Delta \mathbf{A}| = \lim_{\Delta t \to 0} A(\omega_1 \, \Delta t)$$

These results are now combined to obtain

$$\frac{d\mathbf{A}}{dt} = \lim_{\Delta t \to 0} \frac{\Delta \mathbf{A}}{\Delta t} = \lim_{\Delta t \to 0} \mathbf{i}_A \frac{A\omega_1 \, \Delta t}{\Delta t} = (A\omega_1)\mathbf{i}_A \tag{1}$$

$\mathbf{i}_A$ is a unit vector with the direction and sense shown in Fig. 22.41c.

The very important conclusion now arrived at is that the *time derivative* of a vector of *constant magnitude* has *a direction which is normal to the direction of the original vector*.

**22.42** In all of the preceding problems in this text, the axes about which a body rotates is fixed in space.

(*a*) Discuss what happens when an axis of rotation is allowed to *change its direction*.

(*b*) A rotating disk with a symmetrical mass distribution experiences a rotation of its shaft from the original direction of the shaft axis. Find the forces exerted on the shaft by the bearings as a result of this motion.

(*c*) Show the general pattern of the directions and senses of the angular momentum vector $\mathbf{H}_0$, the angular velocity $\omega_1$ of the shaft axis, and the bearing forces.

(*d*) How would the method of solution for part (*b*) change if the rotating mass did not have a symmetrical distribution of mass about the spin axis of the body?

**Fig. 22.42a**

❚ (*a*) As a preliminary consideration, Fig. 22.42*a* shows a piece of equipment which is commonly found in physics laboratories. It consists of a bicycle wheel and tire, with a pair of handles attached at the axis of the wheel. A person grips the handles, and the wheel is set into spinning motion. If the person holding the handles attempts to change the direction of the axis of the handles and of the wheel, a substantial force resists the attempt to rotate the axis of the handles from its original position, and the spinning bicycle wheel appears to be "built in" in space. That physical phenomenon is an example of a *gyroscopic moment*. The theory which explains this effect is now presented.

Figure 22.42*b* shows a symmetrical disk which spins with a constant angular velocity $\omega$ about its axis of symmetry *aa*. The mass moment of inertia of the disk about this axis is designated $I_0$. The

Fig. 22.42b                                                Fig. 22.42c

*angular momentum* $\mathbf{H}_0$ of the disk, using the definition of angular momentum in Prob. 20.51, is

$$\mathbf{H}_0 = I_0 \omega$$

where $I_0$ is the mass moment of inertia about the center axis of the disk, and $\omega$ is the angular velocity. The *angular momentum*, as the product of a scalar quantity $I_0$ and a vector quantity $\omega$, is a *vector quantity*.

If the angular velocity is represented by the double-headed arrow shown in Fig. 22.42c, then the angular momentum $\mathbf{H}_0$ may be represented by the double-headed arrow shown in Fig. 22.42d.

Newton's second law for rotational motion of the disk about the center of mass is

$$M_0 = I_0 \alpha = I_0 \frac{d\omega}{dt} \tag{1}$$

Since $I_0$ is a constant in this problem, the right side of Eq. (1) may be written as

$$I_0 \frac{d\omega}{dt} = \frac{d}{dt}(I_0 \omega) = \frac{dH_0}{dt} \tag{2}$$

Since both $M_0$ and $H_0$ are vector quantities, Eqs. (1) and (2) may be written in boldface vector notation as

$$\mathbf{M}_0 = \frac{d\mathbf{H}_0}{dt} \tag{3}$$

$\mathbf{H}_0$ in the present problem is a vector of constant magnitude, since $\omega$ is constant. It follows that the right-hand side of Eq. (3) represents the operation of finding the time derivative of a vector of constant magnitude. This case was solved in Prob. 22.41, and the results which were obtained there may be used directly.

(b)  A spinning disk, with a yoke shaft, and bearing arrangement, is shown in Fig. 22.42e. The spin axis is collinear with the $x$ axis at the instant shown in the figure, and $l$ is the spacing between bearings $a$ and $b$. The yoke has a constant angular velocity $\omega_1$. It is desired to find the magnitude, direction, and sense of the bearing forces exerted by bearings $a$ and $b$ on the shaft of the disk.

The change in the momentum vector $\mathbf{H}_0$, due to the imposed rotation $\omega_1$, is shown in Fig. 22.42f. The moment $\mathbf{M}_0$ required to produce this change in momentum is found from

$$\mathbf{M}_0 = \frac{d\mathbf{H}_0}{dt}$$

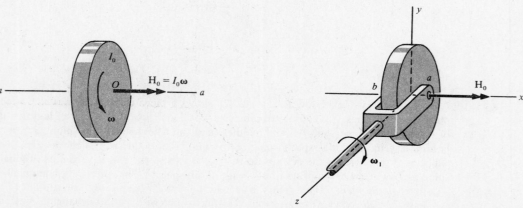

Fig. 22.42d                                                Fig. 22.42e

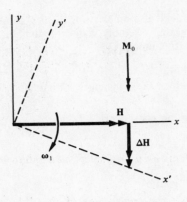

**Fig. 22.42f**

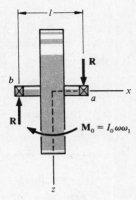

**Fig. 22.42g**

$M_0$ has the direction and sense shown in Fig. 22.42f. From comparison of this equation with Eq. (1) in Prob. 22.41, it follows that

$$M_0 = \frac{dH_0}{dt} = I_0 \omega \omega_1 \qquad (4)$$

Figure 22.42g shows a top view of the disk. The moment $M_0$ given above is the required external moment acting on the disk to give the motion $\omega_1$. This moment must be equal to the couple of the two forces $R$, *of* the bearings *on* the shaft, shown in the figure, so that

$$Rl = M_0 = I_0 \omega \omega_1 \qquad R = \frac{I \omega \omega_1}{l} \qquad (5)$$

The moment $M_0$ and forces $R$ are referred to as the *gyroscopic moment*, and *gyroscopic forces*, respectively.

(c) Figure 22.42h shows the directions and senses of the quantities $\omega_1$ and $H_0$ given in Fig. 22.42e. The vectors $\omega_1$ and $H_0$ lie in the $xz$ plane. The directions of the bearing forces $R$, exerted *on* the shaft *by* the bearings, are parallel to the $z$ axis. The actual senses of the forces $R$ are as shown in Fig. 22.42h, and these senses correspond to the senses of $\omega_1$ and $H_0$ in the figure. If the sense of *either* $\omega_1$ or $H_0$ changes, the sense of $R$ will change. If the senses of *both* $\omega_1$ and $H_0$ change, the sense of $R$ will remain the same.

(d) The disk in part (b) has a symmetrical distribution of mass about the spin axis, and is in a condition of dynamic balance. If the rotating mass did not have a symmetrical mass distribution about the spin axis, the mass moment of inertia about the spin axis would be computed and used in Eq. (4) to find the gyroscopic moment. The additional dynamic bearing forces due to the rotating unbalance would then be found in the usual manner. This case is considered in Prob. 22.47.

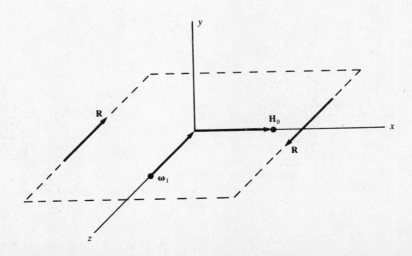

**Fig. 22.42h**

**22.43**  An automobile travels around a horizontal circular track, of diameter 800 ft, at a constant speed of 60 mi/h. The outer diameter of the tires is 30 in. Each wheel-tire assembly weighs 50 lb, has a centroidal radius of gyration of 8 in, and supports 580 lb of the weight of the automobile.

(*a*)  Find the gyroscopic moment acting on each wheel.

(*b*)  Each front wheel is mounted on two bearings which are 6 in apart. Find the values of the gyroscopic forces which act on these bearings.

(*c*)  Find the values of the total reaction forces acting on each bearing.

❙  (*a*)  The translational velocity of the wheel is

$$v = 60 \, \text{mi/h} = 88.0 \, \text{ft/s}$$

The rotational velocity of the wheel is found from

$$v = r\omega \qquad 88.0 = \tfrac{15}{12}\omega \qquad \omega = 70.4 \, \text{rad/s}$$

Figure 22.43*a* shows a front view of the wheel, with the gyroscopic, or dynamic, bearing forces *R*. The assumed sense of motion of the wheel is such as to roll "out" of Fig. 22.43*a*. $\omega_1$ is the angular velocity of the axle, and $\rho$ is the radius of the track. $\omega_1$ is found from

$$v = \rho\omega_1 \qquad 88.0 = 400\omega_1 \qquad \omega_1 = 0.22 \, \text{rad/s}$$

The mass moment of inertia of the wheel about its center axis is given by

$$I_0 = k_0^2 m = \left(\frac{8}{12}\right)^2\left(\frac{50}{32.2}\right) = 0.690 \, \text{lb} \cdot \text{s}^2 \cdot \text{ft}$$

Using Eq. (5) in Prob. 22.42,

$$M_0 = I_0\omega\omega_1 = 0.690(70.4)0.22 = 10.7 \, \text{ft} \cdot \text{lb} = 128 \, \text{in} \cdot \text{lb}$$

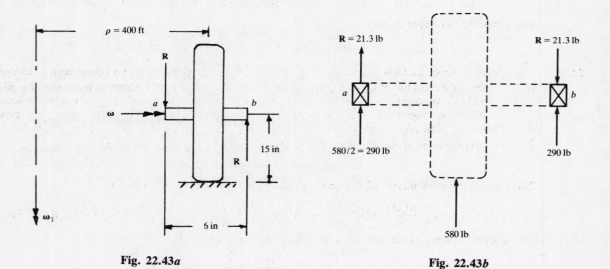

Fig. 22.43*a*                              Fig. 22.43*b*

(*b*)  The gyroscopic, or dynamic, forces which act on the bearings of the wheel are found by using Eq. (5) in Prob. 22.42. The result is

$$Rl = M_0 \qquad R = \frac{M_0}{l} = \frac{128}{6} = 21.3 \, \text{lb}$$

(*c*)  Using Newton's third law, the senses of the gyroscopic forces acting *on* the bearings are as shown in Fig. 22.43*b*. The static force acting on each bearing, due to the weight of the automobile, is 290 lb. This result is shown in Fig. 22.43*b*. The resultant force on each bearing is given by

$$R_a = 290 + 21.3 = 311 \, \text{lb} \qquad \text{(acting up)}$$
$$R_b = 290 - 21.3 = 269 \, \text{lb} \qquad \text{(acting down)}$$

**22.44** Figure 22.44 shows an aircraft landing gear just after the wheel leaves the runway during takeoff. The velocity of the aircraft is 200 km/h, and the retraction angular velocity of the arm is 0.2 rad/s. The wheel may be approximated as a disk of mass 20 kg and diameter 0.7 m. Find the gyroscopic moment, about the axis of the arm, which tends to twist the landing gear arm.

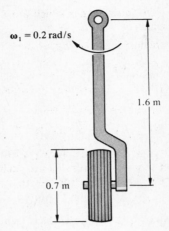

$\omega_1 = 0.2$ rad/s

1.6 m

0.7 m

**Fig. 22.44**

❚ The mass moment of inertia of the wheel is

$$I_0 = \tfrac{1}{8}md^2 = \tfrac{1}{8}(20)(0.7)^2 = 1.23 \text{ kg} \cdot \text{m}^2$$

The angular velocity of the wheel is found from

$$\omega = \frac{v}{r} = \frac{200(1,000)}{3,600(0.35)} = 159 \text{ rad/s}$$

The gyroscopic moment is then

$$M_0 = I_0\omega\omega_1 = 1.23(159)(0.2) = 39.1 \text{ N} \cdot \text{m}$$

**22.45** The system of Prob. 15.58 is repeated in Fig. 22.45a. The fighter plane is in a power dive. The speed of the plane is 960 km/h, and the minimum radius of curvature of the flight path occurs at point $a$ as the pilot pulls out of the dive. A representation of the compressor and turbine rotors of the jet engine is shown in Fig. 22.45b. Find the magnitude of the dynamic bearing forces caused by the gyroscopic moment as the plane passes point $a$. The engine speed is 15,000 r/min, and the specific weight of steel is 0.283 lb/in³.

❚ The mass moment of inertia of each disk about its center axis is of the form

$$I_0 = \tfrac{1}{8}md^2$$

The mass moment of inertia of the entire rotating assembly of the jet engine is

$$I_0 = 8\left[\frac{1}{8}\frac{\pi(15)^2}{4}(2)\left(\frac{0.283}{386}\right)(15)^2\right] + 2\left[\frac{1}{8}\frac{\pi(24)^2}{4}(1.75)\left(\frac{0.283}{386}\right)(24)^2\right] = 142 \text{ lb} \cdot \text{s}^2 \cdot \text{in}$$

The angular velocity of the aircraft as it passes point $a$ is given by

$$\omega_1 = \frac{v}{r} = \frac{960(1,000)}{3,600(2,000)} = 0.133 \text{ rad/s}$$

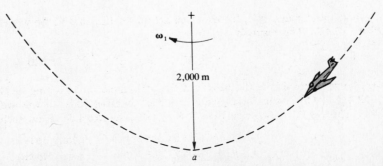

$\omega_1$

2,000 m

$a$

**Fig. 22.45a**

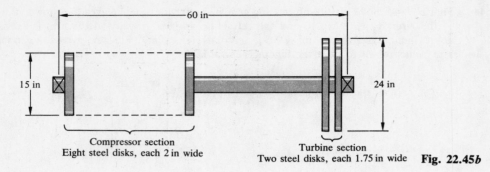

Fig. 22.45b

The speed of rotation of the engine is

$$\omega = 15{,}000\left(\frac{2\pi}{60}\right) = 1{,}570\,\text{rad/s}$$

The gyroscopic moment has the value

$$M_0 = I_0\omega\omega_1 = 142(1{,}570)(0.133) = 29{,}700\,\text{in}\cdot\text{lb}$$

Since the gyroscopic moment is a pure couple effect, the dynamic bearing reaction forces are equal, with the magnitude

$$R = \frac{M_0}{l} = \frac{29{,}700}{60} = 495\,\text{lb}$$

It may be seen that the maneuver of the plane produces large gyroscopic forces on the bearings.

**22.46**  A rotor is mounted on a shaft supported by the yoke arrangement shown in Fig. 22.46a. The weight of the rotor is 4 lb and the centroidal radius of gyration is 1.25 in. The speed of the rotor is $\omega = 2{,}000\,\text{r/min}$ and the speed of the yoke is $\omega_1 = 100\,\text{r/min}$.

*(a)* Find the magnitude, direction, and sense of the dynamic forces exerted on the bearings.

*(b)* Do the same as in part *(a)*, if the rotor has the position shown in Fig. 22.46b. Only bearing a in this figure can resist forces in the x direction.

▌ *(a)* Figure 22.46c shows a top view of Fig. 22.46a, with the momentum vector $\mathbf{H}_0$ and the change in this quantity. $R_{ay}$ and $R_{by}$ are the dynamic forces exerted by the bearings on the shaft. The moment of inertia of the rotor about the shaft axis is

$$I_0 = k_0^2 m = 1.25^2\left(\frac{4}{386}\right) = 0.0162\,\text{lb}\cdot\text{s}^2\cdot\text{in}$$

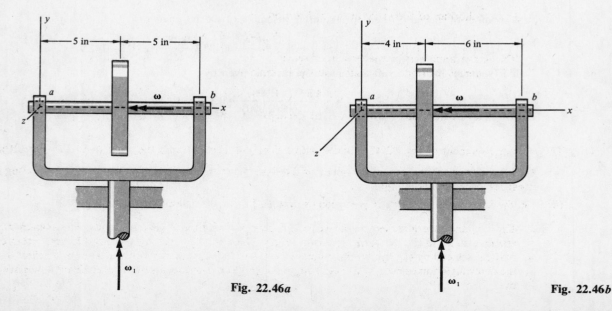

Fig. 22.46a

Fig. 22.46b

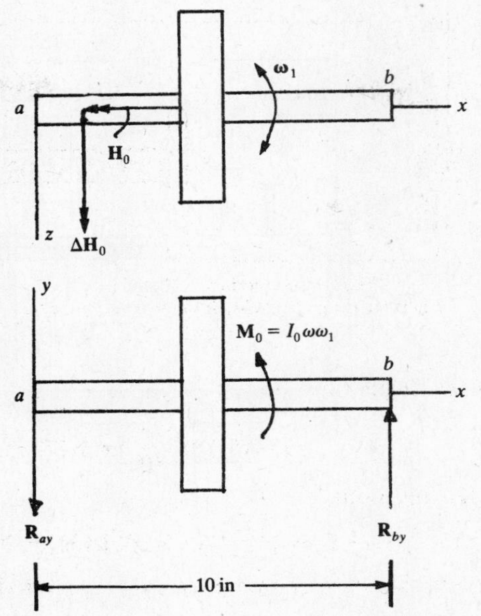

Fig. 22.46c

The angular velocities of the rotor and yoke are

$$\omega = 2{,}000 \, \text{r/min} = 209 \, \text{rad/s} \qquad \omega_1 = 100 \, \text{r/min} = 10.5 \, \text{rad/s}$$

The gyroscopic moment $M_0$, shown in Fig. 22.46c, is

$$M_0 = I_0 \omega \omega_1 = 0.0162(209)10.5 = 35.6 \, \text{in} \cdot \text{lb}$$

The dynamic bearing forces form a couple, and $R_{ay} = R_{by} = R$. These forces have the value

$$R = \frac{M_0}{l} = \frac{35.6}{10} = 3.56 \, \text{lb}$$

The forces exerted by the bearings on the shaft are given by $R = 3.56 \, \text{lb}$, with the senses shown in Fig. 22.46c. The forces exerted by the shaft on the bearings have senses opposite those of the forces $R$ in Fig. 22.46c.

(b)  The gyroscopic forces $R_{ay}$ and $R_{by}$ are the same as in part (a), since the gyroscopic moment is a pure couple effect. The centrifugal force $F_x$ which acts on the rotor, caused by the angular velocity $\omega_1$, is

$$F_x = mr\omega^2 = \tfrac{4}{386}(1)10.5^2 = 1.14 \, \text{lb}$$

For equilibrium of the rotor in the $x$ direction,

$$R_{ax} = 1.14 \, \text{lb}$$

The above force acts in the positive $x$ sense.
The forces exerted by the shaft on the bearings are

$$R_{ax} = -1.14 \, \text{lb} \qquad R_{ay} = 3.56 \, \text{lb} \qquad R_{by} = -3.56 \, \text{lb}$$

The above three forces are positive in the sense of positive $x$, $y$, and $z$.

**22.47**  The yoke arrangement in Fig. 22.47a rotates with a speed of 22 rad/s, and the shaft rotates at 100 rad/s.

(a)  Find the maximum and minimum values of the dynamic forces at bearings $a$ and $b$. Only bearing $b$ can resist forces in the $x$ direction.

(b)  Do the same as in part (a), if $m_a$ becomes detached from the assembly.

❚ (a)  The $xyz$ axes are attached to the yoke and have the motion of this element. The masses are first assumed to lie in the $xy$ plane, as shown in Fig. 22.47b. $R_{a,M}$ and $R_{b,M}$ are the forces, acting on the shaft, produced by the gyroscopic moment $M$. $R_{ay}$ and $R_{by}$ are the forces acting on the shaft to resist the centrifugal forces acting on masses $m_a$ and $m_b$. The components of the centrifugal forces are given by

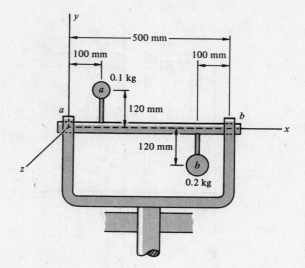

Fig. 22.47a

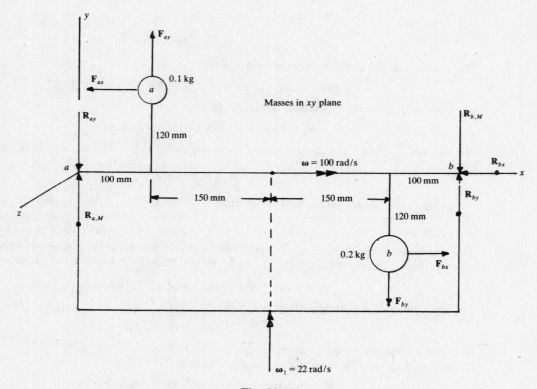

Fig. 22.47b

$$F = mr\omega^2 \qquad F_{ax} = 0.1\left(\frac{150}{1,000}\right)22^2 = 7.26\,\text{N} \qquad F_{ay} = 0.1\left(\frac{120}{1,000}\right)100^2 = 120\,\text{N}$$

$$F_{bx} = 0.2\left(\frac{150}{1,000}\right)22^2 = 14.5\,\text{N} \qquad F_{by} = 0.2\left(\frac{120}{1,000}\right)100^2 = 240\,\text{N}$$

The mass moment of inertia about the $x$ axis is

$$I_x = 0.1\left(\frac{120}{1,000}\right)^2 + 0.2\left(\frac{120}{1,000}\right)^2 = 0.00432\,\text{kg}\cdot\text{m}^2$$

The gyroscopic moment is found as

$$M_0 = I_x\omega\omega_1 = 0.00432(22)100 = 9.50\,\text{N}\cdot\text{m}$$

The gyroscopic bearing forces produced by $M$ are

$$R_{a,M} = R_{b,M} = \frac{9.50}{500/1,000} = 19 \text{ N}$$

Since there is no restriction in this problem on the senses of $\omega$ and $\omega_1$,

$$R_{a,M} = R_{b,M} = \pm 19 \text{ N}$$

The effects of $R_{a,M}$ and $R_{b,M}$ are omitted, and the force components $R_{ay}$ and $R_{by}$ are next found. For equilibrium,

$$\sum M_z = 0$$

$$F_{ax}(120) + F_{ay}(100) + F_{bx}(120) - F_{by}(400) + R_{by}(500)$$

$$= 7.26(120) + 120(100) + 14.5(120) - 240(400) + R_{by}(500) = 0$$

$$R_{by} = 163 \text{ N}$$

$$\sum F_y = 0 \qquad -R_{ay} + F_{ay} - F_{by} + R_{by} = -R_{ay} + 120 - 240 + 163 = 0 \qquad R_{ay} = 43 \text{ N}$$

When the assembly has rotated 180° about the $x$ axis, the senses of $R_{ay}$ and $R_{by}$ change. Thus,

$$R_{ay} = \pm 43 \text{ N} \qquad R_{by} = \pm 163 \text{ N}$$

$R_{an,xy}$ and $R_{bn,xy}$ are the resultant forces, in the $y$ direction, on the shaft when the masses lie in the $xy$ plane.

$$R_{an,xy,max} = |R_{ay}| + |R_{a,M}| = 43 + 19 = 62 \text{ N} \qquad R_{an,xy,min} = |R_{ay}| - |R_{a,M}| = 43 - 19 = 24 \text{ N}$$

$$R_{bn,xy,max} = |R_{by}| + |R_{b,M}| = 163 + 19 = 182 \text{ N} \qquad R_{bn,xy,min} = |R_{by}| - |R_{b,M}| = 163 - 19 = 144 \text{ N}$$

For equilibrium in the $x$ direction,

$$\sum F_x = 0 \qquad -R_{bx} + F_{bx} - F_{az} = -R_{bx} + 14.5 - 7.26 = 0 \qquad R_{bx} = 7.24 \text{ N}$$

The masses are next assumed to lie in the $xz$ plane, as in Fig. 22.47$c$. $R_{a,M}$ and $R_{b,M}$, as before, act in the $y$ direction, and have no components in the $xz$ plane. The centrifugal forces due to $\omega_1$ are

$$F = mr\omega^2 \qquad F_{a,1} = 0.1\left(\frac{192}{1,000}\right)22^2 = 9.29 \text{ N} \qquad F_{b,1} = 0.2\left(\frac{192}{1,000}\right)22^2 = 18.6 \text{ N}$$

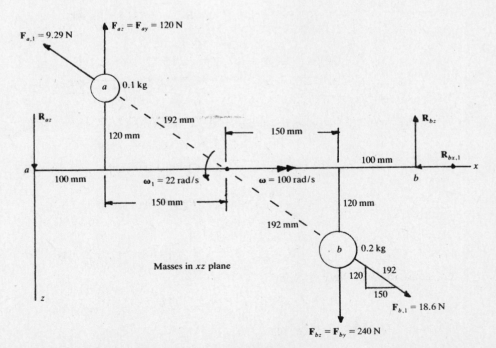

Fig. 22.47$c$

For equilibrium, using Fig. 22.47c,

$$\sum M_a = 0$$

$$\frac{150}{192}(9.29)120 + \frac{120}{192}(9.29)100 + 120(100) + \frac{150}{192}(18.6)120 - \frac{120}{192}(18.6)400 - 240(400) + R_{bz}(500) = 0$$

$$R_{bz} = 171\,\text{N}$$

$$\sum F_z = 0 \qquad R_{az} - \frac{120}{192}(9.29) - 120 + 240 + \frac{120}{192}(18.6) - R_{bz} = 0 \qquad R_{az} = 45.2\,\text{N}$$

$R_{an,xz}$ and $R_{bn,xz}$ are the resultant forces, acting normal to the shaft at $a$ and $b$, when the masses lie in the $xz$ plane. These forces have the values

$$R_{an,xz} = \sqrt{45.2^2 + 19^2} = 49\,\text{N} \qquad R_{bn,xz} = \sqrt{171^2 + 19^2} = 172\,\text{N}$$

For equilibrium in the $x$ direction,

$$\sum F_x = 0 \qquad R_{bx,1} - \frac{150}{192}(9.29) + \frac{150}{192}(18.6) = 0 \qquad R_{bx,1} = -7.27\,\text{N}$$

The final results for the maximum and minimum values of the dynamic forces at bearings $a$ and $b$ are

$$R_{an,\text{max}} = 62\,\text{N} \qquad R_{an,\text{min}} = 24\,\text{N} \qquad R_{bn,\text{max}} = 182\,\text{N} \qquad R_{bn,\text{min}} = 144\,\text{N} \qquad R_{bx,\text{max}} = 7.27\,\text{N}$$

(b) Mass $m_a$ is assumed to become detached from the assembly. The solution follows that in part (a). The mass $m_b$ is first assumed to lie in the $xy$ plane. The centrifugal forces are

$$F_{ax} = F_{ay} = 0 \qquad F_{bx} = 14.5\,\text{N} \qquad F_{by} = 240\,\text{N}$$

The moment of inertia about the $x$ axis is

$$I_x = 0.2\left(\frac{120}{1{,}000}\right)^2 = 0.00288\,\text{kg}\cdot\text{m}^2$$

The gyroscopic moment is given by

$$M_0 = I_x \omega \omega_1 = 0.00288(22)100 = 6.34\,\text{N}\cdot\text{m}$$

Since there is no restriction on the senses of $\omega$ and $\omega_1$,

$$R_{a,M} = R_{b,M} = \pm\frac{6.34}{500/1{,}000} = \pm 12.7\,\text{N}$$

The effects of $R_{a,M}$ and $R_{b,M}$ are omitted, and $R_{ay}$ and $R_{by}$ are found. For equilibrium,

$$\sum M_a = 0$$

$$F_{bx}(120) - F_{by}(400) + R_{by}(500) = 14.5(120) - 240(400) + R_{by}(500) = 0$$

$$R_{by} = 189\,\text{N}$$

$$\sum F_y = 0 \qquad -R_{ay} - F_{by} + R_{by} = -R_{ay} - 240 + 189 = 0 \qquad R_{ay} = -51\,\text{N}$$

When the assembly has rotated 180° about the $x$ axis, the senses of $R_{ay}$ and $R_{by}$ change. Thus,

$$R_{ay} = \pm 51\,\text{N} \qquad R_{by} = \pm 189\,\text{N}$$

$R_{an,xy}$ and $R_{bn,xy}$ are the resultant forces in the $y$ direction on the shaft at $a$ and $b$. When the masses lie in the $xy$ plane,

$$R_{an,xy,\text{max}} = |R_{ay}| + |R_{a,M}| = 51 + 12.7 = 63.7\,\text{N}$$
$$R_{an,xy,\text{min}} = |R_{ay}| - |R_{a,M}| = 51 - 12.7 = 38.3\,\text{N}$$
$$R_{bn,xy,\text{max}} = |R_{by}| + |R_{b,M}| = 189 + 12.7 = 202\,\text{N}$$
$$R_{bn,xy,\text{min}} = |R_{by}| - |R_{b,M}| = 189 - 12.7 = 176\,\text{N}$$

For equilibrium in the $x$ direction,

$$\sum F_x = 0 \qquad -R_{bx} + F_{bx} = 0 \qquad R_{bx} = 14.5\,\text{N}$$

When the masses lie in the $xz$ plane, using Fig. 22.47c,

$$F_{a,1} = F_{az} = 0 \qquad F_{b,1} = 18.6\,\text{N}$$

$$\sum M_a = 0 \qquad \frac{150}{192}(18.6)120 - \frac{120}{192}(18.6)400 - 240(400) + R_{bz}(500) = 0 \qquad R_{bz} = 198\,\text{N}$$

$$\sum F_z = 0 \qquad R_{az} + 240 + \frac{120}{192}(18.6) - R_{bz} = 0 \qquad R_{az} = -53.6\,\text{N}$$

$$\sum F_x = 0 \qquad \frac{150}{192}(18.6) - R_{bx,1} = 0 \qquad R_{bx,1} = 14.5\,\text{N}$$

$R_{an,xz}$ and $R_{bn,xz}$ are the resultant normal forces on the shaft at $a$ and $b$, when the masses lie in the $xz$ plane. These forces have the forms

$$R_{an,xz} = \sqrt{53.6^2 + 12.7^2} = 55.1\,\text{N} \qquad R_{bn,xz} = \sqrt{198^2 + 12.7^2} = 198\,\text{N}$$

$$R_{an,\text{max}} = 63.7\,\text{N} \qquad R_{an,\text{min}} = 38.3\,\text{N} \qquad R_{bn,\text{max}} = 202\,\text{N} \qquad R_{bn,\text{min}} = 176\,\text{N} \qquad R_{bx} = 14.5\,\text{N}$$

It may be seen that the maximum values of the dynamic hearing forces increase when mass $a$ becomes detached.

**22.48** Give a summary of the basic concepts of rigid bodies in three-dimensional motion.

$\blacksquare$ An unbalanced rigid body which rotates about a fixed axis exerts pulsating, or time-varying, forces on the bearings supporting the shaft.

The problem of rotating unbalance may be solved with the direct use of centrifugal forces and the equations of dynamic equilibrium. This problem also may be solved in terms of the centroidal coordinates, and the mass products of inertia, of the rotating mass assembly. The particular problem to be solved usually dictates which method is more convenient.

Mass products of inertia are a measure of the symmetry of placement of a set of coordinate axes on a body. These quantities may be positive or negative, and they have the basic units of mass times length squared. For a thin plane body which lies in the $xy$ plane

$$I_{xy} = \int_V xy\,dm$$

The parallel-axis, or transfer, theorem for the thin plane body is

$$I_{xy} = I_{0,xy} + mx_c y_c$$

where $I_{0,xy}$ is the centroidal product of inertia, $m$ is the mass, and $x_c$ and $y_c$ are the centroidal coordinates of the body, measured in the $xyz$ coordinate system.

The necessary and sufficient conditions for dynamic balance of a rotating mass assembly are that

1. The center of mass of the assembly must lie on the axis of rotation.

2. The axis of rotation and the two rectangular coordinate axes normal to this axis must be principal axes of the body.

If the rotating mass is a thin plane body which lies in the $xy$ plane, the dynamic forces exerted by the bearings $a$ and $b$ on the shaft have the forms

$$R_{by} = \frac{-\omega^2 I_{xy}}{l} \qquad R_{ay} = \frac{\omega^2 I_{xy}}{l} - my_c\omega^2$$

where $\omega$ is the angular velocity, $I_{xy}$ is the mass product of inertia, $y_c$ is the displacement of the center of mass from the axis of rotation, and $l$ is the axial separation distance between bearings $a$ and $b$.

Any rotating mass assembly may be put into a condition of dynamic balance by the addition of correction weights, or by the removal of material, in two parallel planes which are normal to the axis of rotation. The dynamic forces due to rotating unbalance are completely independent of the angular acceleration of the rotating assembly.

A gyroscopic moment occurs when the axis of rotation of a rigid body with angular velocity changes its direction in space. The magnitude of the gyroscopic moment is given by

$$M_0 = I_0\omega\omega_1$$

where $I_0$ is the mass moment of inertia of the body about the spin axis, $\omega$ is the spin angular velocity, and $\omega_1$ is the angular velocity of the axis of rotation.

# CHAPTER 23
# Self-Study Review of the Fundamental Definitions, Concepts, and Techniques of Engineering Mechanics: Dynamics

*Note*: The selected questions below are repeated from the previous chapters in this text. They contain material which is fundamental to the mastery of the subject dynamics. The reader is urged to review these questions carefully to make sure that he or she fully understands the definitions, concepts, and techniques. Reference may be made directly, as required, to the solutions which accompany the original problems. The original problem number is shown in parentheses.

Each question is followed by a set of boxes, for use by the reader to indicate the questions which have been mastered.

## KINEMATICS OF PARTICLES

23.1    What is the definition of the term kinematics? (14.1*a*)

23.2    What is the difference between a particle and a body? (14.1*b*)

23.3    What are the only two possible motions of a particle? (14.2)

23.4    What is the definition of the term displacement? (14.3*a*)

23.5    Show that displacement is a vector quantity. (14.3*b*)

23.6    What is the definition of the term velocity? (14.4a)

23.7    Give the form for the average velocity of a particle. (14.4a)

23.8    What is the relationship between the positive senses of velocity and displacement? (14.4*b*)

23.9    What is meant by the speed of a particle? (14.4*c*)

23.10   What is the definition of the term acceleration? (14.5*a*)

23.11   What is the relationship among the positive senses of acceleration, velocity, and displacement? (14.5*b*)

23.12   What is meant by the term deceleration? (14.5*c*)

23.13   Give the form for the average acceleration of a particle. (14.5*d*)

23.14   What is the graphical interpretation of the definition of velocity as the first time derivative of displacement? (14.15)

23.15   What is the graphical interpretation of the definition of acceleration as the first time derivative of velocity? (14.16*a*)

23.16   What is meant by the term motion diagram? (14.16*b*)

23.17   Show how to identify positive and negative slopes, and increasing and decreasing slopes, in motion diagrams. (14.17*a*)

23.18   Show a simple way to depict positive and negative, and increasing and decreasing, slopes in motion diagrams. (14.17*b*)

23.19   Describe two important physical problems with particles which move in rectilinear motion with constant acceleration. (14.30*a*)

23.20   Show the relationship among velocity, acceleration, and time for the case of a particle which moves in rectilinear motion with constant acceleration. (14.30*b*)

23.21   Show the relationship among displacement, acceleration, and time for the case of a particle which moves in rectilinear motion with constant acceleration. (14.31*a*)

23.22   Show the relationship among displacement, velocity, and acceleration for the case of a particle which moves in rectilinear motion with constant acceleration. (14.31*b*)

23.23   Summarize the three equations used to solve problems of rectilinear motion with constant acceleration. (14.31*c*)

23.24   Give the three equations of motion for the case of a freely falling particle in the earth's gravitational field, in the absence of any frictional retarding effects. (14.43)

23.25   Find the magnitude, direction, and sense of the velocity of a particle which moves in plane curvilinear translation. (14.55)

23.26   Find the magnitude, direction, and sense of the normal component of the acceleration of a particle which moves in plane curvilinear translation. (14.56*a*)

**23.27**    What is the name of the normal component of acceleration of a particle which moves in plane curvilinear translation? (14.56*b*)    ☐☐☐☐

**23.28**    What is the physical effect which produces a normal component of acceleration of a particle which moves in plane curvilinear translation?    ☐☐☐☐

**23.29**    What is the magnitude, direction, and sense of the tangential component of acceleration of a particle which moves in plane curvilinear translation. (14.57)    ☐☐☐☐

**23.30**    Give a summary of the normal and tangential components of acceleration of a particle which moves in plane curvilinear translation. (14.58*a*)    ☐☐☐☐

**23.31**    Show the forms for the magnitude and direction of the resultant of the normal and tangential acceleration of a particle which moves in plane curvilinear translation. (14.58*b*)    ☐☐☐☐

**23.32**    What is the major difference between visualizing the velocity and visualizing the acceleration of a particle which moves in plane curvilinear translation? (14.58*c*)    ☐☐☐☐

**23.33**    What is the expression for the radius of curvature of a plane curve which is defined by $y = f(x)$? (14.63*a*)    ☐☐☐☐

**23.34**    What is a point of inflection on a plane curve? (14.63*b*)    ☐☐☐☐

**23.35**    Show the forms for the rectangular components of the velocity and acceleration of a particle which moves in plane curvilinear translation. (14.66*a*)    ☐☐☐☐

**23.36**    What significant physical problem may be solved by using the rectangular components of the velocity and acceleration of a particle which moves in plane curvilinear translation? (14.66*b*)    ☐☐☐☐

**23.37**    Give the equations of motion of a particle which moves in a gravitational acceleration field. (14.67*a*)    ☐☐☐☐

**23.38**    What is the term used to describe the motion of a particle which moves in a gravitational acceleration field? (14.67*b*)    ☐☐☐☐

**23.39**    Compare the component motions, in the horizontal and vertical directions, of a particle which moves in a gravitational field. (14.67*c*)    ☐☐☐☐

**23.40**    What is an absolute, or inertial, coordinate system? (14.86*a*)    ☐☐☐☐

**23.41**    How may a particle, or point, be referenced to an absolute coordinate system? (14.86*b*)    ☐☐☐☐

**23.42**    What is meant by the terms absolute velocity and absolute acceleration? (14.86*c*)    ☐☐☐☐

**23.43**    Give the form for the relative displacement of two particles. (14.87*a*)    ☐☐☐☐

**23.44**    Show a convention that may be used to distinguish between absolute and relative quantities. (14.87*b*)    ☐☐☐☐

**23.45**    Give the form for the relative velocity of two particles. (14.88*a*)    ☐☐☐☐

**23.46**    Give an interpretation of the relative velocity of two particles. (14.88*b*)    ☐☐☐☐

**23.47**    Give the form for the relative acceleration of two particles. (14.88*c*)    ☐☐☐☐

**23.48**    What is an important useful application of the concepts of relative velocity and acceleration? (14.88*d*)    ☐☐☐☐

## DYNAMICS OF PARTICLES

**23.49**    What is studied in the subjects kinetics and dynamics? (15.1)    ☐☐☐☐

**23.50**    What are the two general types of problems in dynamics? (15.2)    ☐☐☐☐

**23.51**    Summarize the definition, and the possible motions, of a particle. (15.3*a*)    ☐☐☐☐

**23.52**    Under what conditions can motion of a body be considered to be particle motion? (15.3*b*)    ☐☐☐☐

**23.53**    Give the form of Newton's second law for the motion of a particle. (15.4*a*)    ☐☐☐☐

**23.54**    What is the physical interpretation of Newton's second law for the motion of a particle? (15.4*b*)    ☐☐☐☐

**23.55**    What is an alternative way to state Newton's second law? (15.4*c*)    ☐☐☐☐

**23.56**    What type of motion does a particle have if the resultant force which acts on the particle is zero? (15.5)    ☐☐☐☐

**23.57**    Give the USCS and SI units which are used with Newton's second law. (15.6*a*)    ☐☐☐☐

**23.58**    Give the conversion factors between the USCS and SI units which are used with Newton's second law. (15.6*b*)    ☐☐☐☐

**23.59**    Describe the problem where a particle is acted on by an applied force and moves in rectilinear translation. (15.7)    ☐☐☐☐

**23.60**    Give a set of general rules for solving the problem of motion of a particle in rectilinear translation. (15.8*a*)    ☐☐☐☐

**23.61**    Give a short summary of the rules for solving the problem of motion of a particle in rectilinear translation. (15.8*b*)    ☐☐☐☐

**23.62**    What equations are used to find the motion of a particle which is acted on by a constant force and moves in rectilinear translation? (15.9)    ☐☐☐☐

**23.63**    What is the technique for solving the problem of motion of two bodies which are connected by a cable or link and move in rectilinear translation? (15.40)    ☐☐☐☐

**23.64** Give the forms of Newton's second law for the normal and tangential component motions for a particle which moves in plane curvilinear translation. (15.55a) ☐☐☐☐

**23.65** Discuss the normal component of force which acts on the particle which moves in plane curvilinear translation. (15.55b) ☐☐☐☐

**23.66** Under what conditions does a particle move with constant normal acceleration? (15.55c) ☐☐☐☐

**23.67** Discuss the motion which would ensue if the normal force acting on a particle which moves in plane curvilinear translation became equal to zero. (15.56) ☐☐☐☐

**23.68** Give the example of the D'Alembert, or inertia, force. (15.79a) ☐☐☐☐

**23.69** What are the characteristics of an inertia force? (15.79b) ☐☐☐☐

**23.70** Show how the use of the inertia force reduces the form of a dynamics problem to that of a statics problem. (15.79c) ☐☐☐☐

**23.71** Give the general rules for using the inertia force to solve problems in dynamics. (15.80) ☐☐☐☐

## KINEMATICS OF PLANE MOTION OF A RIGID BODY

**23.72** Define the term rigid body (16.1a) ☐☐☐☐

**23.73** What is the significant difference between particle motion and rigid body motion? (16.1b) ☐☐☐☐

**23.74** Describe the motion of a rigid body in rectilinear translation. (16.2) ☐☐☐☐

**23.75** Describe the motion of a rigid body in curvilinear translation. (16.3) ☐☐☐☐

**23.76** What is the relationship between the velocity and acceleration of arbitrary points on a rigid body which moves in rectilinear or curvilinear translation? (16.4a) ☐☐☐☐

**23.77** What is the distinguishing characteristic of rectilinear or curvilinear translation? (16.4b) ☐☐☐☐

**23.78** Describe the possible motions of a link which is hinged to an absolute reference point, such as the earth. (16.5a) ☐☐☐☐

**23.79** What is meant by the term rotation? (16.5b) ☐☐☐☐

**23.80** Define the term angular displacement. (16.6a) ☐☐☐☐

**23.81** Define the term angular velocity. (16.6b) ☐☐☐☐

**23.82** Define the term angular acceleration. (16.6c) ☐☐☐☐

**23.83** What is the graphical interpretation of the definition of angular velocity as the first time derivative of the angular displacement? (16.12) ☐☐☐☐

**23.84** What is the graphical interpretation of the definition of angular acceleration as the first time derivative of the angular velocity? (16.13) ☐☐☐☐

**23.85** What are the relationships among the angular displacement, velocity, and acceleration of a rigid body which rotates about a fixed axis with constant angular acceleration. (16.19) ☐☐☐☐

**23.86** What is the relationship between the angular velocity of a link which rotates about a fixed axis and the translational velocity of points on the link? (16.24a) ☐☐☐☐

**23.87** What is the relationship among the angular velocity and angular acceleration, and the translational acceleration, of points on a link which rotates about a fixed axis? (16.24b) ☐☐☐☐

**23.88** What is the relationship between the translational and rotational motions of connected bodies? (16.30) ☐☐☐☐

**23.89** Show how the most general type of plane motion of a rigid body may be expressed as a combination of translational and rotational motions. (16.36) ☐☐☐☐

**23.90** Give the definition of the instant center of rotation of a body. (16.40) ☐☐☐☐

**23.91** Show the technique for locating the instant center of rotation of a body. (16.41) ☐☐☐☐

**23.92** Describe the possible motions of a cylinder which moves along a straight track. (16.51a) ☐☐☐☐

**23.93** What is meant by the term pure rolling? (16.51b) ☐☐☐☐

**23.94** What is the relationship between the angular velocity of a cylinder in a state of pure rolling and the translational velocity of the center of the cylinder? (16.52a) ☐☐☐☐

**23.95** Give a physical interpretation of the effect of pure rolling of a cylinder. (16.52b) ☐☐☐☐

## CENTROIDS, AND MASS MOMENTS AND PRODUCTS OF INERTIA, OF RIGID BODIES

**23.96** Give the definition of the coordinates of the centroid of a volume which is positioned with respect to a set of $xyz$ coordinate axes. (17.1a) ☐☐☐☐

**23.97** How is the location of the centroid affected by the position of the $xyz$ coordinates with respect to the volume? (17.1b) ☐☐☐☐

**23.98** Show examples of volumes with one, two, and three planes of symmetry. (17.2a) ☐☐☐☐

**23.99** How is the determination of the coordinates of the centroid of a volume simplified if the volume has planes of symmetry? (17.2b) ☐☐☐☐

**23.100** Define the first moments of a volume. (17.3) ☐☐☐☐

**23.101** What is meant by the center of mass of a rigid body? (17.5*a*)

**23.102** What is an alternative term used to describe the center of mass of a rigid body? (17.5*b*)

**23.103** State two practical applications of the concept of the center of mass of a rigid body. (17.5*c*)

**23.104** How is the location of the center of mass of a body made of homogeneous material determined? (17.5*d*)

**23.105** How is the location of the center of mass of a body made of nonhomogeneous material determined? (17.6)

**23.106** Show how the coordinates of the centroid of a homogeneous composite body may be found. (17.12*a*)

**23.107** How is the computation of the centroidal coordinates modified if the body contains holes or cutouts? (17.12*b*)

**23.108** Show a tabular format for the computations for the centroidal coordinates for the case where the body is a complex shape which requires the description of several elementary shapes. (17.17)

**23.109** Give the formal definitions of the mass moments of inertia of a body about the *xyz* coordinate axes. (17.19*a*)

**23.110** What are the units of mass moment of inertia? (17.19*b*)

**23.111** Is the mass moment of inertia a scalar or vector quantity? (17.19*c*)

**23.112** State three significant characteristics of the mass moment of inertia (17.20*a*)

**23.113** Give a physical interpretation of the mass moment of inertia. (17.20*b*)

**23.114** Give the formal definitions of the radii of gyration of a body about the *x*, *y*, and *z* axes. (17.21*a*)

**23.115** What is the physical significance of a radius of gyration? (17.21*b*)

**23.116** Show how the parallel-axis, or transfer, theorem is derived. (17.28*a*)

**23.117** What is a common mistake made when using the parallel-axis theorem? (17.28*b*)

**23.118** What is an inequality relationship between the moment of inertia of a body about a centroidal axis and the moment of inertia about an axis parallel to this centroidal axis? (17.28*c*)

**23.119** How may the mass moment of inertia of a body be found by using the transfer theorem, together with a single integration. (17.39*a*)

**23.120** State the general technique of solution for the mass moment of inertia of a body which is not of simple geometric shape. (17.43)

**23.121** Show a tabular format for the computations for the mass moments of inertia for the case where a body is a complex shape which requires the description of several elementary shapes. (17.57)

**23.122** Show how the forms for the mass moments of inertia may be simplified if a body has the form of a thin, rigid plane body. (17.73)

**23.123** Show the relationship between the moments of inertia of a plane area and the mass moments of inertia of a homogeneous, thin, rigid plane body. (17.74)

**23.124** How is it determined whether a rod used in a rigid body of plane form, made of homogeneous rod material with constant cross-sectional dimensions, is thin? (17.86*a*)

**23.125** How are the coordinates of the center of mass of a rigid body of plane form, made of thin, homogeneous rod material with constant cross-sectional dimensions, found? (17.86*b*)

**23.126** How are the mass moments of inertia of a plane rod shape found? (17.87*a*)

**23.127** What is the general form of the parallel-axis, or transfer, theorem, for a plane rod shape? (17.87*b*)

**23.128** Give the expressions for the mass products of inertia of a rigid body (17.93*a*)

**23.129** Compare the mass moments, and mass products, of inertia. (17.93*b*)

**23.130** What is meant by the term principal axes? (17.93*c*)

**23.131** What is the physical interpretation of the mass products of inertia? (17.93*d*)

**23.132** Give an important example of the use of mass products of inertia. (17.93*e*)

**23.133** Give the general forms of the parallel-axis, or transfer, theorems for mass products of inertia of a rigid body. (17.94*a*)

**23.134** Give the forms of the parallel-axis theorems for mass products of inertia of a thin, rigid plane body. (17.94*b*)

## DYNAMICS OF RIGID BODIES IN PLANE MOTION

**23.135** Give three general classifications of problems in plane motion of a rigid body. (18.1)

**23.136** Develop the equation for the plane motion of a rigid body about a fixed point. (18.2)

**23.137** How is the problem of plane motion of a rigid body about a fixed point simplified if the resultant couple or moment which acts on the body has a constant value? (18.3)

**23.138** What is the usual first step when solving the problem of plane motion of a rigid body about a fixed point? (18.4*a*)

**23.139** Compare the resultant forces which act on bodies in static equilibrium with those which act on bodies in plane motion. (18.4*b*)

**23.140** Which components of the forces which act on a body in plane motion do not contribute to the plane motion? (18.5) ▢▢▢▢

**23.141** What is the equation which describes the translational motion of a body that may have any general type of plane motion? (18.27) ▢▢▢▢

**23.142** Show that the translational acceleration of a rigid body in plane motion is independent of the position of the resultant force that acts on the body. (18.29) ▢▢▢▢

**23.143** Give the equation which describes the rotational motion of a body in general plane motion. (18.30) ▢▢▢▢

**23.144** Summarize the equations which describe the plane motion of a rigid body. (18.31) ▢▢▢▢

**23.145** Find the condition, for a cylindrical body acted on by a horizontal force, to roll without slipping along a straight horizontal track (18.41) ▢▢▢▢

**23.146** Show how the D'Alembert, or inertia, moment may be used to solve problems in plane motion of a rigid body. (18.72) ▢▢▢▢

**23.147** Show how the inertia force may be used to determine whether an object will slide or tip. (18.92) ▢▢▢▢

**23.148** What is meant by the center of percussion? (18.98*a*) ▢▢▢▢

**23.149** How is the location of the center of percussion determined? (18.98*b*) ▢▢▢▢

## WORK-ENERGY METHODS FOR PARTICLES AND RIGID BODIES

**23.150** What is the work done when a force acts on a particle which moves through an infinitesimal translational displacement? (19.1*a*) ▢▢▢▢

**23.151** What are the fundamental units of work? (19.1*b*) ▢▢▢▢

**23.152** Is work a scalar or a vector quantity? (19.1*c*) ▢▢▢▢

**23.153** Find the work done if a force of constant magnitude, direction, and sense acts on a particle which moves through a finite translational displacement. (19.2*a*) ▢▢▢▢

**23.154** What is the difference between positive and negative work? (19.2*b*) ▢▢▢▢

**23.155** What is the work done if the force is normal to the displacement of the particle? (19.2*c*) ▢▢▢▢

**23.156** How is work defined if the magnitude or direction of the force which acts on the particle is not constant? (19.3) ▢▢▢▢

**23.157** Find the work done when a couple or torque, or a moment, acts through an angular displacement. (19.7*a*) ▢▢▢▢

**23.158** Find the work done if the couple or torque, or moment, has constant magnitude, direction, and sense. (19.7*b*) ▢▢▢▢

**23.159** What is the definition of energy? (19.9*a*) ▢▢▢▢

**23.160** In what two forms may energy be stored in a mass particle? (19.9*b*) ▢▢▢▢

**23.161** Find the potential energy of a mass particle which is raised through a vertical displacement. (19.10*a*) ▢▢▢▢

**23.162** What is the fundamental interpretation of potential energy? (19.10*b*) ▢▢▢▢

**23.163** Give the general form for the change in potential energy of a mass particle. (19.11*a*) ▢▢▢▢

**23.164** What is meant by a datum for potential energy? (19.11*b*) ▢▢▢▢

**23.165** Show different choices for a datum for the potential energy of a mass particle. ▢▢▢▢

**23.166** What are two common types of potential energy in a mechanical system? (19.12) ▢▢▢▢

**23.167** Define the kinetic energy of a mass particle. (19.14*a*) ▢▢▢▢

**23.168** What is the fundamental interpretation of kinetic energy? (19.14*b*) ▢▢▢▢

**23.169** What is a significant difference between the datum for potential energy and the datum for kinetic energy? (19.14*c*) ▢▢▢▢

**23.170** What is the relationship between the work done on a mass particle and the change in kinetic energy of this particle? (19.17) ▢▢▢▢

**23.171** What two general types of forces are considered in problems in dynamics? (19.23*a*) ▢▢▢▢

**23.172** Give examples of conservative and nonconservative forces. (19.23*b*) ▢▢▢▢

**23.173** What type of force may be related to the change in energy of a particle? (19.23*c*) ▢▢▢▢

**23.174** Give the definition of the principle of conservation of energy. (19.24) ▢▢▢▢

**23.175** Derive the work-energy method for the motion of a particle. (19.25*a*) ▢▢▢▢

**23.176** Give examples of the exchange of potential and kinetic energies in a physical system. (19.25*b*) ▢▢▢▢

**23.177** What is an important limiting characteristic of the work-energy method of problem solution? (19.25*c*) ▢▢▢▢

**23.178** Find the general expression for the potential energy of a rigid body which moves in a vertical plane. (19.45) ▢▢▢▢

**23.179** Find the general expression for the kinetic energy of a rigid body which rotates about a fixed axis. (19.46*a*) ▢▢▢▢

**23.180** Find the general expression for the kinetic energy of a rigid body which moves in plane motion. (19.46b)

**23.181** Find the relationship between the work done on a body that rotates about a fixed axis and the change in angular velocity of the body. (19.48a)

**23.182** What is the change in angular velocity of a body which rotates about a fixed axis, if the moment effect which acts on the body is constant? (19.48b)

**23.183** Give the form of the work-energy equation for a rigid body which moves in plane motion. (19.53)

**23.184** How may the work-energy method be used to solve for the motion of connected bodies? (19.64a)

**23.185** Give an example of a connected-body problem for which the work-energy method is not a convenient form of solution. (19.64b)

**23.186** Give the equation for the normal acceleration of a particle. (19.80a)

**23.187** How may the work-energy method be used to find the normal acceleration of a particle? (19.80b)

**23.188** What power is expended as a particle moves through an infinitesimal displacement ds in time dt? (19.88a)

**23.189** What is the definition of the term power? (19.88b)

**23.190** What are the basic units of power? (19.88c)

**23.191** What is meant by the term power efficiency? (19.88d)

**23.192** What is the fundamental definition of the power transmitted through a rotating shaft? (19.89)

**23.193** What equation relates the torque, angular speed, and power for a rotating shaft which transmits power? (19.90)

**23.194** Into what two convenient groups may problems in power be divided? (19.91)

## IMPULSE-MOMENTUM METHODS FOR PARTICLES AND RIGID BODIES

**23.195** Summarize the two general methods of solving problems in dynamics (a) by use of Newton's second law and (b) by use of the work-energy method. (20.1)

**23.196** Show that the problem of finding the rebound height of a sphere that is dropped onto a horizontal plate cannot be solved directly by either Newton's second law or the work-energy methods. (20.2)

**23.197** Give the form of the impulse-momentum equation for a particle which moves in rectilinear translation. (20.3a)

**23.198** What is the linear momentum of a particle? (20.3b)

**23.199** What are the units of linear momentum? (20.3c)

**23.200** What is the graphical interpretation of an impulse? (20.3d)

**23.201** What is the form of the impulse-momentum equation for a particle in rectilinear translation if the force which acts on the particle has a constant value? (20.4)

**23.202** What is meant by the term impact? (20.14a)

**23.203** Describe the sequence of events when one body impacts a second body. (20.14b)

**23.204** What three effects characterize the impact problem in dynamics? (20.14c)

**23.205** What is the principal use of solutions to problems in impact? (20.14d)

**23.206** Give the equation for the conservation of linear momentum of two particles which move in rectilinear translation. (20.15a)

**23.207** What is a significant limitation of the equation for the conservation of linear momentum of particles which move in rectilinear translation? (20.15b)

**23.208** Give the definition of the coefficient of restitution. (20.16a)

**23.209** What is plastic impact? (20.16b)

**23.210** What is elastic impact? (20.16c)

**23.211** What is the definition of central impact? (20.40a)

**23.212** What are the definitions of direct central impact and oblique central impact? (20.40b)

**23.213** For the case of oblique central impact what is the significant characteristic of the components of velocity normal to the line of impact? (20.40c)

**23.214** What is the difference between impulsive forces and impact forces? (20.47)

**23.215** Give the form of the angular momentum of a rigid body which rotates about a fixed axis. (20.51a)

**23.216** Is angular momentum a vector or a scalar quantity? (20.51b)

**23.217** Give the form of the impulse-momentum equation for a rigid body which rotates about a fixed axis. (20.52)

**23.218** Give the equations which govern the case of impact of rigid bodies in plane motion. (20.56)

**23.219** Show that, for a certain position where a ball strikes a baseball bat, the person who holds the bat will feel no impulsive force reaction on his or her hands. (20.70) ☐☐☐☐

## RECTILINEAR MOTION OF A BODY WITH RESISTING, OR DRAG, FORCES

**23.220** Summarize the results for the motion of a particle or rigid body which is acted on by a constant force and moves in rectilinear translation. (21.1a) ☐☐☐☐

**23.221** What two general types of resultant forces may act on a particle or rigid body which moves in rectilinear translation? (21.1b) ☐☐☐☐

**23.222** What three common types of drag forces may act on a rigid body which moves in rectilinear translation? (21.1c) ☐☐☐☐

**23.223** Give an example of rectilinear motion of a particle with constant drag force. (21.2) ☐☐☐☐

**23.224** What type of motion results in a drag force which is directly proportional to velocity? (21.3a) ☐☐☐☐

**23.225** Give the general form of a drag force which is directly proportional to velocity. (21.3b) ☐☐☐☐

**23.226** Write the equations of motion for a particle which is acted on by a constant force in the sense of motion, and a viscous drag force, and find the velocity and displacement of the particle. (21.3c) ☐☐☐☐

**23.227** What are the forms of the velocity and displacement of a particle which is acted on by a constant force in the sense of motion and by a viscous drag force (21.13d) ☐☐☐☐

**23.228** Summarize the equations for the velocity and displacement of a particle which is acted on by a constant force in the sense of the motion and by a viscous drag force. (21.3e) ☐☐☐☐

**23.229** What is the viscous drag force constant for a sphere that moves with Stoke's law motion? (21.10) ☐☐☐☐

**23.230** Describe the force effects which resist the motion of a body which moves through a fluid medium. (21.13a) ☐☐☐☐

**23.231** Give the form of a drag force which is caused primarily by pressure effects. (21.13b) ☐☐☐☐

**23.232** What is a quadratic resistance law? (21.13b) ☐☐☐☐

**23.233** Give typical values of the drag coefficient for bodies of simple geometry. (21.13c) ☐☐☐☐

**23.234** Find the velocity and displacement of a body which is acted on by a constant force in the sense of the motion and by a quadratic resisting force. ☐☐☐☐

**23.235** Find the velocity and displacement of a body which is acted on by a constant force with a sense opposite that of the motion and by a quadratic resisting force. ☐☐☐☐

## RIGID BODIES IN THREE-DIMENSIONAL MOTION, DYNAMIC UNBALANCE AND GYROSCOPIC MOMENTS

**23.236** What is the definition of plane motion of a rigid body? (22.1a) ☐☐☐☐

**23.237** State three general types of three-dimensional motion of a rigid body. (22.1b) ☐☐☐☐

**23.238** Review the techniques used to show the D'Alembert, or inertia, force acting on a mass particle. (22.2a) ☐☐☐☐

**23.239** Define the term centrifugal force (22.2b) ☐☐☐☐

**23.240** Show how to find the dynamic forces acting on a rotating shaft with two point masses on the ends of rigid arms attached to the shaft. (22.3a) ☐☐☐☐

**23.241** What is an undesirable effect produced by a rotating unbalanced body? (22.3b) ☐☐☐☐

**23.242** Describe how dynamic bearing forces may be found by integrating the centrifugal forces over the mass elements of a body. (22.14a) ☐☐☐☐

**23.243** What is the general solution for the dynamic bearing forces produced by the rotational motion of a body? (22.25a) ☐☐☐☐

**23.244** What are the necessary and sufficient conditions for perfect dynamic balance of a rotating body? (22.25b) ☐☐☐☐

**23.245** Show the independence of dynamic bearing forces and angular acceleration of rotating, unbalanced bodies. (22.26) ☐☐☐☐

**23.246** How may a rotating mass be put into dynamic balance? (22.36) ☐☐☐☐

**23.247** Give the form of the time derivative of a vector with constant magnitude and changing direction. (22.41) ☐☐☐☐

**23.248** Discuss what happens when the axis of a rotating body is allowed to change its direction. (22.42a) ☐☐☐☐

**23.249** Find the forces exerted on the shaft by the bearings when a rotating disk with a symmetrical mass distribution experiences rotation of its shaft from the original direction of the shaft axis. (22.42b) ☐☐☐☐

**23.250** Show the directions and senses of the angular momentum vector, the angular velocity of the shaft axis, and the bearing forces which act on the shaft of a spinning disk whose axis has an angular velocity. (22.42c) ☐☐☐☐

**23.251** Show how to find the forces exerted on the shaft by the bearings when a rotating disk with an unsymmetrical mass distribution experiences rotation of its shaft from the original direction of the shaft axis. (22.42d) ☐☐☐☐

# Appendix A

# Appendix B

# Appendix C

# Index

*Note*:   Each entry in this index is referenced by problem number. The typical listing of a topic is followed by two descriptions. The first—the term *definition*—is broadly understood to include a basic definition, concept, or technique of solution. The second description—*problems*—identifies the problems which directly use this information. In many cases an initial condition of the problem, such as the initial velocity of a body or the location of the applied loads, is changed to show the effect on the solutions. These types of problems are identified by the statement *problems, with original conditions changed*. All problems in the text are listed in this index, and an asterisk is used to identify problems which have unusually lengthy solutions, or are of a more advanced nature.

The reader is encouraged to review this index and develop facility with its use. It is hoped that this listing has been developed to a sufficient degree of detail to permit rapid identification of specific problems in any desired area of the subject engineering mechanics: dynamics.